P9-DXN-436

---

example 7 **Pollution**

Suppose that for a certain city the cost C of obtaining drinking water that contains $p\%$ impurities (by volume) is given by

$$C = \frac{120{,}000}{p} - 1200$$

a. Determine the domain of this function and graph the function without concern for the context of the problem.

b. Use knowledge of the context of the problem to graph the function in a window that applies to the application.

c. What is the cost of obtaining drinking water that contains 5% impurities?

CHAPTER 4

Pollution

---

45. *Hybrid Vehicle Sales* The total hybrid electric passenger vehicle sales for the years from 1997 to 2006 are shown in the figure below. The sales can be modeled by the equation

$$S(x) = 0.744x^3 - 20.255x^2 + 184.729x - 553.098$$

where $S(x)$ is in thousands of vehicles and x is the number of years after 1990.

a. Graph this model for values of x representing 1990–2006 and positive values of $S(x)$.

b. Find the number of vehicles sold in 2000 and in 2006, according to this model.

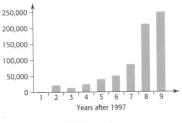

Total Hybrid Electric Passenger Vehicle Sales

Years after 1997

CHAPTER 7

Life Expectancy of Whites and Blacks

---

72. *Global Warming* In an effort to reduce global warming, it has been proposed that a tax be levied based on the emissions of carbon dioxide into the atmosphere. The cost–benefit equation $\ln(1 - P) = -0.0034 - 0.0053t$ estimates the relationship between the percent reduction of emissions of carbon dioxide (as a decimal) and the tax t in dollars per ton of carbon dioxide.

a. Solve the equation for t, giving t as a function of P. Graph the function.

b. Use the equation in part (a) to find what tax will give a 30% reduction in emissions.
(Source: W. Clime, *The Economics of Global Warming*)

CHAPTER 5

Global Warming

CHAPTER 6

Hybrid Vehicle Sales

---

example 3 **Life Expectancy**

Table 7.4 gives the years of life expected at birth for male and female blacks and whites born in the United States in the years 1920, 1940, 1960, 1980, 2000, 2002, and 2004.

Table 7.4

YEARS	WHITES		BLACKS	
	Males	Females	Males	Females
1920	54.4	55.6	45.5	54.9
1940	62.1	66.6	51.1	45.2
1960	67.4	74.1	61.1	67.4
1980	70.7	78.1	63.8	72.5
2000	74.9	80.1	68.3	75.2
2002	75.1	80.3	68.8	75.6
2004	75.7	80.8	69.5	76.3

(Source: National Center for Health Statistics)

a. Make a matrix W containing the life expectancy data for whites and a matrix B for blacks.

b. Use these matrices to find matrix $D = W - B$ that represents the difference between the white and black life expectancy.

c. What does this tell us about race and life expectancy?

College Algebra in Context

with Applications for the Managerial, Life, and Social Sciences

College Algebra in Context

with Applications for the Managerial, Life, and Social Sciences

Third edition

Ronald J. Harshbarger
University of South Carolina—Beaufort

Lisa S. Yocco
Georgia Southern University

Addison-Wesley

Boston San Francisco New York
London Toronto Sydney Tokyo Singapore Madrid
Mexico City Munich Paris Cape Town Hong Kong Montreal

Executive Editor: Anne Kelly
Project Editor: Elizabeth Bernardi
Assistant Editor: Leah Goldberg
Executive Marketing Manager: Becky Anderson
Senior Marketing Manager: Katherine Greig
Marketing Coordinator: Bonnie Gill
Senior Managing Editor: Karen Wernholm
Senior Production Supervisor: Peggy McMahon
Cover Designer: Barbara T. Atkinson
Cover Photo: Ladders on Modern building, low angle view, Australia
Cover Photographer: Oliver Strewe/Getty Images

Photo Researcher: Beth Anderson
Media Producer: Jennifer Thomas
MathXL Project Manager: Eileen Moore
QA Manager, Assessment Content: Marty Wright
Senior Author Support/Technology Specialist: Joe Vetere
Rights and Permissions Advisor: Shannon Barbe
Senior Manufacturing Buyer: Carol Melville
Text Design, Production Coordination, Illustrations, and Composition: Nesbitt Graphics, Inc.

Photo credits: p. 1, Digital Vision; p. 87, Shutterstock; p. 105, Shutterstock; p. 125, Beth Anderson; p. 163, Cultura/Getty RF; p. 187, Photodisc; p. 236, Shutterstock; p. 247, NMHA; p. 269, Shutterstock; p. 306, Shutterstock; p. 315, USGS; p. 408, Blend Images/Getty RF; p. 421, Digital Vision; p. 442, Shutterstock; p. 457, Shutterstock; p. 515, Beth Anderson; p. 587, Shutterstock; p. 591, Riser/Getty Rights Ready; p. 625, Shutterstock; p. 651, Beth Anderson, with permission of Texas Instruments.

Library of Congress Cataloging-in-Publication Data

Harshbarger, Ronald J., 1938-
  College algebra in context with applications for the managerial, life, and social
  sciences/Ronald J. Harshbarger, Lisa S. Yocco.—3rd ed.
      p. cm.
  Includes bibliographical references and index.
  ISBN 0-321-57060-X (student edition)—ISBN 0-321-56360-3 (annotated instructor's edition)
  1. Algebra—Textbooks. 2. Social sciences—Mathematics—Textbooks.
  I. Yocco, Lisa S. II. Title.
  QA152.3.H33 2010
  512.9–dc22

                                                                      2008023587

3 4 5 6 7 8 9 10—WCT—11 10 09

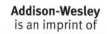

Addison-Wesley
is an imprint of

www.pearsonhighered.com

ISBN-13: 978-0-321-57060-4
ISBN-10: 0-321-57060-X

contents

College Algebra in Context is designed for a course in algebra that is based on data analysis, modeling, and real-life applications from the management, life, and social sciences. The text is designed to show students how to analyze, solve, and interpret problems in this course, in future courses, and in future careers. At the heart of this text is its emphasis on problem solving in meaningful contexts.

The text is application-driven and uses real-data problems that motivate interest in the skills and concepts of algebra. Modeling is introduced early, in the discussion of linear functions and in the discussion of quadratic and power functions. Additional models are introduced as exponential, logarithmic, logistic, cubic, and quartic functions are discussed. Mathematical concepts are introduced informally with an emphasis on applications. Each chapter contains real-data problems and extended application projects that can be solved by students working collaboratively.

The text features a constructive chapter-opening Algebra Toolbox, which reviews previously learned algebra concepts by presenting the prerequisite skills needed for successful completion of the chapter. In addition, the authors emphasize how students can use their new knowledge in a variety of calculus courses by devoting a section to calculus preparation at the end of the text.

Changes in the Third Edition

Many users of the second edition and other reviewers have made suggestions on how to enhance the presentation of material in the text. We have made a number of changes based on their suggestions and from our own classroom experiences.

- To aid instructors and students by providing smaller blocks of topics and more flexibility in scheduling quizzes and exams, Chapters 1 and 2 have each been split into two chapters. This also provides earlier milestones for the students and makes completing a chapter a less daunting task. Chapters 1 and 2, each of which previously had eight sections, and which together comprised more than 130 pages, have now been revised to four chapters with four sections each. The order of the sections has remained, so current users will feel comfortable with the flow of the text.

- The organization and exposition are improved throughout the textbook. In particular,

  ○ More discussion and examples provide justification for techniques and opportunities to graph functions with or without technology.

  ○ The discussion of exact and approximate linear models has been expanded, and the development of linear regression has been improved.

  ○ The relation among cost, revenue, and profit is now discussed earlier in the text.

  ○ The section Quadratic and Power Models is moved one section earlier in the text, so it is in the same chapter where these functions are introduced.

  ○ The discussion of piecewise-defined functions has been expanded in Chapter 3, and the section has been renamed to emphasize the importance of piecewise and power functions.

○ In Chapter 5, Exponential and Logarithmic Functions, properties of logarithms are now discussed in Section 5.2, the introductory section on logarithms.

- Many new Section Previews provide motivation for each section. Examples tied to these previews are now referenced in the Section Preview in the order in which they will appear in the section.

- The overall number of Skills Check and Application Exercises has been increased significantly, especially in the new Chapters 1 to 4.

- Nearly all real data application examples and exercises have been updated, and many new real data applications have been added including applications on global warming and preparing for retirement. In addition, many more applied examples have been added to the exposition.

- Two appendices have been added to the text. One contains a Calculator Guide, and one contains an Excel® Guide. Footnotes throughout the text refer students to these guides for a detailed exposition when a new use of technology is introduced. Additional Excel® solution procedures have been added, but, as before, they can be omitted without loss of continuity in the text.

- A new four-color design improves the detail on graphs and figures and highlights important concepts and skills.

- A new Annotated Instructor's Edition is available.

Continued Features

Features of the text include the following:

- The development of algebra is motivated by and used to find the solutions of real data–based applications.

  Real-life problems demonstrate the need for specific algebraic concepts and techniques. Each section begins with a motivational problem that presents a real life setting. The problem is solved after the necessary skills are presented in that section. The aim is to prepare students to solve problems of all types by first introducing them to various functions and encouraging them to take advantage of available technology. Special business and finance models are included to demonstrate the applications of functions to the business world.

- Technology has been integrated into the text.

  The text discusses the use of graphing calculators and/or computers, but there are no specific technology requirements. As a new calculator or spreadsheet skill becomes useful in a section, the keystrokes or commands required are given in the Graphing Calculator and Excel® Manual that accompanies the text in addition to the appendices mentioned above. The text indicates where calculators and spreadsheets can be used to solve problems. Technology is used to enhance and support learning when appropriate—not to supplant learning.

- Each of the first seven chapters begins with an Algebra Toolbox section that provides the prerequisite skills needed for the successful completion of the chapter.

  Topics discussed in the Toolbox are topics that are prerequisite to a college algebra course (usually found in a chapter R or appendix of a college algebra text).

Key concepts to be learned are listed at the beginning of each Toolbox, and topics are introduced "just in time" to be used in the chapter under consideration.

- Many problems posed in the text are multi-part and multi-level.

  Many problems require thoughtful, real-world answers under varying conditions rather than numerical answers. Questions such as, "When will this model no longer be valid?" "What additional limitations must be placed on your answer?" and "Interpret your answer in the context of the application" are commonplace in the text.

- Each chapter has a Chapter Summary, a Chapter Skills Check, and a Chapter Review.

  The Chapter Summary lists the key terms and formulas discussed in the chapter, with section references. Chapter Skills Check and Chapter Review exercises give additional review problems.

- The text encourages collaborative learning.

  Each chapter ends with one or more Group Activities/Extended Applications that require students to solve multilevel problems involving real data or situations, making it desirable for students to collaborate in their solutions. These activities provide opportunities for students to work together to solve real problems that involve the use of technology and frequently require modeling.

- The text ends with a Preparing for Calculus section that shows how algebra skills from the first four chapters are used in a calculus context.

  Many students have difficulty in calculus because they have trouble applying algebra skills to calculus. This section reviews earlier topics and shows how they apply to the development and application of calculus.

- The text encourages students to improve communication skills and research skills.

  The Group Activities/Extended Applications require written reports and frequently require use of the internet or library. Some Extended Applications require students to use literature or the internet to find a graph or table of discrete data describing an issue. They are then required to make a scatter plot of the data, determine the function type that is the best fit for the data, create the model, discuss how well the model fits the data, and discuss how it can be used to analyze the issue.

- Answers to Selected Exercises includes answers to all Chapter Skills Checks and Chapter Reviews, so students have more feedback regarding the exercises they work.

- Supplements are provided that will help students and instructors use technology to improve the learning and teaching experience. See the supplements list.

Acknowledgments

Many individuals contributed to the development of this textbook. We would like to thank the following reviewers, whose comments and suggestions were invaluable in preparing this text.

Jay Abramson, *Arizona State University*
Khadija Ahmed, *Monroe County Community College*
Jamie Ashby, *Texarkana College*
Sohrab Bakhtyari, *St. Petersburg College*
Jean Bevis, *Georgia State University*
Thomas Bird, *Austin Community College*

Len Brin, *Southern Connecticut State University*
Marc Campbell, *Daytona Beach Community College*
Florence Chambers, *Southern Maine Community College*
Floyd Downs, *Arizona State University*
Aniekan Ebiefung, *University of Tennessee at Chattanooga*
Marjorie Fernandez Karwowski; *Valencia Community College*
Toni W. Fountain, *Chattanooga State Technical Community College*
John Gosselin, *University of Georgia*
David J. Graser, *Yavapai College*
Lee Graubner, *Valencia Community College East*
Linda Green, *Santa Fe Community College*
Richard Brent Griffin, *Georgia Highlands College*
Lee Hanna, *Clemson University*
Deborah Hanus, *Brookhaven College*
Steve Heath, *Southern Utah University*
Todd A. Hendricks, *Georgia Perimeter College*
Suzanne Hill, *New Mexico State University*
Sue Hitchcock, *Palm Beach Community College*
Sandee House, *Georgia Perimeter College*
Mary Hudacheck-Buswell, *Clayton State University*
Arlene Kleinstein, *State University of New York—Farmingdale*
Danny Lau, *Kennesaw State University*
Ann H. Lawrence, *Wake Technical Community College*
Kit Lumley, *Columbus State University*
Antonio Magliaro, *Southern Connecticut State University*
Beverly K. Michael, *University of Pittsburgh*
Nancy R. Moseley, *University of South Carolina Aiken*
Demetria Neal, *Gwinnett Technical College*
Malissa Peery, *University of Tennessee*
Ingrid Peterson, *University of Kansas*
Beverly Reed, *Kent State University*
Jeri Rogers, *Seminole Community College—Oviedo*
Michael Rosenthal, *Florida International University*
Sharon Sanders, *Georgia Perimeter College*
Carolyn Spillman, *Georgia Perimeter College*
Jacqueline Underwood, *Chandler-Gilbert Community College*
Erwin Walker, *Clemson University*
Denise Widup, *University of Wisconsin—Parkside*
Sandi Wilbur, *University of Tennessee—Knoxville*

Our thanks go to Victoria Sapko of Framingham State College, for the "Inconvenient Truth" application. Many thanks to Helen Medley and Patricia Nelson for checking the accuracy of this text. A special thanks goes to the Pearson team for their assistance, encouragement, and direction throughout this project: Greg Tobin, Anne Kelly, Katherine Greig, Elizabeth Bernardi, Peggy McMahon, Barbara Atkinson, Leah Goldberg, Jennifer Thomas, and Bonnie Gill.

Ronald J. Harshbarger
Lisa S. Yocco

list of supplements

Student Supplements	Instructor Supplements

Student's Solutions Manual

- By Lee Graubner, *Valencia Community College*.
- Provides selected solutions to the Skills Check problems and Exercises. Solutions are available to all Chapter Skills Check, Review problems, Algebra Toolbox sets, and Extended Applications.
- ISBN-13: 978-0-321-56968-4; ISBN-10: 0-321-56968-7

Instructor's Solutions Manual

- By Lee Graubner, *Valencia Community College*.
- Provides complete solutions to all of the Algebra Toolbox sets, Skills Check problems, Exercises, Chapter Skills Check and Review problems, and Extended Applications.
- ISBN-13: 978-0-321-56964-6; ISBN-10: 0-321-56964-4

Graphing Calculator and Excel® Manual

- By David J. Graser, *Yavapai College*.
- Provides instructions and keystroke operations for the TI-83, TI-83 Plus, and TI-84, along with EXCEL® commands.
- Also contains worked-out examples correlated to in-text examples.
- ISBN-13: 978-0-321-56969-1; ISBN-10: 0-321-56969-5

Instructor's Testing Manual

- By Mary Hudachek-Buswell, *Clayton State University*.
- Available for download from the Instructor Resource Center at pearsonhighered.com/irc.
- Contains three alternative forms of tests per chapter.
- Answer keys are included with more applications.

Additional Skill & Exercise Manual

- By Melanie Fulton, *High Point University*.
- Provides additional skill and testing preparation for students.
- ISBN-13: 978-0-321-56970-7; ISBN-10: 0-321-56970-9

TestGen®

- Enables instructors to build, edit, print, and administer tests.
- Features a computerized bank of questions developed to cover all text objectives.
- Available for download from the Instructor Resource Center at pearsonhighered.com/irc.

A Review of Algebra

- By Heidi Howard, *Florida Community College at Jacksonville*.
- Provides additional support for those students needing further algebra review.
- ISBN-13: 978-0-201-77347-7; ISBN-10: 0-201-77347-3

Adjunct Support Center

- Offers consultation on suggested syllabi, helpful tips on using the textbook support package, assistance with content, and advice on classroom strategies.
- Available Sunday–Thursday evenings from 5 P.M. to midnight; telephone: EST 1-800-435-4084; e-mail: AdjunctSupport@aw.com; fax: 1-877-262-9774.

Technology Resources

MathXL®

MathXL® is a powerful online homework, tutorial, and assessment system that accompanies Pearson Education's textbooks in mathematics or statistics. With MathXL, instructors can create, edit, and assign online homework and tests using algorithmically generated exercises correlated at the objective level to the textbook. They can also create and assign their own online exercises and import TestGen tests for added flexibility. All student work is tracked in MathXL's online gradebook. Students can take chapter tests in

MathXL and receive personalized study plans based on their test results. The study plan diagnoses weaknesses and links students directly to tutorial exercises for the objectives they need to study and retest. Students can also access supplemental animations and narrated examples directly from selected exercises. MathXL is available to qualified adopters. For more information, visit our Web site at *www.mathxl.com*, or contact your sales representative.

MyMathLab®

MyMathLab is a series of text-specific, easily customizable online courses for Pearson Education's textbooks in mathematics and statistics. Powered by CourseCompass™ (our online teaching and learning environment) and MathXL® (our online homework, tutorial, and assessment system), MyMathLab gives you the tools you need to deliver all or a portion of your course online, whether your students are in a lab setting or working from home. MyMathLab provides a rich and flexible set of course materials, featuring free-response exercises that are algorithmically generated for unlimited practice and mastery. Students can also use online tools, such as narrated examples, animations, and a multimedia textbook, to independently improve their understanding and performance. Instructors can use MyMathLab's homework and test managers to select and assign online exercises correlated directly to the textbook, and they can also create and assign their own online exercises and import TestGen tests for added flexibility. MyMathLab's online gradebook—designed specifically for mathematics and statistics—automatically tracks students' homework and test results and gives the instructor control over how to calculate final grades. Instructors can also add offline (paper-and-pencil) grades to the gradebook. MyMathLab also includes access to Pearson's Tutor Center (www.pearsontutorservices.com) which provides students with tutoring via toll-free phone, fax, e-mail, and interactive Web sessions. MyMathLab is available to qualified adopters. For more information, visit our Web site at *www.mymathlab.com* or contact your sales representative.

InterAct Math Tutorial Web site:
www.interactmath.com

Get practice and tutorial help online! This interactive tutorial Web site provides algorithmically generated practice exercises that correlate directly to the exercises in the textbook. Students can retry an exercise as many times as they like, with new values each time for unlimited practice and mastery. Every exercise is accompanied by an interactive guided solution that provides helpful feedback for incorrect answers, and students can also view a worked-out sample problem that steps them through an exercise similar to the one they're working on.

to the student

College Algebra in Context was written to help you develop the math skills needed to model problems and analyze data—tasks that are required in many jobs in the fields of management, life science, and social science. As you work through this text, you will build upon what you already know about mathematics. You will learn more about algebra in the context of real examples and problems that you may experience in your future studies and career. As you read, you may be surprised by how professionals use algebra—from predicting how a population may vote in an upcoming election to projecting DVD player sales.

There are many ways in which this text will help you succeed in your algebra course, and you can be a partner in that success. Consider the following suggestions that our own students have found helpful:

1. **Take careful notes in an organized notebook**. Good organization is essential in a math course so that you do not fall behind and so that you can quickly reference a topic when you need it.

   - Separate your notebook into 3 sections: **class notes and examples, homework**, and a **problem log** consisting of problems worked in class which will provide a sample test.

   - **Begin each set of notes** with the date, section of the book, page number from the book, and topic.

   - **Write explanations in words**, rather than just the steps to a problem, so that you will understand later what was done in each step. **Use abbreviations and short phrases** rather than complete sentences so that you can keep up with the explanation as you write.

   - **Write step-by-step instructions** for each process.

   - Use a spiral notebook with pockets for **handouts**, or use a loose-leaf notebook. Keep your **tests** in your notebook.

2. **Read the textbook**. Reading a mathematics textbook is different from reading other textbooks. Some suggestions follow.

   - **Skim the material** to get a general idea of the major topics. As you skim the material, **circle** any words that you do not understand. Read the chapter summary and look at the exercises at the end of each section.

   - **Make note cards** for terms, symbols, and formulas. Review these note cards often to retain the information.

   - **Read for explanation and study the steps to work a problem**. It is essential that you learn *how* to work a problem and *why* the process works rather than memorizing sample problems.

   - **Study any illustrations** to work a problem and the sample problems that are given, then cover up the solution and try to work the problem on your own.

   - **Practice the process**. The more problems you do, the more confident you will become in your ability to do math and perform on tests. When doing your homework, don't give in to frustration. Put your homework aside for a while and come back to it later.

- **Recite and review**. You should know and understand the example problems in your text well enough to be able to work similar problems on the test. Make note cards with example problems on one side of the card and the solutions on the other side.

3. **Work through the Algebra Toolbox**. The Algebra Toolbox will give you a great review of the skills needed for success in each chapter. Note the Key Concepts listed at the beginning of the Toolbox; reread them once you have completed the Toolbox exercises.

4. **Practice with Skills Check Exercises**. Skills Check exercises are a way to practice your algebra skills before moving on to more applied problems.

5. **Work the Exercises carefully**. The examples and exercises in this book model ways in which mathematics is used in the world. Look for connections between the examples and what you may be learning in your other classes. The examples will help you work through the applied exercises in each section.

6. **Prepare for your exams**.

   - **Make a study schedule**. Begin to study at least 3 days before the test. You should make a schedule, listing those sections in the book that you will study each day. Schedule a sample test to be taken upon completion of those sections. This sample test should be taken two days before the test date so that you have time to work on areas of difficulty.

   - **Rework problems**. You should actively prepare for a test. Do *more* than read your notes and the textbook. Do *more* than look over your homework. Review the note cards prepared from your class notes and text. Actually *rework* problems from each section of your book. Use the Chapter Summary at the end of each chapter to be sure you know and understand the key concepts and formulas. Then get some more practice with the Chapter Skills Check and Review Exercises. Check your answers!

   - **Get help**. Do not leave questions unanswered. Remember to utilize all resources in getting the help you need. Some resources that you might consider using are your fellow classmates, tutors, MyMathLab, the Student's Solutions Manual, math videotapes, and even your teacher. Do not take the gamble that certain questions will not be on the test!

   - **Make a sample test**. Write a sample test by choosing a variety of problems from each section in the book. Then, write the problems for your sample test in a different order than they appear in the book. (*Hint*: If you write each problem, with directions, on a separate index card and mix them up, you will have a good sample test.)

   - **Review and relax the night before the test**. The night before the test is best used *reviewing* the material. This may include working one problem from each section, reworking problems that have given you difficulty, or thinking about procedures you have used.

   - **Practice taking tests online**. Ask your professor if MathXL or MyMathLab is available at your school. Both provide online homework, tutorial, and assessment systems for unlimited practice exercises correlated to your textbook. (An access card is required to use these products.)

7. **Develop better math test-taking skills**.

   - **Memory download**. As soon as you receive your test, jot down formulas or rules that you will need but are likely to forget. If you get nervous later and forget this information, you only have to refer to the memory cues that you have written down.

- **Skip the difficult questions**. Come back to these later or try to work at least one step for partial credit.

- **Keep a schedule**. The objective is to get the most points. Don't linger over one question very long.

- **Review your work**. Check for careless errors and make sure your answers make sense.

- **Use all the time given**. There are no bonus points for turning your test in early. Use extra time for checking your work.

8. **Have fun!** Look for mathematics all around you. Read the newspaper, look at data on government Web sites, and observe how professionals use mathematics to do their jobs and communicate information to the world.

We have enjoyed teaching this material to our students and watching their understanding grow. We wish you the very best this semester and in your future studies.

Ronald J. Harshbarger
Lisa S. Yocco

Functions, Graphs, and Models; Linear Functions

With digital TV becoming more affordable by the day, the demand for high-definition home entertainment is growing rapidly, with 37 million Americans having digital TV by 2008. Seventy-four percent of European households will have digital TV by 2009, and by 2010 more than 370 million sets should be in use worldwide. Cell phone use is also on the rise, not just in the United States but throughout the world. In the middle of 2006, the number of subscribers to cell phone carriers had dramatically increased, and the number of total users had reached 2.4 billion. If the numbers continue to increase at a steady rate, the number of subscribers is expected to reach into the tens of billions over the next few years. These projections and others are made by collecting real-world data and creating mathematical models. The goal of this chapter and future chapters is to use real data and mathematical models to make predictions and solve meaningful problems.

section	topics	applications
1.1 Functions and Models	Determining graphs, tables, and equations that represent functions; finding domains and ranges; evaluating functions and mathematical models	Body temperature, personal computers, stock market, men in the workforce, federal salaries
1.2 Graphs of Functions	Graphing and evaluating functions with technology; graphing mathematical models; aligning data; graphing data points; scaling data	Cost-benefit, crime rates, bankruptcies, U.S. executions, voting, high school enrollment
1.3 Linear Functions	Identifying and graphing linear functions; finding and interpreting intercepts and slopes; finding constant rates of change; revenue, cost, profit; special linear functions	Loan balances, prescription drugs, marginal profit, marginal revenue
1.4 Equations of Lines	Writing equations of lines; parallel and perpendicular lines; finding average rates of change; approximately linear data	Depreciation, inmate population, service call charges, blood alcohol percent, high school enrollment, hybrid vehicle sales

Algebra Toolbox

The Algebra Toolbox is designed to review prerequisite skills needed for success in each chapter. In this Toolbox we discuss the real numbers, the coordinate system, algebraic expressions, equations, inequalities, absolute values, and subscripts.

Sets

In this chapter we will use sets to write domains and ranges of functions, and in future chapters we will find solution sets to equations and inequalities. A **set** is a well-defined collection of objects, including but not limited to numbers. In this section, we will discuss sets of real numbers, including natural numbers, integers, and rational numbers, and later in the text we will discuss the set of complex numbers. There are two ways to define a set. One way is by listing the **elements** (or **members**) of the set (usually between braces). For example, we may say that a set A contains 2, 3, 5, and 7 by writing $A = \{2, 3, 5, 7\}$. To say that 5 is an element of the set A, we write $5 \in A$. To indicate that 6 is not an element of the set, we write $6 \notin A$. Domains of functions and solutions to equations are sometimes given in sets with the elements listed.

If all the elements of the set can be listed, the set is said to be a **finite set**. If all elements of a set cannot be listed, the set is called an **infinite set**. To indicate that a set continues with the established pattern, we use three dots. For example, $B = \{1, 2, 3, 4, 5, \ldots, 100\}$ describes the finite set of whole numbers from 1 through 100, and the set $N = \{1, 2, 3, 4, 5, \ldots\}$ describes the infinite set of all whole numbers beginning with 1. This set is called the **natural numbers**.

Another way to define a set is to give its description. For example, we may write $\{x \mid x \text{ is a math book}\}$ to define the set of math books. This is read as "the set of all x such that x is a math book." $N = \{x \mid x \text{ is a natural number}\}$ defines the natural numbers, which was also defined by $N = \{1, 2, 3, 4, 5, \ldots\}$ above.

The set that contains no elements is called the **empty set**, and is denoted by $\varnothing$.

example 1

Write the following sets in two ways.

a. The set A containing the natural numbers less than 7.

b. The set B of natural numbers that are at least 7.

Solution

a. $A = \{1, 2, 3, 4, 5, 6\}, A = \{x \mid x \in N, x < 7\}$

b. $B = \{7, 8, 9, 10, \ldots\}, B = \{x \mid x \in N, x \geq 7\}$

The relations that can exist between two sets follow.

Relations Between Sets

1. Sets X and Y are **equal** if they contain exactly the same elements.

2. Set A is called a **subset** of set B if each element of A is an element of B. This is denoted $A \subseteq B$.

3. If set C and D have no elements in common, they are called **disjoint**.

example 2

For the sets $A = \{x \mid x \le 9, x \text{ is a natural number}\}$, $B = \{2, 4, 6\}$, $C = \{3, 5, 8, 10\}$:

a. Which of the sets A, B, and C are subsets of A?

b. Which pairs of sets are disjoint?

c. Are any of these three sets equal?

Solution

a. Every element of B is contained in A. Thus, set B is a subset of A. Because every element of A is contained in A, A is a subset of A.

b. Sets B and C have no elements in common, so they are disjoint.

c. None of these sets have exactly the same elements, so none are equal. ■

The Real Numbers

Because most of the mathematical applications you will encounter in an applied non-technical setting use real numbers, the emphasis in this text is the **real number system**.* Real numbers can be rational or irrational. **Rational numbers** include integers, fractions containing only integers (with no 0 in a denominator), and decimals that either terminate or repeat. Some examples of rational numbers are

$$-9, \quad \frac{1}{2}, \quad 0, \quad 12, \quad -\frac{4}{7}, \quad 6.58, \quad -7.\overline{3}$$

Irrational numbers are real numbers that are not rational. Some examples of irrational numbers are π (a number familiar to us from the study of circles), $\sqrt{2}$, $\sqrt[3]{5}$, and $\sqrt[3]{-10}$.

The types of real numbers are described in Table 1.1.

Table 1.1

Types of Real Numbers	Descriptions
Natural numbers	$1, 2, 3, 4, \ldots$
Integers	Natural numbers, zero, and the negatives of the natural numbers: $\ldots, -3, -2, -1, 0, 1, 2, 3, \ldots$
Rational numbers	All numbers that can be written in the form $\frac{p}{q}$ where p and q are both integers with $q \ne 0$. Rational numbers can be written as terminating or repeating decimals.
Irrational numbers	All real numbers that are not rational numbers. Irrational numbers cannot be written as terminating or repeating decimals.

We can represent real numbers on a **real number line**. Exactly one real number is associated with each point on the line, and we say there is a one-to-one correspondence between the real numbers and the points on the line. That is, the real number line is a graph of the real numbers (see Figure 1.1).

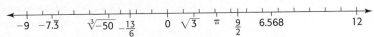

Figure 1.1

---

* The complex number system will be discussed in the Chapter 3 Toolbox.

Notice the number π on the real number line in Figure 1.1. This special number, which can be approximated by 3.14, results when the circumference of (distance around) any circle is divided by the diameter of the circle. Another special real number is e; it is denoted by

$$e \approx 2.71828$$

We will discuss this number, which is important in financial and biological applications, later in the text.

Inequalities and Intervals on the Number Line

In this chapter, we will sometimes use inequalities and interval notation to describe domains and ranges of functions. An **inequality** is a statement that one quantity is greater (or less) than another quantity. We say that a is less than b (written $a < b$) if the point representing a is to the left of the point representing b on the real number line. We may indicate that the number a is greater than or equal to b by writing $a \geq b$. The subset of real numbers x that lie between a and b (excluding a and b) can be denoted by the **double inequality** $a < x < b$ or by the **open interval** (a, b). This is called an open interval because neither of the endpoints is included in the interval. The **closed interval** $[a, b]$ represents the set of all real numbers satisfying $a \leq x \leq b$. Intervals containing one endpoint, such as $[a, b)$ or $(a, b]$, are called **half-open intervals**. We can represent the inequality $x \geq a$ by the interval $[a, \infty)$, and we can represent the inequality $x < a$ by the interval $(-\infty, a)$. Note that ∞ and $-\infty$ are not numbers, but ∞ is used in $[a, \infty)$ to represent the fact that x increases without bound and $-\infty$ is used in $(-\infty, a)$ to indicate that x decreases without bound. Table 1.2 shows the graphs of different types of intervals.

Table 1.2

Interval Notation	Inequality Notation	Verbal Description	Number Line Graph
(a, ∞)	$x > a$	x is greater than a	
$[a, \infty)$	$x \geq a$	x is greater than or equal to a	
$(-\infty, b)$	$x < b$	x is less than b	
$(-\infty, b]$	$x \leq b$	x is less than or equal to b	
(a, b)	$a < x < b$	x is between a and b, not including either a or b	
$[a, b)$	$a \leq x < b$	x is between a and b, including a but not including b	
$(a, b]$	$a < x \leq b$	x is between a and b, not including a but including b	
$[a, b]$	$a \leq x \leq b$	x is between a and b, including both a and b	

Note that open circles may be used instead of parentheses and closed circles may be used instead of brackets in the number line graphs.

example 3	**Intervals**

Write the interval corresponding to each of the inequalities in parts (a)–(e), then graph the inequality.

a. $-1 \le x \le 2$ **b.** $2 < x < 4$ **c.** $-2 < x \le 3$ **d.** $x \ge 3$ **e.** $x < 5$

Solution

a. $[-1, 2]$

b. $(2, 4)$

c. $(-2, 3]$

d. $[3, \infty)$

e. $(-\infty, 5)$

Algebraic Expressions

In algebra we deal with a combination of real numbers and letters. Generally, the letters are symbols used to represent unknown quantities or fixed but unspecified constants. Letters representing unknown quantities are usually called **variables**, and letters representing fixed but unspecified numbers are called **literal constants**. An expression created by performing additions, subtractions, or other arithmetic operations with one or more real numbers and variables is called an **algebraic expression**. Unless otherwise specified, the variables represent real numbers for which the algebraic expression is a real number. Examples of algebraic expressions include

$$5x - 2y, \quad \frac{3x - 5}{12 + 5y}, \quad \text{and} \quad 7z + 2$$

A term of an algebraic expression is the product of one or more variables and a real number; the real number is called a **numerical coefficient**, or simply a **coefficient**. A constant is also considered a term of an algebraic expression and is called a **constant term**. For instance, the term $5yz$ is the product of the factors 5, y, and z; this term has coefficient 5.

Polynomials

An algebraic expression containing a finite number of additions, subtractions, and multiplications of constants and nonnegative integer powers of variables is called a **polynomial**. When simplified, a polynomial cannot contain negative powers of variables, fractional powers of variables, variables in a denominator, or variables inside a radical. The expressions $5x - 2y$ and $7z^3 + 2y$ are polynomials, but $\dfrac{3x - 5}{12 + 5y}$ and $3x^2 - 6\sqrt{x}$ are not polynomials. If the only variable in the polynomial is x, then the polynomial is called a **polynomial in x**. The general form of a polynomial in x is

$$a_n x^n + a_{n-1} x^{n-1} + \cdots + a_1 x + a_0$$

where a_0 and each coefficient $a_n, a_{n-1}, \ldots$ are real numbers and each exponent $n, n - 1, \ldots$ is a positive integer.

For a polynomial in the single variable x, the power of x in each term is the **degree** of that term, with the degree of a constant term equal to 0. The term that has the highest power of x is called the **leading term** of the polynomial, the coefficient of this term is the **leading coefficient**, and the degree of this term is the **degree of the polynomial**. Thus, $5x^4 + 3x^2 - 6$ is a fourth-degree polynomial with leading coefficient 5. Polynomials with one term are called **monomials**, those with two terms are called **binomials**, and those with three terms are called **trinomials**. The right side of the equation $y = 4x + 3$ is a first-degree binomial and the right side of $y = 6x^2 - 5x + 2$ is a second-degree trinomial.

| **example 4** | For each polynomial, state the constant term, the leading coefficient, and the degree of the polynomial. |

a. $5x^2 - 8x + 2x^4 - 3$ **b.** $5x^2 - 6x^3 + 3x^6 + 7$

Solution

a. The constant term is -3; the term of highest degree is $2x^4$, so the leading coefficient is 2 and the degree of the polynomial is 4.

b. The constant term is 7; the term of highest-degree is $3x^6$, so the leading coefficient is 3 and the degree of the polynomial is 6.   ■

Terms that contain exactly the same variables with exactly the same exponents are called **like terms**. For example, $3x^2y$ and $7x^2y$ are like terms, but $3x^2y$ and $3xy$ are not. We can *simplify* an expression by adding or subtracting the coefficients of the like terms. For example, the simplified form of

$$3x + 4y - 8x + 2y \quad \text{is} \quad -5x + 6y$$

and the simplified form of

$$3x^2y + 7xy^2 + 6x^2y - 4xy^2 - 5xy \quad \text{is} \quad 9x^2y + 3xy^2 - 5xy$$

Removing Parentheses

We often need to remove parentheses when simplifying algebraic expressions and when solving equations. Removing parentheses frequently requires use of the **distributive property**, which says that for real numbers a, b, and c, $a(b + c) = ab + ac$. Care must be taken to avoid mistakes with signs when using the distributive property. Multiplying a sum in parentheses by a negative number changes the sign of each term in the parentheses. For example,

$$-3(x - 2y) = -3(x) + (-3)(-2y) = -3x + 6y \quad \text{and} \quad -(3xy - 5x^3) = -3xy + 5x^3$$

We add or subtract (**combine**) algebraic expressions by combining the like terms. For example, the sum of the expressions $5sx - 2y + 7z^3$ and $2y + 5sx - 4z^3$ is

$$(5sx - 2y + 7z^3) + (2y + 5sx - 4z^3) = 5sx - 2y + 7z^3 + 2y + 5sx - 4z^3$$
$$= 10sx + 3z^3$$

and the difference of these two expressions is

$$(5sx - 2y + 7z^3) - (2y + 5sx - 4z^3) = 5sx - 2y + 7z^3 - 2y - 5sx + 4z^3$$
$$= -4y + 11z^3$$

The Coordinate System

Much of our work in algebra involves graphing. To graph in two dimensions, we use a rectangular coordinate system, or **Cartesian coordinate system**. Such a system allows us to assign a unique point in a plane to each ordered pair of real numbers. We construct the coordinate system by drawing a horizontal number line and a vertical number line so that they intersect at their origins (Figure 1.2). The point of intersection is called the **origin** of the system, the number lines are called the coordinate **axes**, and the plane is divided into four parts called **quadrants**. In Figure 1.3, we call the horizontal axis the **x-axis** and the vertical axis the **y-axis**, and we denote any point in the plane as the ordered pair (x, y).

The ordered pair (a, b) represents the point P that is $|a|$ units from the y-axis (right if a is positive, left if a is negative) and $|b|$ units from the x-axis (up if b is positive, down if b is negative). The values of a and b are called the **rectangular coordinates** of the point. Figure 1.3 shows point P with coordinates (a, b). The point is in the second quadrant, where $a < 0$ and $b > 0$.

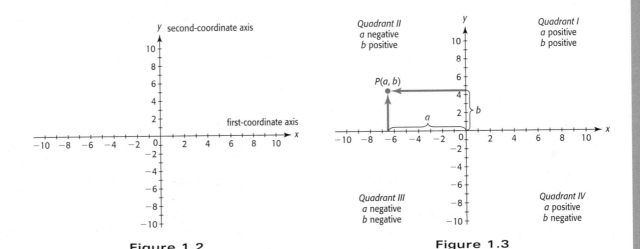

Figure 1.2   **Figure 1.3**

Subscripts

We sometimes need to distinguish between two y-values and/or x-values in the same problem or on the same graph, or to designate literal constants. It is often convenient to do this by using **subscripts**. For example, if we have two fixed but unidentified points on a graph, we can represent one point as (x_1, y_1) and the other as (x_2, y_2). Subscripts also can be used to designate different equations entered in graphing utilities; for example, $y = 2x - 5$ may appear as $y_1 = 2x - 5$ when entered in the equation editor of a graphing calculator.

toolbox exercises

1. Write "the set of all natural numbers less than 9" in two different ways.

2. Is it true that $3 \in \{1, 3, 4, 6, 8, 9, 10\}$?

3. Is A a subset of B if $A = \{2, 3, 5, 7, 8, 9, 10\}$ and $B = \{3, 5, 8, 9\}$?

4. Is it true that $\frac{1}{2} \in N$, if N is the set of natural numbers?

5. Is the set of integers a subset of the set of rational numbers?

6. Are sets of rational numbers and irrational numbers disjoint sets?

Identify the sets of numbers in Exercises 7–9 as containing one or more of the following: integers, rational numbers, and/or irrational numbers.

7. $\{5, 2, 5, 8, -6\}$

8. $\left\{\frac{1}{2}, -4.1, \frac{5}{3}, 1\frac{2}{3}\right\}$

9. $\left\{\sqrt{3}, \pi, \frac{\sqrt[3]{2}}{4}, \sqrt{5}\right\}$

In Exercises 10–12, express each interval or graph as an inequality.

10. x
    -3

11. $[-3, 3]$

12. $(-\infty, 3]$

In Exercises 13–15, express each inequality or graph in interval notation.

13. $x \le 7$

14. $3 < x \le 7$

15. x
    4

In Exercises 16–18, graph the inequality or interval on a real number line.

16. $(-2, \infty)$

17. $5 > x \ge 2$

18. $x < 3$

In Exercises 19–21, plot the points on a coordinate system.

19. $(-1, 3)$

20. $(4, -2)$

21. $(-4, 3)$

22. Plot the points $(-1, 2)$, $(3, -1)$, $(4, 2)$, and $(-2, -3)$ on the same coordinate system.

23. Plot the points (x_1, y_1) and (x_2, y_2) on a coordinate system if

$$x_1 = 2, y_1 = -1, x_2 = -3, y_2 = -5$$

For each algebraic expression in Exercises 24 and 25, give the coefficient of each term and give the constant term.

24. $-3x^2 - 4x + 8$

25. $5x^4 + 7x^3 - 3$

26. Find the sum of $z^4 - 15z^2 + 20z - 6$ and $2z^4 + 4z^3 - 12z^2 - 5$.

27. Simplify the expression

$$3x + 2y^4 - 2x^3y^4 - 119 - 5x - 3y^2 + 5y^4 + 110$$

Remove the parentheses and simplify in Exercises 28–33.

28. $4(p + d)$

29. $-2(3x - 7y)$

30. $-a(b + 8c)$

31. $4(x - y) - (3x + 2y)$

32. $4(2x - y) + 4xy - 5(y - xy) - (2x - 4y)$

33. $2x(4yz - 4) - (5xyz - 3x)$

1.1

Functions and Models

section preview ▪ Body Temperatures

One indication of illness in children is elevated body temperature. There are two common measures of temperature, Fahrenheit (°F) and Celsius (°C), with temperature measured in Fahrenheit degrees in the United States and in Celsius degrees in many other countries of the world. Suppose you know that a child's normal body temperature is 98.6°F and his or her temperature is now 37°C. Does this indicate that the child is ill? To help decide this, we could find the Fahrenheit temperature that corresponds to 37°C. We could do this easily if we knew the relationship between Fahrenheit and Celsius temperature scales. In this section, we will see that the relationship between these measurements can be defined by a **function** and that functions can be defined numerically, graphically, verbally, or by an equation. We also explore how this and other functions can be applied to help solve problems that occur in real-world situations.

Function Definitions

There are several techniques to show how Fahrenheit degree measurements are related to Celsius degree measurements.

One way to show the relationship between Celsius and Fahrenheit degree measurements is by listing some Celsius measurements and the corresponding Fahrenheit measurements. These measurements, and any other real-world information collected in numerical form, are called **data**. These temperature measurements can be shown in a table (Table 1.3).

Table 1.3

Celsius Degrees (°C)	−20	−10	−5	0	25	50	100
Fahrenheit Degrees (°F)	−4	14	23	32	77	122	212

This relationship is also defined by the set of ordered pairs

$$\{(-20, -4), (-10, 14), (-5, 23), (0, 32), (25, 77), (50, 122), (100, 212)\}$$

We can picture the relationship between the measurements with a graph. Figure 1.4 shows a **scatter plot** of the data, that is, a graph of the ordered pairs as points.

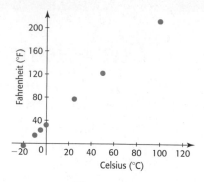

Figure 1.4

Table 1.3, the set of ordered pairs below the table, and the graph in Figure 1.4 define a **function** with a set of Celsius temperature **inputs** (called the **domain** of the function) and a set of corresponding Fahrenheit **outputs** (called the **range** of the function). A function that will give the Fahrenheit temperature measurement F that corresponds to *any* Celsius temperature measurement C between $-20°C$ and $100°C$ is the equation

$$F = \frac{9}{5}C + 32$$

This equation defines F as a function of C because each input C results in exactly one output F. Its graph, shown in Figure 1.5, is a line that contains the points on the scatter plot in Figure 1.4 as well as other points. If we consider only Celsius temperatures from -20 to 100, then the domain of this function is $-20 \leq C \leq 100$ and the resulting range is $-4 \leq F \leq 212$.

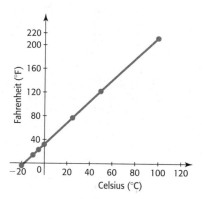

Figure 1.5

Function

A function is a rule or correspondence that assigns to each element of one set (called the domain) exactly one element of a second set (called the range).

The function may be defined by a set of ordered pairs, a table, a graph, an equation, or a verbal description.

example 1

Body Temperature

Suppose a child's normal temperature is 98.6°F and the only thermometer available is Celsius and indicates that the child's temperature is 37°C. Does this reading indicate that the child's temperature is normal?

Solution

We now have a function that will give the output temperature F that corresponds to the input temperature C. Substituting 37 for C in $F = \dfrac{9}{5}C + 32$ gives $F = \dfrac{9}{5}(37) + 32 = 98.6$. This indicates that the child's temperature is normal. ∎

Domains and Ranges

How a function is defined determines its domain and range. For instance, the domain of the function defined by Table 1.3 or by the scatter plot in Figure 1.4 is the finite set $\{-20, -10, -5, 0, 25, 50, 100\}$ with all values measured in degrees Celsius, and the range is the set $\{-4, 14, 23, 32, 77, 122, 212\}$ with all values measured in degrees Fahrenheit. This function has a finite number of inputs in its domain.

The function defined by $F = \dfrac{9}{5}C + 32$ above had the inputs (domain) and outputs (range) restricted to $-20 \leq C \leq 100$ and $-4 \leq F \leq 212$, respectively. Functions defined by equations can also be restricted by the context in which they are used. For example, if the function $F = \dfrac{9}{5}C + 32$ is used in measuring the temperature of water, its domain is limited to real numbers from 0 to 100 and its range is limited to real numbers from 32 to 212, because water changes state with other temperatures.

If x represents any element in the domain, then x is called the **independent variable**, and if y represents an output of the function from an input x, then y is called the **dependent variable**. The figure at left shows a general "function machine" in which the input is called x, the rule is denoted by f, and the output is symbolized by $f(x)$. The symbol $f(x)$ is read "f of x."

If the domain of a function is not specified or restricted by the context in which the function is used, it is assumed that the domain consists of all real number inputs that result in real number outputs in the range, and that the range is a subset of the real numbers. Two special cases where the domain of a function may be limited follow.

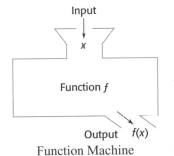

Input
x

Function f

Output $f(x)$

Function Machine

Domains and Ranges

1. Functions with variables in denominators may have input value(s) that give a 0 in the denominator. To find value(s) *not in* the domain:

   Set the denominator equal to 0 and solve for the variable
   (see Example 2(b)).

2. Functions with variables inside even roots may have an input value(s) that give negative value(s) inside the even root. To find value(s) *in* the domain:

   Set the expression inside the even root greater than or equal to 0
   and solve for the variable (see Example 2(c)).

We can use the graph of a function to find or to verify its domain and range. We can usually see the interval(s) on which the graph exists and therefore agree on the subset of the real numbers for which the function is defined. This set is the domain of the function. We can also usually determine if the outputs of the graph form the set of all real numbers or some subset of real numbers. This set is the range of the function.

example 2   Domains and Ranges

For each of the following functions determine the domain. Determine the range of the function in parts (a) and (c).

a. $y = 4x^2$ **b.** $y = 1 + \dfrac{1}{x-2}$ **c.** $y = \sqrt{4-x}$

Solution

a. Because any real number input for x, when squared and multiplied by 4, results in a real number output for y, we conclude that the domain is the set of all real numbers. Because

$$y = 4x^2$$

cannot be negative for any value of x that is input, the range is the set of all nonnegative real numbers ($y \geq 0$). This can be confirmed by looking at the graph of this function, shown in Figure 1.6.

b. Because the denominator of the fractional part of this function will be 0 when $x = 2$, and the outputs for every other value of x are real numbers, the domain of this function contains all real numbers except 2. The graph in Figure 1.7(a) confirms this conclusion.

c. Because

$$y = \sqrt{4-x}$$

cannot be a real number if $4 - x$ is negative, the only values of x that give real outputs to the function are values that satisfy $4 - x \geq 0$, or $4 \geq x$, so the domain is $x \leq 4$. Because $\sqrt{4-x}$ (the principal square root) can never be negative, the range is $y \geq 0$. A table of selected inputs and their outputs, and the graph of this function, are shown in Figure 1.7(b). They confirm that the domain is $(-\infty, 4]$ and the range is $[0, \infty)$.

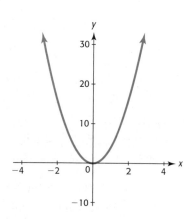

Figure 1.6

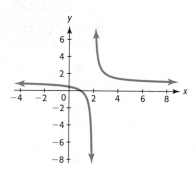

Figure 1.7(a)

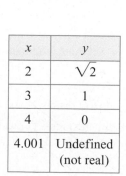

x	y
2	$\sqrt{2}$
3	1
4	0
4.001	Undefined (not real)

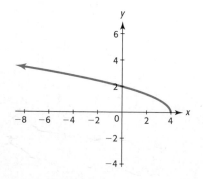

Figure 1.7(b)

Arrow diagrams that show how each individual input results in exactly one output can also represent functions. Each arrow in Figure 1.8(a) and in Figure 1.8(c) goes from an input to exactly one output, so each of these diagrams defines a function. On the other hand, the arrow diagram in Figure 1.8(b) does not define a function because one input, 8, goes to two different outputs, 6 and 9.

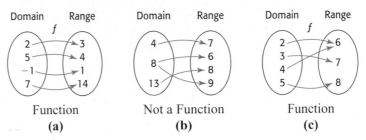

Function   Not a Function   Function
(a)   (b)   (c)

Figure 1.8

Tests for Functions

Functions play an important role in the solution of mathematical problems. Also, to graph a relationship using graphing calculators and computer software, it is usually necessary to express the association between the variables in the form of a function. It is therefore essential for you to recognize when a relationship is a function. Recall that a function is a rule or correspondence that determines exactly one output for each input.

| example 3 |

Recognizing Functions

For each of the following, determine whether or not the indicated relationship represents a function. Explain your reasoning. For each function that is defined, give the domain and range.

a. The number N of personal computers in use worldwide determined by the year x, as defined in Table 1.4. Is N a function of x?

Table 1.4

x Year	N Worldwide Personal Computers (millions)
1991	129.4
1992	150.8
1993	177.4
1994	208.0
1995	245.0
2000	535.6
2005	903.9

(Source: *The Time Almanac*)

b. The daily profit P (in dollars) from the sale of x pounds of candy as shown in Figure 1.9. Is P a function of x?

c. The number of tons x of coal sold determined by the profit P that is made from the sale of the product, as shown in Table 1.5. Is x a function of P?

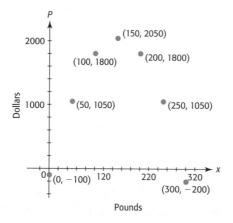

Table 1.5

x (tons)	P ($)
0	$-100{,}000$
500	109,000
1000	480,000
1500	505,000
2000	480,000
2500	109,000
3000	$-100{,}000$

Figure 1.9

d. W is a person's weight in pounds during the nth week of a diet for $n = 1$ and $n = 2$. Is W a function of n?

Solution

a. For each year (input) listed in Table 1.4, only one value is given for the number of computers used worldwide (output), so Table 1.4 represents N as a function of x. The set {1991, 1992, 1993, 1994, 1995, 2000, 2005} is the domain and the range is the set {129.4, 150.8, 177.4, 208, 245, 535.6, 903.9} million.

b. Each input x corresponds to only one daily profit P, so this scatter plot represents P as a function of x. The domain is {0, 50, 100, 150, 200, 250, 300} pounds, and the range is {−100, −200, 1050, 1800, 2050} dollars. For this function, x is the independent variable and P is the dependent variable.

c. The number of tons x of coal sold is not a function of the profit P that is made, because some values of P result in two values of x. For example, a profit of $480,000 corresponds to both 1000 tons of coal and 2000 tons of coal.

d. A person's weight varies during any week; for example, a woman may weigh 122 lb on Tuesday and 121 lb on Friday. Thus there is more than one output (weight) for each input (week), and this relationship is not a function. ∎

| example 4 |

Functions

a. Does the equation $y^2 = 3x - 3$ define y as a function of x?

b. Does the equation $y = -x^2 + 4x$ define y as a function of x?

c. Does the graph in Figure 1.10 give the price of Home Depot, Inc., stock as a function of the day for three months in 2007?

Figure 1.10

Solution

a. This indicated relationship between x and y is not a function because there can be more than one output for each input. For instance, the rule $y^2 = 3x - 3$ determines both $y = 3$ and $y = -3$ for the input $x = 4$. Note that if we solve this equation for y, we get $y = \pm\sqrt{3x - 3}$, so two values of y will result for any value of $x > 1$. In general, if y raised to an even power is contained in an equation, y cannot be solved for uniquely and thus y cannot be a function of another variable.

b. Because each value of x results in exactly one value of y, this equation defines y as a function of x.

c. The graph in Figure 1.10 gives the stock prices for each business day of three months in 2007 for Home Depot, Inc. The graph shows that the price of a share of Home Depot stock during each day of these months in 2007 is not a function. The vertical bar above each day shows that the stock has many prices between its daily high and low. Because of the fluctuation in price during each day of the month, the price of Home Depot stock during these months is not a function of the day. ∎

Vertical Line Test

Another way to determine whether an equation defines a function is to inspect its graph. If y is a function of x, no two distinct points on the graph of $y = f(x)$ can have the same first coordinate. There are two points, $(4, 3)$ and $(4, -3)$, on the graph of $y^2 = 3x - 3$ shown in Figure 1.11(a), so the equation does not represent y as a function of x (as we concluded in Example 4). In general, no two points of the graph can lie on the same vertical line if the relationship is a function.

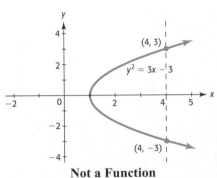

Not a Function

Figure 1.11(a)

Vertical Line Test

A set of points in a coordinate plane is the graph of a function if and only if no vertical line intersects the graph in more than one point.

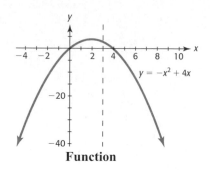

Function

Figure 1.11(b)

When we look at the graph of $y = -x^2 + 4x$ in Figure 1.11(b), we can see that any vertical line will intersect the graph in at most one point for the portion of the graph that is visible. If we know that the graph will extend indefinitely in the same pattern, we can conclude that no vertical line will intersect the graph in two points and so the equation $y = -x^2 + 4x$ represents y as a function of x.

Function Notation

We can use the function notation $y = f(x)$, read "y equals f of x," to indicate that the variable y is a function of the variable x. For specific values of x, $f(x)$ represents the resulting outputs, or y-values. In particular, the point $(a, f(a))$ lies on the graph of $y = f(x)$ for any number a in the domain of the function. We can also say that $f(a)$ is $f(x)$ evaluated at $x = a$.

Thus, if

$$f(x) = 4x^2 - 2x + 3$$

then

$$f(3) = 4(3)^2 - 2(3) + 3 = 33$$
$$f(-1) = 4(-1)^2 - 2(-1) + 3 = 9$$

This means that $(3, 33)$ and $(-1, 9)$ are points on the graph of $f(x) = 4x^2 - 2x + 3$. We can find function values using an equation, values from a table, or points on a graph. For example, because N is a function of x in Table 1.6, we can write $N = f(x)$ and see that $f(1991) = 129.4$ and $f(2000) = 535.6$.

Table 1.6

x Year	N Worldwide Personal Computers (millions)
1991	129.4
1992	150.8
1993	177.4
1994	208.0
1995	245.0
2000	535.6
2005	903.9

(Source: *The Time Almanac*)

| example 5 | **Function Notation** |

Figure 1.12 shows the graph of

$$f(x) = 2x^3 + 5x^2 - 28x - 15$$

a. Use the points shown on the graph to find $f(-2)$ and $f(4)$.

b. Use the equation to find $f(-2)$ and $f(4)$.

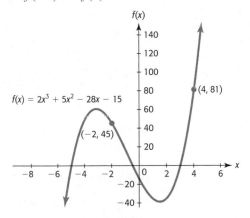

Figure 1.12

Solution

a. By observing Figure 1.12, we see that the point $(-2, 45)$ is on the graph of $f(x) = 2x^3 + 5x^2 - 28x - 15$, so $f(-2) = 45$. We also see that the point $(4, 81)$ is on the graph, so $f(4) = 81$.

b. $f(-2) = 2(-2)^3 + 5(-2)^2 - 28(-2) - 15 = 2(-8) + 5(4) - 28(-2) - 15 = 45$.

$f(4) = 2(4)^3 + 5(4)^2 - 28(4) - 15 = 2(64) + 5(16) - 28(4) - 15 = 81$.

Note that the values found by substitution agree with the y-coordinates of the points on the graph. ■

| example 6 | **Men in the Workforce** |

The points on the graph in Figure 1.13 give the number of men in the workforce (in millions) as a function g of the year for selected years t from 1890 to 2005.

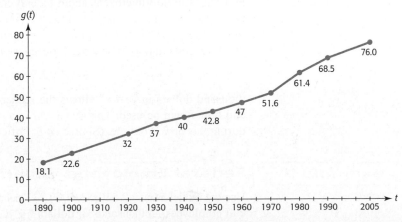

Figure 1.13

(Source: *World Almanac*)

a. Find and interpret $g(1940)$.

b. What is the input t if the output is $g(t) = 51.6$ million men?

c. What can be said about the number of men in the workforce during 1890–2005?

d. What is the maximum number of men in the workforce during the period shown on the graph?

Solution

a. The point above $t = 1940$ has coordinates (1940, 40), so $g(1940) = 40$. This means that there were 40 million men in the workforce in 1940.

b. The point (1970, 51.6) occurs on the graph, so $g(1970) = 51.6$ and $t = 1970$ is the input when the output is $g(t) = 51.6$.

c. The number of men in the workforce increases during the period 1890–2005.

d. The maximum number of men in the workforce during the period is 76.0 million, in 2005.   ∎

Mathematical Models

The process of translating real-world information into a mathematical form so that it can be applied and then interpreted in the real-world setting is called **modeling**. As we most often use it in this text, a **mathematical model** is a functional relationship (usually in the form of an equation) that includes not only the function rule but also descriptions of all involved variables and their units of measure. For example, the function $F = \dfrac{9}{5}C + 32$ was used to describe the relationship between temperature scales at the beginning of the section. The **model** that describes how to convert from one measuring scale to another must include the equation $\left(F = \dfrac{9}{5}C + 32 \right)$ and a description of the variables (F is the temperature measure in degrees Fahrenheit and C is the temperature measure in degrees Celsius). A mathematical model can sometimes provide an exact description of a real situation (such as the Celsius/Fahrenheit model), but a model frequently provides only an approximate description of a real-world situation.

Suppose that the federal government published only the equation

$$S(x) = 2.530x - 5009.231$$

in a report to inform citizens about federal employee salaries. A reader would have no idea what the equation meant or how it could be used; that is, the equation is meaningless without further explanation. A statement such as, "For the years 1998 through 2004, federal employees' average base salary is given by

$$S(x) = 2.530x - 5009.231$$

thousand dollars in year x" allows the reader to understand what the equation means and how it can be used. This statement is a **model** of federal employees' base salary during this period of time. (Source: U.S. Office of Personnel Management)

| example 7 |

Federal Employees' Salaries

Use the model $S(x) = 2.530x - 5009.231$, where $S(x)$ is the average base salary for federal employees, in thousands of dollars for year x from 1998 to 2004, to find the average base salary for federal employees for the years 1999, 2002, and 2004. (Source: U.S. Office of Personnel Management)

Solution

The average base salary for each of these years can be found by evaluating the function at $x = 1999$, 2002, and 2004. Table 1.7 gives the year inputs and the salary outputs that result from substituting each of the years for x in the given equation for $S(x)$.

Table 1.7

Year	1999	2002	2004
Federal Employees Salary ($ thousands)	48.239	55.829	60.889

The table shows that $S(1999) = 48.239$, which gives the base annual salary to be $48,239 in 1999. It also gives base annual salaries $55,829 in 2002 and $60,889 in 2004. ■

skills check

1.1

Use the tables below in Exercises 1–6.

TABLE A

x	-9	-5	-7	6	12	17	20
$y = f(x)$	5	6	7	4	9	9	10

TABLE B

x	-4	-1	0	1	3	7	12
$y = g(x)$	5	7	3	15	8	9	10

1. Table A gives y as a function of x, with $y = f(x)$.

   a. Is -5 an input or an output of this function?

   b. Is $f(-5)$ an input or an output of this function?

   c. State the domain and range of this function.

   d. Explain why this relationship describes y as a function of x.

2. Table B gives y as a function of x, with $y = g(x)$.

   a. Is 0 an input or an output of this function?

   b. Is $g(7)$ an input or an output of this function?

   c. State the domain and range of this function.

   d. Explain why this relationship describes y as a function of x.

3. Use Table A to find $y = f(-9)$ and $y = f(17)$.

4. Use Table B to find $y = g(-4)$ and $y = g(3)$.

5. Does Table A describe x as a function of y? Why or why not?

6. Does Table B describe x as a function of y? Why or why not?

7. For each of the functions $y = f(x)$ described below, find $f(2)$.

   a.

x	-1	0	1	2	3
$f(x)$	5	7	2	-1	-8

   b. $y = 10 - 3x^2$

   c.

   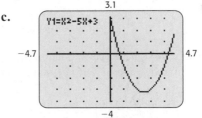

8. For each of the functions $y = f(x)$ described below, find $f(-1)$.

   a.

   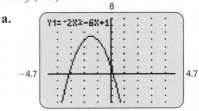

   b.

x	$f(x)$
-3	12
-1	-8
0	5
3	16

   c. $f(x) = x^2 + 3x + 8$

In Exercises 9 and 10, refer to the graph of the function y = f(x) to complete the table.

9.

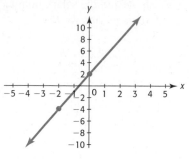

x	y
0	
−2	

10.

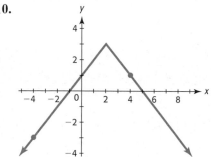

x	y
4	
−4	

11. If $R(x) = 5x + 8$, find (a) $R(-3)$, (b) $R(-1)$, and (c) $R(2)$.

12. If $C(s) = 16 - 2s^2$, find (a) $C(3)$, (b) $C(-2)$, and (c) $C(-1)$.

13. Does the table below define y as a function of x? If so, give the domain and range of f. If not, state why not.

x	−1	0	1	2	3
y	5	7	2	−1	−8

14. Does the table below define y as a function of x? If so, give the domain and range of f. If not, state why not.

x	−3	2	0	3	2
f(x)	12	−3	5	16	4

Determine if each graph in Exercises 15–18 indicates that y is a function of x.

15.

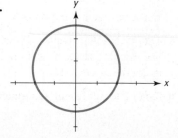

16.

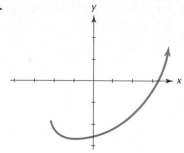

17.

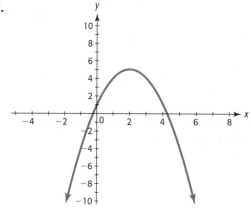

18.

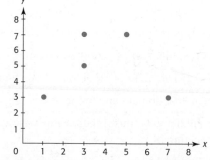

19. Determine if the graph in the figure represents y as a function of x. Explain your reasoning.

20. Determine if the graph below represents y as a function of x. Explain your reasoning.

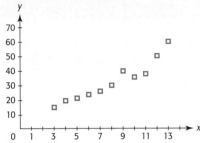

21. Which of the following sets of ordered pairs defines a function?

  a. $\{(1, 6), (4, 12), (4, 8), (3, 3)\}$

  b. $\{(2, 4), (3, -2), (1, -2), (7, 7)\}$

22. Which of the following sets of ordered pairs defines a function?

  a. $\{(1, 3), (-2, 4), (3, 5), (4, 3)\}$

  b. $\{(3, 4), (-2, 5), (4, 6), (3, 6)\}$

23. Which of the following arrow diagrams defines a function?

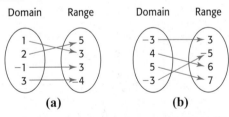

24. Which of the following arrow diagrams defines a function?

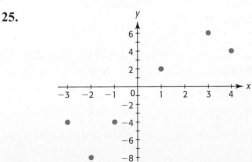

In Exercise 25–28, find the domain and range for the function shown in the graph.

25.

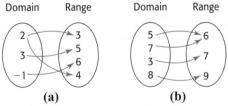

26.

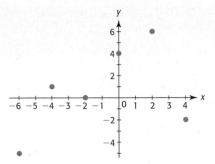

27.

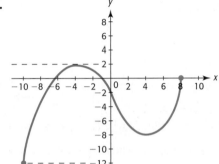

28.

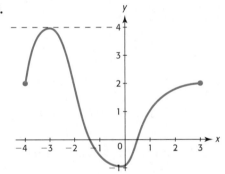

In Exercises 29–32, find the domain of each function.

29. $y = \sqrt{3x - 6}$ **30.** $y = \sqrt{2x - 8}$

31. $y = 2 - \dfrac{5}{x + 4}$ **32.** $y = 4 + \dfrac{8}{2x - 6}$

33. Does $x^2 + y^2 = 4$ describe y as a function of x?

34. Does $x^2 + 2y = 9$ describe y as a function of x?

35. Write an equation to represent the function described by the statement: "The circumference C of a circle is found by multiplying 2π times the radius r."

36. Write a verbal statement to represent the function $D = 3E^2 - 5$.

exercises

1.1

In Exercises 37–41, determine whether the given relationship defines a function. If so, identify the independent and dependent variable, and why the relationship is a function.

37. *Stock Prices*

   a. The price p at which IBM stock can be bought on a given day x.

   b. The closing price p of IBM stock on a given day x.

38. *Odometer*

   a. The odometer reading s when m miles are traveled.

   b. The miles traveled m when the odometer reads s.

39. *Life Insurance*

   a. The monthly premium p for a \$100,000 life insurance policy determined by age a for males aged 26–32 (as shown in the table).

   b. Ages a that get a \$100,000 life insurance policy for a monthly premium of \$11.81 (from the table).

Age (years)	26	27	28	29	30	31	32
Premium (dollars per month)	11.72	11.81	11.81	11.81	11.81	11.81	11.81

40. *Groceries* The average price p of iceberg lettuce in U.S. cities, in cents per pound, for selected months m during 2009, defined by the table.

Month	Average Price (cents per pound)
1	186
2	197
3	195
4	189
5	185

(Source: Bureau of Labor Statistics)

41. *Income* The average income I of male workers with given years y of education, as defined by the graph in the figure.

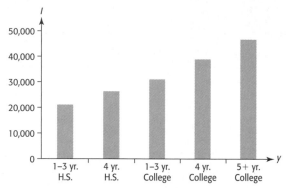

(Source: South Carolina Statistical Abstract)

42. *Temperature* The figure below shows the graph of $T = 0.43m + 76.8$, which gives the temperature T (in degrees Fahrenheit) inside a concert hall m minutes after a 40-minute power outage during a summer rock concert. Stating what the variables m and T represent, explain why T is a function of m.

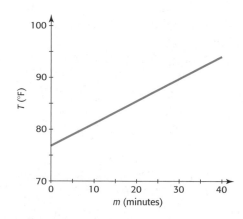

43. *Barcodes* A grocery scanner at Safeway connects the barcode number on a grocery item with the corresponding price.

   a. Is the price a function of the barcode? Explain.

   b. Is the barcode a function of the price? Explain.

44. *Piano* A child's piano has 12 keys, each of which corresponds to a note.

   a. Is the note sounded a function of the key pressed? If it is, how many elements are in the domain of the function?

   b. Is the key pressed a function of the note desired? If it is, how many elements are in the range of the function?

45. *Seawater Pressure* In seawater, the pressure p is related to the depth d according to the model

$$p = \frac{18d + 496}{33}$$

where d is the depth in feet and p is in pounds per square inch. Is p a function of d? Why or why not?

46. *Depreciation* A business property valued at $300,000 is depreciated over 30 years by the straight-line method, so that its value x years after the depreciation began is

$$V = 300{,}000 - 10{,}000x$$

Explain why the value of the property is a function of the number of years.

47. *Weight* During the first two weeks in May, a man weighs himself daily (in pounds) and records the data in the table below.

   a. Does the table define weight as a function of the day in May?

   b. What is the domain of this function?

   c. What is the range?

   d. During what day(s) did he weigh most?

   e. During what day(s) did he weigh least?

   f. He claimed to be on a diet. What is the longest period of time during which his weight decreased?

May	1	2	3	4	5	6	7
Weight (lb)	178	177	178	177	176	176	175

May	8	9	10	11	12	13	14
Weight (lb)	176	175	174	173	173	172	171

48. *Test Scores*
   a. Is the average score on the final exam in a course a function of the average score on a placement test for the course, if the table below defines the relationship?

   b. Is the average score on a placement test a function of the average score on the final exam in a course, if the table below defines the relationship?

Average Score on Math Placement Test (%)	81	75	60	90	75
Average Score on Final Exam in Algebra Course (%)	86	70	58	95	81

49. *Car Financing* A couple wants to buy a $35,000 car and can borrow the money for the purchase at 8%, paying it off in 3, 4, or 5 years. The table below gives the monthly payment and total cost of the purchase (including the loan) for each of the payment plans.

t (Years)	Monthly Payment $	Total Cost $
3	1096.78	39,484.08
4	854.46	41,014.08
5	709.68	42,580.80

(Source: Sky Financial Mortgage Tables)

Suppose that when the payment is over t years, $P(t)$ represents the monthly payment and $C(t)$ represents the total cost for the car and loan.

   a. Find $P(3)$ and write a sentence that explains its meaning.

   b. What is the total cost of the purchase if it is financed over 5 years? Write the answer using function notation.

   c. What is t if $C(t) = 41{,}014.08$?

   d. How much money will the couple save if they finance the car for 3 years rather than 5 years?

50. *Mortgage* A couple can afford $800 per month to purchase a home. As indicated in the table on page 24, if they can get an interest rate of 7.5%, the number of years t that it will take to pay off the mortgage is a function of the dollar amount A of the mortgage for the home they purchase.

Amount A ($)	t (years)
40,000	5
69,000	10
89,000	15
103,000	20
120,000	30

(Source: Comprehensive Mortgage Tables [Publication No. 492]. Financial Publishing Co.)

a. If the couple wishes to finance $103,000, for how long must they make payments? Write this correspondence in function form if $t = f(A)$.

b. What is $f(120,000)$? Write a sentence that explains its meaning.

c. What is $f(3 \cdot 40,000)$?

d. What value of A makes $f(A) = 5$ true?

e. Does $f(3 \cdot 40,000) = 3 \cdot f(40,000)$? Explain your reasoning.

51. *Working Age* The projected ratio of the working-age population (25- to 64-year-olds) to the elderly shown in the figure below defines the ratio as a function of the year shown. If this function is defined as $y = f(t)$ where t is the year, use the graph to answer the following:

a. What is the projected ratio of the working-age population to the elderly population in 2005?

b. Estimate $f(2005)$ and write a sentence that explains its meaning.

c. What is the domain of this function?

d. Is the projected ratio of the working-age population to the elderly increasing or decreasing over the domain shown in the figure?

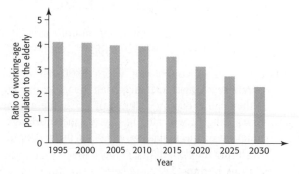

(Source: *Newsweek*)

52. *Women in the Workforce* From 1930 to 2005, the number of women in the workforce increased steadily. The points on the figure below give the number (in millions) of women in the workforce as a function f of the year for selected years.

a. Approximately how many women were in the workforce in 1960?

b. Estimate $f(1930)$ and write a sentence that explains its meaning.

c. What is the domain of this function if we consider only the indicated points?

d. How does the graph reflect the statement that the number of women in the workforce increased?

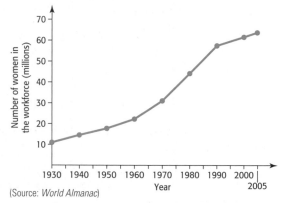

(Source: *World Almanac*)

53. *Gun Crime* The table below gives the total number of nonfatal firearm incidents (crimes) as a function f of the year t.

a. What is $f(2005)$?

b. Interpret the result from (a).

c. What is the maximum number of firearm crimes during this period of time? In what year did it occur?

Year	Firearm Incidents	Year	Firearm Incidents
1993	1,054,820	2000	428,670
1994	1,060,800	2001	467,880
1995	902,680	2002	353,880
1996	845,220	2003	366,840
1997	680,900	2004	280,890
1998	557,200	2005	419,640
1999	457,150		

(Source: National Crime Victimization Survey)

54. *Age at First Marriage* The table below gives the U.S. median age at first marriage for selected years from 1890 to 2004.

a. If the function *f* is the median age of first marriage for men, find $f(1890)$ and $f(2004)$.

b. If the function *g* is the median age of first marriage for women, find $g(1940)$ and $g(2000)$.

c. For what value of *x* is $f(x) = 24.7$? Write the result in a sentence.

d. Did $f(x)$ increase or decrease from 1960 to 2004?

Median Age at First Marriage

Year	Men	Women
1890	26.1	22.0
1900	25.9	21.9
1910	25.1	21.6
1920	24.6	21.2
1930	24.3	21.3
1940	24.3	21.5
1950	22.8	20.3
1960	22.8	20.3
1970	23.2	20.8
1980	24.7	22.0
1990	26.1	23.9
2000	26.8	25.1
2004	27.1	25.8

(Source: U.S. Census Bureau)

55. *Internet Use* The following figure gives the number of millions of U.S. homes using the Internet for the years 1996–2003. If the number of millions of U.S. homes is the function $f(x)$, where *x* is years,

a. How many homes used the Internet during 2000?

b. Find $f(2003)$ and explain its meaning.

c. In what year did 26 million homes use the Internet?

d. Is this function increasing or decreasing? What do you think has happened to home use of the Internet since 2003?

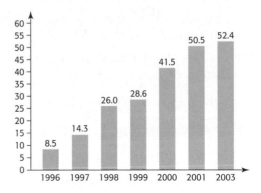

(Source: U.S. Census Bureau)

56. *Farm Workers* The points on the following figure show the percent *p* of U.S. workers in farm occupations for selected years *t*.

a. Is the percent of U.S. workers in farm occupations a function of the year?

b. What is $f(1840)$ if $p = f(t)$?

c. What is *t* if $f(t) = 27$?

d. Write a sentence explaining the meaning of $f(1960) = 6.1$.

e. Describe the change in the percent of U.S. workers in farm occupations.

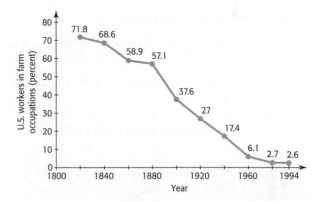

(Source: U.S. Department of Agriculture)

57. *Social Security Funding* Social Security benefits are funded by individuals who are currently employed. The following graph, based on known data and projections into the future, defines a function that gives the number of workers *n* supporting each retiree as a function of time t (given by the calendar year). Denote this function by $n = f(t)$.

a. Find $f(1990)$ and explain its meaning.

b. In what year will the number of workers supporting each retiree equal 2?

c. What does this function tell us about Social Security in the future?

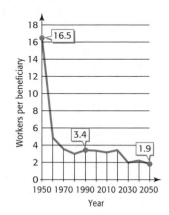

(Source: Social Security Administration)

58. *Teen Pregnancy* The points on the figure below give the number of pregnancies per 1000 U.S. girls (ages 15–19) for the years 1987–2003.

a. Find the output when the input is 1995 and explain its meaning.

b. For what years was the rate 113?

c. During what years did the pregnancy rate increase?

d. For what year is the pregnancy rate at its maximum?

Teen Pregnancy Rate per 1000 Girls

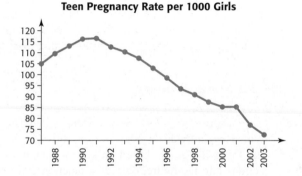

(Source: Guttmacher Institute)

59. *Revenue* The revenue from the sale of specialty golf hats is given by the function $R(x) = 32x$ dollars, where x is the number of hats sold.

a. What is $R(200)$? Interpret this result.

b. What is the revenue from the sale of 2500 hats? Write this in function notation.

60. *Cost* The cost from the production of specialty golf hats is given by the function $C(x) = 4000 + 12x$, where x is the number of hats produced.

a. What is $C(200)$? Interpret this result.

b. What is the cost from the production of 2500 hats? Write this in function notation.

61. *Utilities* An electric utility company determines the monthly bill by charging 76.7 cents per kilowatt-hour (KWh) used plus a base charge of $16.37 per month. Thus the monthly charge is given by the function

$$f(W) = 0.767W + 16.37 \text{ dollars}$$

where W is the number of kilowatt-hours.

a. Find $f(1000)$ and explain what it means.

b. What is the monthly charge if 1500 KWh are used?

62. *Profit* The profit from the production and sale of ipod players is given by the function $P(x) = 450x - 0.1x^2 - 2000$, where x is the number of units produced and sold.

a. What is $P(500)$? Interpret this result.

b. What is the profit from the production of 4000 units? Write this in function notation.

63. *Profit* The daily profit from producing and selling Blue Chief bicycles is given by

$$P(x) = 32x - 0.1x^2 - 1000$$

where x is the number produced and sold and $P(x)$ is in dollars.

a. Find $P(100)$ and explain what it means.

b. Find the daily profit from producing and selling 160 bicycles.

64. *Projectiles* Suppose a ball thrown into the air has its height (in feet) given by the function

$$h(t) = 6 + 96t - 16t^2$$

where t is the number of seconds after the ball is thrown.

a. Find $h(1)$ and explain what it means.

b. Find the height of the ball 3 seconds after it is thrown.

c. Test other values of $h(t)$ to decide if the ball eventually falls. When does the ball stop climbing?

65. *Body-Heat Loss* The description of body-heat loss due to convection involves a coefficient of convection K_c, which depends on wind speed s according to the equation $K_c = 4\sqrt{4s + 1}$.

a. Is K_c a function of s?

b. What is the domain of the function defined by this equation?

c. What restrictions do the physical nature of the model put on the domain?

66. *Test Reliability* If a test that has reliability 0.7 has the number of questions increased by a factor n, the reliability R of the new test is given by

$$R(n) = \frac{0.7n}{0.3 + 0.7n}$$

a. What is the domain of the function defined by this equation?

b. If the application used here requires that the size of the test be increased, what values of n make sense in the application?

67. *Cost-Benefit* Suppose that the cost C (in dollars) of removing $p\%$ of the particulate pollution from the smokestack of a power plant is given by

$$C(p) = \frac{237{,}000p}{100 - p}$$

a. Use the fact that percent is measured between 0 and 100 and find the domain of this function.

b. Evaluate $C(60)$ and $C(90)$.

68. *Demand* Suppose the number of units of a product that is demanded by consumers is given as a function of p by

$$q = \frac{100}{\sqrt{2p + 1}}$$

where p is the price charged per unit.

a. What is the domain of the function defined by this equation?

b. What should the domain and range of this function be to make sense in the application?

69. *Postal Restrictions* Some postal restrictions say that the size of the largest box that can be shipped has its length (longest side) plus its girth (distance around the box in the other two dimensions) equal to 108 inches.

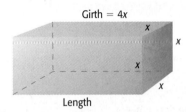

Girth = 4x

Length

If a box has a square cross section that is x inches on each side, then the volume of the box is given by $V = x^2(108 - 4x)$ cubic inches.

a. Find $V(12)$ and $V(18)$.

b. What restrictions must be placed on x to satisfy the conditions of this model?

c. Create a table of function values to investigate the value of x that maximizes the volume. What are the dimensions of the box that has the maximum volume?

70. *Height of a Bullet* The height of a bullet shot into the air is given by $S(t) = -4.9t^2 + 98t + 2$, where t is the number of seconds after it is shot and $S(t)$ is in meters.

a. Find $S(0)$ and interpret it.

b. Find $S(9)$, $S(10)$, and $S(11)$.

c. What appears to be happening to the bullet at 10 seconds? Evaluate the function at some additional times near 10 seconds to confirm your conclusion.

Graphs of Functions

section preview ▪ Personal Savings

Using data from 1960 through 2006, a model can be created that gives the American personal savings rate as a percent. If the data are **aligned** so that the input values (x) represent the number of years after 1960, the personal savings rate can be modeled by the function

$$y = -0.00796x^2 + 0.205x + 7.044$$

(Source: U.S. Census Bureau)

To graph this function for values of x representing 1960 through 2006, we use x-values from 0 through 46 in the function to find the corresponding y-values, which represent percents. These points can be used to sketch the graph of the function. If technology is used to graph the function, the **viewing window** on which the graph is shown can be determined by these values of x and y. (See Example 6.)

In this section, we will graph functions by point plotting and with technology. We will graph application functions on windows determined by the context of the applications, and we will align data so that smaller inputs can be used in models. We will also graph data points.

Graphs of Functions

If an equation defines y as a function of x, we can sketch the graph of the function by plotting enough points to determine the shape of the graph and then drawing a line or curve through the points. This is called the **point-plotting method** of sketching a graph.

| example 1 |

Graphing an Equation by Plotting Points

a. Graph the equation $y = x^2$ by drawing a smooth curve through points determined by integer values of x between 0 and 3.

b. Graph the equation $y = x^2$ by drawing a smooth curve through points determined by integer values of x between -3 and 3.

Solution

a. We use the values in Table 1.8 to determine the points. The graph of the function drawn through these points is shown in Figure 1.14(a).

Table 1.8

x	0	1	2	3
$y = x^2$	0	1	4	9
Points	$(0, 0)$	$(1, 1)$	$(2, 4)$	$(3, 9)$

b. We use Table 1.9 to find the additional points. The graph of the function drawn through these points is shown in Figure 1.14(b).

Table 1.9

x	-3	-2	-1	0	1	2	3
$y = x^2$	9	4	1	0	1	4	9
Points	$(-3, 9)$	$(-2, 4)$	$(-1, 1)$	$(0, 0)$	$(1, 1)$	$(2, 4)$	$(3, 9)$

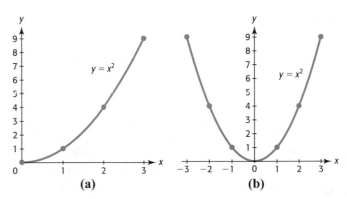

Figure 1.14

Although neither graph in Figure 1.14 shows all of the points satisfying the equation $y = x^2$, the graph in Figure 1.14(b) is a much better representation of the function (as we will learn later). When enough points are connected to determine the shape of the graph, the important parts of the graph, and to suggest what the unseen parts of the graph look like, the graph is called **complete**.

Complete Graph

A graph is a complete graph if it shows the basic shape of the graph and important points on the graph (including points where the graph crosses the axes and points where the graph turns)* and suggests what the unseen portions of the graph will be.

* Points where graphs turn from rising to falling or from falling to rising are called turning points.

example 2

Graphing a Complete Graph

Sketch the complete graph of the equation $f(x) = x^3 - 3x$, using the fact that the graph has at most two "turning points."

Solution

We use the values in Table 1.10 to determine some points. The graph of the function drawn through these points is shown in Figure 1.15.

Table 1.10

x	-3	-2	-1	0	1	2	3
$y = x^3 - 3x$	-18	-2	2	0	-2	2	18
Points	$(-3, -18)$	$(-2, -2)$	$(-1, 2)$	$(0, 0)$	$(1, -2)$	$(2, 2)$	$(3, 18)$

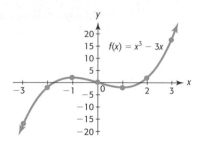

Figure 1.15

Note that this graph has two turning points, so it is a complete graph.

Being able to draw complete graphs, with or without the aid of technology, improves with experience and requires knowledge of the shape of the basic function(s) involved in the graph. As we learn more about types of functions, we will be better able to determine when a graph is complete.

Graphing with Technology

Obtaining a sufficient number of points to sketch a graph by hand can be time consuming. Computers and graphing calculators have graphing utilities that can be used to plot many points quickly, thereby showing the graph with minimal effort. However, graphing with technology involves much more than pushing a few buttons. Unless you know what to enter into the calculator or computer (that is, the domain of the function and sometimes the range), you cannot see or use the graph that is drawn. We suggest the following steps for graphing most functions with a graphing utility.

> ### Using a Graphing Calculator to Draw a Graph
>
> 1. Write the function with x representing the independent variable and y representing the dependent variable. Solve for y, if necessary.
>
> 2. Enter the function in the equation editor of the graphing utility. Use parentheses* as needed to ensure the mathematical correctness of the expression.
>
> 3. Activate the graph by pressing the $\boxed{\text{ZOOM}}$ or $\boxed{\text{GRAPH}}$ key. Most graphing utilities have several preset **viewing windows** under $\boxed{\text{ZOOM}}$, including the **standard viewing window** that gives the graph of a function on a coordinate system in which the x-values range from -10 to 10 and the y-values range from -10 to 10.[†]
>
> 4. To see parts of the graph of a function other than those shown in a standard window, set the x- and y-boundaries of the viewing window before pressing $\boxed{\text{GRAPH}}$. Viewing window boundaries are discussed below and in the technology supplement. As you gain more knowledge of graphs of functions, determining viewing windows that give complete graphs will be less complicated.

---

* Parentheses should be placed around numerators and/or denominators of fractions, fractions that are multiplied by a variable, exponents consisting of more than one symbol, and in other situations where the order of operations needs to be indicated.

[†] For more information on viewing windows see Appendix A, page 654.

Although the standard viewing window is a convenient window to use, it may not show the desired graph. To see parts of the graph of a function other than those that might be shown in a standard window, we change the x- and y-boundaries of the viewing window. The values that define the viewing window can be set manually or by using the ZOOM key. The boundaries of a viewing window are:

x_{min}: the smallest value on the x-axis (the left boundary of the window)

y_{min}: the smallest value on the y-axis (the bottom boundary of the window)

x_{max}: the largest value on the x-axis (the right boundary of the window)

y_{max}: the largest value on the y-axis (the top boundary of the window)

x_{scl}: the distance between ticks on the x-axis (helps visually find x-intercepts and other points)

y_{scl}: the distance between ticks on the y-axis (helps visually find y-intercept and other points)

When showing viewing window boundaries on calculator graphs in this text, we write them in the form:

$$[x_{min}, x_{max}] \text{ by } [y_{min}, y_{max}]$$

example 3

Graphing a Complete Graph

Sketch the graph of $y = x^3 - 3x^2 - 13$

a. using the standard window.

b. using the window $x_{min} = -10, x_{max} = 10, y_{min} = -25, y_{max} = 10$.

Which graph gives a better view of the graph of the function?

Solution

a. A graph of this function in the standard window appears to be a line (Figure 1.16(a)).

b. By setting the window with $x_{min} = -10, x_{max} = 10, y_{min} = -25, y_{max} = 10$, we obtain the graph in Figure 1.16(b). This window gives a better view of the graph of this equation. As we learn more about functions, we will see that the graph in Figure 1.16(b) is a complete graph of this function.

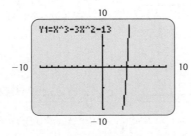

viewing window: $[-10, 10]$ by $[-10, 10]$

This graph looks like a line.

(a)

viewing window: $[-10, 10]$ by $[-25, 10]$

This is a better graph.

(b)

Figure 1.16

Different viewing windows give different views of a graph. There are usually many different viewing windows that give complete graphs for a particular function, but some viewing windows do not show all the important parts of a graph.

| example 4 |

Cost-Benefit

Suppose that the cost C of removing $p\%$ of the pollution from drinking water is given by the model

$$C = \frac{5350p}{100 - p} \text{ dollars}$$

a. Use the restriction on p to determine the limitations on the horizontal-axis values (which are the x-values on a calculator).

b. Graph the function on the viewing window [0, 100] by [0, 50,000]. Why is it reasonable to graph this model on a viewing window with the limitation $C \geq 0$?

c. Find the point on the graph that corresponds to $p = 90$. Interpret the coordinates of this point.

Solution

a. Because p represents the percent of pollution removed, it is limited to values from 0 to 100. However, $p = 100$ makes C undefined in this model, so p is restricted to $0 \leq p < 100$ for this model.

b. The graph of the function is shown in Figure 1.17(a), with x representing p and y representing C. The interval [0, 100) contains all the possible values of p. The value of C is bounded below by 0 because it represents the cost of removing the pollution, which cannot be negative.

c. We can find (or estimate) the output of a function $y = f(x)$ at specific inputs with a graphing utility. We do this with TRACE or TABLE .* Using TRACE with the x-value 90 gives the point (90, 48,150) (see Figure 1.17(b)). Because the values of p are represented by x-values and the values of C are represented by y, the coordinates of the point tell us that the cost of removing 90% of the pollution from the drinking water is $48,150. Figure 1.17(c) shows the value of y for $x = 90$ and other values in a calculator table.

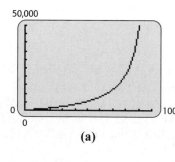

(a)

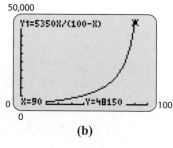

(b)

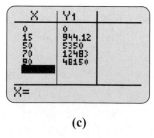

(c)

Figure 1.17

---

* For more details, see Appendix A, page 655.

| spreadsheet solution |

We have briefly described how to use a graphing calculator to graph. Computer software of several types can be used to create more accurate and better-looking graphs. Software such as Scientific Notebook, Maple, and Mathematica can be used to create graphs, and **spreadsheets** like Excel can also be used.

Table 1.11 shows a spreadsheet with inputs and outputs for the function $C = \dfrac{5350p}{100 - p}$ given in Example 4. When evaluating a function with a spreadsheet, we use the cell location of the data to represent the variable. Thus to evaluate $C = \dfrac{5350p}{100 - p}$ at $p = 0$, we type 0 in cell A2 and $= 5350*A2/(100 - A2)$ in cell B2. Typing 10 in cell A3 and using the fill-down capacity of the spreadsheet gives the outputs of this function for $p = 0$ to $p = 90$ in increments of 10 (see Table 1.11). These values can be used to create the graph of this function on Excel (Figure 1.18).*

Table 1.11

	A	B
1	p	C = 5350p/(100 − p)
2	0	0
3	10	594.4444
4	20	1337.5
5	30	2292.857
6	40	3566.667
7	50	5350
8	60	8025
9	70	12483.33
10	80	21400
11	90	48150

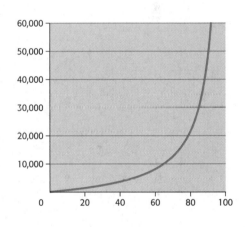

Figure 1.18

Aligning Data

When finding a model to fit a set of data, it is often easier to use aligned inputs rather than the actual data values. **Aligned inputs** are simply input values that have been converted to smaller numbers by subtracting the same number from each input. For instance, instead of using x as the actual year in the following example, it is more convenient to use an aligned input that is the number of years after 1950.

---

* See Excel guide, Appendix B, pages 674–679 for details.

| example 5 | **Voting** |

Between 1950 and 2004, the percent of the voting population who voted in presidential elections (during election years) is given by

$$f(x) = 63.20 - 0.26x$$

where x is the number of years after 1950. That is, the input variable $x = 0$ represents 1950. (Source: Federal Election Commission)

a. What are the values of x that correspond to the years 1960 and 2004?

b. Find $f(10)$ and explain its meaning.

c. If this model is accurate for 2000, find the percent of the voting population who voted in the 2000 presidential election.

Solution

a. Because 1960 is 10 years after 1950, $x = 10$ corresponds to the year 1960 in the aligned data. Using similar reasoning, 2004 corresponds to $x = 54$.

b. An input value of 10 represents 10 years after 1950, which is the year 1960. To find $f(10)$, we substitute 10 for x to obtain $f(10) = 63.20 - 0.26(10) = 60.6$. One possible explanation of this answer is this: Approximately 61% of the voting population voted in the 1960 presidential election.

c. To find the percent in 2000, we first calculate the aligned input for the function. Because 2000 is 50 years after 1950, we use $x = 50$. Thus, we find

$$f(50) = 63.20 - 0.26(50) = 50.2$$

So, if this model is accurate for 2000, approximately 50% of the voting population voted in the 2000 presidential election.   ■

Determining Viewing Windows

Finding the functional values (y-values) for selected inputs (x-values) can be useful when setting viewing windows for graphing utilities.

Technology Note

Once the input values for a viewing window have been selected, TRACE or TABLE can be used to find enough output values to determine a y-view that gives a complete graph.*

| example 6 | **Personal Savings** |

Using data from 1960 to 2006, the personal savings rate (as a percent) of Americans can be modeled by the function

$$y = -0.00796x^2 + 0.205x + 7.044$$

where x is the number of years after 1960. (Source: U.S. Census Bureau)

---

* It is occasionally necessary to make more than one attempt to find a window that gives a complete graph.

a. Choose an appropriate window and graph the function with a graphing calculator.

b. Use the model to estimate the personal savings rate in 2006. Is this possible?

c. Use the graph to estimate the year in which the personal savings rate is a maximum.

Solution

a. The viewing window should include values of x that are equivalent to the years 1960 to 2006, so the x-view should include the aligned values $x = 0$ to $x = 46$. The y-view should include the outputs obtained from the inputs between 0 and 46. We can use $\boxed{\text{TRACE}}$ or $\boxed{\text{TABLE}}$ with some or all of the integers 0 through 46 to find corresponding y-values (regardless of how the y-view is set when we are evaluating). These evaluations indicate that the y-view should include values from about -1 through about 9. Choosing the window $x_{\min} = 0, x_{\max} = 50, y_{\min} = -2$, and $y_{\max} = 10$ gives the graph of the function shown in Figure 1.19(a).

b. The year 2006 is represented by $x = 46$. Evaluating the function with $\boxed{\text{TRACE}}$ on the graph (see Figure 1.19(b)) gives an output of approximately -0.4 when $x = 46$. Figure 1.19(c) shows the values of the function at $x = 0, 40$, and 46 using $\boxed{\text{TABLE}}$. Thus, we estimate that the personal savings rate in 2006 was -0.4%. The personal savings rate of -0.4% is possible. In fact, the actual personal savings rate was -0.4% in 2005 and it was -1.1% in 2006. This means that the average American has borrowed more than he or she has saved in these years.

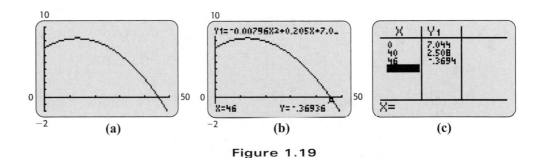

Figure 1.19

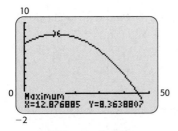

Figure 1.20

c. We can use the "maximum" feature located under $\boxed{\text{2ND}}$ $\boxed{\text{TRACE}}$ to find the maximum point on the graph.* Figure 1.20 shows that the maximum point is (12.88, 8.36). Because x represents the number of years after 1960, an x-value of 12.88 means either the year 1972 or 1973. If we assume data are collected and reported at the end of the year[†], we will round to the year 1973. Thus, the maximum personal savings rate was 8.4% in 1973 (see Figure 1.20). ◼

---

* For more information on computing minimum and maximum points, see the Calculator Guide in Appendix A, page 657.
† This topic will be explored further in Section 2.2.

┌───┐
│ **spreadsheet solution**

Excel can also be used to graph and evaluate the function $y = -0.00796x^2 + 0.205x + 7.044$, discussed in Example 6. The Excel graph is shown in Figure 1.21, and an Excel spreadsheet with values of the function at selected values of x is shown in Table 1.12.
└───┘

Table 1.12

	A	B
1	x	f(x)
2	36	4.10784
3	37	3.73176
4	38	3.33976
5	39	2.93184
6	40	2.508
7	41	2.06824
8	42	1.61256
9	43	1.14096
10	44	0.65344
11	45	0.15
12	46	−0.36936

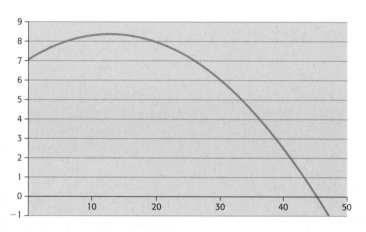

Figure 1.21

Graphing Data Points

Graphing utilities can also be used to create lists of numbers and to create graphs of the data stored in the lists. To see how lists and scatter plots are created, consider the following example.

---

example 7

U.S Executions

Table 1.13 gives the number of executions in the United States for selected years from 1984 to 2005. (Source: "The Death Penalty in the U.S.," www.clarkprosecuter.org)

Table 1.13

Year	Number of Executions
1984	21
1987	25
1990	23
1993	38
1996	45
1999	98
2002	71
2005	60

a. Align the data so that $x =$ the number of years after 1980, and enter these x-values in list L1. Enter the number of executions in L2.

b. Use a graphing command* to create the scatter plot of these data points.

Solution

a. Figure 1.22(a) shows the first 7 aligned inputs in L1 and the first 7 outputs from the data in L2.

b. The scatter plot is shown in Figure 1.22(b), with each entry in L1 represented by an x-coordinate of a point on the graph and the corresponding entry in L2 represented

---

* Many calculators have a $\boxed{\text{STAT PLOT}}$ command, and Excel has an $\boxed{\text{XY}}$ (Scatter) command. See Appendix A, page 653, and Appendix B, page 682.

by the *y*-coordinate of that point on the graph. The window can be set automatically or manually to include *x*-values from L1 and *y*-values from L2.

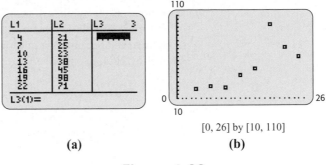

[0, 26] by [10, 110]

(a)   (b)

Figure 1.22

---

spreadsheet solution

Figure 1.23 shows the values from Table 1.13 in an Excel spreadsheet and the scatter plot (graph) of the data.

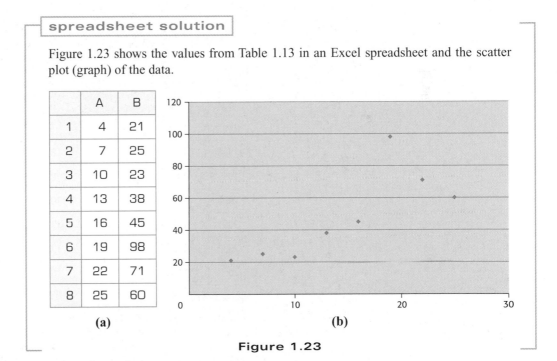

(a)   (b)

Figure 1.23

---

example 8

High School Enrollment

Table 1.14 on the next page shows the enrollment (in thousands) in grades 9–12 of U.S. public and private schools for the years 1990–2003.

a. According to the table, what was the enrollment in 1998?

b. Create a new table with *x* representing the number of years after 1990 and *y* representing the enrollment in millions.

c. According to this new table, what was the enrollment in 1998?

d. Use a graphing utility to graph the new data as a scatter plot.

Solution

a. According to the table, the enrollment in 1998 was 14,428 thousand, or 14,428,000. (Multiply by 1000 to change the number of thousands to a number in standard form.)

b. If x represents the number of years after 1990, we must subtract 1990 from each year in the original table to obtain x. The enrollment in the original table is given in thousands, and because one million is 1000 times one thousand, we must divide each number by 1000 to obtain y. The result is shown in Table 1.15.

Table 1.14

Year	Enrollment (thousands)
1990	12,488
1991	12,703
1992	12,882
1993	13,093
1994	13,376
1995	13,697
1996	14,060
1997	14,272
1998	14,428
1999	14,623
2000	14,802
2001	15,058
2002	15,426
2003	15,723

(Source: U.S. Department of Education)

Table 1.15

Number of Years after 1990, x	Enrollment, y (millions)
0	12.488
1	12.703
2	12.882
3	13.093
4	13.376
5	13.697
6	14.060
7	14.272
8	14.428
9	14.623
10	14.802
11	15.058
12	15.426
13	15.723

c. Because x represents the number of years after 1990, we subtract $1998 - 1990$ to obtain an x-value of 8. Table 1.15 indicated that in year 8 the population was 14.428 million, or (multiplying by 1000) 14,428 thousand. Note that this is the same result as in part (a).

d. A scatter plot of the data is shown in Figure 1.24.

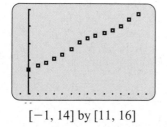

[−1, 14] by [11, 16]

Figure 1.24

Alignment and Scaling of Data

Recall that when we replace the input with x representing the number of years after a certain year, it is called *alignment of the data*. When we replace the outputs by dividing or multiplying each number by a given number, it is called **scaling the data**. In Example 8, we aligned the data by replacing the years with x representing the number of years after 1990, and we scaled the data by replacing the enrollment in thousands with y representing the enrollment in millions.

skills check

1.2

1. a. Complete the table of values for the function $y = x^3$, plot the values of x and y as points in a coordinate plane, and draw a smooth curve through the points.

x	-3	-2	-1	0	1	2	3
y							

b. Graph the function $y = x^3$ with a graphing calculator using the viewing window $x_{min} = -4$, $x_{max} = 4, y_{min} = -30, y_{max} = 30$.

c. Compare the graphs in parts (a) and (b).

2. a. Complete the table of values for the function $y = 2x^2 + 1$, plot the values of x and y as points in a coordinate plane, and draw a smooth curve through the points.

x	-3	-2	-1	0	1	2	3
y							

b. Graph the function $y = 2x^2 + 1$ with a graphing calculator using the viewing window $x_{min} = -5$, $x_{max} = 5, y_{min} = -1, y_{max} = 20$.

c. Compare the graphs in parts (a) and (b).

In Exercises 3–8, make a table of values for each function and then plot the points to graph the function by hand.

3. $f(x) = 3x - 1$ **4.** $f(x) = 2x - 5$

5. $f(x) = \dfrac{1}{2}x^2$ **6.** $f(x) = 3x^2$

7. $f(x) = \dfrac{1}{x - 2}$ **8.** $f(x) = \dfrac{x}{x + 3}$

In Exercises 9–14, graph the functions with a graphing calculator using a standard viewing window.

9. $y = x^2 - 5$ **10.** $y = 4 - x^2$

11. $y = x^3 - 3x^2$ **12.** $y = x^3 - 3x^2 + 4$

13. $y = \dfrac{9}{x^2 + 1}$ **14.** $y = \dfrac{40}{x^2 + 4}$

For Exercises 15–18, graph the given function with a graphing calculator using (a) the standard viewing window and (b) the specified window. Which window gives a better view of the graph of the function?

15. $y = x + 20$ with $x_{min} = -10$, $x_{max} = 10$, $y_{min} = -10, y_{max} = 30$

16. $y = x^3 - 3x + 13$ with $x_{min} = -5$, $x_{max} = 5$, $y_{min} = -10, y_{max} = 30$

17. $y = \dfrac{0.04(x - 0.1)}{x^2 + 300}$ on $[-20, 20]$ by $[-0.002, 0.002]$

18. $y = -x^2 + 20x - 20$ on $[-10, 20]$ by $[-20, 90]$

For Exercises 19–22, find an appropriate viewing window for the function, using the given x-values. Then graph the function.

19. $y = x^2 + 50$, for x-values between -3 and 3.

20. $y = x^2 + 60x + 30$, for values of x between -60 and 0.

21. $y = x^3 + 3x^2 - 45x$, for values of x between -10 and 10.

22. $y = (x - 28)^3$, for x-values between 25 and 31.

23. Find a complete graph of $y = 10x^2 - 90x + 300$. (A complete graph of this function shows one turning point.)

24. Find a complete graph of $y = -x^2 + 34x - 120$. (A complete graph of this function shows one turning point.)

25. Use a calculator or a spreadsheet to find $S(t)$ for the values of t given in the following table.

t	$S(t) = 5.2t - 10.5$
12	
16	
28	
43	

26. Use a calculator or a spreadsheet to find $f(q)$ for the values of q given in the following table.

q	$f(q) = 3q^2 - 5q + 8$
-8	
-5	
24	
43	

27. Enter the data into lists and graph the scatter plot of the data, using the window $[0, 110]$ by $[0, 600]$.

x	20	30	40	50	60	70	80	90	100
y	500	320	276	80	350	270	120	225	250

28. Enter the data below into lists on a graphing utility and graph the scatter plot of the data, using the window $[-20, 120]$ by $[0, 80]$.

x	-15	-1	25	40	71	110	116
y	12	16	10	32	43	62	74

29. Use the table below in parts (a)–(c).

x	-3	-1	1	5	7	9	11
y	-42	-18	6	54	78	102	126

a. Use a graphing utility to graph the points from the table.

b. Use a graphing utility to graph the equation $f(x) = 12x - 6$ on the same set of axes as the data in part (a).

c. Do the points fit on the graph of the equation? Do the respective values of x and y in the table satisfy the equation in part (b)?

30. Use the table below in parts (a)–(c).

x	1	3	6	8	10
y	-10	0	15	25	35

a. Use a graphing utility to graph the points from the table.

b. Use a graphing utility to graph the equation $f(x) = 5x - 15$ on the same set of axes as the data in part (a).

c. Do the points fit on the graph of the equation? Do the respective values of x and y in the table satisfy the equation in part (b)?

31. *Suppose* $f(x) = x^2 - 5x$ million dollars are earned, where x is the number of years after 2000.

a. What is $f(20)$?

b. The answer to part (a) gives the number of millions of dollars earned for what year?

32. *Suppose* $f(x) = 100x^2 - 5x$ thousand units are produced, where x is the number of years after 2000.

a. What is $f(10)$?

b. How many units are produced in 2010, according to this function?

exercises

1.2

33. *Women in the Workforce* The number y (in thousands) of women in the workforce is given by the function

$$y = 5.74x^2 - 17.04x + 600.99$$

where x is the number of years after 1900.

a. Find the value of y when $x = 44$. Explain what this means.

b. Use the model to find the number of women in the workforce in 2010.
(Source: U.S. Census Bureau, U.S. Dept of Commerce)

34. *Welfare Cases* The average number of welfare cases in Niagara, Canada, is given by the model $y = -112x^2 - 107x + 15,056$, where x is the number of years after 1990.

a. What are the values of x that correspond to the years 1994 and 1998?

b. Find the value of y when $x = 8$. Explain what this means.

c. How many welfare cases were there in 1995, according to this model?
(Source: Regional Niagara Social Services Dept, April 25, 2000)

35. *Internet Access* The function $P = 6.9t - 3.18$ gives the percent of households with Internet access as a function of t, the number of years after 1995.

a. What are the values of t that correspond to the years 1996 and 2014?

b. $P = f(10)$ gives the value of P for what year? What is $f(10)$?

c. What x_{min} and x_{max} should be used to set the viewing window so that t represents 1995–2015?

36. *State Lotteries* The cost of prizes and expenses of state lotteries is given by $P = 35t^2 + 740t + 1207$ million dollars, with t equal to the number of years after 1980.

a. What are the values of t that correspond to the years 1982, 1988, and 2000?

b. $P = f(4)$ gives the value of P for what year? What is $f(4)$?

c. What x_{min} and x_{max} should be used to set a viewing window so that t represents 1980–1997?

37. *Height of a Ball* If a ball is thrown into the air at 64 feet per second from the top of a 100-foot-tall building, its height can be modeled by the function $S = 100 + 64t - 16t^2$, where S is in feet and t is in seconds.

a. Graph this function on a viewing window [0, 6] by [0, 200].

b. Find the height of the ball 1 second after it is thrown and 3 seconds after it is thrown. How can these values be equal?

c. Find the maximum height the ball will reach.

38. *Depreciation* A business property valued at $600,000 is depreciated over 30 years by the straight-line method, so that its value x years after the depreciation began is

$$V = 600,000 - 20,000x$$

a. Graph this function on a viewing window [0, 30] by [0, 600,000].

b. What is the value 10 years after the depreciation is started?

39. *Earnings and Gender* A model that relates the median annual salary (in thousands of dollars) of females F, and males M, in the United States is given by $F = 0.78M + 1.32$.

a. Use a graphing utility to graph this function on the viewing window [0, 100] by [0, 80].

b. Use the graphing utility to find the median female salary that corresponds to a male salary of $63,000.
(Source: U.S. Census Bureau)

40. *Education Spending* Federal spending (in billions of dollars) for public education during the years 1996–2001 can be modeled by the function $S = 3.32x + 23.16$, where x is the number of years after 1990.

a. Use a graphing utility to graph this function on the viewing window [0, 11] by [0, 60].

b. Find the federal spending for public education in 2001.
(Source: U.S. Dept of Education)

41. *Medical School* The number of students (in thousands) of osteopathic medicine in the United States can be described by

$$S = 0.027t^2 - 4.85t + 218.93$$

where t is the number of years after 1980.

a. Graph this function on the viewing window [0, 17] by [0, 300].

b. Use technology to find S when t is 15.

c. Use the model to estimate the number of osteopathic students in 2005.
(Source: *Statistical Abstract of the United States*)

42. *State Lotteries* The cost (in millions of dollars) of prizes and expenses for state lotteries can be described by $L = 35.3t^2 + 740.2t + 1207.2$, where t is the number of years after 1980.

a. Graph this function on the viewing window [0, 27] by [1200, 45,000].

b. Use technology to find L when t is 26.

c. What was the cost of prizes and expenses for state lotteries in 2006?
(Source: *Statistical Abstract of the United States*)

43. *Crime* The rate (number per 100,000 people) of juvenile arrests for violent crimes is given by

$$f(x) = -2.01x^2 + 54.84x + 87.31$$

where x is the number of years after 1980.
(Source: Federal Bureau of Investigation)

a. Use technology to graph this model on the viewing window [0, 25] by [0, 550].

b. This viewing window shows the graph for what time period?

c. Did the number of juvenile arrests per 100,000 people increase or decrease after 1994?

d. Use the model to find the number of juvenile arrests per 100,000 in 1994 and in 2005. Is the number increasing or decreasing over this period?

e. Is your answer to part (c) consistent with the answer to part (d)?

44. *Tax Burden* Using data from the Internal Revenue Service, the per capita tax burden B (in hundreds of dollars) can be described by $B(t) = 17.69 + 2.25t$, where t is the number of years after 1980.

a. Graph this function with technology using a viewing window with $t \geq 0$ and $B(t) \geq 0$.

b. Did the tax burden increase or decrease?
(Source: Internal Revenue Service)

45. *Cost* Suppose the cost of the production and sale of x Electra dishwashers is $C(x) = 15,000 + 100x + 0.1x^2$ dollars. Graph this function on a viewing window with x between 0 and 50.

46. *Revenue* Suppose the revenue from the sale of x coffee makers is given by $R(x) = 52x - 0.1x^2$. Graph this function on a viewing window with x between 0 and 100.

47. *Profit* The profit from the production and sale of x laser printers is given by the function $P(x) = 200x - 0.01x^2 - 5000$, where x is the number of units produced and sold. Graph this function on a viewing window with x between 0 and 1000.

48. *Profit* The profit from the production and sale of x digital cameras is given by the function $P(x) = 1500x - 8000 - 0.01x^2$, where x is the number of units produced and sold. Graph this function on a viewing window with x between 0 and 500.

49. *Teacher Salaries* The average U.S. classroom teacher salary is given by $f(t) = 982.06t + 32,903.77$, where t is the number of years from 1990.

a. What inputs correspond to the years 1990–2005?

b. What outputs correspond to the inputs determined for 1990 and 2005?

c. Use the answers to parts (a) and (b) and the fact that the function increases to find an appropriate viewing window and graph this function.
(Source: www.ors2.state.sc.us/abstract)

50. *Cocaine Use* The percent of high school seniors during the years 1975–2005 who have ever used cocaine can be described by

$$y = 0.0035x^3 - 0.1666x^2 + 1.8689x + 9.5782$$

where x is the number of years after 1975.

a. What inputs correspond to the years 1975 through 2005?

b. What outputs for y could be used to estimate the percent of seniors who have ever used cocaine?

c. Based on your answers to parts (a) and (b), choose an appropriate window and graph the equation on a graphing utility.

d. Graph the function again with a new window that gives a graph nearer the center of the screen.

e. Use this function to estimate the percent in 2009.
(Source: monitoringthefuture.org)

51. *U.S. Population* The projected population of the United States for selected years from 2000 to 2060 is shown in the table below, with the population given in millions.

a. According to this table, what will be the U.S. population in 2010?

b. Create a new table with x representing the number of years after 2000 and y representing the number of millions.

c. Use a graphing utility to graph the data from the new table as a scatter plot.

Year	Population (millions)
2000	275.3
2010	299.9
2020	324.9
2030	351.1
2040	377.4
2050	403.7
2060	432.0

(Source: U.S. Census Bureau)

52. *Runway Near-Hits* The number of runway near-hits by airplanes from 1990 to 2000 is shown in the following table. Use a graphing utility to graph the data, with *x* representing the number of years after 1990 and *y* representing the number of near-hits.

Year	Runway Near-Hits	Year	Runway Near-Hits
1990	281	1996	275
1991	242	1997	292
1992	219	1998	325
1993	186	1999	321
1994	200	2000	421
1995	240		

(Source: Federal Aviation Administration)

53. *Hotel Values* The following table gives the annual rental value of an average hotel room for the years 2000–2005.

a. Let *x* represent the number of years from 2000 and *y* represent the value per room in thousands of dollars and sketch the scatter plot of the data.

b. Graph the equation $y = 0.973x^2 - 4.667x + 73.950$ on the same axes as the scatter plot.

Index of Hotel Values	
Year	**Value per Room**
2000	$73,978
2001	$70,358
2002	$68,377
2003	$68,192
2004	$71,691
2005	$74,584

(Source: Pennsylvania State University)

54. *Health Insurance* The table below gives the national monthly health insurance premiums paid by employers and employees for the years 2000–2004.

a. Let *x* represent the number of years from 2000 and *y* represent the monthly premium dollars and sketch the scatter plot of the data.

b. Graph the equation $y = 5.00x^2 + 21.91x + 378.60$ on the same axes as the scatter plot.

Year	Monthly Premiums ($)
2000	379.72
2001	401.20
2002	448.66
2003	485.42
2004	547.29

(Source: www.eip.sc.gov)

55. *Unemployment Rate* The U.S. civilian unemployment rate (as a percent) is given by the table on page 44.

a. According to this table, what was the unemployment rate in 2003?

b. Graph the data from this table as a scatter plot, using the number of years after 1995 as *x*.

c. Graph the equation $y = 0.0583x^4 - 1.315x^3 + 10.636x^2 - 36.622x + 50.019$ on the same axes as the scatter plot.

Year	Unemployment (%)
1998	5.1
1999	4.5
2000	4.9
2001	4.8
2002	3.9
2003	3.5

56. *Dropout Rates* The table to the right gives the dropout rates (as percents) for students in grades 9–12 in Vermont schools during given years.

a. What is the dropout rate in 2004, according to the data?

b. Graph the data from this table as a scatter plot, using the number of years after 1960 as x.

c. On the same axes as the scatter plot, graph the equation $y = 0.0000388x^4 - 0.00336x^3 + 0.0869x^2 - 0.629x + 5.495$.

Year	Dropout (%)
1960	5.5
1970	4.9
1980	7.1
1990	5.6
2000	4
2004	5.6

section 1.3

Linear Functions

key concepts

- Linear functions
- Intercepts
- Slope of a line
- Slope and y-intercept of a line
- Constant rate of change
- Revenue, cost, and profit
- Special linear functions

section preview ▪ Prescription Drug Sales

One of the most serious health care concerns is the rising cost of prescription drugs. The increase in cost is evident by looking at Figure 1.25, which shows the U.S. retail prescription drug sales, in billions of dollars, for the years 1995 through 2005. Because the graph of the function that models the drug sales is a line, the function is called a **linear function**. The function that models the sales for 1995–2005 is

$$y = 16.908x - 20.945$$

where x is the number of years after 1990 and y is the sales in billions of dollars. The rate at which the sales are increasing is constant and is the same as the slope of the line shown in Figure 1.25. (See Example 6.) (Source: National Association of Chain Drug Stores)

In this section, we investigate linear functions and discuss slope, constant rate of change, intercepts, revenue, cost, and profit.

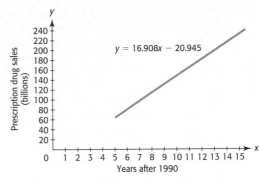

Figure 1.25

Linear Functions

A function whose graph is a line is a **linear function**.

> ### Linear Function
>
> A linear function is a function that can be written in the form $f(x) = ax + b$ where a and b are constants.

If x and y are in separate terms in an equation and each appear to the first power (and not in a denominator), we can rewrite the equation relating them in the form

$$y = ax + b$$

for some constants a and b. So the original equation is a linear equation that represents a linear function. If no restrictions are stated or implied by the context of the problem situation, and if its graph is not a horizontal line, both the domain and range of a linear function consist of the set of all real numbers. Note that an equation of the form $x = d$, where d is a constant, is not a function; its graph is a vertical line.

Recall that in the form $y = ax + b$, a and b represent constants and the variables x and y can represent any variables, with x representing the independent (input) variable and y representing the dependent (output) variable. For example, the function $5q + p = 400$ can be written in the form $p = -5q + 400$, so we can say that p is a linear function of q.

example 1 | ## Linear Functions

Determine whether each equation represents a linear function. If so, give the domain and range.

a. $0 = 2t - s + 1$ **b.** $y = 5$ **c.** $xy = 2$

Solution

a. The equation

$$0 = 2t - s + 1$$

does represent a linear function because each of the variables t and s appear to the first power and each is in a separate term. We can solve this equation for s, getting

$$s = 2t + 1,$$

so s is a linear function of t. Because any real number can be multiplied by 2 and increased by 1, and the result is a real number, both the domain and range consist of the set of all real numbers.

b. The equation $y = 5$ is in the form $y = ax + b$, where $a = 0$ and $b = 5$, so it represents a linear function. (It is in fact a constant function, which is a special linear function.) The domain is the set of all real numbers (because $y = 5$ regardless of what x we choose), and the range is the set containing 5.

c. The equation

$$xy = 2$$

does not represent a linear function because x and y are not in separate terms, and the equation cannot be written in the form $y = ax + b$. ■

Intercepts

The points where a graph crosses or touches the x-axis and the y-axis are called the **x-intercepts** and **y-intercepts**, respectively, of the graph. For example, Figure 1.26 shows that the graph of the linear function $2x - 3y = 12$ crosses the x-axis at $(6, 0)$, so the x-intercept is $(6, 0)$. The graph crosses the y-axis at $(0, -4)$, so the y-intercept is $(0, -4)$.

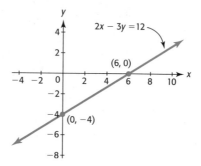

Figure 1.26

In this text, we will use the widely accepted convention that the x-coordinate of the x-intercept may also be called the x-intercept, and that the y-coordinate of the y-intercept may also be called the y-intercept.* The procedure for finding intercepts is a direct result of these definitions.

Finding Intercepts Algebraically

To find the y-intercept of a graph of $y = f(x)$, set $x = 0$ in the equation and solve for y. If the solution is b, we say the y-intercept is b and the graph intersects the y-axis at the point $(0, b)$.

To find the x-intercept(s) of the graph of $y = f(x)$, set $y = 0$ in the equation and solve for x. If the solution is a, we say the x-intercept is a and the graph intersects the x-axis at the point $(a, 0)$.

---

* We usually call the horizontal axis intercept(s) the x-intercept(s) and the vertical axis intercept(s) the y-intercept(s), but realize that symbols other than x and y can be used to represent the input and output. For example, if $p = f(q)$, the vertical intercept is called the p-intercept and the horizontal intercept is called the q-intercept.

example 2	## Finding Intercepts Algebraically

Find the x-intercept and the y-intercept of the graph of $2x - 3y = 12$ algebraically.

Solution

The y-intercept can be found by substituting 0 for x in the equation and solving for y.

$$2(0) - 3y = 12$$
$$-3y = 12$$
$$y = -4$$

Thus the y-intercept is -4 and the graph crosses the y-axis at the point $(0, -4)$.

Similarly, the x-intercept can be found by substituting 0 for y in the equation and solving for x.

$$2x - 3(0) = 12$$
$$2x = 12$$
$$x = 6$$

Thus the x-intercept is 6 and the graph crosses the x-axis at the point $(6, 0)$. ∎

> ## Finding Intercepts Graphically
> To find the intercept(s) of a graph of $y = f(x)$, first graph the function in a window that shows all intercepts.
> To find the y-intercept, $\boxed{\text{TRACE}}$ to $x = 0$ and the y-intercept will be displayed. To find the x-intercept(s) of the graph of $y = f(x)$, use the $\boxed{\text{ZERO}}$ command under the $\boxed{\text{CALC}}$ menu (accessed by $\boxed{\text{2ND}}$ $\boxed{\text{TRACE}}$).*

* For more details, see Appendix A, page 656.

The graph of a linear function has one y-intercept and one x-intercept unless the graph is a horizontal line. The intercepts of the graph of a linear function are often easy to calculate. If the intercepts are distinct, then plotting these two points and connecting them with a line gives the graph.

> ## Technology Note
> When graphing with a graphing utility, finding or estimating the intercepts can help set the viewing window for the graph of a linear equation.

example 3	## Loan Balance

A business property is purchased with a promise to pay off a \$60,000 loan plus the \$16,500 interest on this loan by making 60 monthly payments of \$1275. The amount of money, y, remaining to be paid on \$76,500 (the loan plus interest) is reduced by \$1275 each month. Although the amount of money remaining to be paid changes every month, it can be modeled by the linear function

$$y = 76,500 - 1275x$$

where x is the number of monthly payments made. We recognize that only integer values of x from 0 to 60 apply to this application.

a. Find the x-intercept and the y-intercept of the graph of this linear equation.

b. Interpret the intercepts in the context of this problem situation.

c. How should x and y be limited in this model so that they make sense in the application?

d. Use the intercepts and the results of part (c) to sketch the graph of the given equation.

Solution

a. To find the x-intercept, set $y = 0$ and solve for x.

$$0 = 76,500 - 1275x$$

$$1275x = 76,500$$

$$x = \frac{76,500}{1275} = 60$$

Thus, 60 is the x-intercept.

To find the y-intercept, set $x = 0$ and solve for y.

$$y = 76,500 - 1275(0)$$

$$y = 76,500$$

Thus, 76,500 is the y-intercept.

b. The x-intercept corresponds to the number of months that must pass before the amount owed is \$0. Therefore, a possible interpretation of the x-intercept is "The loan is paid off in 60 months." The y-intercept corresponds to the total (loan plus interest) that must be repaid 0 months after purchase—that is, when the purchase is made. Thus, the y-intercept tells us "A total of \$76,500 must be repaid."

c. We know that the total time to pay the mortgage is 60 months. A value of x larger than 60 will result in a negative value of y, which makes no sense in the application, so x varies from 0 to 60. The output, y, is the total amount owed at any time during the loan. The amount owed cannot be less than 0, and the value of the loan plus interest will be at its maximum, 76,500, when time is 0. Thus, the values of y vary from 0 to 76,500.

d. The graph intersects the horizontal axis at $(60, 0)$ and intersects the y-axis at $(0, 76,500)$, as indicated in Figure 1.27. Because we know that the graph of this equation is a line, we can simply connect the two points to obtain this first-quadrant graph.

Figure 1.27 shows the graph of the function on a viewing window determined by the context of the application.

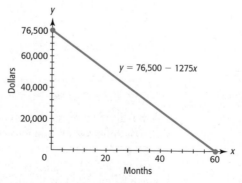

Figure 1.27

Slope of a Line

Consider two stairways: One goes from the park entrance to the shogun shrine at Nikko, Japan, and rises vertically 1 foot for every 1 foot of horizontal increase, and the second goes from the Tokyo subway to the street and rises vertically 20 centimeters for every 25 centimeters of horizontal increase. To find which set of stairs would be easier to climb, we can find the steepness, or **slope**, of each stairway. If a board is placed along the steps of each stairway, its slope is a measure of the incline of the stairway.

$$\text{shrine stairway steepness} = \frac{\text{vertical increase}}{\text{horizontal increase}} = \frac{1 \text{ foot}}{1 \text{ foot}} = 1$$

$$\text{subway stairway steepness} = \frac{\text{vertical increase}}{\text{horizontal increase}} = \frac{20 \text{ cm}}{25 \text{ cm}} = 0.8$$

The shrine stairway has a slope that is larger than that of the subway stairway, so it is steeper than the subway stairway. Thus, the subway steps would be easier to climb. In general, we define the slope of a line as follows.

Slope of a Line

The slope of a line is defined as

$$\text{slope} = \frac{\text{vertical change}}{\text{horizontal change}} = \frac{\text{rise}}{\text{run}}$$

The slope can be found by using any two points on the line (see Figure 1.28). If a nonvertical line passes through the two points, P_1 with coordinates (x_1, y_1) and P_2 with coordinates (x_2, y_2), its slope, denoted by m, is found by using

$$m = \frac{y_2 - y_1}{x_2 - x_1}$$

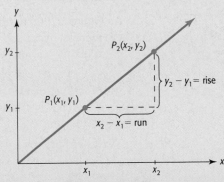

Figure 1.28

The slope of a vertical line is undefined because $x_2 - x_1 = 0$ and division by 0 is undefined.

The slope of any given nonvertical line is a constant. Thus, the same slope will result regardless of which two points on the line are used in its calculation.

example 4 | Calculating the Slope of a Line

a. Find the slope of the line passing through the points $(-3, 2)$ and $(5, -4)$. What does the slope mean?

b. Find the slope of the line joining the x-intercept point and y-intercept point in the mortgage situation of Example 3.

Solution

a. We choose one point as P_1 and the other point as P_2. Although it does not matter which point is chosen to be P_1, it is important to keep the correct order of the terms in the numerator and denominator of the slope formula. Letting $P_1 = (-3, 2)$ and $P_2 = (5, -4)$ and substituting in the slope formula gives

$$m = \frac{-4 - 2}{5 - (-3)} = \frac{-6}{8} = -\frac{3}{4}$$

Note that letting $P_1 = (5, -4)$ and $P_2 = (-3, 2)$ gives the same slope:

$$m = \frac{2 - (-4)}{-3 - 5} = \frac{6}{-8} = -\frac{3}{4}$$

A slope of $-\frac{3}{4}$ means that, from a given point on the line, by moving 3 units down and 4 units to the right, or by moving 3 units up and 4 units to the left, we arrive at another point on the line.

b. Because $x = 60$ is the x-intercept in part (b) of Example 3, $(60, 0)$ is a point on the graph. The y-intercept is $y = 76,500$, so $(0, 76,500)$ is a point on the graph. Recall that x is measured in months and that y has units of dollars in the real-world setting (that is, the *context*) of Example 3. Substituting in the slope formula, we obtain

$$m = \frac{76,500 - 0}{0 - 60} = \frac{76,500}{-60} = -1275$$

This slope means that the amount owed decreases by $1275 each month. ∎

As Figure 1.29(a) to (d) indicates, the slope describes the direction of a line as well as the steepness.

The Relation Between Orientation of a Line and its Slope

1. The slope is *positive* if the line *rises* upward toward the right.

2. The slope is *negative* if the line *falls* downward toward the right.

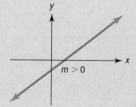

Figure 1.29(a)

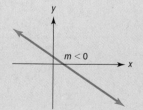

Figure 1.29(b)

3. The slope of a *horizontal line* is 0 because a horizontal line has a vertical change (rise) of 0 between any two points on the line.

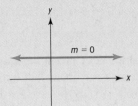

Figure 1.29(c)

4. The slope of a *vertical line* does not exist because a vertical line has a horizontal change (run) of 0 between any two points on the line.

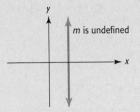

Figure 1.29(d)

Remembering how the orientation of a line is related to its slope can help you check that you have the correct order of the points in the slope formula. For instance, if you find that the slope of a line is positive and you see that the line falls downward from left to right when you observe its graph, you know that there is a mistake either in the slope calculation or in the graph that you are viewing.

Slope and *y*-Intercept of a Line

There is an important connection between the slope of the graph of a linear equation and its equation when it is written in the form $y = f(x)$. To investigate this connection, we can graph the equation

$$y = 3x + 2$$

by plotting points or by using a graphing utility (Figure 1.30(a)). By creating a table for values of x equal to 0, 1, 2, 3, and 4, we can see that each time that x increases by 1, y increases by 3 (see Figure 1.30(b)). Thus the slope is

$$\frac{\text{change in } y}{\text{change in } x} = \frac{3}{1} = 3$$

From this table, we see that the y-intercept is 2, the same value as the constant term of the equation. Observe also that the slope of the graph of $y = 3x + 2$ is the same as the coefficient of x.

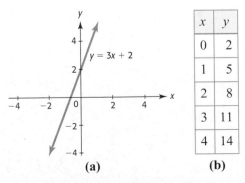

x	y
0	2
1	5
2	8
3	11
4	14

(a) (b)

Figure 1.30

Because we denote the slope of a line by m, we have the following.

> ### Slope and y-Intercept of a Line
> The slope of the graph of the equation $y = mx + b$ is m, and the y-intercept of the graph is b, so the graph crosses the y-axis at $(0, b)$.

Thus, when the equation of a linear function is written in the form $y = mx + b$ or $f(x) = mx + b$, we can "read" the values of the slope and y-intercept of its graph.

example 5

Loan Balance

As we saw in Example 3, the amount of money y remaining to be paid on the loan of $60,000 with $16,500 interest is

$$y = 76,500 - 1275x$$

where x is the number of months the mortgage has been paid.

a. What is the slope and y-intercept of the graph of this function?

b. How does the amount owed on the loan change as the number of months increases?

Solution

a. Writing this equation in the form $y = mx + b$ gives $y = -1275x + 76,500$. The coefficient of x is -1275, so the slope is $m = -1275$; the constant term is 76,500, so the y-intercept is $b = 76,500$.

b. The slope of the line indicates that the amount owed decreases by $1275 each month. This can be verified from the graph of the function and the table of sample inputs and outputs in Figure 1.31.

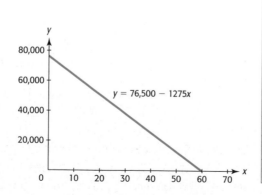

Months (x)	Amount Owed (y)
1	75,225
2	73,950
3	72,675
4	71,400
5	70,125
6	68,850
7	67,575
8	66,300
9	65,025
10	63,750

Figure 1.31

Constant Rate of Change

The function $y = 76,500 - 1275x$, whose graph is shown in Figure 1.31, gives the amount owed as a function of the months remaining on the mortgage. The coefficient of x, -1275, indicates that for each additional month, the value of y changes by -1275 (see the table in Figure 1.31). That is, the amount owed decreases at the **constant rate** of -1275 each month. Note that the **constant rate of change** of this linear function is equal to the **slope** of its graph. This is true for all linear functions.

> ### Constant Rate of Change
> The rate of change of the linear function $y = mx + b$ is the constant m, the slope of the graph of the function.

Note: The rate of change in an applied context should include appropriate units of measure. The rate of change describes by how much the output changes (increases or decreases) for every input unit.

example 6

Prescription Drug Sales

The graph in Figure 1.32 shows retail prescription drug sales for the years 1995–2005 in billions of dollars as a function of years from 1990. The function that models the sales is

$$y = 16.908x - 20.945$$

where x is the number of years after 1990 and y is the sales in billions of dollars. (Source: National Association of Chain Drug Stores)

a. What is the slope of the graph of the function?

b. What is the rate at which the sales grew during this period?

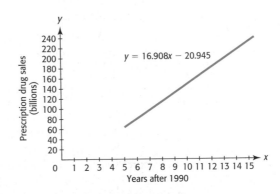

Figure 1.32

Solution

a. The coefficient of x in the linear function is 16.908, so the slope of the line is 16.908.

b. The slope of the line in Figure 1.32 is the rate of change of the function, so the retail sales of prescription drugs increased at a rate of $16.908 billion per year during this period.

Revenue, Cost, and Profit

The profit that a company makes on a product is the difference between the amount that it receives from sales (its **revenue**) and its cost. If we represent the revenue from the sale of x units by $R(x)$ and represent the cost for the production and sale of x units by $C(x)$, then the **profit** from the production and sale of x units is given by the function

$$P(x) = R(x) - C(x)$$

For total cost, total revenue, and profit functions* that are linear, their rates of change are called **marginal cost, marginal revenue**, and **marginal profit**, respectively. Suppose that the cost to produce and sell a product is $C(x) = 54.36x + 6790$ dollars, where x is the number of units produced and sold. This is a linear function, and its graph is a line with slope 54.36. Thus the *rate of change* of this cost function, called the **marginal cost**, is \$54.36 per unit produced and sold. This means that the production and sale of each additional unit will cost an additional \$54.36.

example 7

Marginal Revenue and Marginal Profit

A company produces and sells a product with revenue given by $R(x) = 89.50x$ dollars and cost given by $C(x) = 54.36x + 6790$ dollars, where x is the number of units produced and sold.

a. What is the marginal revenue for this product, and what does it mean?

b. Find the profit function.

c. What is the marginal profit for this product, and what does it mean?

Solution

a. The marginal revenue for this product is the rate of change of the revenue function, which is the slope of its graph. Thus the marginal revenue is \$89.50 per unit sold. This means that the sale of each additional unit will result in additional revenue of \$89.50.

b. To find the profit function, we subtract the cost function from the revenue function.

$$P(x) = 89.50x - (54.36x + 6790) = 35.14x - 6790$$

c. The marginal profit for this product is the rate of change of the profit function, which is the slope of its graph. Thus the marginal profit is \$35.14 per unit sold. This means that the production and sale of each additional unit will result in an additional profit of \$35.14. ■

Special Linear Functions

A special linear function that has the form $y = 0x + b$, or $y = b$, where b is a real number, is called a **constant function**. The graph of the constant function $y = 3$ is shown in Figure 1.33(a). The temperature inside a sealed case containing an Egyptian mummy in a museum is a constant function of time because the temperature inside the case never changes. Notice that even though the range of a constant function consists of a single value, the input of a constant function is any real number or any real number that makes sense in the context of an applied problem. Another special linear function is the **identity function**

$$y = 1x + 0, \quad \text{or} \quad y = x$$

which is a linear function of the form $y = mx + b$ with slope $m = 1$ and y-intercept $b = 0$. For the general identity function $f(x) = x$, the domain and range are each the set of all real numbers. A graph of the identity function f is shown in Figure 1.33(b).

---

* In this text, we frequently use "total cost" and "total revenue" interchangeably with "cost" and "revenue," respectively.

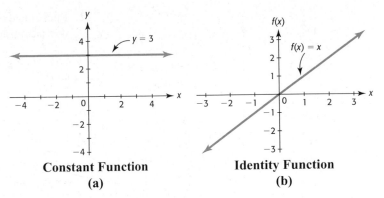

Constant Function
(a)

Identity Function
(b)

Figure 1.33

skills check

1.3

1. Which of the following functions are linear?

   a. $y = 3x^2 + 2$ **b.** $3x + 2y = 12$ **c.** $y = \dfrac{1}{x} + 2$

2. Is the graph in the figure below a function?

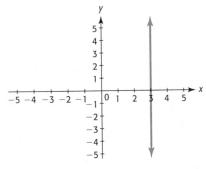

3. Find the slope of the line through $(4, 6)$ and $(28, -6)$.

4. Find the slope of the line through $(8, -10)$ and $(8, 4)$.

5. Find the slope of the line in the graph below.

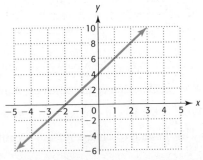

6. Find the slope of the line in the graph below.

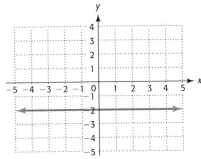

In Exercises 7–10, (a) find the x- and y-intercepts of the graph of the given equation, if they exist, and (b) graph the equation.

7. $5x - 3y = 15$ **8.** $x + 5y = 17$

9. $3y = 9 - 6x$ **10.** $y = 9x$

11. If a line is horizontal, then its slope is _____. If a line is vertical, then its slope is _____.

12. Describe the line whose slope was determined in Exercise 4.

For Exercises 13–14, determine whether the slope of the graph of the line is positive, negative, 0, or undefined.

13. a.

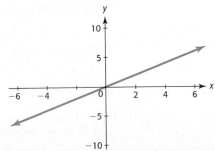

b.

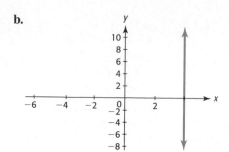

14. a.

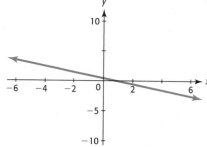

b.

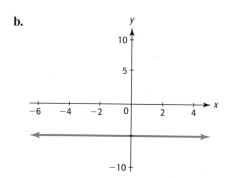

For Exercises 15–18, (a) give the slope of the line (if it exists) and the y-intercept (if it exists) and (b) graph the line.

15. $y = 4x + 8$ **16.** $3x + 2y = 7$

17. $5y = 2$ **18.** $x = 6$

For each of the functions in Exercises 19–21, do the following:

a. Find the slope and y-intercept (if possible) of the graph of the function.

b. Determine if the graph is rising or falling.

c. Graph each function on a window with the given x-range and a y-range that shows a complete graph.

19. $y = 4x + 5; [-5, 5]$

20. $y = 0.001x - 0.03; [-100, 100]$

21. $y = 50,000 - 100x; [0, 500]$

22. Rank the functions in Exercises 19–21 in order of increasing steepness.

For each of the functions in Exercises 23–26, find the rate of change.

23. $y = 4x - 3$ **24.** $y = \dfrac{1}{3}x + 2$

25. $y = 300 - 15x$ **26.** $y = 300x - 15$

27. If a linear function has the points $(-1, 3)$ and $(4, -7)$ on its graph, what is the rate of change of the function?

28. If a linear function has the points $(2, 1)$ and $(6, 3)$ on its graph, what is the rate of change of the function?

29. a. Does graph (i) or graph (ii) represent the identity function?

b. Does graph (i) or graph (ii) represent a constant function?

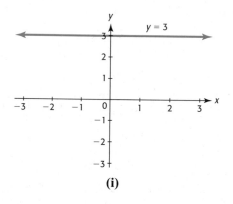

(i)

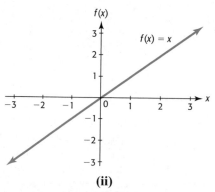

(ii)

30. What is the slope of the identity function?

31. a. What is the slope of the constant function $y = k$?

b. What is the rate of change of a constant function?

32. What is the rate of change of the identity function?

exercises

1.3

33. *Reading Tests* The average reading score of 17-year-olds on the National Assessment of Progress tests is given by $y = 0.155x + 244.37$ points, where x is the number of years past 1970. Is this a linear function? Why or why not?
(Source: U.S. Dept. of Education)

34. *Women in the Workforce* The number y (in thousands) of women in the workforce is given by the function $y = 5.74x^2 - 17.04x + 600.99$, where x is the number of years from 1900. Is this a linear function? Why or why not?
(Source: U.S. Census Bureau, U.S. Dept. of Commerce)

35. *Marriage Rate* When x is the number of years after 1950, the percent of unmarried women who get married each year is given by $M(x) = -0.762x + 85.284$.

a. Why is this a linear function?

b. What is the slope? What does this tell you about the percent of unmarried women who get married?
(Source: Index of Leading Cultural Indicators)

36. *Voters* The percent of the population voting in presidential elections is given by $V(t) = 63.20 - 0.26t$, where t is the number of years after 1950.

a. Why is this a linear function?

b. What is the slope? What does this tell you about the percent of the population voting in presidential elections since 1950?
(Source: Federal Election Commission)

37. *Marijuana Use* The percent p of high school seniors using marijuana daily can be related to x, the number of years after 1990, by the equation $30p - 19x = 30$.

a. Find the x-intercept of the graph of this function.

b. Find and interpret the p-intercept of the graph of this function.

c. Graph the function, using the intercepts. What values of x on the graph represent years 1990 and after?

38. *Depreciation* An \$828,000 building is depreciated for tax purposes by its owner, using the straight-line depreciation method. The value of the building after x months of use is given by $y = 828,000 - 2300x$ dollars.

a. Find and interpret the y-intercept of the graph of this function.

b. Find and interpret the x-intercept of the graph of this function.

c. Use the intercepts to graph the function for non-negative x- and y-values.

39. *Life Insurance* The monthly rates for a \$100,000 life insurance policy for males aged 27–32 are shown in the table.

Age (years), x	27	28	29
Premium (dollars per month), y	11.81	11.81	11.81

Age (years), x	30	31	32
Premium (dollars per month), y	11.81	11.81	11.81

a. Could the data in this table be modeled by a constant function or the identity function?

b. Write an equation whose graph contains the data points in the table.

c. What is the slope of the graph of the function found in part (b)?

d. What is the rate of change of the data in the table?

40. *Eating Asparagus* The per capita consumption of asparagus between 1990 and 1996 is shown in the table.

Year	1990	1991	1992	1993
Asparagus Consumption (pounds per person)	0.6	0.6	0.6	0.6

Year	1994	1995	1996
Asparagus Consumption (pounds per person)	0.6	0.6	0.6

(Source: *Statistical Abstract of the United States*, 1998)

a. Sketch the data as a scatter plot, with y equal to the consumption and x equal to the year.

b. Could the data in the table be modeled by a constant function or the identity function?

c. Write the equation of a function that fits the data points.

d. Sketch a graph of the function you found in part (c) on the same axes as the scatter plot.

41. *Cigarette Use* For the years 1991–2006, the percent p of high school seniors who have tried cigarettes can be modeled by $p = 75.751 - 0.743t$, where t is the number of years after 1991.

a. Is the rate of change of the percent positive or negative?

b. How fast is the percent of seniors who tried cigarettes during this period changing? Use the units in the problem in your answer.
(Source: monitoringthefuture.org)

42. *Internet Recruiting* The percent of Fortune Global 500 firms that actively recruited workers on the Internet from 1998 through 2000 can be modeled by $P(x) = 26.5x - 194.5$ percent, where x is the number of years after 1990.

a. What is the slope of the graph of this function?

b. Interpret the slope as a rate of change.
(Source: *Time*)

43. *Crickets* The number of times per minute n that a cricket chirps can be modeled as a function of the Fahrenheit temperature T. The data can be approximated by the function

$$n = \frac{12T}{7} - \frac{52}{7}$$

a. Is the rate of change of the number of chirps positive or negative?

b. What does this tell us about the relationship between temperature and the number of chirps?

44. *Tax Burden* The per capita tax burden T (in hundreds of dollars) can be described by $T(t) = 17.69 + 2.25t$, where t is the number of years after 1980.

a. What is the slope of the graph of this function?

b. What is the rate of growth of the per capita tax burden per year?
(Source: Internal Revenue Service)

45. *Earnings and Minorities* According to the U.S. Equal Employment Opportunity Commission, the relation between the median annual salaries of minorities and whites can be modeled by the function $M = 0.959W - 1.226$, where M and W represent the median annual salary (in thousands of dollars) for minorities and whites, respectively.

a. Is this function a linear function?

b. What is the slope of the graph of this function?

c. Interpret the slope as a rate of change.
(Source: *Statistical Abstract of the United States*)

46. *Voting* Between 1950 and 2004, the percent of the voting population who voted in presidential elections is given by $p = 63.20 - 0.26x$, where x is the number of years after 1950.

a. What is the slope of the graph of this function?

b. What is the annual rate of change in the percent of the eligible population who voted in presidential elections between 1950 and 2004?
(Source: Federal Election Commission)

47. *Marijuana Use* The percent p of high school seniors using marijuana daily can be modeled by $30p - 19x = 30$, where x is the number of years after 1990.

a. Use this model to determine the slope of the graph of this function if x is the independent variable.

b. What is the rate of change of the percent of high school seniors using marijuana per year?
(Source: Index of Leading Cultural Indicators)

48. *Seawater Pressure* In seawater, the pressure p is related to the depth d according to the model $33p - 18d = 496$, where d is the depth in feet and p is in pounds per square inch.

a. What is the slope of the graph of this function?

b. Interpret the slope as a rate of change.

49. *Advertising Impact* An advertising agency has found that when it promotes a new product in a city the weekly rate of change R of the number of people who are aware of it x weeks after it is introduced is given by $R = 3500 - 70x$. Find the x- and R-intercepts and then graph the function on a viewing window that is meaningful in the application.

50. *ATM Transactions* The dollar volume of transactions at automatic teller machines (ATMs) is modeled by $D(x) = 0.137x - 5.09$ billion dollars, where x is the number of terminals (in thousands).

    a. What was the value of the transactions when 50,000 ATMs were available, according to this model?

    b. This model approximates the dollar value of transactions and does not fit the data exactly. The equation cannot model the data for values of x that give negative dollar values. Test integer values of x to find the values for which this function fails to model the data because it gives negative dollar values. (Source: Electronic Funds Transfer Association, 1993).

51. *Internet Users* The percent of U.S. population with Internet access can be modeled by the function

    $$y = 6.9x - 3.18$$

    where x is the number of years after 1995.

    a. Find the slope and the y-intercept of the graph of this equation.

    b. What interpretation could be given to the slope? (Source: Jupiter Media Metrix)

52. *Wireless Service Spending* The total amount spent in the United States for wireless communication services S (in billions of dollars) can be modeled by the function

    $$S = 6.205 + 11.23t$$

    where t is the number of years after 1995.

    a. Find the slope and the y-intercept of the graph of this equation.

    b. What interpretation could be given to the y-intercept?

    c. What interpretation could be given to the slope? (Source: Cellular Telecommunications and Internet Association)

53. *Depreciation* Suppose the cost of a business property is $1,920,000 and a company depreciates it with the straight-line method. Suppose v is the value of the property after x years, and the line representing the value as a function of years passes through the points (10, 1,310,000) and (20, 700,000).

    a. What is the slope of the line through these points?

    b. What is the annual rate of change of the value of the property?

54. *Men in the Workforce* The number of men in the workforce (in millions) for the years from 1890 to 2005 can be approximated by the linear model determined by connecting the points (1890, 18.1) and (2005, 76.0).

    a. Find the annual rate of change of the model whose graph is the line connecting these points.

    b. What does this tell us about men in the workforce?

55. *Profit* A company charting its profits notices that the relationship between the number of units sold x and the profit P is linear. If 300 units sold results in $4650 profit and 375 units results in $9000 profit, find the marginal profit, which is the rate of change of the profit.

56. *Cost* A company buys and retails baseball caps and the total cost function is linear. The total cost for 200 caps is $2690, and the total cost of 500 caps is $3530. What is the marginal cost, which is the rate of change of the function?

57. *Marginal Cost* Suppose the monthly total cost for the manufacture of golf balls is $C(x) = 3450 + 0.56x$, where x is the number of balls produced each month.

    a. What is the slope of the graph of the total cost function?

    b. What is the marginal cost (rate of change of the cost function) for the product?

    c. What is the cost of each additional ball that is produced in a month?

58. *Marginal Cost* Suppose the monthly total cost for the manufacture of 19-inch television sets is $C(x) = 2546 + 98x$, where x is the number of TVs produced each month.

    a. What is the slope of the graph of the total cost function?

    b. What is the marginal cost for the product?

    c. Interpret the marginal cost for this product.

59. *Marginal Revenue* Suppose the monthly total revenue for the manufacture of golf balls is $R(x) = 1.60x$, where x is the number of balls sold each month.

    a. What is the slope of the graph of the total revenue function?

    b. What is the marginal revenue for the product?

    c. Interpret the marginal revenue for this product.

60. *Marginal Revenue* Suppose the monthly total revenue for the manufacture of television sets is $R(x) = 198x$, where x is the number of TVs sold each month.

   a. What is the slope of the graph of the total revenue function?

   b. What is the marginal revenue for the product?

   c. Interpret the marginal revenue for this product.

61. *Profit* The profit for a product is given by $P(x) = 19x - 5060$, where x is the number of units produced and sold. Find the marginal profit for the product.

62. *Profit* The profit for a product is given by the function $P(x) = 939x - 12{,}207$, where x is the number of units produced and sold. Find the marginal profit for the product.

---

section

1.4

Equations of Lines

key concepts

- Writing equations of lines
- Slope-intercept form
- Point-slope form
- Horizontal line
- Vertical line
- Parallel and perpendicular lines
- General form
- Average rate of change
- Secant line
- Difference quotient
- Approximately linear data

section preview ▪ Blood Alcohol Percent

Suppose that the blood alcohol percent for a 180-lb man is 0.11% if he has 5 drinks and that the percent increases by 0.02% for each additional drink. We can use this information to write the linear equation that models the blood alcohol percent p as a function of the number of drinks he has because the rate of change is constant. (See Example 3.) In this section, we learn how to write the equation of a linear function from information about the line, such as the slope and a point on the line or two points on the line. We also discuss **average rates of change** for data that can be modeled by nonlinear functions, slopes of secant lines, and how to create linear models that approximate data that are nearly linear.

Writing Equations of Lines

Creating a linear model from data involves writing a linear equation that describes the mathematical situation. If we know two points on a line or the slope of the line and one point, we can write the equation of the line.

    Recall that if a linear equation has the form $y = mx + b$, then the coefficient of x is the slope of the line that is the graph of the equation, and the constant b is the y-intercept of the line. Thus if we know the slope and the y-intercept of a line, we can write the equation of this line.

> ### Slope-Intercept Form of the Equation of a Line
>
> The slope-intercept form of the equation of a line with slope m and y-intercept b is
>
> $$y = mx + b$$
>
> In an applied context, m is the rate of change and b is the initial value (when $x = 0$).

| example 1 | **Appliance Repair** |

An appliance repairman charges \$60 for a service call plus \$25 per hour for each hour spent on the repair. Assuming his service call charges can be modeled by a linear function of the number of hours spent on the repair, write the equation of the function.

Solution

Let x represent the number of hours spent on the appliance repair and let y be the service call charge in dollars. The slope of the line is the amount that the charge increases for every hour of work done, so the slope is 25. Because \$60 is the basic charge before any time is spent on the repair, 60 is the y-intercept of the line. Substituting these values in the slope-intercept form of the equation gives the equation of the linear function modeling this situation. When the repairman works x hours on the service call, the charge is

$$y = 25x + 60 \text{ dollars}$$

We next consider a form of an equation of a line that can be written if we know the slope and a point. If the slope of a line is m, then the slope between a fixed point (x_1, y_1) and any other point (x, y) on the line is also m. That is,

$$m = \frac{y - y_1}{x - x_1}$$

Solving for $y - y_1$ (that is, multiplying both sides of this equation by $x - x_1$) gives the **point-slope form** of the equation of a line.

> **Point-Slope Form of the Equation of a Line**
>
> The equation of the line with slope m that passes through a known point (x_1, y_1) is
>
> $$y - y_1 = m(x - x_1)$$

| example 2 | **Using a Point and Slope to Write an Equation of a Line** |

Write an equation for the line that passes through the point $(-1, 5)$ and has slope $\frac{3}{4}$.

Solution

We are given $m = \frac{3}{4}, x_1 = -1,$ and $y_1 = 5$. Substituting in the point-slope form, we obtain

$$y - 5 = \frac{3}{4}(x - (-1))$$

$$y - 5 = \frac{3}{4}(x + 1)$$

$$y - 5 = \frac{3}{4}x + \frac{3}{4}$$

$$y = \frac{3}{4}x + \frac{23}{4}$$

Figure 1.34 shows the graph of this line.

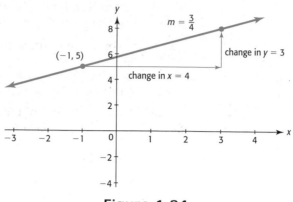

Figure 1.34

| example 3 |

Blood Alcohol Percent

Table 1.16 gives the number of drinks and the resulting blood alcohol percent for a 180-lb man. ("One drink" is equal to 1.25 oz of 80 proof liquor, 12 oz of a regular beer, or 5 oz of table wine, and many states have set 0.08% as the legal limit for driving under the influence.)

a. Is the rate of change of the blood alcohol percent for a 180-lb man a constant? What is it?

b. Write the equation of the function that models the blood alcohol percent as a function of the number of drinks.

Table 1.16

Number of Drinks (x)	3	4	5	6	7	8	9	10
Blood Alcohol Percent (y)	0.07	0.09	0.11	0.13	0.15	0.17	0.19	0.21

Solution

a. Yes. Each additional drink increases the blood alcohol percent by 0.02, so the rate of change is 0.02% per drink.

b. The rate of change is constant, so the function is linear, with its rate of change (and slope of its graph) equal to 0.02. Using this value and any point, like (5, 0.11), on the line gives us the equation we seek, with x equal to the number of drinks and y the blood alcohol percent.

$$y - 0.11 = 0.02(x - 5) \quad \text{or} \quad y = 0.02x + 0.01$$

If we know two points on a line, we can find the slope of the line and use either of the points and this slope to write the equation of the line.

example 4	## Using Two Points to Write an Equation of a Line

Write the equation of the line that passes through the points $(-1, 5)$ and $(2, 4)$.

Solution

Because we know two points on the line, we can find the slope of the line.

$$m = \frac{4 - 5}{2 - (-1)} = \frac{-1}{3} = -\frac{1}{3}$$

We can now substitute one of the points and the slope in the point-slope form to write the equation. Using the point $(-1, 5)$ gives

$$y - 5 = -\frac{1}{3}[x - (-1)], \quad \text{or} \quad y - 5 = -\frac{1}{3}x - \frac{1}{3}, \quad \text{so} \quad y = -\frac{1}{3}x + \frac{14}{3}$$

Using the point $(2, 4)$ gives

$$y - 4 = -\frac{1}{3}(x - 2)$$

$$y - 4 = -\frac{1}{3}x + \frac{2}{3}$$

$$y = -\frac{1}{3}x + \frac{14}{3}$$

Notice that both equations are the same regardless of which of the two given points is used in the point-slope form. ∎

If the rate of change of the outputs with respect to the inputs is a constant, and we know two points that satisfy the conditions of the application, then we can write the equation of the linear function that models the application.

example 5	## Inmate Population

The number of people (in millions) in U.S. prisons or jails grew at a constant rate from 1990 to 2000, with 1.15 million people incarcerated in 1990 and 1.91 million incarcerated in 2000.

a. What is the rate of growth of people incarcerated from 1990 to 2000?

b. Write the linear equation that models the number N of prisoners as a function of the year x.

c. The Bureau of Justice Statistics projected that 2.29 million people would be incarcerated in 2005. Does your model agree with this projection? (Source: Bureau of Justice Statistics)

Solution

a. The rate of growth is constant, so the points fit on a line, and a linear function can be used to find the model for the number of incarcerated people as a function of the years. The rate of change (and slope of the line) is given by

$$m = \frac{1.91 - 1.15}{2000 - 1990} = \frac{0.76}{10} = 0.076$$

b. Substituting in the point-slope form of the equation of a line (with either point) gives the equation of the line that contains the two points and thus models the application.

$$N - 1.91 = 0.076(x - 2000) \quad \text{or} \quad N = 0.076x - 150.09$$

c. Substituting 2005 for x in the function gives

$$N = N(2005) = 0.076(2005) - 150.09 = 2.29$$

Thus the projection that 2.29 million people would be incarcerated in 2005 agrees with this model. ■

| example 6 |

Writing Equations of Horizontal and Vertical Lines

Write equations for the lines that pass through the point $(-1, 5)$ and have

a. slope 0.　　**b.** undefined slope.

Solution

a. If $m = 0$, the point-slope form gives us the equation

$$y - 5 = 0(x - (-1))$$
$$y = 5$$

Because the output is always the same value, the graph of this linear function is a horizontal line. See Figure 1.35(a).

b. Because m is undefined, we cannot use the point-slope form to write the equation of this line. Lines with undefined slope are vertical lines. Every point on the vertical line through $(-1, 5)$ has an x-coordinate of -1. Thus the equation of the line is $x = -1$. Note that this equation does not represent a function. See Figure 1.35(b).

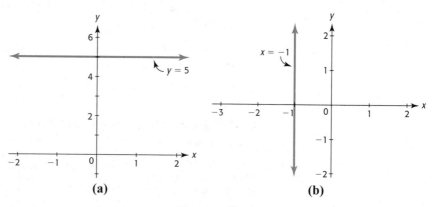

(a)　　　　　　　　　　**(b)**

Figure 1.35 ■

As we saw in Example 6, parts (a) and (b), there are special forms when the lines are horizontal or vertical.

> **Vertical and Horizontal Lines**
>
> A vertical line has the form $x = a$, where a is a constant and a is the x-coordinate of any point on the line.
> A horizontal line has the form $y = b$, where b is a constant and b is the y-coordinate of any point on the line.

Parallel and Perpendicular Lines

Clearly horizontal lines are perpendicular to vertical lines. Two distinct nonvertical lines that have the same slope are **parallel**, and conversely. If a line has slope $m \neq 0$, any line **perpendicular** to it will have slope $-\dfrac{1}{m}$. That is, the slopes of perpendicular lines are negative reciprocals of each other if neither line is horizontal.

example 7

Parallel and Perpendicular Lines

Write the equation of the line through (4, 5) and

a. parallel to the line with equation $3x + 2y = -1$.

b. perpendicular to the line with equation $3x + 2y = -1$.

Solution

a. To find the slope of the line with equation $3x + 2y = -1$, we solve for y.

$$3x + 2y = -1$$
$$2y = -3x - 1$$
$$y = -\frac{3}{2}x - \frac{1}{2}$$

The line through (4, 5) parallel to this line has the same slope, $-\dfrac{3}{2}$, and its equation is

$$y - 5 = -\frac{3}{2}(x - 4)$$

$$y = -\frac{3}{2}x + 11$$

b. The line through (4, 5) perpendicular to $3x + 2y = -1$ has slope $\dfrac{2}{3}$, and its equation is

$$y - 5 = \frac{2}{3}(x - 4)$$

$$y = \frac{2}{3}x + \frac{7}{3}$$

In Example 7(b) we found that the equation of the line was

$$y = \frac{2}{3}x + \frac{7}{3}$$

By multiplying both sides of this equation by 3 and writing the new equation with x and y on the same side of the equation, we have a new form of the equation:

$$3y = 2x + 7 \quad \text{or} \quad 2x - 3y = -7$$

This equation is called the **general form** of the equation of the line.

General Form of the Equation of a Line

The general form of the equation of a line is $ax + by = c$ where a, b, and c are real numbers, with a and b not both equal to 0.

In summary, these are the forms we have discussed for the equation of a line.

Forms of Linear Equations

General form:	$ax + by = c$	where a, b, and c are real numbers, with a and b not both equal to 0.
Point-slope form:	$y - y_1 = m(x - x_1)$	where m is the slope of the line and (x_1, y_1) is a point on the line.
Slope-intercept form:	$y = mx + b$	where m is the slope of the line and b is the y-intercept.
Vertical line:	$x = a$	where a is a constant, and a is the x-coordinate of any point on the line. The slope is undefined.
Horizontal line:	$y = b$	where b is a constant, and b is the y-coordinate of any point on the line. The slope is 0.

Average Rate of Change

For a function whose graph is not a line, the function is nonlinear and the slope of a line joining two points on the curve may change as different points are chosen on the curve. The calculation and interpretation of the slope of a curve are topics you will study if you take a calculus course. However, there is a quantity called the **average rate of change** that can be applied to any function relating two variables.

In general, we can find the average rate of change of a function between two input values if we know how much the function outputs change between the two input values.

Average Rate of Change

The average rate of change of $f(x)$ with respect to x over the interval from $x = a$ to $x = b$ (where $a < b$) is calculated as

$$\text{average rate of change} = \frac{\text{change in } f(x) \text{ values}}{\text{corresponding change in } x \text{ values}} = \frac{f(b) - f(a)}{b - a}$$

How is the average rate of change over some interval of points related to the slope of a line connecting the points? For any function, the average rate of change between two points on its graph is the slope of the line joining the two points. Such a line is called a **secant line**.

example 8

Hybrid Vehicle Sales

The total hybrid electric passenger vehicle sales for the years from 1997 to 2005 are shown in Figure 1.36(a). The number of sales can be modeled by the equation

$$S(x) = 0.744x^3 - 20.255x^2 + 184.729x - 553.098$$

where $S(x)$ is in thousands and x is the number of years after 1990. A graph of $S(x)$ is shown in Figure 1.36(b). $S(10) = 12.692$ and $S(16) = 264.710$, so there were 12,692 units sold in 2000 and 264,710 sold in 2006, according to the model.

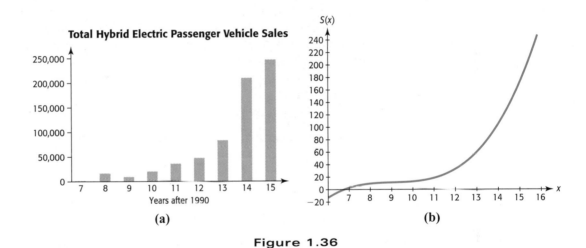

(a)   **(b)**

Figure 1.36

a. Find the average rate of change of hybrid vehicle sales between 2000 and 2006.

b. Interpret your answer to part (a).

c. What is the relationship between the slope of the secant line joining the points (10, 12.692) and (16, 264.710) and the answer to part (a)?

Solution

a. We find the average rate of change between the two points to be

$$\frac{264.710 - 12.692}{16 - 10} = \frac{252.018}{6} = 42.003$$

or 42,003 units per year.

b. A possible interpretation of this is: On average between 2000 and 2006, sales of hybrid electric passenger vehicles increased by 42,003 per year.

c. The slope of this line, shown in Figure 1.37, is numerically the same as the average rate of change of the sales between 2000 and 2006 found in part (a).

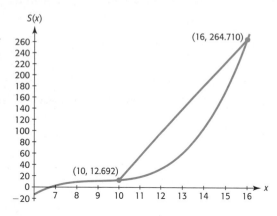

Figure 1.37

If we have points $(x, f(x))$ and $(x + h, f(x + h))$ on the graph of $y = f(x)$, we can find the average rate of change of the function from x to $x + h$, called the **difference quotient**, as follows.

Difference Quotient

The average rate of change of the function $f(x)$ from x to $x + h$ is

$$\frac{f(x + h) - f(x)}{x + h - x} = \frac{f(x + h) - f(x)}{h}$$

| example 9 |

Average Rate of Change

For the function $f(x) = x^2 + 1$, whose graph is shown in Figure 1.38, find:

a. $f(x + h)$ **b.** $f(x + h) - f(x)$

c. the average rate of change $\dfrac{f(x + h) - f(x)}{h}$

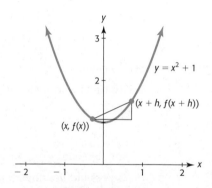

Figure 1.38

Solution

a. $f(x + h)$ is found by substituting $x + h$ for x in $f(x) = x^2 + 1$:

$$f(x + h) = (x + h)^2 + 1 = (x^2 + 2xh + h^2) + 1$$

b. $f(x + h) - f(x) = (x^2 + 2xh + h^2) + 1 - (x^2 + 1)$

$$= x^2 + 2xh + h^2 + 1 - x^2 - 1 = 2xh + h^2$$

c. The average rate of change is $\dfrac{f(x + h) - f(x)}{h} = \dfrac{2xh + h^2}{h} = 2x + h$ ■

Approximately Linear Data

Real-life data are rarely perfectly linear, but some sets of real data points lie sufficiently close to a line so that two points can be used to create a linear function that models the data. Consider the following example.

<table>
<tr><td>

example 10

Table 1.17

Year, x	Enrollment, y (thousands)
1990	12,488
1991	12,703
1992	12,882
1993	13,093
1994	13,376
1995	13,697
1996	14,060
1997	14,272
1998	14,428
1999	14,623
2000	14,802
2001	15,058
2002	15,332
2003	15,722
2004	16,048
2005	16,328
2006	16,498

(Source: U.S. National Center for Education Statistics)

</td><td>

High School Enrollment

Table 1.17 and Figure 1.39(a) show the enrollment (in thousands) in grades 9–12 of U.S. public and private schools for the years 1990–2006.

a. Create a scatter plot of the data. Does a line fit the data points exactly?

b. Find the average rate of change of the high school enrollment between 1990 and 2005.

c. Write the equation of the line determined by this rate and one of the two points.

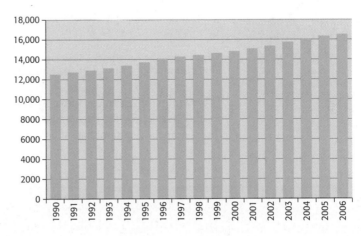

Figure 1.39(a)

Solution

a. A scatter plot is shown in Figure 1.39(b) on the next page. The graph shows that the data do not fit a linear function exactly, but that a linear function could be used as an approximate model for the data.

</td></tr>
</table>

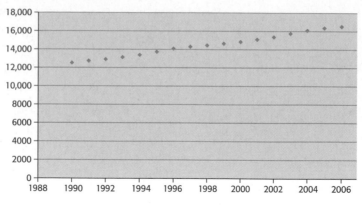

Figure 1.39(b)

b. To find the average rate of change between 1990 and 2005, we use the points (1990, 12,488) and (2005, 16,328).

$$\frac{16{,}328 - 12{,}488}{2005 - 1990} = \frac{3840}{15} = 256$$

This means that the high school enrollments have grown at an average rate of 256 thousand per year (because the outputs are in thousands).

c. Because the enrollment growth is approximately linear, the average rate of change can be used as the slope of the graph of a linear function describing the enrollments from 1990 through 2005. The graph connects the points (1990, 12,488) and (2005, 16,328), so we can use either of these two points to write the equation of the line. Using the point (1990, 12,488) and $m = 256$ gives the equation

$$y - 12{,}488 = 256(x - 1990)$$
$$y - 12{,}488 = 256x - 509{,}440$$
$$y = 256x - 496{,}952$$

This equation is a model of the enrollment, in thousands, as a function of the year.   ∎

The equation in Example 10 approximates a model for the data, but it is not the best possible model for the data. In Section 2.2 we will see how technology can be used to find the linear function that is the best fit for a set of data of this type.

skills check

1.4

For Exercises 1–18, write the equation of the line with the given conditions.

1. slope 4 and y-intercept $\dfrac{1}{2}$

2. slope 5 and y-intercept $\dfrac{1}{3}$

3. slope $\dfrac{1}{3}$ and y-intercept 3

4. slope $-\dfrac{1}{2}$ and y-intercept -8

5. through the point $(4, -6)$ with slope $-\dfrac{3}{4}$

6. through the point $(-4, 3)$ with slope $-\dfrac{1}{2}$

7. vertical line, through the point $(9, -10)$

8. horizontal line, through the point $(9, -10)$

9. passing through $(-2, 1)$ and $(4, 7)$

10. passing through $(-1, 3)$ and $(2, 6)$

11. passing through $(5, 2)$ and $(-3, 2)$

12. passing through $(9, 2)$ and $(9, 5)$

13. x-intercept -5 and y-intercept 4

14. x-intercept 4 and y-intercept -5

15. passing through $(4, -6)$ and parallel to the line with equation $3x + y = 4$

16. passing through $(5, -3)$ and parallel to the line with equation $2x + y = -3$

17. passing through $(-3, 7)$ and perpendicular to the line with equation $2x + 3y = 7$

18. passing through $(-4, 5)$ and perpendicular to the line with equation $3x + 2y = -8$

For Exercises 19 and 20, write the equation of the line whose graph is shown.

19.

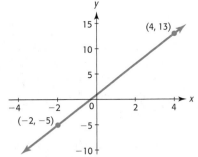

20.

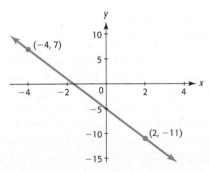

21. Write the equation of a function if its rate of change is -15 and $y = 12$ when $x = 0$.

22. Write the equation of a function if its rate of change is -8 and $y = -7$ when $x = 0$.

23. For the function $y = x^2$, compute the average rate of change between $x = -1$ and $x = 2$.

24. For the function $y = x^3$, compute the average rate of change between $x = -1$ and $x = 2$.

25. For the function shown in the figure, find the average rate of change from $(-2, 7)$ to $(1, -2)$.

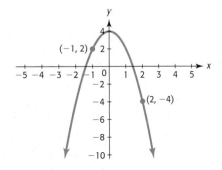

26. For the function shown in the figure, find the average rate of change from $(-1, 2)$ to $(2, -4)$.

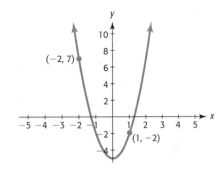

For the functions given in Exercises 27–30, find

$$\frac{f(x + h) - f(x)}{h}$$

27. $f(x) = 45 - 15x$ **28.** $f(x) = 32x + 12$

29. $f(x) = 2x^2 + 4$ **30.** $f(x) = 3x^2 + 1$

In Exercises 31–32, use the table on the next page, which gives a set of input values x and the corresponding outputs y that satisfy a function.

31. a. Would you say that a linear function could be used to model the data? Explain.

b. Write the equation of the linear function that fits the data.

x	10	20	30	40	50
y	585	615	645	675	705

32. a. Verify that the values satisfy a linear function.

b. Write the equation of the linear function that fits the data.

x	1	7	13	19
y	-0.5	8.5	17.5	26.5

exercises

1.4

33. *Utility Charges* Palmetto Electric determines its monthly bills for residential customers by charging a base price of $8.95 plus an energy charge of 9.35 cents for each kilowatt-hour (KWh) used. Write an equation for the monthly charge y (in dollars) as a function of x, the number of KWh used.

34. *Phone Bills* For interstate calls, AT&T charges 7 cents per minute plus a base charge of $4.95 each month. Write an equation for the monthly charge y as a function of the number of minutes of use.

35. *Depreciation* A business uses straight-line depreciation to determine the value y of a piece of machinery over a 10-year period. Suppose the original value (when $t = 0$) is equal to $36,000 and its value is reduced by $3600 each year. Write the linear equation that models the value y of this machinery at the end of year t.

36. *Sleep* Each day a young person should sleep 8 hours plus $\frac{1}{4}$ hour for each year the person is under 18 years of age.

a. Based on this information, how much sleep does a 10-year-old need?

b. Based on this information, how much sleep does a 14-year-old need?

c. Use the answers from parts (a) and (b) to write a linear equation relating hours of sleep y to age x, for $6 \le x \le 18$.

d. Use your equation from part (c) to verify that an 18-year-old needs 8 hours of sleep.

37. *Population Growth* Assume the growth of the population of Del Webb's Sun City Hilton Head community was linear from 1996 to 2000, with a population of 198 in 1996 and a rate of growth of 705 per year.

a. Write an equation for the population P of this community as a function of the number of years x after 1996.

b. Use the function to estimate the population in 2002.
(Source: Island Packet, March 2000)

38. *SAT Scores* The composite SAT score for the Beaufort County School District was 952 points in 1994, and the average rate of increase was 0.51 point per year. Write a linear function that models the SAT scores y as a function of x, the number of years after 1994.
(Source: Island Packet, October 1999)

39. *Depreciation* A business uses straight-line depreciation to determine the value y of an automobile over a 5-year period. Suppose the original value (when $t = 0$) is equal to $26,000 and the salvage value (when $t = 5$) is equal to $1000.

a. By how much has the automobile depreciated over the 5 years?

b. By how much is the value of the automobile reduced at the end of each of the 5 years?

c. Write the linear equation that models the value s of this automobile at the end of year t.

40. *Retirement* For Pennsylvania state employees for whom the average of the three best yearly salaries is $75,000, the retirement plan gives an annual pension of 2.5% times 75,000, multiplied by the number of years of service. Write the linear function that models the pension P in terms of the number of years of service, y.
(Source: Pennsylvania State Retirement Fund)

41. *Patrol Cars* The Beaufort County Sheriff's office assigns a patrol car to each of its deputy sheriffs, who keeps the car 24 hours a day. As the population of the county grew, the number of deputies and the number of patrol cars also grew. The data in the following table could describe how many patrol cars were necessary as the number of deputies increased. Write the function that models the relationship between the number of deputies and patrol cars.

Number of Deputies	10	50	75	125	180	200	250
Number of Patrol Cars in Use	10	50	75	125	180	200	250

42. *Life Insurance* The monthly premiums for a $100,000 life insurance policy for males aged 27–32 are shown in the table below. Write the linear model for the data in the table.

Age (years)	27	28	29
Premium (dollars per month)	11.81	11.81	11.81

Age (years)	30	31	32
Premium (dollars per month)	11.81	11.81	11.81

43. *Profit* A company charting its profits notices that the relationship between the number of units sold, x, and the profit, P, is linear. If 300 units sold results in $4650 profit and 375 units results in $9000 profit, write the equation that models its profit.

44. *Cost* A company buys and retails baseball caps. The total cost function is linear, the total cost for 200 caps is $2680, and the total cost of 500 caps is $3530. Write the equation that models this cost function.

45. *Depreciation* Suppose the cost of a business property is $1,920,000 and a company depreciates it with the straight-line method. If V is the value of the property after x years and the line representing the value as a function of years passes through the points (10, 1,310,000) and (20, 700,000), write the equation that gives the annual value of the property.

46. *Depreciation* Suppose the cost of a business property is $860,000 and a company depreciates it with the straight-line method. Suppose y is the value of the property after t years.

a. What is the value at the beginning of the depreciation (when $t = 0$)?

b. If the property is completely depreciated ($y = 0$) in 25 years, write the equation of the line representing the value as a function of years.

47. *Cigarette Use* The percent of high school seniors who smoke cigarettes can be modeled by a linear function $p = f(t)$, where t is the number of years after 1991. If two points on the graph of this function are (6, 63.5) and (16, 47.1), write the equation of this function.

48. *Earnings and Race* Data from 2003 for various age groups show that for each $100 increase in median weekly income for whites, the median weekly income for blacks increases by $61.90. Also, for these workers, the median weekly income for whites was $676 and for blacks was $527. Write the equation that gives the median weekly income for blacks as a function of the median weekly income for whites. (Source: U.S. Department of Labor)

49. *Blood Alcohol Percent* The table below gives the number of drinks and the resulting blood alcohol percent for a 90-lb woman. ("One drink" is equal to 1.25 oz of 80 proof liquor, 12 oz of regular beer, or 5 oz of table wine, and many states have set 0.08% as the legal limit for driving under the influence.)

a. The rate of change in blood alcohol percent per drink for a 90-lb woman is a constant. What is it?

b. Write the equation of the function that models the blood alcohol percent as a function of the number of drinks.

Number of Drinks	0	1	2	3	4
Blood Alcohol Percent	0	0.05	0.10	0.15	0.20

Number of Drinks	5	6	7	8	9
Blood Alcohol Percent	0.25	0.30	0.35	0.40	0.45

(Source: Pennsylvania Liquor Control Board)

50. *Drinking and Driving* The following table gives the number of drinks and the resulting blood alcohol percent for a 180-lb man legally considered driving under the influence (DUI).

a. The average rate of change of the blood alcohol percent with respect to the number of drinks is a constant. What is it?

b. Use the rate of change and one point determined by a number of drinks and the resulting blood alcohol percent to write the equation of a linear model for this data.

Number of Drinks	5	6	7	8	9	10
Blood Alcohol Percent	0.11	0.13	0.15	0.17	0.19	0.21

(Source: Pennsylvania Liquor Control Board)

51. *Men in the Workforce* The number of men in the workforce (in millions) for selected decades from 1890 to 2010 is shown in the figure. The decade is defined by the year at the beginning of the decade and $g(t)$ is defined by the average number of men (in millions) in the workforce during the decade (indicated by the point on the graph within the decade). This data can be approximated by the linear model determined by the line connecting (1890, 18.1) and (1990, 68.5).

a. Write the equation of the line connecting these two points to find a linear model for this data.

b. Does this line appear to be a reasonable fit to the data points?

c. How does the slope of this line compare with the average rate of change in the function during this period?

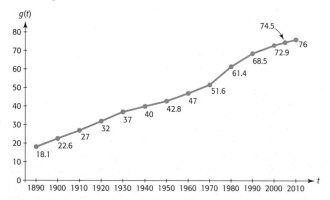

(Source: *World Almanac*)

52. *Farm Workers* The figure below shows the percent p of U.S. workers in farm occupations for selected years t.

a. Write the equation of a line connecting the points (1820, 71.8) and (1994, 2.6), with values rounded to two decimal places.

b. Does this line appear to be a reasonable fit to the data points?

c. Interpret the slope of this line as a rate of change of the percent of farm workers.

d. Can the percent p of U.S. workers in farm occupations continue to fall at this rate for 20 more years? Why or why not?

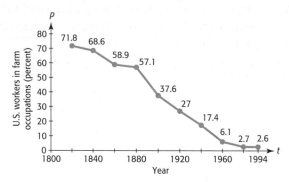

(Source: U.S. Department of Agriculture)

53. *Education Spending* The figure below gives the amount (in billions of dollars) spent on education during the years 1996–2001.

a. If a line were drawn connecting the data points (1996, 23) and (2001, 40.1), what is the slope of the line?

b. What is the average annual rate of change in spending between 1996 and 2001?

c. Is the rate of change in spending for education the same each year?

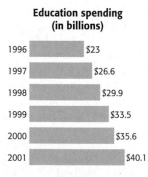

(Source: U.S. Department of Education)

54. *Enrollment Projection* The following figure shows enrollment projections for three schools. (Outputs are measured in students.)

a. What is the slope of the line joining the two points in the figure that show the Beaufort enrollment projections?

b. Find the average rate of change in projected enroll-ment for students in Beaufort schools between 2000 and 2005.

c. How should this information be used when plan-ning future school building in Beaufort?

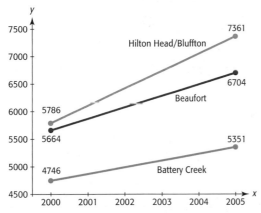

(Source: Beaufort County School District)

55. *Teenage Mothers* The percent of births to teenage mothers that are out-of-wedlock can be approximated by a linear function of the number of years after 1950. The percent was 15 in 1960 and 76 in 1996.

a. What is the slope of the line joining the points (10, 15) and (46, 76)? Round the answer to two decimal places.

b. What is the average rate of change in the percent of teenage out-of-wedlock births over this period?

c. What does this average rate of change tell about the out-of-wedlock births to teenage mothers?

d. Use the slope from part (a) and the number of teenage mothers in 1996 to write the equation of the line. Round the values to two decimal places. (Source: Father Facts)

56. *Voting* The percent of eligible people voting in presi-dential elections can be expressed as a linear function of the years 1960–2004. The percent was 63.1 in 1960 and 55.3 in 2004.

a. What is the slope of the line joining the given points?

b. What is the average rate of change in the percent voting in these elections? Interpret this value.

c. Use the slope from part (a) and the percent of peo-ple voting in the 1960 election to write the equa-tion of the line. (Source: Federal Election Commission)

57. *Prison Population* The graph showing the total num-ber of prisoners in state and federal prisons for the years 1960 through 2005 is shown in the figure. There were 212,953 prisoners in 1960 and 1,512,823 in 2005.
(Source: U.S. Bureau of Justice Statistics)

a. What is the average rate of growth of the prison population from 1960 to 2005?

b. What is the slope of the line connecting the points satisfying the conditions above?

c. Write the equation of the secant line joining these two points on the curve.

d. Can this secant line be used to make a good esti-mate of total number of prisoners in 2008?

e. Points associated with what two years could be used to write a linear equation that will give a better estimate of total number of prisoners in 2008?

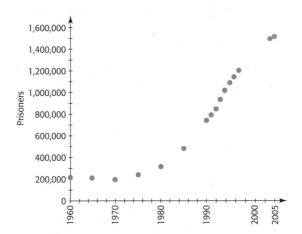

58. *Investment* The graph of the future value of an investment of $1000 for x years earning interest at a rate of 8% compounded continuously is shown in the following figure. The $1000 investment is worth approximately $1083 after 1 year and about $1492 after 5 years.

a. What is the average rate of change of the future value over the 4-year period?

b. Interpret this average rate of change.

c. What is the slope of the line connecting the points satisfying the conditions above?

d. Write the equation of the secant line joining the two given points.

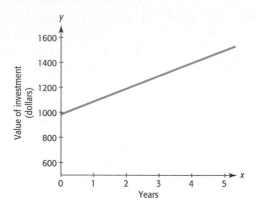

59. *Working Age* The scatter plot below projects the ratio of the working-age population to the elderly.

a. Does the data appear to fit a linear function?

b. The data points shown in the scatter plot from 2010 to 2030 are projections made from a mathematical model. Do those projections appear to be made with a linear model? Explain.

c. If the ratio is projected to be 3.9 in 2010 and 2.2 in 2030, what is the average annual rate of change of the data over this period of time?

d. Write the equation of the line joining these two points.

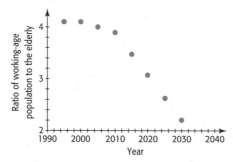

(Source: *Newsweek*, December 1999)

60. *Women in the Workforce* The number of women in the workforce (in millions) for selected years from 1890 to 2005 is shown in the following figure.

a. Would the data in the scatter plot be modeled well by a linear function? Why or why not?

b. The number of women in the workforce was 10.519 million in 1930 and 16.443 million in 1950. What is the average rate of change in the number of women in the workforce during this period?

c. If the number of women in the workforce was 16.443 million in 1950 and 59.531 million in 1990, what is the average rate of change in the number of women in the workforce during this period?

d. Is it reasonable that these two average rates of change are different? How can you tell this from the graph?

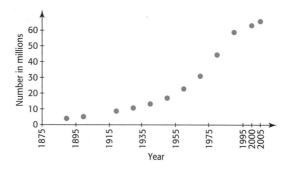

(Source: *Newsweek*)

61. *U.S. Population* The total U.S. population for selected years from 1950 to 2005 is shown in the table below, with the population given in thousands. Using x to represent the number of years from 1950 and y to represent the number of thousands of people:

a. Find the average annual rate of change in population during 1950–2005, with the appropriate units.

b. Use the slope from part (a) and the population in 1950 to write the equation of the line associated with 1950 and 2005.

c. Use the equation to estimate the population in 1975. Does it agree with the population shown in the table?

d. Why might the values be different?

Year	Population (thousands)	Year	Population (thousands)
1950	152,271	1980	227,726
1955	165,931	1985	238,466
1960	180,671	1990	249,948
1965	194,303	1995	263,044
1970	205,052	2000	281,422
1975	215,973	2005	296,410

(Source: U.S. Census Bureau)

62. *Social Agency* A social agency provides emergency food and shelter to two groups of clients. The first group has *x* clients who need an average of $300 for emergencies, and the second group has *y* clients who need an average of $200 for emergencies. The agency has $100,000 to spend for these two groups.

a. Write the equation that gives the number of clients who can be served in each group.

b. Find the *y*-intercept and the slope of the graph of this equation. Interpret each value.

c. If 10 clients are added from the first group, what happens to the number served in the second group?

chapter 1

Summary

In this chapter, we studied the basic concept of a function and how to recognize functions with tables, graphs, and equations. We studied linear functions in particular—graphing, finding slope, model formation, and equation writing. We used graphing utilities to graph functions and to evaluate functions. We solved business and economics problems involving linear functions.

Key Concepts and Formulas

1.1 Functions and Models

Scatter plot of data	A scatter plot is a graph of pairs of values that represent real-world information collected in numerical form.
Function definition	A function is a rule or correspondence that determines exactly one output for each input. The function may be defined by a set of ordered pairs, a table, a graph, or an equation.
Domain and range	The set of possible inputs for a function is called its domain, and the set of possible outputs is called its range. In general, if the domain of a function is not specified, we assume that it includes all real numbers except • Values that result in a denominator of 0 • Values that result in an even root of a negative number
Independent variable and dependent variable	If *x* represents any element in the domain, then *x* is called the independent variable; if *y* represents an output of the function from an input *x*, then *y* is called the dependent variable.
Tests for functions	We can test for a function graphically, numerically, and analytically.
Vertical line test	A set of points in a coordinate plane is the graph of a function if and only if no vertical line intersects the graph in more than one point.
Function notation	We denote that *y* is a function of *x* using function notation when we write $y = f(x)$.
Modeling	The process of translating real-world information into a mathematical form so that it can be applied and then interpreted in the real-world setting is called modeling.

Mathematical model	A mathematical model is a functional relationship (usually in the form of an equation) that includes not only the function rule but also descriptions of all involved variables and their units of measure.

1.2 Graphs of Functions

Point-plotting method	The point-plotting method of sketching a graph means sketching the graph of a function by plotting enough points to determine the shape of the graph and then drawing a smooth curve through the points.
Complete graph	A graph is a complete graph if it shows the basic shape of the graph and important points on the graph (including points where the graph crosses the axes and points where the graph turns), and suggests what the unseen portions of the graph will be.
Using a calculator to draw a graph	After writing the function with x representing the independent variable and y representing the dependent variable, enter the function in the equation editor of the graphing utility. Activate the graph with $\boxed{\text{ZOOM}}$ or $\boxed{\text{GRAPH}}$ (Set the x- and y-boundaries of the viewing window before pressing $\boxed{\text{GRAPH}}$.)
Viewing windows	The values that define the viewing window can be set manually or by using the $\boxed{\text{ZOOM}}$ keys. The boundaries of a viewing window are x_{min}: the smallest value on the x-axis (the left boundary of the window) y_{min}: the smallest value on the y-axis (the bottom boundary of the window) x_{max}: the largest value on the x-axis (the right boundary of the window) y_{max}: the largest value on the y-axis (the top boundary of the window)
Spreadsheets	Spreadsheets like Excel can be used to create accurate graphs, sometimes better-looking graphs than those created with graphing calculators.
Aligning data	Using aligned inputs (input values that have been shifted horizontally to smaller numbers) rather than the actual data values results in smaller coefficients in models and less involved computations.
Evaluating functions with a calculator	We can find (or estimate) the output of a function $y = f(x)$ at specific inputs with a graphing calculator by using $\boxed{\text{TRACE}}$ and moving the cursor to (or close to) the value of the independent variable.
Graphing data points	Graphing utilities can be used to create lists of numbers and to create graphs of the data stored in the lists.
Scaling data	Replacing the outputs of a set of data by dividing or multiplying each number by a given number results in models with less involved computations.

1.3 Linear Functions

Linear functions	A linear function is a function of the form $f(x) = ax + b$ where a and b are constants. The graph of a linear function is a line.
Intercepts	A point (or the x-coordinate of the point) where a graph crosses or touches the horizontal axis is called an x-intercept, and a point (or the y-coordinate of the point) where a graph crosses or touches the vertical axis is called a y-intercept. To find the x-intercept(s) of the graph of an equation, set $y = 0$ in the equation and solve for x. To find the y-intercept(s), set $x = 0$ and solve for y.

Slope of a line	The slope of a line is a measure of its steepness and direction. The slope is defined as $$\text{slope} = \frac{\text{vertical change}}{\text{horizontal change}} = \frac{\text{rise}}{\text{run}}$$ If (x_1, y_1) and (x_2, y_2) are two points on a line, then the slope of the line is $$m = \frac{y_2 - y_1}{x_2 - x_1}$$
Slope and y-intercept of a line	The slope of the graph of the equation $y = mx + b$ is m, and the y-intercept of the graph is b. (This is called slope-intercept form.)
Constant rate of change	If a model is linear, the rate of change of the outputs with respect to the inputs will be constant, and the rate of change equals the slope of the line that is the graph of a linear function. Thus, we can determine if a linear model fits a set of real data by determining if the rate of change remains constant for the data.
Revenue, cost, and profit	If a company sells x units of a product for p dollars per unit, then the total revenue for this product can be modeled by the linear function $R(x) = px$. $$\text{profit} = \text{revenue} - \text{cost} \quad \text{or} \quad P(x) = R(x) - C(x)$$
Special linear functions	
• **Constant function**	A special linear function that has the form $y = c$, where c is a real number, is called a constant function.
• **Identity function**	The identity function $y = x$ is a linear function of the form $y = mx + b$ with slope $m = 1$ and y-intercept $b = 0$.

1.4 Equations of Lines

Writing equations of lines	
• **Slope-intercept form**	The slope-intercept form of the equation of a line with slope m and y-intercept b is $$y = mx + b$$
• **Point-slope form**	The point-slope form of the equation of the line with slope m and passing through a known point (x_1, y_1) is $$y - y_1 = m(x - x_1)$$
• **Horizontal line**	The equation of a horizontal line is $y = b$, where b is a constant.
• **Vertical line**	The equation of a vertical line is $x = a$, where a is a constant.
• **General form**	The general form of the equation of a line is $ax + by = c$, where a, b, and c are constants.
Parallel and perpendicular lines	Two distinct nonvertical lines are parallel if they have the same slope. Two nonvertical and nonhorizontal lines are perpendicular if their slopes are negative reciprocals.
Average rate of change	The average rate of change of a quantity over an interval describes how a change in the output of a function describing that quantity $f(x)$ is related to a change in the input x over that interval. The average rate of change of $f(x)$ with respect to x over the interval from $x = a$ to $x = b$ (where $a < b$) is calculated as $$\text{average rate of change} = \frac{\text{change in } f(x) \text{ values}}{\text{corresponding change in } x \text{ values}} = \frac{f(b) - f(a)}{b - a}$$

Secant line	When a function is not linear, the average rate of change between two points is the slope of the line joining two points on the curve, which is called a secant line.
Difference quotient	The average rate of change of a function from a point $(x, f(x))$ to the point $(x + h, f(x + h))$ is the difference quotient $\dfrac{f(x + h) - f(x)}{h}$.
Approximately linear data	Some sets of real data lie sufficiently close to a line that two points can be used to create a linear function that models the data.

chapter

1

Skills Check

Use the values in the table below in Exercises 1–4.

x	−3	−1	1	3	5	7	9	11	13
y	9	6	3	0	−3	−6	−9	−12	−15

1. Explain why the relationship shown by the table describes y as a function of x.

2. State the domain and range of the function.

3. If the function defined by the table is denoted by f, so that $y = f(x)$, what is $f(3)$?

4. Do the outputs in this table indicate that a linear function fits the data? If so, write the equation of the line.

5. If $C(s) = 16 - 2s^2$, find

   a. $C(3)$ **b.** $C(-2)$ **c.** $C(-1)$

6. For each of the functions $y = f(x)$ described below, find $f(-3)$.

a.   **b.**

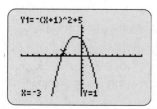

7. Graph the function $f(x) = -2x^3 + 5x$.

8. Graph the function $y = 3x^2$.

9. Graph $y = -10x^2 + 400x + 10$ on a standard window and on a window with $x_{min} = 0$, $x_{max} = 40$, $y_{min} = 0$, $y_{max} = 5000$. Which window gives a better view of the graph of the function?

10. Use a graphing utility to graph the points (x, y) from the table.

x	10	15	20	25	30	35	40
y	−8	−6	−3	0	5	8	10

11. Find the domain of each function.

    a. $y = \sqrt{2x - 8}$ **b.** $f(x) = \dfrac{x + 2}{x - 6}$

12. One line passes through the points $(-12, 16)$ and $(-1, 38)$, and a second line has equation $2y + x = 23$. Are the lines parallel, perpendicular, or neither?

13. A line passes through $(-1, 4)$ and $(5, -3)$. Find the slope of a line parallel to this line and the slope of a line perpendicular to this line.

14. Find the slope of the line through $(-4, 6)$ and $(8, -16)$.

15. Given the equation $2x - 3y = 12$, (a) find the x- and y-intercepts of the graph and (b) graph the equation.

16. What is the slope of the graph of the function given in Exercise 15?

17. Find the slope and y-intercept of the graph of $y = -6x + 3$.

18. Find the rate of change of the function whose equation is given in Exercise 17.

19. Write the equation of the line that has slope $\dfrac{1}{3}$ and y-intercept 3.

20. Write the equation of a line that has slope $\dfrac{-3}{4}$ and passes through $(4, -6)$.

21. Write the equation of the line that passes through $(-1, 3)$ and $(2, 6)$.

22. For the function $y = x^2$, compute the average rate of change between $x = 0$ and $x = 3$.

For the functions given in Exercises 23 and 24, find

a. $f(x + h)$ **b.** $f(x + h) - f(x)$

c. $\dfrac{f(x + h) - f(x)}{h}$

23. $f(x) = 5 - 4x$ **24.** $f(x) = 10x - 50$

---

chapter

1 Review

25. *Voters* The table gives the percent p of African American voters who have supported Democratic candidates for president for the years 1960–1996.

a. Is the percent p a function of the year y ?

b. Let $p = f(y)$ denote that p is a function of y. Find $f(1992)$ and explain what it means.

c. What is y if $f(y) = 94$? What does this mean?

Year	Democrat (%)	GOP (%)
1960	68	32
1964	94	6
1968	85	15
1972	87	13
1976	85	15
1980	86	12
1984	89	9
1992	82	11
1996	84	12

(Source: Joint Center for Political and Economic Studies)

26. *Voters*
a. What is the domain of the function defined by the table in Exercise 25?

b. Is this function defined for 1982? Why is this value not included in the table?

c. Is this function discrete or continuous? Explain.

27. *Voters* Graph the function defined by the table in Exercise 25 on the window [1956, 2000] by [60, 100].

28. *Voters*
a. Find the slope of the line joining points $(1968, 85)$ and $(1996, 84)$.

b. Use the answer from part (a) to find the average annual rate of change of the percent of African American voters who voted for the Democratic candidate for president between the years 1968 and 1996, inclusive.

c. Is the average annual rate of change from 1968 to 1980 equal to the average annual rate of change from 1968 to 1996?

d. Does a linear function model this data exactly?

When money is borrowed to purchase an automobile, the amount borrowed A determines the monthly payment P. In particular, if a dealership offers a 5-year loan at 2.9% interest, then the amount borrowed for the car determines the payment according to the following table. Use the table to define the function P = f(A) in Exercises 29–31.

Amount Borrowed ($)	Monthly Payment ($)
10,000	179.25
15,000	268.87
20,000	358.49
25,000	448.11
30,000	537.73

(Source: Sky Financial)

29. *Car Loans*

   a. Explain why the monthly payment P is a linear function of the amount borrowed A.

   b. Find $f(25,000)$ and interpret its meaning.

   c. If $f(A) = 358.49$, what is A?

30. *Car Loans*

   a. What are the domain and range of the function $P = f(A)$ defined by the table?

   b. Is this function $P = f(A)$ defined for a $12,000 loan?

   c. Is the function defined by the table discrete or continuous?

31. *Car Loans* The equation of the line that fits the data points in the table is $f(A) = 0.017924A + 0.01$.

   a. Use the equation to find $f(28,000)$ and explain what it means.

   b. Can the function f be used to find the monthly payment for any dollar amount A of a loan if the interest rate and length of loan are unchanged?

32. *Life Expectancy* The figure gives the number of years the average woman is estimated to live beyond age 65 during selected years between 1950 and 2030. Let x represent the year and let y represent the expected number of years a woman will live past age 65. Write $y = f(x)$ and answer the following questions:

   a. What is $f(1960)$ and what does it mean?

   b. What is the life expectancy for the average woman in 2010?

c. In what year was the average woman expected to live 19 years past age 65?

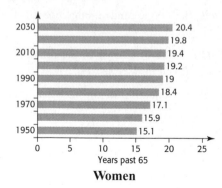

Women

33. *Life Expectancy* The figure gives the number of years the average man is estimated to live beyond age 65 during selected years between 1950 and 2030. Let x represent the year and let y represent the expected number of years a man will live past age 65. Write $y = g(x)$ and answer the following questions:

   a. What is $g(2020)$ and what does it mean?

   b. What was the life expectancy for the average man in 1950?

   c. Write a function expression that indicates that the average man in 1990 has a life expectancy of 80 years.

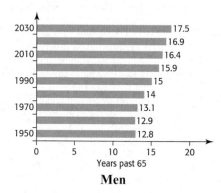

Men

34. *Teacher Salaries* The average classroom teacher salary in the United States is given by $f(t) = 982.06t + 32,903.77$, where t is the number of years from 1990.

   a. What was the average salary in 2000? Write this in function notation.

   b. Find $f(15)$ and interpret it.

   c. Are the salaries increasing or decreasing?
   (Source: www.ors2.state.sc.us/abstract)

35. *Teacher Salaries* The average classroom teacher salary in the United States is given by the function $f(t) = 982.06t + 32,903.77$, where t is the number of years from 1990.

   a. Graph this function for values of t between 0 and 15.

   b. What years do these values represent?
   (Source: www.ors2.state.sc.us/abstract)

36. *Drug Users* The number of current users of illicit drugs increased from 12 million in 1992 to 14.5 million in 1999.

   a. Find the slope of the line connecting the points (1992, 12.0) and (1999, 14.5), rounded to three decimal places.

   b. Find the average annual rate of change of current users between 1992 and 1999.

37. *Fuel* The table below shows data for the number of gallons of gas purchased each day x of a certain week by the 150 taxis owned by the Inner City Transportation Taxi Company. Write the equation that models this data.

Days from first day	0	1	2	3	4	5	6
Gas used (gal)	4500	4500	4500	4500	4500	4500	4500

38. *Work Hours* The average weekly hours worked by production and nonsupervisory workers on private nonfarm payrolls, seasonally adjusted, are given by the data in the table below.

Month and Year	Average Weekly Hours	Month and Year	Average Weekly Hours
June 2006	33.8	Oct. 2006	33.8
July 2006	33.8	Nov. 2006	33.8
Aug. 2006	33.8	Dec. 2006	33.8
Sept. 2006	33.8	Jan. 2006	33.8
		Feb. 2006	33.8

(Source: Bureau of Labor Statistics)

   a. Write the equation of a function that describes the average weekly hours using an input x equal to the number of months past May 2006.

   b. Is this function a constant function?

39. *Revenue* A company has revenue given by $R(x) = 564x$ dollars and total costs given by $C(x) = 40,000 + 64x$ dollars, where x is the number of units produced and sold.

   a. What is the revenue when 120 units are produced?

   b. What is the cost when 120 units are produced?

   c. What is the marginal cost and what is the marginal revenue for this product?

   d. What is the slope of the graph of $C(x) = 40,000 + 64x$?

   e. Graph $R(x)$ and $C(x)$ on the same set of axes.

40. *Profit* A company has revenue given by $R(x) = 564x$ dollars and total cost given by $C(x) = 40,000 + 64x$ dollars, where x is the number of units produced and sold. The profit can be found by forming the function $P(x) = R(x) - C(x)$.

   a. Write the profit function.

   b. Find the profit when 120 units are produced and sold.

   c. How many units give break even?

   d. What is the marginal profit for this product?

   e. How is the marginal profit related to the marginal revenue and the marginal cost?

41. *Depreciation* A business property can be depreciated for tax purposes by using the formula $y + 3000x = 300,000$, where y is the value of the property x years after it was purchased.

   a. Find the y-intercept of the graph of this function. Interpret this value.

   b. Find the x-intercept. Interpret this value.

42. *Marginal Profit* A company has determined that its profit for a product can be described by a linear function. The profit from the production and sale of 150 units is $455 and the profit from 250 units is $895.

   a. What is the average rate of change of the profit for this product when between 150 and 250 units are sold?

   b. What is the slope of the graph of this profit function?

   c. Write the equation of the profit function for this product.

   d. What is the marginal profit for this product?

   e. How many units give break even for this product?

group activities / extended applications

1. Body Mass Index

Obesity is a risk factor for the development of medical problems, including high blood pressure, high cholesterol, heart disease, and diabetes. Of course, how much a person can safely weigh depends on his or her height. One way of comparing weights that account for height is the *body mass index (BMI)*. The table below gives the BMI for a variety of heights and weights of people. Roche Pharmaceuticals, manufacturers of Xenical, state that a BMI of 30 or greater can create an increased risk of developing medical problems associated with obesity.

Describe how to assist a group of people in using the information. Some things you would want to include in your description follow.

a. How a person uses the table to determine his or her BMI.

b. How a person determines the weight that will put him or her at medical risk.

c. How a person whose weight or height is not in the table can determine if his or her BMI is 30. To answer this question, develop a formula to find the weight that would give a BMI of 30 for a person of a given height, including how you would:

1. Pick the points from the table that correspond to a BMI of 30 and create a table of these heights and weights. Change the heights to inches to simplify the data.

2. Create a scatter plot of the data.

3. Using the two points with the smallest and largest heights, write a linear equation that models the data.

4. Graph the linear equation from Question (3) with the scatter plot and discuss the fit.

5. Explain how to use the model to test for obesity.

Body Mass Index for Specified Height (FT./IN.) and Weight (LB)

Height/Weight	120	130	140	150	160	170	180	190	200	210	220	230	240	250
5'0"	23	25	27	29	31	33	35	37	39	41	43	45	47	49
5'1"	23	25	27	28	30	32	34	36	38	40	42	44	45	47
5'2"	22	24	26	27	29	31	33	35	37	38	40	42	44	46
5'3"	21	23	25	27	28	30	32	34	36	37	39	41	43	44
5'4"	21	22	24	26	28	29	31	33	34	36	38	40	41	43
5'5"	20	22	23	25	27	28	30	32	33	35	37	38	40	42
5'6"	19	21	23	24	26	27	29	31	32	34	36	37	39	40
5'7"	19	20	23	24	25	27	28	30	31	33	35	36	38	39
5'8"	18	20	21	23	24	26	27	29	30	32	34	35	37	38
5'9"	18	19	21	22	24	25	27	28	30	31	33	34	36	37
5'10"	17	19	20	22	23	24	25	27	28	29	31	33	35	36
5'11"	17	18	20	21	22	24	25	27	28	29	31	32	34	35
6'0"	16	18	19	20	22	23	24	26	27	29	30	31	33	34
6'1"	16	17	19	20	21	22	24	25	26	28	29	30	32	33

(Source: Roche Pharmaceuticals)

2. Total Revenue, Total Cost, and Profit

The total revenue is the amount a company receives from the sales of its products and can be found by multiplying its selling price per unit times the number of units sold. That is, the revenue is

$$R(x) = p \cdot x$$

where p is the price per unit and x is the number of units sold.

The total cost comprises two costs, the fixed costs and the variable costs. Fixed costs (FC), such as depreciation, rent, utilities, and so on, remain constant regardless of the number of units produced. Variable costs (VC) are those directly related to the number of units produced. The variable cost is the cost per unit (c) times the number of units produced ($VC = c \cdot x$) and the fixed cost is constant ($FC = k$), so the total cost is found by using the equation

$$C(x) = c \cdot x + k$$

where c is the cost per unit and x is the number of units produced.

Also, as discussed in Section 1.3, the profit a company makes on x units of a product is the difference between the revenue and cost from the production and sale of x units.

$$P(x) - R(x) - C(x)$$

Suppose a company manufactures MP3 players and sells them to retailers for $98 each. It has fixed costs of $262,500 related to the production of the MP3 players, and the cost per unit for production is $23.

1. What is the total revenue function?
2. What is the marginal revenue for this product?
3. What is the total cost function?
4. What is the marginal cost for this product? Is the marginal cost equal to the variable cost or the fixed cost?
5. What is the profit function for this product? What is the marginal profit?
6. What is the cost, revenue, and profit if 0 units are produced?
7. Graph the total cost and total revenue functions on the same axes and estimate where the graphs intersect.
8. Graph the profit function and estimate where the graph intersects the x-axis.
9. What do the points of intersection in Question 7 and Question 8 give?

chapter 2

Linear Models, Equations, and Inequalities

Annual data can be used to find the average annual salaries by gender and educational attainment. These salaries can be compared by creating a linear equation that gives the female annual earnings as a function of the male annual earnings. We can then investigate if female salaries are approaching male salaries. We can also find when the weekly market share of Concerta reaches and surpasses the weekly market share of Ritalin, by finding linear models and solving them simultaneously. Market equilibrium occurs when the number of units of a product demanded equals the number of units supplied. In this chapter, we solve these and other problems by creating linear models, solving linear equations and inequalities, and solving systems of equations.

Algebra Toolbox

Properties of Equations

In this chapter we will solve equations. To solve an equation means to find the value(s) of the variable(s) that makes the equation a true statement. For example, the equation $3x = 6$ is true when $x = 2$, so 2 is a solution of this equation. Equations of this type are sometimes called **conditional equations** because they are true only for certain values of the variable. Equations that are true for all values of the variables for which both sides are defined are called **identities**. For example, the equation $7x - 4x = 5x - 2x$ is an identity. Equations that are not true for any value of the variable are called **contradictions**. For example, the equation $2(x + 3) = 2x - 1$ is a contradiction.

We frequently can solve an equation for a given variable by rewriting the equation in an equivalent form whose solution is easy to find. Two equations are **equivalent** if and only if they have the same solutions. The following operations give equivalent equations:

Properties of Equations

1. **Addition Property**   Adding the same number to both sides of an equation gives an equivalent equation. For example, $x - 5 = 2$ is equivalent to $x - 5 + 5 = 2 + 5$, or to $x = 7$.

2. **Subtraction Property**   Subtracting the same number from both sides of an equation gives an equivalent equation. For example, $z + 12 = -9$ is equivalent to $z + 12 - 12 = -9 - 12$, or to $z = -21$.

3. **Multiplication Property**   Multiplying both sides of an equation by the same nonzero number gives an equivalent equation. For example, $\dfrac{y}{6} = 5$ is equivalent to $6\left(\dfrac{y}{6}\right) = 6(5)$, or to $y = 30$.

4. **Division Property**   Dividing both sides of an equation by the same nonzero number gives an equivalent equation. For example, $17x = -34$ is equivalent to $\dfrac{17x}{17} = \dfrac{-34}{17}$, or to $x = -2$.

5. **Substitution Property**   The equation formed by substituting one expression for an equal expression is equivalent to the original equation. For example, if $y = 3x$, then $x + y = 8$ is equivalent to $x + 3x = 8$, so $4x = 8$ and $x = 2$.

---

example 1	**Properties of Equations**

State the property (or properties) of equations that can be used to solve each of the following equations, and then use the property (or properties) to solve the equation.

a. $3x = 6$ **b.** $\dfrac{x}{5} = 12$ **c.** $3x - 5 = 17$ **d.** $4x - 5 = 7 + 2x$

Solution

a. Division Property. Dividing both sides of the equation by 3 gives the solution to the equation.

$$3x = 6$$

$$\frac{3x}{3} = \frac{6}{3}$$

$$x = 2$$

Thus $x = 2$ is the solution to the original equation.

b. Multiplication Property. Multiplying both sides of the equation by 5 gives the solution to the equation.

$$\frac{x}{5} = 12$$

$$5\left(\frac{x}{5}\right) = 5(12)$$

$$x = 60$$

Thus $x = 60$ is the solution to the original equation.

c. Addition Property and Division Property. Adding 5 to both sides of the equation and dividing both sides by 3 gives an equivalent equation.

$$3x - 5 = 17$$

$$3x - 5 + 5 = 17 + 5$$

$$3x = 22$$

$$\frac{3x}{3} = \frac{22}{3}$$

$$x = \frac{22}{3}$$

Thus $x = \dfrac{22}{3}$ is the solution to the original equation.

d. Addition Property, Subtraction Property, and Division Property. Adding 5 to both sides of the equation and subtracting $2x$ from both sides of the equation gives an equivalent equation.

$$4x - 5 = 7 + 2x$$

$$4x - 5 + 5 = 7 + 2x + 5$$

$$4x = 12 + 2x$$

$$4x - 2x = 12 + 2x - 2x$$

$$2x = 12$$

Dividing both sides by 2 gives the solution to the original equation.

$$\frac{2x}{2} = \frac{12}{2}$$

$$x = 6$$

Thus, $x = 6$ is the solution to the original equation.

| example 2 | ## Equations |

Determine whether each equation is a conditional equation, an identity, or a contradiction.

a. $3(x - 5) = 2x - 7$ **b.** $5x - 6(x + 1) = -x - 9$

c. $-3(2x - 4) = 2x + 12 - 8x$

Solution

a. To solve $3(x - 5) = 2x - 7$, we first use the Distributive Property, then the Addition Property.

$$3(x - 5) = 2x - 7$$
$$3x - 15 = 2x - 7$$
$$3x - 15 - 2x = 2x - 7 - 2x$$
$$x - 15 + 15 = -7 + 15$$
$$x = 8$$

Because this equation is true for only the value $x = 8$, it is a conditional equation.

b. To solve $5x - 6(x + 1) = -x - 9$, we first use the Distributive Property, then the Addition Property.

$$5x - 6(x + 1) = -x - 9$$
$$5x - 6x - 6 = -x - 9$$
$$-x - 6 = -x - 9$$
$$-x - 6 + x = -x - 9 + x$$
$$-6 = -9$$

Because $-6 = -9$ is false, the original equation is a contradiction.

c. To solve $-3(2x - 4) = 2x + 12 - 8x$, we first use the Distributive Property, then the Addition Property.

$$-3(2x - 4) = 2x + 12 - 8x$$
$$-6x + 12 = 12 - 6x$$
$$-6x + 12 + 6x = 12 - 6x + 6x$$
$$12 = 12$$

Because $12 = 12$ is a true statement, the equation is true for all real numbers, and thus is an identity.   ◾

Properties of Inequalities

We will solve linear inequalities in this chapter. As with equations, we can find solutions to inequalities by finding equivalent inequalities from which the solutions can be easily seen. We use the following properties to reduce an inequality to a simple equivalent inequality.

Properties of Inequalities	Examples	
Substitution Property The inequality formed by substituting one expression for an equal expression is equivalent to the original inequality.	$7x - 6x \geq 8$ $x \geq 8$	$3(x - 1) < 8$ $3x - 3 < 8$
Addition and Subtraction Properties The inequality formed by adding the same quantity to (or subtracting the same quantity from) both sides of an inequality is equivalent to the original inequality.	$x - 6 < 12$ $x - 6 + 6 < 12 + 6$ $x < 18$	$3x > 13 + 2x$ $3x - 2x > 13 + 2x - 2x$ $x > 13$
Multiplication Property I The inequality formed by multiplying (or dividing) both sides of an inequality by the same *positive* quantity is equivalent to the original inequality.	$\frac{1}{3}x \leq 6$ $3\left(\frac{1}{3}x\right) \leq 3(6)$ $x \leq 18$	$3x > 6$ $\frac{3x}{3} > \frac{6}{3}$ $x > 2$
Multiplication Property II The inequality formed by multiplying (or dividing) both sides of an inequality by the same *negative* number and reversing the inequality symbol is equivalent to the original inequality.	$-x > 7$ $-1(-x) < -1(7)$ $x < -7$	$-4x \geq 12$ $\frac{-4x}{-4} \leq \frac{12}{-4}$ $x \leq -3$

Of course, these properties can be used in combination to solve an inequality. This means that the steps used to solve a linear inequality are the same as those used to solve linear equations, except that the inequality symbol is reversed if both sides are multiplied (or divided) by a negative number.

example 3

Properties of Inequalities

State the property (or properties) of inequalities that can be used to solve each of the following inequalities, and then use the property (or properties) to solve the inequality.

a. $\dfrac{x}{2} \geq 13$ **b.** $-3x < 6$ **c.** $5 - \dfrac{x}{2} \geq 4$ **d.** $-2(x + 3) < 4$

Solution

a. Multiplication Property I. Multiplying both sides by 2 gives the solution to the inequality.

$$\frac{x}{2} \geq 13$$

$$2\left(\frac{x}{2}\right) \geq 2(13)$$

$$x \geq 26$$

The solution to the inequality is $x \geq 26$.

b. Multiplication Property II. Dividing both sides by -3 and reversing the inequality symbol gives the solution to the inequality.

$$-3x < 6$$

$$\frac{-3x}{-3} > \frac{6}{-3}$$

$$x > -2$$

The solution to the inequality is $x > -2$.

c. Subtraction Property, Multiplication Property II. Subtracting 5 from both sides of the inequality gives

$$5 - \frac{x}{3} - 5 \geq -4 - 5$$

$$-\frac{x}{3} \geq -9$$

Multiplying both sides by -3 and reversing the inequality symbol gives the solution to the inequality.

$$-\frac{x}{3} \geq -9$$

$$-3\left(-\frac{x}{3}\right) \leq -3(-9)$$

$$x \leq 27$$

The solution to the inequality is $x \leq 27$.

d. Substitution Property, Addition Property, Multiplication Property II. First we distribute -2 on the left side of the inequality.

$$-2(x + 3) < 4$$

$$-2x - 6 < 4$$

Adding 6 to both sides, then dividing both sides by -2 and reversing the inequality symbol gives the solution to the inequality.

$$-2x - 6 + 6 < 4 + 6$$

$$-2x < 10$$

$$\frac{-2x}{-2} > \frac{10}{-2}$$

$$x > -5$$

The solution to the inequality is $x > -5$. ∎

toolbox exercises

In Exercises 1–8, state the property (or properties) of equations that can be used to solve each of the following equations; then use the property (or properties) to solve the equation.

1. $3x = 6$

2. $x - 7 = 11$

3. $x + 3 = 8$

4. $x - 5 = -2$

5. $\dfrac{x}{3} = 6$

6. $-5x = 10$

7. $2x + 8 = -12$

8. $\dfrac{x}{4} - 3 = 5$

Solve the equations in Exercises 9–16.

9. $4x - 3 = 6 + x$

10. $3x - 2 = 4 - 7x$

11. $\dfrac{3x}{4} = 12$

12. $\dfrac{5x}{2} = -10$

13. $3(x - 5) = -2x - 5$

14. $-2(3x - 1) = 4x - 8$

15. $2x - 7 = -4\left(4x - \dfrac{1}{2}\right)$

16. $-2(2x - 6) = 3\left(3x - \dfrac{1}{3}\right)$

In Exercises 17–20, use the Substitution Property of Equations to solve the equation.

17. Solve for x if $y = 2x$, and $x + y = 12$

18. Solve for x if $y = 4x$, and $x + y = 25$

19. Solve for x if $y = 3x$, and $2x + 4y = 42$

20. Solve for x if $y = 6x$, and $3x + 2y = 75$

In Exercises 21–24, determine whether the equation is a conditional equation, an identity, or a contradiction.

21. $3x - 5x = 2x + 7$ **22.** $3(x + 1) = 3x - 7$

23. $9x - 2(x - 5) = 3x + 10 + 4x$

24. $\dfrac{x}{2} - 5 = \dfrac{x}{4} + 2$

In Exercises 25–32, solve the inequalities.

25. $5x + 1 > -5$ **26.** $1 - 3x \geq 7$

27. $\dfrac{x}{4} > -3$ **28.** $\dfrac{x}{6} > -2$

29. $\dfrac{x}{4} - 2 > 5x$ **30.** $\dfrac{x}{2} + 3 > 6x$

31. $-3(x - 5) < -4$ **32.** $-\dfrac{1}{2}(x + 4) < 6$

section

2.1

Algebraic and Graphical Solution of Linear Equations

key concepts

- Algebraic solution of linear equations
- Solving real-world application problems
- Zero of a function
- Solutions, zeros, and x-intercepts
- Graphical solution of linear equations
  - *x-intercept method*
  - *Intersection method*
- Literal equations
- Solving an equation for a specified variable
- Direct variation

section preview ▪ Prison Sentences

The average sentence length and the average time served in state prisons for various crimes are shown in Table 2.1. These data can be used to create a linear function that is an approximate model for the data:

$$y = 0.55x - 2.886$$

Table 2.1

Average Sentence (months)	Average Time Served (months)
62	30
85	45
180	95
116	66
92	46
61	33
56	26

(Source: U.S. Department of Justice)

This function describes the mean time y served in prison for a crime as a function of the mean sentence length x, where x and y are each measured in months. Assuming that this model remains valid, it predicts that the mean time served on a 5-year (60 months) sentence is $0.55(60) - 2.886 \approx 30$ months, that is, approximately $2\frac{1}{2}$ years.

To find the sentence that would give an expected time served of 10 years (120 months), we use algebraic or graphical methods to solve the equation

$$120 = 0.55x - 2.866$$

for x. (See Example 6.) In this section, we use additional algebraic methods and graphical methods to solve linear equations in one variable.

Algebraic Solution of Linear Equations

We can use algebraic, graphical, or a combination of algebraic and graphical methods to solve linear equations. Sometimes it is more convenient to use algebraic solution methods rather than graphical solution methods, especially if an exact solution is desired. The steps used to solve a linear equation in one variable algebraically follow.

> ### Steps for Solving a Linear Equation in One Variable
>
> 1. If a linear equation contains fractions, multiply both sides of the equation by a number that will remove all denominators from the equation. If there are two or more fractions, use the least common denominator (LCD) of the fractions.
>
> 2. Remove any parentheses or other symbols of grouping.
>
> 3. Perform any additions or subtractions to get all terms containing the variable on one side and all other terms on the other side of the equation. Combine like terms.
>
> 4. Divide both sides of the equation by the coefficient of the variable.
>
> 5. Check the solution by substitution in the original equation. If a real-world solution is desired, check the algebraic solution for reasonableness in the real-world situation.

example 1 | **Algebraic Solutions**

a. Solve for x: $\dfrac{2x - 3}{4} = \dfrac{x}{3} + 1$ **b.** Solve for y: $y - \dfrac{1}{2}\left(\dfrac{y}{2}\right) = -6$

Solution

a.
$$\frac{2x - 3}{4} = \frac{x}{3} + 1$$

$$12\left(\frac{2x - 3}{4}\right) = 12\left(\frac{x}{3} + 1\right) \qquad \text{Multiply both sides by the LCD, 12.}$$

$$3(2x - 3) = 12\left(\frac{x}{3} + 1\right) \qquad \text{Simplify the fraction } \frac{12}{4} \text{ on the left.}$$

$$6x - 9 = 4x + 12$$ Remove parentheses using the distributive property.

$$2x = 21$$ Subtract $4x$ from both sides and add 9 to both sides.

$$x = \frac{21}{2}$$ Divide both sides by 2.

Check the result: $$\frac{2\left(\frac{21}{2}\right) - 3}{4} \overset{?}{=} \frac{\frac{21}{2}}{3} + 1 \Rightarrow \frac{9}{2} = \frac{9}{2}$$

b. $y - \dfrac{1}{2}\left(\dfrac{y}{2}\right) = -6$

$$y - \frac{y}{4} = -6$$ Remove parentheses first to find the LCD.

$$4\left(y - \frac{y}{4}\right) = 4(-6)$$ Multiply both sides by the LCD, 4.

$$4y - y = -24$$ Remove parentheses using the distributive property.

$$3y = -24$$ Combine like terms.

$$y = -8$$ Divide both sides by 3.

Check the result: $$-8 - \frac{1}{2}\left(\frac{-8}{2}\right) \overset{?}{=} -6 \Rightarrow -6 = -6 \qquad \blacksquare$$

As the next example illustrates, we solve an application problem that is set in a real-world context by using the same solution methods. However, you must remember to include units of measure with your answer and check that your answer makes sense in the problem situation.

<table><tr><td>example 2</td></tr></table>

Credit Card Debt

It is hard for some people to pay off credit card debts in a reasonable period of time because of high interest rates. The interest paid on a $10,000 debt over 3 years is approximated by

$$y = 175.393x - 116.287 \text{ dollars}$$

when the interest rate is $x\%$. What is the interest rate if the interest is $1637.60? (Source: Consumer Federation of America)

Solution

To answer this question, we solve the linear equation

$$1637.60 = 175.393x - 116.287$$
$$1753.887 = 175.393x$$
$$x = 9.9998$$

Thus, if the interest rate is approximately 10%, the interest is $1637.60. (Note that if you check the approximate answer, you are checking only for the reasonableness of the estimate.) $\qquad \blacksquare$

| example 3 | **Stock Market** |

For a period of time, a man is very successful speculating on an Internet stock, with its value growing to $100,000. However, the stock value drops rapidly until its value is reduced by 40%. What percent increase will have to occur before the latest value returns to $100,000?

Solution

The value of the stock after the 40% loss is $100,000 - 0.40(100,000) = 60,000$. To find the percent p of increase that is necessary to return the value to 100,000, we solve

$$60,000 + 60,000p = 100,000$$

$$60,000p = 40,000$$

$$p = \frac{40,000}{60,000} = \frac{2}{3} = 66\frac{2}{3}\%$$

Thus, the stock must increase by $66\frac{2}{3}\%$ for it to return to a value of $100,000. ∎

Solutions, Zeros, and x-Intercepts

Because x-intercepts are x values that make the output of the function equal to 0, these intercepts are also called **zeros** of the function.

> ### Zero of a Function
> Any number a for which $f(a) = 0$ is called a **zero** of the function $f(x)$. If a is real, a is an x-intercept of the graph of the function.

The zeros of the function are values that make the function equal to 0, so they are also solutions to the equation $f(x) = 0$.

> The following three concepts are numerically the same:
>
> The x-intercepts of the graph of $y = f(x)$
> The real zeros of the function f
> The real solutions to the equation $f(x) = 0$

The following example illustrates the relationships that exist among x-intercepts of the graph of a function, zeros of the function, and solutions to associated equations.

| example 4 | **Relationships Among x-Intercepts, Zeros, and Solutions** |

For the function $f(x) = 13x - 39$, find:

a. $f(3)$ **b.** The zero of $f(x) = 13x - 39$

c. The x-intercept of the graph of $y = 13x - 39$

d. The solution to the equation $13x - 39 = 0$

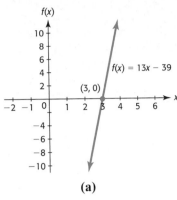

(a)

	A	B
1	x	y = 13x − 39
2	3	0

(b)

Figure 2.1

Solution

a. Evaluate the output $f(3)$ by substituting the input 3 for x in the function:

$$f(3) = 13(3) − 39 = 0$$

b. Because $f(x) = 0$ when $x = 3$, we say that 3 is a zero of the function.

c. The x-intercept of a graph occurs at the value of x where $y = 0$, so the x-intercept is $x = 3$.

The only x-intercept is $x = 3$ because the graph of $f(x) = 13x − 39$ is a line that crosses the x-axis at only one point (Figure 2.1(a)).

d. The solution to the equation $13x − 39 = 0$ is $x = 3$ because $13x − 39 = 0$ gives $13x = 39$ or $x = 3$.

We can use ⬚TRACE⬚ or ⬚TABLE⬚ on a graphing utility to check the reasonableness of a solution. A graph of $y = 13x − 39$ given in Figure 2.1(a) confirms that $x = 3$ is the x-intercept. We can also confirm that $f(3) = 0$ by using a spreadsheet (Figure 2.1(b)). When we enter 3 in cell A2 and the formula "=13*A2–39" in cell B2, the function $f(x) = 13x − 39$ is evaluated at $x = 3$, with a result of 0.*

Graphical Solution of Linear Equations

We can also view the graph or use ⬚TRACE⬚ on a graphing utility to obtain a quick estimate of an answer. ⬚TRACE⬚ may or may not provide the exact solution to an equation, but it will provide an estimate of the solution. If your calculator or computer has a graphical or numerical solver, the solver can be used to obtain the solution to the equation $f(x) = 0$. Recall that an x-intercept of the graph of $y = f(x)$, a real zero of $f(x)$, and the real solution to the equation $f(x) − 0$ are all different names for the same input value. If the graph of the linear function $y = f(x)$ is not a horizontal line, we can find the one solution to the linear equation $f(x) = 0$ as described below. We call this solution method the **x-intercept method**.

> ## Solving a Linear Equation Using the x-Intercept Method with Graphing Utilities
>
> 1. Rewrite the equation to be solved with 0 (and nothing else) on one side of the equation.
>
> 2. Enter the nonzero side of the equation found in the previous step in the equation editor of your graphing utility and graph the line in an appropriate viewing window. Be certain that you can see the line cross the horizontal axis on your graph.
>
> 3. Find the x-intercept by using ⬚ZERO⬚. The x-intercept is the value of x that makes the equation equal to zero, so it is the solution to the equation. The value of x displayed by this method is sometimes a decimal approximation of the exact solution rather than the exact solution. *Note:* Using ⬚MATH⬚ ⬚►FRAC⬚ will often convert a decimal solution of a linear equation (approximated on the display) to the exact solution.†

* See Appendix B, page 676.
† See Appendix A, pages 657–658.

| example 5 | **Graphical Solution** |

Solve $\dfrac{2x - 3}{4} = \dfrac{x}{3} + 1$ for x using the x-intercept method.

Solution

To solve the equation using the x-intercept method, we first rewrite the equation with 0 on one side.

$$0 = \frac{x}{3} + 1 - \frac{2x - 3}{4}$$

Next, enter the right-hand side of the equation,

$$\frac{x}{3} + 1 - \frac{2x - 3}{4},$$

in the equation editor of your graphing utility as

$$y_1 = (x/3) + 1 - (2x - 3)/4$$

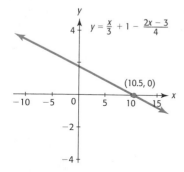

Figure 2.2

and graph this function. Use parentheses as needed to preserve the order of operations. You should obtain a graph similar to the one seen in Figure 2.2, which was obtained using the viewing window $[-10, 15]$ by $[-4, 4]$. However, any graph in which you can see the x-intercept will do. Using $\boxed{\text{ZERO}}$ gives the x-intercept of the graph, 10.5. (See Figure 2.2.) This is the value of x that makes $y = 0$, and thus the original equation true, so it is the solution to this linear equation.

You can determine if $x = 10.5 = \dfrac{21}{2}$, found graphically, is the exact answer by substituting this value into the original equation. Both sides of the original equation are equal when $x = 10.5$, so it is the exact solution.   ■

We can also use the intersection method described below to solve linear equations:

Solving a Linear Equation Using the Intersection Method

1. Enter the left side of the equation as y_1 and the right side as y_2. Graph both of these equations on a window that shows their point of intersection.

2. Find the point of intersection of the two graphs with $\boxed{\text{INTERSECT}}$.* This is the point where $y_1 = y_2$. The x-value of this point is the value of x that makes the two sides of the equation equal, so it is the solution to the original equation.

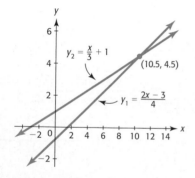

Figure 2.3

* See Appendix A, pages 658–659.

The equation in Example 5,

$$\frac{2x - 3}{4} = \frac{x}{3} + 1,$$

can be solved using the intersection method. Enter the left side as

$$y_1 = (2x - 3)/4 \text{ and the right side as } y_2 = (x/3) + 1$$

Graphing these two functions gives the graphs in Figure 2.3. The point of intersection is found to be $(10.5, 4.5)$. So, the solution is $x = 10.5$ (as we found in Example 5).

| **example 6** | ### Criminal Sentences |

The function $y = 0.55x - 2.886$ describes the mean time y served in prison for a crime as a function of the mean sentence length x, where x and y are each measured in months. To find the sentence for a crime that would give an expected time served of 10 years, write an equation and solve it by using (a) the x-intercept method, and (b) the intersection method.

Solution

a. Note that 10 years would be 120 months. We solve the linear equation $120 = 0.55x - 2.886$ by using the x-intercept method as follows:

Rewrite the equation in a form with 0 on one side:

$$0 = 0.55x - 2.886 - 120$$

Combine the constant terms to obtain

$$0 = 0.55x - 122.886$$

Enter $y_1 = 0.55x - 122.886$ and graph this equation (Figure 2.4). Using $\boxed{\text{ZERO}}$ gives an x-intercept of approximately 223 (Figure 2.4). Thus, if a prisoner receives a sentence of 223 months, we would expect him or her to serve 120 months, or 10 years.

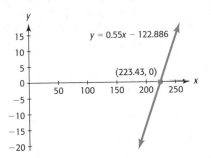

Figure 2.4

b. To use the intersection method, enter $y_1 = 0.55x - 2.886$ and $y_2 = 120$, graph these equations, and find the intersection of the lines with $\boxed{\text{INTERSECT}}$. Figure 2.5 shows the point of intersection, and again we see that x is approximately 223.

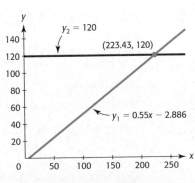

Figure 2.5

---

spreadsheet solution

We can use the **Goal Seek** feature of Excel to find the x-intercept of the graph of a function; that is, the zeros of the function. To solve

$$120 = 0.55x - 2.886$$

we enter the nonzero side of $0 = 0.55x - 122.886$ as shown in Table 2.2. Using **Goal Seek*** with cell B2 set to value 0 gives the value of x that solves the equation as shown in Table 2.3.

Table 2.2

	A	B
1	x	y
2		= 0.55*A2 − 122.886

Table 2.3

	A	B
1	x	y
2	223.43	0

Literal Equations; Solving an Equation for a Specified Linear Variable

An equation that contains two or more letters that represent constants or variables is called a **literal equation**. Formulas are examples of literal equations. If one of two or more of the variables in an equation is present only to the first power, we can solve for that variable by treating the other variables as constants and using the same steps that we used to solve a linear equation in one variable. This is often useful because it is necessary to get equations in the proper form to enter them in a graphing utility.

example 7

Simple Interest

The formula for the future value of an investment of P at simple interest rate r for t years is $A = P(1 + rt)$. Solve the formula for r, the interest rate.

Solution

$$A = P(1 + rt)$$

$$A = P + Prt \qquad \text{Multiply to remove parentheses.}$$

$$A - P = Prt \qquad \text{Get the term containing } r \text{ by itself on one side of the equation.}$$

$$r = \frac{A - P}{Pt} \qquad \text{Divide both sides by } Pt. \qquad ■$$

example 8

Solving an Equation for a Specified Variable

Solve the equation $2(2x - b) = \dfrac{5cx}{3}$ for x.

Solution

We solve the equation for x by treating the other variables as constants:

$$2(2x - b) = \frac{5cx}{3}$$

$$6(2x - b) = 5cx \qquad \text{Clear the equation of fractions by multiplying by the LCD, 3.}$$

---

* See Appendix B, page 685.

$$12x - 6b = 5cx$$ Multiply to remove parentheses.

$$12x - 5cx = 6b$$ Get all terms containing x on one side and all other terms on the other side.

$$x(12 - 5c) = 6b$$ Factor x from the expression. The remaining factor is the coefficient of x.

$$\frac{x(12 - 5c)}{(12 - 5c)} = \frac{6b}{12 - 5c}$$ Divide both sides by the coefficient of x.

$$x = \frac{6b}{12 - 5c}$$

Writing an Equation in Functional Form and Graphing Equations

example 9

Solve each of the following equations for y so that y is expressed as a function of x. Then graph the equation on a graphing utility with a standard viewing window.

a. $2x - 3y = 12$ **b.** $x^2 + 4y = 4$

Solution

a. Because y is to the first power in the equation, we solve the equation for y using linear equation solution methods:

$$2x - 3y = 12$$

$$-3y = -2x + 12$$ Isolate the term involving y by subtracting $2x$ from both sides.

$$y = \frac{2x}{3} - 4$$ Divide both sides by -3, the coefficient of y.

The graph of this equation is shown in Figure 2.6.

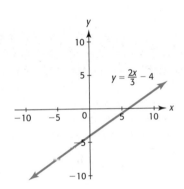

Figure 2.6

b. The variable y is to the first power in the equation

$$x^2 + 4y = 4$$

so we solve for y by using linear solution methods. (Note that to solve this equation for x would be more difficult; this method will be discussed in Chapter 3.)

$$4y = -x^2 + 4$$ Isolate the term containing y by subtracting x^2 from both sides.

$$y = \frac{-x^2}{4} + 1$$ Divide both sides by 4, the coefficient of y.

The graph of this equation is shown in Figure 2.7.

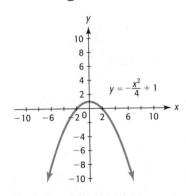

Figure 2.7

Direct Variation

Often in mathematics we need to express relationships between quantities. One relationship that is frequently used in applied mathematics occurs when two quantities are proportional. Two variables x and y are proportional to each other (or vary directly) if their quotient is a constant. That is, y is **directly proportional** to x, or y **varies directly** with x, if y and x are related by the equation

$$\frac{y}{x} = k \text{ or equivalently by } y = kx$$

where k is called the **constant of proportionality** or the **constant of variation**.

| example 10 | Circles |

Does the circumference of a circle vary directly with the radius of the circle?

Solution

The circumference of a circle varies directly with the radius of the circle, because $C = 2\pi r$. In this case, C and r are the variables, and 2π is the constant of variation. We could write an equivalent equation,

$$\frac{C}{r} = 2\pi,$$

which says that the quotient of the circumference divided by the radius is a constant, and that constant is 2π. ■

skills check

2.1

In Exercises 1–12, solve the following equations.

1. $5x - 14 = 23 + 7x$ **2.** $3x - 2 = 7x - 24$

3. $3(x - 7) = 19 - x$ **4.** $5(y - 6) = 18 - 2y$

5. $x - \dfrac{5}{6} = 3x + \dfrac{1}{4}$ **6.** $3x - \dfrac{1}{3} = 5x + \dfrac{3}{4}$

7. $\dfrac{5(x - 3)}{6} - x = 1 - \dfrac{x}{9}$

8. $\dfrac{4(y - 2)}{5} - y = 6 - \dfrac{y}{3}$

9. $5.92t = 1.78t - 4.14$

10. $0.023x + 0.8 = 0.36x - 5.266$

11. $\dfrac{3}{4} + \dfrac{1}{5}x - \dfrac{1}{3} = \dfrac{4}{5}x$ **12.** $\dfrac{2}{3}x - \dfrac{6}{5} = \dfrac{1}{2} + \dfrac{5}{6}x$

For Exercises 13–16, find: (a) the solution to the equation $f(x) = 0$, (b) the x-intercept of the graph of $y = f(x)$, and (c) the zero of $f(x)$.

13. $f(x) = 32 + 1.6x$ **14.** $f(x) = 15x - 60$

15. $f(x) = \dfrac{3}{2}x - 6$ **16.** $f(x) = \dfrac{x - 5}{4}$

In Exercises 17 and 18, you are given a table showing input and output values for a given function $y_1 = f(x)$. Using this table, find (if possible) (a) the x-intercept of the graph of $y = f(x)$, (b) the y-intercept of the graph of $y = f(x)$, and (c) the solution to the equation $f(x) = 0$.

17.

18.

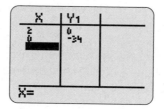

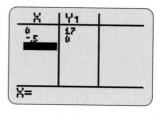

In Exercises 19 and 20, you are given the graph of a certain function $y = f(x)$ and the zero of that function. Using this graph, find (a) the x-intercept of the graph of $y = f(x)$ and (b) the solution to the equation $f(x) = 0$.

19.

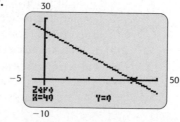

20.

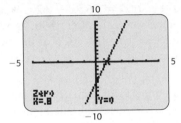

In Exercises 21–24, you are given the equation of a function. For each function, (a) find the zero of the function; (b) find the x-intercept of the graph of the function; and (c) solve the equation $f(x) = 0$.

21. $f(x) = 4x - 100$ **22.** $f(x) = 6x - 120$

23. $f(x) = 330 + 40x$ **24.** $f(x) = 250 + 45x$

In Exercises 25–32, solve the equations using graphical methods.

25. $14x - 24 = 27 - 3x$ **26.** $3x - 8 = 15x + 4$

27. $3(s - 8) = 5(s - 4) + 6$

28. $5(2x + 1) + 5 = 5(x - 2)$

29. $\dfrac{3t}{4} - 2 = \dfrac{5t - 1}{3} + 2$

30. $4 - \dfrac{x}{6} = \dfrac{3(x - 2)}{4}$ **31.** $\dfrac{t}{3} - \dfrac{1}{2} = \dfrac{t + 4}{9}$

32. $\dfrac{x - 5}{4} + x = \dfrac{x}{2} + \dfrac{1}{3}$

33. Solve $A = P(1 + rt)$ (a) for t and (b) for P.

34. Solve $V = \dfrac{1}{3}\pi r^2 h$ for h.

35. Solve $5F - 9C = 160$ for F.

36. Solve $4(a - 2x) = 5x + \dfrac{c}{3}$ for x.

37. Solve $\dfrac{P}{2} + A = 5m - 2n$ for n.

38. Solve $y - y_1 = m(x - x_1)$ for x.

In Exercises 39–42, solve the following equations for y and graph them with a standard window on a graphing utility.

39. $5x - 3y = 5$ **40.** $3x + 2y = 6$

41. $x^2 + 2y = 6$ **42.** $4x^2 + 2y = 8$

exercises

2.1

43. *Depreciation* An $828,000 building is depreciated for tax purposes by its owner using the straight-line depreciation method. The value of the building, y, after x months of use, is given by $y = 828,000 - 2300x$ dollars. After how many years will the value of the building be $690,000?

44. *Temperature Conversion* The equation $5F - 9C = 160$ gives the relationship between Fahrenheit and Celsius temperature measurements. What Fahrenheit measure is equivalent to a Celsius measurement of 20°?

45. *Investments* The future value of a simple interest investment is given by $S = P(1 + rt)$. What principal P must be invested for $t = 5$ years at the simple interest rate $r = 10\%$ so that the future value grows to $9000?

46. *Temperature–Humidity Index* The temperature–humidity index I is $I = t - 0.55(1 - h)(t - 58)$, where t is the air temperature in degrees Fahrenheit and h is the relative humidity expressed as a decimal. If the air temperature is 80°, find the humidity h (as a percent) that gives an index value of 79.

47. *Earnings and Minorities* The median annual salary (in thousands of dollars) of minorities M is related to the median salary W (in thousands of dollars) of whites by $M = 0.959W - 1.226$. What is the median annual salary for whites when the median annual salary for minorities is $50,560? (Source: *Statistical Abstract of the United States*)

48. *Internet Brokerage Accounts* The number of brokerage accounts on the Internet in year t can be modeled by $B(t) = 3.303t - 6591.560$ million accounts. If this model remains valid, in what year were there 14.44 million accounts? (Source: Gomez Advisors)

49. *Game Show Question* The following question was worth $32,000 on the game show *Who Wants to Be a Millionaire?*: "At what temperature are the Fahrenheit and Celsius temperature scales the same?" Answer this question. Recall that Fahrenheit and Celsius temperatures are related by $5F - 9C = 160$.

50. *Out-of-Wedlock Births* The percent of all teen mothers who are unmarried can be given by the equation $y = 1.78x - 3.998$, where x is the number of years from 1950. If this model remains accurate, in what year will the percent have reached 80?
(Source: Father Facts)

51. *Reading Score* The average reading score on the National Assessment of Progress tests is given by $y = 0.155x + 255.37$, where x is the number of years past 1970. In what year would the average reading score be 259.4 if this model is accurate?

52. *Tax Burden* The per capita tax burden B can be described by $B(t) = 17.69 + 2.25t$ hundred dollars, where t is the number of years past 1980. In what year does this model indicate that the per capita tax burden was $4694?
(Source: Internal Revenue Service)

53. *Profit* The profit from the production and sale of specialty golf hats is given by the function $P(x) = 20x - 4000$, where x is the number of hats produced and sold.

   a. Producing and selling how many units will give a profit of $8000?

   b. How many units must be produced and sold to avoid a loss?

54. *Marriage Rate* The percent of unmarried women who get married in any particular year is given by $y = -0.0762x + 8.5284$, where x is the number of years after 1950. During what year does this model estimate the percent was 6.09?
(Source: Index of Leading Cultural Indicators)

55. *U.S. Population* The U.S. population can be modeled for the years 1960–2005 by the function $p = 2490x + 179,280$, where p is in thousands of people and x is in years from 1960. During what year does the model estimate the population to be 303,780,000?
(Source: www.census.gov/statab)

56. *Internet Workers* The percent of Fortune Global 500 firms that actively recruited workers on the Internet can be modeled by $P(x) = 26.5x - 62$ percent, where x is the number of years after 1995. In what year does this model indicate that 44% of the firms recruited on the Internet?
(Source: iLogos.com)

57. *Inmates* The total number of inmates in custody between 1990 and 2005 in state and federal prisons is given approximately by $y = 76x + 115$ thousand prisoners, where x is the number of years after 1990. If the model remains accurate long enough, in what year will the number of inmates be 1,787,000?
(Source: Bureau of Justice Statistics)

58. *Marijuana Use* The percent p of high school seniors using marijuana daily can be related to x, the number of years after 1990, by the equation $30p - 19x = 1$. Assuming the model remains accurate, during what year was the percent using marijuana daily equal to 7%?
(Source: Index of Leading Cultural Indicators)

59. *Presidential Elections* The percent of the population voting in presidential elections has been estimated to be $p = 63.20 - 0.26x$, where x is the number of years after 1950.
(Source: Federal Election Commission)

   a. During what year does this model estimate that the percent voting in a presidential election will be 49.16%?

   b. Of those persons eligible to vote, 53% cast a ballot in the 2000 presidential election. Did this model accurately describe the percent voting in the 2000 election? What does this answer tell you about using models for predictions?

60. *Personal Income* U.S. personal income increased between 1980 and 2005 according to the model $I(x) = 386.17x + 524.32$ billion dollars, where x is the number of years after 1980. In what year does this model predict that U.S. personal income would reach $10,951 billion if this model is accurate beyond 2005?
(Source: *Statistical Abstract of the United States*)

61. *Cigarette Advertising* The total U.S. cigarette advertising and promotional expenditures can be modeled by the equation $y = 277.318x - 1424.766$, where y is measured in millions of dollars and x is the number of years from 1970. If this model remains accurate, in what year did the spending exceed $6 billion?
(Source: Federal Trade Commission)

62. *Cell Phone Subscribers* The number of mobile-phone subscribers (in millions) between 1995 and 2006 can be modeled by $S(x) = 18.13x + 21.03$, where x is the number of years after 1995. In what year does this model indicate that there were 292,980,000 subscribers?
(Source: Semiannual CTIA Wireless Survey)

63. *Grades* To earn an A in a course, a student must get an average score of at least 90 on five tests. If her first four test scores are 92, 86, 79, and 96, what score does she need on the last test to obtain a 90 average?

64. *Grades* To earn an A in a course, a student must get an average score of at least 90 on three tests and a final exam. If the final score is higher than his lowest score, then the lowest score is removed and the final exam score counts double. If his first three test scores are 86, 79, and 96, what is the lowest score he needs on the last test to obtain a 90 average?

65. *Tobacco Judgment* As part of the largest-to-date damage award in a jury trial in history, the July 2000 penalty handed down against the tobacco industry included a $74 billion judgment against Philip Morris. If this amount was 94% of this company's 1999 revenue, how much was Philip Morris's 1999 revenue?
(Source: *Newsweek*, July 24, 2000)

66. *Tobacco Judgment* As part of the largest-to-date damage award in a jury trial in history, the July 2000 penalty handed down against the tobacco industry included a $36 billion judgment against R. J. Reynolds. If this amount was 479% of this company's 1999 revenue, how much was R. J. Reynolds 1999 revenue?
(Source: *Newsweek*, July 24, 2000)

67. *Sales Commission* A salesman earns $50,000 in commission in 1 year and then has his commission reduced by 20% the next year. What percent increase in commission over the second year will give him $50,000 in the third year?

68. *Salaries* A man earning $100,000 per year has his salary reduced by 5% because of a reduction in his company's market. A year later, he receives $104,500 in salary. What percent raise from the reduced salary does this represent?

69. *Sales Tax* The total cost of a new automobile, including a 6% sales tax on the price of the automobile, is $29,998. How much of the total cost of this new automobile is sales tax?

70. *Wildlife Management* In wildlife management, the capture–mark–recapture technique is used to estimate the population size of certain types of fish or animals. To estimate the population, we enter information in the equation

$$\frac{\text{total in population}}{\text{total number marked}} = \frac{\text{total number in second capture}}{\text{number found marked in second capture}}$$

Suppose that 50 sharks are caught along a certain shoreline, marked, and then released. If a second capture of 50 sharks gives 20 sharks that have been marked, what is the resulting population estimate?

71. The formula for the future value A of a simple interest investment is $A = P + Prt$, where P is the original investment, r is the annual interest rate, and t is the time in years. Solve this formula for t.

72. The formula for the future value A of a simple interest investment is $A = P + Prt$, where P is the original investment, r is the annual interest rate, and t is the time in years. Solve this formula for P.

73. If $\$P$ is invested for t years at simple interest rate r, the future value of the investment is $A = P + Prt$. If $2000 invested for 6 years gives a future value of $3200, what is the simple interest rate of this investment?

74. If an investment at 7% simple interest has a future value of $5888 in 12 years, what was the original investment?

75. The simple interest earned in 9 years is directly proportional to the interest rate r. If the interest is $920 when r is 12%, what is the amount of interest earned in 9 years at 8%?

76. The interest earned at 9% simple interest is directly proportional to the number of years the money is invested. If the interest is $4903.65 in 5 years, in how many years will the interest earned at 9% be $7845.84?

Fitting Lines to Data Points: Modeling Linear Functions

section preview ▪ Earnings and Gender

Table 2.4 shows the earnings of year-round full-time workers by gender and educational attainment. Figure 2.8 shows the scatter plot of the data, and it appears that a line would approximately fit along these data points. We can determine the relationship between the two sets of earnings by creating a linear equation that gives the female annual earnings as a function of the male annual earnings. This model can be used to estimate an annual female salary from a given annual male salary. (See Example 3.) In this section, we will learn how to create linear models from data points, how to use graphing utilities to create linear models, and how to use the model to answer questions about the data.

Table 2.4

Educational Attainment	Average Annual Earnings for Males ($ thousand)	Average Annual Earnings for Females ($ thousand)
Less than ninth grade	21.659	17.659
Some high school	26.277	19.162
High school graduate	35.725	26.029
Some college	41.875	30.816
Associate's degree	44.404	33.481
Bachelor's degree	57.220	41.681
Master's degree	71.530	51.316
Doctorate degree	82.401	68.875
Professional degree	100.000	75.036

(Source: U.S. Census Bureau)

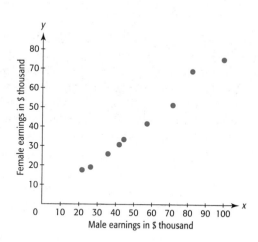

Figure 2.8

Exact and Approximate Linear Models

We have seen that if data points fit exactly on a line, we can use two of the points to model the linear function (write the equation of the line). We can determine that the data points fit exactly on a line by determining that the changes in output values are equal for equal changes in the input values. In this case, we say that the inputs are **uniform** and the **first differences** are constant.

> - If the first differences of data outputs are constant for uniform inputs, the rate of change is constant and a linear function can be found that fits the data exactly.
> - If the first differences are "nearly constant," a linear function can be found that is an approximate fit for the data.

If the first differences of data outputs are constant for uniform inputs, we can use two of the points to write the linear equation that models the data. If the first differences of data outputs are constant for inputs differing by 1, this constant difference is the rate of change of the function, which is the slope of the line fitting the points exactly.

example 1

Retirement

Table 2.5 gives the annual retirement payment of a 62-year-old retiree with 21 or more years of service at Clarion State University as a function of the number of years of service and the first differences of the outputs.

Table 2.5

Year	21	22	23	24	25
Retirement Payment	40,950	42,900	44,850	46,800	48,750
First Differences		1950	1950	1950	1950

Because the first differences of the outputs of this function are constant (equal to 1950) for each unit change of the input (years), the rate of change is the constant 1950. Using this rate of change and a point gives the equation of the line that contains all the points. Representing the annual retirement payment by y and the years of service by x, and using the point (21, 40,950), we obtain the equation

$$y - 40{,}950 = 1950(x - 21)$$
$$y = 1950x$$

Note that the values in Table 2.5 represent points satisfying a **discrete function** (a function with a finite number of inputs) with each input representing the number of years of service. Although only points with integer inputs represent the annual retirement, we can model the application with the **continuous function** $y = 1950x$, whose graph is a line that passes through the 5 data points. Informally, a continuous function has a graph that can be drawn over its domain without lifting the pen from the paper.

In the context of this application, we must give a discrete interpretation to the model. This is because the only inputs of the function that make sense in this case are nonnegative integers representing the number of years of service. The graph of the discrete function defined in Table 2.5 is the scatter plot shown in Figure 2.9(a) on the next page, and the continuous function that fits these data points, $y = 1950x$, is shown in Figure 2.9(b).

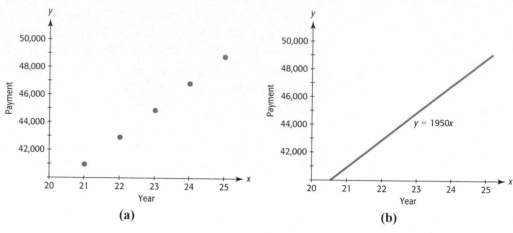

Figure 2.9

When a scatter plot of data can be approximately fitted by a line, we attempt to find a graph that visually gives the best fit for the data and find its equation. For now, we informally define the "best-fit" line as the one that appears to come closest to all the data points.

<table>
<tr><td>example 2</td></tr>
</table>

Health Service Employment

Table 2.6 gives the number of full- and part-time employees in offices and clinics of dentists for selected years between 1990 and 2005. (Source: U.S. Bureau of Labor Statistics)

Table 2.6

Year	1990	1995	2000	2001	2002	2003	2004	2005
Employees (in thousands)	513	592	688	705	725	744	760	771

a. Draw a scatter plot of the data with the x-value of each point representing the number of years after 1990 and the y-value representing the number of dental employees (in thousands) corresponding to that year.

b. Graph the equation $y = 16x + 510$ on the same graph as the scatter plot and determine if the line appears to be a good fit.

c. Draw a "visual fit" line that fits the data well (a piece of spaghetti or pencil lead over your calculator screen works well) and select two points on that line (use the free-moving cursor). Use these two points to write an equation of the "visual fit" line. Determine whether this line or the one from part (b) is the better fit.

Solution

a. By using x as the number of years after 1990, we have aligned the data with $x = 0$ representing 1990, $x = 5$ representing 1995, and so forth. We enter into the lists of a graphing utility (or an Excel spreadsheet) the aligned input data representing the years in Table 2.6 and the output data representing the number of employees shown in the second row of Table 2.6. Figure 2.10(a) shows the lists containing the data. The graph of these data points is shown in Figure 2.10(b). The window for this scatter plot can be set manually or with a command on the graphing utility such as ZoomStat, which is used to automatically set the window and display the graph.

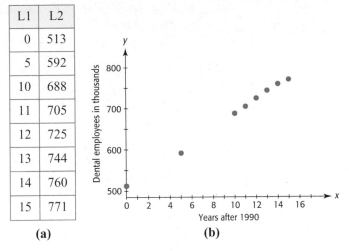

L1	L2
0	513
5	592
10	688
11	705
12	725
13	744
14	760
15	771

(a)   (b)

Figure 2.10

b. The graphs of the equation $y = 16x + 510$ and the scatter plot of the data are shown in Figure 2.11. The line does not appear to be the best possible fit to the data points.

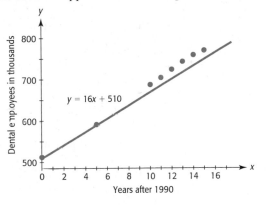

Figure 2.11

c. Figure 2.12(a) shows one example of a visual fit line that fits the data obtained by placing a piece of spaghetti on the calculator screen close to the data points. By using the "free-moving" cursor (press the right, left, up, and down arrows) we obtain two points that lie close to the visual fit line. Two such points that lie on the line are (3.096, 563.934) and (12.479, 731.218) (rounded to three decimal places), and are shown in Figure 2.12(b) and (c).*

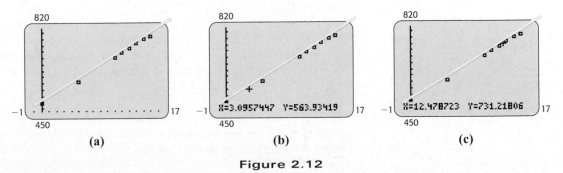

(a)   (b)   (c)

Figure 2.12

---

* Many other points are possible.

To write the equation of the line, we must first find the slope of the line between these two points.

$$m = \frac{731.218 - 563.934}{12.479 - 3.096} \approx 17.83$$

The equation of our visual fit line is

$$y - 563.934 = 17.83(x - 3.096)$$
$$y = 17.830x + 508.732$$

The graphs of the data points and the visual fit line are shown in Figure 2.13.

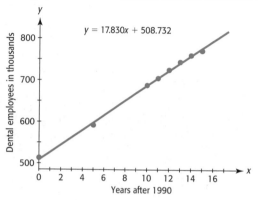

Figure 2.13

Figures 2.11 and 2.13 show that the visual fit line, with equation $y = 17.830x + 508.732$, appears to be a better fit to the data. ∎

Is either of the lines in Figure 2.11 or Figure 2.13 the best-fitting line of all possible lines that could be drawn on the scatter plot? How do we determine the best-fit line? We now discuss the answers to these questions.

Fitting Lines to Data Points; Linear Regression

The points determined by the data in Table 2.6 on page 108 do not all lie on a line, but we can determine the equation of the line that is the "best fit" for these points by using a procedure called **linear regression**. This procedure defines the *best-fit* line as the line for which the sum of the squares of the vertical distances from the data points to the line is a minimum. For this reason, the linear regression procedure is also called the **least squares method**.

The vertical distance between a data point and the corresponding point on a line is simply how much the line misses going through the point—that is, the difference in the y-values of the data point and the point on the line. If we call this difference in outputs d_i (where i takes on the values from 1 to n for n data points), the least squares method* requires that

$$d_1{}^2 + d_2{}^2 + d_3{}^2 + \cdots + d_n{}^2$$

be as small as possible. The line for which this sum of squared differences is as small as possible is called the *linear regression line* or *least squares line* and is the one that we consider to be the **best-fit line** for the data.

---

* The sum of the squared differences is often called SSE, the *sum of squared errors*. The development of the equations that lead to the minimum SSE is a calculus topic.

To illustrate the linear regression process, consider again the last two lines in Figures 2.11 and 2.13, and the data points in Table 2.6. Figures 2.14 and 2.15 indicate the vertical distances for each of these lines. To the right of each figure is the calculation of the sum of squared differences for each line.

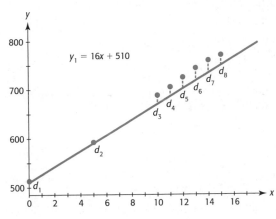

$$d_1^2 + d_2^2 + d_3^2 + \cdots + d_8^2$$
$$= (513 - 510)^2 + (592 - 590)^2 + (688 - 670)^2$$
$$+ (705 - 686)^2 + (725 - 702)^2 + (744 - 718)^2$$
$$+ (760 - 734)^2 + (771 - 750)^2$$
$$= 9 + 4 + 324 + 361 + 529 + 676 + 676 + 441$$
$$= 3020$$

Figure 2.14

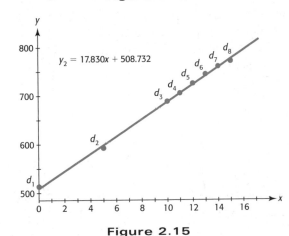

$$d_1^2 + d_2^2 + d_3^2 + \cdots + d_8^2$$
$$= (513 - 508.73)^2 + (592 - 597.88)^2 + (688 - 687.03)^2$$
$$+ (705 - 704.86)^2 + (725 - 722.69)^2 + (744 - 740.52)^2$$
$$+ (760 - 758.35)^2 + (771 - 776.18)^2$$
$$= 100.77$$

Figure 2.15

We see numerically from the sum of squared differences calculations, as well as visually, from Figures 2.14 and 2.15, that the line

$$y_2 = 17.830x + 508.732$$

is the line that is the better model for the data when given a choice of the lines y_1 and y_2. However, is this line the best-fit line for these data? Remember that the best-fit line must have the smallest sum of squared differences of *all* lines that can be drawn through the data! We will see that the line which is the best fit for these data points, with coefficients rounded to three decimal places, is

$$y = 17.733x + 509.917$$

The sum of squared differences for the regression line is approximately 99, which is slightly smaller than the sum 100.77, which was found for $y_2 = 17.830x + 508.732$. ■

Development of the formulas that give the best-fit line for a set of data is beyond the scope of this text, but graphing calculators, computer programs, and spreadsheets have built-in formulas or programs that give the equation of the best-fit line. That is, we can use technology to find the best linear model for the data. The calculation of the sum of squared differences for the dental employees data was given only to illustrate what the best-fit line means. You will not be asked to calculate, nor do we use from this point on, the value of the sum of squared differences in this text.

We illustrate the use of technology to find the regression line by returning to the data of Example 2. To find the equation that gives the number of employees in dental offices as a function of the years after 1990, we use the following steps:

Modeling Data

Step 1:   Enter the data into lists of a graphing utility.

Step 2:   Create a scatter plot of the data to see if a linear model is reasonable. The data should appear to follow a linear pattern with no distinct curvature.

Step 3:   Use the graphing utility to obtain the linear equation that is the best fit for the data. Figure 2.16(a) shows the equation for Example 2, which can be approximated by $y = 17.733x + 509.917$.*

Step 4:   Graph the linear function (unrounded) and the data points on the same graph to see how well the function fits the data. (The equation entered into an equation editor and the graph of the data and the best-fit line for Example 2 are shown in Figures 2.16(b) and Figure 2.16(c), respectively.)

Step 5:   Report the function and/or numerical results in a way that makes sense in the context of the problem, with the appropriate units, and with the variables identified. Unless otherwise indicated, report functions with coefficients rounded to 3 decimal places.

* See Appendix A, pages 661–663.

Figure 2.16(c) shows that

$$y = 17.733x + 509.917$$

is a good model of the number of full- and part-time employees in dentist's offices and clinics, where y is in thousand of employees and x is the number of years after 1990.

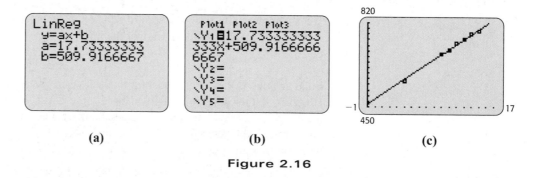

(a)   (b)   (c)

Figure 2.16

The screens for the graphing utility that you use may vary slightly from those given in this text. Also, the regression line you obtain is dependent on your particular technology and may have some decimal places slightly different from those shown here.

When modeling a set of data, it is important to be careful when rounding coefficients in equations and when rounding during calculations. We will use the following guidelines in this text.

Technology Note

After a model for a data set has been found, it can be rounded for reporting purposes. However, do not use a rounded model in calculations, and do not round answers during the calculation process unless otherwise specified. When the model is used to find numerical answers, the answers should be rounded in a way that agrees with the context of the problem.

example 3

Earnings and Gender

Table 2.7 shows the earnings of year-round full-time workers by gender and educational attainment.

a. Let x represent earnings for males, let y represent earnings for females, and create a scatter plot of the data.

b. Create a linear model that expresses females' annual earnings as a function of males' earnings.

c. Graph the linear function and the data points on the same graph, and discuss how well the function models the data.

Table 2.7

Educational Attainment	Average Annual Earnings for Males ($ thousand)	Average Annual Earnings for Females ($ thousand)
Less than ninth grade	21.659	17.659
Some high school	26.277	19.162
High school graduate	35.725	26.029
Some college	41.875	30.816
Associate's degree	44.404	33.481
Bachelor's degree	57.220	41.681
Master's degree	71.530	51.316
Doctorate degree	82.401	68.875
Professional degree	100.000	75.036

(Source: U.S. Census Bureau)

Solution

a. From Table 2.7 enter the data in the lists of a graphing utility. Figure 2.17(a) shows a partial list of the data points. The scatter plot of all the data points from Table 2.7 is shown in Figure 2.17(b).

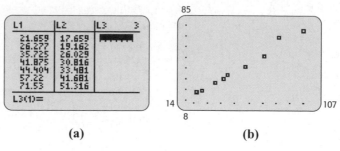

(a)   (b)

Figure 2.17

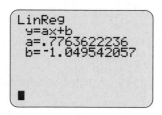

Figure 2.18

b. The points in the scatter plot in Figure 2.17(b) exhibit a nearly linear pattern, so a linear function could be used to model these data. The equation of the line that is the best fit for the data points can be found with a graphing calculator (Figure 2.18). The equation, rounded to three decimal places, is

$$y = 0.776x - 1.050$$

Remember that a *model* gives not only the equation but also a description of the variables and their units of measure. The rounded model

$$y = 0.776x - 1.050 \text{ thousand dollars}$$

where x is in thousands of dollars, expresses women's annual earnings y as a function of men's annual earnings x.

c. Using the unrounded function in the equation editor (Figure 2.19(a)) and graphing it along with the data points,* we observe that the line follows the general trend indicated by the data (Figure 2.19(b)). Note that not all points lie on the graph of the equation, even though this is the line that is the best fit for the data.

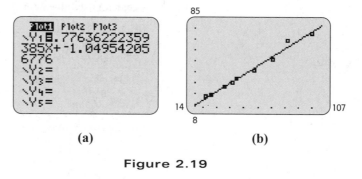

(a)   (b)

Figure 2.19

---

example 4   ## U.S. Population

The total U.S. population for selected years beginning in 1960 and projected to 2050 is shown in Table 2.8, with the population given in millions.

a. Align the data to represent the number of years from 1960, and draw a scatter plot of the data.

b. Create a linear equation that is the best fit for these data, where y is in millions and x is the number of years from 1960.

---

* Even though we write rounded models found from data given in this text, all graphs and calculations use the unrounded model found by the graphing utility.

c. Graph the equation of the linear model on the same graph with the scatter plot and discuss how well the model fits the data.

d. Align the data to represent the years after 1950 and create a linear equation that is the best fit for the data, where y is in millions.

e. How do the x-values for a given year differ?

f. Use both unrounded models to estimate the population in 1997 and in 2000. Are the estimates equal?

Table 2.8

Year	Population (millions)	Year	Population (millions)
1960	180.671	1995	263.044
1965	194.303	1998	270.561
1970	205.052	2000	281.422
1975	215.973	2003	294.043
1980	227.726	2025	358.030
1985	238.466	2050	408.695
1990	249.948		

(Source: U.S. Census Bureau)

Solution

a. The aligned data have $x = 0$ representing 1960, $x = 5$ representing 1965, and so forth. Figure 2.20(a) shows the first seven entries using the aligned data. The scatter plot of these data is shown in Figure 2.20(b). The coordinates of the first two points in the scatter plot are (0, 180.671) and (5, 194.303).

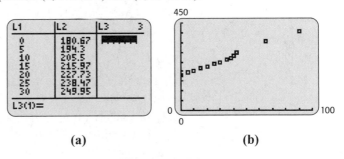

(a)   (b)

Figure 2.20

b. The equation of the best-fit line is found by using linear regression in the graphing calculator. Rounding the decimals to three places, the linear model for the U.S. population is

$$y = 2.607x + 177.195 \text{ million}$$

where x is the number of years from 1960.

c. Using the unrounded function in the equation editor (Figure 2.21(a)) and graphing it along with the scatter plot shows that the graph of the best-fit line is very close to the data points (Figure 2.21(b)). However, the points do not all fit the line because the U.S. population did not increase by exactly the same amount each year.

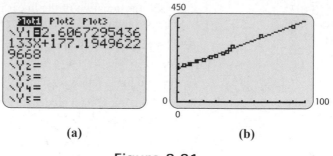

Figure 2.21

d. If we align the data to represent the years after 1950, then $x = 10$ corresponds to 1960, $x = 15$ corresponds to 1965, and so fourth. Figure 2.22(a) shows the first seven entries using the aligned data, and Figure 2.22(b) shows the scatter plot. The equation that best fits the data, found using linear regression with a calculator, is

$$y = 2.607x + 151.128 \text{ million}$$

where x is the number of years from 1950. Figure 2.22(c) shows the regression equation.

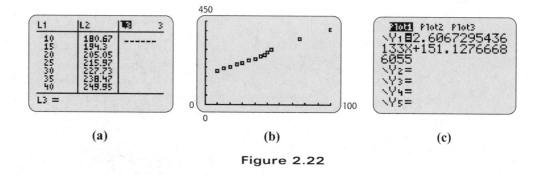

Figure 2.22

e. The x-values for a given year are 10 more for this model than the x-values for the first model.

f. Both models estimate the population to be 273.644 million in 1997 and to be 281.464 million in 2000. They are equal estimates. In fact, if you substitute $x - 10$ for x in the first model, you will get the second model. ∎

| spreadsheet solution |

We can also use spreadsheets to find the linear function that is the best fit for the data. Table 2.9 shows the Excel spreadsheet for the data of Example 4 with x equal to the years after 1960. Selecting the cells containing the data, using **Chart Wizard** to get the scatter plot of the data, and selecting **Add Trendline** gives the equation of the linear function that is the best fit for the data, along with the scatter plot and the graph of the best-fitting line. The equation and graph of the line are shown in Figure 2.23.*

---

* See Appendix B, page 683.

Table 2.9

	A	B
1	Year x	Population (millions) y
2	0	180.671
3	5	194.303
4	10	205.052
5	15	215.973
6	20	227.726
7	25	238.466
8	30	249.948
9	35	263.044
10	38	270.561
11	40	281.422
12	43	294.043
13	65	358.03
14	90	408.695

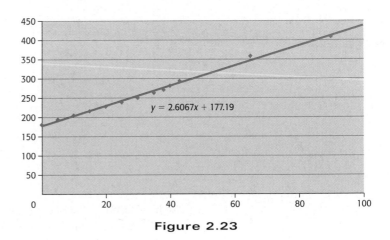

Figure 2.23

Applying Models

Because 1997 is a year between two given values in the table of Example 4, using the model to find the population in 1997 is called **interpolation**. When a model is evaluated for a *prediction* using input(s) outside the given data points, the output is called an **extrapolation**.

Another question arises whenever data that involves time is used. If we label the x-coordinate of a point on the input axis as 1999, what time during 1999 do we mean? Does 1999 refer to the beginning of the year, the middle of the year, its end, or some other time? Because most data from which the functions in applications are derived represent end-of-year totals, we make the following convention when modeling, unless otherwise specified.

A point on the input axis indicating a time refers to the end of the time period.

For instance, the point representing the year 1999 means "at the end of 1999." Notice that this instant in time also represents the beginning of the year 2000. If, for example, data is aligned with x equal to the number of years after 1990, then any x-value greater than 9 and less than or equal to 10 represents some time in the year 2000. If a point represents anything other than the end of the period, this information will be clearly indicated. Also, if a and b are points in time, we use the phrases "from a to b" and "between a and b" to both indicate the same-time interval.

Goodness of Fit

Consider again Example 4, where we modeled the U.S. population for selected years. Looking at Table 2.10, notice that for uniform inputs the first differences of the outputs appear to be relatively close to the same constant, especially compared to the size of the population. (If these differences were closer to a constant, the fit would be better.)

Table 2.10

Uniform Inputs (years)	Outputs Population (millions)	First Differences in Output (difference in population)
1960	180.671	
1965	194.303	**13.632**
1970	205.052	**10.749**
1975	215.973	**10.921**
1980	227.726	**11.753**
1985	238.466	**10.740**
1990	249.948	**11.482**
1995	263.044	**13.096**
2000	281.422	**18.378**

So how "good" is the fit of the linear model $y = 2.6067295436133x + 177.19496229668$ to the data in Example 4? Based on observation of the graph of the line and the data points on the same set of axes (see Figure 2.23), it is reasonable to say that the regression line provides a very good fit, but not an exact fit, to the data.

The goodness of fit of a line to a set of data points can be observed from the graph of the line and the data points on the same set of axes, and it can be measured if your graphing utility computes the **correlation coefficient**. The correlation coefficient is a number r, $-1 \le r \le 1$, that measures the strength of the linear relationship that exists between the two variables. The closer that $|r|$ is to 1, the more closely the data points fit on the linear regression line. (There is no linear relationship between the two variables if $r = 0$.) Positive values of r indicate that the output variable increases as the input variable increases, and negative values of r indicate that the output variable decreases as the input variable increases. But the *strength* of the relationship is indicated by how close $|r|$ is to 1. For the data of Example 4, computing the correlation coefficient gives $r = .997$ (Figure 2.24), which means that the linear relationship is strong and that the linear model is an excellent fit for the data. When a calculator feature or computer program is used to fit a linear model to data, the resulting equation can be considered a best fit for the data. As we will see later in this text, other mathematical models may be better fits to some sets of data, especially if the first differences of the outputs are not close to being constant.

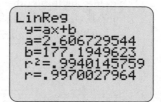

Figure 2.24

skills check

2.2

Report models to three decimal places unless otherwise specified. Use unrounded models to graph and calculate unless otherwise specified.

Discuss whether the data shown in the scatter plots in the figures for Exercises 1 and 2 should be modeled by a linear function.

1.

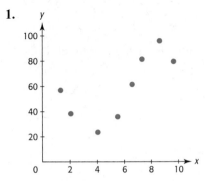

2.

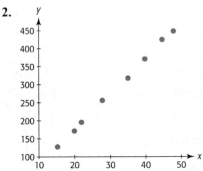

Discuss whether the data shown in the scatter plots in the figures for Exercises 3 and 4 should be modeled by a linear function exactly or approximately.

3.

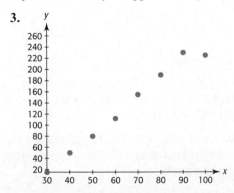

4.

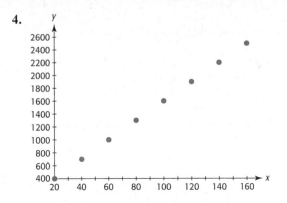

Create a scatter plot for each of the sets of data in Exercises 5 and 6.

5.

x	1	3	5	7	9
y	4	7	10	13	16

6.

x	1	2	3	5	7	9	12
y	1	3	6	1	9	2	6

7. Can the scatter plot in Exercise 5 be fit exactly or only approximately by a linear function? How do you know?

8. Can the scatter plot in Exercise 6 be fit exactly or only approximately by a linear function? How do you know?

9. Find the linear function that is the best fit for the data in Exercise 5.

10. Find the linear function that is the best fit for the data in Exercise 6.

Use the data in the following table for Exercises 11–14.

x	5	8	11	14	17	20
y	7	14	20	28	36	43

11. Construct a scatter plot of the data in the table.

12. Determine if the points plotted in Exercise 11 appear to lie near some line.

13. Create a linear model for the data in the table.

14. Use the function $y = f(x)$ created in Exercise 13 to evaluate $f(3)$ and $f(5)$.

Use the data in the table for Exercises 15–18.

x	2	5	8	9	10	12	16
y	5	10	14	16	18	21	27

15. Construct a scatter plot of the data in the table.

16. Determine if the points plotted in Exercise 15 appear to lie near some line.

17. Create a linear model for the data in the table.

18. Use the rounded function $y = f(x)$ that was reported in Exercise 17 to evaluate $f(3)$ and $f(5)$.

19. Determine which of the equations, $y = -2x + 8$ or $y = -1.5x + 8$, is the better fit for the data points $(0, 8), (1, 6), (2, 5), (3, 3)$.

20. Determine which of the equations, $y = 2.3x + 4$ or $y = 2.1x + 6$, is the better fit for the data points $(20, 50), (30, 73), (40, 96), (50, 119), (60, 142)$.

21. Without graphing, determine which of the following data sets are exactly linear, approximately linear, or nonlinear.

a.

x	y
1	5
2	8
3	11
4	14
5	17

b.

x	y
1	2
2	5
3	10
4	17
5	26

c.

x	y
1	6
2	7
3	12
4	14
5	18

22. Why can't first differences be used to tell if the following data is linear? $(1, 3), (4, 5), (5, 7), (7, 9)$

exercises

2.2

Report models to three decimal places unless otherwise specified. Use unrounded models to graph and calculate unless otherwise specified.

23. *Working Age* The scatter plot in the figure at right projects the ratio of the working-age population to the elderly.

    a. Does the scatter plot in the figure represent discrete or continuous data?

    b. Would a linear function be a good model for the data points shown in the figure? Explain.

    c. Could the data shown for the years beyond 2010 be modeled by a linear function?

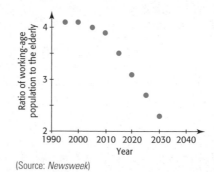

(Source: *Newsweek*)

24. *Women in the Workforce* The number of women in the workforce for selected years from 1890 to 2005 is shown in the following figures.

    a. Does the scatter plot in the figure define the number of working women as a discrete or continuous function of the year?

b. Does the graph of $y = W(x)$ shown in the figure define the number of working women as a discrete or continuous function of the year?

c. Would the data in the scatter plot be better modeled by a linear function rather than by the nonlinear function $y = W(x)$? Why or why not?

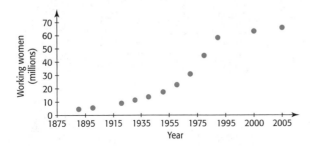

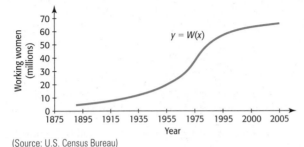

(Source: U.S. Census Bureau)

25. *Future Value of an Investment* If $1000 is invested at 6% simple interest, the initial value and the future value S at the end of each of 5 years is shown in the table below.

a. Can a linear function model exactly the points from the table? Explain.

b. If so, find a linear function $S = f(t)$ that models the points.

c. Use the model to find the future value of this investment at the end of the 7th year. Is this an interpolation or an extrapolation from the data?

d. Should this model be interpreted discretely or continuously?

Year (t)	0	1	2	3	4	5
Future Value (S)	1000	1060	1120	1180	1240	1300

26. *Education Spending* The following figure gives the amount (in billions of dollars) spent on education during the years 1996–2001. Can this data be modeled exactly by a linear function? Explain.

Education spending (in billions)

(Source: U.S. Department of Education)

27. *Taxes* The table below shows some sample incomes and the income tax due for each taxable income.

a. Can a linear function model exactly the points from the table? Explain.

b. If so, find a linear function $T = f(x)$ that models the points.

c. Verify that the model fits the data by evaluating the function at $x = 30{,}100$ and $x = 30{,}300$ and comparing the resulting T-values with the income tax due for these taxable incomes.

d. If the model can be interpreted continuously, use it to find the tax due on taxable income of $30,125. Is this an interpolation or an extrapolation from the data?

e. Can this model be used to compute all tax due on taxable income?

Taxable Income	Income Tax Due
$30,000	$3717.50
30,050	3725
30,100	3732.50
30,150	3740
30,200	3747.50
30,250	3755
30,300	3762.50

(Source: U.S. Federal tax table for 2007)

28. *Farms* The figure on the next page gives the number of farms (in millions) for selected years from 1940 and 2005. Can this data be modeled exactly by a linear function? Explain.

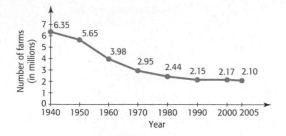

29. *High School Enrollment* The table below gives the enrollment (in thousands) in grades 9–12 of U.S. public and private schools for the years 1990–2006.

a. Create a scatter plot of the data with x equal to the year after 1990 and y equal to enrollment in thousands.

b. Find the linear model that is the best fit for the data.

c. Graph the model and the scatter plot on the same axes. Is the model a "good" fit?

Year, x	Enrollment, y (thousands)	Year, x	Enrollment, y (thousands)
1990	12,488	1999	14,623
1991	12,703	2000	14,802
1992	12,882	2001	15,058
1993	13,093	2002	15,332
1994	13,376	2003	15,721
1995	13,697	2004	16,048
1996	14,060	2005	16,328
1997	14,272	2006	16,498
1998	14,428		

(Source: U.S. Department of Education)

30. *Marriage Rate* The marriage rate (per 1000 unmarried women) is given by the following table for the years 1960 to 1997.

a. Write the linear equation that models the marriage rate as a function of years after 1950.

b. What does the model indicate the marriage rate to be in 1985?

c. For what year does the model indicate that the rate fell below 50 per 1000?

d. Discuss the reliability of your answer to part (c).

Year	Marriage Rate (marriages per 1000 unmarried women)
1960	73.5
1970	76.5
1980	61.4
1990	54.5
1991	54.2
1992	53.3
1993	52.3
1994	51.5
1995	50.8
1996	49.7
1997	49.4

(Source: Index of Leading Cultural Indicators)

31. *U.S. Domestic Travel* The graph below gives the number of millions of persons who took trips of 50 miles or more for the years 1999 through 2004.

a. Find the equation of the line which is the best fit for these data, with x equal to the number of years from 1999 and y equal to the millions of persons.

b. Use the model to estimate the number of persons who took trips in 2008.

c. In what year does the model estimate the number of persons who took trips as 983.23 million?

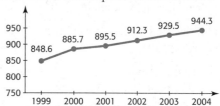

32. *Earnings and Race* The table on the next page gives the median household income for whites and blacks in various years.

a. Let x represent the median household income for whites and y represent the corresponding median household income for blacks, and make a scatter plot of these data.

b. Find a linear model that expresses the median household income for blacks as a function of the median household income for whites.

c. Find the slope of the linear model in (b) and write a sentence that interprets it.

Median Household Income

Year	Whites	Blacks
1981	38,954	21,859
1985	40,614	24,163
1990	42,622	25,488
1995	42,871	26,842
2000	46,910	31,690
2001	46,261	30,625
2002	46,119	29,691
2003	45,631	29,645
2005	48,554	30,858

(Source: U.S. Census Bureau)

33. *Poverty* The table shows the number of millions of people in the United States who lived below the poverty level for selected years.

a. Find a linear model that approximately fits the data, using x as the number of years after 1970.

b. Use a graph of the model and a scatter plot to determine if the model is nearly an exact fit for the data.

Year	Persons Living Below the Poverty Level (millions)
1970	25.4
1975	25.9
1980	29.3
1986	32.4
1990	33.6
1995	36.4
2000	31.1
2004	37.0

(Source: U.S. Census Bureau, U.S. Department of Commerce)

34. *Consumer Price Index* Prices as measured by the U.S. Consumer Price Index (CPI) have risen steadily since World War II. The data in the table give the CPI for selected years between 1985 and 2005. The CPI in this table has 1967 as a reference year; that is, what cost $1 in 1967 cost about $1.13 in 1985 and $3.25 in 2005.

Year	CPI
1985	113.2
1990	162.7
1995	224.2
2000	266.0
2005	324.9

(Source: Bureau of Labor Statistics)

a. Align the input data as the number of years after 1985 and find a linear model for the data rounded to two decimal places.

b. Use the model to estimate when the CPI will be 449.82.

35. *Personal Consumption* The sum of the personal consumption expenditures in the United States, in billions of dollars, for selected years from 1985 through 2005 is shown in the table below.

a. Make a scatter plot of the data, with x equal to the number of years past 1985 and y equal to the billions of dollars spent.

b. Does it appear that a line will be a reasonable fit for this data?

c. Find the linear model which is the best fit for the data.

d. Use the unrounded model to estimate the U.S. personal consumption for 2008.

Year	Personal Consumption ($ billions)
1985	2720.3
1990	3839.9
1995	4975.8
2000	6739.4
2003	7703.6
2004	8211.5
2005	8742.4

(Source: U.S. Department of Commerce)

36. *Gross Domestic Product* The table on the next page gives the gross domestic product (the value of all goods and services, in billions of dollars) of the United States for selected years from 1940 to 2003.

a. Create a scatter plot of the data, with y representing the GDP in billions of dollars and x representing the number of years after 1940.

b. Find the linear function that best fits the data, with x equal to the number of years after 1940.

c. Is a line a good fit for this data?

Year	Gross Domestic Product	Year	Gross Domestic Product
1940	837	1980	3746
1945	1559	1985	4207
1950	1328	1990	4853
1955	1700	1995	5439
1960	1934	2000	9817
1965	2373	2005	12,487
1970	2847		
1975	3173		

(Source: U.S. Bureau of Economic Analysis)

37. *U.S. Population* The following table gives the projections of the U.S. population from 2000 to 2100.

a. Find a linear function that models this data, with x equal to the number of years from 2000 and $f(x)$ equal to the population in millions.

b. Find $f(65)$ and state what it means.

c. What does this model predict the population to be in 2080? How does this compare with the value for 2080 in the table?

Year	Population (millions)	Year	Population (millions)
2000	275.3	2060	432.0
2010	299.9	2070	463.6
2020	324.9	2080	497.8
2030	351.1	2090	533.6
2040	377.4	2100	571.0
2050	403.7		

(Source: www.census.gov/population/projections)

38. *Internet Brokerage Accounts* The number of Internet brokerage accounts is given in the following table.

a. Write the linear equation that models the number of Internet brokerage accounts as a function of the number of years after 1990.

b. According to the model, what is the annual increase in the number of accounts?

c. Use the model to estimate when the number of accounts will be 20 million.

d. What happened in 2001 that may make this model invalid for years after 2001?

Year	Internet Brokerage Accounts (millions)
1996	1.5
1997	4.1
1998	7.1
1999	10.5
2000	14.0
2001	18.0

(Source: *Time*)

39. *ATV Deaths* The table below gives the number of deaths from all-terrain vehicles for selected years from 1995 to 2005.

Years	1995	1997	1999	2001	2003	2005
Number of Deaths	200	241	398	517	636	467

(Source: Consumer Product Safety Commission)

a. Let $x =$ the number of years after 1995 and draw a scatter plot of the data.

b. Find the linear function that is the best fit for the data.

c. Graph the model and the scatter plot on the same axes, and discuss how good the fit is.

40. *Prison Sentences* The National Center for Policy Analysis calculates expected prison time by multiplying the probabilities of being arrested, of being prosecuted, and of being convicted by the length of sentence if convicted. The expected time served (in days) for each serious crime committed is shown in the following table.

a. Use the data to create a linear equation to model expected prison time in days for a serious crime as a function of the number of years after 1960.

b. Use the model to find the expected prison time in 1975.

c. Use the model to determine in what year the expected prison time is 23 days.

Year	Expected Prison Time (days)
1970	10.1
1980	10.6
1985	13.2
1990	18.0
1992	18.5
1993	17.2
1994	19.5
1995	20.2
1996	21.7

(Source: National Center for Policy Analysis)

41. *First Class Postage* The table below gives the U.S. postage rates for first class mail. Each given weight is the largest weight letter that can be mailed for the corresponding postage.

a. Find a linear function $P = f(W)$ that models the postage in the table as a function of the weight in the table.

b. Compare the outputs of the model with the data outputs from the table for several values in the table. How well does the model fit the data?

c. What does the model give as the postage for a 6 ounce letter?

d. Use your knowledge of the U.S. postal service to determine if we should interpret this linear model discretely or continuously?

Weight W (ounces)	First Class Postage P (cents)
1	42
2	59
3	76
4	93
5	110

(Source: USPS)

42. *Parcel Post Postal Rates* The table below gives the local U.S. postage rates for parcel post mail. Each given weight is the largest weight package that can be mailed for the corresponding postage.

a. Find a linear function $P = f(W)$ that models the postage in the table as a function of the weight in the table.

b. Compare the outputs of the model with the data outputs from the table for several values in the table. Is the model a perfect fit?

c. Look at the table heading and decide if we can interpret the linear model discretely or continuously.

Weight not over:	Parcel Post Rate
1 lb	$3.89
2	4.06
3	4.90
4	5.12
5	5.30

(Source: USPS)

43. *Smoking* The table gives the percent of U.S. residents who reported smoking for selected years.

a. Write the equation that is the best fit for the data, with x equal to the number of years after 1985.

b. What does the model estimate the percent to be in 2006?

c. When can we be sure this model no longer applies?

Year	Smoking
1985	38.7
1999	25.8
2000	24.9
2001	24.9
2002	26.0
2003	25.4
2004	24.9
2005	24.9

(Source: *World Almanac*, 2007)

44. *Jail Population* The figure at the right gives the average daily number of inmates in the Beaufort County Detention Center.

  a. Align the data to the number of years after 1990, and create a linear equation that models the data.

  b. Assuming that the model is still appropriate, use the linear function to estimate the population in the Beaufort County jail in 2005.

  c. In what year does the model indicate that the average daily number of inmates is 115?

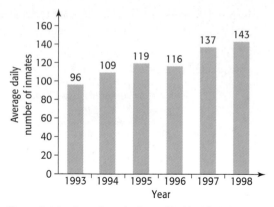

(Source: Beaufort County Detention Center, *The Island Packet*)

45. *U.S. Households with Internet Access* The following table gives the percentage of U.S. households with Internet access in various years.

Year	1996	1997	1998	1999	2000	2001	2003
Percent	8.5	14.3	26.2	28.6	41.5	50.5	52.4

(Source: U.S. Census Bureau)

  a. Create a scatter plot of the data, with x equal to the number of years from 1995.

  b. Create a linear equation that models the data.

  c. Graph the function and the data on the same graph, to see how well the function models (fits) the data.

---

section

2.3

Systems of Linear Equations in Two Variables

key concepts

- System of equations
- Solving a system of linear equations in two variables
  Graphing method
  Substitution method
  Elimination method
- Break-even analysis
- Supply and demand
- Market equilibrium
- Dependent and inconsistent systems
- Modeling systems of equations

section preview ▪ Concerta

The new drug Concerta contains a time-release version of methylphenidate, the main ingredient in Ritalin. Primarily because its effects last 12–14 hours as opposed to Ritalin's 4–5 hours, sales of this drug increased rapidly while Ritalin sales decreased slightly. Looking at the graphs of the market shares of the two drugs in Figure 2.25, we can see that they intersect at a point, and we interpret this to mean that the Concerta market share equaled that of Ritalin in the first few months after it was released. The graphs appear to intersect near the point representing 7 weeks after August 25 (2000) and a 7% market share. (Other drugs, mostly generic, account for the remaining market share.) If we find linear equations that approximately model these graphs, with x representing weeks and y representing market-share percent, the point of intersection of the two graphs would represent the *simultaneous* solution of the two equations because both equations would be satisfied by the coordinates of the point. (See Example 5.)

Weekly Market Share

Figure 2.25

(Source: *Newsweek*, December 4, 2000)

In this section, we solve systems of linear equations in two variables graphically, by substitution, and by the elimination method.

Graphical Solution of Systems

In Section 2.1, we used the intersect method to solve a linear equation by first graphing functions representing the expressions on each side of the equation and then finding the intersection of these graphs. For example, to solve

$$3000x - 7200 = 5800x - 8600$$

we can graph

$$y_1 = 3000x - 7200 \quad \text{and} \quad y_2 = 5800x - 8600$$

and find the point of intersection to be $(0.5, -5700)$. The x-coordinate of the point of intersection of the lines is the value of x that satisfies the original equation, $3000x - 7200 = 5800x - 8600$. Thus, the solution to this equation is $x = 0.5$ (Figure 2.26). In this example, we were actually using a graphical method to solve a **system of two equations in two variables** denoted by

$$\begin{cases} y = 3000x - 7200 \\ y = 5800x - 8600 \end{cases}$$

The coordinates of the point of intersection of the two graphs give the x and y values that satisfy both equations **simultaneously**, and these values are called the **solution** to the system. The following example uses the graphical method to solve a system of equations in two variables.*

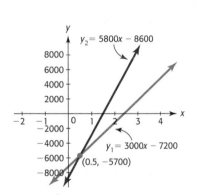

Figure 2.26

| example 1 |

Break Even

A company is said to **break even** from the production and sale of a product if the total revenue equals the total cost—that is, if $R(x) = C(x)$. Because profit $P(x) = R(x) - C(x)$, we can also say that the company breaks even if the profit for the product is zero.

---

* See Appendix A, page 659.

Suppose a company has its total revenue for a product given by $R = 5585x$ and its total cost given by $C = 61{,}740 + 440x$, where x is the number of thousands of tons of the product that are produced and sold per year. The company is said to break even when the total revenue equals the total cost—that is, when $R = C$. Find the number of thousands of tons of the product that gives break even and how much the revenue and cost are at that level of production.

Solution

We graph the revenue function as $y_1 = 5585x$ and the cost function as $y_2 = 61{,}740 + 440x$ (Figure 2.27(a)). We can find break even with the **intersection method** on a calculator by graphing the two equations, $y_1 = 5585x$ and $y_2 = 61{,}740 + 440x$, on a window that contains the point of intersection, then finding the point of intersection, which is the point where the y-values are equal. This point, which gives break even, is $(12, 67{,}020)$ (Figure 2.27(b)).

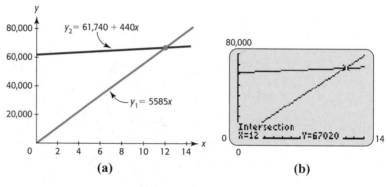

(a)   **(b)**

Figure 2.27

Thus the company will break even on this product if 12 thousand tons of the product are sold, when both the cost and revenue equal \$67,020.   ∎

It is frequently necessary to solve each equation for a variable so that the equation can be graphed with a graphing utility. It is also necessary to find a viewing window that contains the point of intersection. Consider the following example.

| example 2 | **Solving a System of Linear Equations** |

Solve the system

$$\begin{cases} 3x - 4y = 21 \\ 2x + 5y = -9 \end{cases}$$

Solution

To solve this system with a graphing utility, we first solve both equations for y. The solution follows:

$$3x - 4y = 21 \qquad\qquad\qquad 2x + 5y = -9$$

$$-4y = 21 - 3x \qquad\qquad\qquad 5y = -9 - 2x$$

$$y = \frac{21 - 3x}{-4} = \frac{3x - 21}{4} \qquad\qquad y = \frac{-9 - 2x}{5}$$

Graphing these equations with a window that contains the point of intersection (Figure 2.28(a)), and finding the point of intersection (Figure 2.28(b)), gives $x = 3$, $y = -3$, so the solution is $(3, -3)$.

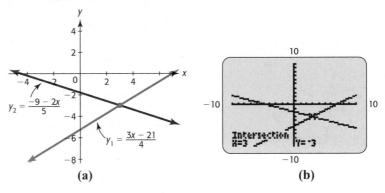

Figure 2.28

Solution by Substitution

Solving systems of equations by graphing with a graphing utility is not always the easiest method to use because the equations must be solved for y to be entered in the utility and an appropriate window must be used. A second solution method for a system of linear equations is called the **substitution method**, where one equation is solved for a variable and that variable is replaced by the equivalent expression in the other equation.

The substitution method is illustrated by the following example, which finds market equilibrium. The quantity of a product that is demanded by consumers is called the **demand** for the product and the quantity that is supplied is called the **supply**. In a free economy, both demand and supply are related to the price, and the price where the number of units demanded equals the number of units supplied is called the **equilibrium price**.

example 3

Market Equilibrium

Suppose the daily demand for a product is given by $p = 200 - 2q$, where q is the number of units demanded and p is the price per unit in dollars, and that the daily supply is given by $p = 60 + 5q$, where q is the number of units supplied and p is the price in dollars. If a price results in more units being supplied than demanded, we say there is a *surplus*, and if the price results in fewer units being supplied than demanded, we say there is a *shortfall*. **Market equilibrium** occurs when the supply quantity equals the demand quantity (and when the prices are equal); that is, when q and p both satisfy the system

$$\begin{cases} p = 200 - 2q \\ p = 60 + 5q \end{cases}$$

a. If the price is \$140, how many units are supplied and how many are demanded?

b. Does this price give a surplus or a shortfall of the product?

c. What price gives market equilibrium?

Solution

a. If the price is \$140, the number of units demanded satisfies $140 = 200 - 2q$, or $q = 30$, and the number of units supplied satisfies $140 = 60 + 5q$, or $q = 16$.

b. At this price, the quantity supplied is less than the quantity demanded, so a shortfall occurs.

c. Because market equilibrium occurs where q and p both satisfy the system

$$\begin{cases} p = 200 - 2q \\ p = 60 + 5q \end{cases}$$

we seek the solution to this system.

   We can solve this system by substitution. Substituting $60 + 5q$ for p in the first equation gives the equation

$$60 + 5q = 200 - 2q$$

Solving this equation gives

$$60 + 5q = 200 - 2q$$
$$7q = 140$$
$$q = 20$$

Thus, market equilibrium occurs when the number of units is 20, and the equilibrium price is

$$p = 200 - 2(20) = 60 + 5(20) = 160 \text{ dollars per unit} \qquad \blacksquare$$

The substitution in Example 3 was not difficult because both equations were solved for p. In general, we use the following steps to solve systems of two equations in two variables by substitution.

Solution of Systems of Equations by Substitution

1. Solve one of the equations for one of the variables in terms of the other variable.

2. Substitute the expression from step 1 into the other equation to give an equation in one variable.

3. Solve the linear equation for the variable.

4. Substitute this solution into the equation from step 1 or into one of the original equations and solve this equation for the second variable.

5. Check the solution in both original equations or check graphically.

| example 4 | ### Solution by Substitution |

Solve the system $\begin{cases} 3x + 4y = 10 \\ 4x - 2y = 6 \end{cases}$ by substitution.

Solution

To solve this system by substitution, we can solve either equation for either variable and substitute the resulting expression into the other equation. Solving the second equation for y gives

$$4x - 2y = 6$$
$$-2y = -4x + 6$$
$$y = 2x - 3$$

Substituting this expression for y in the first equation gives

$$3x + 4(2x - 3) = 10$$

Solving this equation gives

$$3x + 8x - 12 = 10$$
$$11x = 22$$
$$x = 2$$

Substituting $x = 2$ into $y = 2x - 3$ gives $y = 2(2) - 3 = 1$, so the solution to the system is $x = 2, y = 1$, or $(2, 1)$.

Checking shows that this solution satisfies both original equations. ■

example 5

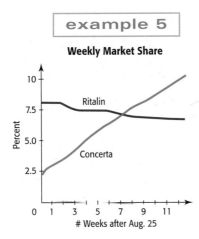

Weekly Market Share

Figure 2.29

(Source: *Newsweek*, December 4, 2000)

Concerta

The graphs that compare the weekly market share of the two drugs Concerta and Ritalin are shown in Figure 2.29.

a. Use the fact that market share for Concerta was $y = 2.4\%$ for August 25 ($x = 0$) and was 10% eleven weeks later to write a linear function representing its market share as a function of time.

b. Use the fact that market share for Ritalin was $y = 7.7\%$ for August 25 ($x = 0$) and was 6.9% eleven weeks later to write a linear function representing its market share as a function of time.

c. Find the number of weeks past the release date that the weekly market share of Concerta reached that of Ritalin.

Solution

a. Using the y-intercept 2.4 and slope $m = \dfrac{10 - 2.4}{11 - 0} = 0.69$, the linear equation is

$$y = 0.69x + 2.4$$

b. Using the y-intercept 7.7 and slope $m = \dfrac{6.9 - 7.7}{11 - 0} = -0.073$, the linear equation is

$$y = -0.073x + 7.7$$

c. To find the number of weeks (x) until the Concerta market share equals the Ritalin share, we solve the system

$$\begin{cases} y = 0.690x + 2.4 \\ y = -0.073x + 7.7 \end{cases}$$

We can solve this system by substitution. From the first equation we see that $0.690x + 2.4$ is equal to y, so we substitute this expression for y in the second equation. This substitution gives the equation

$$0.690x + 2.4 = -0.073x + 7.7$$

Solving this equation for x gives

$$0.763x = 5.3$$
$$x = 6.95 \approx 7 \text{ (weeks)}$$

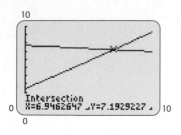

Figure 2.30

Thus, the weekly market share of Concerta reached that of Ritalin 7 weeks after it was released.

We can check the solution graphically by graphing $y = 0.690x + 2.4$ and $y = -0.073x + 7.7$ and finding the point of intersection, as shown in Figure 2.30.

Solution by Elimination

A second analytical method, called the **elimination method**, is frequently an easier method to use to solve a system of linear equations. The elimination method is based on rewriting one or both of the equations in an equivalent form that allows us to eliminate one of the variables by adding or subtracting the equations.

> ### Solving a System of Two Equations in Two Variables by Elimination
>
> 1. If necessary, multiply one or both equations by a nonzero number that will make the coefficients of one of the variables in the equations equal, except perhaps for sign.
>
> 2. Add or subtract the equations to eliminate one of the variables.
>
> 3. Solve for the variable in the resulting equation.
>
> 4. Substitute the solution from step 3 into one of the original equations and solve for the second variable.
>
> 5. Check the solutions in the remaining original equation, or check graphically.

| example 6 | ## Solution by Elimination

Use the elimination method to solve the system

$$\begin{cases} 3x + 4y = 10 \\ 4x - 2y = 6 \end{cases}$$

and check the solution graphically.

Solution

The goal is to convert one of the equations into an equivalent equation of a form so that addition of the two equations will eliminate one of the variables. Notice that the coefficient of y in the second equation, -2, is a factor of the coefficient of y in the first equation, 4. If we multiply both sides of the second equation by 2 and add the two equations, this will eliminate the y-variable.

$$\begin{cases} 3x + 4y = 10 & (1) \\ 4x - 2y = 6 & (2) \end{cases}$$

Multiply 2 times Equation (2),
getting equivalent Equation (3).

$$\begin{cases} 3x + 4y = 10 & (1) \\ 8x - 4y = 12 & (3) \end{cases}$$

Add Equations (1) and (3) to eliminate y. $11x \qquad = 22$

Solve the new equation for x. $x \qquad = 2$

Substituting $x = 2$ in the first equation gives $3(2) + 4y = 10$, or $y = 1$. Thus, the solution to the system is $x = 2, y = 1$, or $(2, 1)$.

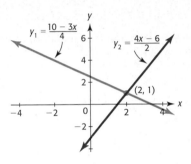

Figure 2.31

To check graphically, we solve both equations for y, getting $y_1 = \dfrac{10 - 3x}{4}$ and $y_2 = \dfrac{4x - 6}{2}$, then graph the equations and find the point of intersection to be $(2, 1)$ (Figure 2.31). ■

Modeling Systems of Linear Equations

Solving some real problems requires us to create two or more equations whose simultaneous solution is the solution to the problem. Consider the following examples.

| example 7 | **Investments** |

An investor has $300,000 to invest, part at 12% and the remainder in a less risky investment at 7%. If her investment goal is to have an annual income of $27,000, how much should she put in each investment?

Solution

If we denote the amount invested at 12% as x and the amount invested at 7% as y, the sum of the investments is $x + y$, so we have the equation

$$x + y = 300{,}000$$

The annual income from the 12% investment is $0.12x$ and the annual income from the 7% investment is $0.07y$. Thus, the desired annual income from the two investments is

$$0.12x + 0.07y = 27{,}000$$

Thus, we can write the given information as a system of equations.

$$\begin{cases} x + y = 300{,}000 \\ 0.12x + 0.07y = 27{,}000 \end{cases}$$

To solve this system, we multiply the first equation by -0.12 and add the two equations. This results in an equation with one variable:

$$\begin{cases} -0.12x - 0.12y = -36{,}000 \\ 0.12x + 0.07y = 27{,}000 \end{cases}$$

$$-0.05y = -9000$$

$$y = 180{,}000$$

Substituting 180,000 for y into the first original equation and solving for x gives $x = 120{,}000$. Thus, $120,000 should be invested at 12%, and $180,000 should be invested at 7%.

To check this solution, we see that the total investment is $120,000 + $180,000, which equals $300,000. The interest earned at 12% is $120,000(0.12) = $14,400, and the interest earned at 7% is $180,000(0.07) = $12,600. The total interest is $14,400 + $12,600, which equals $27,000. This agrees with the given information. ■

| example 8 | **Medication** |

A nurse has two solutions that contain different concentrations of a certain medication. One is a 12% concentration, and the other is an 8% concentration. How many cubic centimeters (cc) of each should she mix together to obtain 20 cc of a 9% solution?

Solution

We begin by denoting the total amount of the first solution by x and the amount of the second solution by y. The total amount of solution is the sum of x and y, so

$$x + y = 20$$

The total medication in the combined solution is 9% of 20 cc, or $0.09(20) = 1.8$ cc, and the mixture is obtained by adding $0.12x$ and $0.08y$, so

$$0.12x + 0.08y = 1.8$$

We can use substitution to solve the system

$$\begin{cases} x + y = 20 \\ 0.12x + 0.08y = 1.8 \end{cases}$$

Substituting $20 - x$ for y in $0.12x + 0.08y = 1.8$ gives $0.12x + 0.08(20 - x) = 1.8$, and solving this equation gives

$$0.12x + 0.08(20 - x) = 1.8$$
$$0.12x + 1.6 - 0.08x = 1.8$$
$$0.04x = 0.2$$
$$x = 5$$

Thus, combining 5 cc of the first solution with $20 - 5 = 15$ cc of the second solution gives 20 cc of the 9% solution.   ■

Dependent and Inconsistent Systems

The system of linear equations discussed in Example 2 has a unique solution, shown as the point of intersection of the graphs. It is possible that two equations in a system of linear equations in two variables describe the same line. When this happens, the equations are equivalent, and the values that satisfy one equation are also solutions to the other equation, and to the system. Such a system is a **dependent system**. If a system contains two equations whose graphs are parallel lines, they have no point in common, and thus the system has no solution. Such a system of equations is **inconsistent**. Figure 2.32(a)–(c) represents these three situations: systems that have a unique solution, many solutions (dependent system), and no solution (inconsistent system), respectively. Note that the slopes of the lines are equal in Figure 2.32(b) and in Figure 2.32(c).

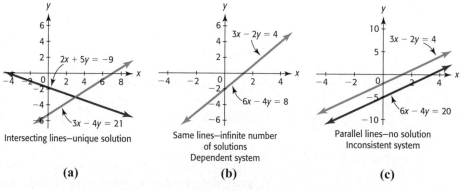

(a)   (b)   (c)

Figure 2.32

| example 9 | **Systems with Nonunique Solutions** |

Use the elimination method to solve each of the following systems, if possible. Verify the solution graphically.

a. $\begin{cases} 2x - 3y = 4 \\ 6x - 9y = 12 \end{cases}$ **b.** $\begin{cases} 2x - 3y = 4 \\ 6x - 9y = 36 \end{cases}$

Solution

a. To solve $\begin{cases} 2x - 3y = 4 \\ 6x - 9y = 12 \end{cases}$, we multiply the first equation by -3 and add the equations, getting the following:

$$\begin{cases} -6x + 9y = -12 \\ 6x - 9y = 12 \end{cases}$$
$$0 = 0$$

This indicates that the graphs of the equations intersect when $0 = 0$, *which is always true*. Thus any value that satisfies one of these equations also satisfies the other, and there are *infinitely many* solutions. Figure 2.33(a) shows that the graphs of the equations lie on the same line. Notice that the second equation is a multiple of the first, so the equations are equivalent. This system is *dependent*.

The infinitely many solutions all satisfy both of the two equations. That is, they are values of x and y that satisfy

$$2x - 3y = 4 \quad \text{or} \quad y = \frac{2}{3}x - \frac{4}{3}$$

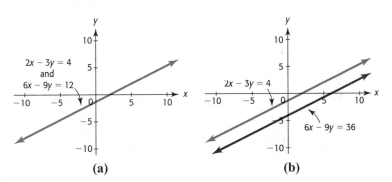

Figure 2.33

b. To solve $\begin{cases} 2x - 3y = 4 \\ 6x - 9y = 36 \end{cases}$, we multiply the first equation by -3 and add the equations, getting the following:

$$\begin{cases} -6x + 9y = -12 \\ 6x - 9y = 36 \end{cases}$$
$$0 = 24$$

This indicates that the equations intersect when $0 = 24$, *which is never true*. Thus, no values of x and y satisfy both of the equations. Figure 2.33(b) shows that the graphs of the equations are parallel. This system is *inconsistent*.

| example 10 | **Investment** |

An investment club has set a goal of earning 15% on the money they invest in stocks. They are considering buying two stocks, with the cost per share and their projected growth per share (both in dollars) summarized in Table 2.11.

Table 2.11

	Utility	Technology
Cost/share	$30	$45
Growth/share	$4.50	$6.75

a. If they have $180,000 to invest, how many shares of each stock should they buy to meet their goal?

b. If they buy 1800 shares of the utility stock, how many of the technology stock should they buy to meet their goal?

Solution

a. The available money to invest in stocks is $180,000, so if x is the number of utility shares and y is the number of technology shares purchased, we have

$$30x + 45y = 180,000$$

A 15% return on their investment would be $0.15(180,000) = 27,000$ dollars, so we have

$$4.50x + 6.75y = 27,000$$

To find x and y, we solve the system

$$\begin{cases} 30x + 45y = 180,000 \\ 4.50x + 6.75y = 27,000 \end{cases}$$

Multiplying 4.5 times both sides of the first equation and -30 times both sides of the second equation gives

$$\begin{cases} 135x + 202.5y = 810,000 \\ -135x - 202.5y = 810,000 \end{cases}$$

Adding the equations gives $0 = 0$, so the system is dependent, with many solutions.

The number of each stock that can be purchased satisfies both of the two original equations. In particular, it satisfies $30x + 45y = 180,000$, so

$$y = \frac{180,000 - 30x}{45} \quad \text{or} \quad y = \frac{12,000 - 2x}{3}$$

with x between 0 and 6000 shares and y between 0 and 4000 shares (because neither x nor y can be negative).

b. Substituting 1800 for x in the equation gives $y = 2800$, so if they buy 1800 shares of the utility stock, they should buy 2800 shares of the technology stock to meet their goal. ∎

skills check

2.3

1. What are the coordinates of the point of intersection of $y = 3x - 2$ and $y = 3 - 2x$?

2. Give the coordinates of the point of intersection of $3x + 2y = 5$ and $5x - 3y = 21$.

In Exercises 3–6, solve the systems of equations graphically.

3. $\begin{cases} y = 3x - 12 \\ y = 4x + 2 \end{cases}$ 4. $\begin{cases} 2x - 4y = 6 \\ 3x + 5y = 20 \end{cases}$

5. $\begin{cases} 4x - 3y = -4 \\ 2x - 5y = -4 \end{cases}$ 6. $\begin{cases} 5x - 6y = 22 \\ 4x - 4y = 16 \end{cases}$

7. Does the system $\begin{cases} 2x + 5y = 6 \\ x + 2.5y = 3 \end{cases}$ have a unique solution, no solution, or many solutions? What does this mean graphically?

8. Does the system $\begin{cases} 6x + 4y = 3 \\ 3x + 2y = 3 \end{cases}$ have a unique solution, no solution, or many solutions? What does this mean graphically?

In Exercises 9–12, solve the systems of equations by substitution.

9. $\begin{cases} x = 5y + 12 \\ 3x + 4y = -2 \end{cases}$ 10. $\begin{cases} 2x - 3y = 2 \\ y = 5x - 18 \end{cases}$

11. $\begin{cases} 2x - 3y = 5 \\ 5x + 4y = 1 \end{cases}$ 12. $\begin{cases} 4x - 5y = -17 \\ 3x + 2y = -7 \end{cases}$

In Exercises 13–22, solve the systems of equations by elimination, if a solution exists.

13. $\begin{cases} x + 3y = 5 \\ 2x + 4y = 8 \end{cases}$ 14. $\begin{cases} 4x - 3y = -13 \\ 5x + 6y = 13 \end{cases}$

15. $\begin{cases} 5x = 8 - 3y \\ 2x + 4y = 8 \end{cases}$ 16. $\begin{cases} 3y = 5 - 3x \\ 2x + 4y = 8 \end{cases}$

17. $\begin{cases} 0.3x + 0.4y = 2.4 \\ 5x - 3y = 11 \end{cases}$ 18. $\begin{cases} 8x - 4y = 0 \\ 0.5x + 0.3y = 2.2 \end{cases}$

19. $\begin{cases} 3x + 6y = 12 \\ 4y - 8 = -2x \end{cases}$ 20. $\begin{cases} 6y - 12 = 4x \\ 10x - 15y = -30 \end{cases}$

21. $\begin{cases} 6x - 9y = 12 \\ 3x - 4.5y = -6 \end{cases}$ 22. $\begin{cases} 4x - 8y = 5 \\ 6x - 12y = 10 \end{cases}$

In Exercises 23–32, solve the systems of equations by any convenient method, if a solution exists.

23. $\begin{cases} y = 3x - 2 \\ y = 5x - 6 \end{cases}$ 24. $\begin{cases} y = 8x - 6 \\ y = 14x - 12 \end{cases}$

25. $\begin{cases} 4x + 6y = 4 \\ x = 4y + 8 \end{cases}$ 26. $\begin{cases} y = 4x - 5 \\ 3x - 4y = 7 \end{cases}$

27. $\begin{cases} 2x - 5y = 16 \\ 6x - 8y = 34 \end{cases}$ 28. $\begin{cases} 4x - y = 4 \\ 6x + 3y = 15 \end{cases}$

29. $\begin{cases} 3x = 7y - 1 \\ 4x = 11 - 3y \end{cases}$ 30. $\begin{cases} 5x = 12 + 3y \\ -5y = 8 - 3x \end{cases}$

31. $\begin{cases} 4x - 3y = 9 \\ 8x - 6y = 16 \end{cases}$ 32. $\begin{cases} 5x - 4y = 8 \\ -15x + 12y = -12 \end{cases}$

exercises

2.3

33. *Break Even* A manufacturer of kitchen sinks has total revenue given by the function $R = 76.50x$ and has total cost given by $C = 2970 + 27x$, where x is

the number of sinks produced and sold. Use graphical methods to find the number of units that gives break even for the product.

34. *Break Even* A jewelry maker has total revenue for her bracelets given by $R = 89.75x$ and incurs a total cost of $C = 23.50x + 1192.50$, where x is the

number of bracelets produced and sold. Use graphical methods to find the number of units that gives break even for the product.

35. *Break Even* A manufacturer of automobile air conditioners has total revenue given by $R = 136.50x$ and total cost given by $C = 9661.60 + 43.60x$, where x is the number of units produced and sold. Use a nongraphical method to find the number of units that gives break even for this product.

36. *Break Even* A manufacturer of reading lamps has total revenue given by $R = 15.80x$ and total cost given by $C = 8593.20 + 3.20x$, where x is the number of units produced and sold. Use a nongraphical method to find the number of units that gives break even for this product.

37. *Supply and Demand* A certain product has supply and demand functions given by $p = 5q + 20$ and $p = 128 - 4q$, respectively.

a. If the price p is $60, how many units q are supplied and how many are demanded?

b. What price gives market equilibrium, and how many units are demanded and supplied at this price?

38. *Market Equilibrium* The demand for a brand of clock radio is given by $p + 2q = 320$, and the supply for these radios is given by $p - 8q = 20$, where p is the price and q is the quantity demanded at price p. Solve the system containing these two equations to find the price at which the quantity demanded equals the quantity supplied and the equilibrium quantity.

39. *Market Equilibrium* Wholesalers' willingness to sell steel framing studs is given by the supply function $p = 1 + 0.02q$, and retailers' willingness to buy the studs is given by $p = 4 - 0.01q$, where p is the price per stud in dollars and q is the number of truckloads of studs. What price will give market equilibrium for the studs?

40. *Market Equilibrium* Wholesalers' willingness to sell laser printers is given by the supply function $p = 50.50 + 0.80q$, and retailers' willingness to buy the printers is given by $p = 400 - 0.70q$, where p is the price per printer in dollars and q is the number of printers. What price will give market equilibrium for the printers?

41. *College Enrollment* Suppose the percent of males who enrolled in college within 12 months of high school graduation is given by $y = -0.126x + 55.72$ and the percent of females enrolled in college within 12 months of high school graduation is given by $y = 0.73x + 39.7$, where x is the number of years after 1960. Use graphical methods to find the year these models indicate that the percent of females equaled the percent of males.
(Source: *Statistical Abstract of the United States*)

42. *Military* The number of active-duty U.S. Navy personnel is given by $y = -5.686x + 676.173$ thousand, and the number of active-duty U.S. Air Force personnel is given by $y = -11.997x + 847.529$ thousand, where x is the number of years from 1960.

a. Use graphical methods to find the year in which the number of Navy personnel reaches the number of Air Force personnel.

b. How many will be in each service when the numbers of personnel are equal?
(Source: *World Almanac*)

43. *Earnings and Race* The median annual earnings for blacks (B) as a function of the median annual earnings for whites (W), both in thousands of dollars, can be modeled by $B = 0.6234W + 0.3785$ using one set of data, and by $B = 1.05W - 18.691$ using more recent data. Use graphical or numerical methods to find what annual earnings by whites will result in both models giving the same median annual earnings for blacks.
(Source: *Statistical Abstract of the United States*)

44. *Snack Sales* The following graphs make it appear that mint sales surpassed gum sales in 1999. Using this data, the best-fitting line for mint sales has equation $y = 24.5x + 93.5$ (million dollars) and the best-fitting line for gum sales is $y = -0.2x + 1007$ (million dollars), where x is the number of years after 1990.

a. For what value of x does mint sales equal gum sales according to these models?

b. In what year do these models predict that mint sales will reach gum sales?

c. Are these graphs misleading even if they are correct? How?

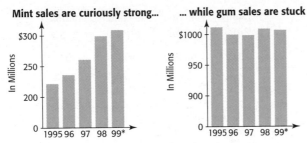

Mint sales are curiously strong...

... while gum sales are stuck

*52 weeks ended July 18, 1999. Source: Information Resources Inc.

(Source: *Newsweek*, November 1, 1999)

45. *Revenue* The sum of the 2009 revenue and twice the 2008 revenue for Mama Joan's International, Inc., is $2144.9 million. The difference in the 2009 and 2008 revenues is $135.5 million. If Mama Joan's revenue between 2000 and 2009 is an increasing linear function, find the 2008 and 2009 revenues.

46. *Stock Prices* The sum of the high and low prices of a share of stock for Johns, Inc., in 2010 is $83.50, and the difference of these two prices in 2010 is $21.88. Find the high and low prices.

47. *Pricing* A concert promoter needs to make $84,000 from the sale of 2400 tickets. The promoter charges $30 for some tickets and $45 for the others.

   a. If there are x of the $30 tickets sold and y of the $45 tickets sold, write an equation that states that the total number of the tickets sold is 2400.

   b. How much money is received from the sale of x tickets for $30 each?

   c. How much money is received from the sale of y tickets for $45 each?

   d. Write an equation that states that the total amount received from the sale is $84,000.

   e. Solve the equations simultaneously to find how many tickets of each type must be sold to yield the $84,000.

48. *Rental Income* A woman has $250,000 invested in two rental properties. One yields an annual return of 10% of her investment, and the other returns 12% per year on her investment. Her total annual return from the two investments is $26,500. Let x represent the amount of the 10% investment and y represent the amount of the 12% investment.

   a. Write an equation that states that the sum of the investments is $250,000.

   b. What is the annual return on the 10% investment?

   c. What is the annual return on the 12% investment?

   d. Write an equation that states that the sum of the annual returns is $26,500.

   e. Solve these two equations simultaneously to find how much is invested in each property.

49. *Investment* One safe investment pays 8% per year, and a more risky investment pays 12% per year.

   a. How much must be invested in each account if the investor of $100,000 would like a return of $9000 per year?

   b. Why might the investor use two accounts rather than put all money in the 12% investment?

50. *Investment* A woman invests $52,000 in two different mutual funds, one that averages 10% per year and another that averages 14% per year. If her average annual return on the two mutual funds is $5720, how much did she invest in each fund?

51. *Investment* Jake has $250,000 to invest. He chooses one money market fund that pays 6.6% and a mutual fund that has more risk but has averaged 8.6% per year. If his goal is to average 7% per year with minimal risk, how much should he invest in each fund?

52. *Investment* Sue chooses one money market fund that pays 6.2% and a mutual fund that has more risk but has averaged 9.2% per year. If she has $300,000 to invest and her goal is to average 7.6% per year with minimal risk, how much should she invest in each fund?

53. *Medication* A pharmacist wants to mix two solutions to obtain 100 cc of a solution that has an 8% concentration of a certain medicine. If one solution has a 10% concentration of the medicine and the second has a 5% concentration, how much of each of these solutions should she mix?

54. *Medication* A pharmacist wants to mix two solutions to obtain 200 cc of a solution that has a 12% concentration of a certain medicine. If one solution has a 16% concentration of the medicine and the second has a 6% concentration, how much of each solution should she mix?

55. *Nutrition* A glass of skim milk supplies 0.1 mg of iron and 8.5 g of protein. A quarter pound of lean meat provides 3.4 mg of iron and 22 g of protein. If a person on a special diet is to have 7.1 mg of iron and 69.5 g of protein, how many glasses of skim milk and how many quarter-pound servings of meat will provide this?

56. *Nutrition* Each ounce of substance A supplies 6% of the nutrient a patient needs, and each ounce of substance B supplies 10% of the required nutrient. If the total number of ounces given to the patient is 14, and 100% of the nutrient is supplied, how many ounces of each substance were given?

57. *Medication* A nurse has two solutions that contain different concentrations of a certain medication. One is a 10% concentration and the other is a 5% concentration. How many cubic centimeters (cc) of each should he mix to obtain 20 cc of an 8% solution?

58. *Medication* A nurse has two solutions that contain different concentrations of a certain medication. One is a 30% concentration and the other is a 15% concentration. How many cubic centimeters (cc) of each should she mix to obtain 45 cc of a 20% solution?

59. *Supply and Demand* The table below gives the quantity of graphing calculators demanded and the quantity supplied for selected prices.

a. Find the linear equation that gives the price as a function of the quantity demanded.

b. Find the linear equation that gives the price as a function of the quantity supplied.

c. Use these equations to find the market equilibrium price.

Price ($)	Quantity Demanded (thousands)	Quantity Supplied (thousands)
50	210	0
60	190	40
70	170	80
80	150	120
100	110	200

60. *Market Analysis* The supply function and the demand function for a product are linear and are determined by the table that follows. Create the supply and demand functions and find the price that gives market equilibrium.

Supply Price	Function Quantity	Demand Price	Function Quantity
200	400	400	400
400	800	200	800
600	1200	0	1200

61. *Alcohol Use* The percent of the U.S. population who admitted to using alcohol at least once during the month prior to being asked about alcohol use is shown in the table below.

Year	1994	1995	1996	1997
18- to 25-Year-Olds (%)	63	61	60	58
26- to 34-Year-Olds (%)	65	63	62	60

(Source: National Household Survey on Drug Abuse)

Letting x represent the number of years after 1990, the models are:

18- to 25-year-olds: $y = -1.6x + 69.3$

26- to 34-year-olds: $y = -1.6x + 71.3$

Solve this system of equations, if possible, to find when the percent of 18- to 25-year-olds who admitted to using alcohol equals the percent of 26- to 34-year-olds.

62. *Medication* Suppose combining x cubic centimeters (cc) of a 20% concentration of a medication and y cc of a 5% concentration of the medication gives $(x + y)$ cc of a 15.5% concentration. If 7 cc of the 20% concentration are added, by how much must the amount of 5% concentration be increased to keep the same concentration?

63. *Social Agency* A social agency provides emergency food and shelter to two groups of clients. The first group has x clients who need an average of $300 for emergencies, and the second group has y clients who need an average of $200 for emergencies. The agency has $100,000 to spend for these two groups.

a. Write an equation that describes the maximum number of clients who can be served with the $100,000.

b. If the first group has twice as many clients as the second group, how many clients are in each group if all the money is spent?

64. *Market Equilibrium* A retail chain will buy 800 televisions if the price is $350 each and 1200 if the price is $300. A wholesaler will supply 700 of these televisions at $280 each and 1400 at $385 each. Assuming that the supply and demand functions are linear, find the market equilibrium point and explain what it means.

65. *Market Equilibrium* A retail chain will buy 900 cordless phones if the price is $10 each and 400 if the price is $60. A wholesaler will supply 700 phones at $30 each and 1400 at $50 each. Assuming that the supply and demand functions are linear, find the market equilibrium point and explain what it means.

section

2.4

Solutions of Linear Inequalities

key concepts

- Linear inequality
- Algebraically solving linear inequalities
- Graphical solution of linear inequalities
- Intersection method
- x-intercept method
- Double inequalities

section preview ▪ Profit

For a certain product, the respective weekly revenue and weekly cost are given by

$$R(x) = 40x \quad \text{and} \quad C(x) = 20x + 1600$$

where x is the number of units produced and sold. For what levels of production will a profit result?

Profit will occur when revenue is greater than cost. So we find the level of production and sale x that gives a profit by solving the **linear inequality**

$$R(x) > C(x), \quad \text{or} \quad 40x > 20x + 1600$$

(See Example 2.) In this section, we will solve linear inequalities of this type algebraically and graphically.

Algebraic Solution of Linear Inequalities

An **inequality** is a statement that one quantity or expression is greater than, less than, greater than or equal to, or less than or equal to another.

> ### Linear Inequality
>
> A linear inequality (or first-degree inequality) in the variable x is an inequality that can be written in the form $ax + b > 0$ where $a \neq 0$.
> (The inequality symbol can be $>$, $\geq$, $<$, or $\leq$.)

The inequality $4x + 3 < 7x - 6$ is a linear inequality (or first-degree inequality) because the highest power of the variable (x) is 1. The values of x that satisfy the inequality form the solution set for the inequality. For example, 5 is in the solution set of this inequality because substituting 5 into the inequality gives

$$4 \cdot 5 + 3 < 7 \cdot 5 - 6 \quad \text{or} \quad 23 < 29$$

which is a true statement. On the other hand, 2 is not in the solution set because

$$4 \cdot 2 + 3 \not< 7 \cdot 2 - 6$$

Solving an inequality means finding its solution set. The solution to an inequality can be written as an inequality or in interval notation. The solution can also be represented by a graph on a real number line.

Two inequalities are *equivalent* if they have the same solution set.

We use the properties of inequalities discussed in the Algebra Toolbox to solve an inequality. In general, the steps used to solve a linear inequality are the same as those used to solve linear equations, except that the inequality symbol is reversed if both sides are multiplied (or divided) by a negative number.

Steps for Solving a Linear Inequality Algebraically

1. If a linear inequality contains fractions with constant denominators, multiply both sides of the inequality by a positive number that will remove all denominators in the inequality. If there are two or more fractions, use the least common denominator (LCD) of the fractions.

2. Remove any parentheses by multiplication.

3. Perform any additions or subtractions to get all terms containing the variable on one side and all other terms on the other side of the inequality. Combine like terms.

4. Divide both sides of the inequality by the coefficient of the variable. *Reverse the inequality symbol if this number is negative.*

5. Check the solution by substitution or with a graphing utility. If a real-world solution is desired, check the algebraic solution for reasonableness in the real-world situation.

| example 1 | ## Solution of a Linear Inequality

Solve the inequality $3x - \dfrac{1}{3} \le -4 + x$.

Solution

$$3x - \frac{1}{3} \le -4 + x$$

Multiplying both sides by 3 gives

$$3\left(3x - \frac{1}{3}\right) \le 3(-4 + x)$$

Removing parentheses gives

$$9x - 1 \le -12 + 3x$$

Performing additions and subtractions to both sides to get the variables on one side and the constants on the other side gives

$$6x \le -11$$

Dividing both sides by the coefficient of the variable gives

$$x \le -\frac{11}{6}$$

The solution set contains all real numbers less than or equal to $-\dfrac{11}{6}$. The graph of the solution set $\left(-\infty, -\dfrac{11}{6}\right]$ is shown in Figure 2.34.

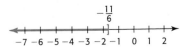

Figure 2.34

| **example 2** | **Profit** |

For a certain product, the respective weekly revenue and weekly cost are given by

$$R(x) = 40x \quad \text{and} \quad C(x) = 20x + 1600$$

where x is the number of units produced and sold. For what levels of production will a profit result?

Solution

Profit will occur when revenue is greater than cost. So we find the level of production and sale that gives a profit by solving the linear inequality $R(x) > C(x)$, or $40x > 20x + 1600$.

$40x > 20x + 1600$	
$20x > 1600$	Subtract 20x from both sides.
$x > 80$	Divide both sides by 20.

Thus, a profit occurs if more than 80 units are produced and sold.

| **example 3** | **Body Temperature** |

A child's health is at risk when his or her body temperature is 103°F or higher. What Celsius temperature reading would indicate that a child's health is at risk?

Solution

A child's health is at risk if $F \ge 103$, and $F = \dfrac{9}{5}C + 32$, where F is the temperature in degrees Fahrenheit and C is the temperature in degrees Celsius. Substituting $\dfrac{9}{5}C + 32$ for F, we have

$$\frac{9}{5}C + 32 \ge 103$$

Now we solve the inequality for C:

$$\frac{9}{5}C + 32 \geq 103$$

$$9C + 160 \geq 515 \qquad \text{Multiply both sides by 5 to clear fractions.}$$

$$9C \geq 355 \qquad \text{Subtract 160 from both sides.}$$

$$C \geq 39.\overline{4} \qquad \text{Divide both sides by 9.}$$

Thus, a child's health is at risk if his or her Celsius temperature is approximately $39.4°$ or higher.   ■

Graphical Solution of Linear Inequalities

In Section 2.1, we used graphical methods to solve linear equations. In a similar manner, graphical methods can be used to solve linear inequalities. We will illustrate both the intersection of graphs method and the x-intercept method.*

Intersection Method

To solve an inequality by the intersection method, we use the following steps.

> ### Steps for Solving a Linear Inequality with the Intersection Method
>
> 1. Set the left side of the inequality equal to y_1, set the right side equal to y_2, and graph the equations using your graphing utility.
>
> 2. Choose a viewing window that contains the point of intersection and find the point of intersection, with x-coordinate a. This is the value of x where $y_1 = y_2$.
>
> 3. The values of x that satisfy the inequality represented by $y_1 < y_2$ are those values for which the graph of y_1 is below the graph of y_2. The values of x that satisfy the inequality represented by $y_1 > y_2$ are those values of x for which the graph of y_1 is above the graph of y_2.

To solve the inequality

$$5x + 2 < 2x + 6$$

by using the intersection method, let

$$y_1 = 5x + 2 \quad \text{and} \quad y_2 = 2x + 6$$

Entering y_1 and y_2 and graphing the equations using a graphing utility (Figure 2.35) shows that the point of intersection occurs at $x = \dfrac{4}{3}$. Some graphing utilities show this answer in the form $x = 1.3333333$. The exact x-value $\left(x = \dfrac{4}{3}\right)$ can also be found by solving the equation $5x + 2 = 2x + 6$ algebraically. Figure 2.35 shows that the graph

---

* See Appendix A, page 660.

of y_1 is below the graph of y_2 when x is less than $\frac{4}{3}$. Thus, the solution to the inequality is $x < \frac{4}{3}$, which can be written in interval notation as $\left(-\infty, \frac{4}{3}\right)$.

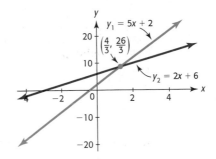

Figure 2.35

example 4

Intersection Method of Solution

Solve $\dfrac{5x + 2}{5} \geq \dfrac{4x - 7}{8}$ using the intersection of graphs method.

Solution

Enter the left side of the inequality as $y_1 = (5x + 2)/5$, enter the right side of the inequality as $y_2 = (4x - 7)/8$, graph these lines, and find their point of intersection. As seen in Figure 2.36, the two lines intersect at the point where $x = -2.55$ and $y = -2.15$.

The solution to the inequality is the x-interval for which the graph of y_1 is above the graph of y_2, or the x-value for which the graph of y_1 intersects the graph of y_2. Figure 2.36 indicates that this is the interval to the right of and including the input value of the point of intersection of the two lines. Thus, the solution is $x \geq -2.55$, or $[-2.55, \infty)$. ∎

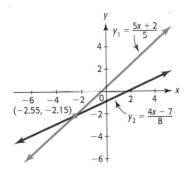

Figure 2.36

x-Intercept Method

To use the x-intercept method to solve a linear inequality, we use the following steps.

> ### Solving Linear Inequalities with the x-Intercept Method
>
> 1. Rewrite the inequality with all nonzero terms on one side of the inequality and combine like terms, getting $f(x) > 0, f(x) < 0, f(x) \leq 0$, or $f(x) \geq 0$.
>
> 2. Graph the nonzero side of this inequality. (Any window in which the x-intercept can be clearly seen is appropriate.)
>
> 3. Find the x-intercept of the graph to find the solution to the equation $f(x) = 0$. (The exact solution can be found algebraically.)
>
> 4. Use the graph to determine where the inequality is satisfied.

To use the x-intercept method to solve the inequality $5x + 2 < 2x + 6$, we rewrite the inequality with all nonzero terms on one side of the inequality and combine like terms:

$$5x + 2 < 2x + 6$$

$$3x - 4 < 0 \qquad \text{Subtract } 2x \text{ and 6 from both sides of the inequality.}$$

Graphing the nonzero side of this inequality as the linear function $f(x) = 3x - 4$ gives the graph in Figure 2.37. Finding the x-intercept of the graph (Figure 2.37) gives the solution to the equation $3x - 4 = 0$. The x-intercept (and zero of the function) is $x = 1.3333\ldots = \dfrac{4}{3}$.

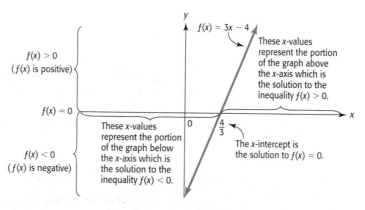

Figure 2.37

We now want to find where $f(x) = 3x - 4$ is less than 0. Notice that the portion of the graph *below* the x-axis gives $3x - 4 < 0$. Thus, the solution to $3x - 4 < 0$ and thus to $5x + 2 < 2x + 6$ is $x < \dfrac{4}{3}$ or $\left(-\infty, \dfrac{4}{3}\right)$.

---

example 5

Apparent Temperature

During the summer of 1998, Dallas, Texas, endured 29 consecutive days where the temperature was at least 100°F. On many of these days, the combination of heat and humidity made it feel even hotter than it was. When the temperature is 100°F, the apparent temperature A (or heat index) depends on the humidity h (expressed as a decimal) according to

$$A = 90.2 + 41.3h$$

For what humidity levels is the apparent temperature at least 110°F? (Source: W. Bosch and C. Cobb, "Temperature-Humidity Indices," *UMAP Journal*, Fall 1989)

Solution

If the apparent temperature is at least 110°F, the inequality to be solved is

$$A \geq 110 \quad \text{or} \quad 90.2 + 41.3h \geq 110$$

Rewriting this inequality with 0 on the right side gives $41.3x - 19.8 \geq 0$. Entering $y_1 = 41.3x - 19.8$ and graphing gives the graph in Figure 2.38. The x-intercept of the graph is (approximately) 0.479.

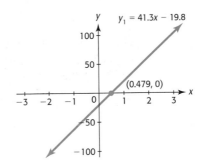

Figure 2.38

The x-interval where the graph is on or above the x-axis is the solution that we seek, so the solution to the inequality is $[0.479, \infty)$. However, humidity is limited to 100%, so the solution is $0.479 \leq h \leq 1.00$, and we say that the apparent temperature is at least 110° when the humidity is between 47.9% and 100%, inclusive. ◼

Double Inequalities

The inequality $0.479 \leq h \leq 1.00$ in Example 5 is a **double inequality**. A double inequality represents two inequalities connected by the word *and* or *or*. The inequality $0.479 \leq h \leq 1.00$ is a compact way of saying $0.479 \leq h$ and $h \leq 1.00$. Double inequalities can be solved algebraically or graphically, as illustrated in the following example. Note that any arithmetic operation is performed to *all three* parts of a double inequality.

---

example 6

Course Grades

A student has taken four tests and has earned grades of 90%, 88%, 93%, and 85%. If all 5 tests count the same, what grade must the student earn on the final test so that his course average is a B (that is, so his average is at least 80% and less than 90%)?

Algebraic Solution

To receive a B, the final test score, represented by x, must satisfy

$$80 \leq \frac{90 + 88 + 93 + 85 + x}{5} < 90$$

Solving this inequality gives

$$80 \leq \frac{356 + x}{5} < 90$$

$$400 \leq 356 + x < 450 \qquad \text{Multiply all three parts by 5.}$$

$$44 \leq x < 94 \qquad \text{Subtract 356 from all three parts.}$$

Thus, he will receive a grade of B if his final test score is at least 44 but less than 94.

Graphical Solution

To solve this inequality graphically, we assign the left side of the inequality to y_1, the middle to y_2, and the right side to y_3, and graph these equations to obtain the graph in Figure 2.39.

$$y_1 = 80$$
$$y_2 = (356 + x)/5$$
$$y_3 = 90$$

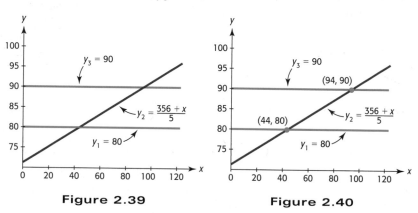

Figure 2.39   **Figure 2.40**

We seek the values of x where the graph of y_2 is above or on the graph of y_1 and below the graph of y_3. The left endpoint of this x-interval occurs at the point of intersection of y_2 and y_1, and the right endpoint of the interval occurs at the intersection of y_2 and y_3. These two points can be found using the intersection method. Figure 2.40 shows the points of intersection of these graphs.

The x-values of the points of intersection are 44 and 94, so the solution to $80 \le \dfrac{356 + x}{5} < 90$ is $44 \le x < 94$, which agrees with our algebraic solution.   ■

| example 7 |

Expected Prison Sentences

The mean (expected) time y served in prison for a serious crime can be approximated by a function of the mean sentence length x, with $y = 0.55x - 2.886$, where x and y are measured in months. According to this model, how many months should a judge sentence a convicted criminal so that the criminal will be expected to serve between 37 and 78 months? (Source: National Center for Policy Analysis)

Solution

We seek values of x that give y-values between 37 and 78, so we solve the inequality $37 \le 0.55x - 2.886 \le 78$ for x:

$$37 \le 0.55x - 2.886 \le 78$$
$$37 + 2.886 \le 0.55x \le 78 + 2.886$$
$$39.886 \le 0.55x \le 80.886$$
$$72.52 \le x \le 147.07$$

Thus, the judge could impose a sentence of 73 to 147 months if she wants the criminal to actually serve between 37 and 78 months.   ■

skills check

2.4

In Exercises 1–12, solve the inequalities both alge-braically and graphically. Draw a number line graph of each solution.

1. $3x - 7 \le 5 - x$

2. $2x + 6 < 4x + 5$

3. $4(3x - 2) \le 5x - 9$

4. $5(2x - 3) > 4x + 6$

5. $4x + 1 < -\dfrac{3}{5}x + 5$

6. $4x - \dfrac{1}{2} \le -2 + \dfrac{x}{3}$

7. $\dfrac{x - 5}{2} < \dfrac{18}{5}$

8. $\dfrac{x - 3}{4} < \dfrac{16}{3}$

9. $\dfrac{3(x - 6)}{2} \ge \dfrac{2x}{5} - 12$

10. $\dfrac{2(x - 4)}{3} \ge \dfrac{3x}{5} - 8$

11. $2.2x - 2.6 \ge 6 - 0.8x$

12. $3.5x - 6.2 \le 8 - 0.5x$

In Exercises 13 and 14, solve graphically by the intersec-tion method. Give the solution in interval notation.

13. $7x + 3 < 2x - 7$

14. $3x + 4 \le 6x - 5$

In Exercises 15 and 16, solve graphically by the x-intercept method. Give the solution in interval notation.

15. $5(2x + 4) \ge 6(x - 2)$

16. $-3(x - 4) < 2(3x - 1)$

17. The graphs of two linear functions f and g are shown in the following figure. (Domains are all real numbers.)

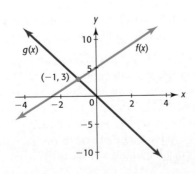

a. Solve the equation $f(x) = g(x)$.

b. Solve the inequality $f(x) < g(x)$.

18. The graphs of three linear functions f, g, and h are shown in the following figure.

a. Solve the equation $f(x) = g(x)$.

b. Solve the inequality $h(x) \le g(x)$.

c. Solve the inequality $f(x) \le g(x) \le h(x)$.

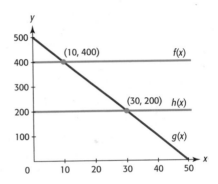

In Exercises 19–28, solve the double inequalities.

19. $17 \le 3x - 5 < 31$

20. $120 < 20x - 40 \le 160$

21. $2x + 1 \ge 6$ and $2x + 1 \le 21$

22. $16x - 8 > 12$ and $16x - 8 < 32$

23. $3x + 1 < -7$ and $2x - 5 > 6$

24. $6x - 2 \le -5$ or $3x + 4 > 9$

25. $\dfrac{3}{4}x - 2 \ge 6 - 2x$ or $\dfrac{2}{3}x - 1 \ge 2x - 2$

26. $\dfrac{1}{2}x - 3 < 5x$ or $\dfrac{2}{5}x - 5 > 6x$

27. $37.002 \le 0.554x - 2.886 \le 77.998$

28. $70 \le \dfrac{60 + 88 + 73 + 65 + x}{5} < 80$

exercises

2.4

29. *Depreciation* Suppose a business purchases equipment for $12,000 and depreciates it over 5 years with the straight-line method until it reaches its salvage value of $2000 (see the figure below). Assuming that the depreciation can be for any part of a year,

   a. Write an equation that represents the depreciated value, V, as a function of the years t.

   b. Write an inequality that indicates that the depreciated value V of the equipment is less than $8000.

   c. Write an inequality that describes the time t during which the depreciated value is at least half of the original value.

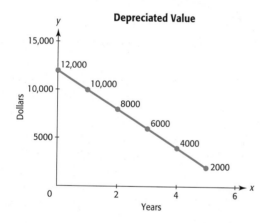

30. *Blood Alcohol Percent* The blood alcohol percent p of a 220-lb male is a function of the number of 12-oz drinks, and the percent at which a person is legally intoxicated (and guilty of DUI if driving) is 0.1% or higher (see the following figure).

   a. Use an inequality to indicate the percent of alcohol in the blood when a person is considered legally intoxicated.

   b. If x is the number of drinks imbibed by a 220-lb male, write an inequality that gives the number of drinks that will cause him to be legally intoxicated.

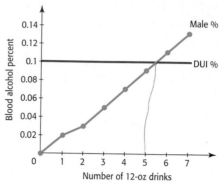

(Source: Pennsylvania Liquor Control Board)

31. *Freezing* The equation $F = \dfrac{9}{5}C + 32$ gives the relationship between temperatures measured in degrees Celsius and degrees Fahrenheit. We know that a temperature at or below 32°F is "freezing." Use an inequality to represent the corresponding "freezing" Celsius temperature.

32. *Boiling* The equation $C = \dfrac{5}{9}(F - 32)$ gives the relationship between temperatures measured in degrees Celsius and degrees Fahrenheit. We know that a temperature at or above 100°C is "boiling." Use an inequality to represent the corresponding "boiling" Fahrenheit temperature.

33. *Job Selection* A job candidate is given the choice of two positions, one paying $3100 per month and the other paying $2000 per month plus a 5% commission on all sales made during the month. What amount must the employee sell in a month for the second position to be more profitable?

34. *Stock Market* Susan Mason purchased 1000 shares of stock for $22 per share, and 3 months later it had dropped by 20%. What is the minimum percent increase required for her to make a profit?

35. *Grades* If Jill Ball has a course average score between 80 and 89, she will earn a grade of B in her algebra course. Suppose she has four exam scores of 78, 69, 92, and 81 and that her teacher said the final exam score has twice the weight of each of the other four exams. What range of scores on the final exam will result in Jill earning a grade of B?

36. *Grades* If John Deal has a course average score between 70 and 79, he will earn a grade of C in his algebra course. Suppose he has three exam scores of 78, 62, and 82 and that his teacher said the final exam score has twice the weight of the other three exams. What range of scores on the final exam will result in John earning a grade of C?

37. *Marijuana Use* The percent p of high school seniors who use marijuana daily is given by $30p - 19x = 1$, where x is the number of years after 1990. If this model is accurate, what is the range of daily use of marijuana for high school seniors between 1996 and 2000?
(Source: Index of Leading Cultural Indicators)

38. *Internet Access* The percent of households in the U.S. with Internet access is given by $y = 6.9x - 3.18$, where x is the number of years from 1995. In what years will the percent be greater than 52?
(Source: U.S. Census Bureau)

39. *SAT Scores* The College Board began reporting SAT scores with a new scale in 1996, with the new scale score y defined as a function of the old scale score x by the equation $y = 0.97x + 128.3829$. Suppose a college requires a new scale score greater than or equal to 1000 to admit a student. To determine what old score values would be equivalent to the new scores that would result in admission to this college,

    a. Write an inequality to represent the problem, and solve it algebraically.

    b. Solve the inequality from part (a) graphically to verify your result.

40. *Cigarettes* World cigarette production can be modeled by $y = 9.3451x + 649.3385$ cigarettes per person per year, where x is the number of years after 1950.
(Source: National Clearinghouse on Tobacco and Health)

    a. If the model is accurate, algebraically determine the year in which the per capita world production of cigarettes is 1000 cigarettes. (Note that the answer should be interpreted discretely.)

    b. Use a graph to verify your answer to part (a).

    c. Use your graph to find when the per capita world production of cigarettes is below 1000 cigarettes.

41. *HID Headlights* The new high-intensity discharge (HID) lights that contain xenon gas and are installed in some new cars have an expected life of 1500 hours. Because a complete system costs $1000, it is hoped that these lights will last for the life of the car. Suppose that the actual life of the lights could be 10% longer or shorter than the advertised expected life. Write an inequality that gives the range of life of these new lights.
(Source: *Automobile,* July 2000)

42. *Prison Sentences* The mean time in prison for a crime y can be found as a function of the mean sentence length x, using the equation $y = 0.554x - 2.886$, where x and y are measured in months. How many months should a judge sentence a convicted criminal if she wants the criminal to actually serve between 4 and 6 years?
(Source: Index of Leading Cultural Indicators)

43. *Marriage Rate* According to data from the Index of Leading Cultural Indicators, the marriage rate (marriages per 1000 unmarried women) can be described by $y = -0.763x + 85.284$, where x is the number of years after 1950. For what years does this model indicate that the marriage rate

    a. Was above 50 marriages per 1000 unmarried women?

    b. Will be below 45 marriages per 1000 unmarried women?

44. *Earnings and Minorities* The relation between the median annual salaries of blacks and whites can be modeled by the function $B = 1.05W - 18.691$, where B and W represent the median annual salary (in thousands of dollars) for blacks and whites, respectively. What is the median salary range for whites that corresponds to a salary range of at least $100,000 for blacks?
(Source: *Statistical Abstract of the United States*)

45. *Home Appraisal* A home purchased in 1996 for $190,000 was appraised at $270,000 in 2000. If the rate of increase in the value of the home is assumed to be constant,

    a. Write an equation for the value of the home as a function of the number of years, x, after 1996.

    b. If the equation in part (a) remains accurate, write an inequality that gives the range of years (until the end of 2010) when the value of the home will be greater than $400,000.

46. *Car Sales Profit* A car dealer purchases 12 new cars for $32,500 each and sells 11 of them at a profit of 5.5%. For how much must he sell the remaining car to average a profit of at least 6% on the 12 cars?

47. *Electrical Components Profit* A company's daily profit from the production and sale of electrical components can be described by the equation $P(x) = 6.45x - 2000$ dollars, where x is the number of units produced and sold. What level of production and sales will give a daily profit of more than $10,900?

48. *Profit* The yearly profit from the production and sale of Plumber's Helper is $P(x) = -40,255 + 9.80x$ dollars, where x is the number of Plumber's Helpers produced and sold. What level of production and sales gives a yearly profit of more than $84,355?

49. *Break Even* A large hardware store's monthly profit from the sale of PVC pipe can be described by the equation $P(x) = 6.45x - 9675$ dollars, where x is the number of feet of PVC pipe sold. What level of monthly sales is necessary to avoid a loss?

50. *Plumber's Helper Break Even* The yearly profit from the production and sale of Plumber's Helper is $P(x) = -40,255 + 9.80x$ dollars, where x is the number of Plumber's Helpers produced and sold. What level of production and sales will result in a loss?

51. *Logic Board Break Even* A company is producing a new logic board for computers. The annual fixed cost for the board is $345,000 and the variable cost is $125 per board. If the logic board sells for $489, write an inequality that gives the number of logic boards that will give a profit for the product.

52. *Temperature* The temperature T (in degrees Fahrenheit) inside a concert hall m minutes after a 40-minute power outage during a summer rock concert is given by $T = 0.43m + 76.8$. Write and solve an inequality that describes when the temperature in the hall is not more than 85°F.

53. *Reading Tests* The average reading score of 17-year-olds on the National Assessment of Progress tests is given by $y = 0.155x + 244.37$ points,

where x is the number of years after 1970. Assuming that this model was valid, write and solve an inequality that describes when the average 17-year-old reading score on this test was between but not including 245 and 248. (Your answer should be interpreted discretely.)
(Source: U.S. Department of Education)

54. *Voting* The percent of the voting population who voted in presidential elections between 1950 and 2004 is given by $p = 63.20 - 0.26x$, where x is the number of years after 1950. Write and solve an inequality that describes in what election years the percent was

a. Less than 52.

b. More than 60.

c. Between and including 50 and 60.

55. *Cigarette Use* The percent p of high school seniors who have tried cigarettes can be modeled by

$$p = 75.751 - 0.743t$$

where t is the number of years after 1990.

a. What percent does this model estimate for the year 2008?

b. Test integer values of x with the TABLE feature of your graphing utility to find the values of x for which $p \geq 59.405$.

c. In what years does this model say the percent is at least 59.405%?

56. *Voting* The percent of the voting population who voted in presidential elections between 1950 and 1996 is given by $p = 63.20 - 0.26x$, where x is the number of years after 1950.

a. What percent does this model estimate for the year 2008?

b. Test integer values of x with the TABLE feature of your graphing utility to find the values of x for which $p \leq 47.6$.

c. In what years does this model say the percent is no more than 47.6%?

chapter

2

Summary

In this chapter, we studied the solution of linear equations and systems of linear equations. We used graphing utilities to solve linear equations. We solved business and economics problems involving linear functions, solved application problems, solved linear inequalities, and used graphing utilities to model linear functions.

Key Concepts and Formulas

2.1 Algebraic and Graphical Solution of Linear Equations

Algebraic solution of linear equations	If a linear equation contains fractions, multiply both sides of the equation by a number that will remove all denominators in the equation. Next, remove any parentheses or other symbols of grouping and then perform any additions or subtractions to get all terms containing the variable on one side and all other terms on the other side of the equation. Combine like terms. Divide both sides of the equation by the coefficient of the variable. Check the solution by substitution in the original equation.
Solving real-world application problems	To solve an application problem that is set in a real-world context, use the same solution methods. However, remember to include units of measure with your answer and check that your answer makes sense in the problem situation.
Zero of a function	Any number a for which $f(a) = 0$ is called a zero of the function $f(x)$.
Solutions, zeros, and x-intercepts	If a is an x-intercept of the graph of a function f, then a is a real zero of the function f, and a is a real solution to the equation $f(x) = 0$.
Graphical solution of linear equations	
• **x-intercept method**	Rewrite the equation with 0 on one side, enter the nonzero side into the equation editor of a graphing calculator, and find the x-intercept of the graph. This is the solution to the equation.
• **Intersection method**	Enter the left side of the equation into y_1, enter the right side of the equation into y_2, and find the point of intersection. The x-coordinate of the point of intersection is the solution to the equation.
Literal equations; solving an equation for a specified linear variable	To solve an equation for one variable if two or more variables are in the equation and if that variable is to the first power in the equation, we can solve for that variable by treating the other variables as constants and using the same steps that we used to solve a linear equation in one variable.

2.2 Fitting Lines to Data Points; Modeling Linear Functions

Fitting lines to data points	When real-world information is collected as numerical information called *data*, technology can be used to determine the pattern exhibited by the data (provided that a recognizable pattern exists). These patterns can often be described by mathematical functions.
Constant first differences	If the first differences of data outputs are constant (for equally spaced inputs), a linear model can be found that fits the data exactly. If the first differences are "nearly constant," a linear model can be found that is an approximate fit for the data.
Linear regression	We can determine the equation of the line that is the best fit for a set of points by using a procedure called linear regression (or the least-squares method), which defines the best-fit line as the line for which the sum of the squares of the vertical distances from the data points to the line is a minimum.
Modeling data	We can model a set of data by entering the data into a graphing utility, obtaining a scatter plot, and using the graphing utility to obtain the linear equation that is the best fit for the data. The equation and/or numerical results should be reported in a way that makes sense in the context of the problem, with the appropriate units and with the variables identified.
Discrete versus continuous	We use the term *discrete* to describe data or a function that is presented in the form of a table or in a scatter plot. We use the term *continuous* to describe a function or graph when the inputs can be any real number or any real number between two specified values.
Applying models	Using a model to find an output for an input between two given data points is called *interpolation*. When a model is used for a prediction using an input outside the given data points, the output is called *extrapolation*.
Goodness of fit	The goodness of fit of a linear model can be observed from a graph of the model and the data points and/or measured with the correlation coefficient.

2.3 Systems of Linear Equations in Two Variables

System of equations	A system of linear equations is a set of equations in two or more variables. A solution of the system must satisfy every equation in the system.
Solving a system of linear equations in two variables	
• **Graphing**	Graph the equations and find their point of intersection.
• **Substitution**	Solve one of the equations for one variable and substitute that expression into the other equation, thus giving an equation in one variable.
• **Elimination**	Rewrite one or both equations in a form that allows us to eliminate one of the variables by adding or subtracting the equations.
Break-even analysis	A company is said to break even from the production and sale of a product if the total revenue equals the total cost—that is, if the profit for that product is zero.
Market equilibrium	*Market equilibrium* is said to occur when the quantity of a commodity demanded is equal to the quantity supplied. The price at this point is called the *equilibrium price*, and the quantity at this point is called the *equilibrium quantity*.

Dependent and inconsistent systems	
• **Unique solution**	Graphs are intersecting lines.
• **No solution**	Graphs are parallel lines; system is *inconsistent.*
• **Many solutions**	Graphs are the same line; system is *dependent.*
Modeling systems of equations	Solution of real problems sometimes requires us to create two or more equations whose simultaneous solution is the solution to the problem.

2.4 Solutions of Linear Inequalities

Linear inequality	A linear inequality (or first-degree inequality) is an inequality that can be written in the form $ax + b > 0$ where $a \neq 0$. (The inequality symbol can be $>$, $\geq$, $<$, or $\leq$.)
Algebraically solving linear inequalities	The steps used to solve a linear inequality are the same as those used to solve linear equations, except that the inequality symbol is reversed if both sides are multiplied (or divided) by a negative number.
Graphical solution of linear inequalities	
• **Intersection method**	Set the left side of the inequality equal to y_1 and set the right side equal to y_2, graph the equations using a graphing utility, and find the x-coordinate of the point of intersection. The values of x that satisfy the inequality represented by $y_1 < y_2$ are those values for which the graph of y_1 is below the graph of y_2.
• **x-intercept method**	To use the x-intercept method to solve an inequality, rewrite the inequality with all nonzero terms on one side and zero on the other side of the inequality and combine like terms. Graph the nonzero side of this inequality and find the x-intercept of the graph. If the inequality to be solved is $f(x) > 0$, the solution will be the interval of x-values representing the portion of the graph above the x-axis. If the inequality to be solved is $f(x) < 0$, the solution will be the interval of x-values representing the portion of the graph below the x-axis.
Double inequalities	A double inequality represents two inequalities connected by the word *and* or *or.* Double inequalities can be solved algebraically or graphically. Any operation performed on a double inequality must be performed on *all three* parts.

chapter

2

Skills Check

In Exercises 1–6, solve the equation for x algebraically and graphically.

1. $3x + 22 = 8x - 12$

2. $2(x - 7) = 5(x + 3) - x$

3. $\dfrac{3(x - 2)}{5} - x = 8 - \dfrac{x}{3}$

4. $\dfrac{6x + 5}{2} = \dfrac{5(2 - x)}{3}$

5. $\dfrac{3x}{4} - \dfrac{1}{3} = 1 - \dfrac{2}{3}\left(x - \dfrac{1}{6}\right)$

6. $3.259x - 198.8546 = -3.8(8.625x + 4.917)$

7. For the function $f(x) = 7x - 105$, (a) find the zero of the function; (b) find the x-intercept of the graph of the function; and (c) solve the equation $f(x) = 0$.

8. Solve $P(a - y) = 1 + \dfrac{m}{3}$ for y.

9. Solve $4x - 3y = 6$ for y and graph it on a graphing utility with a standard window.

Use the table of data below in Exercises 10–13.

x	1	3	6	8	10
y	-9	-1	5	12	18

10. Create a scatter plot of the data.

11. Find the linear function that is the best fit for the data in the table.

12. Use a graphing utility to graph the function found in Exercise 11 on the same set of axes as the scatter plot in

Exercise 10, with $x_{min} = 0$, $x_{max} = 15$, $y_{min} = -12$, and $y_{max} = 20$.

13. Do the data points in the table fit exactly on the graph of the function from Exercise 12?

Solve the systems of linear equations in Exercises 14–19, if possible.

14. $\begin{cases} 3x + 2y = 0 \\ 2x - y = 7 \end{cases}$ **15.** $\begin{cases} 3x + 2y = -3 \\ 2x - 3y = 3 \end{cases}$

16. $\begin{cases} -4x + 2y = -14 \\ 2x - y = 7 \end{cases}$ **17.** $\begin{cases} -6x + 4y = 10 \\ 3x - 2y = 5 \end{cases}$

18. $\begin{cases} 2x + 3y = 9 \\ -x - y = -2 \end{cases}$ **19.** $\begin{cases} 2x + y = -3 \\ 4x - 2y = 10 \end{cases}$

In Exercises 20–22, solve the inequalities both algebraically and graphically.

20. $3x + 8 < 4 - 2x$

21. $3x - \dfrac{1}{2} \le \dfrac{x}{5} + 2$

22. $18 \le 2x + 6 < 42$

chapter

2

Review

When money is borrowed to purchase an automobile, the amount borrowed A determines the monthly payment P. In particular, if a dealership offers a 5-year loan at 2.9% interest, then the amount borrowed for the car determines the payment according to the following table. Use the table to define the function $P = f(A)$ in Exercises 23–24.

Amount Borrowed ($)	Monthly Payment ($)
10,000	179.25
15,000	268.87
20,000	358.49
25,000	448.11
30,000	537.73

(Source: Sky Financial)

23. *Car Loans*
  a. Are the first differences of the outputs in the table constant?
  b. Is there a line on which these data points fit exactly?

24. *Car Loans*
  a. Write the equation $P = f(A)$ of the line that fits the data points in the table.
  b. Use the unrounded linear model found in part (a) to find $P = f(28,000)$ and explain what it means.
  c. Can the function f be used to find the monthly payment for any dollar amount A of a loan if the interest rate and length of loan are unchanged?

d. Determine the amount of a loan that will keep the payment less than or equal to $500, using the unrounded model.

25. *Teacher Salaries* The average classroom teacher salary in the United States is given by the function $f(t) = 982.06t + 32,903.77$, where t is the number of years from 1990. In what year was the average teacher salary $40,760.25, according to this model? (Source: www.ors2.state.sc.us/abstract)

26. *Fuel* The table below shows data for the number of gallons of gas purchased each day of a certain week by the 250 taxis owned by the Inner City Transportation Taxi Company. Write the equation that models this data.

Days from first day	0	1	2	3	4	5	6
Gas used (gal)	4500	4500	4500	4500	4500	4500	4500

27. *Work Hours* The average weekly hours worked by production and nonsupervisory workers on private nonfarm payrolls, seasonally adjusted, are given by the data in the table below.

Month and Year	Average Weekly Hours	Month and Year	Average Weekly Hours
June 1998	34.6	Oct. 1998	34.6
July 1998	34.6	Nov. 1998	34.6
Aug. 1998	34.6	Dec. 1998	34.6
Sept. 1998	34.6	Jan. 1999	34.6
		Feb. 1999	34.6

(Source: Bureau of Labor Statistics)

a. Write the equation of a function that describes the average weekly hours using an input equal to the number of months past May 1998.

b. Is this function a constant function?

28. *Job Selection* A job candidate is given the choice of two positions, one paying $2100 per month and one paying $1000 per month plus a 5% commission on all sales made during the month.

a. How much (in dollars) must the employee sell in a month for the second position to pay as much as the first?

b. To be sure that the second position will pay more than the first, how much (in dollars) must the employee sell each month?

29. *Marketing* A car dealer purchased 12 automobiles for $24,000 each. If she sells 8 of them with an average profit of 12%, for how much must she sell the remaining 4 to obtain an average profit of 10% on all 12?

30. *Investment* A retired couple has $420,000 to invest. They chose one relatively safe investment fund that has an annual yield of 6% and another riskier investment that has a 10% annual yield. How much should they invest in each fund to earn $30,000 per year?

31. *Writing Scores* The average writing scores of 11th-graders on the National Assessment of Educational Progress tests have changed over the years since 1984, with the average score given by $y = -0.629x + 293.871$, where x is the number of years from 1980. For what year does this model give an average score of 285? (Source: U.S. Department of Education)

32. *Profit* A company has revenue given by $R(x) = 500x$ dollars and total costs given by $C(x) = 48,000 + 100x$ dollars, where x is the number of units produced and sold. How many units will give a profit?

33. *Profit* A company has revenue given by $R(x) = 564x$ dollars and total cost given by $C(x) = 40,000 + 64x$ dollars, where x is the number of units produced and sold. The profit can be found by forming the function $P(x) = R(x) - C(x)$.

a. Write the profit function.

b. For what values of x is $P(x) > 0$?

c. For how many units is there a profit?

34. *Depreciation* A business property can be depreciated for tax purposes by using the formula $y + 15,000x = 300,000$, where y is the value of the property x years after it was purchased.

a. For what x-values is the property value below $150,000?

b. After how many years is the property value below $150,000?

35. *Marginal Profit* A company has determined that its profit for a product can be described by a linear function. The profit from the production and sale of 150 units is $455 and the profit from 250 units is $895.

a. Write the equation of the profit function for this product.

b. How many units give profit for this product?

36. *Life Expectancy*

a. Find a linear function $y = f(x)$ that models the data shown in the figure with x equal to the number of years after 1950 and y equal to the number of years the average 65-year-old woman is estimated to live beyond age 65.

b. Graph the data and the model on the same set of axes.

c. Use the model to estimate $f(99)$ and explain what it means.

d. Determine the time period (in years) for which the average 65-year-old woman can expect to live more than 84 years.

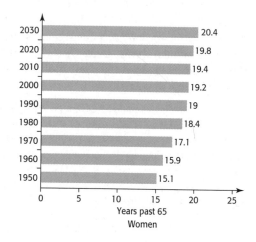

Years past 65
Women

37. *Life Expectancy*

a. Find a linear function $y = g(x)$ that models the data shown in the figure with x equal to the number of years after 1950 and y equal to the number of years the average 65-year-old man is estimated to live beyond age 65.

b. Graph the data and the model on the same set of axes.

c. Use the model to estimate $g(130)$ and explain what it means.

d. In what year would the average 65-year-old man expect to live to age 90?

e. Determine the time period for which the average man could expect to live less than 81 years.

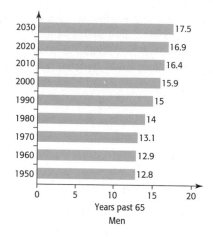

Years past 65
Men

38. *Education Spending* The following table gives the amount (in billions of dollars) spent on education during the years 1996–2001.

a. Graph the data points *or* find the first differences to determine if a linear equation is a reasonable model for this data.

b. If it is reasonable, find the linear model that is the best fit for this data, with x equal to the number of years after 1990.

c. Use the unrounded model to predict the spending in 2002.

Years	Amount Spent ($ billions)
1996	23
1997	26.6
1998	29.9
1999	33.5
2000	35.6
2001	40.1

39. *Population Growth* The resident population of Florida (in thousands) is given in the following table. Let x equal the number of years after 1980 and y equal the number of thousands of residents.

a. Graph the data points to determine if a linear equation is a reasonable model for this data.

b. If it is reasonable, find the linear model that is the best fit for this data, with x equal to the number of years after 1980.

c. What does this unrounded model predict that the population will be in 2002?

Years	1980	1985	1990	1995
Population (thousands)	9746	11,351	12,938	14,180

Years	1996	1997	1998
Population (thousands)	14,425	14,677	14,916

(Source: U.S. Census Bureau)

40. *Earnings per Share* The table below gives the earnings ($ thousands) per share (EPS) for ACS stock for the years 1996–2000. Let x equal the number of years after 1995 and y equal the EPS, in thousands of dollars.

a. Graph the data points to determine if a linear equation is a reasonable model for this data.

b. If it is reasonable, find the linear model that is the best fit for this data.

c. Graph the data points and the function on the same axes and discuss the goodness of fit.

Years	1996	1997	1998	1999	2000
EPS ($ thousands)	0.85	1.05	1.29	1.66	2.05

(Source: 2000 Report of ACS)

41. *Marriage Rate* According to data from the Index of Leading Cultural Indicators, the marriage rate (the number of marriages per 1000 women) can be described by $y = -0.763x + 85.284$, where x is the number of years after 1950. For what years does this model indicate that the rate

a. Was above 48.66 per 1000 women?

b. Will be below 41.03 per 1000 women?

42. *Marijuana Use* The percent p of high school seniors who use marijuana daily is given by the equation $30p - 19x = 1$, where x is the number of years after

1990. If this model is accurate, what is the range of years for which the percent of use is 3.2%–7%? (Source: Index of Leading Cultural Indicators)

43. *Prison Sentences* The mean time in prison y for a crime can be found as a function of the mean sentence length x, using $y = 0.554x - 2.886$, where x and y are in months. If a judge sentences a convicted criminal to serve between 3 and 5 years, how many months would we expect the criminal to serve? (Source: Index of Leading Cultural Indicators)

44. *Investment* A retired couple has $240,000 to invest. They chose one relatively safe investment fund that has an annual yield of 8% and another riskier investment that has a 12% annual yield. How much should they invest in each fund to earn $23,200 per year?

45. *Break Even* A computer manufacturer has a new product with daily total revenue given by $R = 565x$ and daily total cost given by $C = 6000 + 325x$. How many units per day must be produced and sold to give break even for the product?

46. *Medication* Medication A is given six times per day and medication B is given twice per day. For a certain patient, the total intake of the two medications is limited to 25.2 mg per day. If the ratio of the dosage of medication A to the dosage of medication B is 2 to 3, how many milligrams are in each dosage?

47. *Market Equilibrium* The demand for a certain brand of women's shoes is given by $3q + p = 340$, and the supply for these shoes is given by $p - 4q = -220$, where p is the price in dollars and q is the number of pairs demanded at price p. Solve the system containing these two equations to find the equilibrium price and the equilibrium quantity.

48. *Market Analysis* Suppose for a certain product, the supply and demand functions are $p = \dfrac{q}{10} + 8$ and $10p + q = 1500$, respectively, where p is in dollars and q is in units. Find the equilibrium price and quantity.

49. *Pricing* A concert promoter needs to make $120,000 from the sale of 2600 tickets. The promoter charges $40 for some tickets and $60 for the others.

a. If there are x of the $40 tickets and y of the $60 tickets, write an equation that states that the total number of the tickets sold is 2600.

b. How much money is made from the sale of x tickets for $40 each?

c. How much money is made from the sale of y tickets for $60 each?

d. Write an equation that states that the total amount made from the sale is $120,000.

e. Solve the equations simultaneously to find how many tickets of each type must be sold to yield the $120,000.

50. *Rental Income*   A woman has $500,000 invested in two rental properties. One yields an annual return of 12% of her investment, and the other returns 15% per year on her investment. Her total annual return from the two investments is $64,500. If x represents the 12% investment and y represents the 15% investment,

a. Write an equation that states that the sum of the investments is $500,000.

b. What is the annual return on the 12% investment?

c. What is the annual return on the 15% investment?

d. Write an equation that states that the sum of the annual returns is $64,500.

e. Solve these two equations simultaneously to find how much is invested in each property.

group activities /
extended applications

1. Taxes

The table below gives the income tax due, $f(x)$, on each given taxable income, x.

1. What is the domain and range of the function in the table?
2. Create a scatter plot of the data.
3. Do the points appear to lie on a line?
4. Do the inputs change by the same amount? Do the outputs change by the same amount?
5. Is the rate of change in tax per $1 of income constant? What is the rate of change?
6. Will a linear function fit the data points exactly?
7. Write a linear function $y = g(x)$ that fits the data points.
8. Verify that the linear model fits the data points by evaluating the linear function at $x = 63,900$ and $x = 64,100$ and comparing the resulting y-values with the income tax due for these taxable incomes.
9. Is the model a discrete or continuous function?

10. Can the model be used to find the tax due on any taxable income between 63,700 and 64,300?
11. What is the tax due on a taxable income of $64,150, according to this model?

U.S. Federal Taxes

Taxable Income (dollars)	Income Tax Due (dollars)
63,700	8779
63,800	8804
63,900	8829
64,000	8854
64,100	8879
64,200	8904
64,300	8929

(Source: U.S. Internal Revenue Service)

2. Research

Linear functions can be used to model many types of real data. Graphs displaying linear growth are frequently displayed in periodicals such as *Newsweek* and *Time*, in newspapers such as *USA Today* and the *Wall Street Journal*, and on numerous Web sites on the Internet. Tables of data can also be found in these sources, especially in federal and state government Web sites, such as www.census.gov. (This Web site is the source of the Florida resident population data used in the Chapter Review, for example.)

Your mission is to find a company sales, stock price, biological growth, or sociological trend over a period of years (using at least four points) that is linear or "nearly" linear, and to find a linear function that is a model for this data.

A linear model will be a good fit for the data

1. If the data is presented as a graph that is linear or "nearly linear."
2. If the data is presented in a table and the plot of the data points lies near some line.
3. If the data is presented in a table and the first differences of the outputs are nearly constant for equally spaced inputs.

After you have created the model, you should test the goodness of fit of the model to the data and discuss uses that you could make of the model.

Your completed project should include

a. A complete citation of the source of the data you are using.
b. An original copy or photocopy of the data being used.
c. A scatter plot of the data.
d. The equation that you have created.
e. A graph containing the scatter plot and the modeled equation.
f. A statement about how the model could be used to make estimations or predictions about the trend you are observing.

Some helpful hints:

1. If you decide to use a relation determined by a graph that you have found, read the graph very carefully to determine the data points or contact the source of the data to get the data from which the graph was drawn.
2. Align the independent variable by letting x represent the number of years from some convenient year and then enter the data into a graphing utility and create a scatter plot.
3. Use your graphing utility to create the equation of the function that is the best fit for the data. Graph this equation and the data points on the same axes to see if the equation is reasonable.

Quadratic and Other Nonlinear Functions

Revenue and profit from the sale of products frequently cannot be modeled by linear functions because they increase at rates that are not constant. In this chapter, we use nonlinear functions, including quadratic and power functions, to model numerous applications in business, economics, and the life and social sciences.

topics

Graphing quadratic functions; finding vertices of parabolas; increasing and decreasing functions

Solving by factoring; solving graphically; combining graphs and factoring; solving with the root method; completing the square; the quadratic formula; complex solutions; discriminants

Graphing and applying power, root, reciprocal, piecewise-defined, and absolute value functions; direct variation

Modeling with quadratic functions; comparing linear and quadratic models; modeling with power functions; comparing power and quadratic models

applications

Maximum revenue from sales, height of a ball, foreign-born population

Hospital admissions, profit, marijuana use, height of a ball

Windchill factor, service calls, allometric relationships, average cost, postage, residential power costs

Starbucks stores, height of a rocket, aid to dependent children, auto noise, cohabiting households, voting

Algebra Toolbox

In this Toolbox, we discuss absolute value, integer and rational exponents, and radicals. We also discuss the multiplication of monomials and binomials, factoring, and complex numbers.

Integer Exponents

In this chapter and future ones, we will discuss functions and equations containing integer powers of variables. For example, we will discuss the function $y = x^{-1} = \dfrac{1}{x}$.

> If a is a real number and n is a positive integer, then a^n represents a as a factor n times in a product
>
> $$a^n = \underbrace{a \cdot a \cdot a \ldots \cdot a}_{n \text{ times}}$$
>
> In a^n, a is called the base and n is called the exponent.

In particular, $a^2 = a \cdot a$ and $a^1 = a$. Note that for positive integers m and n

$$a^m \cdot a^n = \underbrace{a \cdot a \cdot a \ldots \cdot a}_{m \text{ times}} \underbrace{a \cdot a \ldots \cdot a}_{n \text{ times}} = \underbrace{a \cdot a \cdot a \ldots \cdot a}_{m+n \text{ times}} = a^{m+n}$$

and that, for $m \geq n$,

$$\frac{a^m}{a^n} = \frac{\overbrace{a \cdot a \cdot a \ldots \cdot a}^{m \text{ times}}}{\underbrace{a \cdot a \ldots \cdot a}_{n \text{ times}}} = \overbrace{a \cdot a \cdot a \ldots \cdot a}^{m-n \text{ times}} = a^{m-n} \quad \text{if } a \neq 0$$

These two important properties of exponents can be extended to all integers.

> ## Properties of Exponents
>
> For real numbers a and b, and integers m and n,
>
> 1. $a^m \cdot a^n = a^{m+n}$ (Product Property)
>
> 2. $\dfrac{a^m}{a^n} = a^{m-n}, a \neq 0$ (Quotient Property)

We define an expression raised to zero and to a negative power as follows.

> ## Zero and Negative Exponents
> For $a \neq 0, b \neq 0$
>
> 1. $a^0 = 1$
>
> 2. $a^{-1} = \dfrac{1}{a}$
>
> 3. $a^{-n} = \dfrac{1}{a^n}$
>
> 4. $\left(\dfrac{a}{b}\right)^{-n} = \left(\dfrac{b}{a}\right)^n$

example 1

Zero and Negative Exponents

Simplify the following expressions by removing all zero and negative exponents, for nonzero a, b, and c.

a. $(4c)^0$ **b.** $4c^0$ **c.** $(5b)^{-1}$ **d.** $5b^{-1}$ **e.** $\left(\dfrac{a}{b}\right)^{-3}$ **f.** $6a^{-3}$

Solution

a. $(4c)^0 = 1$ **b.** $4c^0 = 4(1) = 4$ **c.** $(5b)^{-1} = \dfrac{1}{(5b)} = \dfrac{1}{5b}$

d. $5b^{-1} = 5 \cdot \dfrac{1}{b} = \dfrac{5}{b}$ **e.** $\left(\dfrac{a}{b}\right)^{-3} = \left(\dfrac{b}{a}\right)^3 = \dfrac{b^3}{a^3}$ **f.** $6a^{-3} = 6 \cdot \dfrac{1}{a^3} = \dfrac{6}{a^3}$ ∎

Absolute Value

The distance the number a is from 0 on a number line is the **absolute value** of a, denoted by $|a|$. The absolute value of any nonzero number is positive, and the absolute value of 0 is 0. For example, $|5| = 5$ and $|-8| = 8$. Note that if a is a nonnegative number, then $|a| = a$, but if a is negative, then $|a|$ is the positive number $-a$. Formally, we say

$$|a| = \begin{cases} a & \text{if } a \geq 0 \\ -a & \text{if } a < 0 \end{cases}$$

For example, $|5| = 5$ and $|-5| = -(-5) = 5$.

Rational Exponents and Radicals

In this chapter we will study functions involving a variable raised to a rational power (called **power functions**), and we will solve equations involving rational exponents and radicals. This may involve converting expressions involving radicals to expressions involving rational exponents, or vice versa.

Exponential expressions are defined for rational numbers in terms of radicals. Note that for $a \geq 0$ and $b \geq 0$

$$\sqrt{a} = b \text{ only if } a = b^2$$

Thus,

$$\left(\sqrt{a}\right)^2 = b^2 = a, \text{ so}$$

$$\left(\sqrt{a}\right)^2 = a$$

We define $a^{1/2} = \sqrt{a}$, so $(a^{1/2})^2 = a$ for $a \geq 0$.

The following definitions show the connection between rational exponents and radicals.

Rational Exponents

1. If a is a real number, variable, or algebraic expression and n is a positive integer $n \geq 2$, then

$$a^{1/n} = \sqrt[n]{a}$$

provided that $\sqrt[n]{a}$ exists.

2. If a is a real number and if m and n are integers containing no common factor with $n \geq 2$, then

$$a^{m/n} = \sqrt[n]{a^m} = \left(\sqrt[n]{a}\right)^m$$

provided that $\sqrt[n]{a}$ exists.

---

example 2

Write the following expressions with exponents rather than radicals.

a. $\sqrt[3]{x^2}$ **b.** $\sqrt[4]{x^3}$ **c.** $\sqrt{(3xy)^5}$ **d.** $3\sqrt{(xy)^5}$

Solution

a. $\sqrt[3]{x^2} = x^{2/3}$ **b.** $\sqrt[4]{x^3} = x^{3/4}$ **c.** $\sqrt{(3xy)^5} = (3xy)^{5/2}$ **d.** $3(xy)^{5/2}$ ■

example 3

Write the following in radical form.

a. $y^{1/2}$ **b.** $(3x)^{3/7}$ **c.** $12x^{3/5}$

Solution

a. $y^{1/2} = \sqrt{y}$ **b.** $(3x)^{3/7} = \sqrt[7]{(3x)^3} = \sqrt[7]{27x^3}$ **c.** $12x^{3/5} = 12\sqrt[5]{x^3}$ ■

Multiplication of Monomials and Binomials

Polynomials with one term are called monomials, those with two terms are called binomials, and those with three terms are called trinomials. In this chapter, we will factor monomials from polynomials and we will factor trinomials into two binomials. To better see how this factoring is accomplished, we will review multiplying by monomials and binomials.

We multiply two monomials by multiplying the coefficients and adding the exponents of the respective variables that are in both monomials. For example,

$$(3x^3y^2)(4x^2y) = 3 \cdot 4 \cdot x^3 \cdot x^2 \cdot y^2 \cdot y = 3 \cdot 4x^{3+2}y^{2+1} = 12x^5y^3$$

We can multiply more than two monomials in the same manner.

example 4

Find the product:

$$(-2x^4z)(4x^2y^3)(yz^5)$$

Solution

$$(-2x^4z)(4x^2y^3)(yz^5) = -2 \cdot 4x^{4+2}y^{3+1}z^{1+5} = -8x^6y^4z^6 \qquad \blacksquare$$

We can use the **distributive property**,

$$a(b + c) = ab + ac$$

to multiply a monomial times a polynomial. For example,

$$x(3x + y) = x \cdot 3x + x \cdot y = 3x^2 + xy$$

We can extend the property $a(b + c) = ab + ac$ to multiply a monomial times any polynomial. For example,

$$3x(2x + xy + 6) = 3x \cdot 2x + 3x \cdot xy + 3x \cdot 6 = 6x^2 + 3x^2y + 18x$$

The product of two binomials can be found by using the distributive property as follows:

$$(a + b)(c + d) = a(c + d) + b(c + d) = ac + ad + bc + bd$$

Note that this product can be remembered as the sum of the products of the First, Outer, Inner, and Last terms of the binomials, and we use the word **FOIL** to denote this method.

example 5

Find the following products:

a. $(x - 4)(x - 5)$ **b.** $(2x - 3)(3x + 2)$ **c.** $(3x - 5y)(3x + 5y)$

Solution

a. $(x - 4)(x - 5) = x \cdot x + x(-5) + (-4)x + (-4)(-5)$

$$= x^2 - 5x - 4x + 20 = x^2 - 9x + 20$$

b. $(2x - 3)(3x + 2) = (2x)(3x) + (2x)2 + (-3)(3x) + (-3)2$

$$= 6x^2 + 4x - 9x - 6 = 6x^2 - 5x - 6$$

c. $(3x - 5y)(3x + 5y) = (3x)(3x) + (3x)(5y) + (-5y)(3x) + (-5y)(5y)$

$$= 9x^2 + 15xy - 15xy - 25y^2 = 9x^2 - 25y^2 \qquad \blacksquare$$

Certain products and powers involving binomials occur frequently, so the following special products should be remembered.

> ### Special Binomial Products
> 1. $(x + a)(x - a) = x^2 - a^2$ (difference of two squares)
> 2. $(x + a)^2 = x^2 + 2ax + a^2$ (perfect square trinomial)
> 3. $(x - a)^2 = x^2 - 2ax + a^2$ (perfect square trinomial)

example 6

Find the following products by using the special binomial products formulas.

a. $(5x + 1)^2$ **b.** $(2x - 5)(2x + 5)$ **c.** $(3x - 4)^2$

Solution

a. $(5x + 1)^2 = (5x)^2 + 2(5x)(1) + 1^2 = 25x^2 + 10x + 1$

b. $(2x - 5)(2x + 5) = (2x)^2 - 5^2 = 4x^2 - 25$

c. $(3x - 4)^2 = (3x)^2 - 2(3x)(4) + 4^2 = 9x^2 - 24x + 16$ ■

Factoring

Factoring is the process of writing a number or an algebraic expression as the product of two or more numbers or expressions. For example, the distributive property justifies factoring of monomials from polynomials. For example,

$$5x^2 - 10x = 5x(x - 2)$$

Factoring out the **greatest common factor** (gcf) from a polynomial is the first step in factoring.

example 7

Factor out the greatest common factor.

a. $4x^2y^3 - 18xy^4$ **b.** $3x(a - b) - 2y(a - b)$

Solution

a. The gcf of 4 and 18 is 2. The gcf of x^2 and x is the lower power of x, which is x. The gcf of y^3 and y^4 is the lower power of y, which is y^3. Thus, the gcf of $4x^2y^3$ and $18xy^4$ is $2xy^3$. Factoring out the gcf gives

$$4x^2y^3 - 18xy^4 = 2xy^3(2x - 9y)$$

b. The gcf of $3x(a - b)$ and $2y(a - b)$ is the binomial $a - b$. Factoring out $a - b$ from each term gives

$$3x(a - b) - 2y(a - b) = (a - b)(3x - 2y)$$ ■

By recognizing that a polynomial has the form of one of the special products given above, we can factor that polynomial.

| **example 8** | Use knowledge of binomial products to factor the following algebraic expressions. |

a. $9x^2 - 25$ **b.** $4x^2 - 12x + 9$

Solution

a. Both terms are squares, so the polynomial can be recognized as the **difference of two squares**. It will then factor as the product of the sum and the difference of the square roots of the terms (see Special Binomial Products, Formula 1).

$$9x^2 - 25 = (3x + 5)(3x - 5)$$

b. Recognizing that the second-degree term and the constant term are squares leads us to investigate whether $12x$ is twice the product of the square roots of these two terms (see Special Binomial Products, Formula 3). The answer is yes, so the polynomial is a **perfect square**, and it can be factored as follows:

$$4x^2 - 12x + 9 = (2x - 3)^2$$

This can be verified by expanding $(2x - 3)^2$. ■

The first step in factoring is to look for common factors. All applicable factoring techniques should be applied to factor a polynomial completely.

| **example 9** | Factor the following polynomials completely. |

a. $3x^2 - 33x + 72$ **b.** $6x^2 - x - 1$

Solution

a. The number 3 can be factored from all three terms, giving

$$3x^2 - 33x + 72 = 3(x^2 - 11x + 24)$$

If the trinomial can be factored into the product of two binomials, the first term of each binomial must be x, and we seek two numbers whose product is 24 and whose sum is -11. Because -3 and -8 satisfy these requirements, we get

$$3(x^2 - 11x + 24) = 3(x - 3)(x - 8)$$

b. The four possible factorizations of $6x^2 - x - 1$ that give $6x^2$ as the product of the first terms and -1 as the product of the last terms follow:

$$(6x - 1)(x + 1) \qquad (6x + 1)(x - 1)$$
$$(2x - 1)(3x + 1) \qquad (2x + 1)(3x - 1)$$

The factorization that gives a product with middle term $-x$ is the correct factorization.

$$(2x - 1)(3x + 1) = 6x^2 - x - 1$$ ■

Some polynomials, such as $6x^2 + 9x - 8x - 12$, can be factored by **grouping**. To do this, we factor out common factors from pairs of terms and then factor out a common binomial expression if it exists. For example,

$$6x^2 + 9x - 8x - 12 = 3x(2x + 3) - 4(2x + 3) = (2x + 3)(3x - 4)$$

When a second-degree trinomial can be factored but there are many possible factors to test to find the correct one, an alternate method of factoring can be used. The steps used to factor the trinomial using factoring by grouping techniques follow.

Factoring a Trinomial into the Product of Two Binomials Using Grouping

Steps	Example
To factor a quadratic trinomial in the variable x:	Factor $5x - 6 + 6x^2$
1. Arrange the trinomial with the powers of x in descending order.	1. $6x^2 + 5x - 6$
2. Form the product of the second-degree term and the constant term (first and third terms).	2. $6x^2(-6) = -36x^2$
3. Determine if there are two factors of the product in step 2 that will sum to the middle (first-degree) term. (If there are no such factors, the trinomial will not factor into two binomials.)	3. $-36x^2 = (-4x)(9x)$ and $-4x + 9x = 5x$
4. Replace the middle term from step 1 with the *sum* of the two factors from step 3.	4. $6x^2 + 5x - 6 = 6x^2 - 4x + 9x - 6$
5. Factor the four-term polynomial from step 4 by grouping.	5. $6x^2 + 5x - 6 = 2x(3x - 2) + 3(3x - 2) = (3x - 2)(2x + 3)$

example 10 Factor $10x^2 + 23x - 5$, using grouping.

Solution

To use this method to factor $10x^2 + 23x - 5$, we

1. Note that the trinomial has powers of x in descending order. $\qquad 10x^2 + 23x - 5$

2. Multiply the second-degree and constant terms: $\qquad 10x^2(-5) = -50x^2$

3. Factor $-50x^2$ so the sum of factors is $23x$: $\qquad -50x^2 = 25x(-2x),\ 25x + (-2x) = 23x$

4. Rewrite the middle term of the expression as the sum of the factors from step 3. $\qquad 10x^2 + 23x - 5 = 10x^2 + 25x - 2x - 5$

5. Factor by grouping:

$$10x^2 + 25x - 2x - 5$$
$$= 5x(2x + 5) - (2x + 5)$$
$$= (2x + 5)(5x - 1) \qquad \blacksquare$$

Complex Numbers

In this chapter, some equations do not have real number solutions, but do have solutions that are complex numbers.

The numbers discussed up to this point are real numbers (either rational numbers such as $2, -3, \frac{5}{8}$, and $-\frac{2}{3}$ or irrational numbers such as $\sqrt{3}, \sqrt[3]{6}$, and π). But some equations do not have real solutions. For example, if

$$x^2 + 1 = 0, \quad \text{then} \quad x^2 = -1$$

but there is no real number that, when squared, will equal -1. However, we can denote one solution to this equation as the **imaginary unit i**, defined by

$$i = \sqrt{-1}$$

Note that $i^2 = \sqrt{-1} \cdot \sqrt{-1} = -1$.

The set of **complex numbers** is formed by adding real numbers and multiples of *i*.

> ### Complex Number
>
> The number $a + bi$, in which *a* and *b* are real numbers, is said to be a **complex number in standard form**. The *a* is the real part of the number, and *bi* is the imaginary part. If $b = 0$, the number $a + bi = a$ is a real number, and if $b \neq 0$, the number $a + bi$ is an **imaginary number**. If $a = 0$, *bi* is a **pure imaginary number**.

The complex number system includes the real numbers as well as the imaginary numbers (Figure 3.1). Examples of imaginary numbers are $3 + 2i$, $5 - 4i$, and $\sqrt{3} - \frac{1}{2}i$; examples of pure imaginary numbers are $-i$, $2i$, $12i$, $-4i$, $i\sqrt{3}$, and πi.

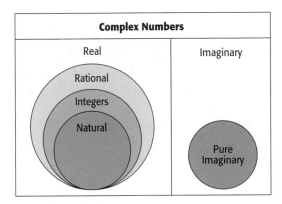

Figure 3.1

The complex number system is an extension of the real numbers that includes imaginary numbers. The term *imaginary number* seems to imply that the numbers do not exist, but in fact they do have important theoretical and technical applications. Complex numbers are used in the design of electrical circuits and airplanes, and they were used in the development of quantum physics. For example, one way of accounting for the amount as well as the phase of the current or voltage in an alternating current electrical circuit involves complex numbers. Special sets of complex numbers can be graphed on the **complex coordinate system** to create pictures called **fractal images**. (Figure 3.2 shows a fractal image called the Mandelbrot set, which can be generated with the complex number *i*.)

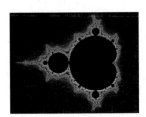

Figure 3.2

From "The Fractal Geometry of Nature" by B.B. Mandelbrot. Copyright 1982. Used with permission of the author.

| example 11 |

Simplifying Complex Numbers

Simplify the following numbers by writing them in the form a, bi, or $a + bi$.

a. $6 + \sqrt{-8}$ **b.** $\dfrac{4 - \sqrt{-6}}{2}$

Solution

a. $6 + \sqrt{-8} = 6 + i\sqrt{8} = 6 + 2i\sqrt{2}$
because $\sqrt{8} = \sqrt{4 \cdot 2} = \sqrt{4} \cdot \sqrt{2} = 2\sqrt{2}$

b. $\dfrac{4 - \sqrt{-6}}{2} = \dfrac{4 - i\sqrt{6}}{2} = \dfrac{4}{2} - \dfrac{i\sqrt{6}}{2} = 2 - \dfrac{i\sqrt{6}}{2}$

example 12 | Complex Numbers

Identify each number as one or more of real, imaginary, or pure imaginary.

a. $6 + 4i$　　**b.** $3i - \sqrt{4}$　　**c.** $3 - 2i^2$　　**d.** $4 - \sqrt{-16}$　　**e.** $\sqrt{-3}$

Solution

a. Imaginary because it contains i.

b. Imaginary because it contains i.

c. Real because $3 - 2i^2 = 3 - 2(-1) = 3 + 2 = 5$.

d. Imaginary because it contains $\sqrt{-16} = 4i$, which gives $4 - 4i$.

e. Imaginary because it contains $\sqrt{-3} = i\sqrt{3}$, and pure imaginary because the real part is 0.

toolbox exercises

In Exercises 1–6, use the rules of exponents to simplify the following expressions and remove all zero and negative exponents. Assume that all variables are nonzero.

1. $\left(\dfrac{2}{3}\right)^{-2}$

2. $\left(\dfrac{3}{2}\right)^{-3}$

3. $10^{-2} \cdot 10^0$

4. $8^{-2} \cdot 8^0$

5. $(2^{-1})^3$

6. $(4^{-2})^2$

Find the absolute values in Exercises 7 and 8.

7. $|-6|$

8. $|7 - 11|$

9. Write each of the following expressions in simplified exponential form.

　a. $\sqrt{x^3}$　**b.** $\sqrt[4]{x^3}$　**c.** $\sqrt[5]{x^3}$

　d. $\sqrt[6]{27y^9}$　**e.** $27\sqrt[6]{y^9}$

10. Write each of the following in radical form.

　a. $a^{3/4}$　　**b.** $-15x^{5/8}$

　c. $(-15x)^{5/8}$

In Exercises 11–15, find the products.

11. $(4x^2y^3)(-3a^2x^3)$

12. $2xy^3(2x^2y + 4xz - 3z^2)$

13. $(x - 7)(2x + 3)$　　**14.** $(k - 3)^2$

15. $(4x - 7y)(4x + 7y)$

In Exercises 16–26, factor each of the polynomials completely.

16. $3x^2 - 12x$　　**17.** $12x^5 - 24x^3$

18. $9x^2 - 25m^2$　　**19.** $x^2 - 8x + 15$

20. $x^2 - 2x - 35$　　**21.** $3x^2 - 5x - 2$

22. $8x^2 - 22x + 5$

23. $6n^2 + 18 + 39n$

24. $3y^4 + 9y^2 - 12y^2 - 36$

25. $18p^2 + 12p - 3p - 2$

26. $5x^2 - 10xy - 3x + 6y$

In Exercises 27 and 28, identify each number as one or more of real, imaginary, or pure imaginary.

27. a. $2 - i\sqrt{2}$　　**b.** $5i$　　**c.** $4 + 0i$

　　d. $2 - 5i^2$

28. a. $3 + i\sqrt{5}$　　**b.** $3 + 0i$　　**c.** $8i$

　　d. $2i^2 - i$

In Exercises 29–31, find values for a and b that make the statement true.

29. $a + bi = 4$　　**30.** $a + 3i = 15 - bi$

31. $a + bi = 2 + 4i$

3.1

Quadratic Functions; Parabolas

section preview ▪ Revenue

Suppose the monthly revenue from the sale of Carlson 42-inch plasma televisions is given by the function

$$R(x) = -0.1x^2 + 600x \text{ dollars}$$

where x is the number of TVs sold. In this case, the revenue of this product is represented by a **second-degree**, or **quadratic function**. A quadratic function is a function that can be written in the form

$$f(x) = ax^2 + bx + c$$

where a, b, and c are real numbers with $a \neq 0$.

The graph of the quadratic function $f(x) = ax^2 + bx + c$ is a **parabola** with a "turning point" called the **vertex**. Figure 3.3 shows the graph of the function $R(x) = -0.1x^2 + 600x$, which is a parabola that opens downward, and the vertex occurs where the function has its maximum value. Notice also that the graph of this revenue function is symmetric about a vertical line through the vertex. This vertical line is called the **axis of symmetry**.

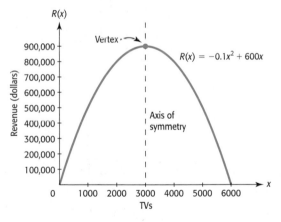

Figure 3.3

Knowing where the maximum value of $R(x)$ occurs can show the company how many units must be sold to obtain the largest revenue. Knowing the maximum output of $R(x)$ helps the company plan its sales campaign. (See Example 1.) In this section, we graph and apply quadratic functions.

Parabolas

The graph of every quadratic function has the distinctive shape known as a parabola. The graph of a quadratic function is determined by the location of the vertex and whether the parabola opens upward or downward.

173

Table 3.1

x	y
-4	16
-3	9
-2	4
-1	1
0	0
1	1
2	4
3	9
4	16

Consider the basic quadratic function $y = x^2$. Each output y is obtained by squaring an input x (Table 3.1). The graph of $y = x^2$, shown in Figure 3.4, is a parabola that opens upward with the **vertex** (turning point) at the origin, (0, 0).

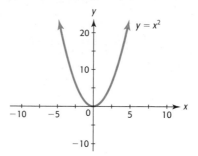

Figure 3.4

Observe that for $x > 0$, the graph of $y = x^2$ rises as it moves from left to right (that is, as the x-values increase), so the function $y = x^2$ is **increasing** for $x > 0$. For values of $x < 0$, the graph falls as it moves from left to right (as the x-values increase), so the function $y = x^2$ is **decreasing** for $x < 0$.

Increasing and Decreasing Functions

A function f is **increasing** on an interval if, for any x_1 and x_2 in the interval, when $x_2 > x_1$, it is true that $f(x_2) > f(x_1)$.

A function f is **decreasing** on an interval if, for any x_1 and x_2 in the interval, when $x_2 > x_1$, it is true that $f(x_2) < f(x_1)$.

The quadratic function $y = -x^2$ has the form $y = ax^2$ with $a < 0$, and its graph is a parabola that opens downward with vertex at (0, 0) (Figure 3.5(a)).

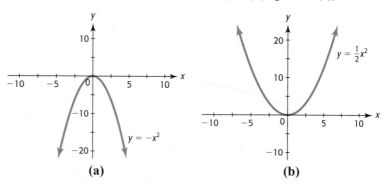

(a)   **(b)**

Figure 3.5

The function $y = \dfrac{1}{2}x^2$ has the form $y = ax^2$ with $a > 0$, and its graph is a parabola opening upward (Figure 3.5(b)). In general, the graph of a quadratic function of the form $y = ax^2$ is a parabola that opens upward (is **concave up**) if a is positive and opens downward (is **concave down**) if a is negative.* The vertex, which is the point

---

*A parabola that is concave up appears that it will "hold water," and a parabola that is concave down will appear to "shed water."

where the parabola turns, is a **minimum point** if a is positive and is a **maximum point** if a is negative. The vertical line through the vertex is called the **axis of symmetry** because this line divides the graph into two halves that are reflections of each other (Figure 3.6).

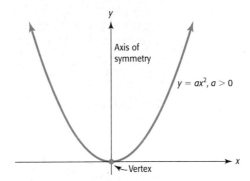

Figure 3.6

We can find the x-coordinate of the vertex of the graph of $y = ax^2 + bx + c$ by using the fact that the axis of symmetry of a parabola passes through the vertex. As Figure 3.7 shows, the y-intercept of the graph of $y = ax^2 + bx + c$ is $(0, c)$, and there is another point on the graph with y-coordinate c.

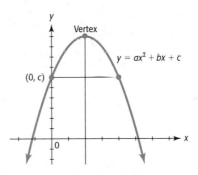

Figure 3.7

The x-coordinates of the points on this graph with y-coordinate c satisfy

$$c = ax^2 + bx + c$$

Solving this equation gives

$$0 = ax^2 + bx$$
$$0 = x(ax + b)$$
$$x = 0 \quad \text{or} \quad x = \frac{-b}{a}$$

The x-coordinate of the vertex is on the axis of symmetry, which is halfway from $x = 0$ to $x = \dfrac{-b}{a}$, so it is at $x = \dfrac{-b}{2a}$. The y-coordinate of the vertex can be found by evaluating the function at the x-coordinate of the vertex.

Graph of a Quadratic Function

The graph of the function

$$f(x) = ax^2 + bx + c$$

is a parabola that opens upward, and the vertex is a minimum, if $a > 0$. The parabola opens downward, and the vertex is a maximum, if $a < 0$.
The larger the value of $|a|$, the more narrow the parabola will be.

Its vertex is at the point $\left(\dfrac{-b}{2a}, f\left(\dfrac{-b}{2a}\right)\right)$ (Figure 3.8).

The axis of symmetry of the parabola has equation $x = \dfrac{-b}{2a}$.

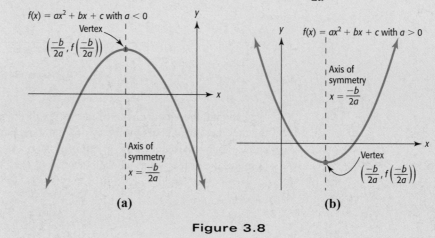

Figure 3.8

Observe that the graph of $y = x^2$ (Figure 3.4) is narrower than the graph of $y = \dfrac{1}{2}x^2$

(Figure 3.5(b)), and that 1 (the coefficient of x^2 in $y = x^2$) is larger than

$\dfrac{1}{2}\left(\text{the coefficient of } x^2 \text{ in } y = \dfrac{1}{2}x^2\right)$.

If we know the location of the vertex and the direction in which the parabola opens, we can make a good sketch of the graph by plotting just a few more points.

---

example 1   ## Maximizing Revenue

Suppose the monthly revenue from the sale of Carlson 42-inch plasma televisions is given by the function

$$R(x) = -0.1x^2 + 600x \text{ dollars}$$

where x is the number of televisions sold.

a. Find the vertex and the axis of symmetry of the graph of this function.

b. Determine if the vertex represents a maximum or minimum point.

c. Interpret the vertex in the context of the application.

d. Graph the function.

Solution

a. The function is a quadratic function with $a = -0.1$, $b = 600$, and $c = 0$. The x-coordinate of the vertex is $\dfrac{-b}{2a} = \dfrac{-600}{2(-0.1)} = 3000$, and the axis of symmetry is the line $x = 3000$. The y-coordinate of the vertex is

$$R(3000) = -0.1(3000)^2 + 600(3000) = 900{,}000$$

so the vertex is $(3000, 900{,}000)$.

b. Because $a < 0$, the parabola opens downward, so the vertex is a maximum point.

c. The x-coordinate of the vertex gives the number of televisions that must be sold to maximize revenue, so selling 3000 sets will result in the maximum revenue. The y-coordinate of the vertex gives the maximum revenue, $\$900{,}000$.

d. A table of some values that satisfy $R(x) = -0.1x^2 + 600x$ and the graph are shown in Figures 3.9(a) and 3.9(b).

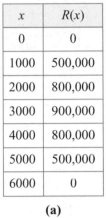

x	$R(x)$
0	0
1000	500,000
2000	800,000
3000	900,000
4000	800,000
5000	500,000
6000	0

(a)

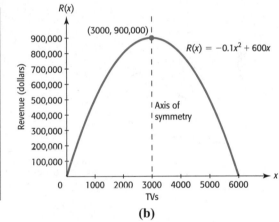

(b)

Figure 3.9

Even when using a graphing utility to graph a quadratic function, it is important to recognize that the graph is a parabola and to locate the vertex. Determining which way the parabola opens and the x-coordinate of the vertex is very useful in setting the viewing window so that a complete graph (that includes the vertex and the intercepts) is shown.

example 2

Foreign-Born Population

Using data from 1900 through 2004, the percent of the U.S. population that was foreign-born can be modeled by the equation

$$y = 0.0025x^2 - 0.138x + 16.602$$

where x is the number of years after 1900.

a. During what year does the model indicate that the percent of foreign-born population was a minimum?

b. What is the minimum percent?

Solution

a. This equation is in the form $f(x) = ax^2 + bx + c$, so $a = 0.0025$. Because $a > 0$, the parabola opens upward, the vertex is a minimum, and the x-coordinate of the vertex is where the minimum percent occurs.

$$x = \frac{-b}{2a} = \frac{-(-0.138)}{2(0.0025)} = 27.6$$

So the percent of the U.S. population that was foreign-born was a minimum in the 28th year from 1900, or 1928.

b. The minimum for the model is found by evaluating the function at $x = 27.6$. This value, 14.698, can be found with TRACE or TABLE, or with direct evaluation (Figure 3.10). So the minimum percent is 14.698%.

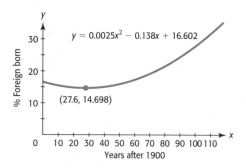

Figure 3.10

Suppose an object is shot or thrown into the air and then falls. If air resistance is ignored, the height in feet of the object after t seconds can be modeled by

$$S(t) = -16t^2 + v_0 t + h_0$$

where -16 ft/sec^2 is the acceleration due to gravity, v_0 ft/sec is the initial velocity (at $t = 0$ sec), and h_0 is the initial height in feet (at $t = 0$).

---

example 3 | ## Height of a Ball

A ball is thrown at 64 feet per second from the top of an 80-foot-high building.

a. Write the quadratic function that models the height (in feet) of the ball as a function of the time t (in seconds).

b. Find the t-coordinate and S-coordinate of the vertex of the graph of this quadratic function.

c. Graph the model.

d. Explain the meaning of the coordinates of the vertex for this model.

Solution

a. The model has the form $S = -16t^2 + v_0 t + h_0$, where $v_0 = 64$ and $h_0 = 80$. Thus the model is

$$S = -16t^2 + 64t + 80 \text{ (feet)}$$

b. The height S is a function of the time t, and the t-coordinate of the vertex is

$$t = \frac{-b}{2a} = \frac{-64}{2(-16)} = 2$$

The S-coordinate of the vertex is the value of S at $t = 2$, so $S = -16(2)^2 + 64(2) + 80 = 144$ is the S-coordinate of the vertex. That is, the vertex is $(2, 144)$.

c. The function is quadratic and the coefficient in the second-degree term is negative, so the graph is a parabola that opens down with vertex $(2, 144)$. To graph the function, we choose a window that includes the vertex $(2, 144)$ near the center top of the screen. Using the window $[0, 6]$ by $[-20, 150]$ gives the graph shown in Figure 3.11(a). Using 2ND CALC MAXIMUM* verifies that the vertex is $(2, 144)$ (see Figure 3.11(b)).

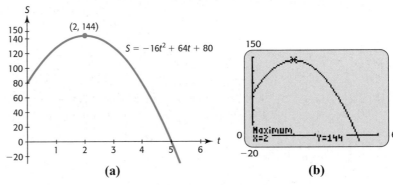

(a)   (b)

Figure 3.11

d. The graph is a parabola that opens down, so the vertex is the highest point on the graph and the function has its maximum there. The t-coordinate of the vertex, 2, is the time (in seconds) at which the ball reaches its maximum height, and the S-coordinate, 144, is the maximum height (in feet) that the ball reaches. ∎

Vertex Form of a Quadratic Function

When a quadratic function is written in the form $f(x) = ax^2 + bx + c$, we can calculate the coordinates of the vertex. But if a quadratic function is written in the form

$$y = a(x - h)^2 + k$$

the vertex of the parabola is at (h, k) (Figure 3.12(a)). For example, the graph of $y = (x - 2)^2 + 3$ is a parabola opening upward with vertex $(2, 3)$ (Figure 3.12(b)).

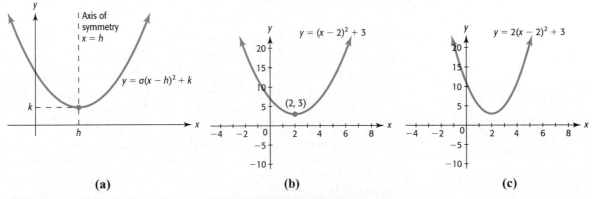

(a)   (b)   (c)

Figure 3.12

*For more details, see Appendix A, page 657.

Graph of a Quadratic Function

In general, the graph of the function

$$y = a(x - h)^2 + k$$

is a parabola with its vertex at the point (h, k).

The parabola opens upward if $a > 0$, and the vertex is a minimum.

The parabola opens downward if $a < 0$, and the vertex is a maximum.

The axis of symmetry of the parabola has equation $x = h$.

The a is the same as the leading coefficient in $y = ax^2 + bx + c$, so the larger the value of $|a|$, the more narrow the parabola will be.

Note that the graph of $y = 2(x - 2)^2 + 3$ in Figure 3.12(c) is more narrow than the graph of $y = (x - 2)^2 + 3$ in Figure 3.12(b).

---

example 4

Minimizing Cost

The cost for producing Champions golf hats is given by the function

$$C(x) = 0.2(x - 40)^2 + 200 \text{ dollars.}$$

a. Find the vertex of this function.

b. Is the vertex a maximum or minimum? Interpret the vertex in the context of the application.

c. Graph the function using a window that includes the vertex.

d. Describe what happens to the function between $x = 0$ and the x-coordinate of the vertex. What does this mean in the context of the application?

Solution

a. This function is in the form $y = a(x - h)^2 + k$ with $h = 40$ and $k = 200$. Thus, the vertex of $C(x) = 0.2(x - 40)^2 + 200$ is $(40, 200)$.

b. Because $a = 0.2$, which is positive, the vertex is a minimum. This means that the cost of producing golf hats is at a minimum of $200 when 40 hats are produced.

c. We know that the vertex of the graph of this function is $(40, 200)$ and that it is a minimum, so we can choose a window with the $x = 40$ near the center of the screen and $y = 200$ near the bottom of the screen. The graph using the window $[0, 100]$ by $[-50, 1000]$ is shown in Figure 3.13.

d. For x-values between 0 and 40, the graph decreases. Thus the cost of producing golf hats is decreasing until 40 hats are produced; after 40 hats are produced, the cost begins to increase. ∎

We can use the vertex form $y = a(x - h)^2 + k$ to write the equation of a quadratic function if we know the vertex and a point on its graph.

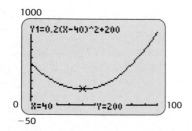

Figure 3.13

| example 5 | **Profit** |

Right Sports Management had its monthly maximum profit, $450,000, when it produced and sold 5500 Waist Trimmers. Its fixed cost is $155,000. If the profit can be modeled by a quadratic function of x, the number of Waist Trimmers produced and sold each month, find this quadratic function $P(x)$.

Solution

When 0 units are produced, the cost is $155,000 and the revenue is $0. Thus the profit is $-\$155,000$ when 0 units are produced and the y-intercept of the graph of the function is $(0, -155,000)$. The vertex of the graph of the quadratic function is $(5500, 450,000)$. Using these points gives

$$P(x) = a(x - 5500)^2 + 450,000$$

and

$$-155,000 = a(0 - 5500)^2 + 450,000$$

which gives

$$a = -0.02$$

Thus the quadratic function that models the profit is $P(x) = -0.02(x - 5500)^2 + 450,000$, or $P(x) = -0.02x^2 + 220x - 155,000$, where $P(x)$ is in dollars and x is the number of units produced and sold. ∎

| example 6 | **Equation of a Quadratic Function** |

If the points in the table lie on a parabola, write the equation whose graph is the parabola.

x	-1	0	1	2
y	13	-2	-7	-2

Solution

The x-values are a uniform distance apart and the symmetry of a parabola indicates that the vertex of this parabola is at $(1, -7)$. (See Figure 3.14.)

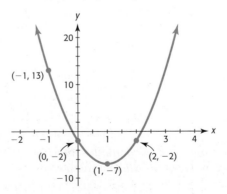

Figure 3.14

Thus the equation of the function is

$$y = a(x - 1)^2 - 7$$

The point $(2, -2)$, or any other point in the table besides $(1, -7)$, can be used to find a.

$$-2 = a(2 - 1)^2 - 7 \quad \text{or} \quad a = 5$$

Thus the equation is

$$y = 5(x - 1)^2 - 7 \quad \text{or} \quad y = 5x^2 - 10x - 2$$ ■

---

example 7 | ### Vertex Form of a Quadratic Function

Write the vertex form of the equation of the quadratic function from the general form $y = 2x^2 - 8x + 5$ by first finding the vertex and a point on the parabola.

Solution

The vertex is at $x = \dfrac{-b}{2a} = \dfrac{-(-8)}{2(2)} = 2$, and $y = 2(2^2) - 8(2) + 5 = -3$. We know a is 2, because a is the same in both forms. Thus

$$y = 2(x - 2)^2 - 3$$

is the vertex form of the equation. ■

---

skills check

3.1

In Exercises 1–6, (a) determine if the function is quadratic. If it is, (b) determine if the graph is concave up or concave down. (c) Determine if the vertex of the graph is a maximum point or a minimum point.

1. $y = 2x^2 - 8x + 6$ **2.** $y = 4x - 3$

3. $y = 2x^3 - 3x^2$ **4.** $f(x) = x^2 + 4x + 4$

5. $g(x) = -5x^2 - 6x + 8$

6. $h(x) = -2x^2 - 4x + 6$

In Exercises 7–14, (a) graph each quadratic function on $[-10, 10]$ by $[-10, 10]$. (b) Does this window give a complete graph?

7. $y = 2x^2 - 8x + 6$ **8.** $f(x) = x^2 + 4x + 4$

9. $g(x) = -5x^2 - 6x + 8$

10. $h(x) = -2x^2 - 4x + 6$

11. $y = x^2 + 8x + 19$

12. $y = x^2 - 4x + 5$

13. $y = 0.01x^2 - 8x$

14. $y = 0.1x^2 + 8x + 2$

15. Write the equation of the quadratic function whose graph is shown.

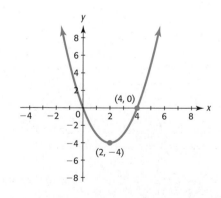

16. Write the equation of the quadratic function whose graph is shown.

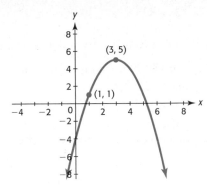

17. The two graphs shown have equations of the form $y = a(x - 2)^2 + 1$. Is the value of a larger for y_1 or y_2?

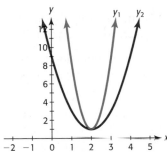

18. The two graphs shown have equations of the form $y = -a(x - 3)^2 + 5$. Is the value of $|a|$ larger for y_1 or y_2?

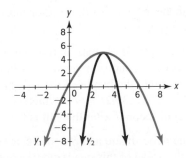

19. If the points in the table lie on a parabola, write the equation whose graph is the parabola.

x	-1	1	3	5
y	-7	13	-7	-67

20. If the points in the table lie on a parabola, write the equation whose graph is the parabola.

x	-6	-5	-4	-3	-2	-1
y	33	12	-3	-12	-15	-12

In Exercises 21–30, (a) give the coordinates of the vertex of the graph of each function. (b) Graph each function on a window that includes the vertex.

21. $y = (x - 1)^2 + 3$ 22. $y = (x + 10)^2 - 6$

23. $y = (x + 8)^2 + 8$ 24. $y = (x - 12)^2 + 1$

25. $f(x) = 2(x - 4)^2 - 6$

26. $f(x) = -0.5(x - 2)^2 + 1$

27. $y = 12x - 3x^2$ 28. $y = 3x + 18x^2$

29. $y = 3x^2 + 18x - 3$ 30. $y = 5x^2 + 75x + 8$

For Exercises 31–34, (a) Find the x-coordinate of the vertex of the graph. (b) Set the viewing window so that the x-coordinate of the vertex is near the center of the window and the vertex is visible, and then graph the given equation.

31. $y = 2x^2 - 40x + 10$

32. $y = -3x^2 - 66x + 12$

33. $y = -0.2x^2 - 32x + 2$

34. $y = 0.3x^2 + 12x - 8$

In Exercises 35–40, sketch complete graphs of the functions.

35. $y = x^2 + 24x + 144$

36. $y = x^2 - 36x + 324$

37. $y = -x^2 - 100x + 1600$

38. $y = -x^2 - 80x - 2000$

39. $y = 2x^2 + 10x - 600$

40. $y = 2x^2 - 75x - 450$

Use the graph of each function in Exercises 41–46 to estimate the x-intercepts.

41. $y = 2x^2 - 8x + 6$ 42. $f(x) = x^2 + 4x + 4$

43. $y = x^2 - x - 110$ 44. $y = x^2 + 9x - 36$

45. $g(x) = -5x^2 - 6x + 8$

46. $h(x) = -2x^2 - 4x + 6$

exercises

3.1

47. *Profit* The daily profit for a product is given by $P = 32x - 0.01x^2 - 1000$, where x is the number of units produced and sold.

   a. Graph this function for x between 0 and 3200.

   b. Describe what happens to the profit for this product when the number of units produced is between 1 and 1600.

   c. What happens after 1600 units are produced?

48. *Profit* The daily profit for a product is given by $P = 420x - 0.1x^2 - 4100$ dollars, where x is the number of units produced and sold.

   a. Graph this function for x between 0 and 4200.

   b. Is the graph of the function concave up or down?

49. *World Population* A low-projection scenario of world population for the years 1995–2150 by the United Nations is given by the function $y = -0.36x^2 + 38.52x + 5822.86$, where x is the number of years after 1990 and the world population is measured in millions of people.

   a. Graph this function for $x = 0$ to $x = 120$.

   b. What will be the world population in 2010 if the projections made using this model are accurate?
   (Source: *World Population Prospects, United Nations*)

50. *Crimes* When x represents the number of years after 1950, the number of arrests of juveniles per 100,000 juveniles for crimes is given by $f(x) = -0.027x^2 + 5.630x + 51.148$.

   a. Graph this function for the years 1950–2010.

   b. What does this model estimate the number of arrests per 100,000 juveniles to be in 2003?
   (Source: Federal Bureau of Investigation)

51. *Tourism Spending* The equation $y = 6.75x^2 - 122.94x + 1009.44$, with x equal to the number of years from 1998 to 2005, models the global spending (in billions of dollars) on travel and tourism.

   a. Graph this function for $x = 0$ to $x = 12$.

   b. If $x = 0$ in 1990, find the spending projected by this model in 2010.

   c. Is the value in part (b) an interpolation or an extrapolation?
   (Source: *Statistical Abstract of the United States*)

52. *Prison Population* The average number of inmates in the Beaufort County, South Carolina, jail is modeled by $y = 0.161x^2 + 7.582x + 82.157$, where x is the number of years after 1991.

   a. Graph this function from $x = 0$ to $x = 20$.

   b. Does this model indicate that the population will increase or decrease in the years 1991–2011?

   c. How would this information be useful to government officials planning for the future?
   (Source: *The Island Packet*)

53. *Flight of a Ball* If a ball is thrown upward at 96 feet per second from the top of a building that is 100 feet high, the height of the ball can be modeled by $S = 100 + 96t - 16t^2$ feet, where t is the number of seconds after the ball is thrown.

   a. Describe the graph of the model.

   b. Find the t-coordinate and S-coordinate of the vertex of the graph of this quadratic function.

   c. Explain the meaning of the coordinates of the vertex for this model.

54. *Flight of a Ball* If a ball is thrown upward at 39.2 meters per second from the top of a building that is 30 meters high, the height of the ball can be modeled by $S = 30 + 39.2t - 9.8t^2$ meters, where t is the number of seconds after the ball is thrown.

   a. Find the t-coordinate and S-coordinate of the vertex of the graph of this quadratic function.

   b. Explain the meaning of the coordinates of the vertex for this function.

   c. Over what time interval is the function increasing? What does this mean in relation to the ball?

55. *Photosynthesis* The rate of photosynthesis R for a certain plant depends on the intensity of light x, in lumens, according to $R(x) = 270x - 90x^2$.

   a. Sketch the graph of this function on a meaningful window.

   b. Determine the intensity x that gives the maximum rate of photosynthesis.

56. *Workers and Output* The weekly output of a certain product is $Q(x) = 200x + 6x^2$. Graph this function for values of x and Q that make sense in this application, if x is the number of weeks, $x \leq 10$.

57. *Profit* The profit for a product can be described by the function $P(x) = 40x - 3000 - 0.01x^2$ dollars, where x is the number of units produced and sold.

   a. To maximize profit, how many units must be produced and sold?

   b. What is the maximum possible profit?

58. *Profit* The profit for a product can be described by the function $P(x) = 840x - 75.6 - 0.4x^2$ dollars, where x is the number of units produced and sold.

   a. To maximize profit, how many units must be produced and sold?

   b. What is the maximum possible profit?

59. *Revenue* The annual total revenue for a product is given by $R(x) = 1500x - 0.02x^2$ dollars, where x is the number of units sold.

   a. To maximize the annual revenue, how many units must be sold?

   b. What is the maximum possible annual revenue?

60. *Revenue* The monthly total revenue for a product is given by $R(x) = 300x - 0.01x^2$ dollars, where x is the number of units sold.

   a. To maximize the monthly revenue, how many units must be sold?

   b. What is the maximum possible monthly revenue?

61. *Area* If 200 feet of fence are used to enclose a rectangular pen, the resulting area of the pen is $A = x(100 - x)$, where x is the width of the pen.

   a. Is A a quadratic function of x?

b. What is the maximum possible area of the pen?

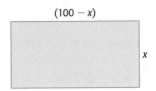

$(100 - x)$

x

62. *Area* If 25,000 feet of fence are used to enclose a rectangular field, the resulting area of the field is $A = (12,500 - x)x$, where x is the width of the pen. What is the maximum possible area of the pen?

63. *Marijuana Use* The percent of U.S. high school seniors who used marijuana during 1990–2006 can be modeled by $y = -0.0584x^2 + 1.096x + 24.3657$ percent, where x is the number of years from 1990. (Source: monitoringthefuture.org)

   a. What is the vertex of the graph of this model?

   b. In what year does the model estimate that the use reached a maximum?

   c. What is the maximum percent of usage, according to the model?

64. *U.S. Visitors* The number of international visitors to the United States, in millions, can be modeled by $y = -0.121x^2 + 4.546x + 5.239$, where x is the number of years after 1980.

   a. Does the model estimate that a maximum or minimum number of visitors will occur during this period? How can you tell without graphing the equation?

   b. Find the input and output at the vertex.

   c. Interpret the results of part (b).
      (Source: *Statistical Abstract of the United States*)

65. *Foreign-Born Population* The percent of the U.S. population that is foreign born can be modeled by the equation $y = 0.0025x^2 - 0.318x + 16.602$, where x is the number of years after 1900.

   a. Is the vertex of the graph of this function a maximum or a minimum?

   b. Assuming that the pattern indicated by this model remains the same, find the input and output at the vertex.

c. Interpret the results of part (b).

d. Use the vertex to set the window and graph the model.
(Source: Census Bureau, U.S. Department of Commerce)

66. *Abortions* Using data from 1975 to 2002, the number of abortions in the United States can be modeled by the function $y = -3.588x^2 + 128.565x + 260.579$, where x is the number of years after 1970 and y is measured in thousands.

  a. In what year does this model indicate the number will be a maximum?

  b. What is the maximum number of abortions in this year if this model is accurate?

  c. For what values of x can we be sure that this model no longer applies?
(Source: Alan Guttmacher Institute)

67. *Wind and Pollution* The amount of particulate pollution p in the air depends on the wind speed s, among other things, with the relationship between p and s approximated by $p = 25 - 0.01s^2$, where p is in ounces per cubic yard and s is in miles per hour.

  a. Sketch the graph of this model with s on the horizontal axis and with nonnegative values of s and p.

  b. Is the function increasing or decreasing on this domain?

  c. What is the p-intercept of the graph?

  d. What does the p-intercept mean in the context of this application?

68. *Drug Sensitivity* The sensitivity S to a drug is related to the dosage size x by $S = 1000x - x^2$.

  a. Sketch the graph of this model using a domain and range with nonnegative x and S.

  b. Is the function increasing or decreasing for x between 0 and 500?

  c. What is the positive x-intercept of the graph?

  d. Why is this x-intercept important in the context of this application?

69. *Falling Object* A tennis ball is thrown into a swimming pool from the top of a tall hotel. The height of the ball from the pool is given by $D(t) = -16t^2 - 4t + 210$ feet where t is the time, in seconds, after the ball was thrown. Graphically find the

t-intercepts for this function. Interpret the value(s) that make sense in this problem context.

70. *Break Even* The profit for a product is given by $P = 1600 - 100x + x^2$, where x is the number of units produced and sold. Graphically find the x-intercepts of the graph of this function to find how many units will give break even (that is, return a profit of zero).

71. *Flight of a Ball* If a softball is hit with upward velocity 32 feet per second when $t = 0$, from a height of 3 feet,

  a. Find the function that models the height of the ball as a function of time.

  b. Find the maximum height of the ball.

72. *Flight of a Ball* If a baseball is hit with upward velocity 48 feet per second when $t = 0$, from a height of 4 feet,

  a. Find the function that models the height of the ball as a function of time.

  b. Find the maximum height of the ball and in how many seconds the ball will reach that height.

73. *Apartment Rental* The owner of an apartment building can rent all 100 apartments if he charges $1200 per apartment per month, but the number of apartments rented is reduced by 2 for every $40 increase in the monthly rent.

  a. Construct a table that gives the revenue if the rent charged is $1240, $1280, and $1320.

  b. Does $R(x) = (1200 + 40x)(100 - 2x)$ model the revenue from these apartments if x represents the number of $40 increases?

  c. What monthly rent gives the maximum revenue for the apartments?

74. *Rink Rental* The owner of a skating rink rents the rink for parties at $720 if 60 or fewer skaters attend, so that the cost is $12 per person if 60 attend. For each 6 skaters above 60, she reduces the price per skater by $.50.

  a. Construct a table that gives the revenue if the number attending is 66, 72, and 78.

  b. Does the function $R(x) = (60 + 6x)(12 - 0.5x)$ model the revenue from the party if x represents the number of increases of 6 people each?

c. How many people should attend for the rink's revenue to be a maximum?

75. *World Population* A low-projection scenario of world population for 1995–2150 by the United Nations is given by the function $y = -0.36x^2 + 38.52x + 5822.86$, where x is the number of years after 1990 and the world population is measured in millions of people.

a. Find the input and output at the vertex of the graph of this model.

b. Interpret the values from part (a).

c. For what years after 1995 does this model predict that the population will increase?
(Source: *World Population Prospects*, United Nations)

section

3.2

Solving Quadratic Equations

key concepts

- Solving quadratic equations
- Factoring; zero product property
- Graphical solution methods
  x-Intercept method
  Intersection method
- Combining graphical and factoring methods
- Combining graphical and numerical methods
- The square root method
- Completing the square
- The quadratic formula
- The discriminant
- Aids for solving quadratic equations
- Equations with complex solutions

section preview ▪ Hospital Admissions

The number of admissions at all hospitals in the United States can be described by the function

$$A(x) = 38.228x^2 - 1069.600x + 40{,}698.547 \text{ thousand people}$$

where x is the number of years after 1985. Viewing and understanding the pattern indicated by this model can help the medical industry, the government, medical training facilities, and consumers to better understand current and future needs for hospital space and facilities and for hospital personnel. (Source: Based on data from the American Hospital Association: *Hospital Statistics*)

To find the year after 1985 that is predicted by this model to give 33,563,000 (that is, 33,563 thousand) admissions, we solve the **quadratic equation**

$$38.228x^2 - 1069.600x + 40{,}698.547 = 33{,}563 \quad \text{or}$$
$$38.228x^2 - 1069.600x + 7135.547 = 0$$

The values of x that satisfy this equation are called the **solutions** of the equation; they are also zeros of the function

$$y = 38.228x^2 - 1069.600x + 7135.547$$

and they are the x-intercepts of the graph of this function. (See Example 9.) In this section, we learn how to solve quadratic equations by using factoring methods, graphical methods, the square root method, completing the square, and the quadratic formula.

Factoring Methods

An equation that can be written in the form $ax^2 + bx + c = 0$, with $a \neq 0$, is called a **quadratic equation**. Solutions to some quadratic equations can be found exactly by factoring; other quadratic equations require different types of solution methods to find or to approximate solutions.

Solution by factoring is based on the following property of real numbers.

> ### Zero Product Property
> For real numbers a and b, the product $ab = 0$ if and only if either $a = 0$ or $b = 0$ or both a and b are zero.

To use this property to solve a quadratic equation by factoring, we must first make sure that the equation is written in a form with zero on one side. If the resulting nonzero expression is factorable, we factor it and use the zero product property to convert the equation into two linear equations that are easily solved.* Before applying this technique to real-world applications, we consider the following example.

| example 1 | ### Solving a Quadratic Equation by Factoring |

Solve the equation $3x^2 + 7x = 6$.

Solution

We first subtract 6 from both sides of the equation to rewrite the equation with 0 on one side:

$$3x^2 + 7x - 6 = 0$$

To begin factoring the trinomial $3x^2 + 7x - 6$, we seek factors of $3x^2$ (that is, $3x$ and x) as the first terms of two binomials and factors of -6 as the last terms of the binomials. The factorization whose inner and outer products combine to $7x$ is $(3x - 2)(x + 3)$, so we have

$$(3x - 2)(x + 3) = 0$$

Using the zero product property gives

$$3x - 2 = 0 \quad \text{or} \quad x + 3 = 0$$

Solving these linear equations gives the two solutions to the original equation.

$$x = \frac{2}{3} \quad \text{or} \quad x = -3 \qquad \blacksquare$$

| example 2 | ### Height of a Ball |

The height above ground of a ball thrown at 64 feet per second from the top of an 80-foot-high building is modeled by $S = 80 + 64t - 16t^2$ feet, where t is the number of seconds after the ball is thrown. How long will the ball be in the air?

---

* For a review of factoring methods, see the Algebra Toolbox.

Solution

The ball will be in the air from $t = 0$ (with the height $S = 80$) until it reaches the ground ($S = 0$). Thus we can find the time in the air by solving

$$0 = -16t^2 + 64t + 80$$

Because 16 is a factor of each of the terms, we can get a simpler but equivalent equation by dividing both sides of the equation by -16.

$$0 = t^2 - 4t - 5$$

This equation can be solved easily by factoring the right side.

$$0 = (t - 5)(t + 1)$$
$$0 = t - 5 \quad \text{or} \quad 0 = t + 1$$
$$t = 5 \quad \text{or} \quad t = -1$$

The time in the air starts at $t = 0$, so $t = -1$ has no meaning in this application. S also equals 0 at $t = 5$, which means that the ball is on the ground 5 seconds after it was thrown; that is, the ball was in the air 5 seconds. Figure 3.15 shows a graph of the function, which confirms that the height of the ball is 0 at $t = 5$.

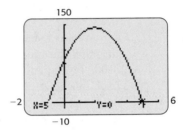

Figure 3.15

Graphical Methods

In cases where factoring $f(x)$ to solve $f(x) = 0$ is difficult or impossible, graphing $y = f(x)$ can be helpful in finding the solution. Recall that if a is a real number, the following three statements are equivalent:

- a is a real solution to the equation $f(x) = 0$.

- a is a real zero of the function f.

- a is an x-intercept of the graph of $y = f(x)$.

It is important to remember that the above three statements are equivalent because connecting these concepts allows us to use different methods for solving equations. Sometimes graphical methods are the easiest way to find or approximate solutions to real data problems. If the x-intercepts of the graph of $y = f(x)$ are easily found, then graphical methods may be helpful in finding the solutions. Note that if the graph of $y = f(x)$ does not cross or touch the x-axis, there are no real solutions to the equation $f(x) = 0$.

We can also find solutions or decimal approximations of solutions to quadratic equations by using the intersection method with a graphing utility.

| example 3 | **Profit** |

Consider the daily profit from the production and sale of x units of a product, given by

$$P(x) = -0.01x^2 + 20x - 500 \text{ dollars}$$

a. Use a graph to find the levels of production and sales that give a daily profit of $1400.

b. Is it possible for the profit to be greater than $1400?

Solution

a. To find the level of production and sales, x, that gives a daily profit of 1400 dollars, we solve

$$1400 = -0.01x^2 + 20x - 500$$

To solve this equation by the intersection method, we graph

$$y_1 = -0.01x^2 + 20x - 500 \quad \text{and} \quad y_2 = 1400$$

To find the appropriate window for this graph, we note that the graph of the function $y_1 = -0.01x^2 + 20x - 500$ is a parabola with the vertex at

$$x = \frac{-b}{2a} = \frac{-20}{2(-0.01)} = 1000 \quad \text{and} \quad y = P(1000) = 9500$$

We use a viewing window containing this point, with $x = 1000$ near the center, to graph the function (Figure 3.16). Using the intersection method, we see that (100, 1400) and (1900, 1400) are the points of intersection of the graphs of the two functions. Thus, $x = 100$ and $x = 1900$ are solutions to the equation $1400 = -0.01x^2 + 20x - 500$, and the profit is $1400 when $x = 100$ units or $x = 1900$ units of the product are produced and sold.

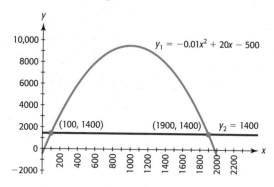

Figure 3.16

b. We can see from the graph in Figure 3.16 that the profit is more than $1400 for many values of x. Because the graph of this profit function is a parabola that opens down, the maximum profit occurs at the vertex of the graph. As we found in part (a), the vertex occurs at $x = 1000$ and the maximum possible profit is $P(1000) = \$9500$, which is more than $1400. ∎

Combining Graphs and Factoring

If the factors of a quadratic function are not easily found, its graph may be helpful in finding the factors. Note once more the important correspondences that exist among solutions, zeros, and x-intercepts: the x-intercepts of the graph of $y = P(x)$ are the real solutions of the equation $0 = P(x)$ and the real zeros of $P(x)$. In addition, the factors of $P(x)$ are also related to the zeros of $P(x)$. The relationship among the factors, solutions, and zeros is true for any polynomial function f and can be generalized by the following theorem.

Factor Theorem

The polynomial function f has a factor $(x - a)$ if and only if $f(a) = 0$. Thus, $(x - a)$ is factor of $f(x)$ if and only if $x = a$ is a solution to $f(x) = 0$.

This means that we can verify our factorization of f and the real solutions to $0 = f(x)$ by graphing $y = f(x)$ and observing where the graph crosses the x-axis. We can also sometimes use our observation of the graph to assist us in the factorization of a quadratic function:

> *If one solution can be found exactly from the graph, it can be used to find one of the factors of the function. The second factor can then be found easily, leading to the second solution.*

example 4

Graphing and Factoring Methods Combined

Solve $0 = 3x^2 - x - 10$ by using the following steps.

a. Graphically find one of the x-intercepts of $y = 3x^2 - x - 10$.

b. Algebraically verify that the zero found in part (a) is an exact solution to $0 = 3x^2 - x - 10$.

c. Use the method of factoring to find the other solution to $0 = 3x^2 - x - 10$.

Solution

a. The vertex of the graph of $y = 3x^2 - x - 10$ is at $x = \dfrac{1}{6}$ and $y \approx -10.08$, and because the parabola opens up, we set a viewing window that includes this point near the bottom center of the window. Graphing the function and using the x-intercept method, we find that the graph crosses the x-axis at $x = 2$ (Figure 3.17). This means that 2 is a zero of $f(x) = 3x^2 - x - 10$.

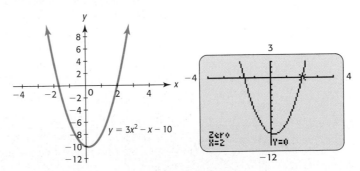

Figure 3.17

b. Because $x = 2$ was obtained graphically, it may be an approximation of the exact solution. To verify that it is exact, we show that $x = 2$ makes the equation $0 = 3x^2 - x - 10$ a true statement:

$$3(2)^2 - 2 - 10 = 12 - 2 - 10 = 0$$

c. Because $x = 2$ is a solution of $0 = 3x^2 - x - 10$, one factor of $3x^2 - x - 10$ is $(x - 2)$. Thus, we use $(x - 2)$ as one factor and seek a second binomial factor so that the product of the two binomials is $3x^2 - x - 10$:

$$3x^2 - x - 10 = 0$$
$$(x - 2)(\quad) = 0$$
$$(x - 2)(3x + 5) = 0$$

The remaining factor is $3x + 5$, which we set equal to 0 to obtain the other solution:

$$3x = -5$$
$$x = -\frac{5}{3}$$

Thus, the two solutions to $0 = 3x^2 - x - 10$ are $x = 2$ and $x = -\frac{5}{3}$. ■

Graphical and Numerical Methods

When approximate solutions to quadratic equations are sufficient, graphical and/or numerical solution methods can be used. These methods of solving are illustrated in Example 5.

example 5

Marijuana Use

For the years 1991 through 2006, the percent p of high school seniors who have tried marijuana can be considered as a function of the time t according to the model

$$p = -0.1967t^2 + 4.0630t + 27.7455$$

where t is the number of years after 1990.

a. Find the year(s) after 1995 during which the percent is predicted to be 40, using a graphical method.

b. Verify the solution(s) numerically.
(Source: National Institute on Drug Abuse)

Solution

a. Note that the output p is measured in percent, so $p = 40$. Thus to find where the percent is 40, we solve

$$40 = -0.1967t^2 + 4.0630t + 27.7455$$

We solve this equation with the intersection method. Figure 3.18(a) shows the graphs of $y_1 = -0.1967x^2 + 4.0630x + 27.7455$ and $y_2 = 40$, and Figure 3.18(b) and Figure 3.18(c) show two points where the graphs intersect.

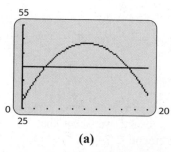

(a)

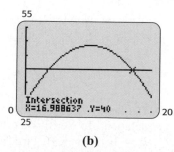

(b)

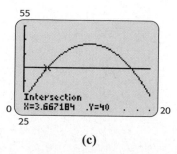

(c)

Figure 3.18

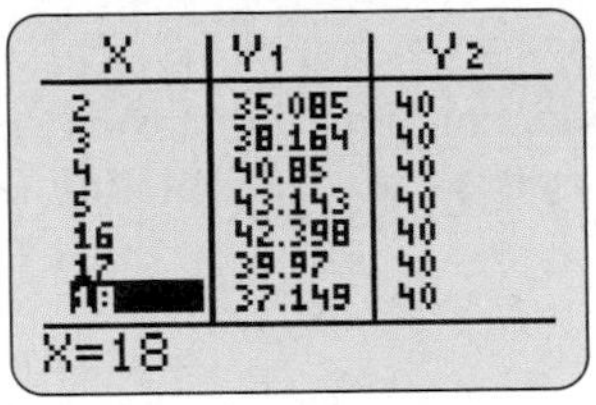

X	Y₁	Y₂
2	35.085	40
3	38.164	40
4	40.85	40
5	43.143	40
16	42.398	40
17	39.97	40
18	37.149	40

X=18

Figure 3.19

One point indicates that the percent is 40 when $t = 17$, so the model predicts that 40% of high school seniors will have tried marijuana 17 years after 1990, in 2007. Another point has $t = 3.7$, which indicates that the percent will be 40 four years after 1990, in 1994. But this is not after 1995, so it is not a solution to the problem.

b. Figure 3.19 shows a table of inputs near 4 and near 17 and the corresponding outputs. The table shows that the percent is 40.85 at the end of the 4th year after 1990, in 1994, and 39.97 in the 17th year after 1990, in 2007. ■

The Square Root Method

We have solved quadratic equations by factoring and by graphical methods. Another method can be used to solve quadratic equations that are written in a particular form. In general, we can find the solutions of quadratic equations of the form $x^2 = C$, where C is a constant, by taking the square root of both sides. For example, to solve $x^2 = 25$, we take the square root of both sides of the equation, getting $x = \pm 5$. Note that there are two solutions because $5^2 = 25$ and $(-5)^2 = 25$.

> **Square Root Method**
>
> The solutions of the quadratic equation $x^2 = C$ are $x = \pm\sqrt{C}$. Note that, when we take the square root of both sides, we use a $\pm$ symbol because there are both a positive and a negative value that, when squared, give C.

Note that this method can also be used to solve equations of the form $(ax + b)^2 = C$.

example 6

Square Root Solution Method

Solve the following equations by using the square root method.

a. $2x^2 - 16 = 0$ b. $(x - 6)^2 = 18$

Solution

a. This equation can be written in the form $x^2 = C$, so the square root method can be used. We want to rewrite the equation so that the x^2-term is isolated and its coefficient is 1 and then take the square root of both sides.

$$2x^2 - 16 = 0$$
$$2x^2 = 16$$
$$x^2 = 8$$
$$x = \pm\sqrt{8} = \pm 2\sqrt{2}$$

The exact solutions to $2x^2 - 16 = 0$ are $x = 2\sqrt{2}$ and $x = -2\sqrt{2}$.

b. The left side of this equation is a square, so we can take the square root of both sides to find x.

$$(x - 6)^2 = 18$$
$$x - 6 = \pm\sqrt{18}$$
$$x = 6 \pm \sqrt{18}$$
$$= 6 \pm \sqrt{9}\sqrt{2}$$
$$= 6 \pm 3\sqrt{2}$$

■

Completing the Square

When we convert one side of a quadratic equation to a perfect binomial square and use the square root method to solve the equation, the method used is called **completing the square**.

example 7   ### Completing the Square

Solve the equation $x^2 - 12x + 7 = 0$ by completing the square.

Solution

Because the left side of this equation is not a perfect square, we rewrite it in a form where we can more easily get a perfect square. We subtract 7 from both sides, getting

$$x^2 - 12x = -7$$

We complete the square on the left side of this equation by adding the appropriate number to both sides of the equation to make the left side a perfect square trinomial. Because $(x + a)^2 = x^2 + 2ax + a^2$, the constant term of a perfect square trinomial will be the *square of half the coefficient of x*. Using this rule with the equation $x^2 - 12x = -7$, we would need to take half the coefficient of x and add the square of this number to get a perfect square trinomial. Half of -12 is -6, so adding $(-6)^2 = 36$ to $x^2 - 12x$ gives the perfect square $x^2 - 12x + 36$. We also have to add 36 to the other side of the equation (to preserve the equality), giving

$$x^2 - 12x + 36 = -7 + 36$$

Factoring this perfect square trinomial gives

$$(x - 6)^2 = 29$$

We can now solve the equation with the square root method.

$$(x - 6)^2 = 29$$
$$x - 6 = \pm\sqrt{29}$$
$$x = 6 \pm \sqrt{29}$$

The Quadratic Formula

We can generalize the method of completing the square to derive a general formula that gives the solution to any quadratic equation. We use this method to find the general solution to $ax^2 + bx + c = 0, a \neq 0$.

$$ax^2 + bx + c = 0 \qquad \text{Standard form}$$
$$ax^2 + bx = -c \qquad \text{Subtract } c \text{ from both sides.}$$
$$x^2 + \frac{b}{a}x = -\frac{c}{a} \qquad \text{Divide both sides by } a.$$

We would like to make the left side of the last equation a perfect square trinomial. Half the coefficient of x is $\dfrac{b}{2a}$, and squaring this gives $\dfrac{b^2}{4a^2}$. Hence, adding $\dfrac{b^2}{4a^2}$ to both sides of the equation gives a perfect square trinomial on the left side, and we can continue with the solution.

$$x^2 + \frac{b}{a}x + \frac{b^2}{4a^2} = \frac{b^2}{4a^2} - \frac{c}{a}$$ Add $\dfrac{b^2}{4a^2}$ to both sides of the equation.

$$\left(x + \frac{b}{2a}\right)^2 = \frac{b^2 - 4ac}{4a^2}$$ Factor the left side and combine the fractions on the right side.

$$x + \frac{b}{2a} = \pm\sqrt{\frac{b^2 - 4ac}{4a^2}}$$ Take the square root of both sides.

$$x = -\frac{b}{2a} \pm \frac{\sqrt{b^2 - 4ac}}{2|a|}$$ Subtract $\dfrac{b}{2a}$ from both sides and simplify.

$$x = -\frac{b}{2a} \pm \frac{\sqrt{b^2 - 4ac}}{2a}$$ $\pm 2|a| = \pm 2a$

$$x = \frac{-b \pm \sqrt{b^2 - 4ac}}{2a}$$ Combine the fractions.

The formula we have developed is called the **quadratic formula**.

Quadratic Formula

The solutions of the quadratic equation $ax^2 + bx + c = 0$ are given by the formula

$$x = \frac{-b \pm \sqrt{b^2 - 4ac}}{2a}$$

Note that a is the coefficient of x^2, b is the coefficient of x, and c is the constant term.

Because of the $\pm$ sign, the solutions can be written as:

$$x = \frac{-b + \sqrt{b^2 - 4ac}}{2a} \quad \text{and} \quad x = \frac{-b - \sqrt{b^2 - 4ac}}{2a}$$

We can use the quadratic formula to solve all quadratic equations exactly, but it is especially useful for finding exact solutions to those equations for which factorization is difficult or impossible. For example, the solutions to $40 = -0.1967t^2 + 4.0630t + 27.7455$ were found approximately by graphical methods in Example 5. If we need to find the exact solutions, we could use the quadratic formula.

example 8

Solving Using the Quadratic Formula

Solve $6 - 3x^2 + 4x = 0$ using the quadratic formula.

Solution

The equation $6 - 3x^2 + 4x = 0$ can be rewritten as $-3x^2 + 4x + 6 = 0$, so $a = -3$, $b = 4$, and $c = 6$. The two solutions to this equation are

$$x = \frac{-4 \pm \sqrt{4^2 - 4(-3)(6)}}{2(-3)} = \frac{-4 \pm \sqrt{88}}{-6} = \frac{-4 \pm 2\sqrt{22}}{-6} = \frac{2 \pm \sqrt{22}}{3}$$

Thus, the exact solutions are the irrational numbers $x = \dfrac{2 + \sqrt{22}}{3}$ and $x = \dfrac{2 - \sqrt{22}}{3}$.
Three-place decimal approximations for these solutions are $x \approx -0.897$ and $x \approx 2.230$.

◼

Decimal approximations of irrational solutions found with the quadratic formula will often suffice as answers to an applied problem. The quadratic formula is especially useful when the coefficients of a quadratic equation are decimal values that make factorization impractical. This occurs in many applied problems, such as the one discussed at the beginning of this section. If the graph of the quadratic function $y = f(x)$ does not intersect the x-axis at "nice" values of x, the solutions may be irrational numbers, and using the quadratic formula allows us to find these solutions exactly.

example 9 | Hospital Admissions

The annual number of admissions at all hospitals in the United States can be described by the function $A(x) = 38.228x^2 - 1069.600x + 40{,}698.547$ thousand people, where x is the number of years after 1985. If the model is valid until 2010, find the years after 1985 that this model estimates the admissions to be 33,563,000.
(Source: American Hospital Association: *Hospital Statistics*)

Solution

Note that the output of A is measured in thousands of admissions, so we must first write 33,563,000 admissions as 33,563 thousand admissions. Then, to answer the given question, we solve (using an algebraic, numerical, or graphical method) the quadratic equation

$$38.228x^2 - 1069.600x + 40{,}698.547 = 33{,}563$$

We choose to use the quadratic formula (even though an approximate answer is all that is needed). We first write $38.228x^2 - 1069.600x + 40{,}698.547 = 33{,}563$ with 0 on one side:

$$38.228x^2 - 1069.600x + 40{,}698.547 - 33{,}563 = 0$$
$$38.228x^2 - 1069.600x + 7135.547 = 0$$

This gives $a = 38.228$, $b = -1069.6$, and $c = 7135.547$. Substituting in the quadratic formula gives

$$x = \frac{-(-1069.6) \pm \sqrt{(-1069.6)^2 - 4(38.228)(7135.547)}}{2(38.228)}$$
$$= \frac{1069.6 \pm \sqrt{52{,}933.39714}}{76.456}$$
$$x \approx 16.999 \quad \text{or} \quad x \approx 10.981$$

Thus, 33,563,000 admissions occur in the 17th year from the end of 1985, in 2002, and in $1985 + 11 = 1996$. Figure 3.20 shows the outputs of

$$y_1 = 38.228x^2 - 1069.600x + 40{,}698.547$$

at integer values near the x-values that are the results from the quadratic formula, which confirms our conclusions above.

◼

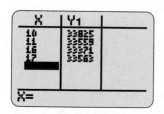

Figure 3.20

The Discriminant

We can also determine the type of solutions a quadratic equation has by looking at the expression $b^2 - 4ac$, which is called the **discriminant** of the quadratic equation $ax^2 + bx + c = 0$. The discriminant is the expression inside the radical in the quadratic formula $x = \dfrac{-b \pm \sqrt{b^2 - 4ac}}{2a}$, so it determines if the quantity inside the radical is positive, zero, or negative. Thus, we have

- If $b^2 - 4ac > 0$, there are two different real solutions.
- If $b^2 - 4ac = 0$, there is one real solution.
- If $b^2 - 4ac < 0$, there is no real solution.

For example, the equation $3x^2 + 4x + 2 = 0$ has no real solution because $4^2 - 4(3)(2) = -8 < 0$, and the equation $x^2 + 4x + 2 = 0$ has two different real solutions because $4^2 - 4(1)(2) = 8 > 0$.

Aids for Solving Quadratic Equations

The x-intercepts of the graph of the quadratic function $y = f(x)$ can be used to determine how to solve the quadratic equation $f(x) = 0$. Table 3.2 summarizes these ideas with suggested methods for solving, and the graphical representations of the solutions.

Table 3.2 Connections Between Graphs of Quadratic Functions and Solution Methods

Graph			
Type of x-Intercepts	Graph crosses x-axis twice.	Graph touches but does not cross x-axis.	Graph does not cross the x-axis.
Type of Solutions	Equation has two real solutions.	Equation has one real solution.	Equation has no real solutions.
Suggested Solution Methods*	Use factoring, graphing, or quadratic formula.	Use factoring, graphing, or the square root method.	Use quadratic formula to verify that solutions are not real.

*Solution methods other than the suggested methods may also be successful.

Equations with Complex Solutions

Recall that the solutions of the quadratic equation $x^2 = C$ are $x = \pm\sqrt{C}$, so the solutions of the equation $x^2 = -a$ for $a > 0$ are

$$x = \pm\sqrt{-a} = \pm\sqrt{-1}\sqrt{a} = \pm i\sqrt{a}$$

example 10

Solution Using the Root Method

Solve the equations:

a. $x^2 = -9$ **b.** $3x^2 + 24 = 0$

Solution

a. Taking the square root of both sides of the equation gives the solution of $x^2 = -9$:

$$x = \pm\sqrt{-9} = \pm\sqrt{-1}\sqrt{9} = \pm 3i$$

b. We solve $3x^2 + 24 = 0$ using the root method, as follows:

$$3x^2 = -24$$

$$x^2 = -8$$

$$x = \pm\sqrt{-8} = \pm\sqrt{-1}\sqrt{4 \cdot 2} = \pm 2i\sqrt{2}$$ ∎

We used the root method to solve the equations in Example 10 because neither equation contained a first-degree term (that is, a term containing x to the first power). We can also find complex solutions by using the quadratic formula.* Recall that the solutions of the quadratic equation $ax^2 + bx + c = 0$ are given by the formula

$$x = \frac{-b \pm \sqrt{b^2 - 4ac}}{2a}$$

example 11	**Solution with the Quadratic Formula**

Solve the equations:

a. $x^2 - 3x + 5 = 0$ **b.** $3x^2 + 4x = -3$

Solution

a. Using the quadratic formula, with $a = 1$, $b = -3$, and $c = 5$, gives

$$x = \frac{-(-3) \pm \sqrt{(-3)^2 - 4(1)(5)}}{2(1)} = \frac{3 \pm \sqrt{-11}}{2} = \frac{3 \pm i\sqrt{11}}{2}$$

Note that the solutions can also be written in the form $\dfrac{3}{2} \pm \dfrac{\sqrt{11}}{2}i$.

Thus, the solutions are the complex numbers $\dfrac{3}{2} + \dfrac{\sqrt{11}}{2}i$ and $\dfrac{3}{2} - \dfrac{\sqrt{11}}{2}i$.

b. Writing $3x^2 + 4x = -3$ in the form $3x^2 + 4x + 3 = 0$ gives $a = 3$, $b = 4$, and $c = 3$, so the solutions are

$$x = \frac{-4 \pm \sqrt{(4)^2 - 4(3)(3)}}{2(3)} = \frac{-4 \pm \sqrt{-20}}{6} = \frac{-4 \pm \sqrt{-1}\sqrt{4}\sqrt{5}}{6}$$

$$= \frac{-4 \pm 2i\sqrt{5}}{6} = \frac{2(-2 \pm i\sqrt{5})}{2 \cdot 3} = \frac{-2 \pm i\sqrt{5}}{3}$$

Thus,

$$x = -\frac{2}{3} + \frac{\sqrt{5}}{3}i \quad \text{and} \quad x = -\frac{2}{3} - \frac{\sqrt{5}}{3}i$$ ∎

---

* The solutions could also be found by completing the square. Recall that the quadratic formula was developed by completing the square on the quadratic equation $ax^2 + bx + c = 0$.

skills check

3.2

In Exercises 1–10, use factoring to solve the equations.

1. $x^2 - 3x - 10 = 0$ **2.** $x^2 - 9x + 18 = 0$

3. $x^2 - 11x + 24 = 0$ **4.** $x^2 + 3x - 10 = 0$

5. $2x^2 + 2x - 12 = 0$ **6.** $2s^2 + s - 6 = 0$

7. $0 = 2t^2 - 11t + 12$ **8.** $6x^2 - 13x + 6 = 0$

9. $6x^2 + 10x = 4$ **10.** $10x^2 + 11x = 6$

Use a graphing utility to find or to approximate the x-intercepts of the graph of each function in Exercises 11–16.

11. $y = x^2 - 3x - 10$ **12.** $y = x^2 + 4x - 32$

13. $y = 3x^2 - 8x + 4$ **14.** $y = 2x^2 + 8x - 10$

15. $y = 2x^2 + 7x - 4$ **16.** $y = 5x^2 - 17x + 6$

Use a graphing utility as an aid in factoring to solve the equations in Exercises 17–22.

17. $2w^2 - 5w - 3 = 0$ **18.** $3x^2 - 4x - 4 = 0$

19. $x^2 - 40x + 256 = 0$ **20.** $x^2 - 32x + 112 = 0$

21. $2s^2 - 70s = 1500$ **22.** $3s^2 - 130s = -1000$

In Exercises 23–26, use the square root method to solve the quadratic equations.

23. $4x^2 - 9 = 0$ **24.** $x^2 - 20 = 0$

25. $x^2 - 32 = 0$ **26.** $5x^2 - 25 = 0$

In Exercises 27–30, complete the square to solve the quadratic equations.

27. $x^2 - 4x - 9 = 0$ **28.** $x^2 - 6x + 1 = 0$

29. $x^2 - 3x + 2 = 0$ **30.** $2x^2 - 9x + 8 = 0$

In Exercises 31–34, use the quadratic formula to solve the equations.

31. $x^2 - 5x + 2 = 0$ **32.** $3x^2 - 6x - 12 = 0$

33. $5x + 3x^2 = 8$ **34.** $3x^2 - 30x - 180 = 0$

In Exercises 35–40, use a graphing utility to find or approximate solutions of the equations.

35. $2x^2 + 2x - 12 = 0$ **36.** $2x^2 + x - 6 = 0$

37. $0 = 6x^2 + 5x - 6$ **38.** $10x^2 = 22x - 4$

39. $4x + 2 = 6x^2 + 3x$ **40.** $(x - 3)(x + 2) = -4$

In Exercises 41–46, find the exact solutions to $f(x) = 0$ in the complex numbers and confirm that the solutions are not real by showing that the graph of $y = f(x)$ does not cross the x-axis.

41. $x^2 + 25 = 0$ **42.** $2x^2 + 40 = 0$

43. $(x - 1)^2 = -4$ **44.** $(2x + 1)^2 + 7 = 0$

45. $x^2 + 4x + 8 = 0$ **46.** $x^2 - 5x + 7 = 0$

In Exercises 47 and 48, you are given the graphs of several functions of the form $f(x) = ax^2 + bx + c$ for different values of a, b, and c. For each function,

a. Determine if the discriminant is positive, negative, or zero.

b. Determine if there are 0, 1, or 2 real solutions to $f(x) = 0$.

c. Solve the equation $f(x) = 0$.

47.

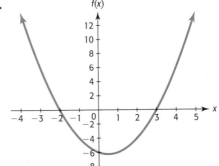

48.

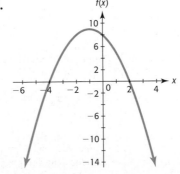

exercises

3.2

In Exercises 49–60, solve analytically and then check graphically.

49. *Flight of a Ball* If a ball is thrown upward at 96 feet per second from the top of a building that is 100 feet high, the height of the ball can be modeled by $S = 100 + 96t - 16t^2$ feet, where t is the number of seconds after the ball is thrown. How long after the ball is thrown is the height 228 feet?

50. *Falling Object* A tennis ball is thrown into a swimming pool from the top of a tall hotel. The height of the ball above the pool is modeled by $D(t) = -16t^2 - 4t + 200$ feet, where t is the time, in seconds, after the ball is thrown. How long after the ball is thrown is it 44 feet above the pool?

51. *Break Even* The profit for a product is given by $P(x) = -12x^2 + 1320x - 21,600$, where x is the number of units produced and sold. How many units give break even (that is, give zero profit) for this product?

52. *Break Even* The profit for a product is given by $P(x) = -15x^2 + 180x - 405$ thousand dollars, where x is the number of tons of product produced and sold. How many tons give break even (that is, give zero profit) for this product?

53. *Break Even* The total revenue function for a product is given by $R = 550x$ dollars, and the total cost function for this same product is given by $C = 10,000 + 30x + x^2$, where C is measured in dollars. For both functions, the input x is the number of units produced and sold.

   a. Form the profit function for this product from the two given functions.

   b. What is the profit when 18 units are produced and sold?

   c. What is the profit when 32 units are produced and sold?

   d. How many units must be sold to break even on this product?

54. *Break Even* The total revenue function for a product is given by $R = 266x$, and the total cost function for this same product is $C = 2000 + 46x + 2x^2$, where R and C are each measured in thousands of dollars and x is the number of units produced and sold.

   a. Form the profit function for this product from the two given functions.

   b. What is the profit when 55 units are produced and sold?

   c. How many units must be sold to break even on this product?

55. *Wind and Pollution* The amount of particulate pollution p from a power plant in the air above the plant depends on the wind speed s, among other things, with the relationship between p and s approximated by $p = 25 - 0.01s^2$, with s in miles per hour.

   a. Find the value(s) of s that will make $p = 0$.

   b. What does $p = 0$ mean in this application?

   c. What solution to $0 = 25 - 0.01s^2$ makes sense in the context of this application?

56. *Drug Sensitivity* The sensitivity S to a drug is related to the dosage size x by $S = 100x - x^2$, where x is the dosage size in milliliters.

   a. What dosage(s) give zero sensitivity?

   b. Explain what your answers in part (a) might mean.

57. *Velocity of Blood* Because of friction from the walls of an artery, the velocity of a blood corpuscle in an artery is greatest at the center of the artery and decreases as the distance r from the center increases. The velocity of the blood in the artery can be modeled by the function

$$v = k(R^2 - r^2)$$

where R is the radius of the artery and k is a constant that is determined by the pressure, viscosity of the blood, and the length of the artery. In the case where $k = 2$ and $R = 0.1$ centimeter, the velocity is $v = 2(0.01 - r^2)$ centimeters per second (cm/sec).

   a. What distance r would give a velocity of 0.02 cm/sec?

b. What distance r would give a velocity of 0.015 cm/sec?

c. What distance r would give a velocity of 0 cm/sec? Where is the blood corpuscle?

58. *Body-Heat Loss* The model for body-heat loss depends on the coefficient of convection K, which depends on wind speed s according to the equation $K^2 = 16s + 4$, where s is in miles per hour. Find the positive coefficient of convection when the wind speed is

a. 20 mph. **b.** 60 mph.

c. What is the change in K for a change in speed from 20 mph to 60 mph?

59. *Market Equilibrium* Suppose the demand for artificial Christmas trees is given by the function

$$p = 109.70 - 0.10q$$

and that the supply of these trees is given by

$$p = 0.01q^2 + 5.91$$

where p is the price of a tree in dollars and q is the quantity of trees that are demanded/supplied in hundreds. Find the price that gives market equilibrium price and the number of trees that will be sold/bought at this price.

60. *Market Equilibrium* The demand for a product is given by $p = 7000 - 2x$ dollars, and the supply for this product is given by $p = 0.01x^2 + 2x + 1000$ dollars, where x is the number of units demanded and supplied when the price per unit is p dollars. Find the equilibrium quantity and equilibrium price.

61. *Foreign-Born Population* Suppose the percent of the U.S. population that is foreign born can be modeled by the equation $y = 0.003x^2 - 0.390x + 18.8$, where x is the number of years after 1900.

a. Verify graphically that one solution to $8 = 0.003x^2 - 0.390x + 18.8$ is $x = 40$.

b. Use this information and factoring to find the year or years when the percent of the U.S. population that is foreign born is 8.
(Source: U.S. Census Bureau)

62. *Smoking Cessation* Using data from 1965 to 1990, the percent of people over 19 years of age who have ever smoked and quit is given by the equation $y = 0.010x^2 + 0.344x + 28.74$, where x is the number of years after 1960.

a. One solution to the equation $58.5 = 0.010x^2 + 0.344x + 28.74$ is $x = 40$. What does this mean?

b. To find when after 1960 the percent of people over 19 years of age who have ever smoked and quit is 58.5, do we need to find the second solution to this equation? Why or why not?

c. Graphically verify that $x = 40$ is a solution to $58.5 = 0.010x^2 + 0.344x + 28.74$.
(Source: *Substance Abuse*, Princeton, N.J.)

63. *World Population* A low-projection scenario of world population for the years 1995–2150 by the United Nations is given by the function $y = -0.36x^2 + 38.52x + 5822.86$, where x is the number of years after 1990 and the world population is measured in millions of people. In what years will the world population reach 6581 million?

64. *Internet Use* An equation that models the number of worldwide users of the Internet is

$$y = 3.604x^2 + 27.626x - 26.714 \text{ million users}$$

where x is the number of years after 1990. If the pattern indicated by the model is based on data collected between 1990 and 2007, when does this model predict there will be 1638.25 million users?
(Source: U.S. Census Bureau)

65. *Tourism Spending* The global spending on travel and tourism (in billions of dollars) can be described by the equation $y = 0.502x^2 + 15.719x + 282.667$, where x equals the number of years after 1990. Graphically find the year in which spending is projected to reach $820.5 billion.
(Source: *Statistical Abstract of the United States*)

66. *Prison Population* The average daily number of inmates in the Beaufort County, South Carolina, prison can be modeled by

$$I(x) = 0.161x^2 + 7.582x + 82.157$$

where x is the number of years after 1991.

a. How would you set the viewing window to show a graph of the model for the years 1991 to 2011? Graph the function.

b. Graphically estimate when this model predicts that the average daily inmate population will be 185.
(Source: *The Island Packet*)

67. *Hospital Admissions* The number of admissions to all types of hospitals in the U.S. can be described by the function

$$A(t) = 38.228t^2 - 1069.60t + 40,698.547$$

thousand people where t is the number of years after 1980.

a. By how much did the number of hospital admissions change between 1990 and 1997?

b. Use this model and graphical methods to find in what year after 1980 the number of admissions was equal to 33,217,000.

c. Use the model to estimate the number of hospital admissions in 2005. What assumptions are you making when you make this estimation?
(Source: American Hospital Association, *Hospital Statistics*)

68. *Smoking Cessation* The percent of people over 19 years of age who have ever smoked and quit is given by the equation $y = 0.010x^2 + 0.344x + 28.74$, where x is the number of years after 1960. In what year will this model predict y to be over 100 percent, making it certain that the model is invalid?
(Source: *Substance Abuse*, Princeton, N.J.)

69. *Student Aid* Federal funds providing for student financial assistance through programs funded by congressional appropriations between 1980 and 2000 can be described by the model

$$f(x) = 9.032x^2 + 99.970x + 3645.90$$

million dollars where x is the number of years after 1980.

a. Use the model to compare the federal aid available to students in 2000 and 2008.

b. When did the available federal aid reach and exceed $14,000,000,000?
(Source: U.S. Department of Education, National Center for Educational Statistics)

70. *Bomb Threats* The number of bomb threats received against U.S. aircraft between 1980 and 1996 can be approximated by the function $B(t) = 3.268t^2 - 45.733t + 277.910$ threats, where t is the number of years after 1980.

a. What does this model estimate the number of bomb threats to be in 1997?

b. Use graphical or numerical methods with this model to estimate when the number of bomb threats will reach 1177 if this model remained valid.
(Source: *Statistical Abstract of the United States*)

71. *Quitting Smoking* The percent of people over 19 years of age who have ever smoked and quit is given by the equation $y = 0.010x^2 + 0.344x + 28.74$, where x is the number of years since 1960.

a. Graph this function for values of x representing 1960–2010.

b. Assuming that the pattern indicated by this model continued through 2010, what would be the percent in 2010?

c. When does this model indicate that the percent reached 40%?
(Source: *Substance Abuse*, Princeton, N.J.)

72. *Cell Phones* The number of U.S. cell phone subscribers (in thousands) is given by

$$y = 732.216x^2 + 2687.049x + 3056.333$$

where x is the number of years after 1990.

a. Use the model to estimate when the number of U.S. subscribers is 288,661,199.

b. When will the model estimate that the number of U.S. subscribers reaches 349,683,713?

c. What does the answer to (b) tell about this model?
(Source: U.S. Census Bureau)

73. *World Population* One projection of the world population by the United Nations (a low-projection scenario) is given in the table below. These data can be modeled by $y = -0.36x^2 + 38.52x + 5822.86$ million people, where x is the number of years from 1990. In what year past 1990 does this model predict the world population will first reach 6,702,000,000?

Year	Projected Population (millions)
1995	5666
2000	6028
2025	7275
2050	7343
2075	6402
2100	5153
2125	4074
2150	3236

(Source: *World Population Prospects*, United Nations)

Piecewise-Defined Functions and Power Functions

section preview ▪ Residential Power Costs

The data in Table 3.3 give the charges for electricity that Georgia Power Company charges its residential power customers for electricity during the months of June through September, excluding fuel adjustment costs and taxes. The monthly charges can be modeled by a **piecewise-defined function**, as we will see in Example 3. Piecewise-defined functions are used in applications in the life, social, and physical sciences, as well as in business, when there is not a single function that accurately represents the application.

Power functions are another type of function that is a basic building block for many mathematical applications. In this section, we see how to evaluate, graph, and apply piecewise-defined and power functions.

Table 3.3

Monthly Kilowatt-hours (kWh)	Monthly Charge
0 to 650	$7.50 plus $0.04657 per kWh
More than 650, up to 1000	$37.77 plus $0.07738 per kWh above 650
More than 1000	$64.85 plus $0.07976 per kWh above 1000

Piecewise-Defined Functions

It is possible that a set of data cannot be modeled with a single equation. We can use a **piecewise-defined function** when there is not a single function that accurately represents the situation. A piecewise-defined function is so named because it is defined with different pieces for different parts of its domain rather than one equation.

example 1

Postage

The postage paid for a first-class letter "jumps" by 17 cents for each ounce after the first ounce but does not increase until the weight increases by 1 ounce. Table 3.4 gives the postage for letters up to 3.5 ounces. Write a piecewise-defined function that models the price of postage and graph the function.

Table 3.4

Weight x (oz)	Postage y (cents)
$0 < x \leq 1$	42
$1 < x \leq 2$	59
$2 < x \leq 3$	76
$3 < x \leq 3.5$	93

Solution

The function that models the price P of postage for x ounces, where x is between 0 and 3.5, is

$$P(x) = \begin{cases} 42 & \text{if } 0 < x \leq 1 \\ 59 & \text{if } 1 < x \leq 2 \\ 76 & \text{if } 2 < x \leq 3 \\ 93 & \text{if } 3 < x \leq 3.5 \end{cases}$$

The graph of this function (Figure 3.21) shows that each output is a constant with discontinuous "steps" at $x = 1, 2$, and 3. This is a special piecewise-defined function, called a **step function**. Other piecewise-defined functions can be defined by polynomial functions over limited domains.

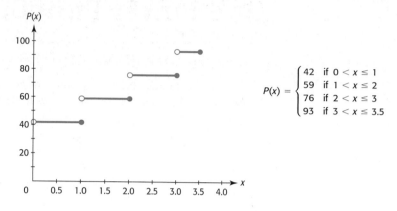

$$P(x) = \begin{cases} 42 & \text{if } 0 < x \le 1 \\ 59 & \text{if } 1 < x \le 2 \\ 76 & \text{if } 2 < x \le 3 \\ 93 & \text{if } 3 < x \le 3.5 \end{cases}$$

Figure 3.21

example 2

Piecewise-Defined Function

Graph the function

$$f(x) = \begin{cases} 5x + 2 & \text{if } 0 \le x < 3 \\ x^3 & \text{if } 3 \le x \le 5 \end{cases}$$

Solution

Table 3.5(a)

x	0	1	2	2.99
$f(x)$	2	7	12	16.95

Table 3.5(b)

x	3	4	5
$f(x)$	27	64	125

To graph this function, we can construct tables of values for each of the pieces. Table 3.5(a) gives outputs of the function for some sample inputs x on the interval $[0, 3)$; on this interval, $f(x)$ is defined by $f(x) = 5x + 2$. Table 3.5(b) gives outputs for some sample inputs on the interval $[3, 5]$; on this interval $f(x)$ is defined by $f(x) = x^3$.

Plotting the points from Table 3.5(a) and connecting them with a smooth curve gives the graph of $y = f(x)$ on the x-interval $[0, 3)$ (Figure 3.22). The open circle on this piece of the graph indicates that this piece of the function is not defined for $x = 3$. Plotting the points from Table 3.5(b) and connecting them with a smooth curve gives the graph of $y = f(x)$ on the x-interval $[3, 5]$ (Figure 3.22). The closed circles on this piece of the graph indicate that this piece of the function is defined for $x = 3$ and $x = 5$. The graph of $y = f(x)$, shown in Figure 3.22, consists of these two pieces.

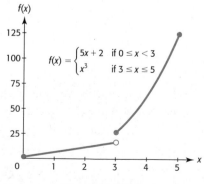

Figure 3.22

| example 3 | ## Residential Power Costs |

Excluding fuel adjustment costs and taxes, Georgia Power Company charges its residential power customers for electricity during the months of June through September according to Table 3.6.

Table 3.6

Monthly Kilowatt-hours (kWh)	Monthly Charge
0 to 650	$7.50 plus $0.04657 per kWh
More than 650, up to 1000	$37.77 plus $0.07738 per kWh above 650
More than 1000	$64.85 plus $0.07976 per kWh above 1000

a. Write the piecewise-defined function C that gives the monthly charge for residential electricity, with input x equal to the monthly number of kilowatt-hours.

b. Find $C(950)$ and explain what it means.

c. Find the charge for using 1560 kWh in a month.

Solution

a. The monthly charge is given by the function

$$C(x) = \begin{cases} 7.50 + 0.04657x & \text{if } 0 \le x \le 650 \\ 37.77 + 0.07738(x - 650) & \text{if } 650 < x \le 1000 \\ 64.85 + 0.07976(x - 1000) & \text{if } x > 1000 \end{cases}$$

where $C(x)$ is the charge in dollars for x kWh of electricity.

b. To evaluate $C(950)$, we must determine which "piece" defines the function when $x = 950$. Because 950 is between 650 and 1000, we use the "middle piece" of the function:

$$C(950) = 37.77 + 0.07738(950 - 650) = 60.984$$

Companies regularly round charges *up* to the next cent if any part of a cent is due. This means that if 950 kWh are used in a month, the bill is $60.99.

c. To find the charge for 1560 kWh, we evaluate $C(1560)$, using the "bottom piece" of the function because $1560 > 1000$. Evaluating this gives

$$C(1560) = 64.85 + 0.07976(1560 - 1000) = 109.5156$$

so the charge for the month is $109.52.

| example 4 | ## Wind Chill Factor |

Wind chill factors are used to measure the effect of the combination of temperature and wind speed on human comfort, by providing equivalent air temperatures with no wind blowing. One formula that gives the wind chill factor for a 30°F temperature and a wind with velocity V in miles per hour is

$$W = \begin{cases} 30 & \text{if } 0 \le V < 4 \\ 1.259V - 18.611\sqrt{V} + 62.255 & \text{if } 4 \le V \le 55.9 \\ -6.5 & \text{if } V > 55.9 \end{cases}$$

a. Find the wind chill factor for the 30°F temperature if the wind is 40 mph.

b. Find the wind chill factor for the 30°F temperature if the wind is 65 mph.

c. Graph this function for $0 \leq V \leq 70$.

d. What are the domain and range of the function graphed in (c)?
(Source: The National Weather Service)

Solution

a. Because $V = 40$ is in the interval $4 \leq V \leq 55.9$, the wind chill factor is

$$1.259(40) - 18.611\sqrt{40} + 62.255 \approx -5.091 \approx -5$$

This means that, if the wind is 40 mph, a temperature of 30°F would actually feel like −5°F.

b. Because $V = 65$ is in the interval $V > 55.9$, the wind chill factor is −6.5.

This means that, if the wind is 65 mph, a temperature of 30°F would actually feel like −6.5°F.

c. The graph is shown in Figure 3.23.

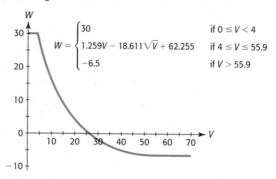

Figure 3.23

d. The domain is specified to be $0 \leq V \leq 70$, and the range, which is the set of outputs for W, is $-6.5 \leq W \leq 30$. That is, for wind speeds between 0 and 70 mph, the wind chill factor is no lower than −6.5°F and no higher than 30°F. ∎

---

example 5

Service Calls

The cost of weekend service calls by Airtech Services is shown by the graph in Figure 3.24. Write a piecewise-defined function that represents the cost for service during the first five hours, as defined by the graph.

Solution

The graph indicates that there are three pieces to the function, and each piece is a constant function. For t-values between 0 and 1, $C = 330$; for t-values between 1 and 2, $C = 550$; and for t-values between 2 and 5, $C = 770$. The open and closed circles on the graph tell us where the endpoints of each interval of the domain are defined. Thus, the cost function, with C in dollars and t in hours, is

$$C = \begin{cases} 330 & \text{if } 0 < t \leq 1 \\ 550 & \text{if } 1 < t \leq 2 \\ 770 & \text{if } 2 < t \leq 5 \end{cases}$$

∎

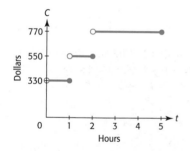

Figure 3.24

Absolute Value Function

We can construct a new function by combining two functions into a special piecewise-defined function. For example, we can write a function with the form

$$f(x) = \begin{cases} x & \text{if } x \geq 0 \\ -x & \text{if } x < 0 \end{cases}$$

This function is called the **absolute value function**, which is denoted by $f(x) = |x|$ and is derived from the definition of the absolute value of a number. Recall that the definition of the absolute value of a number is

$$|x| = \begin{cases} x & \text{if } x \geq 0 \\ -x & \text{if } x < 0 \end{cases}$$

To graph $f(x) = |x|$, we graph the portion of the line $y_1 = x$ for $x \geq 0$ (Figure 3.25(a)) and the portion of the line $y_2 = -x$ for $x < 0$ (Figure 3.25(b)). When these pieces are joined on the same graph (Figure 3.25(c)), we have the graph of $y = |x|$.

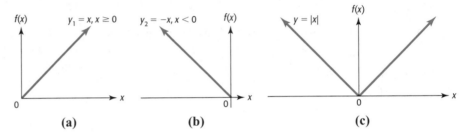

Figure 3.25

Solving Absolute Value Equations

We now consider the solution of equations that contain absolute value symbols, called **absolute value equations**. Consider the equation $|x| = 5$. To solve this equation, we know that $|5| = 5$ and $|-5| = 5$. Therefore, the solution to $|x| = 5$ is $x = 5$ or $x = -5$. Also, if $|x| = 0$, then x must be 0. Finally, because the absolute value of a number is never negative, we cannot solve $|x| = a$ if a is negative. We generalize this as follows.

> **Absolute Value Equation**
>
> If $|x| = a$ and $a > 0$, then $x = a$ or $x = -a$.
> There is no solution to $|x| = a$ if $a < 0$; $|x| = 0$ has solution $x = 0$.

If one side of an equation is a function contained in an absolute value and the other side is a nonnegative constant, we can solve the equation by using the method above.

---

example 6 | **Absolute Value Equations**

Solve the following equations:

a. $|x - 3| = 9$ **b.** $|2x - 4| = 8$ **c.** $|x^2 - 5x| = 6$

Solution

a. If $|x - 3| = 9$, then $x - 3 = 9$ or $x - 3 = -9$. Thus, the solution is $x = 12$ or $x = -6$.

b. If $|2x - 4| = 8$, then

$$2x - 4 = 8 \quad \text{or} \quad 2x - 4 = -8$$

$$
\begin{array}{c|c}
2x = 12 & 2x = -4 \\
x = 6 & x = -2
\end{array}
$$

Thus, the solution is $x = 6$ or $x = -2$.

c. If $|x^2 - 5x| = 6$, then

$$x^2 - 5x = 6 \quad \text{or} \quad x^2 - 5x = -6$$

$$
\begin{array}{c|c}
x^2 - 5x - 6 = 0 & x^2 - 5x + 6 = 0 \\
(x - 6)(x + 1) = 0 & (x - 3)(x - 2) = 0 \\
x = 6 \quad \text{or} \quad x = -1 & x = 3 \quad \text{or} \quad x = 2
\end{array}
$$

Thus, four values of x satisfy the equation. We can check these solutions by graphing or by substitution. Using the intersection method, we graph $y_1 = |x^2 - 5x|$ and $y_2 = 6$ and find the points of intersection. The solutions to the equation, $x = 6, -1, 3$, and 2 found above, are also the x-values of the points of intersection (Figure 3.26).

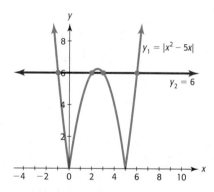

Figure 3.26

Power Functions

We define power functions as follows.

> ## Power Functions
>
> A **power function** is a function of the form $y = ax^b$, where a and b are real numbers, $b \neq 0$.

We know that the area of a rectangle is $A = lw$ square units, where l and w are the length and the width, respectively, of the rectangle. Because $l = w$ when the rectangle is a square, the area of a square can be found with the formula $A = x^2$ if each side is x units long.

This is the function $A = x^2$ for $x \geq 0$. A related function is $y = x^2$, defined on the set of all real numbers; this is a power function with power 2. This function is also a basic quadratic function, called the **squaring function**, and its complete graph is a parabola (see Figure 3.27(a)).

If we fill a larger cube with cubes that are 1 unit on each edge and have a volume of 1 cubic unit, the number of these small cubes that fit gives the volume of the larger cube. It is easily seen that if the length of an edge of a cube is x units, its volume is x^3 cubic units for any nonnegative value of x. The related function $y = x^3$ defined for all real numbers x is called the **cubing function**. A complete graph of this function is shown in Figure 3.27(b).

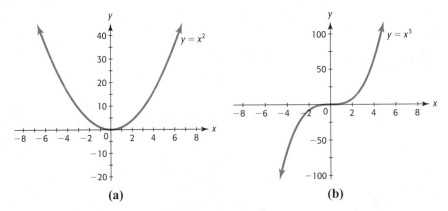

Figure 3.27

The functions $y = x$, $y = x^2$, and $y = x^3$ are examples of **power functions** with positive integer powers. Additional examples of power functions include functions with noninteger powers of x. For example, the graphs of the functions $y = x^{2/3}$ and $y = x^{3/2}$ are shown in Figures 3.28(a) and 3.28(b).

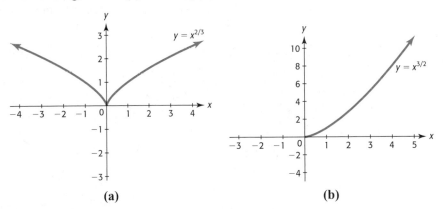

Figure 3.28

Note that $y = x^{2/3}$ is a power function with power less than 1, and its graph increases in the first quadrant but not as fast as $y = x$ does for values of x greater than 1. We say that the graph is *concave down* in the first quadrant. Also, $y = x^{3/2}$ is a power function with power greater than 1, and its graph increases in the first quadrant faster than $y = x$ does for values of x greater than 1. We say that the graph is *concave up* in the first quadrant.

In general, if $a > 0$ and $x > 0$, the graph of $y = ax^b$ is concave up if $b > 1$ and concave down if $0 < b < 1$. Figures 3.29(a) and 3.29(b) show the first-quadrant portion of typical graphs of $y = ax^b$ for $a > 0$ and different values of b.

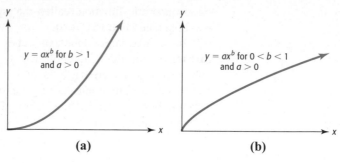

Figure 3.29

| example 7 | **Allometric Relationships** |

For fiddler crabs, data show that the relationship between the weight C of the claw and the weight W of the body is given by

$$C = 0.11W^{1.54}$$

with W and C in grams.

(Source: d'Arcy Thompson, *On Growth and Form*, Cambridge University Press, 1961)

a. Graph this function in the context of this application.

b. Use this function to compute the weight of a claw if the weight of the body of a crab is 5 grams.

Solution

a. It is appropriate to graph this model only in the first quadrant because $C > 0$ and $W > 0$ in the problem context. The graph is shown in Figure 3.30.

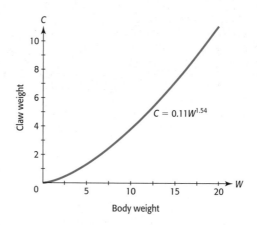

Figure 3.30

b. We find the weight of the claw by evaluating the function with an input of 5, giving an output of $C = 0.11(5)^{1.54} \approx 1.312$. Thus, the weight of the claw is approximately 1.3 grams. ∎

Note that the power in $C = 0.11W^{1.54}$ is 1.54, which can be written in the form $\dfrac{154}{100}$ and is therefore a rational number.

Functions with rational powers can also be written with radicals. For example, $y = x^{1/3}$ can be written in the form $y = \sqrt[3]{x}$, and $y = x^{1/2}$ can be written in the form $y = \sqrt{x}$. Functions like $y = \sqrt[3]{x}$ and $y = \sqrt{x}$ are special power functions called **root functions**.

Root Functions

A root function is a function of the form $y = ax^{1/n}$, or $y = a\sqrt[n]{x}$, where n is an integer, $n \geq 2$.

The graphs of $y = \sqrt{x}$ and $y = \sqrt[3]{x}$ are shown in Figures 3.31 and 3.32. Note that the domain of $y = \sqrt{x}$ is $x \geq 0$ because $\sqrt{x}$ is undefined for negative values of x. Note also that these root functions can be written as power functions.

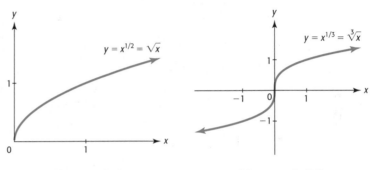

Figure 3.31   **Figure 3.32**

Reciprocal Function

One special power function is the function

$$y = x^{-1} = \frac{1}{x}$$

This function is usually called the **reciprocal function**. Its graph, shown in Figure 3.33, is called a rectangular hyperbola. Note that x cannot equal 0, so 0 is not in the domain of this function. As the graph of $y = \dfrac{1}{x}$ shows, as x gets close to 0, $|y|$ gets large and the graph approaches but does not touch the y-axis. We say that the y-axis is a **vertical asymptote** of this graph.

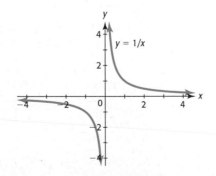

Figure 3.33

Note also that the values of $y = \dfrac{1}{x}$ get very small as $|x|$ gets large, and the graph approaches but does not touch the x-axis. We say that the x-axis is a **horizontal asymptote** for this graph.

In the following example, we consider an application of a reciprocal function.

| example 8 | **Average Cost** |

The average cost per hat for producing baseball hats is $\overline{C}(x) = 4.50 + \dfrac{1}{x}$ dollars, where x is the number of units produced.

a. Determine the domain of this function without concern for the context of the application and graph the function on a standard window.

b. Use knowledge of the context of the problem to graph the function on a window that applies to the application.

c. What is the horizontal asymptote of the graph? What does this tell you about how low the average cost can be?

Solution

a. All values except $x = 0$ result in real values for the function. Thus, the domain of $\overline{C}(x)$ is all real numbers except 0. The graph, shown in Figure 3.34(a), has the same shape as the graph of $y = \dfrac{1}{x}$, shown in Figure 3.33.

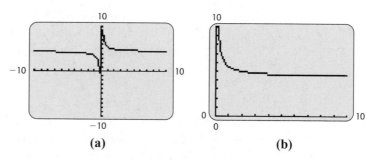

(a)   (b)

Figure 3.34

b. Because x represents the number of units produced, the viewing window is set for values of $x \geq 0$. The average cost cannot be negative, so the window is set with $\overline{C}(x) \geq 0$. Setting x-max and y-max at 10 gives the graph shown in Figure 3.34(b).

c. The horizontal asymptote is $y = 4.50$. This means that the average cost approaches \$4.50 as the number of hats produced increases. ■

Direct Variation as the *n*th Power

In Section 2.1 we discussed direct variation. If a quantity y varies directly as a power of x, we say that y is directly proportional to the *n*th power of x. That is, y varies directly as the *n*th power ($n > 0$) of x if there is a constant k such that

$$y = kx^n$$

The number k is called the *constant of variation* or the *constant of proportionality*.

example 9	Production

In the production of an item, the number of units of one raw material required varies as the cube of the number of units of a second raw material that is required. Suppose 500 units of the first and 5 units of the second raw material are required to produce 100 units of the item. How many units of the first raw material are required if the number of units produced requires 10 units of the second raw material?

Solution

If x is the number of units of the second raw material required and y is the number of units of the first raw material required, then y varies as the cube of x, or

$$y = kx^3$$

Because $y = 500$ when $x = 5$, we have

$$500 = k \cdot 5^3 \quad \text{or} \quad k = 4$$

Then $y = 4x^3$ and $y = 4(10^3) = 4000$. Thus 4000 units of the first raw material are required if the number of units produced requires 10 units of the second raw material. ∎

skills check

3.3

In Exercises 1–10, sketch the graph of each function using a window that gives a complete graph.

1. $y = x^3$

2. $y = x^4$

3. $y = x^{1/3}$

4. $y = x^{4/3}$

5. $y = \sqrt{x} + 2$

6. $y = \sqrt[3]{x} - 2$

7. $y = \dfrac{1}{x} - 3$

8. $y = 4 - \dfrac{1}{x}$

9. $y = \begin{cases} -1 & \text{if } x < 0 \\ 1 & \text{if } x \geq 0 \end{cases}$

10. $y = \begin{cases} 2 & \text{if } x \geq 2 \\ 6 & \text{if } x < 2 \end{cases}$

11. a. Graph the function $f(x) = \begin{cases} 5 & \text{if } 0 \leq x < 2 \\ 10 & \text{if } 2 \leq x < 4 \\ 15 & \text{if } 4 \leq x < 6 \\ 20 & \text{if } 6 \leq x < 8 \end{cases}$

b. What type of function is this?

12. a. Graph the function

$$f(x) = \begin{cases} 100 & \text{if } \ 0 \leq x < 20 \\ 200 & \text{if } 20 \leq x < 40 \\ 300 & \text{if } 40 \leq x < 60 \\ 400 & \text{if } 60 \leq x < 80 \end{cases}$$

b. What type of function is this?

13. a. Graph $f(x) = \begin{cases} 4x - 3 & \text{if } x \leq 3 \\ x^2 & \text{if } x > 3 \end{cases}$.

b. Find $f(2)$ and $f(4)$.

c. State the domain of the function.

14. a. Graph $f(x) = \begin{cases} 3 - x & \text{if } x \leq 2 \\ x^2 & \text{if } x > 2 \end{cases}$.

b. Find $f(2)$ and $f(3)$.

c. State the domain of the function.

15. a. Graph $f(x) = |x|$.

b. Find $f(-2)$ and $f(5)$.

c. State the domain of the function.

16. a. Graph $f(x) = |x - 4|$.

   b. Find $f(-2)$ and $f(5)$.

   c. State the domain of the function.

For each of the functions in Exercises 17–20, find the value of (a) $f(-1)$ and (b) $f(3)$, if possible.

17. $y = \begin{cases} 5 & \text{if } x \leq 1 \\ 6 & \text{if } x > 1 \end{cases}$ **18.** $y = \begin{cases} -2 & \text{if } x < -1 \\ 4 & \text{if } x \geq -1 \end{cases}$

19. $y = \begin{cases} x^2 - 1 & \text{if } x \leq 0 \\ x^3 + 2 & \text{if } x > 0 \end{cases}$

20. $y = \begin{cases} 3x + 1 & \text{if } x < 3 \\ x^2 & \text{if } x \geq 3 \end{cases}$

21. Determine if the function $y = 4x^3$ is increasing or decreasing for

   a. $x < 0$.   **b.** $x > 0$.

22. Determine if the function $y = -3x^4$ is increasing or decreasing for

   a. $x < 0$.   **b.** $x > 0$.

For each of the functions in Exercises 23–26, determine if the function is concave up or concave down in the first quadrant.

23. $y = x^{1/2}$ **24.** $y = x^{3/2}$

25. $y = x^{1.4}$ **26.** $y = x^{0.6}$

27. Graph $f(x) = \begin{cases} x & \text{if } x \geq 0 \\ -x & \text{if } x < 0 \end{cases}$

28. Graph $f(x) = \begin{cases} x - 4 & \text{if } x \geq 4 \\ 4 - x & \text{if } x < 4 \end{cases}$

29. Graph $f(x) = \begin{cases} x & \text{if } x < 0 \\ -x & \text{if } x \geq 0 \end{cases}$

30. Compare the graph in Exercise 27 with the graph in Exercise 15(a).

31. Compare the graph in Exercise 28 with the graph in Exercise 16(a).

In Exercises 32–36, solve the equations and check graphically.

32. $|2x - 5| = 3$ **33.** $\left|x - \dfrac{1}{2}\right| = 3$

34. $|x| = x^2 + 4x$ **35.** $|3x - 1| = 4x$

36. $|x - 5| = x^2 - 5x$

37. Suppose that S varies directly as the 2/3 power of T, and that $S = 64$ when $T = 64$. Find S when $T = 8$.

38. Suppose that y varies directly as the square root of x, and that $y = 16$ when $x = 4$. Find x when $y = 24$.

exercises

3.3

39. For the nonextreme weather months, Palmetto Electric charges $7.10 plus 6.747 cents per kilowatt-hour (kWh) for the first 1200 kWh and $88.06 plus 5.788 cents for all kilowatt-hours above 1200.

   a. Write the function that gives the monthly charge in dollars as a function of the kilowatt-hours used.

   b. What is the monthly charge if 960 kWh are used?

   c. What is the monthly charge if 1580 kWh are used?

40. *Postal Rates* The table below gives the local postal rates as a function of the weight of the printed matter that is mailed. Write a step function that gives the postage P as a function of the weight in pounds x for $1.5 \leq x \leq 5$.

Weight (lb)	Postal Rate ($)
$1.5 \leq x \leq 2$	1.97
$2 < x \leq 3$	2.04
$3 < x \leq 4$	2.14
$4 < x \leq 5$	2.34

(Source: USPS, 2008)

41. *First-Class Postage* The postage charged for first-class mail is a function of its weight. The U.S. Postal Service uses the following table to describe the rates for 2008.

Weight Increment x (oz)	First-Class Postage P(x)
First ounce or fraction of an ounce	42¢
Each additional ounce or fraction	17¢

(Source: pe.usps.gov/text)

a. Convert this table to a piecewise-defined function that represents first-class postage for letters weighing up to 4 ounces, using x as the weight in ounces and P the postage in cents.

b. Find $P(1.2)$ and explain what it means.

c. Give the domain of P as it is defined above.

d. Find $P(2)$ and $P(2.01)$.

e. Find the postage for a 2-ounce letter and for a 2.01-ounce letter.

42. *Federal Support for Education* The federal on-budget funds for all educational programs (in millions of constant 2000 dollars) between 1965 and 2000 can be modeled by the function

$$P(t) = \begin{cases} 1.965t - 5.65 & \text{when } 5 \le t \le 20 \\ 0.095t^2 - 2.925t + 54.429 & \text{when } 20 < t \le 40 \end{cases}$$

where t is the number of years after 1960.

a. Graph the function P for $5 \le t \le 40$. Describe how the funding for educational programs changed between 1965 and 1980.

b. What was the amount of funding for educational programs in 1980?

c. How much federal funding was allotted for educational programs in 1998?
(Source: U.S. Department of Education, National Center for Educational Statistics)

43. *Income Tax* The 2004 U.S. federal income tax owed by a married couple filing jointly can be found from the following table.
(Source: Internal Revenue Service, 2004, Form 1040 Instructions)

Filing Status: Married Filing Jointly

If Taxable Income is between	Tax Due is	of the amount over
$0 — $15,650	$0.00 + 10.0%	$0
$15,650 — $63,700	$1,565.00 + 15.0%	$15,650
$63,700 — $128,500	$8,772.50 + 25.0%	$63,700
$128,500 — $195,850	$24,872.50 + 28.0%	$128,500
$195,850 — $349,700	$43,830.50 + 33.0%	$195,850
$349,700 — up	$86,328.00 + 35.0%	$349,700

a. Write the piecewise-defined function T with input x that models the federal tax dollars owed as a function of x, the taxable income dollars earned, with $0 < x \le 128,500$.

b. Use the function to find $T(42,000)$.

c. Find the tax owed on a taxable income of $65,000.

d. A friend tells Jack Waddell not to earn any money over $63,700 because it would raise his tax rate to 25% on all of his taxable income. Test this statement by finding the tax on $63,700 and $63,700 + $1. What do you conclude?

44. *Wind Chill* The formula that gives the wind chill factor for a 60°F temperature and a wind with velocity V in miles per hour is

$$W = \begin{cases} 60 & \text{if } 0 \le V < 4 \\ 0.644V - 9.518\sqrt{V} + 76.495 & \text{if } 4 \le V \le 55.9 \\ 41 & \text{if } V > 55.9 \end{cases}$$

a. Find the wind chill factor for the 60° temperature if the wind is 20 mph.

b. Find the wind chill factor for the 60° temperature if the wind is 65 mph.

c. Graph the function for $0 \le V \le 80$.

d. What are the domain and range of the function graphed in part (c)?

45. *Suicides* The number of suicides can be modeled by the function $f(x) = 20,000x^{0.11}$, where x is the number of years after 1960.

a. What type of function is this?

b. What is $f(5)$? What does this mean?

c. How many suicides does this model estimate for 1982?
(Source: National Center for Health Statistics, for selected years)

46. *Single-Parent Families* The percent of all families that were single-parent families after 1960 was found to be modeled by $y = 1.053x^{0.888}$, with $x = 0$ in 1950.

   a. What is $f(45)$? What does this mean?

   b. Does this model indicate that the percent of single-parent families increased or decreased during the period after 1960?

47. *Taxi Miles* The Inner City Taxi Company estimated, on the basis of collected data, that the number of taxi miles driven each day can be modeled by the function $Q = 489L^{0.6}$, when they employ L drivers per day.

   a. Graph this function for $0 \leq L \leq 35$.

   b. How many taxi miles are driven each day if there are 32 drivers employed?

   c. Does this model indicate that the number of taxi miles increases or decreases as the number of drivers increases? Is this reasonable?

48. *Single-Parent Families* The percent of all families that were single-parent families after 1960 can be modeled by $y = 1.053x^{0.888}$, with $x = 0$ in 1950.

   a. According to the model, what percent of all families were estimated to have been single-parent families in 1975?

   b. What percent of all families did the model predict would be single-parent families in 2005? Comment on the reliability of your answer.

   c. Is this function concave up or concave down during this period of time?

   d. Use numerical or graphical methods to find when the model predicts that the percent will be 40.
(Source: U.S. Census Bureau)

49. *Mutual Funds* The household assets in mutual funds, measured in billions of dollars, can be modeled by $f(x) = 105.095x^{1.5307}$ billion dollars, where x is the number of years after 1990.

   a. Is this function increasing or decreasing from 1995 to 1999?

   b. Is this function concave up or concave down during this period of time?

   c. Use numerical or graphical methods to find when the model predicts that the assets will reach $4,000,000,000,000.
(Source: Federal Reserve, *Time*, April 3, 2000)

50. *U.S. Population* The U.S. population can be modeled by the function $y = 165.6x^{1.345}$, where y is in thousands and x is the number of years after 1800.

   a. What was the population in 1960, according to this model?

   b. Is the graph of this function concave up or concave down?

   c. Use numerical or graphical methods to find when the model estimates the population to be 92,370,000.

51. *Production Output* The monthly output of a product (in units) is given by $P = 1200x^{5/2}$, where x is the capital investment in thousands of dollars.

   a. Graph this function for x from 0 to 10 and P from 0 to 200,000.

   b. Is the graph concave up or concave down?

52. *Harvesting* A farmer's main cash crop is tomatoes, and the tomato harvest begins in the month of May. The number of bushels of tomatoes harvested on the xth day of May is given by the equation $B(x) = 6(x + 1)^{3/2}$. How many bushels did the farmer harvest on May 8?

53. *Voter Turnout* The function $y = 86.17x^{-0.135}$ gives the percent of voter turnout during presidential election years, with x representing the number of years after 1950.

   a. Graph the function for $x > 0$.

   b. Does this model indicate that the percent of voter turnout is increasing or decreasing?

   c. Use the graph to determine the percent in 2000.

   d. What percent voter turnout did this model estimate for the 2004 election?

   e. The voter turnout in 2004 was 55.3%. Does the estimate from the model agree with this turnout?
(Source: Federal Election Commission)

54. *Trust in the Government* The percent of people who say they trust the government in Washington always or most of the time is given by $y = 154.131x^{-0.492}$, with x equal to the number of years after 1960.

   a. Graph the function.

   b. Does this model indicate that trust in the government is increasing or decreasing?

   c. Use the graph to estimate the percent of people who trust the government in 1998.
   (Source: Pew Research Center)

55. *Average Cost* The monthly average cost of producing 27-inch television sets is $C(x) = 105 + \dfrac{50,000}{x}$ dollars, where x is the number of sets produced per month. What is the average cost per set if 2000 sets are produced?

56. *Concentration of Body Substances* The concentration C of a substance in the body depends on the quantity of substance Q and the volume V through which it is distributed. For a static substance, the concentration is given by

$$C = \frac{Q}{V}$$

   a. For $Q = 1000$ milliliters (mL) of a substance, graph the concentration as a function of the volume on the interval from $V = 1000$ mL to $V = 5000$ mL.

   b. For a fixed quantity of a substance, does the concentration of the substance in the body increase or decrease as the volume through which it is distributed increases?

57. *Investing* If money is invested for 3 years with interest compounded annually, the future value of the investment varies directly as the cube of $1 + r$, where r is the annual interest rate. If the future value of the investment is $6298.56 when the interest rate is 8%, what rate gives a future value of $5955.08?

58. *Investing* If money is invested for 4 years with interest compounded annually, the future value of the investment varies directly as the fourth power of $1 + r$, where r is the annual interest rate. If the future value of the investment is $17,569.20 when the interest rate is 10%, what rate gives a future value of $24,883.20?

section

3.4 Quadratic and Power Models

key concepts

- Modeling with quadratic functions
- Comparison of models; first and second differences
- Modeling with power functions
- Comparison of quadratic and power models

section preview ▪ Voting

Table 3.7 on the next page shows that the percent of registered voters who voted in Presidential elections mostly declined for the years 1960 through 2004. The scatter plot of these data, shown in Figure 3.35, shows that a line would not fit these data well, so a linear function is not a good model for the data. (See Example 6.) The use of graphing calculators permits us to model nonlinear data with other types of functions by using steps similar to those that we used to model data with linear functions. In this section, we model sets of data with quadratic and power functions.

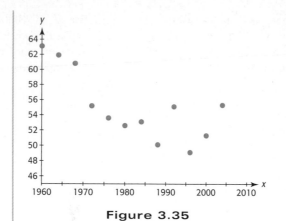

Figure 3.35

Table 3.7   Percent Voting in Presidential Elections

Year	Percent	Year	Percent
1960	63.1	1984	53.1
1964	61.9	1988	50.1
1968	60.8	1992	55.1
1972	55.2	1996	49.1
1976	53.6	2000	51.3
1980	52.6	2004	55.3

(Source: Federal Election Commission)

Modeling a Quadratic Function from Three Points on Its Graph

If we know three (or more) points that fit exactly on a parabola, we can find the quadratic function whose graph is the parabola.

example 1   ### Equation of a Quadratic Function

Find the equation of the quadratic function whose graph is a parabola containing the points $(-1, 9)$, $(2, 6)$, and $(3, 13)$.

Solution

Using the three points $(-1, 9)$, $(2, 6)$, and $(3, 13)$, we substitute the values for x and y in the general equation $y = ax^2 + bx + c$, getting three equations.

$$\begin{cases} 9 = a(-1)^2 + b(-1) + c \\ 6 = a(2)^2 + b(2) + c \\ 13 = a(3)^2 + b(3) + c \end{cases} \quad \text{or} \quad \begin{cases} 9 = a - b + c \\ 6 = 4a + 2b + c \\ 13 = 9a + 3b + c \end{cases}$$

We use these three equations to solve for a, b, and c, using techniques similar to those used to solve two equations in two variables.*

Subtracting the third equation, $13 = 9a + 3b + c$, from each of the first and second equations, gives a system of two equations in two variables.

$$\begin{cases} -4 = -8a - 4b \\ -7 = -5a - b \end{cases}$$

Multiplying the second equation by -4 and adding gives

$$\begin{cases} -4 = -8a - 4b \\ 28 = 20a + 4b \end{cases}$$
$$\overline{\hspace{0.5cm} 24 = 12a \quad \Rightarrow a = 2}$$

---

* We will discuss solution of systems of three equations in three variables further in Chapter 7.

Substituting $a = 2$ in $-4 = -8a - 4b$ gives

$$-4 = -16 - 4b \quad \text{or} \quad b = -3$$

and substituting $a = 2$ and $b = -3$ in the original third equation, $9 = a - b + c$, gives

$$9 = 2 - (-3) + c \quad \text{or} \quad c = 4$$

Thus $a = 2, b = -3, c = 4$, and the quadratic function whose graph contains the points is

$$y = 2x^2 - 3x + 4$$

■

Modeling with Quadratic Functions

If the graph of a set of data has a pattern that approximates the shape of a parabola or part of a parabola, a quadratic function may be appropriate to model the data. In Example 2, we will use technology to model the number of Starbucks stores as a function of the number of years after 1990.

example 2

Starbucks Stores

Table 3.8 gives the number of Starbucks stores in the United States for the years 1992 through 2007.

Table 3.8

Year	Starbucks Stores	Year	Starbucks Stores
1992	113	2000	2119
1993	163	2001	2925
1994	264	2002	3756
1995	430	2003	4453
1996	663	2004	5452
1997	974	2005	6423
1998	1321	2006	7715
1999	1657	2007	9401

a. Create a scatter plot of the data points, with x equal to the number of years after 1990.

b. Create a quadratic function that models the data, using the number of years after 1990 as the input x.

c. Graph the aligned data and the quadratic function on the same axes. Does this model seem like a reasonable fit?

d. Use the model to estimate the number of stores in 2006. Is the estimate close to the actual number?

e. Use the model to estimate the number of stores in 2010. Discuss the reliability of this estimate.

Solution

a. Figure 3.36(a) shows a scatter plot of the data. The shape looks as though it could be part of a parabola, so it is reasonable to find a quadratic function using the data points.

b. Enter the aligned inputs 2 through 17 in one list of a graphing utility and the corresponding outputs in a second list. Using quadratic regression in a graphing calculator gives a quadratic function that models the data. The function, with the coefficients rounded to three decimal places, is

$$y = 48.278x^2 - 334.702x + 785.961$$

where x is the number of years from 1990. Note that we will use the unrounded model in graphing and performing calculations.

c. The scatter plot of the aligned data and the graph of the function are shown in Figure 3.36(b). The function appears to be an excellent fit to the data.

d. Evaluating the function at $x = 16$ gives the number of U.S. stores in 2006 as 7790 (Figure 3.36(c)). This is a reasonable estimate of the actual number of stores, which is 7715.

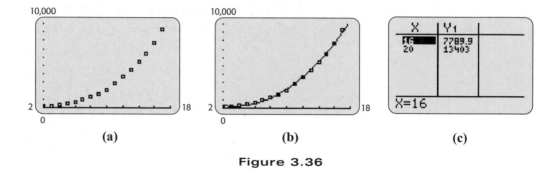

(a)   (b)   (c)

Figure 3.36

e. Figure 3.36(c) also gives the prediction for 2010, which is 13,403. This extrapolation assumes that the number of stores will grow very rapidly, and many things may change during the 3 years after this data has been collected. Even for Starbucks, this prediction seems optimistic. ▪

spreadsheet solution

We can use graphing utilities, software programs, and spreadsheets to find the quadratic function that is the best fit for data. Table 3.9 shows a partial Excel spreadsheet for the aligned data of Example 2. Selecting the cells containing the data, using Chart Wizard to get the scatter plot of the data, selecting Add Trendline, and picking Polynomial with order 2 gives the equation of the quadratic function that is the best fit for the data, along with the scatter plot and the graph of the best-fitting parabola (see Figure 3.37).*

---

*See Appendix B, page 683.

Table 3.9

	A	B
1	Year x	Starbucks Stores y
2	2	113
3	3	163
4	4	264
5	5	430
6	6	663
7	7	974
8	8	1321

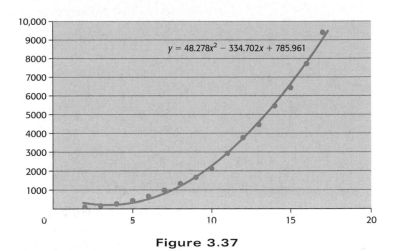

$y = 48.278x^2 - 334.702x + 785.961$

Figure 3.37

Comparison of Linear and Quadratic Models

Recall that when the changes in inputs are constant and the (first) differences of the outputs are constant or nearly constant, a linear model will give a good fit for the data. In a similar manner, we can compare the differences of the first differences, which are called the **second differences**. If the second differences are constant for equally spaced inputs, the data can be modeled exactly by a quadratic function.

Consider the data in Table 3.10, which gives the measured height y of a toy rocket x seconds after it has been shot into the air from the ground.

Table 3.10 Height of a Rocket

Time x (seconds)	Height (meters)
1	68.6
2	117.6
3	147
4	156.8
5	147
6	117.6
7	68.6

Table 3.11 gives the first differences and second differences for the equally spaced inputs for the rocket height data in Table 3.10.

Table 3.11

Outputs	68.6		117.6		147		156.8		147		117.6		68.6
First Differences		49		29.4		9.8		−9.8		−29.4		−49	
Second Differences			−19.6		−19.6		−19.6		−19.6		−19.6			

The first differences are not constant, but each of the second differences is -19.6, which indicates that the data can be fit exactly by a quadratic function. We can find that the quadratic model for the height of the toy rocket is $y = 78.4x - 9.8x^2$ meters, where x is the time in seconds.

example 3   ## Aid to Dependent Children

Table 3.12 gives the number of families (in thousands) who were recipients under the Federal Aid to Families with Dependent Children program for the years 1990 to 1996.

Table 3.12

Year	1990	1991	1992	1993	1994	1995	1996
Recipient Families (thousands)	4218	4708	4936	5050	4979	4641	4166

(Source: U.S. Census Bureau, *Statistical Abstract of the United States*)

a. Find the first and second differences for the data to justify that a quadratic function will be a better model for the data than a linear function.

b. Align the data so that the input is the number of years from 1990 and create a quadratic function that models the data.

c. Graph the aligned data and the quadratic function on the same axes.

d. What does the model give as the maximum number of families who were recipients of federal aid during the period 1990–1996?

Solution

a. The inputs are equally spaced, and the first and second differences for the data are shown in Table 3.13. The second differences are closer to being constant than the first differences, so a quadratic model will be a better fit for the data. In practice, it is easier to create a scatter plot to determine which models may be good fits for data points.

Table 3.13

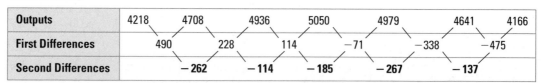

Outputs	4218	4708	4936	5050	4979	4641	4166
First Differences		490	228	114	-71	-338	-475
Second Differences			-262	-114	-185	-267	-137

b. Enter the aligned inputs 0, 1, 2, 3, 4, 5, and 6 in a list of a graphing utility and enter the corresponding output values from the second row of Table 3.12. Using quadratic regression in a graphing utility gives a quadratic function that models the data. The function, with the coefficients rounded to four decimal places, is

$$y = -95.5357x^2 + 564.3929x + 4219.9286$$

c. The scatter plot of the aligned data and the graph of the function are shown in Figure 3.38(a). We see that a quadratic model fits the data well even though the second differences are not constant.

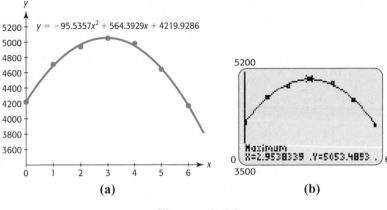

Figure 3.38

d. The maximum value of the rounded function occurs at the vertex of the parabola shown in Figure 3.38(a). The x-coordinate of this vertex is $x = \dfrac{-564.3929}{2(-95.5357)} = 2.9538 \approx 3$, and the y-coordinate is $y \approx 5053$. Thus, the model gives 5053 thousand, or 5,053,000 families as the maximum number of families who received federal aid under this program in $1990 + 3 = 1993$. This can be verified by finding the maximum point on the graph with technology (Figure 3.38(b)). ■

Modeling with Power Functions

We can model some experimental data by observing patterns. For example, by observing how the area of each of the (square) faces of a cube is found and that a cube has six faces, we can deduce that the surface area of a cube that is x units on each edge is

$$S = 6x^2 \text{ square units}$$

We can also measure and record the surface areas for cubes of different sizes to investigate the relationship between the edge length and the surface areas for cubes. Table 3.14 contains selected measures of edges and the resulting surface areas.

Table 3.14

Edge length x (units)	Surface Area of Cube (square units)
1	6
2	24
3	54
4	96
5	150

We can enter the lengths from the table as the independent (x) variable and the corresponding surface areas as the dependent (y) variable in a graphing utility, and then have the utility create the **power function** that is the best model for the data (Figure 3.39). This model also has the equation $y = 6x^2$ square units.

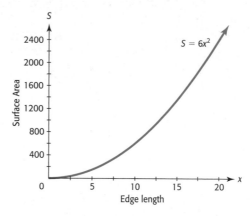

Figure 3.39

example 4

Auto Noise

The noise level of a Vauxhall VX220 increases as the speed of the car increases. Table 3.15 gives the noise, in decibels (db), at different speeds.

Table 3.15

Speed (mph)	Noise Level (db)
10	50
30	68
50	75
70	79
100	84

(Source: *Auto Car Magazine,* November 2001)

a. Fit a power function model to the data.

b. Graph the data points and the model on the same axes.

c. Use the result from part (a) to estimate the noise level at 80 mph.

Solution

a. Entering the data in a graphing utility and creating the power regression model with the utility gives the equation. The model is

$$y = 30.414x^{0.226}$$

where x is in miles per hour and y is in decibels.

b. The graphs of the data points and the model are shown in Figure 3.40.

c. Evaluating the model at $x = 80$ gives $y = 81.88$, so the noise level at 80 mph is 81.88 decibels.

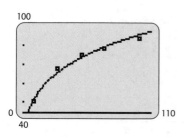

Figure 3.40

| example 5 | **Cohabiting Households** |

The data in Table 3.16 give the number of cohabiting (without marriage) households (in thousands) for selected years from 1960 to 2004. The scatter plot of these data, with x representing the number of years from 1950 and y representing thousands of households, is shown in Figure 3.41.

Table 3.16   Cohabiting Households

Year	Cohabiting Households (thousands)	Year	Cohabiting Households (thousands)
1960	439	1993	3510
1970	523	1994	3661
1980	1589	1995	3668
1985	1983	1996	3958
1990	2856	1997	4130
1991	3039	1998	4236
1992	3308	2000	5476
		2004	5841

(Source: Index of Leading Cultural Indicators)

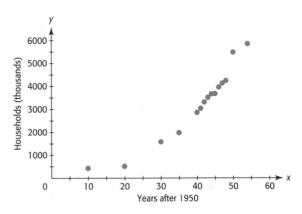

Figure 3.41

a. Find the power function that models these data.

b. Graph the data and the model on the same axes.

Solution

a. Using power regression in a graphing utility gives the power function that models these data. The power function that is the best fit is

$$y = 5.817x^{1.694}$$

where y is in thousands and x is the number of years from 1950.

b. The graphs of the data and the power function that models it are shown in Figure 3.42.

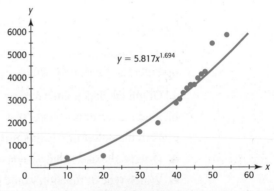

Figure 3.42

Power models can also be found with Excel. The Excel spreadsheet in Table 3.17 shows the household assets in mutual funds, measured in billions of dollars, for 1995 to 1999, listed as years after 1990. Selecting the cells containing the data, using Chart Wizard to get the scatter plot of the data, selecting Add Trendline, and picking Power gives the equation of the power function that is the best fit for the data, along with the scatter plot and the graph of the best-fitting power function. The equation and graph of the function are shown in Figure 3.43.

Table 3.17

	A	B
1	Year x	Assets in mutual funds y ($ dollars)
2	5	1265
3	6	1586
4	7	2057
5	8	2501
6	9	3104

(Source: "Federal Reserve," *Time*, April 3, 2000)

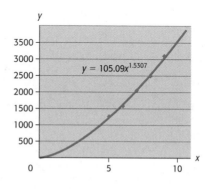

Figure 3.43

Comparison of Power and Quadratic Models

We found a power function that is a good fit for the data in Table 3.17, but a linear function or a quadratic function may also be a good fit for this data. A quadratic function may fit data points even if there is no obvious "turning point" in the graph of the data points. If the data points appear to rise (or fall) more rapidly than a line, then a quadratic model or a power model may fit the data well. In some cases it may be necessary to find both models to determine which is the better fit for the data.

example 6 | **Voting**

Table 3.18 shows the percent of voting-age population who voted in presidential elections for the years 1960–2004.

a. Graph the data points, representing the years from 1950 as x and the percent as y.

b. Find the quadratic model that is the best fit for the data.

c. Find the power model that is the best fit for the data.

d. Discuss the use of the two models to predict the percent voting after 2004.

e. Which type of function would give the better fit if a point were added giving the percent voting as 58.1 in 2008?

Table 3.18 Percent Voting in Presidential Elections

Year	Percent	Year	Percent
1960	63.1	1984	53.1
1964	61.9	1988	50.1
1968	60.8	1992	55.1
1972	55.2	1996	49.1
1976	53.6	2000	51.3
1980	52.6	2004	55.3

(Source: Federal Election Commission)

Solution

a. The scatter plot of the data is shown in Figure 3.44(a). It has no obvious "turning point," so either a quadratic or a power model may possibly be a good fit.

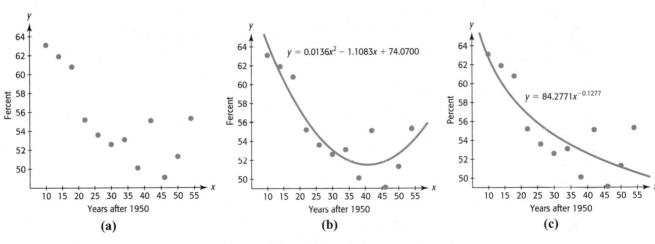

Figure 3.44

b. The quadratic model that is the best fit for the data is

$$y = 0.0136x^2 - 1.1083x + 74.0700$$

The graphs of this model and the data points are shown in Figure 3.44(b).

c. The power model that is the best fit for the data is

$$y = 84.2771x^{-0.1277}$$

The graphs of this model and the data points are shown in Figure 3.44(c).

d. The graph of the quadratic model turns and begins to rise, as did the voter turnout in the very close elections in 2000 and 2004.

e. The quadratic model would give a better fit because its graph turns and begins to rise after 1990.

Note that the quadratic model for the percent of voter turnout found in Example 6 is a continuous function, but it can only be meaningfully interpreted discretely, for election years.

skills check
3.4

1. Find the equation of the quadratic function whose graph is a parabola containing the points $(-1, 6)$, $(2, 3)$ and $(3, 10)$.

2. Find the equation of the quadratic function whose graph is a parabola containing the points $(-2, -4)$, $(3, 1)$ and $(2, 4)$.

3. Find the quadratic function that models the data in the table below.

x	-2	-1	0	1	2	3	4
y	16	5	0	1	8	21	40

x	5	6	7	8	9	10
y	65	96	133	176	225	280

4. The following table has the inputs, x, and the outputs for three functions, f, g, and h. Use second differences to determine which function is exactly quadratic, which is approximately quadratic, and which is not quadratic.

x	$f(x)$	$g(x)$	$h(x)$
0	0	2	0
2	399	0.8	110
4	1601	1.2	300
6	3600	3.2	195
8	6402	6.8	230
10	9998	12	290

5. As you can verify, the following data points give constant second differences, but the points do not fit on the graph of a quadratic function. How can this be?

x	1	2	4	8	16	32	64
y	1	6	15	28	45	66	91

6. **a.** Make a scatter plot of the data in the table below.

   b. Does it appear that a linear model or a power model is the best fit for the data?

x	y
1	4
2	9
3	11
4	21
5	32
6	45

7. Find the quadratic function that is the best fit for $f(x)$ defined by the table in Exercise 4.

8. Find the quadratic function that is the best fit for $g(x)$ defined by the table in Exercise 4.

9. **a.** Find a power function that models the data in the table in Exercise 6.

   b. Find a linear function that models the data.

   c. Visually determine which model is the better fit for the data.

10. **a.** Make a scatter plot of the data in the table below.

    b. Does it appear that a linear model or a power model is the better fit for the data?

x	y
3	4.7
5	8.6
7	13
9	17

11. **a.** Find a power function that models the data in the table in Exercise 10.

    b. Find a linear function that models the data.

    c. Visually determine if each model is a good fit.

12. Find the quadratic function that models the data in the table below.

x	−2	−1	0	1	2	3
y	15	5	2	1	3	10

x	4	5	6	7	8
y	20	35	55	75	176

13. Find the power function that models the data in the table below.

x	1	2	3	4	5	6	7	8
y	3	4.5	5.8	7	8	9	10	10.5

14. Is a power function or a quadratic function the better model for the data below?

x	0.5	1	2	3	4	5	6
y	1	7	17	32	49	70	90

exercises

3.4

Calculate numerical results with the unrounded models, unless otherwise instructed. Report models to 3 decimal places unless otherwise stated.

15. *Income* The median annual incomes of males in the United States for selected years between 1960 and 2004 are shown in the table below.

Year	Median Annual Income ($)	Year	Median Annual Income ($)
1960	4080	2001	29,101
1970	6670	2002	29,238
1980	12,530	2003	29,931
1990	20,293	2004	30,513
2000	28,343		

(Source: U.S. Census Bureau)

a. Find the quadratic function that models the median income as a function of x, the number of years after 1960.

b. Use the model from part (a) to estimate the median annual income of males in 1997 and in 2010.

c. Do you feel that these estimates are reliable? Explain.

16. *Age and Income* The median annual income of men for certain average ages is given in the table at right.

a. Find the quadratic function that models the annual income as a function of the average age.

b. Does the model appear to be a good fit?

c. Find and interpret the vertex of the graph of the *rounded* function.

Average Age (yr)	Median Income ($)	Average Age (yr)	Median Income ($)
19.5	6960	49.5	36,526
29.5	25,179	59.5	29,526
39.5	32,167	69.5	16,684

(Source: *Statistical Abstract of the United States*)

17. *National Health Care* The table on the next page shows the national expenditures for health care in the United States for selected years, with projections to 2015.

a. Use a scatter plot with x as the number of years past 1950 and y as the total expenditures for health care (in billions) to identify what type (or types) of function(s) would make a good model for these data.

b. Find a power model and a quadratic model for the data.

c. Which model from (b) more accurately estimates the 2010 expenditures for national health care?

d. Use the better model from (c) to estimate the 2020 expenditures for national health care.

Year	National Expenditures for Health Care (in billions)
1960	$28
1970	75
1980	255
1990	717
1995	1020
2000	1359
2005	2016
2010	2879
2015	4032

(Source: U.S. Centers for Medicare and Medicaid Services)

18. *Unemployment* The percent of unemployment in the United States for the years 2000–2007 is given by the data in the table below.

a. Create a scatter plot for the data, with x equal to the number of years after 2000.

b. Does it appear that a quadratic model will fit the data? If so, find the best-fitting quadratic model.

c. Does the y-intercept of the function in part (b) have meaning in the context of this problem? If so, interpret the value.

Year	Percent Unemployment	Year	Percent Unemployment
2000	4.0	2004	5.5
2001	4.7	2005	5.1
2002	5.4	2006	4.8
2003	6.0	2007	4.6

(Source: South Carolina Employment Security Commission)

19. *Foreign-Born Population* The following table gives the percent of U.S. population that is foreign born.

a. Create a scatter plot for the data, with x equal to the number of years from 1900 and y equal to the percent.

b. Does it appear that the data could be modeled with a quadratic function?

c. Find the best-fitting quadratic function for the data. Report your answer to 4 decimal places.

d. Use the function to estimate the percent in 2010.

Year	Foreign Born (%)	Year	Foreign Born (%)
1900	13.6	1960	5.4
1910	14.7	1970	4.7
1920	13.2	1980	6.2
1930	11.6	1990	8.0
1940	8.8	2000	10.4
1950	6.9	2005	11.7

(Source: U.S. Census Bureau)

20. *Modeling Homicide Rate* The numbers of homicides per 100,000 people for selected years from 1950 to 2004 are given in the following table.

Year	Homicides per 100,000	Year	Homicides per 100,000
1950	4.6	1995	8.2
1955	4.1	1997	6.8
1960	5.1	1998	6.3
1965	5.1	1999	5.7
1970	7.9	2000	5.5
1975	9.6	2001	5.6
1980	10.2	2002	5.6
1985	7.9	2003	5.7
1990	9.4	2004	5.5

(Source: FBI, Uniform Crime Statistics)

a. Find the quadratic function that models the data, with $x = 0$ in 1950.

b. Graph the data and the model on the same axes.

c. Find the year in which the maximum number of homicides occurs, according to this model.

21. *Accidental Deaths* The table below gives the rate of accidental deaths (per 100,000 residents) in the United States, as a function of the number years from 1999.

Years from 1999	Accidental Deaths per 100,000
1	35.6
2	35.6
3	37.1
4	37.4
5	37.8

   a. What is the best-fitting power model for these data?

   b. What is the accidental death rate in 2012, according to this model?

22. *U.S. Population* The table gives the U.S. population, in millions, for selected years, with projections to 2050.

   a. Create a scatter plot for the data in the table, with x equal to the number of years past 1960.

   b. Use the scatter plot to determine the type of function that can be used to model the data, and create a function that best fits the data, with x equal to the number of years past 1960.

Year	U.S. Population (millions)	Year	U.S. Population (millions)
1960	180.671	1995	263.044
1965	194.303	1998	270.561
1970	205.052	2000	281.422
1975	215.973	2003	294.043
1980	227.726	2025	358.030
1985	238.466	2050	408.695
1990	249.948		

(Source: U.S. Census Bureau)

23. *Violent Crime* The following table gives the rate of violent crimes (per 100,000 residents) in the United State, as a function of the year x.

   a. Find the power model that is the best fit for the data, with x equal to the number of years after 1999.

   b. What rate of violent crimes does the model predict for 2012?

Year	Violent Crimes per 100,000	Year	Violent Crimes per 100,000
2000	506.5	2003	475.8
2001	504.5	2004	463.2
2002	494.4	2005	469.2

(Source: U.S. Census Bureau)

24. *Volume* The measured volume of a pyramid with each edge of the base equal to x units and with its altitude (height) equal to x units is given in the table below.

   a. Determine if the second differences of the outputs are constant.

   b. If the answer is yes, find the quadratic model that is the best fit for the data. Otherwise, find the power function that is the best fit.

Edge Length x (units)	Volume of Pyramid (cubic units)
1	1/3
2	8/3
3	9
4	64/3
5	125/3
6	72

25. *Cell Phones* The table on the next page gives the number of millions of U.S. cellular telephone subscribers.

   a. Create a scatter plot for the data with x equal to the number of years from 1985. Does it appear that the data could be modeled with a quadratic function?

   b. Find the quadratic function that is the best fit for these data, with x equal to the number of years from 1985 and y equal to the number of subscribers in millions.

   c. Use the model to estimate the number in 2010.

   d. What part of the U.S. population does this estimate equal if the U.S. population is 300 million?

Year	Subscribers (millions)	Year	Subscribers (millions)
1985	0.340	1996	44.043
1986	0.682	1997	55.312
1987	1.231	1998	69.209
1988	2.069	1999	86.047
1989	3.509	2000	109.478
1990	5.283	2001	128.375
1991	7.557	2002	140.767
1992	11.033	2003	158.722
1993	16.009	2004	182.140
1994	24.134	2005	207.896
1995	33.786	2006	233.000

(Source: Semiannual CTIA Wireless Industry Survey)

26. *World Population* One projection of the world population by the United Nations for selected years (a low projection scenario) is given in the table below.

Year	Projected Population (millions)	Year	Projected Population (millions)
1995	5666	2075	6402
2000	6028	2100	5153
2025	7275	2125	4074
2050	7343	2150	3236

(Source: *World Population Prospects, United Nations*)

a. Find a quadratic function that fits these data, using the number of years after 1990 as the input.

b. Find the positive *x*-intercept of this graph, to the nearest year.

c. When can we be certain that this model no longer applies?

27. *Personal Savings* The following table gives Americans' personal-savings rate for selected years from 1960 to 2005.

a. Find a quadratic function that models the personal-savings rate as a function of the number of years from 1960.

b. Find and interpret the vertex of the graph of the *rounded* model from part (a).

c. Find the positive *x*-intercept of this graph, to the nearest tenth.

d. When can we be certain that this model no longer applies?

Year	Personal-Savings Rate (%)	Year	Personal-Savings Rate (%)
1960	7.3	1985	7.0
1965	7.8	1990	7.0
1970	9.4	1995	4.6
1975	10.0	2000	2.3
1980	10.0	2005	−0.4

(Source: Index of Leading Cultural Indicators)

28. *Mortgages* The balance owed *y* on a $50,000 mortgage after *x* monthly payments is shown in the table below. Graph the data points with each of the equations below to determine which of them is the better model for the data, if *x* is the number of months that payments have been made.

a. $y = 338{,}111.278x^{-0.676}$

b. $y = 4700\sqrt{110 - x}$

Monthly Payments	Balance Owed ($)
12	47,243
24	44,136
48	36,693
72	27,241
96	15,239
108	8074

29. *International Visitors* The number of international visitors to the U.S. for selected years 1986–2005 is given in the table below.

Year	U.S. Visitors (millions)	Year	U.S. Visitors (millions)
1986	26	1996	46.3
1987	29.5	1997	48.9
1988	34.1	2000	50.9
1989	36.6	2001	44.9
1990	39.5	2002	41.9
1991	43	2003	41.2
1992	47.3	2004	46.1
1994	45.5	2005	49.4
1995	44		

(Source: *World Tourism Organization*)

a. Using an input equal to the number of years after 1980, graph the aligned data points and both of the given equations to determine which equation is the better model for the aligned data.

   i. $y = 17\sqrt[3]{x}$

   ii. $y = -0.125x^2 + 4.680x + 4.420$

b. If you had to pick one of these models to predict the number of international visitors in the year 2010, which model would be the more reasonable choice?

30. *Consumer Price Index* The effect of inflation is described by the Consumer Price Index, which tells how much it takes to buy an item in a given year if it cost $1 in an earlier year. The following table gives the prices of all goods and services for urban households, based on a price of $1 for them in 1913.

   a. Find the quadratic function that is the best fit for the data, with x equal to the number of years after 1900.

   b. The model gives the minimum index in what year?

   c. What event in history explains why the index is lower after 1920?

Year	Amount It Took to Equal $1 in 1913	Year	Amount It Took to Equal $1 in 1913
1920	2.02	1970	3.92
1925	1.77	1975	5.43
1930	1.69	1980	8.32
1935	1.38	1985	10.87
1940	1.41	1990	13.20
1945	1.82	1995	15.39
1950	2.43	2000	17.39
1955	2.71	2005	19.73
1960	2.99	2006	20.18
1965	3.18		

31. *Modeling Personal Income* Total personal income in the United States (in billions of dollars) for selected years from 1960 to 2005 is given in the following table.

Year	Personal Income
1960	411.5
1970	830.8
1980	2307.9
1990	4878.6
2000	8429.7
2005	10,239.2

(Source: Bureau of Economic Analysis, U.S. Department of Commerce)

a. These data can be modeled by a power function. Write the equation of this function, with x as the number of years past 1950.

b. If this model is accurate, what will be the total U.S. personal income in 2010?

c. Find the quadratic function that is the best fit for the data.

d. Which model is the best fit for the data?

32. *U.S. Population* The data in the table below give the U.S. population for selected years from 1790 to 2000.

   a. Create a quadratic function that models these data, with y equal to the population in millions and x equal to the number of years from 1700. Report your answer to 4 decimal places.

   b. During what year after 1700 does this model indicate that the population was 140.2 million?

Year	Population (millions)
1790	3.929
1870	38.558
1930	123.203
1990	248.794
2000	281.422
2005	296.410

(Source: U.S. Census Bureau)

33. *Educational Funding* Data that give the amount of federal on-budget funds for research programs at universities and related institutions appear in the table below.

Year	1965	1970	1975	1980	1985
Federal Funds ($ billions)	1.82	2.28	3.42	5.80	8.84

Year	1990	1995	1998	1999	2000
Federal Funds ($ billions)	12.61	15.68	18.48	20.24	21.02

(Source: U.S. Department of Education, National Center for Educational Statistics)

   a. Using an input equal to the number of years after 1960, find a quadratic function that models the data.

   b. Using an input equal to the number of years after 1965, find a quadratic function that models the data.

   c. Use the model in part (a) to find when (between 1965 and 2000) federal funds for research first exceeded $10 billion.

   d. If you had used the model in part (b) instead of the one in part (a) to answer part (c), how would your results have differed?

34. *Medicare Trust Fund Balance* The year 1994 marked the 30th anniversary of Medicare. A 1994 pamphlet by Representative Lindsey O. Graham gave projections for the Medicare Trust Fund Balance as shown in the table below.

   a. Representative Graham stated, "The fact of the matter is that Medicare is going broke." Use the data in the table to find when he predicted that this would happen.

   b. Find a quadratic model to fit the data, with x equal to the number of years after 1990. When does this model predict that the Medicare Trust Fund balance will be 0?

   c. The pamphlet reported the Medicare trustees as saying "the present financing schedule for the hospital insurance program is sufficient to ensure the payment of benefits only over the next 7 years." Does the function in part (b) confirm or refute the value 7 in this statement?

   d. Find and interpret the vertex of the quadratic function in part (b).

Year	Medicare Trust Fund Balance ($ billions)	Year	Medicare Trust Fund Balance ($ billions)
1993	128	1999	98
1994	133	2000	72
1995	136	2001	37
1996	135	2002	−7
1997	129	2003	−61
1998	117	2004	−126

35. *Travel and Tourism Spending* The global spending on travel and tourism (in billions of dollars) for the years 1991–2005 is given in the table on the next page.

   a. Write the equation of a power function that models the data, letting your input represent the number of years after 1990.

   b. Use the model to estimate the global spending for 2010.

   c. When did the global spending reach $300 billion, according to this model?

Year	Spending ($ billions)	Year	Spending ($ billions)
1991	278	1999	455
1992	317	2000	483
1993	323	2001	472
1994	356	2002	487
1995	413	2003	533
1996	439	2004	633
1997	443	2005	682
1998	445		

(Source: *World Almanac*)

36. *Insurance Rates* The following table gives the monthly insurance rates for a $100,000 life insurance policy for smokers 35–50 years of age.

 a. Create a scatter plot for the data.

 b. Does it appear that a quadratic function can be used to model the data? If so, find the best-fitting quadratic model.

 c. Find the power model that is the best fit for the data.

 d. Compare the two models by graphing each model on the same axes with the data points. Which model appears to be the better fit?

Age (yr)	Monthly Insurance Rate ($)	Age (yr)	Monthly Insurance Rate ($)
35	17.32	43	23.71
36	17.67	44	25.11
37	18.02	45	26.60
38	18.46	46	28.00
39	19.07	47	29.40
40	19.95	48	30.80
41	21.00	49	32.55
42	22.22	50	34.47

(Source: American General Life Insurance Company)

37. *U.S. Gross Domestic Product* The table gives the U.S. gross domestic product (in billions of dollars) for selected years from 1940 through 2005.

 a. Find the best-fitting quadratic model for the data, with x equal to the number of years from 1900.

 b. Find the power model that is the best fit for the data, with x equal to the number of years from 1900.

 c. Compare the two models by graphing each model on the same axes with the data points. Which model appears to be the better fit?

Year	Domestic Gross Product	Year	Domestic Gross Product
1940	837	1985	4,207
1945	1,559	1990	4,853
1950	1,328	1995	5,439
1955	1,700	2000	9,817
1960	1,934	2001	10,128
1965	2,373	2002	10,470
1970	2,847	2003	10,976
1975	3,173	2004	11,713
1980	3,746	2005	12,456

(Source: U.S. Bureau of Economic Analysis)

38. *Abortions* The total numbers of abortions, in thousands, in the United States for selected years are given in the table below.

 a. Use the data to find a quadratic function that models the number of abortions as a function of years from 1970.

 b. Does the function reported in part (a) yield a maximum or minimum? When is this value obtained, and what does the value represent?

Year	Number of Abortions (thousands)	Year	Number of Abortions (thousands)
1970	193.4	1995	1210.8
1975	854.9	1998	884.3
1980	1297.6	2000	857.5
1985	1328.6	2002	854.1
1990	1429.2		

(Source: Centers for Disease Control)

39. *Banks* The table on the next page gives the number of banks in the United States for selected years from 1935 to 2005.

a. Create a scatterplot of the data, with x equal to the number of years after 1900.

b. Find a quadratic function that models the data.

c. Use the model to estimate the number of banks in 1990.

d. In what year does the model indicate that the number of banks is a maximum?

U.S. Banks

Year	Number of Banks
1935	15,295
1940	15,772
1950	16,500
1960	17,549
1970	18,205
1980	18,763
1990	15,158
2000	9905
2005	8832

40. *Box-Office Revenues* The data in the table below give the box-office revenues, in billions of dollars, for movies released in selected years between 1980 and 1998.

Year	Revenue ($ billions)	Year	Revenue ($ billions)
1980	2.75	1994	5.39
1985	3.75	1995	5.49
1990	5.02	1996	5.91
1991	4.80	1997	6.37
1992	4.87	1998	6.95
1993	5.15		

(Source: Index of Leading Cultural Indicators)

a. Find the power function that best fits the revenues as a function of the number of years from 1970.

b. What does the unrounded model estimate as the revenue in 2005?

c. Discuss the reliability of this estimation.

---

chapter 3

Summary

In this chapter, we discussed in depth quadratic functions, including realistic applications that involve the vertex and x-intercepts of parabolas and solving quadratic equations. We then studied piecewise-defined functions, power functions, and other nonlinear functions. Real-world data are provided throughout the chapter, and we learned how to fit power and quadratic functions to some of these data.

Key Concepts and Formulas

3.1 Quadratic Functions; Parabolas

Quadratic function	Also called a *second-degree polynomial function*, this function can be written in the form $f(x) = ax^2 + bx + c$ where $a \neq 0$.
Parabola	A parabola is the graph of a quadratic function.

Vertex	The turning point on a parabola is the vertex.
Maximum point	If a quadratic function has $a < 0$, the vertex of the parabola is the maximum point, and the parabola opens downward.
Minimum point	If a quadratic function has $a > 0$, the vertex of the parabola is the minimum point, and the parabola opens upward.
Axis of symmetry	The axis of symmetry is the vertical line through the vertex of the parabola.
Forms of quadratic functions	$y = a(x - h)^2 + k$ $y = ax^2 + bx + c$ Vertex at (h, k) x-coordinate of vertex at $x = -\dfrac{b}{2a}$ Axis of symmetry is the line $x = h$. Axis of symmetry is the line $x = -\dfrac{b}{2a}$ Parabola opens up if a is positive and down if a is negative. Parabola opens up if a is positive and down if a is negative.

3.2 Solving Quadratic Equations

Solving quadratic equations	An equation that can be written in the form $ax^2 + bx + c = 0$, $a \neq 0$, is called a *quadratic equation*.
Zero product property	For real numbers a and b, the product $ab = 0$ if and only if either $a = 0$ or $b = 0$ or both a and b are zero.
Solving by factoring	To solve a quadratic equation by factoring, write the equation in a form with 0 on one side. Then factor the nonzero side of the equation, if possible, and use the zero product property to convert the equation into two linear equations that are easily solved.
Solving and checking graphically	Solutions or decimal approximations of solutions to quadratic equations can be found by using TRACE, ZERO, or INTERSECT with a graphing utility.

Factor Theorem	The factorization of a polynomial function $f(x)$ and the real solutions to $f(x) = 0$ can be verified by graphing $y = f(x)$ and observing where the graph crosses the x-axis. If $x = a$ is a solution to $f(x) = 0$, then $x - a$ is a factor of f.
Solving using the square root method	When a quadratic equation has the simplified form $x^2 = C$, the solutions are $x = \pm\sqrt{C}$. This method can also be used to solve equations of the form $(ax + b)^2 = C$.
Completing the square	A quadratic equation can be solved by converting one side to a perfect binomial square and taking the square root of both sides.
Solving using the quadratic formula	The solutions of the quadratic equation $ax^2 + bx + c = 0$, with $a \neq 0$, are given by the formula $$x = \frac{-b \pm \sqrt{b^2 - 4ac}}{2a}$$
Solutions, zeros, x-intercepts, and factors	If a is a real number, the following three statements are equivalent: • a is a real solution to the equation $f(x) = 0$. • a is a real zero of the function $f(x)$. • a is an x-intercept of the graph of $y = f(x)$.

3.3 Piecewise-Defined Functions and Power Functions

Piecewise-defined functions	This is a function that is created by combining two or more functions. Piecewise-defined functions can be graphed by graphing their pieces on the same axes. A familiar example of a piecewise-defined function is the *absolute value function*.
Absolute value function	$$|x| = \begin{cases} x & \text{if } x \geq 0 \\ -x & \text{if } x < 0 \end{cases}$$
Absolute value equations	The solution of the absolute value equation $|x| = a$ is $x = a$ or $x = -a$ when $a \geq 0$. There is no solution to $|x| = a$ if $a < 0$.
Power functions	A power function is a function of the form $y = ax^b$, where a and b are real numbers, $b \neq 0$.
Cubing function	A special power function, $y = x^3$, is called the cubing function.
Root functions	A root function is a function of the form $y = ax^{1/n}$, or $y = a\sqrt[n]{x}$, where n is an integer, $n \geq 2$.
Reciprocal function	This function is formed by the quotient of a constant function and the identity function, $f(x) = \dfrac{1}{x}$
Direct variation as the nth power	If a quantity y varies directly as the nth power of x, then y is directly proportional to the nth power of x and $$y = kx^n$$ The number k is called the constant of variation or the constant of proportionality.

3.4 Quadratic and Power Models

Modeling a quadratic function from three points on its graph	If we know three (or more) points that fit exactly on a parabola, we can find the quadratic function whose graph is the parabola.
Quadratic modeling	The use of graphing utilities permits us to fit quadratic functions to data by using technology.
Second differences	If the second differences of data are constant for equally spaced inputs, a quadratic function is the exact fit for the data.
Power modeling	The use of graphing utilities permits us to fit power functions to nonlinear data by using technology.

chapter

3

Skills Check

In Exercises 1–8, (a) give the coordinates of the vertex of the graph of each quadratic function and (b) graph each function on a window that includes the vertex and all intercepts.

1. $y = (x - 5)^2 + 3$

2. $y = (x + 7)^2 - 2$

3. $y = 3x^2 - 6x - 24$

4. $y = 2x^2 + 8x - 10$

5. $y = -x^2 + 30x - 145$

6. $y = -2x^2 + 120x - 2200$

7. $y = x^2 - 0.1x - 59.998$

8. $y = x^2 + 0.4x - 99.96$

In Exercises 9 and 10, use factoring to solve the equations.

9. $x^2 - 5x + 4 = 0$

10. $6x^2 + x - 2 = 0$

Use a graphing utility as an aid in factoring to solve the equations in Exercises 11 and 12.

11. $5x^2 - x - 4 = 0$

12. $3x^2 + 4x - 4 = 0$

In Exercises 13 and 14, use the quadratic formula to solve the equations.

13. $x^2 - 4x + 3 = 0$

14. $4x^2 + 4x - 3 = 0$

15. a. Use graphical and algebraic methods to find the x-intercepts of the graph of $f(x) = 3x^2 - 6x - 24$.

  b. Find the solutions to $f(x) = 0$ if $f(x) = 3x^2 - 6x - 24$.

16. a. Use graphical and algebraic methods to find the x-intercepts of graph of $f(x) = 2x^2 + 8x - 10$.

  b. Find the solutions to $f(x) = 0$ if $f(x) = 2x^2 + 8x - 10$.

In Exercises 17 and 18, use the square-root method to solve the equations.

17. $5x^2 - 20 = 0$

18. $(x - 4)^2 = 25$

In Exercises 19–22, find the exact solutions to the equations in the complex numbers.

19. $z^2 - 4z + 6 = 0$

20. $w^2 - 4w + 5 = 0$

21. $4x^2 - 5x + 3 = 0$

22. $4x^2 + 2x + 1 = 0$

In Exercises 23–30, graph each function.

23. $f(x) = \begin{cases} 3x - 2 & \text{if } x < -1 \\ 4 - x^2 & \text{if } x \geq -1 \end{cases}$

24. $f(x) = \begin{cases} 4 - x & \text{if } x \le 3 \\ x^2 - 5 & \text{if } x > 3 \end{cases}$

25. $f(x) = 2x^3$ **26.** $f(x) = x^{3/2}$

27. $f(x) = \sqrt{x - 4}$ **28.** $f(x) = \dfrac{1}{x} - 2$

29. $y = x^{4/5}$ **30.** $y = \sqrt[3]{x + 2}$

31. Determine if the function $y = -3x^2$ is increasing or decreasing

    a. For $x < 0$.   **b.** For $x > 0$.

32. For each of the functions, determine if the function is concave up or concave down.

    a. $y = x^{3/2}$ **b.** $y = x^{1/2}$

33. Solve $|3x - 6| = 24$. **34.** Solve $|2x + 3| = 13$.

35. Find a power function that models the data below.

x	1	2	3	4	5	6
y	4	9	11	21	32	45

36. Find a quadratic function that models the data below.

x	1	3	4	6	8
y	2	8	15	37	63

37. Suppose that q varies directly as the 3/2 power of p and that $q = 16$ when $p = 4$. Find q when $p = 16$.

38. If $f(x) = \begin{cases} 3x - 2 & \text{if } -8 \le x < 0 \\ x^2 - 4 & \text{if } 0 \le x < 3 \\ -5 & \text{if } x \ge 3 \end{cases}$, find

$f(-8), f(0),$ and $f(4)$.

---

chapter

3

Review

39. *Maximizing Profit* The monthly profit from producing and selling x units of a product is given by the function $P(x) = -0.01x^2 + 62x - 12{,}000$.

    a. Producing and selling how many units will result in the maximum profit for this product?

    b. What is the maximum possible profit for the product?

40. *Profit* The revenue from sales of x units of a product is given by $R(x) = 200x - 0.01x^2$, and the cost of producing and selling the product can be described by $C(x) = 38x + 0.01x^2 + 16{,}000$.

    a. Producing and selling how many units will give maximum profit?

    b. What is the maximum possible profit for the product?

41. *Height of a Ball* If a ball is thrown into the air at 64 feet per second from a height of 192 feet, its height (in feet) is given by $S = 192 + 64t - 16t^2$, where t is in seconds.

    a. In how many seconds will the ball reach its maximum height?

    b. What is the maximum possible height for the ball?

42. *Height of a Ball* If a ball is thrown into the air at 29.4 meters per second from a height of 60 meters, its height (in meters) is given by $S = 60 + 29.4t - 9.8t^2$, where t is in seconds.

    a. In how many seconds will the ball reach its maximum height?

    b. What is the maximum possible height for the ball?

43. *Visas* The number of skilled workers' visas issued in the United States can be modeled by $y = -1.48x^2 + 38.901x - 118.429$, where x is the number of years after 1990 and y is the number of visas in thousands.

    a. Use numerical methods to determine the year when the number of visas issued was a maximum.

    b. What was the maximum number of visas issued?

c. During what year after 2000 does the model indicate that the number of visas issued will be 100 thousand?
(Source: Department of Homeland Security)

44. *Break Even* The profit for a product is given by $P = -3600 + 150x - x^2$, where x is the number of units produced and sold. How many units will give break even (that is, return a profit of 0)?

45. *Falling Ball* If a ball is dropped from the top of a 400-foot-high building, its height S in feet is given by $S = 400 - 16t^2$, where t is in seconds. In how many seconds will it hit the ground?

46. *Profit* The profit from producing and selling x units of a product is given by the function $P(x) = -0.3x^2 + 1230x - 120{,}000$ dollars. Producing and selling how many units will result in a profit of $324{,}000 for this product?

47. *Federal Support for Education* The federal on-budget funds for all educational agencies between 1965 and 2000 can be modeled by the function

$$f(t) = \begin{cases} 0.011t^3 - 0.696t^2 + 14.380t - 29.370 \\ \qquad\qquad\text{when } 5 \le t < 35 \\ 0.931t^2 - 67.374t + 1295.785 \\ \qquad\qquad\text{when } 35 \le t \le 40 \end{cases}$$

where t is the number of years after 1960 and $f(t)$ is in millions of constant 2000 dollars.

a. Graph the function f for $5 \le t \le 40$.

b. What was the amount of funding given to educational agencies in 1989?

c. How much federal funding was given to educational agencies in 1997?
(Source: U.S. Department of Education, National Center for Educational Statistics)

48. The number of deaths from all-terrain vehicles for selected years from 1986 to 2004 can be modeled by $y = 2.554x^2 - 34.252x + 324.794$, where x is the number of years after 1985. During what year after 2000 does this model indicate that the number of deaths will be 596?
(Source: Consumer Product Safety Commission)

49. *Firefighters* The number of firefighters killed while on duty has decreased and can be modeled by $y = 269.34x^{-0.29}$, where x is the number of years after 1970.

a. What does this model predict the number of deaths to be in 2010?

b. Graph the function on a window that makes sense for the application.

c. Use the graph to determine in what year the model indicates that the number of deaths is 115.

50. *Indiana Population* The population of Indiana (in thousands) between 1980 and 1997 can be described by

$$P(x) = \begin{cases} 2.320x^2 - 389x + 21762 & \text{where } 80 \le x \le 90 \\ 46.06x + 1462.133 & \text{where } x > 90 \end{cases}$$

and x is the number of years after 1900.

a. What was the Indiana population in 1990?

b. Use P to predict when the population of Indiana reached 6 million people. What assumptions are made if this prediction is to be considered valid?
(Source: U.S. Census Bureau 2007)

51. *Internet Usage* The worldwide Internet usage from 1997 through 2007 is shown in the table below.

a. Find the quadratic function that models this data, with x equal to the number of years after 1990.

b. Graph the data and the function that models the data.

c. Does this model appear to be a good fit for the data?

d. Use the result of part (a) to estimate when the Internet usage is 2 billion.

Year	Internet Usage (millions)	Year	Internet Usage (millions)
1997	75	2001	340
1998	105	2002	580
1999	175	2005	1018
2000	255	2007	1215

(Source: Infoplane.com)

52. *Personal Income* The income received by persons from all sources minus their personal contributions for Social Security insurance is called *personal income*. The table on the next page lists the personal

income, in billions of dollars, of persons living in the United States for the indicated years.

Year	Personal Income ($ billions)
1980	2307.9
1990	4878.6
2000	8429.7
2002	8881.9
2003	9163.6
2004	9731.4
2005	10,239.2

(Source: *Statistical Abstract of the United States*)

a. Using an input equal to the number of years after 1980, find a quadratic function that models these data.

b. Use your unrounded model to estimate when before 2030 the personal income was $5500 billion.

c. When does the model estimate that the personal income will be double its actual 1990 value?

53. *Resident Population* The U.S. resident population from 15 to 19 years of age is given for selected years in the table below. Write the quadratic function (to three decimal places) that models the 15- to 19-year-old population as a function of the years from 1980. Include the description of this population and its units of measure.

Year	Age 15–19 Population (thousands)	Year	Age 15–19 Population (thousands)
1980	21,168	2000	20,262
1985	18,727	2002	20,366
1990	17,890	2004	20,724
1995	18,152	2005	21,039
1997	19,068		

(Source: U.S. Bureau of the Census, *Current Population Reports*)

54. *Insurance Premiums* The following table gives the monthly premiums required for a $250,000 term-life insurance policy on a 35-year-old female nonsmoker for different guaranteed term periods.

a. Find a quadratic function that models the monthly premium as a function of the length of term for a 35-year-old female nonsmoking policyholder. Report your answer to 5 decimal places.

b. Assuming that the domain of the function contains integer values between 10 years and 30 years, what term in years could a 35-year-old nonsmoking female purchase for $130 a month?

Term Period (years)	Monthly Premium for 35-Year-Old Female (dollars)
10	103
15	125
20	145
25	183
30	205

(Source: Quotesmith.com)

55. *Computer Usage* The number of students per computer in U.S. public schools for selected years from the 1983–1984 through the 2003–2004 school years is shown in the table.

School Year	Students per Computer
1983–1984	125
1985–1986	50
1987–1988	32
1989–1990	22
1991–1992	18
1993–1994	14
1995–1996	10
1997–1998	6.1
2003–2004	4.4

(Source: U.S. Census Bureau)

a. Align the input data as the number of years after the beginning of the 1980–1981 school year and find a power function f to fit the data. Graph the data and the function $y = f(t)$ that models the data. Does this model appear to be a good fit?

b. If we assume that the function in part (a) is valid for all school years after the 2003–2004 one, will the function in part (a) ever indicate that there will be one student per computer? Explain.

56. *Hawaii Population* The data in the table below give the 1990 through 2006 population of Hawaii.

Year	Population (thousands)	Year	Population (thousands)
1990	1108	1995	1179
1991	1131	1996	1183
1992	1150	1997	1187
1993	1160	2000	1212
1994	1173	2006	1285

(Source: *Statistical Abstract of the United States*)

a. Align the data with $x =$ the number of years after 1985 and find a power function to fit the aligned data.

b. If the pattern indicated by the model remains valid, estimate in what year Hawaii's population will rise to 1.3 million people.

57. *Hospital Utilization* The average length of stay in noninstitutional, short-stay hospitals (exclusive of federal hospitals and not counting newborn infants) for selected years is given in the table below. Align the years to be the number of years after 1980.

Year	Average Stay (days)	Year	Average Stay (days)
1980	7.3	1994	5.7
1985	6.5	1995	5.4
1990	6.4	1996	5.2
1991	6.4	2000	4.9
1992	6.2	2004	4.8
1993	6.0		

a. Fit a quadratic model to the 1980–1991 aligned data.

b. Fit a power model to the 1992–2004 aligned data.

c. Combine the results of parts (a) and (b) to form a piecewise model for the 1980–2004 data. The function should be defined for all years between 1980 and 2004.

d. Use the piecewise-defined model to answer the following:

   i. Find and interpret the output in 1987.

   ii. When was the average stay 6.1 days?

   iii. What was the average stay in 2002?

58. *Consumer Price Index* Prices as measured by the U.S. Consumer Price Index (CPI) have risen steadily since World War II. The data in the table gives the CPI for selected years between 1985 and 2005. The CPI in this table has 1967 as a reference year; that is, what cost $100 in 1967 cost about $322.20 in 1985 and $585.00 in 2005.

Year	CPI ($)
1985	322.2
1990	391.4
1995	456.5
2000	515.8
2005	585.0

(Source: Bureau of Labor Statistics)

Align the input data as the number of years after 1985 and report any models to three decimal places.

a. Find a linear model for the data. Discuss the fit to the data.

b. Find a quadratic model for the data. How good is the fit?

c. Use the models to estimate when the CPI is 696.

group activities / extended applications

1. Modeling

Graphs displaying linear and nonlinear growth are frequently displayed in periodicals such as *Newsweek* and *Time*, in newspapers such as *USA Today* and the *Wall Street Journal*, and on numerous Web sites on the Internet. Tables of data can also be found in these sources, especially in federal and state government Web sites, such as www.census.gov.

Your mission is to find a company sales, stock price, biological growth, or sociological trend over a period of years that is nonlinear, to determine which type of nonlinear function is the best fit for the data, and to find a nonlinear function that is a model for this data.

A quadratic model will be a good fit for the data:

1. If the data are presented as a graph that resembles a parabola or part of a parabola.
2. If the data are presented in a table and the plot of the data points lie near some parabola or part of a parabola.
3. If the data are presented in a table and the second differences of the outputs are nearly constant for equally spaced inputs.

A power model may be a good fit if the shape of the graph or graph of the data points resembles part of a parabola, but a parabola is not a good fit for the data.

After creating the model, you should test the goodness of fit of the model to the data and discuss uses that you could make of the model.

Your completed project should include

a. A complete citation of the source of the data that you are using.
b. An original copy or photocopy of the data being used.
c. A scatter plot of the data.
d. The equation that you have created.
e. A graph containing the scatter plot and the modeled equation.
f. A statement about how the model could be used to make estimations or predictions about the trend you are observing.

Some helpful hints:

1. If you decide to use a relation determined by a graph that you have found, read the graph very carefully to determine the data points or contact the source of the data to get the data from which the graph was drawn.
2. Align the independent variable by letting x represent the number of years from some convenient year and then enter the data into a graphing utility and create a scatter plot.
3. Use your graphing utility to create the equation of the function that is the best fit for the data. Graph this equation and the data points on the same axes to see if the equation is reasonable.

2. Gender and Earnings

The table below gives the number of male and female employees in the United States and the median earnings (in thousands of 2006 dollars) of full-time, year-round workers for males and females for selected years from 1970 to 2006. To investigate the equity of employment for females:

1. Find the quadratic function $y = f(x)$ that models the number of employed males as a function of the number of years from 1970.
2. Find the linear function $y = g(x)$ that models the number of employed females as a function of the number of years from 1970.
3. Graph $y = f(x)$ and $y = g(x)$ to determine if the number of female workers will equal the number of male workers. If so, estimate the year when the number of female workers will equal the number of male workers.

4. Calculate the female-to-male earnings ratio for each of the given years, and enter the ratios in the table.
5. Use each of the given years and the corresponding female-to-male earnings ratio to create a quadratic function that models the female-to-male earnings ratio as a function of number of years after 1970.
6. Write an equation to find an estimate of the year when the median female earnings reaches the median male earnings, and solve the equation algebraically.
7. Confirm your solution graphically.
8. One possible reason that it will take a long time for median female earnings to reach the median male earnings is gender discrimination. What other reasons are possible?

Median Earnings of Full-Time, Year-Round Workers by Gender

	Males		Females		Female-to-Male Earnings Ratio
Year	Number (millions)	Earnings ($ thousands)	Number (millions)	Earnings ($ thousands)	
1970	36.132	40.656	15.476	24.137	
1975	37.267	42.493	17.452	24.994	
1980	41.881	43.360	22.859	26.085	
1985	44.943	43.236	27.383	27.920	
1990	49.171	41.391	31.682	29.643	
1995	52.667	41.375	35.482	29.554	
2000	59.602	43.615	41.719	32.153	
2005	61.500	42.743	43.351	32.903	
2006	63.055	42.261	44.663	32.515	

(Source: U.S. Census Bureau)

Additional Topics with Functions

A profit function can be formed by algebraically combining cost and revenue functions. An average cost function can be constructed by finding the quotient of two functions. Other new functions can be created using function composition and inverse functions. In this chapter we study additional topics with functions to create functions and solve applied problems.

Algebra Toolbox

key concepts

- Symmetry
- Relations with symmetric graphs
- One-to-one functions

In this Toolbox, we discuss graphs that are symmetric about the y-axis and graphs that are symmetric about the x-axis. We discuss graphing relations whose graphs are symmetric about the x-axis. We also discuss one-to-one functions.

Symmetry About the y-axis

In Chapter 3, we observed that any parabola that is a graph of a quadratic function is symmetric about a vertical line called the axis of symmetry. This means that the two halves of the parabola are reflections of each other (Figure 4.1(a)). In particular, the graph of $y = ax^2$, for any real number a, is symmetric about the y-axis (Figure 4.1(b)). We see that for any point (m, n) on the graph of this function, there will also be a point $(-m, n)$ on the graph. In this chapter, we will see how to use the equation of a function to determine if its graph is symmetric about the y-axis.

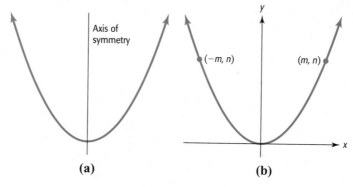

Figure 4.1

Relations with Symmetric Graphs

The graphs of some equations can be symmetric with respect to the x-axis, as shown in Figure 4.2.

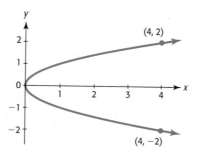

Figure 4.2

Notice that the graph in Figure 4.2 does not pass the vertical line test and thus is not the graph of a function. We can graph this relation and many others by constructing tables of values and plotting points. For example, Table 4.1 gives points on the graph of $y^2 = x$. These points can be found by substituting values for y and solving for x.

Table 4.1 $y^2 = x$

y	-2	-1	0	1	2
x	4	1	0	1	4

The equation could also be graphed by solving it for y, getting $y = \pm\sqrt{x}$, and then substituting values for x to get points on the graph. We can also write this equation as $y = \sqrt{x}$ or $y = -\sqrt{x}$. Because each of these two equations is actually a function, we can enter them individually in a graphing calculator. Graphing both of them at the same time gives the graph (Figure 4.3). This graph is called a **horizontal parabola**.

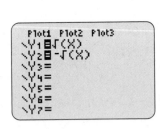
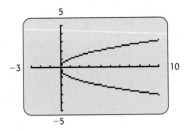

Figure 4.3

---

example 1

Symmetry

Determine visually whether each of the graphs in Figure 4.4 is symmetric about the x-axis, y-axis, or neither.

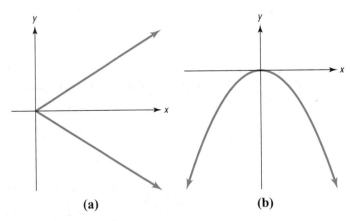

(a)   (b)

Figure 4.4

Solution

a. The two halves of the graph in Figure 4.4(a) appear to be reflections of each other about the x-axis, so we conclude that the graph is symmetric about the x-axis.

b. Because the two halves of the graph in Figure 4.4(b) appear to be reflections of each other about the y-axis, we conclude that the graph is symmetric about the y-axis.

| example 2 | ## Symmetry |

Based on the ordered pairs shown in the pair of tables in a graphing calculator, determine whether the graph of the function defined by these points is symmetric about the y-axis.

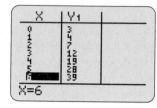 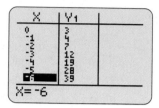

Solution

The pair of tables illustrates that whenever a point (x, y) is on the graph, the point $(-x, y)$ is also on the graph. Thus we conclude that the graph of this function is symmetric with respect to the y-axis.   ∎

One-to-One Functions

In this chapter we will discuss inverse functions. A function can have another function as its inverse if it is a one-to one function.

We define a one-to-one function as follows.

> ### One-to-One Function
>
> A function f is a one-to-one function if each output of the function corresponds to exactly one input in the domain of the function. This means there is a one-to-one correspondence between the elements of the domain and the elements of the range.

This statement means that for a one-to-one function f, $f(a) \neq f(b)$ if $a \neq b$.

| example 3 | ## One-to-One Functions |

Determine if each of the functions is a one-to-one function.

a. $f(x) = 3x^4$ **b.** $f(x) = x^3 - 1$

Solution

a. Clearly $a = -2$ and $b = 2$ are different inputs, but they both give the same output for $f(x) = 3x^4$.

$$f(-2) = 3(-2)^4 = 48 \quad \text{and} \quad f(2) = 3(2)^4 = 48$$

Thus $y = 3x^4$ is not a one-to-one function.

b. Suppose that $a \neq b$. Then $a^3 \neq b^3$ and $a^3 - 1 \neq b^3 - 1$, so if $f(x) = x^3 - 1$, $f(a) \neq f(b)$. This satisfies the condition $f(a) \neq f(b)$ if $a \neq b$, so the function $f(x) = x^3 - 1$ is one-to-one.   ∎

Recall that no vertical line can intersect the graph of a function in more than one point. The definition of a one-to-one function means that if a function is one-to-one, a horizontal line will intersect its graph in at most one point.

> ### Horizontal Line Test
>
> A function is one-to-one if no horizontal line can intersect the graph of the function in more than one point.

---

example 4

Horizontal Line Test

Determine if the functions (a) $y = 3x^4$ and (b) $y = x^3 - 1$ are one-to-one by using the horizontal line test.

Solution

We can see that $y = 3x^4$ is not a one-to-one function because we can observe that the graph of this function does not pass the horizontal line test (Figure 4.5(a)). Note that when $x = 2$, $y = 48$, and when $x = -2$, $y = 48$.

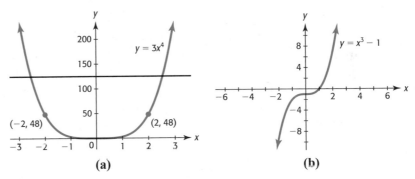

(a)   **(b)**

Figure 4.5

The graph in Figure 4.5(b) and the horizontal line test also indicate that the function $y = x^3 - 1$ is one-to-one. That is, no horizontal line will intersect this graph in more than one point. ∎

---

toolbox exercises

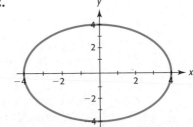

In Exercises 1–4, determine visually whether each of the graphs is symmetric about the x-axis, y-axis, or neither.

1.

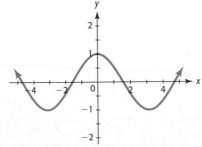

2.

3.

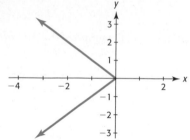

4.

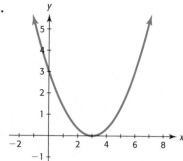

In Exercises 5 and 6, determine whether the graph of the equation in Y_1 is symmetric about the y-axis, based on the ordered pairs shown in each pair of tables.

5.

X	Y₁
0	0
1	2
2	8
3	18
4	32
5	50

X=0

X	Y₁
0	0
-1	2
-2	8
-3	18
-4	32
-5	50

X=

6.

X	Y₁
0	.5
1	2
2	
3	4.5
4	8
5	12.5

X=

X	Y₁
0	.5
-1	2
-2	
-3	4.5
-4	8
-5	12.5

X=0

In Exercises 7–12, determine if each of the functions is a one-to-one function.

7. $\{(1, 5), (2, 6), (3, 7), (4, 5)\}$

8. $\{(2, -4), (5, -8), (8, -12), (11, -16)\}$

9. $y = -3x^2$ **10.** $f(x) = \sqrt{x - 5}$

11. $f(x) = (x - 3)^3$ **12.** $f(x) = \dfrac{1}{x}$

In Exercises 13 and 14, determine whether each graph is the graph of a one-to-one function.

13.

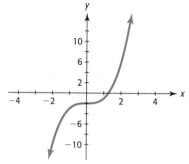

14.

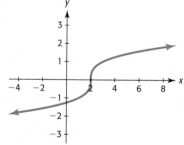

In Exercises 15–18, determine whether the function is one-to-one.

15. $y = -2x^4$ **16.** $y = -2x + 5$

17. $y = \sqrt{x + 3}$ **18.** $y = x^2 - 8$

4.1 Transformations of Graphs and Symmetry

section preview ▪ Voting

Suppose that the percent of voting-age population that voted in presidential elections from 1960 through 2004 can be modeled by the quadratic function

$$y = 0.0135x^2 - 1.0990x + 73.9727$$

where x is the number of years from 1950. To find the quadratic model that will give the percent where x is the election year and y is the percent, we can rewrite the equation in a "shifted" form. (See Example 5.)

In this section, we discuss shifting, stretching, compressing, and reflecting the graph of a function. This is useful in obtaining graphs of additional functions and also in finding windows in which to graph them. We also discuss symmetry of graphs.

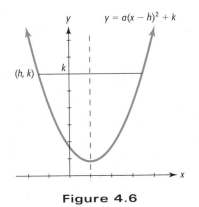

Figure 4.6

Shifts of Graphs of Functions

Consider the quadratic function in the form $y = a(x - h)^2 + k$. Substituting h for x in this equation gives $y = k$, so the graph of this function contains the point (h, k). Because the graph is symmetric about the vertical line through the vertex, there is a second point on the graph with y-coordinate k, *unless* the point (h, k) is the vertex of the parabola (Figure 4.6). To see if the second point exists or if (h, k) is the vertex, we solve

$$k = a(x - h)^2 + k$$

This gives

$$0 = a(x - h)^2$$
$$0 = (x - h)^2$$
$$0 = x - h$$
$$x = h$$

Thus, only one value of x corresponds to $y = k$, so the point (h, k) is the vertex of any parabola with an equation of the form $y = a(x - h)^2 + k$.* We can say that the vertex of $y = a(x - h)^2 + k$ has been **shifted** from $(0, 0)$ to the point (h, k), and this entire graph is in fact the graph of $y = x^2$ shifted h units horizontally and k units vertically. We will see that graphs of other functions can be shifted similarly.

In Section 3.3, we studied absolute value functions. In the following example, we investigate shifts of the absolute value function.

---

* This confirms use of the vertex form of a quadratic function in Section 3.1.

| example 1 | **Shifts of Functions** |

Graph the function $f(x) = |x|$ and $g(x) = |x| + 3$ on the same axes. What relationship do you notice between the two graphs?

Solution

Table 4.2 shows inputs for x and outputs for both $f(x) = |x|$ and $g(x) = |x| + 3$. Observe that for a given input, each output value of $g(x) = |x| + 3$ is 3 more than the corresponding output value for $f(x) = |x|$. Thus the y-coordinate of each point on the graph of $g(x) = |x| + 3$ is 3 more than the y-coordinate of the point on the graph of $f(x) = |x|$ with the same x-coordinate. We say that the graph of $g(x) = |x| + 3$ is the graph of $f(x) = |x|$ shifted up 3 units (Figure 4.7).

Table 4.2

x	$f(x) = |x|$	$g(x) = |x| + 3$
-3	3	6
-2	2	5
-1	1	4
0	0	3
1	1	4
2	2	5
3	3	6
4	4	7

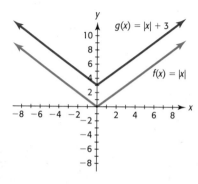

Figure 4.7

In Example 1, the equation of the shifted graph is $g(x) = |x| + 3 = f(x) + 3$. In general, we have the following.

> **Vertical Shifts of Graphs**
>
> If k is a positive real number:
>
> The graph of $g(x) = f(x) + k$ can be obtained by shifting the graph of $f(x)$ upward k units.
> The graph of $g(x) = f(x) - k$ can be obtained by shifting the graph of $f(x)$ downward k units.

| example 2 | **Shifts of Functions** |

Graph the function $f(x) = |x|$ and $g(x) = |x - 5|$ on the same axes. What relationship do you notice between the two graphs?

Solution

Table 4.3 shows inputs for x and outputs for both $f(x) = |x|$ and $g(x) = |x - 5|$. Observe that the output values for both functions are equal *if* the input value for $g(x) = |x - 5|$ is 5 more than the corresponding input value for $f(x) = |x|$. Thus the x-coordinate of each point on the graph of $g(x) = |x - 5|$ is 5 more than the x-coordinate of the point on the graph of $f(x) = |x|$ with the same y-coordinate. We say that the graph of $g(x) = |x - 5|$ is the graph of $f(x) = |x|$ shifted 5 units to the right (Figure 4.8).

Table 4.3

x	$f(x) = |x|$	x	$g(x) = |x - 5|$
-3	3	2	3
-2	2	3	2
-1	1	4	1
0	0	5	0
1	1	6	1
2	2	7	2
3	3	8	3
4	4	9	4

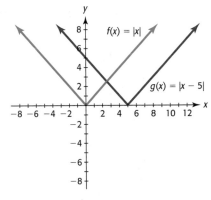

Figure 4.8

In Example 2, the equation of the shifted graph is $g(x) = |x - 5| = f(x - 5)$. In general, we have the following.

Horizontal Shifts of Graphs

If h is a positive real number:

The graph of $g(x) = f(x - h)$ can be obtained by shifting the graph of $f(x)$ to the right h units.

The graph of $g(x) = f(x + h)$ can be obtained by shifting the graph of $f(x)$ to the left h units.

---

example 3

Shifts of Functions

Graph the function $f(x) = \sqrt{x}$ and $g(x) = \sqrt{x + 3} - 4$ on the same axes. What relationship do you notice between the two graphs?

Solution

The graphs of $f(x) = \sqrt{x}$ and $g(x) = \sqrt{x + 3} - 4$ are shown in Figure 4.9. We see that the graph of $g(x) = \sqrt{x + 3} - 4$ can be obtained by shifting the graph of $f(x) = \sqrt{x}$ to the left 3 units and down 4 units. To verify this, consider a few points from each graph in Table 4.4 on the next page.

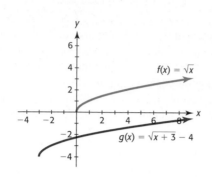

Figure 4.9

Table 4.4

$f(x) = \sqrt{x}$		$g(x) = \sqrt{x + 3} - 4$	
x	$f(x) = \sqrt{x}$	x	$g(x) = \sqrt{x + 3} - 4$
0	0	-3	-4
1	1	-2	-3
2	$\sqrt{2}$	-1	$\sqrt{2} - 4$
3	$\sqrt{3}$	0	$\sqrt{3} - 4$
4	2	1	-2

example 4 Profit

Use the graph of $y = -x^2$ and the appropriate transformation to set the window and graph the profit function

$$P = -(x - 10)^2 + 30$$

Solution

Shifting the graph of $y = -x^2$ (Figure 4.10(a)) 10 units to the right and up 30 units gives the graph of $P = -(x - 10)^2 + 30$ (Figure 4.10(b)).

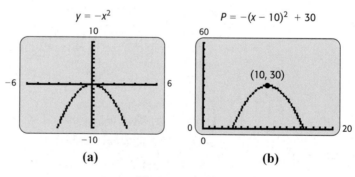

(a)　　　　**(b)**

Figure 4.10

example 5 Voter Turnout

Suppose that the percent of voting-age population that voted in presidential elections from 1960 through 2004 can be modeled by the quadratic function

$$y = 0.0135x^2 - 1.0990x + 73.9727$$

where x is the number of years from 1950. Find the quadratic model that will give the percent, where x is the election year and y is the percent.

Solution

Because y represents the percent in both models, there is no vertical shift from one function to the other. To find the quadratic model that will give the percent where x is the election year rather than the number of years from 1950, we rewrite the equation in a new form where $x = 1950$ represents the same input as $x = 0$ does in the original model. This can be accomplished by replacing x in the original model with $x - 1950$. This gives the new model

$$y = 0.0135(x - 1950)^2 - 1.0990(x - 1950) + 73.9727$$

where x is the election year and y is the percent.

   To see that these two models are equivalent, observe the inputs and outputs for selected election years in Table 4.5 and the graphs in Figure 4.11.

Table 4.5

Original Model		New Model	
Years from 1950	Percent	Election Year	Percent
14	61.2	1964	61.2
26	54.5	1976	54.5
30	53.2	1980	53.2
50	52.8	2000	52.8

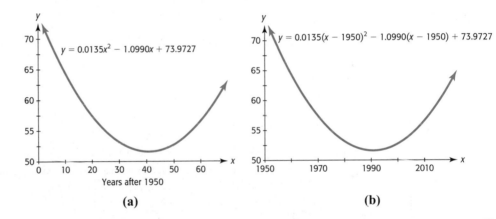

(a)   (b)

Figure 4.11

Stretching and Compressing Graphs

example 6

Stimulus-Response

One of the early results in psychology relates the magnitude of a stimulus x to the magnitude of the response y with the model $y = kx^2$, where k is an experimental constant. Compare the graphs of $y = kx^2$ for $k = 1, 2,$ and $\frac{1}{2}$.

Solution

Table 4.6 shows the values of y for $y = f(x) = x^2, y = 2f(x) = 2x^2$, and $y = \frac{1}{2}f(x) = \frac{1}{2}x^2$ for selected values of x. Observe that for these x-values, the y-values for $y = 2x^2$ are 2 times the y-values for $y = x^2$, and the points on the graph of $y = 2x^2$ have y-values that are 2 times the y-values on the graph of $y = x^2$ for equal x-values. We say that the graph of $y = 2x^2$ is a **vertical stretch** of $y = x^2$ by a factor of 2. (Compare Figure 4.12(a) and Figure 4.12 (b) on the next page.)

Observe also that for these x-values, the y-values for $y = \frac{1}{2}x^2$ are $\frac{1}{2}$ of the y-values for $y = x^2$, and the points on the graph of $y = \frac{1}{2}x^2$ have y-values that are $\frac{1}{2}$ of the y-values on the graph of $y = x^2$ for equal x-values. We say that the graph of $y = \frac{1}{2}x^2$ is a **vertical compression** of $y = x^2$ by a factor of $\frac{1}{2}$ (Figure 4.12(c)).

Table 4.6

x	0	1	2
$f(x)$	0	1	4

x	0	1	2
$2f(x)$	0	2	8

x	0	1	2
$\frac{1}{2}f(x)$	0	$\frac{1}{2}$	2

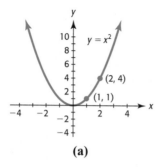 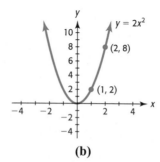 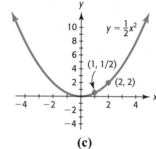

(a) (b) (c)

Figure 4.12

Stretching and Compressing Graphs

The graph of $y = af(x)$ is obtained by vertically stretching the graph of $f(x)$ by a factor of $|a|$ if $|a| > 1$, and vertically compressing the graph of $f(x)$ by a factor of $|a|$ if $0 < |a| < 1$.

Reflections of Graphs

In Figure 4.13(a), we see that the graph of $y = -x^2$ is a parabola that *opens down*. It can be obtained by reflecting the graph of $f(x) = x^2$, which *opens up*, across the x-axis. We can compare the y-coordinates of the graphs of these two functions by looking at Table 4.7 and Figure 4.13(b). Notice that for a given value of x, the y-coordinates of $y = x^2$ and $y = -x^2$ are negatives of each other.

Table 4.7

x	$y = x^2$	$y = -x^2$
-2	4	-4
-1	1	-1
0	0	0
1	1	-1
2	4	-4

(a) (b)

Figure 4.13

We can also see that the graph of $y = (-x)^3$ is a reflection of the graph of $y = x^3$ across the y-axis by looking at Table 4.8 and Figure 4.14. Notice that for a given value of y, the x-coordinates of $y = x^3$ and $y = (-x)^3$ are negatives of each other.

Table 4.8

x	$y = x^3$	x	$y = (-x)^3$
2	8	−2	8
1	1	−1	1
0	0	0	0
−1	−1	1	−1
−2	−8	2	−8

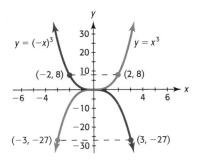

Figure 4.14

In general, we have the following:

Reflections of Graphs Across the Coordinate Axes

1. The graph of $y = -f(x)$ can be obtained by reflecting the graph of $y = f(x)$ across the x-axis.

2. The graph of $y = f(-x)$ can be obtained by reflecting the graph of $y = f(x)$ across the y-axis.

A summary of the transformations of a graph follows.

Graph Transformations

For a given function $y = f(x)$,

Vertical Shift	$y = f(x) + k$	Graph is shifted k units up if $k > 0$ and k units down if $k < 0$.
Horizontal Shift	$y = f(x - h)$	Graph is shifted h units right if $h > 0$ and h units left if $h < 0$.
Stretch/Compress	$y = a f(x)$	Graph is vertically stretched by a factor of $\|a\|$ if $\|a\| > 1$.
		Graph is compressed by a factor of $\|a\|$ if $\|a\| < 1$.
Reflection	$y = -f(x)$	Graph is reflected across the x-axis.
	$y = f(-x)$	Graph is reflected across the y-axis.

| example 7 |

Pollution

Suppose that for a certain city the cost C of obtaining drinking water that contains $p\%$ impurities (by volume) is given by

$$C = \frac{120,000}{p} - 1200$$

a. Determine the domain of this function and graph the function without concern for the context of the problem.

b. Use knowledge of the context of the problem to graph the function in a window that applies to the application.

c. What is the cost of obtaining drinking water that contains 5% impurities?

Solution

a. All values of p except $p = 0$ result in real values for the function. Thus, the domain of C is all real numbers except 0. The graph of this function is a transformation of the graph of $C = \dfrac{1}{p}$, stretched by a factor of 120,000, then shifted downward 1200 units. The viewing window should have its center near $p = 0$ (horizontally) and $C = -1200$ (vertically). We increase the vertical view to allow for the large stretching factor, using the viewing window $[-100, 100]$ by $[-20,000, 20,000]$. The graph, shown in Figure 4.15(a), has the same shape as the graph of $y = \dfrac{1}{x}$.

b. Because p represents the percent of impurities, the viewing window is set for values of p from 0 to 100. (Recall from part (a) that p cannot be 0.) The cost of reducing the impurities cannot be negative, so the vertical view is set from 0 to 20,000. The graph is shown in Figure 4.15(b).

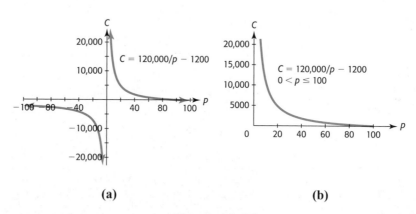

(a)   (b)

Figure 4.15

c. To determine the cost of obtaining drinking water that contains 5% impurities, we substitute 5 for p in $C = \dfrac{120,000}{p} - 1200$, giving

$$C = \frac{120,000}{5} - 1200 = 22,800$$

Thus, the cost is $22,800 to obtain drinking water that contains 5% impurities. ■

example 8

Velocity of Blood

Because of friction from the walls of an artery, the velocity of blood is greatest at the center of the artery and decreases as the distance r from the center increases. The velocity of the blood in the artery can be modeled by the function

$$v = k(R^2 - r^2)$$

where R is the radius of the artery and k is a constant that is determined by the pressure, viscosity of the blood, and the length of the artery. In the case where $k = 2$ and $R = 0.1$ centimeter, the velocity is

$$v = 2(0.01 - r^2) \text{ centimeter per second}$$

a. Graph this function and the functions $v = r^2$ and $v = -2r^2$ on the viewing window $[-0.1, 0.1]$ by $[-0.05, 0.05]$.

b. How is the function $v = 2(0.01 - r^2)$ related to the second-degree power function $v = r^2$?

Solution

a. The graph of the function $v = 2(0.01 - r^2)$ is shown in Figure 4.16(a), the graph of $v = r^2$ is shown in Figure 4.16(b), and the graph of $v = -2r^2$ is shown in Figure 4.16(c).

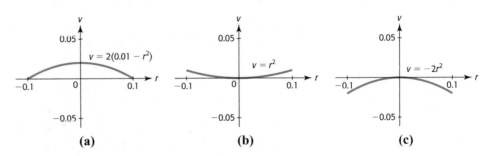

(a)   (b)   (c)

Figure 4.16

b. The function $v = 2(0.01 - r^2)$ can also be written in the form $v = -2r^2 + 0.02$, so it should be no surprise that the graph is a reflected, stretched, and shifted form of the graph of $v = r^2$. If the graph of $v = r^2$ is reflected about the r-axis, has each r^2-value doubled, and then is shifted upward 0.02 unit, the graph of $v = -2r^2 + 0.02$ results. ∎

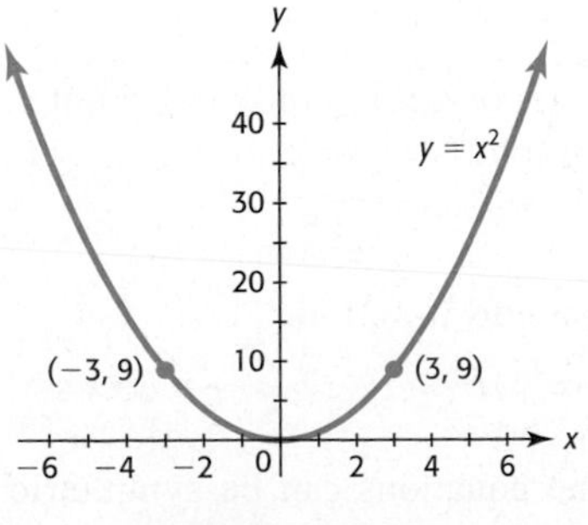

Symmetry; Even and Odd Functions

Notice that the complete graph of $y = x^2$ in Figure 4.17 is symmetric (that is, is a reflection of itself) about the y-axis. For example, the point $(3, 9)$ is on the graph, and its mirror image across the y-axis is the point $(-3, 9)$. This suggests an algebraic way to determine whether the graph of an equation is symmetric with respect to the y-axis. That is, replace x with $-x$ and simplify. If the resulting equation is equivalent to the original equation, then the graph is symmetric with respect to the y-axis.

Figure 4.17

Symmetry with Respect to the y-axis

The graph of $y = f(x)$ is symmetric with respect to the y-axis if, for every point (x, y) on the graph, the point $(-x, y)$ is also on the graph. That is,

$$f(-x) = f(x)$$

for all x in the domain of f. Such a function is called an **even function**.

All power functions with even integer exponents will have graphs that are symmetric about the y-axis. This is why all functions satisfying this condition are called even functions, although functions other than power functions exist that are even functions.

A graph of $y = x^3$ is shown in Figure 4.18(a). Notice in Figure 4.18(b) that if we draw a line through the origin and any point on the graph, it will intersect the graph at a second point, and the distances from the two points to the origin will be equal.

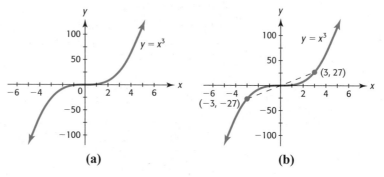

(a)   **(b)**

Figure 4.18

For example, the point $(3, 27)$ is on the graph of $y = x^3$, and the line through $(3, 27)$ and the origin intersects the graph at the point $(-3, -27)$, which is the same distance from the origin as $(3, 27)$. In general, the graph of a function is called symmetric with respect to the origin if for every point (x, y) on the graph, the point $(-x, -y)$ is also on the graph.

To determine algebraically whether the graph of an equation is symmetric about the origin, replace x with $-x$, y with $-y$, and simplify. If the resulting equation is equivalent to the original equation, then the graph is symmetric with respect to the origin.

Symmetry with Respect to the Origin

The graph of $y = f(x)$ is symmetric with respect to the origin if, for every point (x, y) on the graph, the point $(-x, -y)$ is also on the graph. That is,

$$f(-x) = -f(x)$$

for all x in the domain of f. Such a function is called an **odd function**.

As noted in the Chapter 4 Toolbox, the graphs of some equations can be symmetric with respect to the x-axis. Such graphs do not represent y as a function of x (because they do not pass the vertical line test). For example, we can graph the equation $x^2 + y^2 = 9$ by hand or with technology after solving it for y. Solving gives

$$y^2 = 9 - x^2$$
$$y = \pm\sqrt{9 - x^2}$$

Entering values for x in the equation gives points on the **circle** graphed in Figure 4.19(a) and entering the two functions in a calculator gives the graph. Note that the calculator graph will not look like a circle unless the calculator window is square. Figure 4.19(b) gives the graph with the standard window, Figure 4.19(c) gives the graph after Zoom Square,* and Figure 4.19(d) gives the graph with Zoom Decimal.*

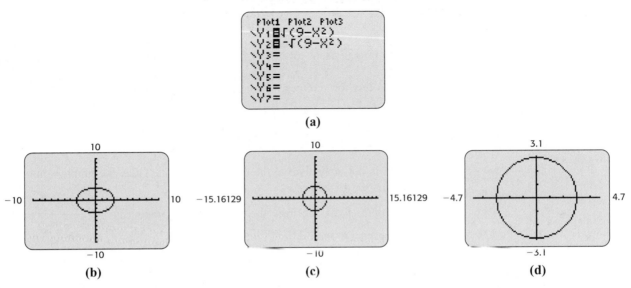

Figure 4.19

In general, if the equation contains one and only one even power of y, its graph will be symmetric about the x-axis.

To determine algebraically whether the graph of an equation is symmetric about the x-axis, replace y with $-y$ and simplify. If the resulting equation is equivalent to the original equation, then the graph is symmetric with respect to the x-axis.

Symmetry with Respect to the x-axis

The graph of an equation is symmetric with respect to the x-axis if, for every point (x, y) on the graph, the point $(x, -y)$ is also on the graph.

| example 9 | **Symmetry** |

Determine algebraically whether the graph of each equation is symmetric with respect to the x-axis, y-axis, or origin. Confirm your conclusion graphically.

a. $y = \dfrac{2x^2}{x^2 + 1}$ **b.** $y = x^3 - 3x$ **c.** $x^2 + y^2 = 16$

---

* See Appendix A, page 653.

Solution

a. To test the graph of $y = \dfrac{2x^2}{x^2 + 1}$ for symmetry with respect to the x-axis, we replace y with $-y$. Because the result,

$$-y = \frac{2x^2}{x^2 + 1}$$

is not equivalent to the original function, the graph is not symmetric with respect to the x-axis.

To test for y-axis symmetry, we replace x with $-x$:

$$y = \frac{2(-x)^2}{(-x)^2 + 1} = \frac{2x^2}{x^2 + 1}$$

which is equivalent to the original equation. Thus, the graph is symmetric to the y-axis.

To test for symmetry with respect to the origin, we replace x with $-x$ and y with $-y$:

$$-y = \frac{2(-x)^2}{(-x)^2 + 1} = \frac{2x^2}{x^2 + 1}$$

which is not equivalent to the original equation. Thus, the graph is not symmetric with respect to the origin.

Figure 4.20 confirms the fact that the graph of $y = \dfrac{2x^2}{x^2 + 1}$ is symmetric with respect to the y-axis.

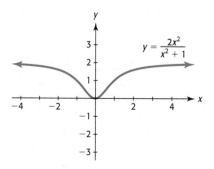

$$y = \frac{2x^2}{x^2 + 1}$$

Figure 4.20

b. To test the graph of $y = x^3 - 3x$ for symmetry with respect to the x-axis, we replace y with $-y$. Because the result

$$-y = x^3 - 3x$$

is not equivalent to the original function, the graph is not symmetric with respect to the x-axis.

To test for y-axis symmetry, we replace x with $-x$:

$$y = (-x)^3 - 3(-x) = -x^3 + 3x$$

which is not equivalent to the original equation. Thus, the graph is not symmetric to the y-axis.

To test for symmetry with respect to the origin, we replace x with $-x$ and y with $-y$:

$$-y = (-x)^3 - 3(-x) = -x^3 + 3x, \text{ or } y = x^3 - 3x$$

which is equivalent to the original equation. Thus, the graph is symmetric with respect to the origin.

Figure 4.21 confirms this.

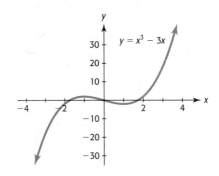

Figure 4.21

c. To test the graph of $x^2 + y^2 = 16$ for symmetry with respect to the x-axis, we replace y with $-y$:

$$x^2 + (-y)^2 = 16 \quad \text{or} \quad x^2 + y^2 = 16$$

This is equivalent to the original equation, so the graph is symmetric with respect to the x-axis.

To test for y-axis symmetry, we replace x with $-x$:

$$(-x)^2 + y^2 = 16 \quad \text{or} \quad x^2 + y^2 = 16$$

This is equivalent to the original equation, so the graph is symmetric to the y-axis.

To test for symmetry with respect to the origin, we replace x with $-x$ and y with $-y$:

$$(-x)^2 + (-y)^2 = 16 \quad \text{or} \quad x^2 + y^2 = 16$$

which is equivalent to the original equation. Thus, the graph is symmetric with respect to the origin.

Figure 4.22 confirms the fact that the graph of $x^2 + y^2 = 16$ is symmetric with respect to the x-axis, y-axis, and origin. Note that this equation does not define a function.

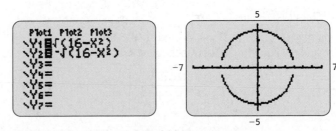

Figure 4.22

skills check

4.1

In Exercises 1–16, (a) sketch the graph of each pair of functions using a standard window, and (b) describe the transformations used to obtain the graph of the second function from the first function.

1. $y = x^3, y = x^3 + 5$ 2. $y = x^2, y = x^2 + 3$

3. $y = \sqrt{x}, y = \sqrt{x} - 4$

4. $y = x^2, y = (x + 2)^2$

5. $y = \sqrt[3]{x}, y = \sqrt[3]{x + 2} - 1$

6. $y = x^3, y = (x - 5)^3 - 3$

7. $y = |x|, y = |x - 2| + 1$

8. $y = |x|, y = |x + 3| - 4$

9. $y = x^2, y = -x^2 + 5$

10. $y = \sqrt{x}, y = -\sqrt{x - 2}$

11. $y = \dfrac{1}{x}, y = \dfrac{1}{x} - 3$

12. $y = \dfrac{1}{x}, y = \dfrac{2}{x - 1}$

13. $f(x) = x^2, g(x) = \dfrac{1}{3}x^2$

14. $f(x) = x^3, g(x) = 0.4x^3$

15. $f(x) = |x|, g(x) = 3|x|$

16. $f(x) = \sqrt{x}, g(x) = 4\sqrt{x}$

17. How is the graph of $y = (x - 2)^2 + 3$ transformed from the graph of $y = x^2$?

18. How is the graph of $y = (x + 4)^3 - 2$ transformed from the graph of $y = x^3$?

19. Suppose the graph of $y = x^{3/2}$ is shifted to the left 4 units. What is the equation that gives the new graph?

20. Suppose the graph of $y = x^{3/2}$ is shifted down 5 units and to the right 4 units. What is the equation that gives the new graph?

21. Suppose the graph of $y = x^{3/2}$ is stretched by a factor of 3 and then shifted up 5 units. What is the equation that gives the new graph?

22. Suppose the graph of $y = x^{2/3}$ is compressed by a factor of $\dfrac{1}{5}$ and then shifted right 6 units. What is the equation that gives the new graph?

In Exercises 23–26, write the equation of the function g(x) that is transformed from the given function f(x), and whose graph is shown.

23. $f(x) = x^2$

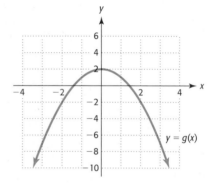

24. $f(x) = x^2$

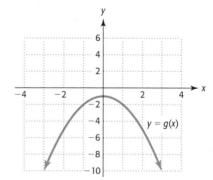

25. $f(x) = |x|$

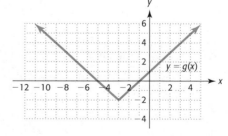

26. $f(x) = x^2$

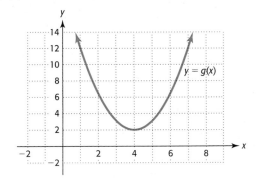

In Exercises 27–34, determine algebraically whether the graph of the given equation is symmetric with respect to the x-axis, the y-axis, and/or the origin. Confirm graphically.

27. $y = 2x^2 - 3$
28. $y = -x^2 + 4$

29. $y = x^3 - x$
30. $y = -x^3 + 5x$

31. $y = \dfrac{6}{x}$
32. $x = 3y^2$

33. $x^2 + y^2 = 25$
34. $x^2 - y^2 = 25$

In Exercises 35–40, determine whether the function is even, odd, or neither.

35. $f(x) = |x| - 5$
36. $f(x) = |x - 2|$

37. $g(x) = \sqrt{x^2 + 3}$
38. $f(x) = \dfrac{1}{2}x^3 - x$

39. $g(x) = \dfrac{5}{x}$
40. $g(x) = 4x + x^2$

Determine if each of the complete graphs in Exercises 41–42 represents functions that are even, odd, or neither.

41.

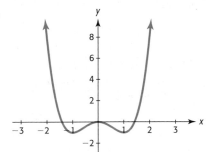

42.

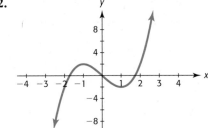

43. Graph $(y + 1)^2 = x + 4$.

44. About what line is the function $y = 3 \pm \sqrt{x - 2}$ symmetric?

exercises

4.1

45. *Ballistics* Ballistic experts are able to identify the weapon that fired a certain bullet by studying the markings on the bullet. If a test is conducted by firing the bullet into a water tank, the distance that the bullet will travel is given by $s = 27 - (3 - 10t)^3$ inches, for $0 \le t \le 0.3$, t in seconds.

    a. The graph of this function is a shifted graph of which basic function?

    b. Graph this function for $0 \le t \le 0.3$.

    c. How far does the bullet travel during this time period?

46. *Marijuana Use* The percent of persons 12 years of age and over in the United States who said that they had used marijuana at least once within the month prior to being asked during selected years is described by the function

$$M(x) = \frac{-1}{12}\left(x - \frac{789}{10}\right)^2 + \frac{15,541}{1200}$$

when $79 \le x \le 88$, where x is the number of years after 1900.
(Source: National Household Survey on Drug Abuse, Series H6 and H7)

    a. The graph of this function is a shifted graph of which basic function?

b. Find and interpret $M(79)$, $M(80)$, and $M(88)$.

c. Sketch a graph of M.

47. *Supply and Demand* The price per unit of a product is $\$p$, and the number of units of the product is denoted by q. The supply function for a product is given by $p = \dfrac{180 + q}{6}$, and the demand for the product is given by $p = \dfrac{30,000}{q} - 20$.

a. Is the supply function a linear function or a shifted reciprocal function?

b. Is the demand function a shifted linear function or a shifted reciprocal function? Describe the transformations needed to obtain the specific function from the basic function.

48. *Supply and Demand* The supply function for a commodity is given by $p = 58 + \dfrac{q}{2}$, and the demand function for this commodity is given by $p = \dfrac{2555}{q + 5}$.

a. Is the supply function a linear function or a shifted reciprocal function?

b. Is the demand function a shifted linear function or a shifted reciprocal function? Describe the transformations needed to obtain the specific function from the basic function.

49. *Mob Behavior* In a study of lynchings between 1899 and 1946, psychologist Brian Mullin concluded that the size of the mob relative to the number of victims predicted the level of brutality. The formula he developed gives y, the other-total ratio that predicts the level of self-attentiveness of people in a crowd of size x with one victim. Mullin's formula is $y = \dfrac{1}{x + 1}$, and the lower the value of y, the more likely an individual is to be influenced by "mob psychology."

a. This function is a shifted version of what basic function, and how is it shifted?

b. Graph this function for $x > 0$.

c. Will the amount of self-attentiveness of a person increase or decrease as the crowd size becomes larger?

50. *Pollution* The daily cost C (in dollars) of removing pollution from the smokestack of a coal-fired electric power plant is related to the percent of pollution p

being removed according to the equation

$$C = \frac{10,500}{100 - p}$$

a. Describe the transformations needed to obtain this function from the function $C = \dfrac{1}{p}$.

b. Graph the function for $0 \leq p < 100$.

c. What is the daily cost of removing 80% of the pollution?

51. *Population Growth* Suppose the population of a certain microorganism at time t (in minutes) is given by

$$P = -1000\left(\frac{1}{t + 10} - 1\right)$$

a. Describe the transformations needed to obtain this function from the function $f(t) = \dfrac{1}{t}$.

b. Graph this function for values of t representing 0 to 100 minutes.

52. *Mortgages* The balance owed y on a $\$50,000$ mortgage after x monthly payments is shown in the table below. The function that models this data is

$$y = 4700\sqrt{110 - x}$$

Monthly Payments	Balance Owed ($)
12	47,243
24	44,136
48	36,693
72	27,241
96	15,239
108	8074

a. Is this a shifted root function?

b. What is the domain of the function in the context of this application?

c. Describe the transformations needed to obtain the graph from the graph of $y = \sqrt{x}$.

53. *Mutual Funds* The household assets in mutual funds, measured in billions of dollars, for the years 1995–1999 can be modeled by $f(x) = 105.095x^{1.5307}$ billion dollars, where x is the number of years from 1990. Rewrite the model with x equal to the number

of years from 1995. (*Hint:* Shift the *x*-value by the difference in years.)

54. *Single-Parent Families* The percent of all families that were single-parent families between 1960 and 1998 can be modeled by $y = 1.053x^{0.888}$, with $x = 0$ in 1950. Rewrite the model with *x* equal to the number of years from 1960.

55. *Tobacco Sales* The domestic sales of tobacco (in millions of kilograms) in Canada can be described by $y = 0.084x^2 + 1.124x + 4.028$, where *x* is the number of years after 1986. Convert the equation so that *x* represents the number of years after 1990.

56. *Aircraft Accidents* The number of U.S. aircraft accidents, for all military services, can be modeled by $y = 0.184x^2 - 5.437x + 58.427$, with $x = 0$ in 1970. Rewrite the model with *x* equal to the number of years from 1960.

57. *U.S. Cellular Subscribers* The following table shows the number of U.S. cellular subscribers from 1985 to 2006.

Year	Subscribers (millions)	Year	Subscribers (millions)
1985	0.340	1996	44.043
1986	0.682	1997	55.312
1987	1.231	1998	69.209
1988	2.069	1999	86.047
1989	3.509	2000	109.478
1990	5.283	2001	128.375
1991	7.557	2002	140.767
1992	11.033	2003	158.722
1993	16.009	2004	182.140
1994	24.134	2005	207.896
1995	33.786	2006	233.000

(Source: Semiannual CTIA Wireless Survey)

The number of U.S. cellular subscribers, in millions, can be modeled by the function $S(t) = 0.000442t^{4.103}$, where *t* is the number of years after 1980.

a. Use the model to find the number of subscribers in 2005.

b. Rewrite a new model $C(t)$ with *t* equal to the number of years after 1985.

c. Using the model in part (b), what value of *t* should be used to determine the number of subscribers in 2005? Does the value of $C(t)$ agree with the answer to part (a)?

58. *U.S. Poverty Threshold* The table below shows the yearly income poverty thresholds for a single person for selected years from 1990 to 2002.

Year	Income ($)	Year	Income ($)
1990	6652	2000	8794
1995	7763	2001	9039
1998	8316	2002	9182
1999	8501		

(Source: U.S. Census Bureau)

The poverty threshold income for a single person can be modeled by the function $P(t) = 850.36t^{0.686}$, where *t* is the number of years after 1970.

a. Assuming that the model is accurate after 2002, find the poverty threshold income in 2008.

b. Rewrite a new model $S(t)$ with *t* equal to the number of years after 1990.

c. Using the model in part (b), what value of *t* should be used to determine the poverty threshold income in 2008? Does the value of $S(t)$ agree with the answer to part (a)?

59. *Average Cost* The monthly average cost of producing 32-inch plasma television sets is given by $\overline{C}(x) = \dfrac{100,000}{x} + 150$ dollars, where *x* is the number of units produced.

a. Graph this function for $x > 0$.

b. Will the average cost function decrease or increase as the number of units produced increases?

c. What transformations of the graph of the reciprocal function give the graph of this function?

60. *Average Cost* The monthly average cost of producing x sofas is $\overline{C}(x) = \dfrac{50{,}000}{x} + 120$ dollars.

a. Graph this function for $x > 0$.

b. Will the average cost function decrease or increase as the number of units produced increases?

c. What transformations of the graph of the reciprocal function give the graph of this function?

61. *Cost-Benefit* Suppose for a certain city the cost C of obtaining drinking water that contains $p\%$ impurities (by volume) is given by

$$C = \frac{120{,}000}{p} - 1200$$

a. What is the cost of drinking water that is 100% impure?

b. What is the cost of drinking water that is 50% impure?

c. What transformations of the graph of the reciprocal function give the graph of this function?

62. *Pollution* The daily cost C (in dollars) of removing pollution from the smokestack of a coal-fired electric power plant is related to the percent of pollution p being removed according to the equation $100C - Cp = 10{,}500$.

a. Solve this equation for C to write the daily cost as a function of p, the percent of pollution removed.

b. What is the daily cost of removing 50% of the pollution?

c. Why would this company probably resist removing 99% of the pollution?

section

4.2

Combining Functions; Composite Functions

key concepts

- Operations with functions
  Sum
  Difference
  Product
  Quotient
- Revenue, cost, and profit
- Average cost
- Composition of functions

section preview ▪ Profit

If the daily total cost to produce x units of a product is

$$C(x) = 360 + 40x + 0.1x^2 \text{ thousand dollars}$$

and the daily revenue from the sale of x units of this product is

$$R(x) = 60x \text{ thousand dollars,}$$

we can model the profit function as $P(x) = R(x) - C(x)$, giving

$$P(x) = 60x - (360 + 40x + 0.1x^2)$$

or

$$P(x) = -0.1x^2 + 20x - 360$$

In a manner similar to the one that formed this function, we can construct new functions by performing algebraic operations with two or more functions. For example, we can build an average cost function by finding the quotient of two functions. In addition to using arithmetic operations with functions, we can create new functions using function composition.

Operations with Functions

New functions that are the sum, difference, product, and quotient of two functions are defined as follows:

Operation	Formula	Example with $f(x) = \sqrt{x}$ and $g(x) = x^3$
Sum	$(f + g)(x) = f(x) + g(x)$	$(f + g)(x) = \sqrt{x} + x^3$
Difference	$(f - g)(x) = f(x) - g(x)$	$(f - g)(x) = \sqrt{x} - x^3$
Product	$(f \cdot g)(x) = f(x) \cdot g(x)$	$(f \cdot g)(x) = \sqrt{x} \cdot x^3 = x^3 \sqrt{x}$
Quotient	$\left(\dfrac{f}{g}\right)(x) = \dfrac{f(x)}{g(x)} \quad (g(x) \neq 0)$	$\left(\dfrac{f}{g}\right)(x) = \dfrac{\sqrt{x}}{x^3} \quad (x \neq 0)$

The domain of the sum, difference, and product of f and g consists of all real numbers of the input variable for which f and g are defined. The domain of the quotient function consists of all real numbers for which f and g are defined and $g \neq 0$.

example 1

Operations with Functions

If $f(x) = x^3$ and $g(x) = x - 1$, find the following functions and give their domains:

a. $(f + g)(x)$ **b.** $(f - g)(x)$ **c.** $(f \cdot g)(x)$ **d.** $\left(\dfrac{f}{g}\right)(x)$

Solution

a. $(f + g)(x) = f(x) + g(x) = x^3 + x - 1$; all real numbers

b. $(f - g)(x) = f(x) - g(x) = x^3 - (x - 1) = x^3 - x + 1$; all real numbers

c. $(f \cdot g)(x) = f(x) \cdot g(x) = x^3(x - 1) = x^4 - x^3$; all real numbers

d. $\left(\dfrac{f}{g}\right)(x) = \dfrac{f(x)}{g(x)} = \dfrac{x^3}{x - 1}$; all real numbers except 1

Revenue, Cost, and Profit

Because a number of business applications occur frequently, functions have been developed to model them. Examples include models for total revenue, total cost, and profit.

If a company sells x units of a product for p dollars per unit, then the total revenue for this product can be modeled by the linear function

$$R(x) = px$$

For example, if the Telstar computer is sold for \$899, then the revenue for x computers can be modeled by the function

$$R(x) = 899x \text{ dollars}$$

The **total cost** of producing and selling a product involves two parts, the fixed costs and the variable costs. **Fixed costs** include such things as rent, utilities, and equipment, and they remain constant regardless of the number of units produced. **Variable costs** are those directly related to the number of units produced. Thus, the total cost (frequently called the cost) is found by using the formula

$$\text{total cost} = \text{variable costs} + \text{fixed costs}$$

For example, if the monthly cost of producing the Telstar computer is $249 per unit, with a fixed cost of $13,000, then the monthly cost of producing x computers is

$$C(x) = 249x + 13{,}000 \text{ dollars}$$

The profit that a company makes on its product is found by subtracting the total cost of production from the total revenue for the product:

$$P(x) = R(x) - C(x)$$

Thus, the monthly profit for x Telstar computers is given by

$$P(x) = 899x - (249x + 13{,}000) \quad \text{or} \quad P(x) = 650x - 13{,}000 \text{ dollars}$$

Note that each of these three quantities, total revenue, total cost, and profit, is a function of the number of units that are produced and sold.

| example 2 | **Revenue, Cost, and Profit** |

The demand for a certain electronic component is given by $p(x) = 1000 - 2x$. Producing and selling x units of this component involve a monthly fixed cost of $1999, and the cost of producing each component is $4.

a. Write the equations that model total revenue and total cost as functions of the units produced for this electronic component in a month.

b. Write the equation that models the profit as a function of the units produced for this component during a month.

c. Find the maximum possible monthly profit.

Solution

a. The revenue for the components is given by the product of $p(x) = 1000 - 2x$ and x, the number of units sold.

$$R(x) = p(x) \cdot x = (1000 - 2x)x = 1000x - 2x^2 \text{ dollars}$$

The monthly total cost is the sum of the variable cost, $4x$, and the fixed cost, 1999.

$$C(x) = 4x + 1999 \text{ dollars}$$

b. The monthly profit for the production and sale of the electronic components is the difference between the revenue and cost functions.

$$P(x) = R(x) - C(x) = (1000x - 2x^2) - (4x + 1999) = -2x^2 + 996x - 1999$$

c. The maximum monthly profit occurs where $x = \dfrac{-996}{2(-2)} = 249$. The maximum profit is $P(249) = 122{,}003$ dollars.

example 3

Peanut Production

Georgia's production of peanuts has increased moderately in the last 10 years, but profit margins have been reduced by lower prices, declining yields, and increasing costs. Table 4.9 gives the annual revenue, variable costs, and fixed costs for peanut production in Georgia for the years 1998–2002.

Table 4.9

	1998	1999	2000	2001	2002
Revenue ($)	57,250	58,395	59,563	60,754	61,969
Variable Costs ($)	28,064	28,905	29,773	30,666	31,586
Fixed Costs ($)	11,382	11,724	12,076	12,438	12,811

(Source: University of Georgia College of Agricultural and Environment Sciences Cooperative Extension Agency)

The function that models the revenue from peanut production in Georgia for the years 1998–2002 is given by

$$R(x) = 1.1797x + 47.7892 \text{ thousand dollars}$$

where x is number of years after 1990. The function that models the total cost (variable plus fixed) of peanut production in Georgia is given by

$$C(x) = 1.2377x + 29.508 \text{ thousand dollars}$$

where x is the number of years after 1990.

a. Write a function $P(x)$ that models the profit from peanut production.

b. Graph the functions R, C, and P on the same set of axes.

c. What is the slope of the graph of $P(x)$? Interpret the slope as a rate of change.

d. If the model remains accurate, what will be the profit for peanut production in 2010?

Solution

a. We obtain the profit by subtracting the total cost from the revenue:

$$P(x) = R(x) - C(x) = 1.1797x + 47.7892 - (1.2377x + 29.508)$$
$$= -0.058x + 18.2812$$

b. The graphs of $R(x)$, $C(x)$, and $P(x) = R(x) - C(x)$ are shown in Figure 4.23.

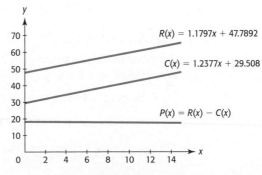

Figure 4.23

c. The slope of $P(x)$ is -0.058. This means that profit from peanut production in Georgia has decreased by 0.058 thousand dollars, or $58, per year during the years 1998–2002.

d. If the model remains accurate, the profit for peanut production in 2010 can be found by substituting $x = 20$ in $P(x)$:

$$P(20) = -0.058(20) + 18.2812 = 17.1212$$

This means that the profit will be approximately $17,121 in 2010.   ∎

The quotient of the function $C(x)$, which gives total production cost, and the identity function, which gives the number of units produced, gives a new function, the **average cost** function, $\overline{C}$.

$$\overline{C}(x) = \frac{C(x)}{x}$$

| example 4 | **Average Cost** |

a. Form the average cost function if the total cost function for the production of x units of a product is

$$C(x) = 50x + 5000$$

b. For which input values is $\overline{C}$ defined? Give a real-world explanation of this answer.

Solution

a. The average cost function is the quotient of the cost function $C(x) = 50x + 5000$, and the identity function $I(x) = x$.

$$\overline{C}(x) = \frac{50x + 5000}{x}$$

b. The average cost function is defined for all real numbers such that $x > 0$ because producing negative units is not possible and the function is undefined for $x = 0$. (This is reasonable, because if nothing is produced, it does not make sense to discuss average cost per unit of product.)   ∎

Composition of Functions

We have seen that we can convert a temperature of $x°$ Celsius to a Fahrenheit temperature by using the formula $F(x) = \frac{9}{5}x + 32$. There are also temperature scales in which $0°$ is absolute zero, which is set at $-273.15°C$. One of these, called the Kelvin temperature scale (K), uses the same degree size as Celsius until it reaches absolute zero, which is $0°K$. To convert a Celsius temperature to a Kelvin temperature, we can add 273.15 to the Celsius temperature, and to convert $x°$ Kelvin to a Celsius temperature we can use the formula

$$C(x) = x - 273.15$$

To convert from $x°$ Kelvin to a Fahrenheit temperature, we find $F(C(x))$ by substituting $C(x) = x - 273.15$ for x in $F(x) = \frac{9}{5}x + 32$, getting

$$F(C(x)) = F(x - 273.15) = \frac{9}{5}(x - 273.15) + 32 = \frac{9}{5}x - 459.67$$

That is, we can convert from the Kelvin scale to the Fahrenheit scale by using

$$F(K) = \frac{9}{5}K - 459.67$$

The process we have used to get a new function from these two functions is called **composition of functions**. The function $F(K(x))$ is called a **composite function**.

Composite Function

The composite function, f of g, is denoted by $f \circ g$ and defined by

$$(f \circ g)(x) = f(g(x))$$

The domain of $f \circ g$ is the subset of the domain of g for which $f \circ g$ is defined.
The composite function $g \circ f$ is defined by $(g \circ f)(x) = g(f(x))$.
The domain of $g \circ f$ is the subset of the domain of f for which $g \circ f$ is defined.

When computing the composite function $f \circ g$, keep in mind that the output of g becomes the input for f and that the rule for f is applied to this new input.

A composite function machine can be thought of as a machine within a machine. Figure 4.24 shows a composite "function machine" in which the input is denoted by x, the inside function is g, the outside function is f, the composite function rule is denoted by $f \circ g$, and the composite function output is symbolized by $f(g(x))$. Note that for most functions f and g, $f \circ g \neq g \circ f$.

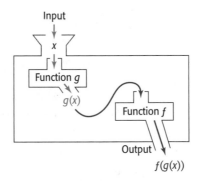

Figure 4.24

| example 5 | **Function Composition and Orange Juice** |

There are two machines in a room; one of the machines squeezes oranges to make orange juice, and the other is a canning machine that puts the orange juice into cans. Thinking of these two processes as functions, we name the squeezing function g and the canning function f.

a. Describe the composite function $(f \circ g)(\text{orange}) = f(g(\text{orange}))$.

b. Describe the composite function $(g \circ f)(\text{orange}) = g(f(\text{orange}))$.

c. Which function, $f(g(\text{orange}))$, $g(f(\text{orange}))$, neither, or both, makes sense in context?

Solution

a. The process by which this output $f(g(\text{orange}))$ is obtained is easiest to understand thinking "from the inside out." Think of the input as an orange. The orange first goes into the g machine, so it is squeezed. The output of g, liquid orange juice, is then put into the f machine, which puts it into a can. The result is a can containing orange juice (Figure 4.25).

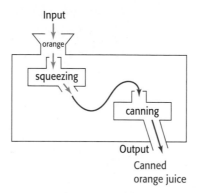

Input

orange

squeezing

canning

Output

Canned
orange juice

Figure 4.25

b. Again thinking from the inside out, the output $g(f(\text{orange}))$ is obtained by first putting an orange into the f machine, which puts it into a can. The output of f, the canned orange, is then put into the squeezing machine, g. The result is a compacted can, containing an orange (Figure 4.26).

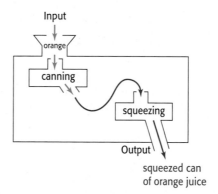

Input

orange

canning

squeezing

Output

squeezed can
of orange juice

Figure 4.26

c. As seen from the results of parts (a) and (b), the order in which the functions appear in the composite function symbol does make a difference. The process that makes sense is the one in part (a), $f(g(\text{orange}))$.

| example 6 | **Finding Composite Function Outputs** |

Find the following composite function outputs, using $f(x) = 2x - 5$, $g(x) = 6 - x^2$, and $h(x) = \dfrac{1}{x}$. Give the domain of each new function formed.

a. $(h \circ f)(x) = h(f(x))$ **b.** $(f \circ g)(x) = f(g(x))$ **c.** $(g \circ f)(x) = g(f(x))$

Solution

a. First, the output of f becomes the input for h.

$$(h \circ f)(x) = h(\,f(x)) = h(\overbrace{2x - 5}) = \frac{1}{2x - 5}$$

Next, the rule for h is applied to this input.

The domain of this function is all $x \neq \dfrac{5}{2}$ because $\dfrac{1}{2x - 5}$ is undefined if $x = \dfrac{5}{2}$.

b. $(f \circ g)(x) = f(g(x)) = f(6 - x^2) = 2(6 - x^2) - 5 = 12 - 2x^2 - 5$

$$= -2x^2 + 7$$

The domain of this function is the set of all real numbers.

c. $(g \circ f)(x) = g(\,f(x)) = g(2x - 5) = 6 - (2x - 5)^2 = 6 - (4x^2 - 20x + 25)$

$$= 6 - 4x^2 + 20x - 25 = -4x^2 + 20x - 19$$

The domain of this function is the set of all real numbers. ■

Technology Note

If you have two functions input as Y_1 and Y_2, you can enter the sum, difference, product, quotient, or composition of the two functions in Y_3. These combinations of functions can then be graphed or evaluated for input values of x.

---

example 7

Combinations with Functions

If $f(x) = \sqrt{x - 5}$ and $g(x) = 2x^2 - 4$,

a. Graph $(f + g)(x)$ using the window $[-1, 10]$ by $[0, 200]$. What is the domain of $f + g$?

b. Graph $\left(\dfrac{f}{g}\right)(x)$ using the window $[5, 15]$ by $[0, 0.016]$. What is the domain of $\dfrac{f}{g}$?

c. Compute $(f \circ g)(-3)$ and $(g \circ f)(9)$.

Solution

a. Functions f and g are entered as Y_1 and Y_2 in a graphing calculator and the sum of f and g is entered in Y_3, as shown in Figure 4.27(a). The graph of $(f + g)(x)$ is shown in Figure 4.27(b).

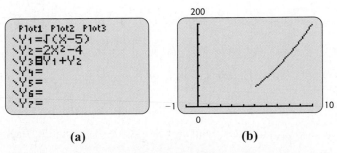

(a)   (b)

Figure 4.27

The domain of $(f + g)(x)$ is $x \geq 5$ because values of x less than 5 make $f(x) + g(x)$ undefined. This can also be seen in the graph in Figure 4.27(b).

b. The quotient of f and g is entered in Y_3, as shown in Figure 4.28(a). The graph of $\left(\dfrac{f}{g}\right)(x)$ is shown in Figure 4.28(b). The domain of $\left(\dfrac{f}{g}\right)(x)$ is $x \geq 5$ because values of x less than 5 make $\dfrac{f(x)}{g(x)}$ undefined. This can also be seen in the graph in Figure 4.28(b).

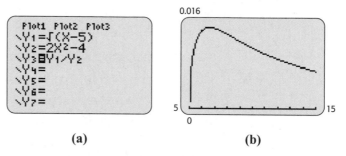

(a)   (b)

Figure 4.28

c. To compute $(f \circ g)(-3)$, we enter $Y_1(Y_2(-3))$, obtaining 3 (Figure 4.29(a)). To compute $(g \circ f)(9)$, we enter $Y_2(Y_1(9))$, obtaining 4 (Figure 4.29(b)).

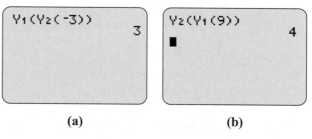

(a)   (b)

Figure 4.29

skills check

4.2

In Exercises 1–8, find the following: (a) $(f + g)(x)$, (b) $(f - g)(x)$, (c) $(f \cdot g)(x)$, (d) $\left(\dfrac{f}{g}\right)(x)$, and (e) the domain of $\dfrac{f}{g}$.

1. $f(x) = 3x - 5$; $g(x) = 4 - x$

2. $f(x) = 2x - 3$; $g(x) = 5 - x$

3. $f(x) = x^2 - 2x$; $g(x) = 1 + x$

4. $f(x) = 2x^2 - x$; $g(x) = 2x + 1$

5. $f(x) = \dfrac{1}{x}$; $g(x) = \dfrac{x + 1}{5}$

6. $f(x) = \dfrac{x - 2}{3}$; $g(x) = \dfrac{1}{x}$

7. $f(x) = \sqrt{x}$; $g(x) = 1 - x^2$

8. $f(x) = x^3$; $g(x) = \sqrt{x + 3}$

9. If $f(x) = x^2 - 5x$ and $g(x) = 6 - x^3$, evaluate

    a. $(f + g)(2)$ **b.** $(g - f)(-1)$

    c. $(f \cdot g)(-2)$ **d.** $\left(\dfrac{g}{f}\right)(3)$

10. If $f(x) = 4 - x^2$ and $g(x) = x^3 + x$, evaluate

    a. $(f + g)(1)$ **b.** $(f - g)(-2)$

    c. $(f \cdot g)(-3)$ **d.** $\left(\dfrac{g}{f}\right)(2)$

In Exercises 11–20, find (a) $(f \circ g)(x)$ *and (b)* $(g \circ f)(x)$.

11. $f(x) = 2x - 6; g(x) = 3x - 1$

12. $f(x) = 3x - 2; g(x) = 2x - 2$

13. $f(x) = x^2; g(x) = \dfrac{1}{x}$ **14.** $f(x) = x^3; g(x) = \dfrac{2}{x}$

15. $f(x) = \sqrt{x - 1}; g(x) = 2x - 7$

16. $f(x) = \sqrt{3 - x}; g(x) = x - 5$

17. $f(x) = |x - 3|; g(x) = 4x$

18. $f(x) = |4 - x|; g(x) = 2x + 1$

19. $f(x) = \dfrac{3x + 1}{2}; g(x) = \dfrac{2x - 1}{3}$

20. $f(x) = \sqrt[3]{x + 1}; g(x) = x^3 + 1$

In Exercises 21 and 22, use $f(x)$ and $g(x)$ to evaluate each expression.

21. $f(x) = 2x^2; g(x) = \dfrac{x - 5}{3}$

    a. $(f \circ g)(2)$ **b.** $(g \circ f)(-2)$

22. $f(x) = (x - 1)^2; g(x) = 3x - 1$

    a. $(f \circ g)(2)$ **b.** $(g \circ f)(-2)$

In Exercises 23 and 24, use the graphs of f and g below to evaluate the functions.

23. a. $(f + g)(2)$ **b.** $(f \circ g)(-1)$

    c. $\left(\dfrac{f}{g}\right)(4)$ **d.** $(f \circ g)(1)$

    e. $(g \circ f)(-2)$

24. a. $(g - f)(-2)$ **b.** $(f \circ g)(3)$

    c. $\left(\dfrac{f}{g}\right)(0)$ **d.** $(f \circ g)(-2)$

    e. $(g \circ f)(2)$

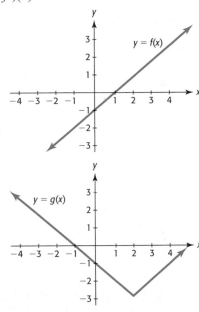

exercises

4.2

25. *Profit* Suppose the total weekly cost for the production and sale of x bicycles is $C(x) = 23x + 3420$ dollars and that the total revenue is given by $R(x) = 89x$ dollars, where x is the number of bicycles.

    a. Write the equation of the function that models the weekly profit from the production and sale of x bicycles.

    b. What is the profit on the production and sale of 150 bicycles?

26. *Profit* Suppose the total weekly cost for the production and sale of television sets is $C(x) = 189x + 5460$

and that the total revenue is given by $R(x) = 988x$, where x is the number of televisions and $C(x)$ and $R(x)$ are in dollars.

a. Write the equation of the function that models the weekly profit from the production and sale of x television sets.

b. What is the profit on the production and sale of 80 television sets in a given week?

27. *Revenue and Cost* The total revenue function for a certain product is given by $R = 1050x$ dollars, and the total cost function for this product is $C = 10,000 + 30x + x^2$ dollars, where x is the number of units of the product that are produced and sold.

a. Which function is quadratic, and which is linear?

b. Form the profit function for this product from these two functions.

c. Is the profit function a linear function, a quadratic function, or neither of these?

28. *Revenue and Cost* The total revenue function for a product is given by $R = 26,600x$ dollars, and the total cost function for this product is $C = 200,000 + 4600x + 2x^2$ dollars, where x is the number of units of the product that are produced and sold.

a. Which function is quadratic, and which is linear?

b. Form the profit function for this product from these two functions.

c. Is the profit function a linear function, a quadratic function, or neither of these?

29. *Revenue and Cost* The total revenue function for a certain product is given by $R = 550x$ dollars, and the total cost function for this product is $C = 10,000 + 30x + x^2$ dollars, where x is the number of units of the product that are produced and sold.

a. Find the profit function.

b. Find the number of units that gives maximum profit.

c. Find the maximum possible profit.

30. *Revenue and Cost* The total revenue function for a product is given by $R = 6600x$ dollars, and the total cost function for this product is $C = 2000 + 4800x + 2x^2$ thousand dollars, where x is the number of units of the product that are produced and sold.

a. Find the profit function.

b. Find the number of units that gives maximum profit.

c. Find the maximum possible profit.

31. *Average Cost* If the monthly total cost of producing 27-inch television sets is given by $C(x) = 50,000 + 105x$, where x is the number of sets produced per month, then the average cost per unit is given by

$$\overline{C}(x) = \frac{50,000 + 105x}{x}$$

a. Explain how $C(x)$ and another function can be combined to obtain the average cost function.

b. What is the average cost per set if 3000 sets are produced?

32. *Cost-Benefit* Suppose that for a certain city the cost C of obtaining drinking water that contains $p\%$ impurities (by volume) is given by

$$C = \frac{120,000}{p} - 1200$$

a. This function can be considered as the difference of what two functions?

b. What is the cost of drinking water that is 80% impure?

33. *Printers* The weekly total cost function for producing a dot matrix printer is $C(x) = 3000 + 72x$, where x is the number of printers produced per week.

a. Form the weekly average cost function for this product.

b. Find the average cost for the production of 100 printers.

34. *Electronic Components* The monthly cost of producing x electronic components is $C(x) = 2.15x + 2350$.

a. Find the monthly average cost function.

b. Find the average cost for the production of 100 components.

35. *Football Tickets* At a certain school, the number of student tickets sold for a home football game can be modeled by $S(p) = 62p + 8500$, where p is the winning percent of the home team. The number of non-student tickets sold for these home games is given by $N(p) = 0.5p^2 + 16p + 4400$ tickets.

a. Write an equation for the total number of tickets sold for a home football game at this school as a function of the winning percent p.

b. What is the domain for the function in part (a) in this context?

c. Assuming that the football stadium is filled to capacity when the team wins 90% of its home games, what is the capacity of the school's stadium?

36. *T-Shirt Sales* Let $T(c)$ be the number of T-shirts that are sold when the shirts have c colors and $P(c)$ be the price, in dollars, of a T-shirt that has c colors. Write a sentence explaining the meaning of the function $(T \cdot P)(c)$.

37. *Harvesting* A farmer's main cash crop is tomatoes, and the tomato harvest begins in the month of May. The number of bushels of tomatoes harvested on the xth day of May is given by the equation $B(x) = 6(x + 1)^{3/2}$. The market price in dollars of 1 bushel of tomatoes on the xth day of May is given by the formula $P(x) = 8.5 - 0.12x$.

a. How many bushels did the farmer harvest on May 8?

b. What was the market price of tomatoes on May 8?

c. How much was the farmer's tomato harvest worth on May 8?

d. Write a model for the worth W of the tomato harvest on the xth day of May.

38. *Total Cost* If the fixed cost for producing a product is $3000 and the variable cost for the production is $3.30x^2$ dollars, where x is the number of units produced, form the total cost function $C(x)$.

39. *Profit* A manufacturer of satellite systems has monthly fixed costs of $32,000 and variable costs of $432 per system, and it sells the systems for $592 per unit.

a. Write the function that models the profit P from the production and sale of x units of the system.

b. What is the profit if 600 satellite systems are produced and sold in 1 month?

c. At what rate does the profit grow as the number of units increases?

40. *Profit* A manufacturer of computers has monthly fixed costs of $87,500 and variable costs of $87 per computer, and it sells the computer for $295 per unit.

a. Write the function that models the profit P from the production and sale of x computers.

b. What is the profit if 700 computers are produced and sold in 1 month?

c. What is the y-intercept of the graph of the profit function? What does it mean?

41. *Population of Children* The following table gives the estimated population (in millions) of U.S. boys age 5 and under and the estimated U.S. population (in millions) of girls age 5 and under in selected years.

Year	1995	2000	2005	2010
Boys	10.02	9.71	9.79	10.24
Girls	9.57	9.27	9.43	9.77

(Source: U.S. Department of Commerce)

A function that models the population (in millions) of U.S. boys age 5 and under t years after 1990 is $B(t) = 0.0076t^2 - 0.1752t + 10.705$, and a function that models the population (in millions) of U.S. girls age 5 and under t years after 1990 is $G(t) = 0.0064t^2 - 0.1448t + 10.12$.

a. Find the equation of a function that models the estimated U.S. population (in millions) of children age 5 and under t years after 1990.

b. Use the result of part (a) to estimate the U.S. population of children age 5 and under in 2003.

42. *Educational Attainment* Data collected in March of the indicated years given in the table below show the percentage of 25- to 29-year-old male and female high school graduates who also have a bachelor's degree or higher.

Year	1975	1980	1985	1990
Males (%)	29.7	28.1	26.9	28.0
Females (%)	22.9	24.5	24.6	26.2

Year	1993	1995	1997	1999
Males (%)	27.2	28.4	30.7	31.2
Females (%)	27.4	28.5	32.9	33.0

(Source: U.S. Census Bureau)

a. Will adding the percents for males and females create a new function that gives the total percent of all 25- to 29-year-old high school graduates who also have a bachelor's degree or higher? Why or why not?

b. A function that approximately models the total percent of all 25- to 29-year-old high school graduates who also have a bachelor's degree or higher is $h(t) = 0.024t^2 - 0.595t + 29.121$, where t is the number of years after 1970. Use this model to estimate the total percent in 1990 and 1999.

c. Add the percent of males and the percent of females for the 1990 data and the 1999 data in the table and divide by 2. Do these values conflict with your answers in part (b)?

43. *Function Composition* Think of each of the following processes as a function designated by the indicated letter: f, placing in a styrofoam container; g, grinding. Describe each of the functions in parts (a) to (e), then answer part (f).

a. f(meat)   b. g(meat)   c. $(g \circ g)$(meat)

d. $f(g(\text{meat}))$ e. $g(f(\text{meat}))$

f. Which of the functions in parts (d) and (e) gives a sensible operation?

44. *Function Composition* Think of each of the following processes as a function designated by the indicated letter: f, putting on a sock; g, taking off a sock. Describe each of the functions in parts (a)–(c).

a. f(left foot)   b. $f(f(\text{left foot}))$

c. $(g \circ f)$(right foot)

45. *Shoe Sizes* A woman's shoe that is size x in Japan is size $s(x)$ in the United States, where $s(x) = x - 17$. A woman's shoe that is size x in the United States is size $p(x)$ in Britain, where $p(x) = x - 1.5$. Find a function that will convert Japanese shoe size to British shoe size.
(Source: Kuru International Exchange Association)

46. *Shoe Sizes* A man's shoe that is size x in Britain is size $d(x)$ in the United States, where $d(x) = x + 0.5$. A man's shoe that is size x in the United States is size $t(x)$ in Continental size, where $t(x) = x + 34.5$. Find a function that will convert British shoe size to Continental shoe size.
(Source: Kuru International Exchange Association)

47. *Exchange Rates* On October 9, 2007, each Austrian schilling was worth 2.56504 Russian rubles and each Chilean peso was worth 0.019367 Austrian schillings. Find the value of 1000 Chilean pesos in Russian rubles on October 9, 2007. Round the answer to two decimal places.
(Source: Expedia.com)

48. *Exchange Rates* On October 9, 2007, each Euro was worth 1.41400 U.S. dollars and each Australian dollar was worth 0.63496 Euros. Find the value of 100 Australian dollars in U.S. dollars.

49. *AOL* If $f(x)$ represents the number of AOL subscribers x years after 1992 and $g(x)$ represents the number of Internet users x years after 1992, what function represents the percent of Internet users who are AOL subscribers x years after 1992?

50. *Home Computers* If $f(x)$ represents the percent of American homes with computers and $g(x)$ represents the number of American homes, with x equal to the number of years after 1990, then what function represents the number of American homes with computers, with x equal to the number of years after 1990?

51. *Education* If the function $f(x)$ gives the number of female PhDs produced by American universities x years after 1990 and the function $g(x)$ gives the number of male PhDs produced by American universities x years after 1990, what function gives the total number of PhDs produced by American universities x years after 1990?

52. *Wind Chill* If the air temperature is 25°F, the wind chill temperature C is given by $C = 59.914 - 2.35s - 20.14\sqrt{s}$, where s is the wind speed in miles per hour.

a. State two functions whose difference gives this function.

b. Graph this function for $3 \le s \le 12$.

c. Is the function increasing or decreasing on this domain?

53. *Discount Prices* Half-Price Books has a sale with an additional 20% off the regular (1/2) price of their books. What percent of the retail price is charged during this sale?
(Source: Half-Price Books, Cleveland, Ohio)

4.3

Inverse Functions

section preview ▪ Loans

A business property is purchased with a promise to pay off a $60,000 loan plus the $16,500 interest on this loan by making 60 monthly payments of $1275. The amount of money remaining to be paid on the loan plus interest is given by the function

$$f(x) = 76,500 - 1275x$$

where x is the number of months for which payments have been made.

If we know how much remains to be paid and want to find how many months remain to make payments, we can find the **inverse** of the above function.

In this section, we determine if two functions are inverses of each other and find the inverse of a function if it has one. We also solve applied problems by using inverse functions.

Inverse Functions

We have seen that we can convert a temperature of $x°$ Celsius to a Fahrenheit temperature by using the formula

$$F(x) = \frac{9}{5}x + 32$$

The function that can be used to convert a temperature of $x°$ Fahrenheit back to a Celsius temperature is

$$C(x) = \frac{5x - 160}{9}$$

To see that one of these functions "undoes" what the other one does, see Table 4.10, which gives selected inputs and the resulting outputs for both functions. Observe that if we input the number 50 into F, the output after it is operated on by F is 122, and if 122 is then operated on by C the output is 50, which means that $C(F(50)) = 50$. In addition, if 122 is operated on by C the output is 50, and if 50 is operated on by F the output is 122, so $F(C(122)) = 122$.

Table 4.10

$x°$ C	−20	0	50	100
$F(x)$	−4	32	122	212

$x°$ F	−4	32	122	212
$C(x)$	−20	0	50	100

In fact, $C(F(x)) = x$ and $F(C(x)) = x$, for any input x, as we will see in Example 1. This means that both of these composite functions act as identity functions because their outputs are the same as their inputs. This happens because the second function performs the inverse operations of the first function. Because of this, we say that $C(x)$ and $F(x)$ are **inverse functions**.

> ### Inverse Functions
>
> Functions f and g for which $f(g(x)) = x$ for all x in the domain of g, and $g(f(x)) = x$ for all x in the domain of f, are called inverse functions. In this case, we denote g by f^{-1}, read as "f inverse."

example 1   ## Temperature Measurement

The function that can be used to convert a temperature of $x°$ Celsius to a Fahrenheit temperature is

$$F(x) = \frac{9}{5}x + 32$$

The function that can be used to convert a temperature of $x°$ Fahrenheit back to a Celsius temperature is

$$C(x) = \frac{5x - 160}{9}$$

To see how the two conversion formulas for temperature are related, find $C(F(x))$ and $F(C(x))$ and determine if the functions are inverse functions.

Solution

We compute $C(F(x))$ and $F(C(x))$. Evaluating C at $F(x)$ gives

$$C(F(x)) = C\left(\frac{9}{5}x + 32\right) = \frac{5\left(\frac{9}{5}x + 32\right) - 160}{9} = \frac{9x + 160 - 160}{9} = x$$

and evaluating F at $C(x)$ gives

$$F(C(x)) = F\left(\frac{5x - 160}{9}\right) = \frac{9}{5} \cdot \left(\frac{5x - 160}{9}\right) + 32 = \frac{45x}{45} - \frac{1440}{45} + 32 = x$$

Because $C(F(x)) = x$ and $F(C(x)) = x$, the two functions F and C are inverses.   ■

Consider a function that doubles each input. Its inverse takes half of each input. For example, the discrete function f with equation $f(x) = 2x$ and domain $\{1, 4, 5, 7\}$ has the range $\{2, 8, 10, 14\}$. The inverse of this function has the equation $f^{-1}(x) = \frac{x}{2}$. The domain of the inverse function is the set of outputs of the original function, $\{2, 8, 10, 14\}$. The outputs of the inverse function form the set $\{1, 4, 5, 7\}$, which is the domain of the original function. Figure 4.30 illustrates the relationship between the domains and ranges of the function f and its inverse f^{-1}. In this case and in every case, the domain of the inverse function is the range of the original function, and the range of the inverse function is the domain of the original function.

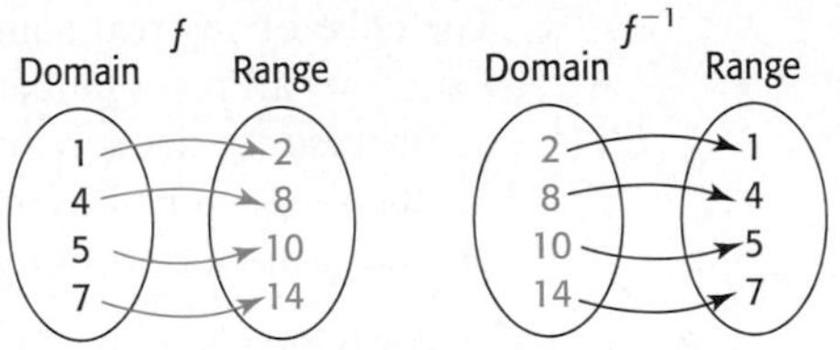

Figure 4.30

We summarize the information about inverse functions as follows.

> ## Inverse Functions
>
> The functions f and g are inverse functions if, whenever the pair (a, b) satisfies $y = f(x)$, the pair (b, a) satisfies $y = g(x)$. Note that when this happens,
>
> $$f(g(x)) = x \quad \text{and} \quad g(f(x)) = x$$
>
> for all x in the domain of g and f, respectively.

It is very important to note that the "-1" used in f^{-1} and g^{-1} is *not* an exponent, but rather a symbol used to denote the inverse of the function. The expression f^{-1} *always* refers to the inverse function of f and *never* to the reciprocal $\dfrac{1}{f}$ of f; that is, $f^{-1}(x) \neq \dfrac{1}{f(x)}$.

In general, we can show that a function has an inverse if it is a one-to-one function. Recall from the Algebra Toolbox that a one-to-one function has exactly one output for each input and exactly one input for each output. There is a one-to-one correspondence between the independent and dependent variables defining the function.

Also, recall that a function is one-to-one if no horizontal line can intersect the graph of the function in more than one point.

| example 2 | ## Inverse Functions |

a. Determine if $f(x) = x^3 - 1$ has an inverse function.

b. Verify that $g(x) = \sqrt[3]{x + 1}$ is the inverse function of $f(x) = x^3 - 1$.

c. Find the domain and range of each function.

Solution

a. Because each output of the function $f(x) = x^3 - 1$ corresponds to exactly one input, the function is one-to-one. Thus it has an inverse function.

b. To verify that the functions $f(x) = x^3 - 1$ and $g(x) = \sqrt[3]{x + 1}$ are inverse functions, we show that the compositions $f(g(x))$ and $g(f(x))$ are each equal to the identity function, x.

$$f(g(x)) = f(\sqrt[3]{x + 1}) = (\sqrt[3]{x + 1})^3 - 1 = (x + 1) - 1 = x \quad \text{and}$$
$$g(f(x)) = g(x^3 - 1) = \sqrt[3]{(x^3 - 1) + 1} = \sqrt[3]{x^3} = x$$

Thus, we have verified that these functions are inverses.

c. The cube of any real number decreased by 1 is a real number, so the domain of f is the set of all real numbers. The cube root of any real number plus 1 is also a real number, so the domain of g is the set of all real numbers. The domain of the function f is the range of its inverse g, and the domain of g is the range of f. Thus, the ranges of f and g are the set of real numbers. ■

By using the definition of inverse functions, we can find the equation for the inverse function of f by interchanging x and y in the equation $y = f(x)$ and solving the new equation for y.

Finding the Inverse of a Function

To find the inverse of the function f that is defined by the equation $y = f(x)$:

1. Rewrite the equation replacing $f(x)$ with y.

2. Interchange x and y in the equation defining the function.

3. Solve the new equation for y. If this equation cannot be solved uniquely for y, the original function has no inverse function.

4. Replace y with $f^{-1}(x)$.

---

example 3   ### Finding an Inverse Function

a. Find the inverse function of $f(x) = \dfrac{2x - 1}{3}$.

b. Graph $f(x) = \dfrac{2x - 1}{3}$ and its inverse function on the same axes.

Solution

a. Using the steps for finding the inverse of a function, we have

$$y = \frac{2x - 1}{3} \qquad \text{Replace } f(x) \text{ with } y.$$

$$x = \frac{2y - 1}{3} \qquad \text{Interchange } x \text{ and } y.$$

$$3x = 2y - 1 \qquad \text{Solve for } y.$$

$$3x + 1 = 2y$$

$$\frac{3x + 1}{2} = y$$

$$f^{-1}(x) = \frac{3x + 1}{2} \qquad \text{Replace } y \text{ with } f^{-1}(x).$$

b. The graphs of $f(x) = \dfrac{2x - 1}{3}$ and its inverse $f^{-1}(x) = \dfrac{3x + 1}{2}$ are shown in Figure 4.31.

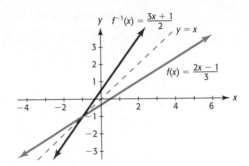

Figure 4.31

Notice that the graphs of $y = f(x)$ and $y = f^{-1}(x)$ in Figure 4.31 appear to be reflections of each other about the line $y = x$. This occurs because, if you were to choose any point (a, b) on the graph of the function f and interchange the x- and y-coordinates, the new point (b, a) will be on the graph of the inverse function f^{-1}. This should make sense because the inverse function is formed by interchanging x and y in the equation defining the original function. In fact, this relationship occurs for every function and its inverse.

Graphs of Inverse Functions

The graphs of a function and its inverse are symmetric with respect to the line $y = x$.

We again illustrate the symmetry of graphs of inverse functions in Figure 4.32, which shows the graphs of the inverse functions $f(x) = x^3 - 2$ and $f^{-1}(x) = \sqrt[3]{x + 2}$. Note that the point $(0, -2)$ is on the graph of $f(x) = x^3 - 2$ and the point $(-2, 0)$ is on the graph of $f^{-1}(x) = \sqrt[3]{x + 2}$.

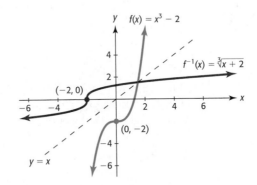

Figure 4.32

---

example 4

Loan Repayment

A business property is purchased with a promise to pay off a $60,000 loan plus the $16,500 interest on this loan by making 60 monthly payments of $1275. The amount of money remaining to be paid on the loan plus interest is given by the function

$$f(x) = 76{,}500 - 1275x$$

where x is the number of months remaining for payments to be made.

a. Find the inverse of this function.

b. Use the inverse to determine how many months remain to make payments if $35,700 remains to be paid.

Solution

a. Replacing $f(x)$ with y gives $y = 76{,}500 - 1275x$.

Interchanging x and y gives the equation $x = 76{,}500 - 1275y$.

Solving this new equation for y gives $y = \dfrac{76{,}500 - x}{1275}$.

Replacing y with $f^{-1}(x)$ gives the inverse function $f^{-1}(x) = \dfrac{76{,}500 - x}{1275}$.

b. The inverse function gives the number of months remaining to make payments if x dollars remain to be paid.

$$f^{-1}(35{,}700) = \frac{76{,}500 - 35{,}700}{1275} = 32$$

so 32 months remain to make payments.

Inverse Functions on Limited Domains

As we have stated, a function cannot have an inverse function if it is not a one-to-one function. However, if there is a limited domain over which such a function is a one-to-one function, then it has an inverse function for this domain. Consider the function $f(x) = x^2$. The horizontal line test on the graph of this function (Figure 4.33(a)) indicates that this function is not a one-to-one function and that the function f does not have an inverse function. To see why, consider the attempt to find the inverse function:

$$y = x^2$$
$$x = y^2 \qquad \text{Interchange } x \text{ and } y.$$
$$\pm\sqrt{x} = y \qquad \text{Solve for } y.$$

This equation is not the inverse function because it is not a function (one value of x gives two values for y). Because we cannot solve $x = y^2$ uniquely for y, $f(x) = x^2$ does not have an inverse function.

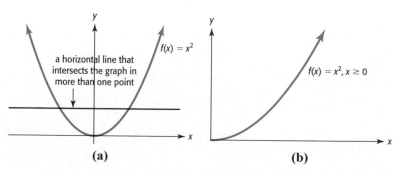

(a)   **(b)**

Figure 4.33

However, if we limit the domain of the function to $x \geq 0$, no horizontal line intersects the graph of $f(x) = x^2$ more than once, so the function is one-to-one on this limited domain (Figure 4.33(b)), and the function has an inverse. If we restrict the domain of the original function by requiring that $x \geq 0$, then the range of the inverse function is restricted to $y \geq 0$, and the equation $y = \sqrt{x}$ or $f^{-1}(x) = \sqrt{x}$ defines the inverse function. Figure 4.34 shows the graph of $f(x) = x^2$ for $x \geq 0$ and its inverse $f^{-1}(x) = \sqrt{x}$. Note that the graphs are symmetric about the line $y = x$.

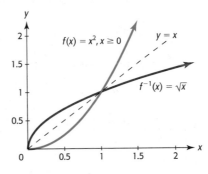

Figure 4.34

| example 5 | **Velocity of Blood** |

Because of friction from the walls of an artery, the velocity of blood is greatest at the center of the artery and decreases as the distance x from the center increases. The (non-negative) velocity in centimeters per second of a blood corpuscle in the artery can be modeled by the function

$$v = k(R^2 - x^2), k > 0, 0 < r \leq R$$

where R is the radius of the artery and k is a constant that is determined by the pressure, viscosity of the blood, and the length of the artery. In the case where $k = 2$ and $R = 0.1$ centimeter, the velocity can be written as a function of the distance x from the center as

$$v(x) = 2(0.01 - x^2) \text{ centimeter/second}$$

a. Assuming the velocity is nonnegative, does the inverse of this function exist on a domain limited by the context of the application?

b. What is the inverse function?

c. What does the inverse function mean in the context of this application?

Solution

a. Because x represents distance, it is nonnegative, and because $R = 0.1$ and x equals the distance of blood in an artery from the center of the artery, $0 \leq x \leq 0.1$. Hence, the function is one-to-one for $0 \leq x \leq 0.1$ and we can find its inverse.

b. We find the inverse function as follows.

$$f(x) = 2(0.01 - x^2)$$
$$y = 2(0.01 - x^2) \qquad \text{Replace } f(x) \text{ with } y.$$
$$x = 2(0.01 - y^2) \qquad \text{Interchange } y \text{ and } x.$$
$$x = 0.02 - 2y^2 \qquad \text{Solve for } y.$$

$$2y^2 = 0.02 - x$$

$$y^2 = \frac{0.02 - x}{2}$$

$$y = \sqrt{\frac{0.02 - x}{2}} \qquad \text{Only nonnegative values of } y \text{ are possible.}$$

$$f^{-1}(x) = \sqrt{\frac{0.02 - x}{2}} \qquad \text{Replace } y \text{ with } f^{-1}(x).$$

c. The inverse function gives the distance from the center of the artery as a function of the velocity of a blood corpuscle.   ■

skills check

4.3

In each of Exercises 1 and 2, determine if the function f defined by the arrow diagram has an inverse. If it does, create an arrow diagram that defines the inverse. If not, explain why not.

1.
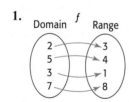

2.
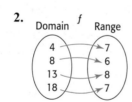

In each of Exercises 3 and 4, determine if the function f defined by the set of ordered pairs has an inverse. If so, find the inverse.

3. $\{(5, 2), (4, 1), (3, 7), (6, 2)\}$

4. $\{(2, 8), (3, 9), (4, 10), (5, 11)\}$

5. If $f(x) = 3x$ and $g(x) = \dfrac{x}{3}$,

   a. What are $f(g(x))$ and $g(f(x))$?

   b. Are $f(x)$ and $g(x)$ inverse functions?

6. If $f(x) = 4x - 1$ and $g(x) = \dfrac{x + 1}{4}$,

   a. What are $f(g(x))$ and $g(f(x))$?

   b. Are $f(x)$ and $g(x)$ inverse functions?

7. If $f(x) = x^3 + 1$ and $g(x) = \sqrt[3]{x} - 1$, are $f(x)$ and $g(x)$ inverse functions?

8. If $f(x) = (x - 2)^3$ and $g(x) = \sqrt[3]{x} + 2$, are $f(x)$ and $g(x)$ inverse functions?

9. For the function f defined by $f(x) = 3x - 4$, complete the tables below for f and f^{-1}.

x	$f(x)$		x	$f^{-1}(x)$
-1	-7		-7	-1
0			-4	
1			-1	
2			2	
3			5	

10. For the function g defined by $g(x) = 2x^3 - 1$, complete the tables below for g and g^{-1}.

x	$g(x)$		x	$g^{-1}(x)$
-2	-17		-17	-2
-1			-3	
0			-1	
1			1	
2			15	

11. a. Write the inverse of $f(x) = 3x - 4$.

   b. Do the values for f^{-1} in the table of Exercise 9 fit the equation for f^{-1}?

12. a. Write the inverse of $g(x) = 2x^3 - 1$.

   b. Do the values for g^{-1} in the table of Exercise 10 fit the equation for g^{-1}?

13. If the graph of $y = f(x)$ has the point (a, b) on its graph, name a point on the graph of $y = f^{-1}(x)$.

14. If function h has an inverse and $h^{-1}(-2) = 3$, find $h(3)$.

15. Find the inverse of $f(x) = \dfrac{1}{x}$.

16. Find the inverse of $g(x) = 4x + 1$.

17. Find the inverse of $f(x) = 4x^2$ for $x \geq 0$.

18. Find the inverse of $g(x) = x^2 - 3$ for $x \geq 0$.

19. Graph $g(x) = \sqrt{x}$ and its inverse $g^{-1}(x)$ for $x \geq 0$ on the same axes.

20. Graph $g(x) = \sqrt[3]{x}$ and its inverse $g^{-1}(x)$ on the same axes.

21. $f(x) = (x - 2)^2$ and $g(x) = \sqrt{x} + 2$ are inverse functions for what values of x?

22. $f(x) = x^2 + 5$ and $g(x) = \sqrt{x - 5}$ are inverse functions for what values of x?

23. Is the function $f(x) = 2x^3 + 1$ a one-to-one function? Does it have an inverse?

24. Is the function $f(x) = 3x^2 + 1$ a one-to-one function? Does it have an inverse?

25. Is the function defined by $\{(1, 3), (2, 2), (3, 4)\}$ a one-to-one function?

26. Is the function defined by $\{(2, 4), (6, 7), (3, 5)\}$ a one-to-one function?

27. Is the function with the graph below one-to-one?

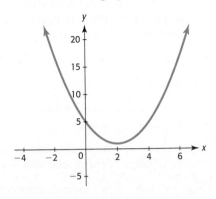

28. Is the function with the graph below one-to-one?

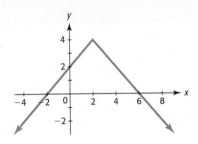

29. Sketch the graph of $y = f^{-1}(x)$ on the axes with the graph of $y = f(x)$, shown below.

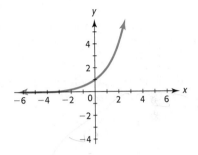

30. Sketch the graph of $y = f^{-1}(x)$ on the axes with the graph of $y = f(x)$, shown below.

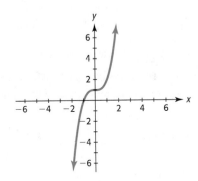

31. *Shoe Sizes* A man's shoe that is size x in Britain is equivalent to size $d(x)$ in the United States, where $d(x) = x + 0.5$.

a. Find the inverse of the function.

b. Use the inverse function to find the British size of a shoe if it is U.S. size $8\frac{1}{2}$.
(Source: Kuru International Exchange Association)

32. *Shoe Sizes* A man's shoe that is size x in the United States is size $t(x)$ in Continental size, where $t(x) = x + 34.5$.

a. Find a function that will convert Continental shoe size to U.S. shoe size.

b. Use the inverse function to find the U.S. size if the Continental size of a shoe is 43.
(Source: Kuru International Exchange Association)

33. *Cigarettes* For the years 1975–1991, the percent of high school seniors who have tried cigarettes is given by $f(t) = 75.451 - 0.707t$, where t is the number of years after 1975. Find the inverse of this function and use it to find the year in which the percent fell below 65%.
(Source: National Institute on Drug Abuse)

34. *Investments* If x dollars are invested at 10% for 6 years, the future value of the investment is given by $S(x) = x + 0.6x$.

a. Find the inverse of this function.

b. What do the outputs of the inverse function represent?

c. Use this function to find the amount of money that must be invested for 6 years at 10% to have a future value of $24,000.

35. *Ritalin* The function that models the consumption of Ritalin over the years 1990–1996 is $f(x) = 225.304x + 493.432$, where x is the number of years after 1990 and y is measured in grams per 100,000 people.

a. Find the inverse of this function. What do the outputs of the inverse function represent?

b. Use the inverse function to find when consumption equals 1170 grams per 100,000 people.

36. *Apparent Temperature* If the outside temperature is 90°F, the apparent temperature is given by $A(x) = 82.35 + 29.3x$, where x is the humidity written as a decimal. Find the inverse of this function and use it to find the percent humidity that will give an apparent temperature of 97° if the temperature is 90°F.
(Source: "Temperature-Humidity Indices," *The UMAP Journal*, Fall 1989)

37. *Body-Heat Loss* The model for body-heat loss depends on the coefficient of convection $K = f(x)$, which depends on wind speed x according to the equation $f(x) = 4\sqrt{4x + 1}$.

a. What are the domain and range of this function without regard to the context of this application?

b. Find the inverse of this function.

c. What are the domain and range of the inverse function?

d. In the context of the application, what are the domain and range of the inverse function?

38. *Algorithmic Relationship* For many species of fish, the weight W is a function of the length x, given by $W = kx^3$, where k is a constant depending on the species. Suppose $k = 0.002$, W is in pounds, and x is in inches, so that the weight is $W(x) = 0.002x^3$.

a. Find the inverse function of this function.

b. What does the inverse function give?

c. Use the inverse function to find the length of a fish that weighs 2 pounds.

d. In the context of the application, what are the domain and range of the inverse function?

39. *Decoding Messages* If we assign numbers to the letters of the alphabet as follows and assign 27 to a blank space, we can convert a message to a numerical sequence. We can "encode" a message by adding 3 to each number that represents a letter in a message.

A	B	C	D	E	F	G	H	I	J	K	L	M
1	2	3	4	5	6	7	8	9	10	11	12	13

N	O	P	Q	R	S	T	U	V	W	X	Y	Z
14	15	16	17	18	19	20	21	22	23	24	25	26

Thus, the message "Go for it" can be encoded by using the numbers to represent the letters and further encoded by using the function $C(x) = x + 3$. The coded message would be 10 18 30 9 18 21 30 12 23. Find the inverse of the function and use it to decode 23 11 8 30 21 8 4 15 30 23 11 12 17 10.

40. *Decoding Messages* Use the numerical representation from Exercise 39 and the inverse of the encoding function $C(x) = 3x + 2$ to decode 41 5 35 17 83 41 77 83 14 5 77.

41. *Social Security Numbers and Income Taxes* Consider the function that assigns each person who pays federal income tax his or her Social Security number. Is this a one-to-one function? Explain.

42. *Checkbook Balance* Consider the function with the check number in your checkbook as input and the dollar amount of the check as the output. Is this a one-to-one function? Explain.

43. *Volume of a Cube* The volume of a cube is $f(x) = x^3$ cubic inches, where x is the length of the edge of the cube in inches.

  a. Is this function one-to-one?

  b. Find the inverse of this function.

  c. What are the domain and range of this inverse function in the context of the application?

  d. How could this inverse function be used?

44. *Volume of a Sphere* The volume of a sphere is $f(x) = \frac{4}{3}\pi x^3$ cubic inches, where x is the radius of the sphere in inches.

  a. Is this function one-to-one?

  b. Find the inverse of this function.

  c. What are the domain and range of the inverse function for this application?

  d. How could this inverse function be used?

  e. What is the radius of a sphere if its volume is 65,450 cubic inches?

45. *Currency Conversion* The function that converts Canadian dollars to U.S. dollars according to the January 19, 2008, values is $f(x) = 0.99295x$, where x is the number of Canadian dollars and $f(x)$ is the number of U.S. dollars.

  a. Find the inverse function for f and interpret its meaning.

  b. Use f and f^{-1} to determine the money you will have if you take 500 U.S. dollars to Canada, convert it to Canadian dollars, don't spend any, and then convert it back to U.S. dollars. (Assume that there is no fee for conversion and the conversion rate remains the same.)
  (Source: Expedia.com)

46. *Surface Area* The surface area of a cube is $y = 6x^2$ cm^2, where x is the length of the edge of the cube in centimeters.

  a. For what values of x does this model make sense? Is the model a one-to-one function for these values of x?

  b. What is the inverse of this function on this interval?

  c. How could the inverse function be used?

47. *Illumination* The intensity of illumination of a light is a function of the distance from the light. For a given light, the intensity is given by $I(x) = \dfrac{300,000}{x^2}$ candle power, where x is the distance in feet from the light.

  a. Is this function a one-to-one function if it is not limited by the context of the application?

  b. What is the domain of this function in the context of the application?

  c. Is the function one-to-one for the domain in part (b)?

  d. Find the inverse of this function on the domain from part (b) and use it to find the distance at which the intensity of the light is 75,000 candlepower.

48. *Supply* The supply function for a product is $p(x) = \frac{1}{4}x^2 + 20$, where x is the number of thousands of units a manufacturer will supply if the price is $p(x)$ dollars.

a. Is this function a one-to-one function?

b. What is the domain of this function in the context of the application?

c. Is the function one-to-one for the domain in part (b)?

d. Find the inverse of this function and use it to find how many units the manufacturer is willing to supply if the price is $101.

49. Suppose the function that converts United Kingdom (U.K.) pounds to U.S. dollars is $f(x) = 1.9733x$, where x is the number of pounds and $f(x)$ is the number of U.S. dollars.

a. Find the inverse function for f and interpret its meaning.

b. Use f and f^{-1} to determine the money you will have if you take 1000 U.S. dollars to the United Kingdom, convert it to pounds, don't spend any, and then convert it back to U.S. dollars. (Assume that there is no fee for conversion and the conversion rate remains the same.)
(Source: International Monetary Fund)

50. *First-Class Postage* The postage charged for first-class mail is a function of its weight. The U.S. Postal Service uses the following table to describe the rates for 2008.

Weight Increment, *x*	Postal Rate
First ounce or fraction of an ounce	$0.42
Each additional ounce or fraction	$0.17

(Source: pe.usps.gov/text)

a. Convert this table to a piecewise-defined function $P(x)$ that represents postage for letters weighing more than 0 and no more than 3 ounces, using x as the weight in ounces and $P(x)$ as the postage in cents.

b. Does P have an inverse function? Why or why not?

51. *Path of a Ball* If a ball is thrown into the air at a velocity of 96 feet per second from a building that is 256 feet high, the height of the ball after x seconds is $f(x) = 256 + 96x - 16x^2$ feet.

a. For how many seconds will the ball be in the air?

b. Is this function a one-to-one function over this time interval?

c. Give an interval over which the function is one-to-one.

d. Find the inverse of the function over the interval $0 \le x \le 3$. What does it give?

---

section

4.4

Additional Equations and Inequalities

key concepts

- Radical equations
- Equations with rational powers
- Quadratic inequalities
- Power inequalities
- Absolute value inequalities

section preview ▪ Profit

The daily profit from the production and sale of x units of a product is given by

$$P(x) = -0.01x^2 + 20.25x - 500$$

Because a profit occurs when $P(x)$ is positive, there is a profit for those values of x that make $P(x) > 0$. Thus, the values of x that give a profit are the solutions to

$$-0.01x^2 + 20.25x - 500 > 0$$

In this section, we solve quadratic inequalities by using both analytical and graphical methods. We also solve equations and inequalities involving radicals, rational powers, and absolute values.

Radical Equations; Equations Involving Rational Powers

An equation containing a radical can frequently be converted to an equation that does not contain radicals by raising both sides of the equation to an appropriate power. For example, an equation containing a square root radical can usually be converted by squaring both sides of the equation. It is possible, however, that raising both sides of an equation to a power may produce an *extraneous* solution (a value that does not satisfy the original equation). For this reason, we must check solutions when using this technique. To solve an equation containing a radical, we use the following steps.

Solving Radical Equations

1. Isolate a single radical on one side of the equation.

2. Square both sides of the equation. (Or raise both sides to a power that is equal to the index of the radical.)

3. If a radical remains, repeat steps 1 and 2.

4. Solve the resulting equation.

5. All solutions must be checked in the original equation, and only those that satisfy the original equation are actual solutions.

---

example 1

Radical Equation

Solve $\sqrt{x + 5} + 1 = x$.

Solution

To eliminate the radical, we isolate it on one side and square both sides:

$$\sqrt{x + 5} = x - 1$$
$$x + 5 = x^2 - 2x + 1$$
$$0 = x^2 - 3x - 4$$
$$0 = (x - 4)(x + 1)$$
$$x = 4 \quad \text{or} \quad x = -1$$

Checking these values shows that 4 is a solution and -1 is not.

$x = 4$	$x = -1$
$\sqrt{4 + 5} + 1 \stackrel{?}{=} 4$	$\sqrt{-1 + 5} + 1 \stackrel{?}{=} -1$
$\sqrt{9} + 1 \stackrel{?}{=} 4$	$\sqrt{4} + 1 \stackrel{?}{=} -1$
$3 + 1 \stackrel{?}{=} 4$	$2 + 1 \stackrel{?}{=} -1$
$4 = 4$	$3 \neq -1$

We can also check graphically by the intersection method. Graphing $y = \sqrt{x + 5} + 1$ and $y = x$, we find the point of intersection to be $(4, 4)$ (Figure 4.35). Note that there is *not* a point of intersection at $x = -1$.

Thus, the solution to the equation $\sqrt{x + 5} + 1 = x$ is $x = 4$.

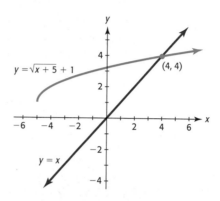

Figure 4.35

---

example 2 | **Radical Equation**

Solve $\sqrt{4x - 8} - 1 = \sqrt{2x - 5}$.

Solution

We begin by confirming that one of the radicals is isolated. Then we square both sides of the equation:

$$\left(\sqrt{4x - 8} - 1\right)^2 = \left(\sqrt{2x - 5}\right)^2$$
$$4x - 8 - 2\sqrt{4x - 8} + 1 = 2x - 5$$
$$4x - 7 - 2\sqrt{4x - 8} = 2x - 5$$

Now we must isolate the remaining radical and square both sides of the equation again:

$$2x - 2 = 2\sqrt{4x - 8}$$
$$(2x - 2)^2 = \left(2\sqrt{4x - 8}\right)^2$$
$$4x^2 - 8x + 4 = 4(4x - 8)$$
$$4x^2 - 8x + 4 = 16x - 32$$
$$4x^2 - 24x + 36 = 0$$
$$4(x^2 - 6x + 9) = 0$$
$$4(x - 3)^2 = 0$$
$$x = 3$$

We can check by substituting or graphing. Graphing $y = \sqrt{4x - 8} - 1$ and $y = \sqrt{2x - 5}$, we find the point of intersection at $x = 3$ (Figure 4.36). Thus $x = 3$ is the solution.

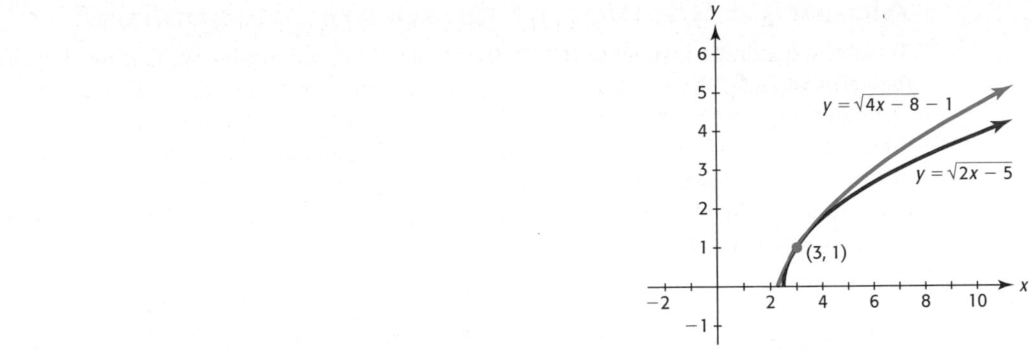

$$y = \sqrt{4x - 8} - 1$$

$$y = \sqrt{2x - 5}$$

$(3, 1)$

Figure 4.36

Equations Containing Rational Powers

Some equations containing rational powers can be solved by writing the equation as a radical equation.

example 3

Equation with Rational Powers

Solve the equation $(x - 3)^{2/3} - 4 = 0$.

Solution

Rewriting the equation as a radical equation and with the radical isolated gives

$$(x - 3)^{2/3} = 4$$

$$\sqrt[3]{(x - 3)^2} = 4$$

Cubing both sides of the equation and solving give

$$(x - 3)^2 = 64$$

$$\sqrt{(x - 3)^2} = \pm\sqrt{64}$$

$$x - 3 = \pm 8$$

$$x = 3 \pm 8$$

$$x = 11 \quad \text{or} \quad x = -5$$

Entering each solution in the original equation shows that each checks there.

$x = 11$	$x = -5$
$(11 - 3)^{2/3} - 4 \overset{?}{=} 0$	$(-5 - 3)^{2/3} - 4 \overset{?}{=} 0$
$8^{2/3} - 4 \overset{?}{=} 0$	$(-8)^{2/3} - 4 \overset{?}{=} 0$
$(\sqrt[3]{8})^2 - 4 \overset{?}{=} 0$	$(\sqrt[3]{-8})^2 - 4 \overset{?}{=} 0$
$2^2 - 4 \overset{?}{=} 0$	$(-2)^2 - 4 \overset{?}{=} 0$
$0 = 0$	$0 = 0$

Quadratic Inequalities

A **quadratic inequality** is an inequality that can be written in the form

$$ax^2 + bx + c > 0$$

where a, b, and c are real numbers and $a \neq 0$ (or with $>$ replaced by $<$, $\geq$, or $\leq$).

Algebraic Solution of Quadratic Inequalities

To solve a quadratic inequality $f(x) > 0$ or $f(x) < 0$ algebraically, we first need to find the zeros of $f(x)$. The zeros can be found by factoring or the quadratic formula. If the quadratic function has two real zeros, these two zeros divide the real number line into three intervals. Within each interval, the value of the quadratic function is either always positive or always negative, so we can use one value in each interval to test the function on the interval. We can use the following steps, which summarize these ideas, to solve quadratic inequalities.

Solving a Quadratic Inequality Algebraically

1. Write an equivalent inequality with 0 on one side and with the function $f(x)$ on the other side.

2. Solve $f(x) = 0$.

3. Create a sign diagram that uses the solutions from step 2 to divide the number line into intervals. Pick a test value in each interval and determine whether $f(x)$ is positive or negative in that interval to create a sign diagram.*

4. Identify the intervals that satisfy the inequality in step 1. The values of x that define these intervals are solutions to the original inequality.

* The numerical feature of your graphing utility can be used to test the x-values.

example 4

Solve the inequality $x^2 - 3x > 3 - 5x$.

Solution

Rewriting the inequality with 0 on the right side of the inequality gives $f(x) > 0$ with $f(x) = x^2 + 2x - 3$:

$$x^2 + 2x - 3 > 0$$

Writing the equation $f(x) = 0$ and solving for x gives

$$x^2 + 2x - 3 = 0$$
$$(x + 3)(x - 1) = 0$$
$$x = -3 \quad \text{or} \quad x = 1$$

The two values of x, -3 and 1, divide the number line into three intervals, the numbers less than -3, the numbers between -3 and 1, and the numbers greater than 1. We need only find the sign of $(x + 3)$ and $(x - 1)$ in each interval and then find the sign of their product to find the solution to the original inequality. Testing a value in each interval determines the sign of each factor in each interval. (See the sign diagram in Figure 4.37.)

```
sign of (x + 3)(x − 1)  +++++++++++   − − − − − − − − − −   +++++++++++
      sign of (x − 1)   − − − − − − − − − −   − − − − − − − − − −   +++++++++++
      sign of (x + 3)   − − − − − − − − − −   +++++++++++   +++++++++++
                      ──────────────────┼──────────────────┼──────────
                                       −3                  1
```

Figure 4.37

The function $f(x) = (x + 3)(x - 1)$ is positive on the intervals $(-\infty, -3)$ and $(1, \infty)$, so the solution to $x^2 + 2x - 3 > 0$ and thus the original inequality is

$$x < -3 \quad \text{or} \quad x > 1 \qquad \blacksquare$$

Graphical Solution of Quadratic Inequalities

Recall that we solved a linear inequality $f(x) > 0$ (or $f(x) < 0$) graphically by graphing the related equation $f(x) = 0$ and observing where the graph is above (or below) the x-axis—that is, where $f(x)$ is positive (or negative).

Similarly, we can solve a quadratic inequality $f(x) > 0$ (or $f(x) < 0$) graphically by graphing the related equation $f(x) = 0$ and observing the x-values of the intervals where $f(x)$ is positive (or negative).

For example, we can graphically solve the inequality

$$x^2 - 3x - 4 \leq 0$$

by graphing

$$y = x^2 - 3x - 4$$

and observing the part of the graph that is below or on the x-axis. Using the x-intercept method, we find that $x = -1$ and $x = 4$ give $y = 0$, and we see that the graph is below or on the x-axis for values of x satisfying $-1 \leq x \leq 4$ (Figure 4.38). Thus, the solution to the inequality is $-1 \leq x \leq 4$.

Table 4.11 shows the possible graphs of quadratic functions and the solutions to the related inequalities. Note that if the graph of $y = f(x)$ lies entirely above the x-axis, the solution to $f(x) > 0$ is the set of all real numbers, and there is no solution to $f(x) < 0$. (The graph of $y = f(x)$ can also touch the x-axis in one point.)

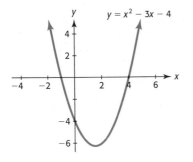

Figure 4.38

Table 4.11

Orientation of Graph of $y = f(x)$	Inequality to Be Solved	Part of Graph That Satisfies the Inequality	Solution to Inequality
	$f(x) > 0$	where the graph is above the x-axis	$x < a$ or $x > b$
	$f(x) < 0$	where the graph is below the x-axis	$a < x < b$
	$f(x) > 0$	where the graph is above the x-axis	$a < x < b$
	$f(x) < 0$	where the graph is below the x-axis	$x < a$ or $x > b$
	$f(x) > 0$	the entire graph	all real numbers
	$f(x) < 0$	none of the graph	no solution

(continued)

Table 4.11 (continued)

Orientation of Graph of $y = f(x)$	Inequality to Be Solved	Part of Graph That Satisfies the Inequality	Solution to Inequality
(graph opening downward with x-axis above)	$f(x) > 0$	none of the graph	no solution
	$f(x) < 0$	the entire graph	all real numbers

---

example 5

Height of a Model Rocket

A model rocket is projected straight upward from ground level according to the equation

$$h = -16t^2 + 192t, \; t \geq 0$$

where h is the height in feet and t is the time in seconds. During what time interval will the height of the rocket exceed 320 feet?

Algebraic Solution

To find the time interval when the rocket is higher than 320 feet, we find the values of t for which

$$-16t^2 + 192t > 320$$

Getting 0 on the right side of the inequality, we have

$$-16t^2 + 192t - 320 > 0$$

To determine when the height is 0, we solve

$$-16t^2 + 192t - 320 = 0$$
$$-16(t^2 - 12t + 20) = 0$$
$$-16(t - 2)(t - 10) = 0$$
$$t = 2 \quad \text{or} \quad t = 10$$

The values of $t = 2$ and $t = 10$ divide the number line into three intervals—the numbers less than 2, the numbers between 2 and 10, and the numbers greater than 10 (Figure 4.39). To solve the inequality, we find the sign of the product of -16, $(t - 2)$, and $(t - 10)$ in each interval. We do this by testing a value in each interval.

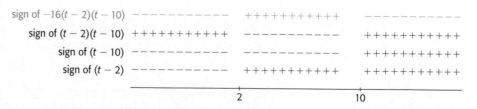

Figure 4.39

This shows that the product $-16(t - 2)(t - 10)$ is positive only for $2 < t < 10$, so the solution to $-16t^2 + 192t - 320 > 0$ is $2 < t < 10$. This solution is also a solution to the original inequality, so the rocket exceeds 320 feet between 2 and 10 seconds after launch.

Note that if the inequality problem is applied, we must check that the solution makes sense in the context of the problem.

Graphical Solution

To find the interval when the height exceeds 320 feet, we can also solve $-16t^2 + 192t > 320$ using a graphing utility. To use the x-intercept method, we rewrite the inequality with 0 on the right side:

$$-16t^2 + 192t - 320 > 0$$

Using the variable x in place of the variable t, we enter $y_1 = -16x^2 + 192x - 320$ and graph the function (Figure 4.40). The x-intercepts of the graph are $x = 2$ and $x = 10$.

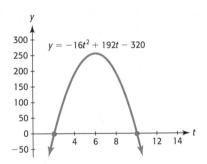

Figure 4.40

The solution to $-16t^2 + 192t - 320 > 0$ is the interval where the graph is above the t-axis, which is

$$2 < t < 10$$

Thus, the height of the rocket exceeds 320 feet between 2 and 10 seconds. ∎

---

example 6

Internet Use

According to Internet World Stats, the number of worldwide Internet users, in millions, during the years 1995 to 2007 can be modeled by the equation $y = 3.604x^2 + 27.626x - 261.714$, where x is the number of years after 1990. For what years from 1995 through 2015 does this model indicate that the number of worldwide Internet users is at least 1103 million?

Solution

To find the years from 1995 through 2015 when the number of Internet users is at least 1103 million, we solve the inequality

$$3.604x^2 + 27.626x - 261.714 \geq 1103$$

We begin solving this inequality by graphing $y = 3.604x^2 + 27.626x - 261.714$. We set the viewing window from $x = 5$ (1995) to $x = 25$ (2015). The graph of this function is shown in Figure 4.41(a) on the next page.

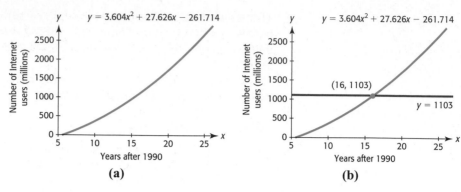

Figure 4.41

We seek the point(s) where the graph of $y = 1103$ intersects the graph of $y = 3.604x^2 + 27.626x - 261.714$. Graphing the line $y = 1103$, and intersecting it with the graph of the quadratic function in the given window gives the point of intersection $(16, 1103)$ as shown in Figure 4.41(b). This graph shows that the number of users is at or above 1103 million for $16 \leq x \leq 25$. Because x represents the number of years after 1990, the number of Internet users is at least 1103 million for the years 2006 through 2015. ∎

Power Inequalities

To solve a power inequality, we will use a combination of analytical and graphical methods.

> ### Power Inequalities
>
> To solve a power inequality, first solve the related equation. Then use graphical methods to find the values of the variable that satisfy the inequality.

example 7   **Investment**

The future value of $3000 invested for 3 years at rate r, compounded annually, is given by $S = 3000(1 + r)^3$. What interest rate will give a future value of at least $3630?

Solution

To solve this problem, we solve the inequality

$$3000(1 + r)^3 \geq 3630$$

We begin by solving the related equation using the root method.

$$3000(1 + r)^3 = 3630$$

$$(1 + r)^3 = 1.21 \qquad \text{Divide both sides by 3000.}$$

$$1 + r = \sqrt[3]{1.21} \qquad \text{Take the cube root } \left(\frac{1}{3} \text{ power}\right) \text{ of both sides.}$$

$$1 + r \approx 1.0656$$

$$r \approx 0.0656$$

This tells us that the investment will have a future value of $3630 in 3 years if the interest rate is approximately 6.56%, and helps us set the window to solve the inequality graphically. Graphing $y_1 = 3000(1 + r)^3$ and $y_2 = 3630$ on the same axes shows that $3000(1 + r)^3 \geq 3630$ if $r \geq 0.0656$ (Figure 4.42). Thus, the investment will result in at least $3630 if the interest rate is 6.56% or higher. Note that in the context of this problem, only values of r from 0 (0%) to 1 (100%) make sense.

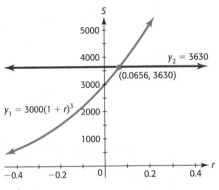

Figure 4.42

Inequalities Involving Absolute Values

Recall that if u is an algebraic expression, then for $a \geq 0$, $|u| = a$ means that $u = a$ or $u = -a$. We can represent inequalities involving absolute values as follows.

> For $a \geq 0$:
>
> $|u| < a$ means that $-a < u < a$.
>
> $|u| \leq a$ means that $-a \leq u \leq a$.
>
> $|u| > a$ means that $u < -a$ or $u > a$.
>
> $|u| \geq a$ means that $u \leq -a$ or $u \geq a$.

We can solve inequalities involving absolute values algebraically or graphically.

example 8

Solve the following inequalities and verify the solutions graphically:

a. $|2x - 3| \leq 5$ **b.** $|3x + 4| - 5 > 0$

Solution

a. This inequality is equivalent to $-5 \leq 2x - 3 \leq 5$. If we add 3 to all three parts of this inequality and then divide all parts by 2, we get the solution.

$$-5 \leq 2x - 3 \leq 5$$
$$-2 \leq 2x \leq 8$$
$$-1 \leq x \leq 4$$

Figure 4.43(a) on the next page shows that the graph of $y = |2x - 3|$ is on or below the graph of $y = 5$ for $-1 \leq x \leq 4$.

b. This inequality is equivalent to $|3x + 4| > 5$, which is equivalent to $3x + 4 < -5$ or $3x + 4 > 5$. We solve each of the two inequalities as follows:

$$3x + 4 < -5 \quad \text{or} \quad 3x + 4 > 5$$
$$3x < -9 \qquad\qquad 3x > 1$$
$$x < -3 \qquad\qquad x > \frac{1}{3}$$

Figure 4.43(b) shows that the graph of $y = |3x + 4| - 5$ is above the x-axis for $x < -3$ or $x > \frac{1}{3}$.

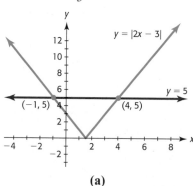

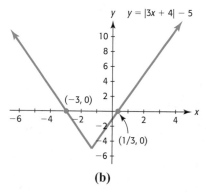

(a)   (b)

Figure 4.43

skills check

4.4

In Exercises 1–12, solve the following equations algebraically and check graphically or by substitution.

1. $\sqrt{2x^2 - 1} - x = 0$ 2. $\sqrt{3x^2 + 4} - 2x = 0$

3. $\sqrt[3]{x - 1} = -2$ 4. $\sqrt[3]{4 - x} = 3$

5. $\sqrt{3x - 2} + 2 = x$ 6. $\sqrt{x - 2} + 2 = x$

7. $\sqrt[3]{4x + 5} = \sqrt[3]{x^2 - 7}$

8. $\sqrt{5x - 6} = \sqrt{x^2 - 2x}$

9. $\sqrt{x - 1} = \sqrt{x - 5}$

10. $\sqrt{x - 10} = -\sqrt{x - 20}$

11. $(x + 4)^{2/3} = 9$ 12. $(x - 5)^{3/2} = 64$

In Exercises 13–22, use algebraic methods to solve the inequalities.

13. $x^2 + 4x < 0$ 14. $x^2 - 25x < 0$

15. $9 - x^2 \geq 0$ 16. $x > x^2$

17. $-x^2 + 9x - 20 > 0$

18. $2x^2 - 8x < 0$

19. $2x^2 - 8x \geq 24$ 20. $t^2 + 17t \leq 8t - 14$

21. $x^2 - 6x < 7$ 22. $4x^2 - 4x + 1 > 0$

Use graphical methods to solve the inequalities in Exercises 23–26.

23. $2x^2 - 7x + 2 \geq 0$ 24. $w^2 - 5w + 4 > 0$

25. $5x^2 \geq 2x + 6$ 26. $2x^2 \leq 5x + 6$

In Exercises 27–34, solve the inequality by using algebraic and graphical methods.

27. $(x + 1)^3 < 4$ 28. $(x - 2)^3 \geq -2$

29. $(x - 3)^5 < 32$ 30. $(x + 5)^4 > 16$

31. $|2x - 1| < 3$ 32. $|3x + 1| \leq 5$

33. $|x - 6| \geq 2$ 34. $|x + 8| > 7$

In Exercises 35–38, you are given the graphs of several functions of the form $f(x) = ax^2 + bx + c$ for different values of a, b, and c. For each function,

a. Solve $f(x) \geq 0$ **b.** Solve $f(x) < 0$

35.

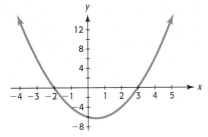

36.

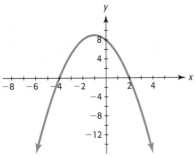

37.

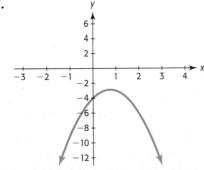

38.

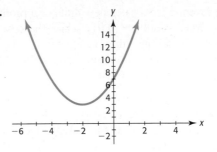

In Exercises 39 and 40, you are given the graphs of two functions f(x) and g(x). Solve $f(x) \leq g(x)$.

39.

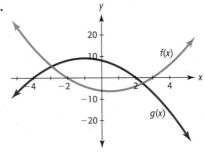

40.

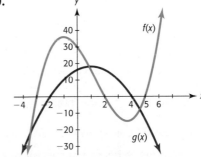

exercises

4.4

Use algebraic and/or graphical methods to solve Exercises 41–50.

41. *Profit* The monthly profit from producing and selling x units of a product is given by

$$P(x) = -0.3x^2 + 1230x - 120,000$$

Producing and selling how many units will result in a profit for this product?

42. *Profit* The monthly profit from producing and selling x units of a product is given by

$$P(x) = -0.01x^2 + 62x - 12,000$$

Producing and selling how many units will result in a profit for this product?

43. *Profit* The revenue from sales of x units of a product is given by $R(x) = 200x - 0.01x^2$, and the cost of producing and selling the product is $C(x) = 38x + .01x^2 + 16,000$. Producing and selling how many units will result in a profit?

44. *Profit* The revenue from sales of x units of a product is given by $R(x) = 600x - 0.01x^2$, and the cost of producing and selling the product is $C(x) = 77x + 0.02x^2 + 52,000$. Producing and selling how many units will result in a profit?

45. *Projectiles* Two projectiles are fired into the air over a lake, with the height of the first projectile given by $y = 100 + 130t - 16t^2$ and the height of the second projectile given by $y = -16t^2 + 180t$, where y is in feet and t is in seconds. Over what time interval, before the lower one hits the lake, is the second projectile above the first?

46. *Projectile* A rocket shot into the air has height $s = 128t - 16t^2$ feet, where t is the number of seconds after the rocket is shot. During what time after the rocket is shot is it at least 240 feet high?

47. *Tobacco Sales* The domestic sales of tobacco (in millions of kilograms) in Canada is given by $y = -0.084x^2 + 1.124x + 4.028$, where x is the number of years after 1986. During what years does this model indicate that the sales will be at least 5,940,000 kilograms?
(Source: Canadian Council on Smoking and Health)

48. *World Population* The low long-range world population numbers and projections for the years 1995–2150 are given by the equation $y = -0.00036x^2 + 0.0385x + 5.823$, where x is the number of years after 1990 and y is in billions. During what years does this model estimate that the population was above 6 billion?
(Source: U.N. Department of Economic and Social Affairs)

49. *Foreign-Born Population* Suppose the percent of the U.S. population that is foreign born is given by $y = 0.003x^2 - 0.390x + 18.800$, where x is the number of years after 1900. During what years does this model indicate that the percent is at most 18.8?
(Source: U.S. Census Bureau)

50. *Gross Domestic Product* The U.S. gross domestic product (in billions of constant dollars) can be modeled by the equation $y = 3.99x^2 - 432.50x + 12,862.21$, where x is the number of years after 1900. During what years prior to 2010 is the gross domestic product less than $4 trillion?
(Source: Microsoft Encarta 98)

Use graphical and/or numerical methods to solve Exercises 51–56.

51. *Homicide Rate* The homicide rate (number per 100,000 people) is given by the function $r = -0.00965x^2 +$

$0.629x - 0.847$, where x is the number of years after 1950. During what years after 1950 does this model indicate that the rate will be above 8 per 100,000 people?
(Source: Bureau of Justice Statistics)

52. *Hotel Room Supply* The percentage change p (from the previous year) in the hotel room supply is given by $p = 0.025x^2 - 0.645x + 4.005$, where x is the number of years after 1998. If this model is accurate, during what years after 1998 will the percentage change be less than 3%?
(Source: Hotel and Motel Management)

53. *Drug Use* The percent of high school students who have ever used marijuana for the years 1990 through 2006 is given by $y = -0.1940x^2 + 4.0142x + 27.861$, where x is the number of years from 1990. Use the model to estimate the years when the percent who ever used marijuana is greater than 42.4%.

54. *Tobacco Sales* The domestic sales of tobacco (in millions of kilograms) in Canada is given by $y = -0.084x^2 + 1.124x + 4.028$, where x is the number of years after 1986. During what years from 1986 does this model indicate that the sales will exceed 6,643,000 kilograms?
(Source: Canadian Council on Smoking and Health)

55. *Airplane Crashes* The number of all airplane crashes (in thousands) is given by the equation $y = 0.0057x^2 - 0.197x + 3.613$, where x is the number of years after 1980. During what years from 1980 through 2015 does the model indicate the number of crashes as being below 3146?

56. *Marriage Age* Based on data appearing in the *World Almanac*, the median age at first marriage for women in the United States can be described by $A = 0.0024x^2 + 0.0418x + 20.24$ years of age, where x is the number of years after 1960. During what years does this model indicate that the median age at first marriage for women is at least 23?

57. *Car Design* Sports cars are designed so that the driver's seat is comfortable for persons with height 5 feet 8 inches, plus or minus 8 inches.

a. Write an absolute value inequality that gives the height x of a person who will be uncomfortable.

b. Solve this inequality for x to identify the heights of people who are uncomfortable.

58. *Voltage* Required voltage for an electric oven is 220 volts, but it will function normally if the voltage varies from 220 by 10 volts.

a. Write an absolute value inequality that gives the voltage x for which the oven will work normally.

b. Solve this inequality for x.

chapter

4

Summary

In this chapter, we presented the building blocks for function construction. One way to construct new functions is by horizontal or vertical transformations, stretching or compressing, and reflection. We can also construct new functions by combining functions algebraically or with function composition. Inverse functions can be used to "undo" the operations in a function. We finished the chapter by solving additional equations and inequalities.

Key Concepts and Formulas

4.1 Transformations of Graphs and Symmetry

Vertical shifts of graphs	If k is a positive real number, • The graph of $g(x) = f(x) + k$ can be obtained by shifting the graph of $f(x)$ upward k units. • The graph of $g(x) = f(x) - k$ can be obtained by shifting the graph of $f(x)$ downward k units.
Horizontal shifts of graphs	If h is a positive real number, • The graph of $g(x) = f(x - h)$ can be obtained by shifting the graph of $f(x)$ to the right h units. • The graph of $g(x) = f(x + h)$ can be obtained by shifting the graph of $f(x)$ to the left h units.
Stretching and compressing graphs	The graph of $y = a f(x)$ is obtained by stretching the graph of $f(x)$ by a factor of $\|a\|$ if $\|a\| > 1$ and compressing the graph of $f(x)$ by a factor of $\|a\|$ if $\|a\| < 1$.
Reflections of graphs across the coordinate axes	The graph of $y = -f(x)$ can be obtained by reflecting the graph of $y = f(x)$ across the x-axis. The graph of $y = f(-x)$ can be obtained by reflecting the graph of $y = f(x)$ across the y-axis.

Symmetry with respect to the y-axis	The graph of $y = f(x)$ is symmetric with respect to the y-axis if $f(-x) = f(x)$.
Symmetry with respect to the origin	The graph of $y = f(x)$ is symmetric with respect to the origin if $f(-x) = -f(x)$.
Symmetry with respect to the x-axis	The graph of an equation is symmetric with respect to the x-axis if, for every point (x, y) on the graph, the point $(x, -y)$ is also on the graph.

4.2 Combining Functions; Composite Functions

Operations with functions	*Sum*: $(f + g)(x) = f(x) + g(x)$ *Difference*: $(f - g)(x) = f(x) - g(x)$ *Product*: $(f \cdot g)(x) = f(x) \cdot g(x)$ Quotient: $\left(\dfrac{f}{g}\right)(x) = \dfrac{f(x)}{g(x)}, g(x) \neq 0$ The domain of the sum, difference, and product of f and g consists of all real numbers of the input variable for which f and g are defined. The domain of the quotient function consists of all real numbers for which f and g are defined and $g \neq 0$.
Revenue, cost, and profit	If a company sells x units of a product for p dollars per unit, then the total revenue for this product can be modeled by the linear function $R(x) = px$. The total cost is: cost = variable costs + fixed costs (Note: Variable costs depend on the number of units produced.) The profit is: profit = revenue − cost or $P(x) = R(x) - C(x)$
Composite functions	The notation for the composite function "f of g" is $(f \circ g)(x) = f(g(x))$. The domain of $f \circ g$ is the subset of the domain of g for which $f \circ g$ is defined.

4.3 Inverse Functions

Inverse functions	If $f(g(x)) = x$ and $g(f(x)) = x$ for all x in the domain of g and f, then f and g are inverse functions. In addition, the functions f and g are inverse functions if, whenever the pair (a, b) satisfies $y = f(x)$, the pair (b, a) satisfies $y = g(x)$. We denote g by f^{-1}, read as "f inverse."
One-to-one functions	For a function f to have an inverse, f must be one-to-one. A function f is one-to-one if each output of f corresponds to exactly one input of f.
Horizontal line test	If no horizontal line can intersect the graph of a function in more than one point, then the function is one-to-one.
Finding the equation of the inverse function	To find the inverse of the function f that is defined by the equation $y = f(x)$: 1. Rewrite the equation with y replacing $f(x)$. 2. Interchange x and y in the equation defining the function. 3. Solve the new equation for y. If this equation cannot be solved uniquely for y, the function has no inverse. 4. Replace y by $f^{-1}(x)$.

Graphs of inverse functions	The graphs of a function and its inverse are symmetric with respect to the line $y = x$.
Inverse functions on limited domains	If a function is one-to-one on a limited domain, then it has an inverse function on that domain.

4.4 Additional Equations and Inequalities

Radical equations	An equation containing radicals can frequently be converted to an equation that does not contain radicals by raising both sides of the equation to a power that is equal to the index of the radical. The solutions must be checked when using this technique.
Equations with rational powers	Some equations with rational powers can be solved by writing the equation as a radical equation.
Quadratic inequality	A quadratic inequality is an inequality that can be written in the form $ax^2 + bx + c > 0$, where a, b, c are real numbers and $a \neq 0$ (or with $>$ replaced by $<, \geq,$ or $\leq$).
Solving a quadratic inequality algebraically	To solve a quadratic inequality $f(x) < 0$ or $f(x) > 0$ algebraically, solve $f(x) = 0$ and use the solutions to divide the number line into intervals. Pick a test value in each interval to determine whether $f(x)$ is positive or negative in that interval and identify the intervals that satisfy the original inequality.
Solving a quadratic inequality graphically	To solve a quadratic inequality $f(x) < 0$ (or $f(x) > 0$) graphically, graph the related equation $y = f(x)$ and observe the x-values of the intervals where the graph of $f(x)$ is below (or above) the x-axis.
Power inequalities	To solve a power inequality, first solve the related equation. Then use graphical methods to find the values of the variable that satisfy the inequality.
Absolute value inequalities	If $a > 0$, the solution to $|u| < a$ is $-a < u < a$. If $a > 0$, the solution to $|u| > a$ is $u < -a$ or $u > a$.

chapter

4

Skills Check

1. How is the graph of $g(x) = (x - 8)^2 + 7$ transformed from the graph of $f(x) = x^2$?

2. How is the graph of $g(x) = -2(x + 1)^3$ transformed from the graph of $f(x) = x^3$?

3. **a.** Graph the functions $f(x) = \sqrt{x}$ and $g(x) = \sqrt{x + 2} - 3$.

   b. How are the graphs related?

4. What is the domain of the function $g(x) = \sqrt{x + 2} - 3$?

5. Suppose the graph of $f(x) = x^{1/3}$ is shifted up 4 units and to the right 6 units. What is the equation that gives the new graph?

6. Suppose the graph of $f(x) = x^{1/3}$ is vertically stretched by a factor of 3 and then shifted down 5 units. What is the equation that gives the new graph?

For Exercises 7–9, match each graph with the correct equation.

a. $y = |x| + 2$

b. $y = x^3$

c. $y = \dfrac{3}{x - 1} + 1$

d. $y = \dfrac{-3}{x + 1} + 1$

e. $y = 2x^3$

f. $y = |x + 2|$

7.

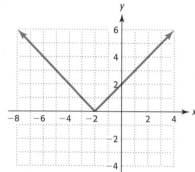

8.

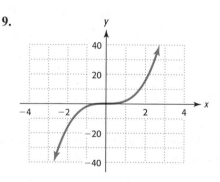

9.

In Exercises 10 and 11, determine algebraically whether the graph of the function is symmetric with respect to the x-axis, y-axis, and/or the origin. Confirm graphically.

10. $f(x) = x^3 - 4x$ **11.** $f(x) = -x^2 + 5$

12. Determine whether the function $f(x) = -\dfrac{2}{x}$ is even, odd, or neither.

For Exercises 13–20, use the functions $f(x) = 3x^2 - 5x$; $g(x) = 6x - 4$; $h(x) = 5 - x^3$ to find the following.

13. $(f + g)(x)$ **14.** $(h - g)(x)$

15. $(g \cdot f)(x)$ **16.** $\left(\dfrac{h}{g}\right)(x)$

17. $(f - g)(-2)$ **18.** $(f \circ g)(x)$

19. $(g \circ f)(x)$ **20.** $(g \circ h)(-3)$

21. For the functions $f(x) = 2x - 5$ and $g(x) = \dfrac{x + 5}{2}$,

    a. Find $f(g(x))$ and $g(f(x))$.

    b. What is the relationship between $f(x)$ and $g(x)$?

22. Find the inverse of $f(x) = 3x - 2$.

23. Find the inverse of $g(x) = \sqrt[3]{x - 1}$.

24. Graph $f(x) = (x + 1)^2$ and its inverse $f^{-1}(x)$ on the domain $[-1, 10]$.

25. Is the function $f(x) = \dfrac{x}{x - 1}$ a one-to-one function?

26. Does the function $f(x) = x^2 - 3x$ have an inverse?

27. Solve $\sqrt{4x^2 + 1} = 2x + 2$.

28. Solve $\sqrt{3x^2 - 8} + x = 0$.

29. Solve the inequality $x^2 - 7x \le 18$.

30. Solve the inequality $2x^2 + 5x \ge 3$.

31. Solve $|2x - 4| \le 8$.

32. Solve $|4x - 3| \ge 15$.

33. Solve $(x - 4)^3 < 4096$.

34. Solve $(x + 2)^2 \ge 512$.

chapter

4 Review

35. *Ballistics* Ballistic experts are able to identify the weapon that fired a certain bullet by studying the markings on the bullet. If a test is conducted by firing the bullet into a bale of paper, the distance that the bullet will travel is given by $s = 64 - (4 - 10t)^3$ inches, for $0 \le t \le 0.4$, t in seconds.

   a. Graph this function for $0 \le t \le 0.4$.

   b. How far does the bullet travel during this time period?

36. *Projectile* A toy rocket is fired into the air from the top of a building, with its height given by $S = -16(t - 4)^2 + 380$ feet, with t in seconds.

   a. In how many seconds will the rocket reach its maximum height?

   b. What is the maximum possible height?

   c. Describe the transformations needed to obtain the graph of $S(t)$ from the graph of $y = t^2$.

37. *Aircraft Accidents* According to www.1001crash.com, your chances of being involved in an aircraft accident during a flight are about 1 in 2 million, considering that the average flight time is 5 hours. The number of airplane crashes during the years 2000–2006 can be modeled by the equation $y = 0.589x^2 - 4.161x + 27.464$, where x is the number of years after 2000.

   a. During what year does this model indicate that the number of crashes was at a minimum?

   b. Use a graphing utility to determine the years during which the annual number of crashes is below 25.

38. *Homicide Rate* The homicide rate (the number of homicides per 100,000 people) is given by the function $y = -0.00965x^2 + 0.629x - 0.847$, where x is the number of years after 1950. Use a graphing utility to find during what years this model indicates that the rate will be above 8 per 100,000 people.
(Source: Bureau of Justice Statistics)

39. *Hotel Room Supply* The percentage change p (from the previous year) in the hotel room supply is given by $p = 0.025x^2 - 0.645x + 4.005$, where x is the number of years after 1998. If this model is accurate, use a graphing utility to find during what years between 1998 and 2010 the percentage change will fall below 3%.
(Source: Hotel and Motel Management)

40. *Average Cost* Suppose the total cost function for a product is determined to be $C(x) = 30x + 3150$, where x is the number of units produced and sold.

   a. Write the average cost function, which is the quotient of two functions.

   b. Graph this function for $0 < x \le 20$.

41. *Supply and Demand* The price per unit of a product is $\$p$ and the number of units of the product is denoted by q. Suppose the supply function for a product is given by $p = \dfrac{180 + q}{6}$, and the demand for the product is given by $p = \dfrac{300}{q} - 20$.

   a. Which of these functions is a shifted reciprocal function?

   b. Graph the supply and demand functions on the same axes.

42. *Supply and Demand* The supply function for a commodity is given by $p = 58 + \dfrac{q}{2}$, and the demand function for this commodity is given by $p = \dfrac{2555}{q + 5}$.

   a. Which of these functions is a shifted reciprocal function?

   b. Graph the supply and demand functions on the same axes.

43. *Marijuana Use* For the years 1991 through 2006, the percent M of high school seniors who have tried marijuana can be considered as a function of the time x according to the model

$$M(x) = -0.2(x - 10.3)^2 + 48.968$$

where x is the number of years after 1990.

a. Describe the transformations of the graph of $f(x) = x^2$ needed to obtain the graph of M.

b. Find and interpret $M(1)$, $M(2)$, $M(4)$, and $M(6)$.

c. Using the information in parts (a) and (b), sketch a graph of M.

d. The survey on which M is based was taken once during each of the selected years. State the domain of the related discrete function.

44. *Southwest Airlines* The number of employees of Southwest Airlines can be modeled by the function $E(t) = 0.017t^2 + 2.164t + 8.061$ thousand employees, where t is the number of years after 1990. Southwest Airlines' revenue during the same time period is given by the model $R(E) = 0.165E - 0.226$ billion dollars when there are E thousand employees of the company.

a. Find and interpret the meaning of the function $R(E(t))$.

b. Use the result of part *a* to find $R(E(3))$. Interpret this result.

c. How many employees worked for Southwest Airlines in 1997?

d. What was the 1997 revenue for this airline?
   (Source: Hoover's Online Capsules)

45. *Prison Sentences* The mean time in prison y for certain crimes can be found as a function of the mean sentence length x, using $f(x) = 0.554x - 2.886$, where x and y are measured in months.

a. Find the inverse of this function.

b. Interpret the inverse function from part (a).
   (Source: Index of Leading Cultural Indicators)

46. *Dairy Cows* The number of cows and heifers kept for milk production in the United States can be approximated by the function $C(x) = -0.093x + 9.929$ million head, where x is the number of years after 1990.

a. Find a formula for the inverse of the function C.

b. Interpret your answer to part (a).

c. If the inverse function is called C^{-1}, find $C^{-1}(C(8))$.
   (Source: *Statistical Abstract of the United States, 1998*)

47. *Computer Usage* The number of students per computer in U.S. public schools from the 1983–1984 school year through the 2003–2004 school year can be modeled by the function $f(t) = 750.487t^{-1.619}$, where t is the number of years after the beginning of the 1980–1981 school year.

a. Another function that might be used to model this data is

$$C(t) = \frac{380}{t + 0.3} - 15 \text{ students per computer}$$

where t is the number of years after the beginning of the 1980–1981 school year. What is the basic function that can be transformed to obtain C? Describe the transformations.

b. Do you feel that the function $y = f(t)$ above or $y = C(t)$ better fits the data? Why?

c. There were 5.7 students per computer in 1998–1999. Which function, $y = f(t)$ or $y = C(t)$, comes closer to estimating the actual value? Does this result agree with your thoughts in part (b)?

48. *Profit* The monthly profit from producing and selling x units of a product is given by

$$P(x) = -0.01x^2 + 62x - 12{,}000$$

Producing and selling how many units will result in profit for this product ($P(x) > 0$)?

49. *Personal Savings Rate* The personal savings rate for Americans for certain years from 1960 to 2006 can be modeled by the equation

$$y = -0.008x^2 + 0.21x + 7.04$$

where x is the number of years after 1960 and y is the rate as a percent. Use graphical or numerical methods to find the years after 1960 during which the model indicates that the personal savings rate is above 4%.

group activities / extended applications

1. Cost, Revenue, and Profit

The following table gives the weekly revenue and cost, respectively, for a selected number of units of production and sale of a product by the Quest Manufacturing Company.

Number of Units	Revenue ($)	Number of Units	Cost ($)
100	6,800	100	32,900
300	20,400	300	39,300
500	34,000	500	46,500
900	61,200	900	63,300
1,400	95,200	1,400	88,800
1,800	122,400	1,800	112,800
2,500	170,000	2,500	162,500

Provide the information requested and answer the questions.

1. Use technology to determine the functions that model revenue and cost functions for this product, using x as the number of units produced and sold.
2. **a.** Combine the revenue and cost functions with the correct operation to create the profit function for this product.
   b. Use the profit function to complete the following table:

x (units)	Profit, P(x) ($)
0	
100	
600	
1600	
2000	
2500	

3. Find the number of units of this product that must be produced and sold to break even.
4. Find the maximum possible profit and the number of units that gives the maximum profit.
5. **a.** Use operations with functions to create the average cost function for the product.
   b. Complete the following table.

x (units)	Average Cost, C(x) ($/unit)
1	
100	
300	
1400	
2000	
2500	

   c. Graph this function using the viewing window [0, 2500] by [0, 400].
6. Graph the average cost function using the viewing window [0, 4000] by [0, 100]. Determine the number of units that should be produced to minimize the average cost and the minimum average cost.
7. Compare the number of units that produced the minimum average cost with the number of units that produced the maximum profit. Are they the same number of units? Discuss which of these values is more important to the manufacturer and why.

2. Cell Phone Revenue

The table in the right column gives the number of U.S. cell phone subscribers and the average monthly bill for the years 1998–2005. To investigate the average monthly revenue for cell phone companies over the time period from 1998–2005, answer the following.

1. Find a quadratic function $S = f(t)$ that models the number of subscribers, in millions, as a function of the number of years from 1995. Write the model with 3 decimal places.
2. Graph the data and the model to visually determine if the model is a good fit for the subscriber data.
3. Find a quadratic function $D = g(t)$ that models the average monthly cell phone bill, in dollars, as a function of the number of years from 1995. Write the model with 3 decimal places.
4. Graph the data and the model to visually determine if the model is a good fit for the billing data.
5. Using models from steps 3 and 4 with coefficients rounded to one decimal place, find the product of these two functions to find a function that models the average monthly revenue for the cell phone companies.
6. Graph the data points representing the years and average revenue on the same axes as the function found in step 5. Is it a good fit? Explain.
7. Create a new column of revenue data by multiplying the number of subscribers in each year times the average monthly bill for that year.

8. Find a quadratic function that is a good fit for the data representing the years after 1995 and average revenue. Is your model a better fit than the function found in step 5?
9. Find a 4th degree (quartic on a graphing calculator) function that models the data representing the years after 1995 and average revenue. Does this model or the model found in step 5 fit the data better?

Year	Number of Subscribers (Millions)	Average Monthly Bill (Dollars)
1998	69.209	39.43
1999	86.047	41.24
2000	109.478	45.27
2001	128.375	47.37
2002	140.766	48.40
2003	158.722	49.91
2004	182.140	50.64
2005	207.896	49.98

(CTIA–The Wireless Association)

Exponential and Logarithmic Functions

The intensities of earthquakes like the one in San Francisco in 1989 are measured with the Richter scale, which uses logarithmic functions. Logarithmic functions are also used to measure loudness (in decibels) and stellar magnitude and in calculating pH values. Exponential functions are used in many real-world settings, such as the growth of investments, the growth of populations, sales decay, and radioactive decay.

Algebra Toolbox

key concepts

- Properties of exponents
- Real number exponents
- Exponential expressions
- Scientific notation

Additional Properties of Exponents

In this toolbox, we discuss properties of integer and real exponents, and exponential expressions. These properties are useful in the discussion of exponential functions and of logarithmic functions, which are related to exponential functions. Integer exponents and rational exponents were discussed in Chapter 3, as well as the Product Property and Quotient Property.

For real numbers a and b, and integers m and n,

1. $a^m \cdot a^n = a^{m+n}$ (Product Property)

2. $\dfrac{a^m}{a^n} = a^{m-n}, a \neq 0$ (Quotient Property)

Additional properties of exponents, which can be developed using the properties above, follow:

For real numbers a and b, and integers m and n,

3. $(ab)^m = a^m b^m$ (Power of a Product Property)

4. For $b \neq 0$, $\left(\dfrac{a}{b}\right)^m = \dfrac{a^m}{b^m}$ (Power of a Quotient Property)

5. $(a^m)^n = a^{mn}$ (Power of a Power Property)

6. $a^{-m} = \dfrac{1}{a^m}$, for $a \neq 0$

For example, we can prove Property 3 for positive integer m as follows:

$$(ab)^m = \underbrace{(ab)(ab)\cdots(ab)}_{m \text{ times}} = \underbrace{(a \cdot a \cdots a)}_{m \text{ times}}\underbrace{(b \cdot b \cdots b)}_{m \text{ times}} = a^m b^m$$

example 1

Using the Properties of Exponents

Use properties of exponents to simplify each of the following. Assume that denominators are nonzero.

a. $\dfrac{5^6}{5^4}$ b. $\dfrac{y^2}{y^5}$ c. $(3xy)^3$ d. $\left(\dfrac{y}{z}\right)^4$ e. $3^{15-2m} \cdot 3^{2m}$

Solution

a. $\dfrac{5^6}{5^4} = 5^{6-4} = 5^2 = 25$ b. $\dfrac{y^2}{y^5} = y^{2-5} = y^{-3} = \dfrac{1}{y^3}$ c. $(3xy)^3 = 3^3 x^3 y^3 = 27x^3 y^3$

d. $\left(\dfrac{y}{z}\right)^4 = \dfrac{y^4}{z^4}$ e. $3^{15-2m} \cdot 3^{2m} = 3^{(15-2m)+2m} = 3^{15-2m+2m} = 3^{15}$ ∎

We can also simplify and evaluate expressions involving powers of powers.

example 2 | Powers of Powers

a. Simplify $(x^4)^5$.   **b.** Evaluate $(2^3)^5$.   **c.** Simplify $(x^2y^3)^5$.   **d.** Evaluate 2^{3^2}.

Solution

a. $(x^4)^5 = x^{4 \cdot 5} = x^{20}$

b. We can evaluate $(2^3)^5$ in two ways:

$$(2^3)^5 = 2^{15} = 32{,}768 \quad \text{or} \quad (2^3)^5 = (8)^5 = 32{,}768$$

c. $(x^2y^3)^5 = (x^2)^5(y^3)^5 = x^{10}y^{15}$ **d.** $2^{3^2} = 2^9 = 512$

example 3 | Applying Exponent Properties

Compute the following products and quotients and write the answer with positive exponents.

a. $(-2x^{-2}y)(5x^{-2}y^{-3})$ **b.** $\dfrac{8xy^{-2}}{2x^4y^{-6}}$ **c.** $\dfrac{\dfrac{2x^{-1}y}{3a}}{\dfrac{6xy^{-2}}{5a}}$ **d.** $(4^0x^3y^{-2})^{-3}$

Solution

a. $(-2x^{-2}y)(5x^{-2}y^{-3}) = -10x^{-2+(-2)}y^{1+(-3)} = -10x^{-4}y^{-2}$

$$= -10 \cdot \frac{1}{x^4} \cdot \frac{1}{y^2} = \frac{-10}{x^4y^2}$$

b. $\dfrac{8xy^{-2}}{2x^4y^{-6}} = 4x^{1-4}y^{-2-(-6)} = 4x^{-3}y^4 = 4 \cdot \dfrac{1}{x^3} \cdot y^4 = \dfrac{4y^4}{x^3}$

c. To perform this division, we invert and multiply.

$$\frac{\dfrac{2x^{-1}y}{3a}}{\dfrac{6xy^{-2}}{5a}} = \frac{2x^{-1}y}{3a} \cdot \frac{5a}{6xy^{-2}} = \frac{10ax^{-1}y}{18axy^{-2}} = \frac{5x^{-1-1}y^{1-(-2)}}{9} = \frac{5x^{-2}y^3}{9} = \frac{5y^3}{9x^2}$$

d. $(4^0x^3y^{-2})^{-3} = 4^0x^{-9}y^6 = 1 \cdot \dfrac{1}{x^9} \cdot y^6 = \dfrac{y^6}{x^9}$

Real Number Exponents

All the above properties also hold for rational exponents so long as no negative numbers in even powered roots result, and they also apply for all real numbers that give expressions that are real numbers.

example 4 | Operations with Real Exponents

Perform the indicated operations.

a. $(x^{1/3})(x^{3/4})$ **b.** $\dfrac{y^{1/2}}{y^{2/5}}$ **c.** $(c^{1/3})^{3/4}$

Solution

a. $(x^{1/3})(x^{3/4}) = x^{1/3+3/4} = x^{13/12}$

b. $\dfrac{y^{1/2}}{y^{2/5}} = y^{1/2-2/5} = y^{1/10}$

c. $(c^{1/3})^{3/4} = c^{1/3 \cdot 3/4} = c^{1/4}$ ■

Scientific Notation

Evaluating exponential and logarithmic functions in this chapter may result in outputs that are very large or very close to 0. **Scientific notation** is a convenient way to write very large (positive or negative) numbers or numbers close to 0. Numbers in scientific notation have the form

$$N \times 10^p \text{ (where } 1 \leq N < 10 \text{ and } p \text{ is an integer)}$$

For instance, the scientific notation form of 2,654,000 is 2.654×10^6. When evaluating functions or solving equations with a calculator, we may see an expression that looks like 5.122E-8 appear on the screen. This calculator expression represents 5.122×10^{-8}, which is scientific notation for $0.00000005122 \approx 0$. Figure 5.1 shows two numbers expressed in standard notation and in scientific notation, and the calculator display of scientific notation.

Standard Notation	Scientific Notation	Calculator Display of Scientific Notation
62256	6.2256×10^4	62256
0.000235	2.35×10^{-4}	6.2256E4
		0.000235
		2.35E-4
		■

Figure 5.1

A number written in scientific notation can be converted to standard notation by multiplying (when the exponent on 10 is positive) or by dividing (when the exponent on 10 is negative.) For example, we convert 7.983×10^5 to standard notation by multiplying 7.983 by $10^5 = 100,000$:

$$7.983 \times 10^5 = 7.983 \cdot 100,000 = 798,300$$

and we convert 4.563×10^{-7} to standard notation by dividing 4.563 by $10^7 = 10,000,000$:

$$4.563 \times 10^{-7} = \frac{4.563}{10,000,000} = 0.0000004563$$

Multiplying two numbers in scientific notation involves adding the powers of 10, and dividing them involves subtracting the powers.

example 5 | ## Scientific Notation

Compute the following and write the answers in scientific notation.

a. $(7.983 \times 10^5)(4.563 \times 10^{-7})$ b. $(7.983 \times 10^5)/(4.563 \times 10^{-7})$

Solution

a. $(7.983 \times 10^5)(4.563 \times 10^{-7}) = 36.426429 \times 10^{5+(-7)}$

$= 36.426429 \times 10^{-2} = (3.6426429 \times 10^1) \times 10^{-2} = 3.6426429 \times 10^{-1}$

The calculation using technology is shown in Figure 5.2(a).

b. $(7.983 \times 10^5)/(4.563 \times 10^{-7}) = 1.749506903 \times 10^{5-(-7)}$

$= 1.749506903 \times 10^{12}$. Figure 5.2(b) shows the calculation using technology.

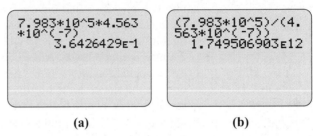

(a)   (b)

Figure 5.2

toolbox exercises

In Exercises 1 and 2, use the properties of exponents to simplify.

1. a. $x^4 \cdot x^3$ **b.** $\dfrac{x^{12}}{x^7}$ **c.** $(4ay)^4$ **d.** $\left(\dfrac{3}{z}\right)^4$

e. $2^3 \cdot 2^2$ **f.** $(x^4)^2$

2. a. $y^5 \cdot y$ **b.** $\dfrac{w^{10}}{w^4}$ **c.** $(6bx)^3$ **d.** $\left(\dfrac{5z}{2}\right)^3$

e. $3^2 \cdot 3^3$ **f.** $(2y^3)^4$

In Exercises 3–18, use the properties of exponents to simplify the following expressions and remove all zero and negative exponents. Assume that all variables are nonzero.

3. 10^{5^0}

4. 4^{2^2}

5. $x^{-4} \cdot x^{-3}$

6. $y^{-5} \cdot y^{-3}$

7. $(c^{-6})^3$

8. $(x^{-2})^4$

9. $\dfrac{a^{-4}}{a^{-5}}$

10. $\dfrac{b^{-6}}{b^{-8}}$

11. $(x^{-1/2})(x^{2/3})$

12. $(y^{-1/3})(y^{2/5})$

13. $(3a^{-3}b^2)(2a^2b^{-4})$

14. $(4a^{-2}b^3)(-2a^4b^{-5})$

15. $\left(\dfrac{2x^{-3}}{x^2}\right)^{-2}$

16. $\left(\dfrac{3y^{-4}}{2y^2}\right)^{-3}$

17. $\dfrac{28a^4b^{-3}}{-4a^6b^{-2}}$

18. $\dfrac{36x^5y^{-2}}{-6x^6y^{-4}}$

In Exercises 19–22, write the following numbers in scientific notation.

19. 46,000,000

20. 862,000,000,000

21. 0.000094

22. 0.00000278

In Exercises 23–26, write each of the following numbers in standard form.

23. 4.372×10^5

24. 7.91×10^6

25. 5.6294×10^{-4}

26. 6.3478×10^{-3}

In Exercises 27 and 28, multiply or divide, as indicated, and write the result in scientific notation.

27. $(6.250 \times 10^7)(5.933 \times 10^{-2})$

28. $\dfrac{2.961 \times 10^{-2}}{4.583 \times 10^{-4}}$

In Exercises 29–34, use the properties of exponents to simplify the following expressions. Assume that all variables are nonzero.

29. $x^{1/2} \cdot x^{5/6}$ **30.** $y^{2/5} \cdot y^{1/4}$

31. $(c^{2/3})^{5/2}$ **32.** $(x^{3/2})^{3/4}$

33. $\dfrac{x^{3/4}}{x^{1/2}}$ **34.** $\dfrac{y^{3/8}}{y^{1/4}}$

5.1

Exponential Functions

key concepts

- Exponential functions
- Horizontal asymptotes
- Exponential growth function
- Exponential decay function
- The number e
- Transformations of graphs of exponential functions

section preview ▪ Paramecia

The primitive single-cell animal called a paramecium reproduces by splitting into two pieces (called binary fission) so that the population doubles each time there is a split. If we assume that the population begins with 1 paramecium and doubles each hour, then there will be 2 paramecia after 1 hour, 4 paramecia after 2 hours, 8 after 3 hours, and so on. If we let y represent the number of paramecia in the population after x hours have passed, the points (x, y) that satisfy this function for the first 9 hours of growth are described by the data in Table 5.1 and the scatter plot in Figure 5.3(a). We can show that each of these points and the data points for $x = 10, 11$, and so forth lie on the graph of the function $y = 2^x$, with $x \geq 0$ (Figure 5.3(b)).

Functions like $y = 2^x$, which have a constant base raised to a variable power, are called **exponential functions**. We discuss exponential functions and their applications in this section.

Table 5.1

x (hours)	0	1	2	3	4	5	6	7	8	9
y (paramecia)	1	2	4	8	16	32	64	128	256	512

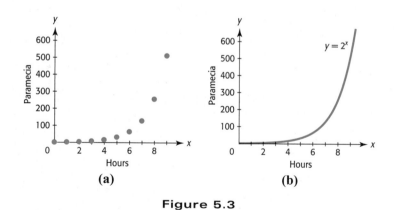

(a)   **(b)**

Figure 5.3

Exponential Functions

Recall that a linear function has a constant rate of change; that is, the outputs change by the same amount for each unit increase in the inputs. An exponential function has outputs that are *multiplied* by a fixed number for each unit increase in the inputs.

Note that in Example 1, the number y of paramecia is multiplied by 2 for each hour increase in time x, so the number is an exponential function of time.

example 1

Paramecia

As discussed in the Section Preview, the primitive single-cell animal called paramecium reproduces by splitting into two pieces so that the population doubles each time there is a split. The number y of paramecia in the population after x hours have passed fits the graph of $y = 2^x$.

a. In the context of the paramecium application, what inputs can be used?

b. Is the function $y = 2^x$ discrete or continuous?

c. Graph the function $y = 2^x$, without regard to the restriction to the inputs that make sense for the paramecium application.

d. What are the domain and range of this function?

Solution

a. Not every point on the graph of $y = 2^x$ in Figure 5.3(b) describes a number of paramecia. For example, the point $(7.1, 2^{7.1})$ is on this graph, but $2^{7.1} \approx 137.187$ does not represent a number of paramecia because it is not a whole number. Since only whole number inputs will give whole number outputs, only whole number inputs can be used for this application.

b. Every point describing a number of paramecia is on the graph of $y = 2^x$, so the function $y = 2^x$ is a continuous model for this discrete paramecium growth function.

c. Some values satisfying the function $y = 2^x$ are shown in Table 5.2, and the graph of the function is shown in Figure 5.4. The function that models the growth of paramecia was restricted because of the physical setting, but if the domain of the function $y = 2^x$ is not restricted, it is the set of real numbers. Note that the range of this function is the set of all positive real numbers because there is no power of 2 that results in a value of 0 or a negative number.

Table 5.2

x	y
-3	$2^{-3} = 0.125$
-1.5	$2^{-1.5} \approx 0.35$
-0.5	$2^{-0.5} \approx 0.71$
0	$2^0 = 1$
0.5	$2^{0.5} \approx 1.41$
1.5	$2^{1.5} \approx 2.83$
3	$2^3 = 8$
5	$2^5 = 32$

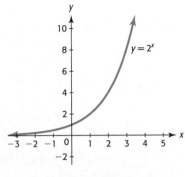

Figure 5.4

The graph in Figure 5.4 approaches but never touches the x-axis as x becomes more negative (approaching $-\infty$), so the x-axis is a **horizontal asymptote** for the graph. As previously mentioned, the function $y = 2^x$ is an example of a special class of functions called **exponential functions**. In general, we define an exponential function as follows.

Exponential Function

If b is a positive real number, $b \neq 1$, then the function $f(x) = b^x$ is an exponential function. The constant b is called the *base* of the function and the variable x is the *exponent*.

If we graph another exponential function $y = b^x$ that has a base b greater than 1, the graph will have the same basic shape as the graph of $y = 2^x$. (See the graphs of $y = 1.5^x$ and $y = 12^x$ in Figure 5.5.)

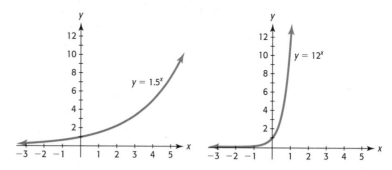

Figure 5.5

Exponential Growth

Exponential functions can be used to model growth in many different applications. Whenever the base b of $y = a(b^{kx})$ is greater than 1 and $a > 0$, $k > 0$, the exponential function is increasing and can be used to model growth.

Exponential Growth Function

Equation:	$y = a(b^{kx})$, $b > 1$, $a > 0$, $k > 0$
x-intercept:	none
y-intercept:	$(0, a)$
Domain:	all real numbers
Range:	all real numbers $y > 0$
Horizontal asymptote:	x-axis (the line $y = 0$)
Shape:	increasing on domain and concave up

As we will see in Section 5.5, the future value S of an investment of P dollars invested for t years at interest rate r, compounded annually, is given by the exponential growth function

$$S = P(1 + r)^t$$

---

example 2 | ## Future Value of an Investment

a. Graph the function that gives the future value of $24,000 invested at 8%, compounded annually, for t years, $0 \leq t \leq 10$.

b. What is the future value of $24,000 invested for 5 years at 8%, compounded annually?

Solution

a. The function that gives this future value is

$$S = 24{,}000(1 + 0.08)^t \quad (0 \le t \le 10)$$

and is graphed in Figure 5.6.

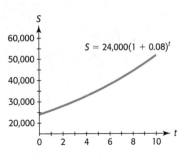

Figure 5.6

b. The future value of $24,000 invested for 5 years at 8%, compounded annually, is
$S = 24{,}000(1 + 0.08)^5 = 35{,}263.87$ dollars.

---

example 3

Head Start Program Growth

Federal funds for the Head Start Program in the U.S. Department of Health and Human Services increased dramatically between 1975 and 2000. The on-budget (federal) funds that were allocated for Head Start during this time period can be described by the model

$$f(x) = 400(1.107)^x \text{ million dollars}$$

where x is the number of years after 1975.

a. Graph this function. Find and interpret the vertical-axis intercept.

b. According to the model, how much was allocated for Head Start in 2000?
(Source: *Federal Support for Education, Fiscal Years 1980–2000*, National Center for Educational Statistics, U.S. Department of Education, September 2000)

Solution

a. The graph is shown in Figure 5.7(a). The vertical-axis intercept is found by evaluating the function at $x = 0$, obtaining

$$f(0) = 400(1.107)^0 = 400$$

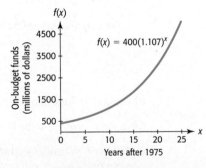

Figure 5.7(a)

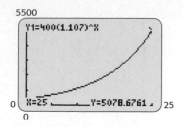

Figure 5.7(b)

Because our data are aligned so that $x = 0$ at the end of 1975, one possible interpretation is this: $400 million in federal funds were allocated for the Head Start Program in 1975.

b. The year 2000 is 25 years after 1975, so we find the allocation for 2000 by evaluating $f(25)$:

$$f(25) = 400(1.107)^{25} \approx 5078.68 \text{ million dollars}$$

so federal funds allocated for Head Start in 2000 amounted to about $5.079 billion. Note that we could find the value of the function at $x = 25$ graphically or numerically by using the table feature of a graphing utility (Figure 5.7(b)). ∎

---

spreadsheet solution

As with linear and quadratic functions, computer software of several types can be used to create exponential graphs, and spreadsheets like Excel can also be used. Figure 5.8 shows the output of $y = 400(1.107)^x$ at several values of x, including $x = 25$, and the graph of this function with Excel.

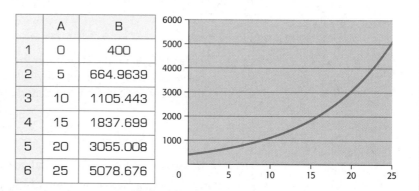

Figure 5.8

Exponential Decay

Replacing x with $-x$ in $y = f(x)$ gives a function whose graph is a reflection of the graph of $y = f(x)$ about the y-axis. For example, the graph of $y = 3^{-x}$ is the reflection of the graph of $y = 3^x$ about the y-axis, and the function $y = 3^{-x}$ decreases rather than increases (Figure 5.9).

We can also write the function $y = 3^{-x}$ in the form $y = \left(\dfrac{1}{3}\right)^x$ because

$$3^{-x} = (3^{-1})^x = \left(\frac{1}{3}\right)^x$$

In general, a function of the form $f(x) = a(b^{-x})$ with $b > 1$ can also be written in the form $f(x) = a(c^x)$ with $0 < c < 1$ and $c = \dfrac{1}{b}$. Functions with equations of this form can be used to model **exponential decay**.

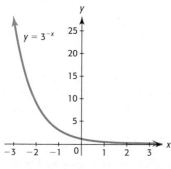

Figure 5.9

Exponential Decay Function

Equation:	$y = a(b^{-kx}), b > 1, a > 0, k > 0$
	or
	$y = a(c^x), 0 < c < 1, a > 0$
x-intercept:	none
y-intercept:	$(0, a)$
Domain:	all real numbers
Range:	$y > 0$
Horizontal asymptote:	*x*-axis (the line $y = 0$)
Shape:	decreasing on domain and concave up

---

example 4

Sales Decay

It pays to advertise, and it is frequently true that weekly sales will drop rapidly for certain products after an advertising campaign ends. This decline in sales is called *sales decay*. Suppose that the decay in the sales of a product is given by

$$S = 1000(2^{-0.5x}) \text{ dollars}$$

where x is the number of weeks after the end of a sales campaign. Use this function to answer the following.

a. What is the level of sales when the advertising campaign ends?

b. What is the level of sales 1 week after the end of the campaign?

c. Use a graph of the function to estimate the week in which sales equal $500.

d. According to this model, will sales ever fall to zero?

Solution

a. The campaign ends when $x = 0$, so $S = 1000(2^{-0.5(0)}) = 1000(2^0) = 1000(1) = \$1000$.

b. At 1 week after the end of the campaign, $x = 1$. Thus, $S = 1000(2^{-0.5(1)}) = \$707.11$.

c. The graph of this sales decay function is shown in Figure 5.10(a) on the next page. One way to find the x-value for which $S = 500$ is to graph $y_1 = 1000(2^{-0.5x})$ and $y_2 = 500$ and find the point of intersection of the two graphs. See Figure 5.10(b), which shows that $y = 500$ when $x = 2$.

Thus, sales fall to half their original amount after 2 weeks.

d. The graph of this sales decay function approaches the positive x-axis as x gets large, but it never reaches the x-axis. Thus, sales will never reach a value of $0.

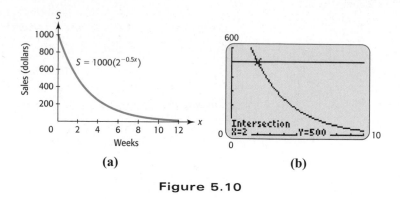

Figure 5.10

The equations of growth and decay functions are related as follows.

Growth and Decay

For initial amount $a > 0$, the equation $y = ab^{kx}$ for $b > 1$ defines a growth function if $k > 0$ and a decay function if $k < 0$.

The Number *e*

Many real applications involve exponential functions with the base e. The number e is an irrational number with decimal approximation 2.718281828 (to nine decimal places).

$$e \approx 2.718281828$$

The exponential function with base e occurs frequently in biology and in finance. We discuss the derivation of e and how it is used in finance in Section 5.5. Because e is close in value to 3, the graph of the exponential function $y = e^x$ has the same basic shape as $y = 3^x$. (See Figure 5.11, which compares the graphs of $y = e^x$, $y = 3^x$, and $y = 2^x$.)

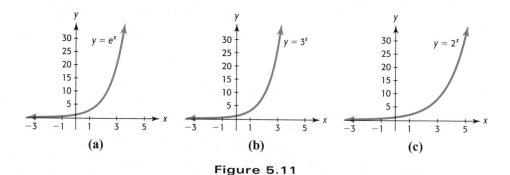

Figure 5.11

The growth in the value of an investment is another important application that can be modeled by an exponential function. As we prove in Section 5.5, the future value S of an investment of P dollars for t years at interest rate r, compounded continuously, is given by

$$S = Pe^{rt} \text{ dollars}$$

| example 5 | ## Growth of an Investment |

If $10,000 is invested for 15 years at 10%, compounded continuously, what is the future value of the investment?

Solution

The future value of this investment is

$$S = Pe^{rt} = 10,000e^{0.10(15)} = 10,000e^{1.5} = 44,816.89 \text{ dollars}$$

| example 6 | ## Comparing Exponential Functions |

Compare the functions $f(x) = 5e^{2x}$ and $g(x) = 5e^{-2x}$ algebraically and graphically.

Solution

To compare these functions algebraically, we see that the base in each function is e and that the function g can be created by replacing x with $-x$ in $f(x) = 5e^{2x}$. Thus, the graph of $g(x) = 5e^{-2x}$ is a reflection of the graph of $f(x) = 5e^{2x}$ about the y-axis. The function $f(x) = 5e^{2x}$ is an exponential growth function, and the function $g(x) = 5e^{-2x}$ is an exponential decay function.

The graphs of $y = f(x)$ in Figure 5.12(a) and $y = g(x)$ in Figure 5.12(b) confirm the results of the algebraic investigation. In addition, both graphs have $(0, 5)$ as the y-intercept and the x-axis as a horizontal asymptote.

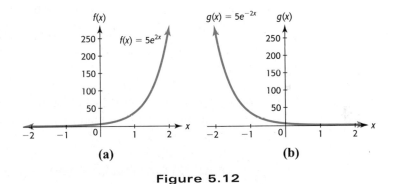

(a)   (b)

Figure 5.12

Note in Example 6 that the exponent of $f(x) = 5e^{2x}$ is $2x$. To see the effect of the 2 in this exponent, we can use a property of exponents* to rewrite $f(x) = 5e^{2x}$ as $f(x) = 5(e^x)^2$, so multiplying the exponent by 2 has the effect of squaring the exponential e^x. In general,

$$y = e^{rx}$$

will increase more rapidly than $y = e^x$ if $r > 1$, will increase more slowly than $y = e^x$ if $0 < r < 1$, and will decrease if $r < 0$.

An exponential decay function can be used to model the amount of a radioactive material that remains after a period of time.

* See the Algebra Toolbox at the beginning of this chapter.

| example 7 | **Carbon-14 Dating** |

During the life of an organism, the ratio of carbon-14 (C-14) to carbon-12 (C-12) atoms remains about the same. When the organism dies, the number of C-14 atoms within its carcass begins to gradually decrease, but the C-12 does not, so they can be measured to see how many C-14 atoms were present at death. Comparing the number of C-14 atoms when it is discovered with the number at death can be used to find how long ago the organism lived. The ratio of the present amount of C-14 to the original amount t years ago is given by

$$\frac{y}{y_0} = e^{-0.00012097t}$$

This gives a function

$$y = f(t) = y_0 e^{-0.00012097t}$$

that gives the amount remaining after t years.
(Source: Thomas Jefferson National Accelerator Facility, http://www.jlab.org/)

a. If the original amount of carbon-14 present in an artifact is 100 grams, how much remains after 2000 years?

b. What percent of the original amount of carbon-14 remains after 4500 years?

Solution

a. The initial amount y_0 is 100 grams, and the time t is 2000 years, so we substitute these values in $f(t) = y_0 e^{-0.00012097t}$ and calculate:

$$f(2000) = 100e^{-0.00012097(2000)} \approx 78.51$$

Thus the amount of carbon-14 present after 2000 years is approximately 78.5 grams.

b. If y_0 is the original amount, the amount that will remain after 4500 years is

$$f(4500) = y_0 e^{-0.00012097(4500)} = 0.58y_0$$

Thus, approximately 58% of the original amount will remain after 4500 years. (Note that you do not need to know the original amount of carbon-14 in order to answer this question.) ∎

Transformations of Graphs of Exponential Functions

Graphs of exponential functions, like other types of functions, can be shifted, reflected, or stretched. Transformations, which were introduced in Chapter 4, can be applied to graphs of exponential functions.

| example 8 | **Transformations of Graphs of Exponential Functions** |

a. Explain how the graph of $y = 3^{x-4}$ compares to the graph of $y = 3^x$ and sketch the graphs.

b. Explain how the graph of $y = 3^x - 4$ compares to the graph of $y = 3^x$ and sketch the graphs.

c. Explain how the graph of $y = 2 + 3^{x-4}$ compares to the graph of $y = 3^x$ and sketch the graphs.

d. Compare the graph of $y = 5(3^x)$ with the graph of $y = 3^x$.

Solution

a. The graph of $y = 3^{x-4}$ has the same shape as $y = 3^x$, but it is shifted 4 units to the right. The graph of $y = 3^x$ is shown in Figure 5.13(a) and the graph of $y = 3^{x-4}$ is shown in Figure 5.13(b).

b. The graph of $y = 3^x - 4$ has the same shape as $y = 3^x$, but it is shifted 4 units down. The graph of $y = 3^x - 4$ is shown in Figure 5.13(c). Note that the horizontal asymptote shifted from the x-axis ($y = 0$) to $y = -4$.

c. The graph of $y = 2 + 3^{x-4}$ has the same shape as $y = 3^x$, but it is shifted 4 units to the right and 2 units up. The graph of $y = 2 + 3^{x-4}$ is shown in Figure 5.13(d). The horizontal asymptote is $y = 2$.

d. As we saw in Chapter 4, multiplication of a function by a constant greater than 1 stretches the graph of the function by a factor equal to that constant. Each of the y-values of $y = 5(3^x)$ is 5 times the corresponding y-value of $y = 3^x$. The graph of $y = 5(3^x)$ is shown in Figure 5.13(e).

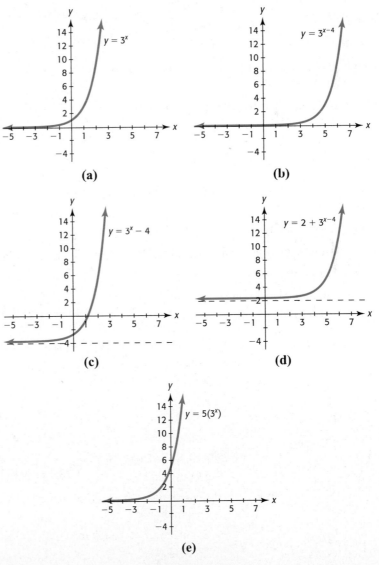

Figure 5.13

skills check

5.1

1. Which of the following functions are exponential functions?

   a. $y = 3x + 3$ **b.** $y = x^3$

   c. $y = 3^x$ **d.** $y = x^2 - 3x$

   e. $y = e^{x^2 - 3x}$ **f.** $y = x^e$

2. Determine if each of the following functions is a growth exponential or a decay exponential.

   a. $y = 2^{0.1x}$ **b.** $y = 3^{-1.4x}$

   c. $y = 4e^{-5x}$ **d.** $y = 0.8^{3x}$

3. **a.** Graph the function $f(x) = e^x$ on the window $[-5, 5]$ by $[-10, 30]$.

   b. Find $f(1), f(-1),$ and $f(4)$, rounded to three decimal places.

   c. What is the horizontal asymptote of the graph?

   d. What is the y-intercept?

4. **a.** Graph the function $f(x) = 5^x$ on the window $[-5, 5]$ by $[-10, 30]$.

   b. Find $f(1), f(3),$ and $f(-2)$.

   c. What is the horizontal asymptote of the graph?

   d. What is the y-intercept?

5. Graph the function $y = 3^x$ on $[-5, 5]$ by $[-10, 30]$.

6. Graph $y = 5e^{-x}$ on $[-4, 4]$ by $[-1, 20]$.

In Exercises 7–12, graph each function.

7. $y = 5^{-x}$

8. $y = 3^{-2x}$

9. $y = 3^x + 5$

10. $y = 2(1.5)^{-x}$

11. $y = 3^{(x-2)} - 4$

12. $y = 3^{(x-1)} + 2$

In Exercises 13–18, match the graph with its equation.

13. $y = 4^{x+3}$ **A.**
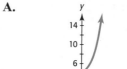

14. $y = 2 \cdot 4^x$ **B.**
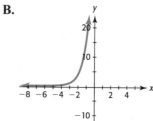

15. $y = 4^x + 3$ **C.**
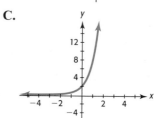

16. $y = 4^{-x}$ **D.**
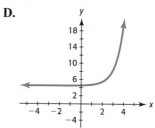

17. $y = -4^x$ **E.**
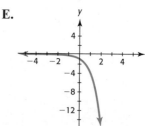

18. $y = 4^{x-2} + 4$ **F.**
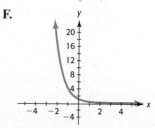

In Exercises 19–24, use your knowledge of transformations to compare the graph of the function with the graph of $f(x) = 4^x$. Then graph each function for $-4 \leq x \leq 4$.

19. $y = 4^x + 2$

20. $y = 4^{x-1}$

21. $y = 4^{-x}$

22. $y = -4^x$

23. $y = 3(4^x)$

24. $y = 3 \cdot 4^{(x-2)} - 3$

25. How does the graph of Exercise 24 compare with the graph of Exercise 23?

26. Which of the functions in Exercises 19–24 are increasing functions, and which are decreasing functions?

27. a. Graph $f(x) = 12e^{-0.2x}$ for $-5 \leq x \leq 15$.

   b. Find $f(10)$ and $f(-10)$.

   c. Does this function represent growth or decay?

28. If $y = 200(2^{-0.01x})$,

   a. Find the value of y (to two decimal places) when $x = 20$.

   b. Use graphical or numerical methods to determine the value of x that gives $y = 100$.

exercises

5.1

29. *Sales Decay* At the end of an advertising campaign, weekly sales declined according to the equation $y = 12{,}000(2^{-0.08x})$ dollars, where x is the number of weeks after the end of the campaign.

   a. Determine the sales at the end of the ad campaign.

   b. Determine the sales 6 weeks after the end of the campaign.

   c. Does this model indicate that sales will eventually reach $0?

30. *Sales Decay* At the end of an advertising campaign, weekly sales declined according to the equation $y = 10{,}000(3^{-0.05x})$ dollars, where x is the number of weeks after the campaign ended.

   a. Determine the sales at the end of the ad campaign.

   b. Determine the sales 8 weeks after the end of the campaign.

   c. How do we know, by inspecting the equation, that this function is decreasing?

31. *Investment* Use the formula $S = P(1 + r)^t$ to:

   a. Graph the function that gives the future value of $80,000 invested at 5%, compounded annually, for t years, $0 \leq t \leq 12$.

   b. Find the future value of $80,000 invested for 10 years at 5%, compounded annually.

32. *Investment* Use the formula $S = P(1 + r)^t$ to:

   a. Graph the function that gives the future value of $56,000 invested at 9%, compounded annually, for t years, $0 \leq t \leq 15$.

   b. Find the future value of $56,000 invested for 13 years at 9%, compounded annually.

33. *Continuous Compounding* If $8000 is invested for t years at 8% interest compounded continuously, the future value is given by $S = 8000e^{0.08t}$ dollars.

   a. Graph this function for $0 \leq t \leq 15$.

   b. Use the graph to estimate when the future value will be $20,000.

   c. Complete the following table.

t (Year)	S ($)
10	
20	
	46,449.50

34. *Continuous Compounding* If $35,000 is invested for t years at 9% interest compounded continuously, the future value is given by $S = 35{,}000e^{0.09t}$ dollars.

a. Graph this function for $0 \leq t \leq 8$.

b. Use the graph to estimate when the future value will be $60,000.

c. Complete the following table.

t (Year)	S ($)
10	
15	
	253,496

35. *Radioactive Decay* The amount of radioactive isotope thorium-234 present at time t is given by $A(t) = 500e^{-0.02828t}$ grams, where t is the time in years that the isotope decays. The initial amount present is 500 grams.

   a. How many grams remain after 10 years?

   b. Graph this function for $0 \leq t \leq 100$.

   c. If the half-life is the time it takes for half of the initial amount to decay, use graphical methods to estimate the half-life of this isotope.

36. *Radioactive Decay* A breeder reactor converts stable uranium-238 into the isotope plutonium-239. The decay of this isotope is given by $A(t) = 100e^{-0.00002876t}$, where $A(t)$ is the amount of the isotope at time t (in years) and 100 grams is the original amount.

   a. How many grams remain after 100 years?

   b. Graph this function for $0 \leq t \leq 50,000$.

   c. The half-life is the time it takes for half of the initial amount to decay; use graphical methods to estimate the half-life of this isotope.

37. *Sales Decay* At the end of an advertising campaign, weekly sales at an electronics store declined according to the equation $y = 2000(2^{-0.1x})$ dollars, where x is the number of weeks after the end of the campaign.

   a. Graph this function for $0 \leq x \leq 60$.

   b. Use the graph to find the weekly sales 10 weeks after the campaign ended.

   c. Comment on "It pays to advertise" for this store.

38. *Sales Decay* At the end of an advertising campaign, weekly retail sales of a product declined according to the equation $y = 40,000(3^{-0.1x})$ dollars, where x is the number of weeks after the campaign ended.

   a. Graph this function for $0 \leq x \leq 50$.

   b. Find the weekly sales 10 weeks after the campaign ended.

   c. Should the retailers consider another advertising campaign even if it costs $5000?

39. *Purchasing Power* The purchasing power (real value of money) decreases if inflation is present in the economy. For example, the purchasing power of R after t years of 5% inflation is given by the model

$$P = R(0.95^t) \text{ dollars}$$

   a. What will be the purchasing power of $40,000 after 20 years of 5% inflation?

   b. How does this impact people planning to retire at age 50?
   (*Source: Viewpoints*, VALIC, 1993)

40. *Purchasing Power* If a retired couple has a fixed income of $60,000 per year, the purchasing power (real value of money) after t years of 5% inflation is given by the equation $P = 60,000(0.95^t)$ dollars.

   a. What is the purchasing power of their income after 4 years?

   b. Test numerical values in this function to determine the years when their purchasing power will be less than $30,000.

41. *Real Estate Inflation* During a 5-year period of constant inflation, the value of a $100,000 property will increase according to the equation $v = 100,000e^{0.05t}$.

   a. What will be the value of this property in 4 years?

   b. Use a table or graph to estimate when this property will double in value.

42. *Inflation* An antique table increases in value according to the function $v(x) = 850(1.04^x)$ dollars, where x is the number of years after 1990.

   a. How much was the table worth in 1990?

b. If the pattern indicated by the function remains valid, what was the value of the table in 2005?

c. Use a table or graph to estimate the year when this table will reach double its 1990 value.

43. *Population* The population in a certain city was 53,000 in 2000, and its future size is predicted to be $P(t) = 53,000e^{0.015t}$ people, where t is the number of years after 2000.

    a. Does this model indicate that the population is increasing or decreasing?

    b. Use this function to estimate the population of the city in 2005.

    c. Use this function to predict the population of the city in 2010.

    d. What is the average rate of growth between 2000 and 2010?

44. *Population* The population in a certain city was 800,000 in 2003, and its future size is predicted to be $P = 800,000e^{-0.020t}$ people, where t is the number of years after 2003.

    a. Does this model indicate that the population is increasing or decreasing?

    b. Use this model to predict the population of the city in 2010.

    c. Use this model to predict the population of the city in 2020.

    d. What is the average rate of change in population between 2010 and 2020?

45. *Carbon-14 Dating* An exponential decay function can be used to model the number of grams of a radioactive material that remain after a period of time. Carbon-14 decays over time, with the amount remaining after t years given by $y = 100e^{-0.00012097t}$ if 100 grams is the original amount.

    a. How much remains after 1000 years?

    b. Use graphical methods to estimate the number of years until 10 grams of carbon-14 remain.

46. *Drugs in the Bloodstream* If a drug is injected into the bloodstream, the percent of the maximum dosage that is present at time t is given by

$$y = 100(1 - e^{-0.35(10-t)})$$

where t is in hours, with $0 \leq t \leq 10$.

    a. What percent of the drug is present after 2 hours?

    b. Graph this function.

    c. When is the drug totally gone from the bloodstream?

47. *Normal Curve* The "curve" on which many students like to be graded is the bell-shaped normal curve. The equation $y = \dfrac{1}{\sqrt{2\pi}}e^{-(x-50)^2/2}$ describes the normal curve for a standardized test, where x is the test score before curving.

    a. Graph this function for x between 47 and 53 and for y between 0 and 0.5.

    b. The average score for the test is the score that gives the largest output y. Use the graph to find the average score.

48. *IQ Measure* The frequency of IQ measures x follows a bell-shaped normal curve with equation $y = \dfrac{1}{\sqrt{2\pi}}e^{-(x-100)^2/20}$. Graph this function for x between 80 and 120 and for y between 0 and 0.4.

5.2

Logarithmic Functions; Properties of Logarithms

section preview ▪ Investing

Suppose $2500 is invested in an account earning 10% annual interest compounded continuously. How long will it take for the amount to grow to $5000 (that is, to double its value)? To find the time t in which this investment will double, we can solve

$$5000 = 2500e^{0.10t} \quad \text{or} \quad 2 = e^{0.10t}$$

We can solve this equation graphically as we did in the last section (Figure 5.14). In this section, we learn how to rewrite this equation in a new form, called the **logarithmic form** of the equation, to find the value of t algebraically. We also introduce logarithmic functions, discuss the relationship between exponential and logarithmic functions, and apply logarithmic functions in real situations.

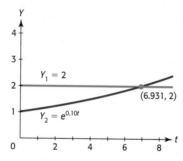

Figure 5.14

Logarithmic Functions

Recall that we discussed inverse functions in Section 4.3. The inverse of a function is a second function that "undoes" what the original function does. For example, if the original function doubles each input, its inverse takes half of each input. If a function is defined by $y = f(x)$, the inverse of this function can be found by interchanging x and y in $y = f(x)$ and solving the new equation for y. As we saw in Section 4.3, a function has an inverse if it is a one-to-one function (that is, there is a one-to-one correspondence between the inputs and outputs defining the function).

Every exponential function of the form $y = b^x$, with $b > 0$ and $b \neq 1$, is one-to-one, so every exponential function of this form has an inverse function. The inverse function of

$$y = b^x$$

is found by interchanging x and y and solving the new equation for y. Interchanging x and y in $y = b^x$ gives

$$x = b^y$$

In this new function, y is the power to which we raise b to get the number x. To solve an expression like this for the exponent, we need new notation. We define y, the power to which we raise the base b to get the number x, as

$$y = \log_b x$$

This inverse function is called a **logarithmic function** with base b.

Logarithmic Function

For $x > 0$, $b > 0$, and $b \neq 1$, the logarithmic function to the base b is

$$y = \log_b x$$

which is defined by $x = b^y$.

That is, $\log_b x$ is the exponent to which we must raise the base b to get the number x.

This function is the inverse function of the exponential function $y = b^x$.

Note that, according to the definition of the logarithmic function, $y = \log_b x$ and $x = b^y$ are two different forms of the same equation. We call $y = \log_b x$ the *logarithmic form* of the equation and $x = b^y$ the *exponential form* of the equation. The number b is called the **base** in both $y = \log_b x$ and $x = b^y$, and y is the **logarithm** in $y = \log_b x$ and the **exponent** in $x = b^y$. Thus, a logarithm is an exponent.

| example 1 |

Exponential and Logarithmic Forms

a. Write the equation $y - \log_3 x$ in exponential form.

b. Write $x = 3^y$ in logarithmic form.

Solution

a. The base of $y = \log_3 x$ is 3, and y, which equals the logarithm, is the exponent in the equivalent exponential form. Thus, the exponential form is $x = 3^y$.

b. To write $x = 3^y$ in logarithmic form, note that the base of the logarithm will be 3, the same as the base of the exponential expression. Also note that y is the exponent in $x = 3^y$, so y is the value of the logarithm. That is, the exponential form of $x = 3^y$ is $y = \log_3 x$. ∎

Table 5.3 shows some logarithmic equations and their equivalent exponential forms.

Table 5.3

Logarithmic Form	Exponential Form
$\log_2 x = y$	$2^y = x$
$\log_{10} 100 = 2$	$10^2 = 100$
$\log_a 1 = 0 \,(a > 0)$	$a^0 = 1$
$\log_a a = 1 \,(a > 0)$	$a^1 = a$

We can sometimes more easily evaluate logarithmic functions for different inputs by changing the function from logarithmic form to exponential form.

| example 2 | **Evaluating Logarithms** |

a. Write $y = \log_2 x$ in exponential form.

b. Use the exponential form from part (a) to find y when $x = 8$.

c. Evaluate $\log_2 8$. **d.** Evaluate $\log_4 \dfrac{1}{16}$. **e.** Evaluate $\log_{10} 0.001$.

Solution

a. To rewrite $y = \log_2 x$ in exponential form, note that the base of the logarithm, 2, becomes the base of the exponential expression, and y equals the exponent in the equivalent exponential form. That is, the exponential form is

$$2^y = x$$

b. To find y when x is 8, we solve $2^y = 8$. We can easily see that $y = 3$ because $2^3 = 8$.

c. To evaluate $\log_2 8$, we write $\log_2 8 = y$ in its equivalent exponential form $2^y = 8$. From part (b), $y = 3$. Thus, $\log_2 8 = 3$.

d. To evaluate $\log_4 \dfrac{1}{16}$, we write $\log_4 \dfrac{1}{16} = y$ in its equivalent exponential form $4^y = \dfrac{1}{16}$. Because $4^{-2} = \dfrac{1}{16}, y = -2$. Thus, $\log_4 \dfrac{1}{16} = -2$.

e. To evaluate $\log_{10} 0.001$, we write $\log_{10} 0.001 = y$ in its equivalent exponential form $10^y = 0.001$. Because $10^{-3} = 0.001, y = -3$. Thus, $\log_{10} 0.001 = -3$. ∎

The relationship between the logarithmic and exponential forms of an equation is used in the following example.

| example 3 | **Graphing a Logarithmic Function** |

Graph $y = \log_2 x$.

Solution

We can graph $y = \log_2 x$ by graphing the equivalent equation $x = 2^y$. The table of values (found by substituting values of y and calculating x) and the graph are shown in Figure 5.15.

$x = 2^y$	y
$\dfrac{1}{8}$	-3
$\dfrac{1}{4}$	-2
$\dfrac{1}{2}$	-1
1	0
2	1
4	2
8	3

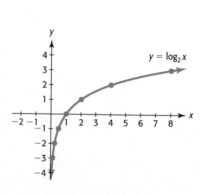

Figure 5.15

Figures 5.16(a) and (b) show graphs of the logarithmic function $y = \log_b x$ for a base $b > 1$ and a base $0 < b < 1$.

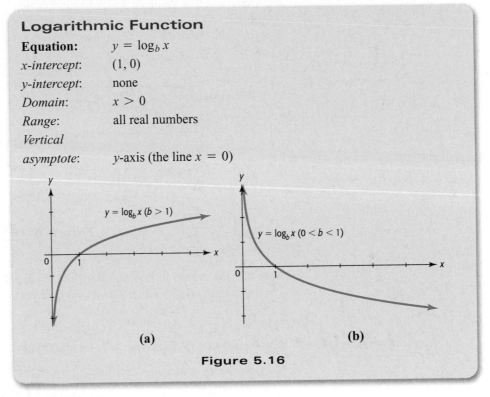

Logarithmic Function

Equation:	$y = \log_b x$
x-intercept:	$(1, 0)$
y-intercept:	none
Domain:	$x > 0$
Range:	all real numbers
Vertical asymptote:	*y*-axis (the line $x = 0$)

$y = \log_b x \ (b > 1)$

$y = \log_b x \ (0 < b < 1)$

(a)   (b)

Figure 5.16

Note that the graph of the function in Figure 5.16(a) is increasing, and the graph of the function in Figure 5.16(b) is decreasing.

Common Logarithms

We can use technology to evaluate logarithmic functions if the base of the logarithm is 10 or *e*. Because our number system uses base 10, logarithms with a base of 10 are called **common logarithms**, and, by convention, $\log_{10} x$ is simply denoted $\log x$. The graph of $y = \log x$ can be drawn by plotting points satisfying $x = 10^y$. Figure 5.17(a) shows some points on the graph of $y = \log x$, and Figure 5.17(b) shows the graph.

x	$y = \log x$
1000	3
100	2
10	1
1	0
0.1	-1
0.01	-2
0.001	-3

$y = \log x$

(a)   (b)

Figure 5.17

Graphing utilities and spreadsheets can be used to evaluate and graph $f(x) = \log x$, using the key or command $\boxed{\log}$. Figure 5.18(a) shows the graph of $y = \log x$ on a graphing calculator. The graph of a logarithmic function with any base can be created with a spreadsheet such as Excel by entering the formula $= \log(x, base)$. The graph in Figure 5.18(b) results from the Excel command "$= \log(x, 10)$."

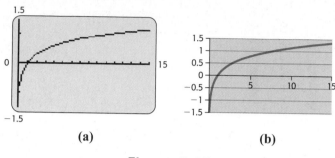

(a)   (b)

Figure 5.18

Note that the graphs in Figure 5.18(a) and (b) appear to touch the y-axis, although we know that the y-axis is actually a vertical asymptote for the graph.

| example 4 |

Evaluating Base 10 Logarithms

Using each of the indicated methods, find $f(10,000)$ if $f(x) = \log x$.

a. Write the equation $y = \log x$ in exponential form and find y when $x = 10,000$.

b. Use technology to evaluate $\log 10,000$.

```
log(10000)
                    4
```

Figure 5.19

Solution

a. The exponential form of $y = \log x = \log_{10} x$ is $10^y = x$. Substituting 10,000 for x in this equation gives $10^y = 10,000$. Because $10^4 = 10,000$, we see that $y = 4$. Thus, $f(10,000) = 4$.

b. The value of $\log 10,000$ is shown to be 4 on the calculator screen in Figure 5.19, so $f(10,000) = 4$.   ■

Richter Scale

Because the intensities of earthquakes are so large, the Richter scale was developed to provide a "scaled down" measuring system that permits earthquakes to be more easily measured and compared. The Richter scale gives the magnitude R of an earthquake using the formula

$$R = \log\left(\frac{I}{I_0}\right)$$

where I is the intensity of the earthquake and I_0 is a certain minimum intensity used for comparison.

The following table describes the typical effects of earthquakes of various magnitudes near the epicenter.

Description	Richter Magnitudes	Earthquake Effects
Micro	Less than 2.0	Microearthquakes, not felt
Very minor	2.0–2.9	Generally not felt, but recorded
Minor	3.0–3.9	Often felt, but rarely causes damage
Light	4.0–4.9	Noticeable shaking of indoor items, rattling noises. Significant damage unlikely
Moderate	5.0–5.9	Can cause damage to poorly constructed buildings over small regions
Strong	6.0–6.9	Can be destructive in areas up to about 100 miles across in populated areas
Major	7.0–7.9	Can cause serious damage over larger areas
Great	8.0–8.9	Can cause serious damage in areas several hundred miles across
Rarely, great	9.0–9.9	Devastating in areas several thousand miles across
Meteoric	10.0+	Never recorded

(Source: Wikipedia)

example 5

Richter Scale

a. If an earthquake has an intensity of 10,000 times I_0, what is the magnitude of the earthquake? Describe the effects of this earthquake.

b. Show that if the Richter scale reading of an earthquake is k, the intensity of the earthquake is $I = 10^k I_0$.

c. An earthquake that measured 9.0 on the Richter scale occurred in the Indian Ocean in December 2004, causing a devastating tsunami that killed thousands of people. Express the intensity of this earthquake in terms of I_0.

d. If an earthquake measures 7.0 on the Richter scale, give the intensity of this earthquake in terms of I_0. How much more is the intensity of the earthquake in part (c) than the one with Richter scale measurement of 7.0?

Solution

a. If the intensity of the earthquake is $10{,}000 I_0$, we substitute $10{,}000 I_0$ for I in the equation $R = \log\left(\dfrac{I}{I_0}\right)$, getting $R = \log\dfrac{10{,}000 I_0}{I_0} = \log 10{,}000 = 4$. Thus, the magnitude of the earthquake is 4, a light earthquake that would not likely cause significant damage.

b. If $R = k$, $k = \log\left(\dfrac{I}{I_0}\right)$, and the exponential form of this equation is $10^k = \dfrac{I}{I_0}$, so $I = 10^k I_0$.

c. The Richter scale reading for this earthquake is 9.0, so its intensity is
$$I = 10^{9.0} I_0 = 1{,}000{,}000{,}000\ I_0$$
Thus the earthquake had an intensity of 1 billion times I_0.

d. The Richter scale reading for this earthquake is 7.0, so its intensity is $I = 10^{7.0}I_0$. Thus the ratio of intensities of the two earthquakes is

$$\frac{10^{9.0}I_0}{10^{7.0}I_0} = \frac{10^{9.0}}{10^{7.0}} = 10^2 = 100$$

Therefore, the Indian Ocean earthquake is 100 times as intense. ■

Note from Example 5(d) that the Richter scale measurement of the Indian Ocean earthquake is 2 larger than the one with the Richter scale measurement 7.0, and that the Indian Ocean earthquake is 10^2 times as intense. In general, we have the following.

> ### Richter Scale
>
> 1. If the intensity of an earthquake is I, its Richter scale measurement is
>
> $$R = \log\frac{I}{I_0}$$
>
> 2. If the Richter scale reading of an earthquake is k, the intensity of the earthquake is
>
> $$I = 10^k I_0$$
>
> 3. If the difference of the Richter scale measurements of two earthquakes is the positive number d, the intensity of the larger earthquake is 10^d times that of the smaller earthquake.

Natural Logarithms

As we have seen in the previous section, the number e is important in many business and life science applications. Likewise, logarithms to base e are important in many applications. Because they are used so frequently, especially in science, they are called **natural logarithms**, and we use the special notation $\ln x$ to denote $\log_e x$.

> ### Natural Logarithms
>
> The logarithmic function with base e is $y = \log_e x$, defined by $x = e^y$ for all positive numbers x and denoted by
>
> $$\ln x = \log_e x$$

Most graphing utilities and spreadsheets have the key or command $\boxed{\ln}$ to use in evaluating functions involving natural logarithms. Figure 5.20(a) shows the graph of $y = \ln x$, which is similar to the graph of $y = \log_3 x$ (Figure 5.20(b)) because the value of e, the base of $\ln x$, is close to 3.

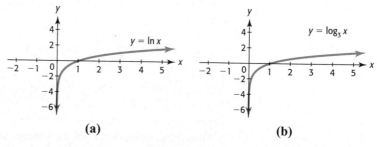

(a)   (b)

Figure 5.20

| example 6 | ## Mothers in the Workforce |

Between the years 1976 and 2004, the percent of moms who returned to the workforce within 1 year after having a child is given by $w(x) = 5.200 + 15.268 \ln x$, where x is the number of years after 1970.
(Source: Associated Press)

a. Graph this function.

b. What does this model estimate the percent of mothers returning to the workforce to be for 2006?

c. Use the graph drawn in part (a) to estimate the year in which the percent reached 53.

Solution

a. We are given that the model is valid between 1976 and 2004, so an appropriate domain is $6 \le x \le 34$. This logarithmic function increases as the input becomes larger, so a reasonable range includes the interval $w(6) \le y \le w(34)$, or $32.56 \le y \le 59.04$. We choose to use the viewing window [0, 34] by [0, 60], but there are many other appropriate windows. The graph of $y = w(x) = 5.200 + 15.268 \ln x$ appears in Figure 5.21(a).

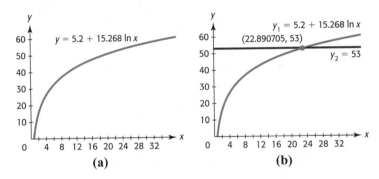

(a)   **(b)**

Figure 5.21

b. The year 2006 is 36 years after 1970, so we find

$$w(36) = 5.200 + 15.268 \ln 36 = 59.913 \approx 60$$

Thus the model estimates that in 2006 about 60% of moms returned to the workforce within 1 year after having a child.

c. To find the x-value for which $y = 53$, we graph $y_1 = 5.200 + 15.268 \ln x$ and $y_2 = 53$ on the same window and find the point of intersection of the graphs (Figure 5.21(b)). We see that the input of this point is $x \approx 23$, which corresponds to the year 1993. So we conclude that approximately 53% of mothers returned to the workforce within 1 year after having a child in 1993. ■

Logarithmic Properties

We have learned that logarithms are exponents and that a logarithmic function with base a is the inverse of an exponential function with base a. Thus, the properties of logarithms can be derived from the properties of exponents. These properties are useful in solving equations involving exponents and logarithms.

The basic properties of logarithms are easy to derive from the definition of a logarithm.

Basic Properties of Logarithms

For $b > 0, b \neq 1$,

1. $\log_b b = 1$
2. $\log_b 1 = 0$
3. $\log_b b^x = x$
4. $b^{\log_b x} = x$
5. For positive real numbers M and N, if $M = N$, then $\log_b M = \log_b N$.

These properties are easily justified by rewriting the logarithmic form as an exponential form or vice versa. In particular, $\log_b 1 = 0$ because $b^0 = 1$, $\log_b b = 1$ because $b^1 = b$, $\log_b b^x = x$ because $b^x = b^x$, and $b^{\log_b x} = x$ because $\log_b x = \log_b x$.

example 7 ## Logarithmic Properties

Use the basic properties of logarithms to simplify the following:

a. $\log_5 5^{10}$ **b.** $\log_4 4$ **c.** $\log_4 1$ **d.** $\log 10^7$ **e.** $\ln e^3$ **f.** $\log\left(\dfrac{1}{10^3}\right)$

Solution

a. $\log_5 5^{10} = 10$ by Property 3   **b.** $\log_4 4 = 1$ by Property 1
c. $\log_4 1 = 0$ by Property 2   **d.** $\log 10^7 = \log_{10} 10^7 = 7$ by Property 3

e. $\ln e^3 = \log_e e^3 = 3$ by Property 3   **f.** $\log\left(\dfrac{1}{10^3}\right) = \log 10^{-3} = -3$ ∎

example 8 ## Earthquakes

Show that if the intensity of an earthquake is $I = 10^k I_0$, then the magnitude is $R = k$.

Solution

If the intensity of an earthquake is $I = 10^k I_0$,

then $R = \log\left(\dfrac{I}{I_0}\right) = \log\dfrac{10^k I_0}{I_0} = \log 10^k = \log_{10} 10^k = k$ by Property 3. Thus $R = k$. ∎

Other logarithmic properties are helpful in simplifying logarithmic expressions, which in turn help us solve some logarithmic equations. These properties are also directly related to the properties of exponents.

Additional Logarithmic Properties

For $b > 0, b \neq 1$, k a real number, and M and N positive real numbers,

Product Property   6. $\log_b(MN) = \log_b M + \log_b N$

Quotient Property   7. $\log_b\left(\dfrac{M}{N}\right) = \log_b M - \log_b N$

Power Property   8. $\log_b M^k = k\log_b M$

To prove Property 6, we let $x = \log_b M$ and $y = \log_b N$, so that $M = b^x$ and $N = b^y$. Then

$$\log_b(MN) = \log_b(b^x \cdot b^y) = \log_b(b^{x+y}) = x + y = \log_b M + \log_b N$$

To prove Property 8, we let $x = \log_b M$, so that $M = b^x$. Then

$$\log_b M^k = \log_b(b^x)^k = \log_b b^{kx} = kx = k \log_b M$$

example 9 ## Rewriting Logarithms

Rewrite the following expressions as a sum, difference, or product of logarithms, and simplify if possible.

a. $\log_4 5(x - 7)$ **b.** $\ln(e^2(e + 3))$ **c.** $\log\left(\dfrac{x - 8}{x}\right)$

d. $\log_4 y^6$ **e.** $\ln\left(\dfrac{1}{x^5}\right)$

Solution

a. By Property 6, $\log_4 5(x - 7) = \log_4 5 + \log_4(x - 7)$.

b. By Property 6, $\ln(e^2(e + 3)) = \ln e^2 + \ln(e + 3)$.
   By Property 3, this equals $2 + \ln(e + 3)$.

c. By Property 7, $\log\left(\dfrac{x - 8}{x}\right) = \log(x - 8) - \log x$.

d. By Property 8, $\log_4 y^6 = 6 \log_4 y$.

e. By Property 7, Property 2, and Property 8,

$$\ln\left(\frac{1}{x^5}\right) = \ln 1 - \ln x^5 = 0 - 5 \ln x = -5 \ln x \qquad \blacksquare$$

Caution: There is no property of logarithms to rewrite a logarithm of a sum or difference. That is why in Example 9(a) the expression $\log_4(x - 7)$ was **not** written as $\log_4 x - \log_4 7$.

example 10 ## Rewriting Logarithms

Rewrite the following expressions as a single logarithm.

a. $\log_3 x + 4 \log_3 y$ **b.** $\dfrac{1}{2} \log a - 3 \log b$ **c.** $\ln(5x) - 3 \ln z$

Solution

a. $\log_3 x + 4 \log_3 y = \log_3 x + \log_3 y^4$ Use Logarithmic Property 8.

   $= \log_3 xy^4$ Use Logarithmic Property 6.

b. $\dfrac{1}{2} \log a - 3 \log b = \log a^{1/2} - \log b^3$ Use Logarithmic Property 8.

   $= \log\left(\dfrac{a^{1/2}}{b^3}\right)$ Use Logarithmic Property 7.

c. $\ln(5x) - 3 \ln z = \ln(5x) - \ln z^3$ Use Logarithmic Property 8.

$$= \ln\left(\frac{5x}{z^3}\right)$$ Use Logarithmic Property 7. ∎

---

example 11   ## Applying the Properties of Logarithms

a. Use logarithmic properties to estimate $\ln(5e)$ if $\ln 5 \approx 1.61$.

b. Find $\log_a\left(\dfrac{15}{8}\right)$ if $\log_a 15 = 2.71$ and $\log_a 8 = 2.079$.

c. Find $\log_a\left(\dfrac{15^2}{\sqrt{8}}\right)^3$ if $\log_a 15 = 2.71$ and $\log_a 8 = 2.079$.

Solution

a. $\ln(5e) = \ln 5 + \ln e \approx 1.61 + 1 = 2.61$

b. $\log_a\left(\dfrac{15}{8}\right) = \log_a 15 - \log_a 8 = 2.71 - 2.079 = 0.631$

c. $\log_a\left(\dfrac{15^2}{\sqrt{8}}\right)^3 = 3 \log_a\left(\dfrac{15^2}{\sqrt{8}}\right) = 3[\log_a 15^2 - \log_a 8^{1/2}] = 3\left[2 \log_a 15 - \dfrac{1}{2} \log_a 8\right]$

$$= 3\left[2(2.71) - \dfrac{1}{2}(2.079)\right] = 13.1415$$ ∎

---

example 12   ## Earthquakes

If one earthquake has intensity 320,000 times I_0 and a second has intensity 3,200,000 times I_0, what is the difference in their Richter scale measurements?

Solution

Recall that the Richter scale measurement is given by $R = \log\left(\dfrac{I}{I_0}\right)$. The difference of the Richter scale measurements is

$$R_2 - R_1 = \log\frac{3,200,000 I_0}{I_0} - \log\frac{320,000 I_0}{I_0} = \log 3,200,000 - \log 320,000$$

Using Property 6 gives us

$$\log 3,200,000 - \log 320,000 = \log\frac{3,200,000}{320,000} = \log 10 = 1$$

Thus the difference in the Richter scale measurements is 1.   ∎

In Exercises 1–4, write the logarithmic equations in exponential form.

1. $y = \log_3 x$ **2.** $2y = \log_5 x$

3. $y = \ln(2x)$ **4.** $y = \log(-x)$

In Exercises 5–8, write the exponential equations in logarithmic form.

5. $x = 4^y$ **6.** $m = 3^p$

7. $32 = 2^5$ **8.** $9^{2x} = y$

In Exercises 9 and 10 evaluate the logarithms, if possible. Round each answer to three decimal places.

9. a. $\log 7$ **10. a.** $\log 456$

   b. $\ln 86$ **b.** $\log(-12)$

   c. $\log 63{,}980$ **c.** $\ln 10$

In Exercises 11–13, find the value of the logarithms without using a calculator.

11. a. $\log_2 32$ **b.** $\log_9 81$

   c. $\log_3 27$ **d.** $\log_4 64$

   e. $\log_5 625$

12. a. $\log_2 64$ **b.** $\log_9 27$

   c. $\log_4 2$ **d.** $\ln(e^3)$

   e. $\log 100$

13. a. $\log_3 \dfrac{1}{27}$ **b.** $\ln 1$

   c. $\ln e$ **d.** $\log 0.0001$

In Exercises 14, graph the functions by changing to exponential form.

14. a. $y = \log_3 x$ **b.** $y = \log_5 x$

In Exercises 15–18, graph the functions with technology.

15. $y = 2 \ln x$ **16.** $y = 4 \log x$

17. $y = \log(x + 1) + 2$ **18.** $y = \ln(x - 2) - 3$

19. a. Write the inverse of $y = 4^x$ in logarithmic form.

   b. Graph $y = 4^x$ and its inverse and discuss the symmetry of their graphs.

20. a. Write the inverse of $y = 3^x$ in logarithmic form.

   b. Graph $y = 3^x$ and its inverse and discuss the symmetry of their graphs.

21. Write $\log_a a = x$ in exponential form and find x to evaluate $\log_a a$ for any $a > 0, a \neq 1$.

22. Write $\log_a 1 = x$ in exponential form and find x to evaluate $\log_a 1$ for any $a > 0, a \neq 1$.

In Exercises 23–26, use the properties of logarithms to evaluate the expressions.

23. $\log 10^{14}$ **24.** $\ln e^5$

25. $10^{\log_{10} 12}$ **26.** $6^{\log_6 25}$

In Exercises 27–30, use $\log_a(20) = 1.4406$ and $\log_a(5) = 0.7740$ to evaluate each expression.

27. $\log_a(100)$ **28.** $\log_a(4)$

29. $\log_a 5^3$ **30.** $\log_a \sqrt{20}$

In Exercises 31–34, rewrite each expression as a sum, difference, or product of logarithms, and simplify if possible.

31. $\ln \dfrac{3x - 2}{x + 1}$ **32.** $\log [x^3(3x - 4)^5]$

33. $\log_3 \dfrac{\sqrt[4]{4x + 1}}{4x^2}$ **34.** $\log_3 \dfrac{\sqrt[3]{3x - 1}}{5x^2}$

In Exercises 35–38, rewrite each expression as a single logarithm.

35. $3 \log_2 x + \log_2 y$ **36.** $\log x - \dfrac{1}{3} \log y$

37. $4 \ln(2a) - \ln b$

38. $6 \ln(5y) + 2 \ln x$

exercises
5.2

39. *Life Span* On the basis of data for the years 1910 through 2007, the expected life span of people in the United States can be described by the function $f(x) = 10.963 + 14.321 \ln x$ years, where x is the number of years from 1900 to the person's birth year.

a. What does this model estimate the life span to be for people born in 1925? In 2007? (Give each answer to the nearest year.)

b. Explain why these numbers are so different.
(Source: National Center for Health Statistics)

40. *Japan's Population* The population of Japan for the years 1984–2004 is approximated by the logarithmic function $y = 113.885 + 4.327 \ln x$ million people, with x equal to the number of years after 1980. According to the model, what is the estimated population in 2000? In 2009?
(Source: www.jinjapan.org/stat/)

41. *Poverty Threshold* A single person is considered as living in poverty if his or her income falls below the federal government's official poverty level, which is adjusted every year for inflation. The function that models the poverty threshold for the years 1987–2005 is $f(x) = -110.073 + 3358.105 \ln x$ dollars, where x is the number of years after 1980.

a. What does this model give as the poverty threshold in 1990? In 2005?

b. Is this function increasing or decreasing?

c. What factors have led to this change in the threshold?
(Source: U.S. Census Bureau)

42. *Loan Repayment* The number of years t that it takes to pay off a $100,000 loan at 10% interest by making annual payments of R is

$$t = \frac{\ln R - \ln(R - 10,000)}{\ln 1.1}, R > 10,000.$$

If the annual payment is $16,274.54, in how many years will the loan be paid off?

43. *Supply* Suppose that the supply of a product is given by $p = 20 + 6 \ln(2q + 1)$ where p is the price per

unit and q is the number of units supplied. What price will give a supply of 5200 units?

44. *Demand* Suppose that the demand function for a product is $p = \dfrac{500}{\ln(q + 1)}$, where p is the price per unit and q is the number of units demanded. What price will give a demand for 6400 units?

45. *Female Workers* For the years 1970–2006, the percent of females in the workforce is given by $y = 11.101 + 8.090 \ln x$, where x is the number of years from 1960.

a. What does the model predict the percent to be in 2011? In 2015?

b. Is the percent of female workers increasing or decreasing?

46. *Nonmarital Childbearing* The percent of live births to unwed mothers for the years 1970–2003 can be modeled by the function $y = -32.422 + 17.664 \ln x$, where x is the number of years from 1960.

a. What does the model predict the percent to be in 2010? In 2015?

b. Is the percent increasing or decreasing?

47. *Doubling Time* If $4000 is invested in an account earning 10% annual interest compounded continuously, then the number of years that it takes for the amount to grow to $8000 is $n = \dfrac{\ln 2}{0.10}$. Find the number of years.

48. *Doubling Time* If $5400 is invested in an account earning 7% annual interest compounded continuously, then the number of years that it takes for the amount to grow to $10,800 is $n = \dfrac{\ln 2}{0.07}$. Find the number of years.

49. *Doubling Time* The number of quarters (a quarter equals 3 months) needed to double an investment when a lump sum is invested at 8% compounded quarterly is given by $n = \dfrac{\log 2}{0.0086}$. In how many years will the investment double?

50. *Doubling Time* The number of periods needed to double an investment when a lump sum is invested at

12% compounded semiannually is $n = \dfrac{\log 2}{0.0253}$. How many years pass before the investment doubles in value?

The number of years it takes for an investment to double if it earns r percent (as a decimal), compounded annually, is $t = \dfrac{\ln 2}{\ln(1 + r)}$. Use this formula in Exercises 51 and 52.

51. *Investing* In how many years will an investment double if it is invested at 8%, compounded annually?

52. *Investing* In how many years will an investment double if it is invested at 12.3%, compounded annually?

53. *Earthquakes* If an earthquake has an intensity of 25,000 times I_0, what is the magnitude of the earthquake?

54. *Earthquakes*
  a. If an earthquake has an intensity of 250,000 times I_0, what is the magnitude of the earthquake?

  b. Compare the magnitudes of two earthquakes when the intensity of one is 10 times the intensity of the other. (See Exercise 53.)

55. *Earthquakes* The Richter scale measurement for the southern Sumatra earthquake of 2007 was 6.4. Find the intensity of this earthquake as a multiple of I_0.

56. *Earthquakes* The Richter scale measurement for the San Francisco earthquake that occurred in 1906 was 8.25. Find the intensity of this earthquake as a multiple of I_0.

57. *Earthquakes* The Richter scale reading for the San Francisco earthquake of 1989 was 7.1. Find the intensity I of this earthquake as a multiple of I_0.

58. *Earthquakes* On January 9, 2008, an earthquake in northern Algeria measuring 4.81 on the Richter scale killed 1 person, and on May 12, 2008, an earthquake in China measuring 7.9 on the Richter scale killed 67,180 people. How many times more intense was the earthquake in China than the one in northern Algeria?

59. *Earthquakes* The largest earthquake ever to strike San Francisco (in 1906) measured 8.25 on the Richter scale, and the second largest (in 1989) measured 7.1. How many times more intense was the 1906 earthquake than the 1989 earthquake?

60. *Earthquakes* The largest earthquake ever recorded measured 8.9 on the Richter scale, and an earthquake measuring 8.25 occurred in Japan in 1983. Calculate the ratio of the intensities of these earthquakes.

Use the following information to answer the questions in Exercises 61–66. To quantify the intensity of sound, the decibel scale (named for Alexander Graham Bell) was developed. The formula for loudness L on the decibel scale is $L = 10 \log\left(\dfrac{I}{I_0}\right)$, where I_0 is the intensity of sound just below the threshold of hearing, which is approximately 10^{-16} watt per square centimeter.

61. *Decibel Scale* Find the decibel reading for a sound with intensity 20,000 times I_0.

62. *Decibel Scale* Compare the decibel readings of two sounds if the intensity of the first sound is 100 times the intensity of the second sound.

63. *Decibel Scale* Use the exponential form of the function $L = 10 \log\left(\dfrac{I}{I_0}\right)$ to find the intensity of a sound if its decibel reading is 40.

64. *Decibel Scale* Use the exponential form of $L = 10 \log\left(\dfrac{I}{I_0}\right)$ to find the intensity of a sound if its decibel reading is 140. (Sound at this level causes pain.)

65. *Decibel Scale* The intensity of a whisper is $115I_0$, and the intensity of a busy street is $9,500,000I_0$. How do their decibel levels compare?

66. *Decibel Scale* If the decibel reading of a sound that is painful is 140 and the decibel reading of rock music is 120, how much more intense is the painful sound than the rock concert?

Use the following information to answer the questions in Exercises 67–69. To simplify the measurement of the acidity or basicity of a solution, the pH (hydrogen potential) measure was developed. The pH is given by the formula $pH = -\log[H^+]$, where $[H^+]$ is the concentration of hydrogen ions in moles per liter in the solution.

67. *pH Levels* Find the pH value of beer for which $[H^+] = 0.0000631$.

68. *pH Levels* The pH value of eggs is 7.79. Find the hydrogen-ion concentration for eggs.

69. *pH Levels* The most common solutions have a pH range between 1 and 14. Use an exponential form of the pH formula to find the values of $[H^+]$ associated with these pH levels.

5.3

Exponential and Logarithmic Equations

section preview ▪ Carbon-14 Dating

Carbon-14 dating is a process by which scientists can tell the age of many fossils. To find the age of a fossil that originally contained 1000 grams of carbon-14 and now contains 1 gram, we solve the equation

$$1 = 1000e^{-0.00012097t}$$

An approximate solution of this equation can be found graphically, but sometimes a window that shows the solution is hard to find. If this is the case, it may be easier to solve an exponential equation like this algebraically. In this section, we consider two algebraic methods of solving exponential equations. The first method involves converting the equation to logarithmic form and then solving the logarithmic equation for the variable. The second method involves taking the logarithm of both sides of the equation and using the properties of logarithms to solve for the variable.

Solving Exponential Equations Using Logarithmic Forms

When we wish to solve an equation for a variable that is contained in an exponent, we can remove the variable from the exponent by converting the equation to its logarithmic form. The steps in this solution method follow.

Solving Exponential Equations Using Logarithmic Forms

To solve an exponential equation using logarithmic form:

1. Rewrite the equation with the term containing the exponent by itself on one side.
2. Divide both sides by the coefficient of the term containing the exponent.
3. Change the new equation to logarithmic form.
4. Solve for the variable.

This method is illustrated in the following example.

example 1 — Solving a Base 10 Exponential Equation

a. Solve the equation $3000 = 150(10^{4t})$ for t by converting it to logarithmic form.

b. Solve the equation graphically to confirm the solution.

Solution

a. 1. The term containing the exponent is by itself on one side of the equation.

2. We divide both sides of the equation by 150.

$$3000 = 150(10^{4t})$$
$$20 = 10^{4t}$$

3. We rewrite this equation in logarithmic form.

$$4t = \log_{10} 20 \quad \text{or} \quad 4t = \log 20$$

4. Solving for t gives the solution:

$$t = \frac{\log 20}{4} \approx 0.32526$$

b. We solve the equation graphically by using the intersection method. Entering $y_1 = 3000$ and $y_2 = 150(10^{4t})$, we graph using the window $[0, 0.4]$ by $[0, 3500]$ and find the point of intersection to be about $(0.32526, 3000)$ (Figure 5.22). Thus, the solution to the equation $3000 = 150(10^{4t})$ is $t \approx 0.32526$, as we found in part (a).

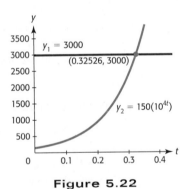

Figure 5.22

<table>
<tr><td>example 2</td></tr>
</table>

Doubling Time

a. Prove that the time it takes for an investment to double its value is $t = \dfrac{\ln 2}{r}$, if the interest rate is r, compounded continuously,

b. Suppose \$2500 is invested in an account earning 6% annual interest compounded continuously. How long will it take for the amount to grow to \$5000?

Solution

a. If \$$P$ are invested for t years at an annual interest rate r compounded continuously, then the investment will grow to a future value S according to the equation

$$S = Pe^{rt}$$

and the investment will be doubled when $S = 2P$, giving

$$2P = Pe^{rt} \quad \text{or} \quad 2 = e^{rt}$$

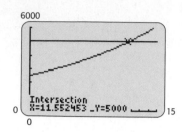

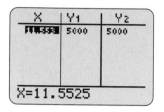

Figure 5.23

example 3

Converting this exponential equation to its equivalent logarithmic form gives

$$\log_e 2 = rt$$

and solving this equation for t gives the doubling time,

$$t = \frac{\ln 2}{r}$$

b. We seek the time it takes to double the investment, and the doubling time is given by

$$t = \frac{\ln 2}{0.06} \approx 11.5525$$

Thus, it will take approximately 11.6 years for an investment of $2500 to grow to $5000 if it is invested at an annual rate of 6% compounded continuously. We can confirm this solution graphically or numerically (Figure 5.23). Note that this is the same time it would take any investment at 6%, compounded continuously, to double. ∎

Carbon-14 Dating

Radioactive carbon-14 decays according to the equation

$$y = y_0 e^{-0.00012097t}$$

where y_0 is the original amount and y is the amount of carbon-14 at time t years. To find the age of a fossil if the original amount of carbon-14 was 1000 grams and the present amount is 1 gram, we solve the equation

$$1 = 1000e^{-0.00012097t}$$

a. Find the age of the fossil by converting the equation to logarithmic form.

b. Find the age of the fossil using graphical methods.

Solution

a. To write the equation so that the coefficient of the exponential term is 1, we divide both sides by 1000, obtaining

$$0.001 = e^{-0.00012097t}$$

Writing this equation in logarithmic form gives the equation

$$-0.00012097t = \log_e 0.001 \quad \text{or} \quad -0.00012097t = \ln 0.001$$

Solving the equation for t gives

$$t = \frac{\ln 0.001}{-0.00012097} \approx 57{,}103$$

Thus, the age of the fossil is calculated to be about 57,103 years.

We can confirm that this is the solution to $1 = 1000e^{-0.00012097t}$ by using a table of values with $Y_1 = 1000e^{-0.00012097t}$. The result of entering 57,103 for x is shown in Figure 5.24.

b. We can solve the equation graphically by using the x-intercept method. Entering

$$y_1 = 1000e^{-0.00012097t} - 1$$

and graphing using the window [0, 80000] by [−3, 3], we find the x-intercept to be 57,103.044 (see Figure 5.25.) Thus, as we found in part (a), the age of the fossil is about 57,103 years.

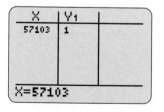

Figure 5.24

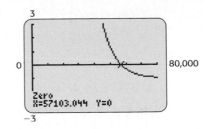

Figure 5.25

While finding this solution from a graph is easy, actually finding a window that contains the x-intercept is not. It may be easier to solve the equation algebraically, as we did in part (a), and then use graphical or numerical methods to confirm the solution.

■

Change of Base

Most calculators and other graphing utilities can be used to evaluate logarithms and graph logarithmic functions if the base is 10 or e. Graphing logarithmic functions with other bases usually requires converting to exponential form to determine outputs and then graphing the functions by plotting points by hand. However, we can use a special formula called the **change of base formula** to rewrite logarithms so that the base is 10 or e. The formula for changing from base a to base b is developed below.

Suppose $y = \log_a x$. Then

$a^y = x$

$\log_b a^y = \log_b x$ Take logarithm, base b, of both sides. (Property 5 of Logarithms)

$y \log_b a = \log_b x$ Use the Power Property of Logarithms.

$y = \dfrac{\log_b x}{\log_b a}$ Solve for y.

$\log_a x = \dfrac{\log_b x}{\log_b a}$ Substitute $\log_a x$ for y.

The general change of base formula is summarized below.

Change of Base Formula

If $b > 0$, $b \neq 1$, $a > 0$, $a \neq 1$, and $x > 0$, then

$$\log_a x = \frac{\log_b x}{\log_b a}$$

In particular, for base 10 and base e,

$$\log_a x = \frac{\log x}{\log a} \quad \text{and} \quad \log_a x = \frac{\ln x}{\ln a}$$

---

example 4

Applying the Change of Base Formula

a. Evaluate $\log_8 124$.

Graph the functions in parts (b) and (c) by changing each logarithm to a common logarithm and then by changing the logarithm to a natural logarithm.

b. $y = \log_3 x$ **c.** $y = \log_2(-3x)$

Solution

a. The change of base formula (to base 10) can be used to evaluate this logarithm:

$$\log_8 124 = \frac{\log 124}{\log 8} = 2.318 \text{ approximately}$$

b. Changing $y = \log_3 x$ to base 10 gives $\log_3 x = \dfrac{\log x}{\log 3}$, so we graph $y = \dfrac{\log x}{\log 3}$ (Figure 5.26(a)). Changing to base e gives $\log_3 x = \dfrac{\ln x}{\ln 3}$, so we graph $y = \dfrac{\ln x}{\ln 3}$ (Figure 5.26(b)). Note that the graphs appear to be identical (as they should be because they both represent $y = \log_3 x$).

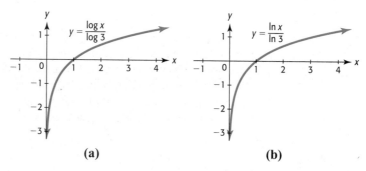

(a)   **(b)**

Figure 5.26

c. Changing $y = \log_2(-3x)$ to base 10 gives $\log_2(-3x) = \dfrac{\log(-3x)}{\log 2}$, so we graph $y = \dfrac{\log(-3x)}{\log 2}$ (Figure 5.27(a)). Changing to base e gives $\log_2(-3x) = \dfrac{\ln(-3x)}{\ln 2}$, so we graph $y = \dfrac{\ln(-3x)}{\ln 2}$ (Figure 5.27(b)). Note that the graphs are identical. Also notice that because logarithms are defined only for positive inputs, x must be negative so that $-3x$ is positive. Thus, the domain of $y = \log_2(-3x)$ is the interval $(-\infty, 0)$.

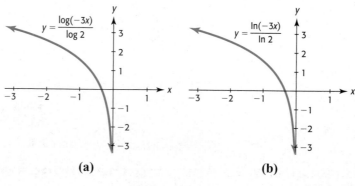

(a)   **(b)**

Figure 5.27

The change of base formula is also useful in solving exponential equations whose base is neither 10 nor e.

example 5	## Investment

If \$10,000 is invested for t years at 10% compounded annually, the future value is given by

$$S = 10,000(1.10^t)$$

In how many years will the investment grow to \$45,950?

Solution

We seek to solve the equation

$$45,950 = 10,000(1.10^t)$$

Dividing both sides by 10,000 gives

$$4.5950 = 1.10^t$$

Writing the equation in logarithmic form solves the equation for t:

$$t = \log_{1.10} 4.5950$$

Using the change of base formula gives the number of years:

$$t = \log_{1.10} 4.5950 = \frac{\log 4.5950}{\log 1.10} = 16 \text{ years}$$

Solving Exponential Equations Using Logarithmic Properties

With the properties of logarithms at our disposal, we consider a second method for solving exponential equations, which may be easier to use. This method of solving an exponential equation involves taking the logarithm of both sides of the equation (Property 5 of Logarithms) and then using properties of logarithms to write the equation in a form that we can solve.

> ### Solving Exponential Equations Using Logarithmic Properties
>
> To solve an exponential equation using logarithmic properties:
>
> 1. Rewrite the equation with a base raised to a power on one side.
> 2. Take the logarithm, base e or 10, of both sides of the equation.
> 3. Use a logarithmic property to remove the variable from the exponent.
> 4. Solve for the variable.

This method is illustrated in the following example.

example 6	## Solution of Exponential Equations

Solve the following exponential equations:

a. $4096 = 8^{2x}$ **b.** $6(4^{3x-2}) = 120$

Solution

a. **1.** This equation has the base 8 raised to a variable power on one side.

2. Taking the logarithm, base 10, of both sides of the equation $4096 = 8^{2x}$ gives

$$\log 4096 = \log 8^{2x}$$

3. Using the Power Property of Logarithms removes the variable x from the exponent:

$$\log 4096 = 2x \log 8$$

4. Solving for x gives the solution.

$$\frac{\log 4096}{2 \log 8} = x$$

$$x = 2$$

b. We first isolate the exponential expression on one side of the equation by dividing both sides by 6.

$$\frac{6(4^{3x-2})}{6} = \frac{120}{6}$$

$$4^{3x-2} = 20$$

Taking the natural logarithm of both sides leads to the solution.

$$\ln 4^{3x-2} = \ln 20$$

$$(3x - 2) \ln 4 = \ln 20$$

$$3x - 2 = \frac{\ln 20}{\ln 4}$$

$$x = \frac{1}{3}\left(\frac{\ln 20}{\ln 4} + 2 \right) \approx 1.387$$

An alternate method of solving the equation is to write $4^{3x-2} = 20$ in logarithmic form.

$$\log_4 20 = 3x - 2$$

$$x = \frac{\log_4 20 + 2}{3}$$

The change of base formula can be used to compute this value and verify that this solution is the same as above.

$$x = \frac{\dfrac{\ln 20}{\ln 4} + 2}{3} \approx 1.387 \qquad \blacksquare$$

Solution of Logarithmic Equations

Some logarithmic equations can be solved by converting them to exponential form.

example 7	**Solving a Logarithmic Equation**

Solve $4 \log_3 x = -8$ by converting to exponential form and verify the solution graphically.

Solution

We first isolate the logarithm by dividing both sides of the equation by 4.

$$\log_3 x = -2$$

Writing $\log_3 x = -2$ in exponential form gives the solution

$$3^{-2} = x$$

$$x = \frac{1}{9}$$

Graphing $y_1 = 4\log_3 x$ and $y_2 = -8$ with a window that contains $x = \frac{1}{9}$ and $y = -8$, we can find the point of intersection to be $\left(\dfrac{1}{9}, -8\right)$. See Figure 5.28(a) and (b).

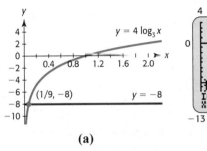

(a)

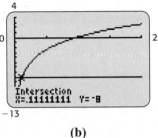

(b)

Figure 5.28

example 8 | ## Solving a Logarithmic Equation

a. Solve $6 + 3\ln x = 12$ by writing the equation in exponential form.

b. Solve the equation graphically.

Solution

a. We first solve the equation for $\ln x$:

$$6 + 3\ln x = 12$$

$$3\ln x = 6$$

$$\ln x = 2$$

Writing $\log_e x = 2$ in exponential form gives

$$x = e^2$$

b. We solve $6 + 3\ln x = 12$ graphically by the intersection method. Entering $y_1 = 6 + 3\ln x$ and $y_2 = 12$, we graph using the window $[-2, 10]$ by $[-10, 15]$ and find the point of intersection to be about $(7.38906, 12)$ (Figure 5.29). Thus, the solution to the equation $6 + 3\ln x = 12$ is $x \approx 7.38906$. Because $e^2 \approx 7.38906$, the solutions agree.

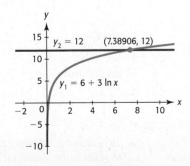

Figure 5.29

The properties of logarithms are frequently useful in solving logarithmic equations. Consider the following example.

| example 9 | **Solving a Logarithmic Equation** |

Solve $\ln x + 3 = \ln(x + 4)$ by converting the equation to exponential form and then using algebraic methods.

Solution

We first write the equation with the logarithmic expressions on one side:

$$\ln x + 3 = \ln(x + 4)$$
$$3 = \ln(x + 4) - \ln x$$

Using the Quotient Property of Logarithms gives

$$3 = \ln \frac{x + 4}{x}$$

Writing the equation in exponential form gives

$$e^3 = \frac{x + 4}{x}$$

We can now solve for x:

$$e^3 x = x + 4$$
$$e^3 x - x = 4$$
$$x(e^3 - 1) = 4$$
$$x = \frac{4}{e^3 - 1} \approx 0.21 \qquad \blacksquare$$

| example 10 | **Global Warming** |

In an effort to reduce global warming, it has been proposed that a tax be levied based on the emissions of carbon dioxide into the atmosphere. The cost–benefit equation $\ln(1 - P) = -0.0034 - 0.0053t$ estimates the relationship between the percent reduction of emissions of carbon dioxide (as a decimal) and the tax t in dollars per ton of carbon.

a. Solve the equation for P, the estimated percent reduction in emissions.

b. Determine the estimated percent reduction in emissions if a tax of \$100 per ton is levied.

(Source: W. Clime, *The Economics of Global Warming*)

Solution

a. We solve $\ln(1 - P) = -0.0034 - 0.0053t$ for P by writing the equation in exponential form.

$$(1 - P) = e^{-0.0034 - 0.0053t}$$
$$P = 1 - e^{-0.0034 - 0.0053t}$$

b. Substituting 100 for t gives $P = 0.4134$, so a \$100 tax per ton of carbon is estimated to reduce the carbon dioxide emissions by 41.3%. $\qquad \blacksquare$

Some exponential and logarithmic equations are difficult or impossible to solve algebraically, and finding approximate solutions to real data problems is frequently easier with graphical methods.

Exponential and Logarithmic Inequalities

Inequalities involving exponential and logarithmic functions can be solved by solving the related equation algebraically and then investigating the inequality graphically. Consider the following example.

---

example 11

Sales Decay

After the end of an advertising campaign, the daily sales of Genapet fell rapidly, with daily sales given by $S = 3200e^{-0.08x}$ dollars, where x is the number of days from the end of the campaign. For how many days after the campaign ended were sales at least $1980?

Solution

To solve this problem, we find the solution to the inequality $3200e^{-0.08x} > 1980$. We begin our solution by solving the related equation.

$$3200e^{-0.08x} = 1980$$
$$e^{-0.08x} = .61875 \qquad \text{Divide both sides by 3200.}$$
$$\ln e^{-0.08x} = \ln .61875 \qquad \text{Take the logarithm, base } e, \text{ of both sides.}$$
$$-0.08x \approx -.4801 \qquad \text{Use the Power Property of Logarithms.}$$
$$x \approx 6$$

We now investigate the inequality graphically.

   To solve this inequality graphically, we graph $y_1 = 3200e^{-0.08x}$ and $y_2 = 1980$ on the same axes, with nonnegative x-values since x represents the number of days. The graph shows that $y_1 = 3200e^{-0.08x}$ is above $y_2 = 1980$ for $x < 6$ (Figure 5.30). Thus, the daily revenue is at least $1980 for each of the first 6 days after the end of the advertising campaign.

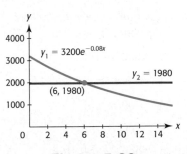

Figure 5.30

---

example 12

Cost–Benefit

The cost–benefit equation $\ln(1 - P) = -0.0034 - 0.0053t$ estimates the relationship between the percent reduction of emissions of carbon dioxide P (as a decimal) and the tax t in dollars per ton of carbon. What tax will give a reduction of at least 50%?

Solution

To find the tax, we find the value of t that gives $P = 0.50$.

$$\ln(1 - 0.50) = -0.0034 - 0.0053t$$
$$-0.6931 = -0.0034 - 0.0053t$$
$$0.0053t = 0.6897$$
$$t = 130.14$$

In Example 10, we solved this cost–benefit for P, getting $P = 1 - e^{-0.0034-0.0053t}$. The graphs of this equation and $P = 0.50$ on the same axes are shown in Figure 5.31. The point of intersection of the graphs is $(130.14, 0.5)$, and the graph shows that the percent reduction of emissions is more than 50% if the tax is above \$130.14 per ton of carbon.

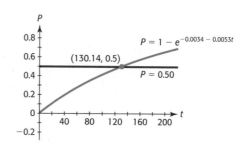

Figure 5.31

skills check
5.3

In Exercises 1–10, solve the equations algebraically and check graphically. Round to 3 decimal places.

1. $1600 = 10^x$ **2.** $4600 = 10^x$

3. $2500 = e^x$ **4.** $54.6 = e^x$

5. $8900 = e^{5x}$ **6.** $2400 = 10^{8x}$

7. $4000 = 200e^{8x}$ **8.** $5200 = 13e^{12x}$

9. $8000 = 500(10^x)$ **10.** $9000 = 400(10^x)$

In Exercises 11–14, use a change of base formula to evaluate each logarithm. Give your answers rounded to four decimal places.

11. $\log_6(18)$ **12.** $\log_7(215)$

13. $\log_8(\sqrt{2})$ **14.** $\log_4(\sqrt[3]{10})$

In Exercises 15–22, solve the equations.

15. $8^x = 1024$ **16.** $9^x = 2187$

17. $2(5^{3x}) = 31{,}250$ **18.** $2(6^{2x}) = 2592$

19. $5^{x-2} = 11.18$ **20.** $3^{x-4} = 140.3$

21. $18{,}000 = 30(2^{12x})$ **22.** $5880 = 21(2^{3x})$

In Exercises 23–35, solve the logarithmic equations.

23. $\log_2 x = 3$ **24.** $\log_4 x = -2$

25. $5 + 2\ln x = 8$ **26.** $4 + 3\log x = 10$

27. $5 + \ln(8x) = 23 - 2\ln x$

28. $3\ln x + 8 = \ln(3x) + 12.18$

29. $2\log x - 2 = \log(x - 25)$

30. $\ln(x - 6) + 4 = \ln x + 3$

31. $\log_3 x + \log_3 9 = 1$

32. $\log_2 x + \log_2(x - 6) = 4$

33. $\log_2 x = \log_2 5 + 3$

34. $\log_2 x = 3 - \log_2 2x$

35. $\log 3x + \log 2x = \log 150$

36. $\ln(x + 2) + \ln x = \ln(x + 12)$

In Exercises 37–40, solve the inequalities.

37. $3^x < 243$

38. $7^x \geq 2401$

39. $5(2^x) \geq 2560$

40. $15(4^x) \leq 15,360$

exercises

5.3

41. *Supply* The supply function for a certain size boat is given by $p = 340(2^q)$ boats, where p dollars is the price per boat and q is the quantity of boats supplied at that price. What quantity will be supplied if the price is $10,880 per boat?

42. *Demand* The demand function for a dining room table is given by $p = 4000(3^{-q})$ dollars per table, where p is the price and q is the quantity, in thousands of tables, demanded at that price. What quantity will be demanded if the price per table is $256.60?

43. *Sales Decay* After a television advertising campaign ended, the weekly sales of Korbel champagne fell rapidly. Weekly sales in a city were given by $S = 25{,}000e^{-0.072x}$ dollars, where x is the number of weeks after the campaign ended.

   a. Write the logarithmic form of this function.

   b. Use the logarithmic form of this function to find the number of weeks after the end of the campaign before weekly sales fell to $16,230.

44. *Sales Decay* After a television advertising campaign ended, sales of Genapet fell rapidly, with daily sales given by $S = 3200e^{-0.08x}$ dollars, where x is the number of days after the campaign ended.

   a. Write the logarithmic form of this function.

   b. Use the logarithmic form of this function to find the number of days after the end of the campaign that passed before daily sales fell to $2145.

45. *Sales Decay* After the end of an advertising campaign, the daily sales of Genapet fell rapidly, with daily sales given by $S = 3200e^{-0.08x}$ dollars, where x is the number of days from the end of the campaign.

   a. What were daily sales when the campaign ended?

   b. How many days passed after the campaign ended before daily sales were below half of what they were at the end of the campaign?

46. *Sales Decay* After the end of a television advertising campaign, the weekly sales of Korbel champagne fell rapidly, with weekly sales given by $S = 25{,}000e^{-0.072x}$ dollars, where x is the number of weeks from the end of the campaign.

   a. What were weekly sales when the campaign ended?

   b. How many weeks passed after the campaign ended before weekly sales were below half of what they were at the end of the campaign?

47. *Snapple Beverage Revenues* Prior to the November 1994 $1.7 billion takeover proposal by Quaker Oats, Snapple Beverage Corporation's revenues were given by the function $B(t) = 1.337e^{0.718t}$ million dollars, where t is the number of years after 1985.

   a. According to the model, what was Snapple's 1995 revenue?

   b. If the revenue continued to increase as described by this model, when did it reach $3599 million? (Source: *U.S. News and World Report*, November 14, 1994)

48. *Population Growth* The population in a certain city was 53,000 in 2000, and its future size is predicted to be $P = 53{,}000e^{0.015t}$ people, where t is the number of years after 2000. Determine algebraically when the population reaches 60,000, and verify your solution graphically.

49. *Purchasing Power* The purchasing power (real value of money) decreases if inflation is present in the economy. For example, the purchasing power of $40,000 after t years of 5% inflation is given by the model

$$P = 40{,}000e^{-0.05t} \text{ dollars}$$

How long will it take for the value of a $40,000 pension to have a purchasing power of $20,000 under 5% inflation?

(Source: *Viewpoints*, VALIC)

50. *Purchasing Power* If a retired couple has a fixed income of $60,000 per year, the purchasing power (adjusted value of the money) after t years of 5% inflation is given by the equation $P = 60,000e^{-0.05t}$. In how many years will the purchasing power of their income be half of their current income?

51. *Real Estate Inflation* During a 5-year period of constant inflation, the value of a $100,000 property increases according to the equation $v = 100,000e^{0.03t}$ dollars. In how many years will the value of this building be double its current value?

52. *Real Estate Inflation* During a 10-year period of constant inflation, the value of a $200,000 property is given by the equation $v = 200,000e^{0.05t}$ dollars. In how many years will the value of this building be $254,250?

53. *Radioactive Decay* The amount of radioactive isotope thorium-234 present in a certain sample at time t is given by $A(t) = 500e^{-0.02828t}$ grams, where t years is the time since the initial amount was measured.

   a. Find the initial amount of the isotope that is present in the sample.

   b. Find the half-life of this isotope. That is, find the number of years until half of the original amount of the isotope remains.

54. *Radioactive Decay* The amount of radioactive isotope thorium-234 present in a certain sample at time t is given by $A(t) = 500e^{-0.02828t}$ grams, where t years is the time since the initial amount was measured. How long will it take for the amount of isotope to equal 318 grams?

55. *Drugs in the Bloodstream* The concentration of a drug in the bloodstream from the time the drug is injected until 8 hours later is given by

$$y = 100(1 - e^{-0.312(8-t)}) \text{ percent, } t \text{ in hours,}$$

where the drug is administered at time $t = 0$. In how many hours will the drug concentration be 79% of the initial dose?

56. *Drugs in the Bloodstream* If a drug is injected into the bloodstream, the percent of the maximum dosage that is present at time t is given by

$y = 100(1 - e^{-0.35(10-t)})$, where t is in hours, with $0 \le t \le 10$. In how many hours will the percent reach 65%?

57. *Radioactive Decay* A breeder reactor converts stable uranium-238 into the isotope plutonium-239. The decay of this isotope is given by $A(t) = A_0 e^{-0.00002876t}$, where $A(t)$ is the amount of the isotope at time t (in years) and A_0 is the original amount. If the original amount is 100 pounds, find the half-life of this isotope.

58. *Recording Media Sales* On the basis of data collected between 1982 and 1997, manufacturer's shipments of vinyl record albums (LPs and EPs) declined according to the function $A(t) = 482.794e^{-0.351t}$ million albums, where t is the number of years after 1980.

   a. Graph the function A for $0 \le t \le 20$.

   b. If album sales continue according to the pattern indicated in the graph, what sales level are vinyl record albums approaching?

   c. According to the function, when did vinyl record album sales fall to 5 million?

   d. Why are the sales decreasing?
   (Source: *Statistical Abstract of the United States*)

59. *Cost* Suppose the weekly cost for the production of x units of a product is given by $C(x) = 3452 + 50 \ln(x + 1)$ dollars. Use graphical methods to estimate the number of units produced if the total cost is $3556.

60. *Supply* Suppose the daily supply function for a product is $p = 31 + \ln(x + 2)$, where p is in dollars and x is the number of units supplied. Use graphical methods to estimate the number of units that will be supplied if the price is $35.70.

61. *Doubling Time* If P is invested at an annual interest rate r compounded annually for t years, the future value of the investment is given by $S = P(1.07)^t$. Find a formula for the number of years it will take to double the initial investment.

62. *Doubling Time* The future value of a lump sum P that is invested for n years at 10% compounded annually is $S = P(1.10)^n$. Show that the number of years it would take for this investment to double is $n = \log_{1.10} 2$.

63. *Investment* At the end of t years, the future value of an investment of $20,000 at 7%, compounded

annually, is given by $S = 20,000(1 + 0.07)^t$. In how many years will the investment grow to $48,196.90?

64. *Investment* At the end of t years, the future value of an investment of $30,000 at 9%, compounded annually, is given by $S = 30,000(1 + 0.09)^t$. In how many years will the investment grow to $129,829?

65. *Investing* Find in how many years $40,000 invested at 10%, compounded annually, will grow to $64,420.40.

66. *Investing* In how many years will $40,000 invested at 8%, compounded annually, grow to $86,357?

67. *Life Span* Based on data for the years 1910–2007, the expected life span of people in the United States can be described by the function $f(x) = 10.963 + 14.321 \ln x$, where x is the number of years from 1900 to the person's birth year.
(Source: National Center for Health Statistics)

   a. Estimate the birth year for which the expected life span is 78 years.

   b. Use graphical methods to determine the birth year for which the expected life span is 78 years. Does this agree with the solution in part (a)?

68. *Mothers in Workforce* Between the years 1976 and 2004, the percent of moms who returned to the workforce within 1 year after they had a child is given by $w(x) = 5.200 + 15.268 \ln x$, where x is the number of years past 1970.

   a. Use algebraic methods to estimate the year in which the percent is 50.

   b. Use graphical methods to determine the birth year for which the percent is 50. Does this agree with the solution in part (a)?
   (Source: Associated Press)

69. *Supply* Suppose that the supply of a product is given by $p = 20 + 6 \ln(2q + 1)$, where p is the price per unit and q is the number of units supplied. How many units will be supplied if the price per unit is $68.04?

70. *Demand* Suppose that the demand function for a product is $p = \dfrac{500}{\ln(q + 1)}$, where p is the price per unit and q is the number of units demanded. How many units will be demanded if the price is $61.71 per unit?

71. *Deforestation* The number of square miles per year of rain forest destroyed in Brazil is given by

$y = 4899.7601(1.0468^x)$, where x is the number of years from 1990. In what year did 6447 square miles get destroyed, according to the model?
(Source: National Institute of Space Research [INPE] data)

72. *Global Warming* In an effort to reduce global warming, it has been proposed that a tax be levied based on the emissions of carbon dioxide into the atmosphere. The cost–benefit equation $\ln(1 - P) = -0.0034 - 0.0053t$ estimates the relationship between the percent reduction of emissions of carbon dioxide (as a decimal) and the tax t in dollars per ton of carbon dioxide.

   a. Solve the equation for t, giving t as a function of P. Graph the function.

   b. Use the equation in part (a) to find what tax will give a 30% reduction in emissions.
   (Source: W. Clime, *The Economics of Global Warming*)

73. *Doubling Time* The number of quarters needed to double an investment when a lump sum is invested at 8% compounded quarterly is given by $n = \log_{1.02} 2$.

   a. Use the change of base formula to find n.

   b. In how many years will the investment double?

74. *Doubling Time* The number of periods needed to double an investment when a lump sum is invested at 12% compounded semiannually is given by $n = \log_{1.06} 2$.

   a. Use the change of base formula to find n.

   b. How many years pass before the investment doubles in value?

75. *Annuities* If $2000 is invested at the end of each year in an annuity that pays 5% compounded annually, the number of years it takes for the future value to amount to $40,000 is given by $t = \log_{1.05} 2$. Use the change of base formula to find the number of years until the future value is $40,000.

76. *Annuities* If $1000 is invested at the end of each year in an annuity that pays 8% compounded annually, the number of years it takes for the future value to amount to $30,000 is given by $t = \log_{1.08} 3.4$. Use the change of base formula to find the number of years until the future value is $30,000.

77. *Deforestation* One of the major causes of rain forest deforestation is agricultural and residential development. The number of hectares (2.47 acres) destroyed in a particular year t can be modeled by $y = -3.91435 + 2.62196 \ln t$, where $t = 0$ in 1950. When will more than 6 hectares be destroyed per year?

78. *Life Span* The life span of people in the United States is given by $L(x) = 11.307 + 14.229 \ln x$, where x is the number of years after 1900. After what year is the life span greater than 67?

79. *Market Share* Suppose that after a company introduces a new product, the number of months before its market share is x percent is given by

$$m = 20 \ln\left(\frac{50}{50 - x}\right), x < 50$$

After how many months is the market share more than 45%, according to this model?

80. *Drugs in the Bloodstream* The concentration of a drug in the bloodstream from the time the drug is administered until 8 hours later is given by $y = 100(1 - e^{-0.312(8-t)})$ percent, where the drug is administered at time $t = 0$. For what time period is the amount of drug present more than 60%?

81. *Carbon-14 Dating* An exponential decay function can be used to model the number of grams of a radioactive material that remain after a period of time. Carbon-14 decays over time, with the amount remaining after t years given by $y = y_0 e^{-0.00012097t}$, where y_0 is the original amount. If the original

amount of carbon-14 is 200 grams, find the number of years until 155.6 grams of carbon-14 remain.

82. *Sales Decay* After a television advertising campaign ended, the weekly sales of Korbel champagne fell rapidly. Weekly sales in a city were given by $S = 25{,}000e^{-0.072x}$ dollars, where x is the number of weeks after the campaign ended.

  a. Use the logarithmic form of this function to find the number of weeks after the end of the campaign that passed before weekly sales fell below $16,230.

  b. Check your solution by graphical or numerical methods.

83. *Sales Decay* After a television advertising campaign ended, the weekly sales of Turtledove bars fell rapidly. Weekly sales are given by $S = 600e^{-0.05x}$ thousand dollars, where x is the number of days after the campaign ended.

  a. Use the logarithmic form of this function to find the number of weeks after the end of the campaign before weekly sales fell below $269.60.

  b. Check your solution by graphical or numerical methods.

section

5.4

Exponential and Logarithmic Models

key concepts

- Modeling with exponential functions
- Constant percent change
- Comparing quadratic and exponential models
- Logarithmic modeling
- Exponents, logarithms, and linear regression

section preview ▪ Mothers in the Workforce

In Example 6 of Section 5.2, we solved problems about mothers in the workforce by using the fact that the percent of moms who returned to work within one year after having a child is given by

$$w(x) = 5.200 + 15.268 \ln x$$

where x is the number of years after 1970. In this section we will use real data to create this logarithmic function with technology. (See Example 5.)

Many sources provide data that can be modeled by exponential growth and decay functions, and technology can be used to find exponential functions that model data. In this section we model real data with exponential functions and determine when this type of model is appropriate. We also create exponential functions that model phenomena characterized by constant percent change, and we model real data with logarithmic functions when appropriate.

Modeling with Exponential Functions

Exponential functions can be used to model real data if the data exhibited rapid growth or decay. When the scatter plot of data shows a very rapid increase or decrease, it is possible that an exponential function can be used to model the data. Consider the following examples.

| example 1 | **Insurance Premiums** |

The monthly premiums for $250,000 in term-life insurance over a 10-year term period increase with the age of the men purchasing the insurance. The monthly premiums for nonsmoking males are shown in Table 5.4

Table 5.4

Age (years)	Monthly Premium for 10-Year Term Insurance (dollars)	Age (years)	Monthly Premium for 10-Year Term Insurance (dollars)
35	123	60	783
40	148	65	1330
45	225	70	2448
50	338	75	4400
55	500		

(Source: Quotesmith.com)

a. Graph the data in the table with x as age and y in dollars.

b. Create an exponential function that models these premiums as a function of age.

c. Graph the data and the exponential function that models the data on the same axes.

Solution

a. The scatter plot of the data is shown in Figure 5.32(a) on the next page. It shows that the outputs rise rapidly as the inputs increase.

b. Using technology gives the exponential model, rounded to four decimal places, as

$$y = 4.0389(1.0946^x) \text{ dollars}$$

as the monthly premium where x is the age in years.

c. Figure 5.32(b) shows a scatter plot of the data and a graph of the (unrounded) exponential equation used to model it. The graph of the model appears to be a good, but not perfect, fit for the data.

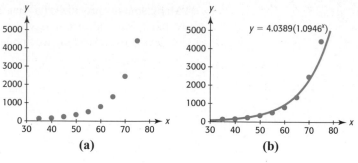

Figure 5.32

---

example 2   Smokers

The percents of U.S. males 18 years of age and older who smoked in selected years from 1965 to 2004 are given in Table 5.5.

Table 5.5

Year	Percent	Year	Percent	Year	Percent
1965	51.6	1990	28.0	2000	25.7
1974	42.9	1991	27.5	2002	25.2
1979	37.2	1992	28.2	2003	24.1
1983	34.7	1993	27.5	2004	23.4
1985	32.1	1994	27.8		
1987	31.0	1995	26.7		

(Source: Centers for Disease Control, www.cdc.gov/nchs/datawh/statabl/)

a. Graph the data, with x equal to the number of years after 1960.

b. Find an exponential function that models the data, using as input the number of years after 1960.

c. Graph the data and the exponential function that models the data on the same axes.

d. Use the model to estimate the percent of U.S. male smokers age 18 and over in 2008.

Solution

a. The graph of the data is shown in Figure 5.33(a).

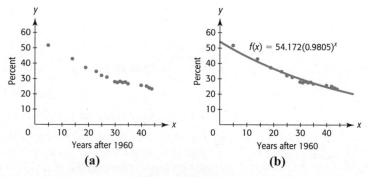

Figure 5.33

b. Exponential regression gives the function $f(x) = 54.172(0.9805)^x$, with three-decimal-place accuracy.

c. The graphs of the (unrounded) model and the data are shown in Figure 5.33(b).

d. Evaluating the function at $x = 48$ gives $f(48) \approx 21.1$, so the estimated percent for 2008 is 21.

---

spreadsheet solution

We can use software programs and spreadsheets to find the exponential function that is the best fit for the data. Table 5.6 shows the Excel spreadsheet for the aligned data of Example 2. Selecting the cells containing the data, using Chart Wizard to get the scatter plot of the data, selecting Add Trendline, and picking Exponential give the equation of the exponential function that is the best fit for the data, along with the scatter plot and the graph of the best-fitting function. The equation and graph are shown in Figure 5.34.

Table 5.6

	A	B
1	Year x	Percent y
2	5	51.6
3	14	42.9
4	19	37.2
5	23	34.7
6	25	32.1
7	27	31
8	30	28
9	31	27.5
10	32	28.2
11	33	27.5
12	34	27.8
13	35	26.7
14	40	25.7
15	42	25.2
16	43	24.1
17	44	23.4

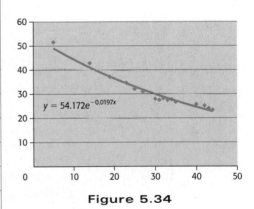

$y = 54.172e^{-0.0197x}$

Figure 5.34

The function created by Excel,

$$y = 54.172e^{-0.0197x}$$

is in a different form than the model found in Example 2, which is

$$f(x) = 54.172(0.9805)^x$$

However, they are equivalent because $e^{-0.0197} \approx 0.9805$, which means that

$$y = 54.172(e^{-0.0197})^x = 54.172(0.9805)^x$$

Constant Percent Change in Exponential Models

How can we determine if a set of data can be modeled by an exponential function? To investigate this, we look at Table 5.7, which gives the growth of paramecia for each hour up to 9 hours.

Table 5.7

x (hours)	0	1	2	3	4	5	6	7	8	9
y (paramecia)	1	2	4	8	16	32	64	128	256	512

If we look at the first differences in the output, we see that they are not constant (Table 5.8). But if we calculate the percent change of the outputs for equally spaced inputs, we see that they are constant. For example, from hour 2 to hour 3, the population grew by 4 units; this represents a 100% increase from the hour-2 population. From hour 3 to hour 4, the population grew by 8, which is a 100% increase over the hour-3 population. In fact, this population increases by 100% each hour. This means that the population 1 hour from now will be the present population plus 100% of the present population.

Table 5.8

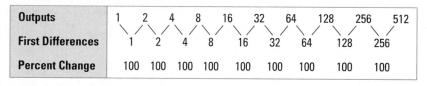

Because the percent change in the outputs is constant for equally spaced inputs in this example, an exponential model ($y = 2^x$) fits the data perfectly.

Constant Percent Changes

If the percent change of the outputs of a set of data is constant for equally spaced inputs, an exponential function will be a perfect fit for the data.

If the percent change of the outputs is approximately constant for equally spaced inputs, an exponential function will be an approximate fit for the data.

example 3

Sales Decay

Suppose a company develops a product that is released with great expectations and extensive advertising, but sales suffer due to bad word of mouth from dissatisfied customers.

a. Use the monthly sales data shown in Table 5.9 to determine the percent change for each of the months given.

Table 5.9

Month	1	2	3	4	5	6	7	8
Sales ($ thousands)	780	608	475	370	289	225	176	137

b. Find the exponential function that models the data.

c. Graph the data and the model on the same axes.

Solution

a. The inputs are equally spaced; the differences of outputs and the percent changes are shown in Table 5.10.

Table 5.10

Outputs	780		608		475		370		289		225		176		137
First Differences		-172		-133		-105		-81		-64		-49		-39	
Percent of Change		-22.1		-21.9		-22.1		-21.9		-22.1		-21.8		-22.2	

The percent change of the outputs is approximately -22%. This means that the sales 1 month from now will be approximately 22% less than the sales now.

b. Because the percent change is nearly constant, an exponential function should fit these data well. Technology gives the model

$$y = 999.781(0.780^x)$$

where x is the month and y is the sales in thousands of dollars.

c. The graphs of the data and the (unrounded) model are shown in Figure 5.35.

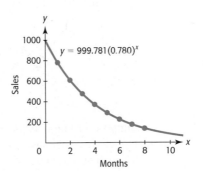

Figure 5.35

Exponential Models

Most graphing utilities give the best-fitting exponential model in the form

$$y = a \cdot b^x$$

It can be shown that the constant percent change of this function is

$$(b - 1) \cdot 100\%$$

Thus, if $b > 1$, the function is increasing (growing), and if $0 < b < 1$, the function is decreasing (decaying).

For the function $y = 2^x$, $b = 2$, so the constant percent change is $r = (2 - 1) \cdot 100\% = 100\%$ (which we found in Table 5.8). For the function $y = 1000(0.78^x)$, the base is $b = 0.78$, and the percent change is $r = (0.78 - 1) \cdot 100\% = -22\%$.

Because $a \cdot b^0 = a \cdot 1 = a$, the value of $y = a \cdot b^x$ is a when $x = 0$, so a is the y-intercept of the graph of the function. Because $y = a$ when $x = 0$, we say that the **initial value** of the function is a. If the constant percent change (as a decimal) is $r = b - 1$, then $b = 1 + r$, and we can write $y = a \cdot b^x$ as

$$y = a(1 + r)^x$$

Exponential Model

If a set of data has initial value a and a constant percent change r (written as a decimal) for equally spaced inputs x, the data can be modeled by the exponential function

$$y = a(1 + r)^x$$

for exponential growth, and

$$y = a(1 - r)^x$$

for exponential decay.

To illustrate this, suppose we know that the present population of a city is 100,000 and that it will grow at a constant percent of 10% per year. Then we know that the future population can be modeled by an exponential function with initial value $a = 100,000$ and constant percent change $r = 10\% = 0.10$ per year. That is, the population can be modeled by

$$y = 100,000(1 + 0.10)^x = 100,000(1.10)^x$$

where x is the number of years from the present.

| example 4 | **Inflation** |

Suppose inflation averages 4% per year for each year from 2000 to 2010. This means that an item that costs \$1 one year will cost \$1.04 one year later. In the second year, the \$1.04 cost will increase by a factor of 1.04, to $(1.04)(1.04) = (1.04)^2$.

a. Write an expression that gives the cost t years after 2000 of an item costing \$1 in 2000.

b. Write an exponential function that models the cost of an item t years from 2000 if its cost was \$100 in 2000.

c. Use the model to find the cost of the item from part (b) in 2010.

Solution

a. The cost of an item is \$1 in 2000, and the cost increases at a constant percent of $4\% = 0.04$ per year, so the cost after t years will be

$$1(1 + 0.04)^t = (1.04)^t \text{ dollars}$$

b. The inflation rate is 4% = 0.04. Thus, if an item costs $100 in 2000, the function that gives the cost after t years is

$$f(t) = 100(1 + 0.04)^t = 100(1.04)^t \text{ dollars}$$

c. The year 2010 is 10 years after 2000, so the cost of an item costing $100 in 2000 is

$$f(10) = 100(1.04)^{10} = 148.02 \text{ dollars} \qquad \blacksquare$$

Comparison of Models

Sometimes it is hard to determine what type of model is the best fit for a set of data. When a scatter plot exhibits a single curvature without a visible high or low point, it is sometimes difficult to determine whether a power, quadratic, or exponential function should be used to model the data. Suppose after graphing the data points we are unsure which of these models is more appropriate. Then we can find the three models, graph them on the same axes as the data points, and inspect the graphs to see which is a better visual fit.*

Consider the exponential model for insurance premiums as a function of age that we found in Example 1. The curvature indicated by the scatter plot suggests that either a power, quadratic, or exponential model may fit the data. To compare the goodness of fit of the models, we find each model and compare its graph with the scatter plot (Figure 5.36).

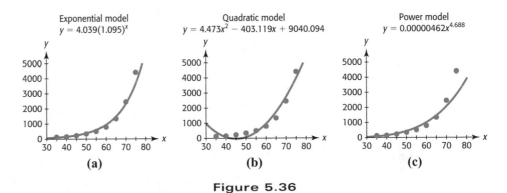

Figure 5.36

The exponential function appears to provide a better fit than the quadratic function because the data points seem to be approaching the x-axis as the input becomes closer to zero from the right. (Recall that the x-axis is a horizontal asymptote for the basic exponential function.) The exponential function also appears to be a better fit than the power function, especially for larger values of x.

When fitting exponential functions to data using technology, it is important to know that problems arise if the input values are large. For instance, when the input values are years (like 2000 and 2010), a technology-determined model may appear to be of the form $y = 0 \cdot b^x$, a function that certainly does not make sense as a model for exponential data. At other times when large inputs are used, your calculator or computer software may give an error message and not even produce an equation. To avoid these

---

*Statistical measures of goodness of fit exist, but caution must be used in applying these measures. For example, the correlation coefficient r found for linear fits to data is different from the coefficient of determination R^2 found for quadratic fits to data.

problems, it is helpful to align the inputs by converting years to the number of years from a specified year (for example, inputting years from 1990 reduces the size of the inputs).

> **Technology Note**
>
> When using technology to fit an exponential model to data, you should align the inputs to reasonably small values.

Logarithmic Models

As with other functions we have studied, we can create models involving logarithms. Data that exhibit an initial rapid increase and then have a slow rate of growth can often be described by the function

$$f(x) = a + b \ln x \qquad (\text{for } b > 0)$$

Note that the parameter a is the vertical shift of the graph of $y = \ln x$ and the parameter b affects how much the graph of $y = \ln x$ is stretched. If $b < 0$, the graph of $f(x) = a + b \ln x$ decreases rather than increases. Most graphing utilities will create logarithmic models in the form $y = a + b \ln x$.

example 5 | **Mothers in the Workforce**

Table 5.11 gives the percent of mothers who returned to the workforce within 1 year after having a child (for the years 1976–2004). Find a logarithmic model for these data, with x equal to the number of years after 1970.

Table 5.11

Year	Returning Mothers (%)	Year	Returning Mothers (%)
1976	31	1990	52
1978	36	1992	53
1980	39	1994	52
1982	43	1995	55
1984	48	1998	59
1986	50	2000	55
1988	51	2004	55

(Source: Associated Press)

Solution

Entering the aligned input data (the number of years after 1970) as the values of x and the percents as the values of y, we use logarithmic regression on a graphing utility to find the function that models the data. This function, rounded to three decimal places, is

$$y = 5.200 + 15.268 \ln x$$

where x is the number of years after 1970. The graphs of this function and the data are shown in Figure 5.37.

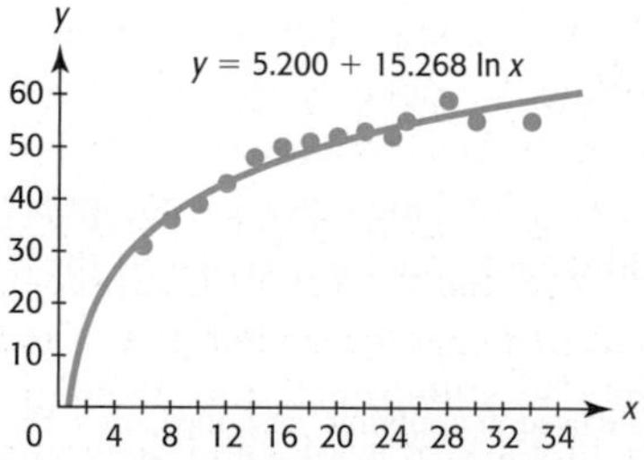

Figure 5.37

Note that the input data in Example 5 were not aligned as the number of years after 1976 because the first aligned input value would be 0, and the logarithm of 0 does not exist.

> ### Technology Note
> When using technology to fit a logarithmic model to data, you must align the data so that all input values are positive.

example 6

Female Workers

The percent of all workers who are female during selected years from 1970 to 2006 is given in Table 5.12.

Table 5.12

Year	Percent of Workers Who Are Female
1970	30.0
1975	31.9
1980	35.3
1985	37.9
1990	37.2
1995	40.3
2000	41.2
2005	41.3
2006	41.5

(Source: U.S. Census Bureau)

a. Find a logarithmic function that models these data. Align the input to be the number of years after 1960.

b. Graph the equation and the aligned data points. Comment on how the model fits the data.

c. Assuming that the model is valid in 2010, use it to estimate the percent of female workers in 2010.

Solution

a. A logarithmic function that models these data is

$$f(x) = 11.244 + 7.978 \ln x$$

percent, where x is the number of years after 1960.

b. The scatter plot of the data and the graph of the logarithmic function that models the data are shown in Figure 5.38. The model appears to be a good fit for the data.

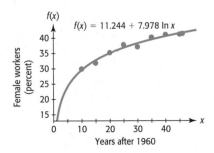

Figure 5.38

c. The percent of all workers who are female in 2010 can be estimated by evaluating $f(50) = 42.5$.

spreadsheet solution

We can use software programs and spreadsheets to find the logarithmic function that is the best fit for data. Table 5.13 shows the Excel spreadsheet for the data of Example 6. Selecting the cells containing the data, using Chart Wizard to get the scatter plot of the data, selecting Add Trendline, and picking Logarithmic gives the equation of the logarithmic function that is the best fit for the data, along with the scatter plot and the graph of the best-fitting function. The equation and graph are shown in Figure 5.39.

Table 5.13

	A	B
	Years After 1960 x	Percent of Female Workers y
1		
2	10	30.0
3	15	31.9
4	20	35.3
5	25	37.9
6	30	37.2
7	35	40.3
8	40	41.2
9	45	41.3
10	46	41.5

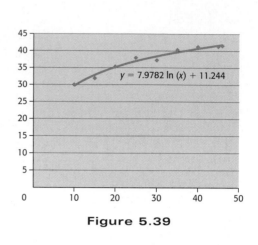

$y = 7.9782 \ln(x) + 11.244$

Figure 5.39

Exponents, Logarithms, and Linear Regression

Now that we have knowledge of the properties of logarithms and how exponential functions are related to logarithmic functions, we can show that linear regression can be used to create exponential models that fit data. (The development of linear regression is discussed in Section 2.2.)

Logarithmic Property 3 states that $\log_b b^x = x$ and in particular that $\ln e^x = x$, so if we have data that can be approximated by an exponential function, we can convert the data to a linear form by taking the logarithm, base e, of the outputs. We can use linear regression to find the linear function that is the best fit for the converted data and then use the linear function as the exponent of e, which gives the exponential function that is the best fit for the original data. Consider Table 5.14, which gives the number y of paramecia in a population after x hours. We have shown that the population can be modeled by $y = 2^x$ in Section 5.1.

Table 5.14

x (hours)	0	1	2	3	4	5	6	7	8	9
y (paramecia)	1	2	4	8	16	32	64	128	256	512
ln y	0	.6931	1.3863	2.0794	2.7726	3.4657	4.1589	4.8520	5.5452	6.2383

The third row of Table 5.14 has the logarithms, base e, of the numbers (y-values) in the second row rounded to 4 decimal places, and the relationship between x and $\ln y$ is linear. The first differences of the $\ln y$ values are constant, with the difference approximately equal to 0.6931. Using linear regression on the x and $\ln y$ values (to 10 decimal places) gives the equation of the linear model:*

$$\ln y = 0.6931471806x$$

---

* This equation could be found directly, in the form $\ln y = ax + b$, where $b = 0$ and a equals the constant difference. This is because the exponential function is a perfect fit for the data in this application.

Because we seek the equation solved for y, we can write this equation in its exponential form, getting

$$y = e^{0.6931471806x}$$

To show that this model for the data is equal to $y = 2^x$, which we found in Section 5.1, we use properties of exponents:

$$y = e^{0.6931471806x} = (e^{0.6931471806})^x \approx 2^x$$

Fortunately, most graphing utilities have combined these steps to give the exponential model directly from the data.

skills check

5.4

1. Find the exponential function that models the data in the table below.

x	-2	-1	0	1	2	3	4	5
y	2/9	2/3	2	6	18	54	162	486

2. The following table has input x and output $f(x)$. Test the percent change of the outputs to determine if the function is exactly exponential, approximately exponential, or not exponential.

x	$f(x)$
1	4
2	16
3	64
4	256
5	1024
6	4096

3. The following table has input x and output $g(x)$. Test the percent change of the outputs to determine if the function is exactly exponential, approximately exponential, or not exponential.

x	$g(x)$
1	2.5
2	6
3	8.5
4	10
5	8
6	6

4. The following table has input x and output $h(x)$. Test the percent change of the outputs to determine if the function is exactly exponential, approximately exponential, or not exponential.

x	$h(x)$
1	1.5
2	2.25
3	3.8
4	5
5	11
6	17

5. Find the exponential function that is the best fit for $f(x)$ defined by the table in Exercise 2.

6. Find the exponential function that is the best fit for $h(x)$ defined by the table in Exercise 4.

7. **a.** Make a scatter plot of the data in the table below.

   b. Does it appear that a linear model or an exponential model is the better fit for the data?

x	1	2	3	4	5	6
y	2	3.1	4.3	5.4	6.5	7.6

8. Find the exponential function that models the data in the table below. Round the model with three-decimal place accuracy.

x	-2	-1	0	1	2	3
y	5	30	150	1000	4000	20,000

9. Compare the first differences and the percent change of the outputs to determine if the data in the table below should be modeled by a linear or an exponential function.

x	1	2	3	4	5
y	2	6	14	34	81

10. Use a scatter plot to determine if a linear or exponential function is the better fit for the data in Exercise 9.

11. Find the linear *or* exponential function that is the better fit for the data in Exercise 9. Round the model to three-decimal-place accuracy.

12. Find the logarithmic function that models the data in the table below. Round the model to two-decimal-place accuracy.

x	1	2	3	4	5	6	7
y	2	4.08	5.3	6.16	6.83	7.38	7.84

13. **a.** Make a scatter plot of the data in the table below.

    b. Does it appear that a linear model or a logarithmic model is the better fit for the data?

x	1	3	5	7	9
y	-2	1	3	4	5

14. **a.** Find a logarithmic function that models the data in the table in Exercise 13. Round the model to three-decimal-place accuracy.

    b. Find a linear function that models the data.

    c. Visually determine which model is the better fit for the data.

15. **a.** Make a scatter plot of the data in the table below.

    b. Find a power function that models the data. Round to 3 decimal places.

    c. Find a quadratic function that models the data. Round to 3 decimal places.

    d. Find a logarithmic function that models the data. Round to 3 decimal places.

x	y
1	3.5
2	5.5
3	6.8
4	7.2
5	8
6	9

16. Let $y = f(x)$ represent the power model found in Exercise 15(b), $y = g(x)$ represent the quadratic model found in Exercise 15(c), and $y = h(x)$ represent the logarithmic model found in Exercise 15(d). Graph each function on the same axes as the scatter plot using the window [0, 12] by [0, 15]. Which model appears to be the best fit?

exercises
5.4

Report models accurate to three decimal places unless otherwise specified. Use the unrounded function to calculate and to graph the function.

Use the exponential form $y = a(1 + r)^x$ to model the information in Exercises 17–20.

17. *Inflation* Suppose the retail price of an automobile is $30,000 in 2000 and that it increases at 4% per year.

    a. Write the equation of the exponential function that models the retail price of the automobile t years after 2000.

    b. Use the model to predict the retail price of the automobile in 2015.

18. *Population* Suppose the population of a city is 190,000 in 2000 and that it grows at 3% per year.

    a. Write the equation of the exponential function that models the annual growth.

    b. Use the model to find the population of this city in 2010.

19. *Sales Decay* At the end of an advertising campaign, weekly sales amounted to $20,000 and then decreased by 2% each week after the end of the campaign.

    a. Write the equation of the exponential function that models the weekly sales.

    b. Find the sales 5 weeks after the end of the advertising campaign.

20. *Inflation* The average price of a house in a certain city is $220,000 in 2008, and it increases at 3% per year.

    a. Write the equation of the exponential function that models the average price of a house t years after 2008.

    b. Use the model to predict the average price of a house in 2013.

21. *Personal Income* Total personal income in the United States (in billions of dollars) for selected years from 1960 to 2005 is given in the following table.

Year	Personal Income
1960	411.5
1970	838.8
1980	2307.9
1990	4878.6
2000	8429.7
2005	10,239.2

(Source: Bureau of Economic Analysis, U.S. Department of Commerce)

    a. These data can be modeled by an exponential function. Write the equation of this function, with x as the number of years after 1960.

    b. If this model is accurate, what will be the total U.S. personal income in 2010?

    c. In what year does the model predict the total personal income will reach $22 trillion?

22. *Cohabiting Households* The table below gives the number of cohabiting without marriage households (in thousands) for selected years from 1960 to 2004.

Year (x)	Cohabiting Households (thousands)	Year (x)	Cohabiting Households (thousands)
1960	439	1994	3661
1970	523	1995	3668
1980	1589	1996	3958
1985	1983	1997	4130
1990	2856	1998	4236
1991	3039	2000	5457
1992	3308	2004	5841
1993	3510		

a. Find the exponential function $y = f(x)$ that models the data, with $x = 0$ in 1960.

b. Estimate the number of cohabiting households in 2010 using the model.

23. *Insurance Premiums* The table below gives the annual premiums required for a $250,000 20-year term-life insurance policy on female nonsmokers of different ages.

a. Find an exponential function that models the monthly premium as a function of the age of the female nonsmoking policyholder.

b. Find the quadratic function that is the best fit for this data.

c. Graph each function on the same axes with the data points to determine visually which model is the better fit for the data. Use the window [30, 80] by [−10, 6800].

Age	Monthly Premium for a 20-Year Policy ($)	Age	Monthly Premium for a 20-Year Policy ($)
35	145	60	845
40	185	65	1593
45	253	70	2970
50	363	75	5820
55	550		

(Source: Quotesmith.com)

24. *National Debt* The table below gives the U.S. national debt for selected years from 1900 to 2006.

Year	U.S. Debt ($ billions)	Year	U.S. Debt ($ billions)	Year	U.S. Debt ($ billions)
1900	1.2	1975	533.2	2000	5674.2
1910	1.1	1985	1823.1	2001	5807.5
1920	24.2	1990	3233.3	2002	6228.2
1930	16.1	1992	4064.6	2003	6783.2
1940	43.0	1994	4692.8	2004	7379.1
1945	258.7	1996	5224.8	2005	7932.7
1955	272.8	1998	5526.2	2006	8507.0
1965	313.8	1999	5656.3		

(Source: Bureau of Public Debt, U.S. Treasury)

a. Using a function of the form $y = a(b^x)$, with $x = 0$ in 1900 and y equal to the national debt in billions, model the data.

b. Use the model to predict the debt in 2010.

c. Predict when the debt will be $25 trillion ($25,000 billion).

d. Look at a graph of both the data and the model. What events may affect the accuracy of this model as a predictor of future public debt? Explain.

25. *Consumer Price Index* The consumer price index (CPI) is calculated by averaging the prices of various items after assigning a weight to each item. The following table gives the consumer price indexes for selected years from 1940 through 2005, reflecting buying patterns of all urban consumers, with x representing years past 1900.

a. Find an equation that models these data.

b. Use the model to predict the consumer price index in 2013.

c. According to the model, during what year will the consumer price index pass 300?

Year	Consumer Price Index	Year	Consumer Price Index
1940	14	1990	130.7
1950	24.1	2000	172.2
1960	29.6	2002	179.9
1970	38.8	2004	188.9
1980	82.4	2005	195.3

(Source: U.S. Census Bureau)

26. *Political Contributions* The figure below gives the total contributions, in millions of dollars, to federal political campaigns from the venture-capital industry for selected years from 1990–2000.

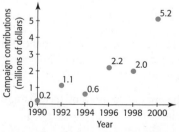

(Source: *The Wall Street Journal*, October 31, 2000)

a. Find an exponential function that models the taxes paid for these years. Use the number of years after 1990 as the input.

b. Find the quadratic function that is the best fit for these data.

c. Graph each function on the same axes with the data points and discuss which model is the better fit for the data.

Year	Income ($)	Year	Income ($)	Year	Income ($)
1990	6652	2000	8794	2003	9573
1995	7763	2001	9039	2004	9827
1998	8316	2002	9182	2005	10,160
1999	8501				

(Source: U.S. Census Bureau)

27. *Life Span* The table below gives the life expectancy for people in the United States for the birth years 1920–2007.

a. Find the logarithmic function that models these data, with x equal to 0 in 1900.

b. Find the quadratic function that is the best fit for the data. Round the quadratic coefficient to five decimal places.

c. Graph each of these functions on the same axes with the data points to determine visually which function is the better model for the data for the years 1920–2007.

d. Evaluate both models for the birth year 2010.

Year	Life Span (years)	Year	Life Span (years)	Year	Life Span (years)
1920	54.1	1987	75.0	1999	76.7
1930	59.7	1988	74.9	2000	77.0
1940	62.9	1989	75.2	2001	77.2
1950	68.2	1990	75.4	2002	77.3
1960	69.7	1992	75.8	2003	77.5
1970	70.8	1994	75.7	2004	77.8
1975	72.6	1996	76.1	2005	77.9
1980	73.7	1998	76.7	2007	77.9

(Source: National Center for Health Statistics)

28. *Poverty Threshold* The following table gives the average poverty thresholds for one person for selected years from 1990 to 2005.

a. Use a logarithmic equation to model these data, with x as the number of years after 1980.

b. Find an exponential model for these data with $x = 0$ in 1980.

c. Which model is the better fit for the data?

29. *Sales Abroad* Because of the weakening of the U.S. dollar, U.S.-based corporations are generating a growing share of their sales overseas. The figure shows the percent of sales made abroad.

a. Find a logarithmic function that is the best fit for the data, with x equal to the number of years from 2000 and y equal to the percent.

b. If the model found in part (a) is accurate in 2007, what would you expect the percent of sales made abroad to be in 2007?

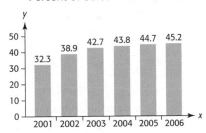

Percent of Sales Made Abroad

(Source: Standard & Poor's)

30. *Female Workers* The percents of females in the workforce for the selected years 1970–2006 are shown in the table below.

a. Find a logarithmic function that models the data, with y the percent and x the number of years from 1960.

b. When will the percent reach 43%, according to the model?

Year	Percent	Year	Percent
1970	30.0	1995	40.3
1975	31.9	2000	41.3
1980	35.3	2005	41.3
1985	37.9	2006	41.5
1990	39.2		

31. *CO_2 Emissions* The global CO_2 fossil fuel emissions for selected years from 1850–2004 are given in the table below.

    a. Find an exponential function that models the data, with y equal to the number of metric tons of CO_2 and x equal to the number of years after 1850.

    b. Predict the number of metric tons of CO_2 in 2015, if the model still applies.

Year	CO$_2$ (metric tons)	Year	CO$_2$ (metric tons)
1850	54	1940	1299
1860	91	1950	1630
1870	147	1960	2577
1880	236	1970	4077
1890	356	1980	5348
1900	534	1990	6196
1910	819	2000	6981
1920	932	2004	7910
1930	1053		

32. *SAT Scores* The composite SAT scores for entering freshmen at Georgia Southern University are shown in the table below.

Year (x)	1996	1997	1998	1999	2000	2001
SAT Score (y)	967	973	983	987	1008	1028

Year (x)	2002	2003	2004	2005	2006
SAT Score (y)	1052	1056	1080	1098	1104

(Source: Student Information Report System [SIRS])

    a. Find an exponential function that models the data, with x equal to the number of years from 1995.

    b. Use the model to estimate the SAT score for this school in 2010.

    c. If this model applies beyond 2006, when does it predict a composite score of 1300?

33. *Sexually Active Girls* The percents of girls of age x or younger who have been sexually active are given in the table below.

    a. Create a logarithmic function that models the data, using an input equal to the age of the girls.

    b. Use the model to estimate the percent for girls age 17 or younger who have been sexually active.

    c. Find the quadratic function that is the best fit for the data.

    d. Graph each of these functions on the same axes with the data points to determine which function is the better model for the data.

Age	Cumulative Percent Sexually Active Girls	Cumulative Percent Sexually Active Boys
15	5.4	16.6
16	12.6	28.7
17	27.1	47.9
18	44.0	64.0
19	62.9	77.6
20	73.6	83.0

(Source: "National Longitudinal Survey of Youth," *Risking the Future.* Washington D.C: National Academy Press)

34. *Sexually Active Boys* The percents of boys age x or younger who have been sexually active are given in the table in Exercise 33.

    a. Create a logarithmic function that models the data, using an input equal to the age of the boys.

    b. Use the model to estimate the percent for boys age 17 or younger who have been sexually active.

    c. Compare the percent that are sexually active for the two genders (see Exercise 33). What do you conclude?

Exponential Functions and Investing

section preview ▪ Investments

If $1000 is invested in an account that earns 6% interest, with the interest added to the account at the end of each year, the interest is said to be **compounded annually**. The amount to which an investment grows over a period of time is called its **future value**. In this section, we use general formulas to find the future value of money invested when the interest is compounded at regular intervals of time and when it is compounded continuously. We also find the lump sum that must be invested to have it grow to a specified amount in the future. This is called the **present value** of the investment.

Compound Interest

In Section 5.4, we found that if a set of data has initial value a and a constant percent change r for equally spaced inputs x, the data can be modeled by the exponential function

$$y = a(1 + r)^x$$

Thus, if an investment of $1000 earns 6% interest compounded annually, the future value at the end of t years will be

$$S = \$1000(1.06)^t \text{ dollars}$$

---

example 1

Future Value of an Account

To numerically and graphically view how $1000 grows at 6% interest compounded annually, do the following:

a. Construct a table that gives the future values $S = 1000(1 + 0.06)^t$ for $t = 0, 1, 2, 3, 4,$ and 5 years after the $1000 was invested.

b. Graph $S = 1000(1 + 0.06)^t$ for the values of t given in part (a).

c. Graph the function $S = 1000(1 + 0.06)^t$ as a continuous function of t for $0 \le t \le 5$.

d. Is there another graph that more accurately represents the amount that would be in the account at any time during the first 5 years that the money is invested?

Solution

a. Substitute the values of t into the equation $S = 1000(1 + 0.06)^t$. Because the output units for S are dollars, we round to the nearest cent to obtain the values in Table 5.15.

b. Figure 5.40 on the next page shows the scatter plot of the data in Table 5.15.

Table 5.15

t (years)	Future Value S ($)
0	1000.00
1	1060.00
2	1123.60
3	1191.02
4	1262.48
5	1338.23

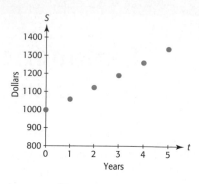

Figure 5.40

c. Figure 5.41(a) shows a continuous graph of the function $S = 1000(1 + 0.06)^t$ for $0 \leq t \leq 5$.

d. The scatter plot in Figure 5.40 shows the amount in the account at the end of each year, but not at any other time during the 5-year period. The continuous graph in Figure 5.41(a) shows the amount continually increasing, but interest is paid only at the end of each year and the amount in the account is constant at the previous level until more interest is added. So the graph in Figure 5.41(b) is a more accurate graph of the amount that would be in this account because it shows the amount remaining constant until the end of each year, when interest is added.

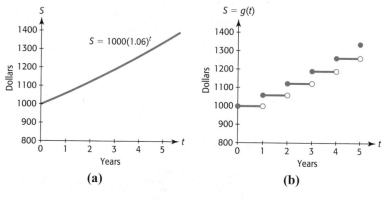

(a)   **(b)**

Figure 5.41

The function g graphed in Figure 5.41(b) is the step function defined by

$$g(t) = \begin{cases} 1000.00 & \text{if } 0 \leq t < 1 \\ 1060.00 & \text{if } 1 \leq t < 2 \\ 1123.60 & \text{if } 2 \leq t < 3 \\ 1191.02 & \text{if } 3 \leq t < 4 \\ 1262.48 & \text{if } 4 \leq t < 5 \\ 1338.23 & \text{if } t = 5 \end{cases} \quad \text{dollars in } t \text{ years}$$

This step function is rather cumbersome to work with, so we will find the future value by using the graph in Figure 5.41(a) and the function $S = 1000(1 + 0.06)^t$ *interpreted discretely.* That is, we draw the graph of this function as it appears in Figure 5.41(a) and evaluate the formula for S with the understanding that the only inputs that make sense in this investment example are nonnegative integers. In general, we have the following.

Future Value of an Investment with Annual Compounding

If $P is invested at an interest rate r per year, compounded annually, the future value S at the end of t years is

$$S = P(1 + r)^t$$

The **annual interest rate** r is also called the **nominal interest rate**, or simply the **rate**. Remember that the interest rate is usually stated as a percent and that it is converted to a decimal when computing future values of investments.

If the interest is compounded more than once per year, then the additional compounding will result in a larger future value. For example, if the investment is compounded twice per year (semiannually), there would be twice as many compounding periods with the interest rate in each period equal to

$$r\left(\frac{1}{2}\right) = \frac{r}{2}$$

Thus, the future value is found by doubling the number of periods and halving the interest rate if the compounding is done twice per year. In general, we use the following model, interpreted discretely, to find the amount of money that results when P dollars are invested and earn compound interest.

Future Value of an Investment with Periodic Compounding

If $P is invested for t years at the annual interest rate r, where the interest is compounded k times per year, then the interest rate per period is $\frac{r}{k}$, the number of compounding periods is kt, and the future value that results is given by

$$S = P\left(1 + \frac{r}{k}\right)^{kt} \text{ dollars}$$

For example, if $1000 is placed in an account that earns 6% interest per year, with the interest added to the account at the end of every month (compounded monthly), the future value of this investment in 5 years is given by

$$S = 1000\left(1 + \frac{0.06}{12}\right)^{12(5)} = \$1348.85$$

Recall from Table 5.15 that compounding annually gives $1338.23, so compounding monthly rather than yearly results in an additional $1348.85 − $1338.23 = $10.62 from the 5-year investment.

example 2

Daily Versus Annual Compounding of Interest

a. Write the equation that gives the future value of $1000 invested for t years at 8% compounded annually.

b. Write the equation that gives the future value of $1000 invested for t years at 8% compounded daily.

c. Graph the equations from parts (a) and (b) on the same axes with t between 0 and 30.

d. What is the additional amount of interest earned in 30 years from compounding daily rather than annually?

Solution

a. Substituting $P = 1000$ and $r = 0.08$ in $S = P(1 + r)^t$ gives

$$S = 1000(1 + 0.08)^t \quad \text{or} \quad S = 1000(1.08)^t$$

b. Substituting $P = 1000$, $r = 0.08$, and $k = 365$ in $S = P\left(1 + \dfrac{r}{k}\right)^{kt}$ gives

$$S = 1000\left(1 + \frac{0.08}{365}\right)^{365t}$$

c. The graphs of both functions are shown in Figure 5.42.

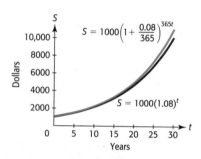

Figure 5.42

d. Evaluating the two functions in parts (a) and (b) at $t = 30$, we calculate the future value of $1000 compounded annually at 8% to be $10,062.66 and the future value of $1000 at 8%, compounded daily, to be $11,020.28. The additional amount earned in 30 years from compounding daily rather than annually is the difference between these two amounts:

$$\$11{,}020.28 - \$10{,}062.66 = \$957.62.$$

Continuous Compounding and the Number *e*

In Section 5.1, we stated that if $P is invested for t years at interest rate r compounded continuously, the future value of the investment is

$$S = Pe^{rt}$$

To see how the number e becomes part of this formula, we consider the future value of $1 invested at an annual rate of 100%, compounded for 1 year with different compounding periods. If we denote the number of periods per year by k, the model that gives the future value is

$$S = \left(1 + \frac{1}{k}\right)^k \text{dollars}$$

Table 5.16 shows the future value of this investment for different compounding periods.

Table 5.16

Type of Compounding	Number of Compounding Periods per Year	Future Value ($)
Annually	1	$\left(1 + \dfrac{1}{1}\right)^1 = 2$
Quarterly	4	$\left(1 + \dfrac{1}{4}\right)^4 = 2.44140625$
Monthly	12	$\left(1 + \dfrac{1}{12}\right)^{12} \approx 2.61303529022$
Daily	365	$\left(1 + \dfrac{1}{365}\right)^{365} \approx 2.71456748202$
Hourly	8760	$\left(1 + \dfrac{1}{8760}\right)^{8760} \approx 2.71812669063$
Each minute	525,600	$\left(1 + \dfrac{1}{525,600}\right)^{525,600} \approx 2.7182792154$
x times per year	x	$\left(1 + \dfrac{1}{x}\right)^x$

As Table 5.16 indicates, the future value increases (but not very rapidly) as the number of compounding periods during the year increases. As x gets very large, the future value approaches the number e, which is 2.718281828 to nine decimal places. In general, the outputs that result from larger and larger inputs in the function

$$f(x) = \left(1 + \frac{1}{x}\right)^x$$

approach the number e. (In calculus, e is defined as the *limit* of $\left(1 + \dfrac{1}{x}\right)^x$ as x approaches ∞.)

This definition of e permits us to create a new model for the future value of an investment when interest is compounded continuously. We can let $m = \dfrac{k}{r}$ to rewrite the formula

$$S = P\left(1 + \frac{r}{k}\right)^{kt} \quad \text{as} \quad S = P\left(1 + \frac{1}{m}\right)^{mrt} \quad \text{or} \quad S = P\left[\left(1 + \frac{1}{m}\right)^m\right]^{rt}$$

Now as the compounding periods k increase without bound, $m = \dfrac{k}{r}$ increases without bound and

$$\left(1 + \frac{1}{m}\right)^m$$

approaches e, so $S = P\left[\left(1 + \dfrac{1}{m}\right)^m\right]^{rt}$ approaches Pe^{rt}.

> ## Future Value of an Investment with Continuous Compounding
>
> If P is invested for t years at an annual interest rate r compounded continuously, then the future value S is given by
>
> $$S = Pe^{rt} \text{ dollars}$$

For a given principal and interest rate, the function $S = Pe^{rt}$ is a continuous function whose domain consists of real numbers greater than or equal to 0. Unlike the other compound interest functions, this one is not discretely interpreted because of the continuous compounding.

example 3 — Future Value and Continuous Compounding

a. What is the future value of $2650 invested for 8 years at 12% compounded continuously?

b. How much interest will be earned on this investment?

Solution

a. The future value of this investment is $S = 2650e^{0.12(8)} = 6921.00$

b. The interest earned on this investment is the future value minus the original investment:

$$\$6921 - \$2650 = \$4271$$ ∎

example 4 — Continuous Versus Annual Compounding of Interest

a. For each of 9 years, compare the future value of an investment of $1000 at 8% compounded annually and of $1000 at 8% compounded continuously.

b. Graph the functions for annual compounding and for continuous compounding for $t = 30$ years on the same axes.

c. What conclusion can be made regarding compounding annually and compounding continuously?

Solution

a. The future value of $1000 compounded annually is given by the function $S = 1000(1 + 0.08)^t$, and the future value of $1000 compounded continuously is given by $S = 1000e^{0.08t}$. By entering these formulas in an Excel spreadsheet and evaluating them for each of 9 years (Table 5.17), we can compare the future value for each of these 9 years. At the end of the 9 years, we see that compounding continuously results in $2054.43 - \$1999.00 = \$55.43$ more than compounding annually yields. ·

b. The graphs are shown in Figure 5.43.

c. The value of the investment increases more under continuous compounding than under annual compounding.

Table 5.17

	A	B	C
1	YEAR	$S = 1000*(1.08)^t$	$S = 1000*e^{(.08t)}$
2	1	1080	1083.287068
3	2	1166.4	1173.510871
4	3	1259.712	1271.249150
5	4	1360.48896	1377.127764
6	5	1469.328077	1491.824698
7	6	1586.874323	1616.07440
8	7	1713.824269	1750.672500
9	8	1850.93021	1896.480879
10	9	1999.004627	2054.433211

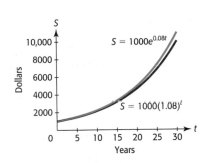

Figure 5.43

Present Value of an Investment

We can write the formula for the future value S when a lump sum P is invested for $n = kt$ periods at interest rate $i = \dfrac{r}{k}$ per compounding period as

$$S = P(1 + i)^n$$

We sometimes need to know the lump sum investment P that will give an amount S in the future. This is called the **present value** P, and it can be found by solving $S = P(1 + i)^n$ for P. Solving for P gives the present value as follows.

Present Value

The lump sum that will give future value S in n compounding periods at rate i per period is the present value

$$P = \frac{S}{(1 + i)^n} \quad \text{or} \quad P = S(1 + i)^{-n}$$

---

example 5

Present Value

What lump sum must be invested at 10% compounded semiannually for the investment to grow to $15,000 in 7 years?

Solution

We have the future value $S = 15,000$, $i = \dfrac{0.10}{2} = 0.05$, and $n = 2 \cdot 7 = 14$, so the present value of this investment is

$$P = \frac{15,000}{(1 + 0.05)^{14}} = 7576.02 \text{ dollars}$$

Investment Models

We have developed and used investment formulas. We can also use graphing utilities to model actual investment data. Consider the following example.

| example 6 | **Mutual Fund Growth** |

Table 5.18

Time (years)	Average Annual Total Return ($)
1	1.23
3	2.24
5	3.11
10	6.82

(Source: *FundStation*)

The data in Table 5.18 give the annual return on an investment of $1.00 made on March 31, 1990, in the AIM Value Fund, Class A Shares. The values reflect reinvestment of all distributions and changes in net asset value but exclude sales charges.

a. Use an exponential function to model these data.

b. Use the model to find what the fund will amount to on March 31, 2001 if $10,000 is invested March 31, 1990 and the fund continues to follow this model after 2000.

c. Is it likely that this fund continued to grow at this rate in 2002?

Solution

a. Using technology to create an exponential model for the data gives

$$A(x) = 1.1609(1.2004)^x$$

Thus, $1.00 invested in this fund grows to $A(x) = 1.1609(1.2004)^x$ dollars x years after March 31, 1990.

b. March 31, 2001 is 11 years after March 31, 1990, so using the equation from part (a) with $x = 11$ gives $A(11) = 1.1609(1.2004)^{11}$ as the future value of an investment of $1.00. Thus, the future value of an investment of $10,000 in 11 years is

$$\$10,000[1.1609(1.2004)^{11}] = \$86,572.64$$

c. No. Most stocks and mutual funds decreased in value in 2002 because of terrorist attacks in 2001. ∎

skills check

5.5

Evaluate the expressions in Exercises 1–10. Write approximate answers rounded to two decimal places.

1. $15,000e^{0.06(20)}$

2. $8000e^{0.05(10)}$

3. $3000(1.06^x)$ for $x = 30$

4. $20,000(1.07^x)$ for $x = 20$

5. $12,000\left(1 + \dfrac{0.10}{k}\right)^{kn}$ for $k = 4$ and $n = 8$

6. $23,000\left(1 + \dfrac{0.08}{k}\right)^{kn}$ for $k = 12$ and $n = 20$

7. $P\left(1 + \dfrac{r}{k}\right)^{kn}$ for $P = 3000, r = 8\%, k = 2, n = 18$

8. $P\left(1 + \dfrac{r}{k}\right)^{kn}$ for $P = 8000, r = 12\%, k = 12, n = 8$

9. $300\left[\dfrac{1.02^n - 1}{0.02}\right]$ for $n = 240$

10. $2000\left[\dfrac{1.10^n - 1}{0.10}\right]$ for $n = 12$

11. Find $g(2.5)$, $g(3)$, and $g(3.5)$ if

$$g(x) = \begin{cases} 1000.00 & \text{if } 0 \le x < 1 \\ 1060.00 & \text{if } 1 \le x < 2 \\ 1123.60 & \text{if } 2 \le x < 3 \\ 1191.00 & \text{if } 3 \le x < 4 \\ 1262.50 & \text{if } 4 \le x < 5 \\ 1338.20 & \text{if } x = 5 \end{cases}$$

12. Find $f(2)$, $f(1.99)$, and $f(2.1)$ if

$$f(x) = \begin{cases} 100 & \text{if } 0 \le x \le 1 \\ 200 & \text{if } 1 < x < 2 \\ 300 & \text{if } 2 \le x \le 3 \end{cases}$$

13. Solve $S = P\left(1 + \dfrac{r}{k}\right)^{kn}$ for P.

14. Solve $S = P(1 + i)^n$ for P.

exercises

5.5

15. *Future Value* If $8800 is invested for x years at 8% interest compounded annually, find the future value that results in

  a. 8 years.

  b. 30 years.

16. *Investments* Suppose $6400 is invested for x years at 7% interest compounded annually. Find the future value of this investment at the end of

  a. 10 years.

  b. 30 years.

17. *Future Value* If $3300 is invested for x years at 10% interest compounded annually, the future value that results is $S = 3300(1.10)^x$ dollars.

  a. Graph the function for $x = 0$ to $x = 8$.

  b. Use the graph to estimate when the money in the account will double.

18. *Investments* If $5500 is invested for x years at 12% interest compounded quarterly, the future value that results is $S = 5500(1.03)^{4x}$ dollars.

  a. Graph this function for $0 \le x \le 8$.

  b. Use the graph to estimate when the money in the account will double.

19. *Future Value* If $10,000 is invested at 12% interest compounded quarterly, find the future value in 10 years.

20. *Future Value* If $8800 is invested at 6% interest compounded semiannually, find the future value in 10 years.

21. *Future Value* An amount of $10,000 is invested at 12% interest compounded daily.

  a. Find the future value in 10 years.

  b. How does this future value compare with the future value in Exercise 19? Why are they different?

22. *Future Value* A total of $8800 is invested at 6% interest compounded daily.

  a. Find the future value in 10 years.

  b. How does this future value compare with the future value in Exercise 20? Why are they different?

23. *Compound Interest* If $10,000 is invested at 12% interest compounded monthly, find the interest earned in 15 years.

24. *Compound Interest* If $20,000 is invested at 8% interest compounded quarterly, find the interest earned in 25 years.

25. *Continuous Compounding* Suppose $10,000 is invested for t years at 6% interest compounded continuously. Give the future value at the end of

  a. 12 years.

  b. 18 years.

26. *Continuous Compounding* If $42,000 is invested for t years at 7% interest compounded continuously, find the future value in

a. 10 years.

b. 20 years.

27. *Continuous versus Annual Compounding*
   a. If $10,000 is invested at 6% compounded annually, find the future value in 18 years.

   b. How does this compare with the result from Exercise 25(b)?

28. *Continuous versus Annual Compounding*
   a. If $42,000 is invested at 7% compounded annually, find the future value in 20 years.

   b. How does this compare with the result from Exercise 26(b)?

29. *Doubling Time* Use a spreadsheet, a table, or a graph to estimate how long it takes for an investment to double if it is invested at 10% interest

a. Compounded annually.

b. Compounded continuously.

30. *Doubling Time* Use a spreadsheet, a table, or a graph to estimate how long it takes for an investment to double if it is invested at 6% interest

a. Compounded annually.

b. Compounded continuously.

31. *Future Value* Suppose $2000 is invested in an account paying 5% interest compounded annually. What is the future value of this investment

a. After 8 years?

b. After 18 years?

32. *Future Value* If $12,000 is invested in an account that pays 8% interest compounded quarterly, find the future value of this investment

a. After 2 quarters.

b. After 10 years.

33. *Investments* Suppose $3000 is invested in an account that pays 6% interest compounded monthly. What is the future value of this investment after 12 years?

34. *Investments* If $9000 is invested in an account that pays 8% interest compounded quarterly, find the future value of this investment

a. After 0.5 year.

b. After 15 years.

35. *Doubling Time* If money is invested at 10% interest compounded quarterly, the future value of the investment doubles approximately every 7 years.

a. Use this information to complete the table below for an investment of $1000 at 10% interest compounded quarterly.

b. Create an exponential function, rounded to three decimal places, that models the discrete function defined by the table.

c. Because the interest is compounded quarterly, this model must be interpreted discretely. Use the rounded function to find the value of the investment in 5 years and in $10\frac{1}{2}$ years after the money was invested.

Years	0	7	14	21	28
Future Value ($)	1000				

36. *Doubling Time* If money is invested at 11.6% interest compounded monthly, the future value of the investment doubles approximately every 6 years.

a. Use this information to complete the table below for an investment of $1000 at 11.6% interest compounded monthly.

b. Create an exponential function, rounded to three decimal places, that models the discrete function defined by the table.

c. Because the interest is compounded monthly, this model must be interpreted discretely. Use the rounded model to find the value of the investment in 2 months, in 4 years, and in $12\frac{1}{2}$ years.

Years	0	6	12	18	24
Future Value ($)	1000				

37. *Present Value* What lump sum must be invested at 10%, compounded monthly, for the investment to grow to $65,000 in 8 years?

38. *Present Value* What lump sum must be invested at 8%, compounded quarterly, for the investment to grow to $30,000 in 12 years?

39. *Present Value* What lump sum investment will grow to $10,000 in 10 years if it is invested at 6% compounded annually?

40. *Present Value* What lump sum investment will grow to $30,000 if it is invested for 15 years at 7% compounded annually?

41. *College Tuition* New parents want to put a lump sum into a money market fund to provide $30,000 in 18 years, to help pay for college tuition for their child. If the fund averages 10% per year compounded monthly, how much should they invest?

42. *Trust Fund* Grandparents decide to put a lump sum of money into a trust fund on their granddaughter's 10th birthday so that she will have $1,000,000 on her 60th birthday. If the fund pays 11% compounded monthly, how much money must they put in the account?

43. *College Tuition* The Toshes need to have $80,000 in 12 years for their son's college tuition. What amount must they invest to meet this goal if they can invest money at 10%, compounded monthly?

44. *Retirement* Hennie and Bob inherit $100,000 and plan to invest part of it for 25 years at 10%, compounded monthly. If they want it to grow to $1 million dollars for their retirement, how much should they invest?

45. *Investment* At the end of t years, the future value of an investment of $10,000 in an account that pays 8% APR compounded monthly is

$$S = 10{,}000\left(1 + \frac{0.08}{12}\right)^{12t} \text{ dollars}$$

Assuming no withdrawals or additional deposits, how long will it take for the investment to amount to $40,000?

46. *Investment* At the end of t years, the future value of an investment of $25,000 in an account that pays 12% annual percentage rate (APR) compounded quarterly is

$$S = 25{,}000\left(1 + \frac{0.12}{4}\right)^{4t} \text{ dollars}$$

In how many years will the investment amount to $60,000?

47. *Investment* At the end of t years, the future value of an investment of $38,500 in an account that pays 8% compounded monthly is

$$S = 38{,}500\left(1 + \frac{0.08}{12}\right)^{12t} \text{ dollars}$$

For what time period (in years) will the future value of the investment be more than $100,230?

48. *Investment* At the end of t years, the future value of an investment of $12,000 in an account that pays 8% compounded monthly is

$$S = 12{,}000\left(1 + \frac{0.08}{12}\right)^{12t} \text{ dollars}$$

Assuming no withdrawals or additional deposits, for what time period (in years) will the future value of the investment be less than $48,000?

49. *Investing* If an investment of $P earns r percent (as a decimal), compounded m times per year, its future value in t years is $A = P\left(1 + \dfrac{r}{m}\right)^{mt}$. Prove that the number of years it takes for this investment to double is $t = \dfrac{\ln 2}{m \ln(1 + r/m)}$.

Annuities; Loan Repayment

section preview ▪ Annuities

To prepare for future retirement, people frequently invest a fixed amount of money at the end of each month into an account that pays interest that is compounded monthly. Such an investment plan, or any other characterized by regular payments, is called an annuity. In this section, we develop a model to find future values of annuities.

A person planning retirement may also want to know how much money is needed to provide regular payments to him or her during retirement, and a couple may want to know what lump sum to put in an investment to pay future college expenses for their child. The amount of money they need to invest to receive a series of payments in the future is the present value of an annuity.

We can also use the present value concept to determine the payments that are necessary to repay a loan with equal payments made on a regular schedule, which is called amortization. The formulas used to determine the future value and the present value of annuities and the payments necessary to repay loans are applications of exponential functions.

Future Value of an Annuity

An **annuity** is a financial plan characterized by regular payments. We can view an annuity as a savings plan where regular payments are made to an account, and we can use an exponential function to determine what the future value of the account will be. One type of annuity, called an **ordinary annuity**, is a financial plan where equal payments are contributed at the end of each period to an account that pays a fixed rate of interest compounded at the same time as payments are made. For example, suppose $1000 is invested at the end of each year for 5 years in an account that pays interest at 10% compounded annually. To find the future value of this annuity, we can think of it as the sum of 5 investments of $1000, one that draws interest for 4 years (from the end of the first year to the end of the fifth year), one that draws interest for 3 years, and so on (Table 5.19).

Table 5.19

Investment	Future Value
Invested at end of 1st year	$1000(1.10)^4$
Invested at end of 2nd year	$1000(1.10)^3$
Invested at end of 3rd year	$1000(1.10)^2$
Invested at end of 4th year	$1000(1.10)^1$
Invested at end of 5th year	1000

Note that the last investment is at the end of the fifth year, so it earned no interest. The future value of the annuity is the sum of the separate future values. It is

$$S = 1000 + 1000(1.10)^1 + 1000(1.10)^2 + 1000(1.10)^3 + 1000(1.10)^4$$

$$= 6105.10 \text{ dollars}$$

In Section 8.4, we will develop the following formula to find the future value of an annuity.

> ### Future Value of an Ordinary Annuity
>
> If R dollars are contributed at the end of each period for n periods into an annuity that pays interest at rate i at the end of the period, the future value of the annuity is
>
> $$S = R\left[\frac{(1 + i)^n - 1}{i}\right] \text{ dollars}$$

Note that the interest rate used in the above formula is the rate per period, not the rate for a year. We can use this formula to find the future value of an annuity in which $1000 is invested at the end of each year for 5 years into an account that pays interest at 10% compounded annually. This future value, which we found earlier without the formula, is

$$S = R\left[\frac{(1 + i)^n - 1}{i}\right] = 1000\left[\frac{(1 + 0.10)^5 - 1}{0.10}\right] = 6105.10 \text{ dollars}$$

example 1

Future Value of an Ordinary Annuity

Find the 5-year future value of an ordinary annuity with a contribution of $500 per quarter into an account that pays 8% per year compounded quarterly.

Solution

The payments and interest compounding occur quarterly, so the interest rate per period is $\dfrac{0.08}{4} = 0.02$ and the number of compounding periods is $4(5) = 20$. Substituting the information into the formula for the future value of an annuity gives

$$S = 500\left[\frac{(1 + 0.02)^{20} - 1}{0.02}\right] = 12{,}148.68$$

Thus, the future value of this investment is $12,148.68. ∎

example 2

Future Value of an Annuity

Harry deposits $200 at the end of each month into an account that pays interest 12% per year, compounded monthly. Find the future value for every 4-month period, for up to 36 months.

Solution

To find the future value of this annuity, we note that the interest rate per period (month) is $\dfrac{12\%}{12} = 0.01$, so the future value at the end of n months is given by the function

$$S = 200\left(\frac{1.01^n - 1}{0.01}\right) \text{ dollars}$$

This model for the future value of an ordinary annuity is a continuous function with discrete interpretation because the future value changes only at the end of each period. A graph of the function and the table of future values for every 4 months up to 36 months are shown in Figure 5.44 and Table 5.20, respectively. Keep in mind that the graph should be interpreted discretely; that is, only positive integer inputs and corresponding outputs make sense.

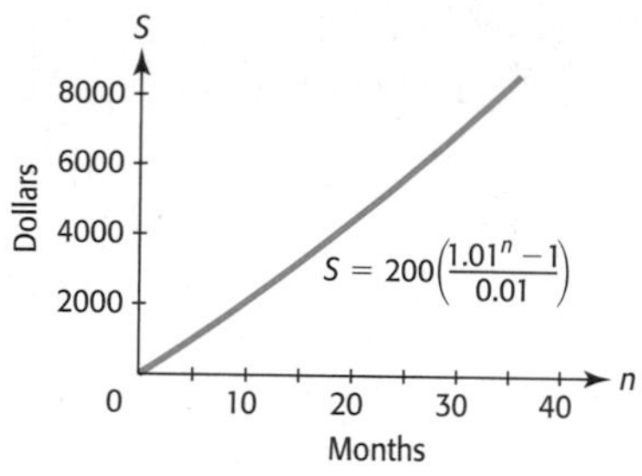

Table 5.20

Number of Months	Value of Annuity ($)
4	812.08
8	1657.13
12	2536.50
16	3451.57
20	4403.80
24	5394.69
28	6425.82
32	7498.81
36	8615.38

$$S = 200\left(\frac{1.01^n - 1}{0.01}\right)$$

Figure 5.44

Present Value of an Annuity

Many retirees purchase annuities that give them regular payments through a period of years. Suppose a lump sum of money is invested to return payments of $R at the end of each of n periods, with the investment earning interest at rate i per period, after which $0 will remain in the account. This lump sum is the **present value** of this ordinary annuity, and it is equal to the sum of the present values of each of the future payments. Table 5.21 shows the present value of each payment of this annuity.

Table 5.21

Present	Present Value of This Payment ($)
Paid at end of 1st period	$R(1 + i)^{-1}$
Paid at end of 2nd period	$R(1 + i)^{-2}$
Paid at end of 3rd period	$R(1 + i)^{-3}$
.	.
.	.
Paid at end of $(n - 1)$st period	$R(1 + i)^{-(n-1)}$
Paid at end of nth period	$R(1 + i)^{-n}$

We must have enough money in the account to provide the present value of each of these payments, so the present value of the annuity is

$$A = R(1 + i)^{-1} + R(1 + i)^{-2} + R(1 + i)^{-3} + \cdots + R(1 + i)^{-(n-1)} + R(1 + i)^{-n}$$

We will show in Section 8.4 that this sum can be written in the form

$$A = R\left[\frac{1 - (1 + i)^{-n}}{i}\right]$$

> ## Present Value of an Ordinary Annuity
>
> If a payment of R is to be made at the end of each period for n periods from an account that earns interest at a rate of i per period, then the account is an **ordinary annuity**, and the **present value** is
>
> $$A = R \left[\frac{1 - (1 + i)^{-n}}{i} \right]$$

| example 3 | ## Present Value of an Annuity |

Suppose a retiring couple wants to establish an annuity that will provide $2000 at the end of each month for 20 years. If the annuity earns 6%, compounded monthly, how much must the couple put in the account to establish the annuity?

Solution

We seek the present value of an annuity that pays $2000 at the end of each month for $12(20) = 240$ months, with interest at $\dfrac{6\%}{12} = \dfrac{0.06}{12} = 0.005$ per month. The present value is

$$A = 2000 \left[\frac{1 - (1 + 0.005)^{-240}}{0.005} \right] = 279{,}161.54 \text{ dollars}$$

Thus, the couple can receive a payment of $2000 at the end of each month for 20 years if they put a lump sum of $279,161.54 in an annuity. (Note that the sum of the money they will receive over the 20 years is $2000 \cdot 20 \cdot 12 = \$480{,}000$.) ∎

| example 4 | ## Home Mortgage |

A couple who wants to purchase a home has $30,000 for a down payment and wants to make monthly payments of $2200. If the interest rate for a 25-year mortgage is 6% per year on the unpaid balance, what is the price of a house they can buy?

Solution

The amount of money that they can pay for the house is the sum of the down payment and the present value of the payments that they can afford to pay. The present value of the $12(25) = 300$ monthly payments of $2200 with interest at $\dfrac{6\%}{12} = \dfrac{0.06}{12} = 0.005$ per month is

$$A = 2200 \left[\frac{1 - (1 + 0.005)^{-300}}{0.005} \right] = 341{,}455.10$$

They can buy a house costing $341{,}455 + 30{,}000 = \$371{,}455.10$. ∎

Loan Repayment

When money is borrowed, the borrower must repay the total amount that was borrowed (the debt) plus interest on that debt. Most loans (including those for houses, for cars, and for other consumer goods) require regular payments on the debt plus payment of interest on the unpaid balance of the loan. That is, loans are paid off by a series of partial payments with interest charged on the unpaid balance at the end of each period. The stated interest rate (the **nominal rate**) is the *annual rate*. The rate per period is the nominal rate divided by the number of payment periods per year.

There are two popular repayment plans for these loans. One plan applies an equal amount to the debt each payment period plus the interest for the period, which is the interest on the unpaid balance. For example, a loan of $240,000 for 10 years at 12% could be repaid with 120 monthly payments of $2000 plus 1% $\left(\dfrac{1}{12} \text{ of } 12\%\right)$ of the unpaid balance each month. When this payment method is used, the payments will decrease as the unpaid balance decreases. A few sample payments are shown in Table 5.22.

Table 5.22

Payment Number (month)	Unpaid Balance ($)	Interest on the Unpaid Balance ($)	Balance Reduction ($)	Monthly Payment ($)	New Balance ($)
1	240,000	2400	2000	4400	238,000
2	238,000	2380	2000	4380	236,000
.					
51	140,000	1400	2000	3400	138,000
.					
101	40,000	400	2000	2400	38,000
.					
118	6000	60	2000	2060	4000

A loan of this type can also be repaid by making all payments (including the payment on the balance and the interest) of equal size. The process of repaying the loan in this way is called **amortization**. If a bank makes a loan of this type, it uses an amortization formula that gives the size of the equal payments. This formula is developed by considering the loan as an annuity purchased from the borrower by the bank, with the annuity paying a fixed return to the bank each payment period. The lump sum that the bank gives to the borrower (the principal of the loan) is the present value of this annuity, and each payment that the bank receives from the borrower is a payment from this annuity. To find the size of these equal payments, we solve the formula for the present value of an annuity, $A = R\left[\dfrac{1 - (1 + i)^{-n}}{i}\right]$, for R. This gives the following formula.

Amortization Formula

If a debt of A, with interest at a rate of i per period, is amortized by n equal periodic payments made at the end of each period, then the size of each payment is

$$R = A\left[\dfrac{i}{1 - (1 + i)^{-n}}\right]$$

We can use this formula to find the *equal* monthly payments that will amortize the loan of $240,000 for 10 years at 12%. The interest rate per month is 0.01, and there are 120 periods.

$$R = A\left[\frac{i}{1 - (1 + i)^{-n}}\right] = 240{,}000\left[\frac{0.01}{1 - (1 + 0.01)^{-120}}\right] \approx 3443.303$$

Thus, repaying this loan requires 120 equal payments of $3443.31. If we compare this payment with the selected payments in Table 5.22, we see that the first payment plan in Table 5.22 is nearly $1000 more than the equal amortization payment and that every payment after the 101st is $1000 less than the amortization payment. The advantage of the amortization method of repaying the loan is that the borrower can budget better by knowing what the payment will be each month. Of course, the balance of the loan will not be reduced as fast with this payment method, because most of the monthly payment will be needed to pay interest in the first few months. The partial spreadsheet in Table 5.23 shows how the outstanding balance is reduced.

Table 5.23

	A	B	C	D	E	F
1	Payment	Unpaid		Monthly	Balance	New
2	Number	Balance	Interest	Payment	Reduction	Balance
3	1	240,000	2400	3443.31	1043.31	238,956.69
4	2	238,956.69	2389.57	3443.31	1053.74	237,902.95
	.					
53	51	172,745.38	1727.45	3443.31	1715.86	171,029.52
	.					
103	101	63,136.30	631.36	3443.31	2811.95	60,324.35

---

example 5

Home Mortgage

A couple that wants to purchase a home with a price of $230,000 has $50,000 for a down payment. If they can get a 25-year mortgage at 9% per year on the unpaid balance,

a. What will be their equal monthly payments?

b. What is the total amount they will pay before they own the house outright?

c. How much interest will they pay?

Solution

a. The amount of money that they must borrow is $230,000 - $50,000 = $180,000. The number of monthly payments is 12(25) = 300, and the interest rate is $\frac{9\%}{12} = \frac{0.09}{12} = 0.0075$ per month. The monthly payment is

$$R = 180{,}000\left[\frac{0.0075}{1 - (1.0075)^{-300}}\right] \approx 1510.553$$

so the required payment would be $1510.56.

b. The amount that they must pay before owning the house is the down payment plus the total of the 300 payments:

$$\$50,000 + 300(\$1510.56) = \$503,168$$

c. The total interest is the total amount paid minus the price paid for the house. The interest is

$$\$503,168 - \$230,000 = \$273,168 \qquad \blacksquare$$

When the interest paid is based on the periodic interest rate on the unpaid balance of the loan, the nominal rate is also called the **annual percentage rate** (APR). You will occasionally see an advertisement for a loan with a stated interest rate and a different, larger APR. This can happen if a lending institution charges fees (sometimes called points) in addition to the stated interest rate. Federal law states that all these charges must be included in computing the APR, which is the true interest rate that is being charged on the loan.

skills check

5.6

Give answers to two decimal places.

1. Solve $S = P(1 + i)^n$ for P with positive exponent n.

2. Solve $S = P(1 + i)^n$ for P, with no denominator.

3. Solve $Ai = R[1 - (1 + i)^{-n}]$ for A.

4. Evaluate $2000\left[\dfrac{1 - (1 + 0.01)^{-240}}{0.01}\right]$.

5. Solve $A = R\left[\dfrac{1 - (1 + i)^{-n}}{i}\right]$ for R.

6. Evaluate $240,000\left[\dfrac{0.01}{1 - (1 + 0.01)^{-120}}\right]$.

exercises

5.6

7. *IRA* Ruth Wright decides to invest $4000 in an IRA CD at the end of each year for 10 years. If she makes these payments and the certificates all pay 6%, compounded annually, how much will she have at the end of the 10 years?

8. *Future Value* Find the future value of an annuity of $5000 paid at the end of each year for 20 years, if interest is earned at 9%, compounded annually.

9. *Down Payment* To start a new business Beth deposits $1000 at the end of each six-month period in an account that pays 8%, compounded semiannually. How much will she have at the end of 8 years?

10. *Future Value* Find the future value of an annuity of $2600 paid at the end of each 3-month period for 5 years, if interest is earned at 6%, compounded quarterly.

11. *Retirement* Mr. Lawrence invests $600 at the end of each month in an account that pays 7%, compounded monthly. How much will be in the account in 25 years?

12. *Down Payment* To accumulate money for the down payment on a house, the Kings deposited $800 at the end of each month into an account paying 7%, compounded monthly. How much will they have at the end of 5 years?

13. *Dean's List* Parents agree to invest $1000 (at 10%, compounded semiannually) for their daughter on the December 31 or June 30 following each semester that she makes the Dean's List during her 4 years in college. If she makes the Dean's List in each of the 8 semesters, how much will the parents have to give her when she graduates?

14. *College Tuition* To help pay for tuition, parents deposited $1000 into an account on each of their child's birthdays, starting with the first birthday. If the account pays 6%, compounded annually, how much will they have for tuition on her 19th birthday?

15. *Annuities* Find the present value of an annuity that will pay $1000 at the end of each year for 10 years if the interest rate is 7% compounded annually.

16. *Annuities* Find the present value of an annuity that will pay $500 at the end of each year for 20 years if the interest rate is 9% compounded annually.

17. *Lottery Winnings* The winner of a "million dollar" lottery is to receive $50,000 plus $50,000 at the end of each year for 19 years, or the present value of this annuity in cash. How much cash would she receive if money is worth 8% compounded annually?

18. *College Tuition* A couple wants to establish a fund that will provide $3000 for tuition at the end of each 6-month period for 4 years. If a lump sum can be placed in an account that pays 8% compounded semiannually, what lump sum is required?

19. *Insurance Payment* A man is disabled in an accident and wants to receive an insurance payment that will provide him with $3000 at the end of each month for 30 years. If the payment can be placed in an account that pays 9% compounded monthly, what size payment should he seek?

20. *Auto Leasing* A woman wants to lease rather than buy a car but does not want to make monthly payments. A dealer has the car she wants, which leases for $400 at the end of each month for 48 months. If money is worth 8% compounded monthly, what lump sum should she offer the dealer to keep the car for 48 months?

21. *Business Sale* A man can sell his Thrifty Electronics business for $800,000 cash or for $100,000 plus $122,000 at the end of each year for 9 years.
    a. Find the present value of the annuity that is offered if money is worth 10% compounded annually.
    b. If he takes the $800,000, spends $100,000 of it, and invests the rest in a 9-year annuity at 10% compounded annually, what size annuity payment will he receive at the end of each year?
    c. Which is better, taking the $100,000 and the annuity or taking the cash settlement? Discuss the advantages of your choice.

22. *Sale of a Practice* A physician can sell her practice for $1,200,000 cash or for $200,000 plus $250,000 at the end of each year for 5 years.
    a. Find the present value of the annuity that is offered if money is worth 7%, compounded annually.
    b. If she takes the $1,200,000, spends $200,000 of it, and invests the rest in a 5-year annuity at 7% compounded annually, what size annuity payment will she receive at the end of each year?
    c. Which is better, taking the $200,000 and the annuity or taking the cash settlement? Discuss the advantages of your choice.

23. *Home Mortgage* A couple wants to buy a house and can afford to pay $1600 per month.
    a. If they can get a loan for 30 years with interest at 9% per year on the unpaid balance and make monthly payments, how much can they pay for a house?
    b. What is the total amount paid over the life of the loan?
    c. What is the total interest paid on the loan?

24. *Auto Loan* A man wants to buy a car and can afford to pay $400 per month.
    a. If he can get a loan for 48 months with interest at 12% per year on the unpaid balance and makes monthly payments, how much can he pay for a car?
    b. What is the total amount paid over the life of the loan?
    c. What is the total interest paid on the loan?

25. *Loan Repayment* A loan of $10,000 is to be amortized with quarterly payments over 4 years. If the interest on the loan is 8% per year, paid on the unpaid balance,

  a. What is the interest rate charged each quarter on the unpaid balance?

  b. How many payments are made to repay the loan?

  c. What payment is required quarterly to amortize the loan?

26. *Loan Repayment* A loan of $36,000 is to be amortized with monthly payments over 6 years. If the interest on the loan is 6% per year, paid on the unpaid balance,

  a. What is the interest rate charged each month on the unpaid balance?

  b. How many payments are made to repay the loan?

  c. What payment is required each month to amortize the loan?

27. *Home Mortgage* A couple who wants to purchase a home with a price of $350,000 has $100,000 for a down payment. If they can get a 30-year mortgage at 6% per year on the unpaid balance,

  a. What will be their monthly payments?

  b. What is the total amount they will pay before they own the house outright?

  c. How much interest will they pay over the life of the loan?

28. *Business Loan* Business partners want to purchase a restaurant that costs $750,000. They have $300,000 for a down payment, and they can get a 25-year business loan for the remaining funds at 8% per year on the unpaid balance, with quarterly payments.

  a. What will be the payments?

  b. What is the total amount that they will pay over the 25-year period?

  c. How much interest will they pay over the life of the loan?

section

5.7 Logistic and Gompertz Functions

key concepts

- Logistic function
- Limiting value of logistic function
- Modeling logistic functions
- Gompertz function

section preview ▪ DVD Players

DVD players entered U.S. households faster than any other piece of home-electronics equipment in history. The past and projected sales from 1999 to 2004 are shown in Figure 5.45. Notice that the graph is an S-shaped curve with a relatively slow start, then a steep climb followed by a leveling off. It is reasonable that sales would eventually level off, even if everyone in the United States bought one. Very seldom will exponential growth continue indefinitely, so functions of this type better represent many types of sales and organizational growth. Two functions that are characterized by rapid growth followed by leveling off are **logistic functions** and **Gompertz functions**. We consider applications of these functions in this section.

Boffo Sales!

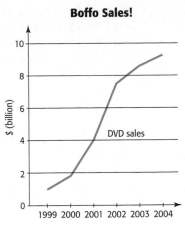

Figure 5.45

(Source: *Newsweek*, August 20, 2001)

Logistic Functions

When growth begins slowly, then increases at a rapid rate and finally slows over time to a rate that is almost zero, the amount (or number) present at any given time frequently fits on an S-shaped curve. Growth of business organizations and the spread of a virus or disease sometimes occur according to this pattern. For example, the total number y of people on a college campus infected by a virus can be modeled by

$$y = \frac{10,000}{1 + 9999e^{-0.99t}}$$

where t is the number of days after an infected student arrives on campus. Table 5.24 shows the number infected on selected days. Note that the number infected grows rapidly, and then levels off near 10,000. The graph of this function is shown in Figure 5.46.

Table 5.24

Days	1	3	5	7	9	11	13	15	17
Number Infected	3	19	139	928	4255	8429	9749	9965	9995

A function that can be used to model growth of this type is called a **logistic function**; its graph is similar to the one shown in Figure 5.46.

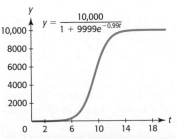

Figure 5.46

Logistic Function

For real numbers a, b, and c, the function

$$f(x) = \frac{c}{1 + ae^{-bx}}$$

is a logistic function.

If $a > 0$, a logistic function increases when $b > 0$ and decreases when $b < 0$.

The parameter c is often called the **limiting value**,* or **upper limit**, because the line $y = c$ is a horizontal asymptote for the logistic function. As seen in Figure 5.47, the line $y = 0$ is also a horizontal asymptote. Logistic growth begins at a rate that is near zero, rapidly increases in an exponential pattern, and then slows to a rate that is again near zero.

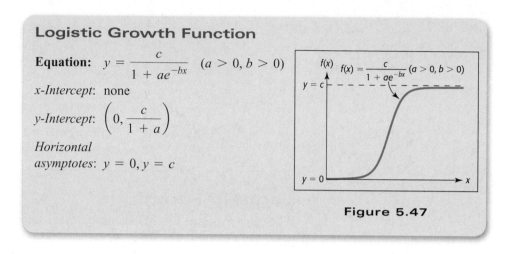

Logistic Growth Function

Equation: $y = \dfrac{c}{1 + ae^{-bx}}$ $(a > 0, b > 0)$

x-Intercept: none

y-Intercept: $\left(0, \dfrac{c}{1 + a}\right)$

Horizontal asymptotes: $y = 0, y = c$

$f(x) = \dfrac{c}{1 + ae^{-bx}}$ $(a > 0, b > 0)$

Figure 5.47

Even though the graph of an increasing logistic function has the elongated S-shape indicated in Figure 5.47, it may be the case that we have data showing only a portion of the S. Consider the following example, which illustrates this.

---

example 1   ## Life Span

The data in Table 5.25 show the expected life span of people for certain birth years in the United States. A technology-determined logistic function† for the life-span data is

$$y = \frac{81.837}{1 + 0.8213e^{-0.0255x}} \text{ years}$$

where x is the number of birth years after 1900. The graph of this logistic function is shown in Figure 5.48. Use this model to

a. Estimate the expected life span of a person born in the United States in 1955 and a person born in 1980.

b. Find an upper limit for a person's expected life span in the United States according to this model.

---

*The logistic function fit by most technologies is a best-fit (least-squares) logistic equation. There may therefore be data values that are larger than the value of c. To avoid confusion, we talk about the limiting value of the logistic function or the upper limit in the problem context.

† Note that because of the exponential term in a logistic model, you must align input data to reasonably small values before using technology to fit a logistic model to data.

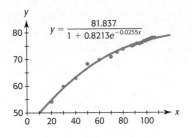

$$y = \frac{81.837}{1 + 0.8213e^{-0.0255x}}$$

Figure 5.48

Table 5.25

Birth Year	Life Span (years)	Year	Life Span (years)	Year	Life Span (years)
1920	54.1	1987	75.0	1999	76.7
1930	59.7	1988	74.9	2000	77.0
1940	62.9	1989	75.2	2001	77.2
1950	68.2	1990	75.4	2002	77.3
1960	69.7	1992	75.8	2003	77.5
1970	70.8	1994	75.7	2004	77.8
1975	72.6	1996	76.1	2005	77.9
1980	73.7	1998	76.7	2007	77.9

(Source: National Center for Health Statistics)

Solution

a. To find the expected life span for people born in 1955, we evaluate the function for $x = 55$. This gives an expected life span of $y = \dfrac{81.837}{1 + 0.8213e^{-0.0255(55)}} = 68.1$, or approximately 68 years. The expected life span for people born in 1980 is found by evaluating y when $x = 80$. This evaluation gives a life span of $y = \dfrac{81.837}{1 + 0.8213e^{-0.0255(80)}} = 73.9$, or approximately 74 years.

b. By substituting larger values for x in the equation $y = \dfrac{81.837}{1 + 0.8213e^{-0.0255x}}$, we see that the outputs increase very slowly and approach the value 81.837 (Figure 5.49). Thus, the upper limit of expected life span, according to this model, is about 82 years.

	A	B
1	100	76.905
2	150	80.396
3	200	81.429
4	250	81.723
5	300	81.805
6	350	81.828
7	400	81.835

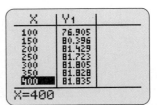

Figure 5.49

A logistic function can also decrease; this behavior occurs when $b < 0$. Demand functions for products often follow such a pattern. Decreasing logistic functions can also be used to represent decay over time. Figure 5.50 shows a decreasing logistic function.

Logistic Decay Function

Equation: $y = \dfrac{c}{1 + ae^{-bx}}$ $(a > 0, b < 0)$

x-Intercept: none

y-Intercept: $\left(0, \dfrac{c}{1 + a}\right)$

Horizontal asymptotes: $y = 0, y = c$

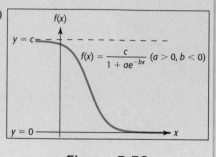

$f(x) = \dfrac{c}{1 + ae^{-bx}}$ $(a > 0, b < 0)$

Figure 5.50

The properties of logarithms can also be used to solve for a variable in an exponent in a logistic function.

---

example 2

Expected Life Span

The expected life span at birth of people born in the United States can be modeled by the equation

$$y = \frac{81.837}{1 + 0.8213e^{-0.0255x}}$$

where x is the number of years from 1900.

Use this model to estimate the birth year after 1900 that gives an expected life span of 65 years with

a. Graphical methods.

b. Algebraic methods.

Solution

a. To solve the equation

$$65 = \frac{81.837}{1 + 0.8213e^{-0.0255x}}$$

with the *x*-intercept method, we solve

$$0 = \frac{81.837}{1 + 0.8213e^{-0.0255x}} - 65$$

Graphing the function

$$y = \frac{81.837}{1 + 0.8213e^{-0.0255x}} - 65$$

on the window [30, 55] by [−2, 3] shows a point where the graph crosses the *x*-axis (Figure 5.51(a)). Finding the *x*-intercept gives the approximate solution

$$x \approx 45.25$$

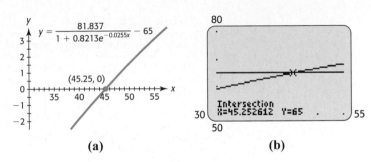

Figure 5.51

We round $x = 45.25$ to 46 years after 1900, so we conclude that people born in 1946 have an expected life span of 65 years.

We can also use the intersection method to solve the equation graphically. To do this, we graph the equations $y_1 = 65$ and $y_2 = \dfrac{81.837}{1 + 0.8213e^{-0.0255x}}$ and find the point of intersection of the graphs (Figure 5.51(b)).

b. We can also solve the equation

$$65 = \frac{81.837}{1 + 0.8213e^{-0.0255x}}$$

algebraically. To do this, we first multiply both sides by the denominator and then solve for $e^{-0.0255x}$:

$$65(1 + 0.8213e^{-0.0255x}) = 81.837$$

$$65 + 53.3845e^{-0.0255x} = 81.837$$

$$53.3845e^{-0.0255x} = 16.837$$

$$e^{-0.0255x} = \frac{16.837}{53.3845}$$

Taking the natural logarithm of both sides of the equation and using a logarithmic property are the next steps in the solution:

$$\ln e^{-0.0255x} = \ln \frac{16.837}{53.3845}$$

$$-0.0255x = \ln \frac{16.837}{53.3845}$$

$$x = \frac{\ln \dfrac{16.837}{53.3845}}{-0.0255} \approx 45.252612$$

This solution is the same solution that was found (more easily) by using the graphical method in Figure 5.51(a); 1946 is the birth year for people with an expected life span of 65 years. ■

<table>
<tr><td>example 3</td><td>Miles per Gallon</td></tr>
</table>

The average miles per gallon of gasoline for passenger cars for selected years from 1975 through 2006 are shown in Table 5.26 on the next page. Find a logistic function that models these data, with x equal to the number of years after 1970.

Table 5.26

Year	Miles per Gallon	Year	Miles per Gallon	Year	Miles per Gallon
1975	13.5	1997	24.3	2002	24.5
1980	20.0	1998	24.4	2003	24.7
1985	23.0	1999	24.1	2004	24.7
1990	23.7	2000	24.1	2005	25.0
1995	24.2	2001	24.3	2006	24.6
1996	24.2				

(Source: Environmental Protection Agency)

Solution

The scatter plot of the data is shown in Figure 5.52(a), with y representing miles per gallon and x equal to the number of years after 1970. Technology gives the equation of the best-fitting logistic function as

$$y = \frac{24.458}{1 + 2.805e^{-0.250x}}$$

Figure 5.52(b) shows the graph of this model and the scatter plot on the same axes.

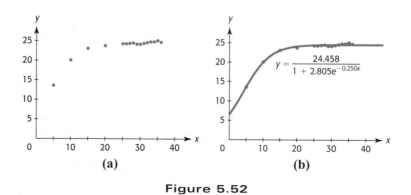

(a) (b)

Figure 5.52

Gompertz Functions

Another type of function that models rapid growth that eventually levels off is called a **Gompertz function**. This type of function can also be used to describe human growth and development and the growth of organizations in a limited environment. These functions have equations of the form

$$N = Ca^{R^t}$$

where t represents the time, R $(0 < R < 1)$ is the expected rate of growth of the population, C is the maximum possible number of individuals, and a represents the proportion of C present when $t = 0$.

For example, the equation

$$N = 3000(0.2)^{0.6^t}$$

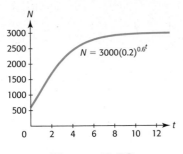

Figure 5.53

could be used to predict the number of employees t years after the opening of a new facility. Here the maximum number of employees C would be 3000, the proportion of 3000 present when $t = 0$ is $a = 0.2$, and R is 0.6. The graph of this function is shown in Figure 5.53. Observe that the initial number of employees, at $t = 0$, is

$$N = 3000(0.2)^{0.6^0} = 3000(0.2)^1 = 3000(0.2) = 600$$

To verify that the maximum possible number of employees is 3000, observe the following. Because $0.6 < 1$, higher powers of t make 0.6^t smaller, with 0.6^t approaching 0 as t approaches ∞. Thus, $N \to 3000(0.2)^0 = 3000(1) = 3000$ as $t \to \infty$.

---

example 4

Deer Population

The Gompertz equation

$$N = 1000(0.06)^{0.2^t}$$

predicts the size of a deer herd on a small island t decades from now.

a. What is the number of deer on the island now ($t = 0$)?

b. How many deer are predicted to be on the island 1 decade from now ($t = 1$)?

c. Graph the function.

d. What is the maximum number of deer predicted by this model?

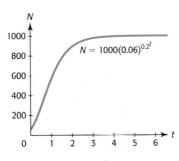

Figure 5.54

Solution

a. If $t = 0$, $N = 1000(0.06)^{0.2^0} = 1000(0.06)^1 = 60$.

b. If $t = 1$, $N = 1000(0.06)^{0.2^1} = 1000(0.06)^{0.2} = 570$ (approximately).

c. The graph is shown in Figure 5.54.

d. According to the model, the maximum number of deer is predicted to be 1000. Evaluating N as t gets large, we see the values of N approach, but never reach, 1000. We say that $N = 1000$ is an asymptote for this graph. ∎

---

example 5

Company Growth

A new dotcom company starts with 3 owners and 5 employees but tells investors that it will grow rapidly, with the total number of people in the company given by the model

$$N = 2000(0.004)^{0.5^t}$$

where t is the number of years from the present. Use graphical methods to determine the year in which they predict that the number of employees will be 1000.

Solution

If the company has 1000 employees plus the 3 owners, the total number in the company will be 1003; so we seek to solve the equation

$$1003 = 2000(0.004)^{0.5^t}$$

Using the equations $y_1 = 1003$ and $y_2 = 2000(0.004)^{0.5^t}$ and the intersection method gives $t = 3$ (Figure 5.55 on the next page). Because t represents the number of years, the owners predict that they will have 1000 employees in 3 years.

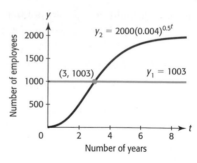

Figure 5.55

Give approximate answers to two decimal places.

1. Evaluate $\dfrac{79.514}{1 + 0.835e^{-0.0298(80)}}$.

2. If $y = \dfrac{79.514}{1 + 0.835e^{-0.0298x}}$ find

   a. y when $x = 10$.

   b. y when $x = 50$.

3. Evaluate $1000(0.06)^{0.2^t}$ for $t = 4$ and for $t = 6$.

4. Evaluate $2000(.004)^{0.5^t}$ for $t = 5$ and for $t = 10$.

5. a. Graph the function $f(x) = \dfrac{100}{1 + 3e^{-x}}$ for $x = 0$ to $x = 15$.

   b. Find $f(0)$ and $f(10)$.

   c. Is this function increasing or decreasing?

   d. What is the limiting value of this function?

6. a. Graph $f(t) = \dfrac{1000}{1 + 9e^{-0.9t}}$ for $t = 0$ to $t = 15$.

   b. Find $f(2)$ and $f(5)$.

   c. What is the limiting value of this function?

7. a. Graph $y = 100(0.05)^{0.3^x}$ for $x = 0$ to $x = 10$.

   b. What is the initial value of this function (the y-value when $x = 0$)?

   c. What is the limiting value of this function?

8. a. Graph $N = 2000(0.004)^{0.5^t}$ for $t = 0$ to $t = 10$.

   b. What is the initial value of this function (the y-value when $x = 0$)?

   c. What is the limiting value of this function?

9. *Spread of a Disease* The spread of a highly contagious virus in a high school can be described by the logistic function

$$y = \frac{5000}{1 + 999e^{-0.8x}}$$

where x is the number of days after the virus is identified in the school and y is the total number of people who are infected by the virus.

   a. Graph the function for $0 \le x \le 15$.

   b. How many students had the virus when it was first discovered?

   c. What is the upper limit of the number infected by the virus during this period?

10. *Population Growth* Suppose that the size of a population of an island is given by

$$p(t) = \frac{98}{1 + 4e^{-0.1t}} \text{ thousand people}$$

where t is the number of years after 1988.

 a. Graph this function for $0 \le x \le 30$.

 b. Find and interpret $p(10)$.

 c. Find and interpret $p(100)$.

 d. What appears to be an upper limit for the size of this population?

11. *Sexually Active Boys* The percent of boys between ages 15 and 20 that have been sexually active at some time (the cumulative percent) can be modeled by the logistic function

$$y = \frac{89.786}{1 + 4.6531e^{-0.8256x}}$$

where x is the number of years after age 15.

 a. Graph the function for $0 \le x \le 5$.

 b. What does the model estimate the cumulative percent to be for boys whose age is 16?

 c. What cumulative percent does the model estimate for boys of age 21, if it is valid after age 20?

 d. What is the limiting value implied by this model?
 (Source: "National Longitudinal Survey of Youth," risking thefuture.com)

12. *Sexually Active Girls* The percent of girls between ages 15 and 20 that have been sexually active at some time (the cumulative percent) can be modeled by the logistic function

$$y = \frac{83.84}{1 + 13.9233e^{-0.9248x}}$$

where x is the number of years after age 15.

 a. Graph the function for $0 \le x \le 5$.

 b. What does the model estimate the cumulative percent to be for girls of age 16?

 c. What cumulative percent does the model estimate for girls of age 20?

 d. What is the upper limit implied by the given logistic model?
 (Source: "National Longitudinal Survey of Youth," risking the future.com)

13. *Spread of a Rumor* The number of people in a small town who are reached by a rumor about the mayor and an intern is given by

$$N = \frac{10,000}{1 + 100e^{-0.8t}}$$

where t is the number of days after the rumor begins.

 a. How many people will have heard the rumor by the end of the first day?

 b. How many will have heard the rumor by the end of the fourth day?

 c. Use graphical or numerical methods to find the day in which 7300 people in town have heard the rumor.

14. *Sexually Active Girls* The cumulative percent of sexually active girls ages 15 to 20 is given in the table below.

Age (years)	Cumulative Percent Sexually Active
15	5.4
16	12.6
17	27.1
18	44.0
19	62.9
20	73.6

(Source: "National Longitudinal Survey of Youth," riskingthefuture.com)

 a. A logarithmic function was used to model this data in Exercises 5.4, problem 33. Find the logistic function that models the data, with x equal to the number of years after age 15.

 b. Does this model agree with the model used in Exercise 12?

 c. Find the linear function that is the best fit for the data.

 d. Graph each function on the same axes with the data points to determine which model appears to be the better fit for the data.

15. *Sexually Active Boys* The cumulative percent of sexually active boys ages 15 to 20 is given in the table on the next page.

a. A logarithmic function was used to model this data in Exercises 5.4, problem 34. Create the logistic function that models the data, with *x* equal to the number of years after age 15.

b. Does this agree with the model used in Exercise 11?

c. Find the linear function that is the best fit for the data.

d. Determine which model appears to be the better fit for the data and graph this function on the same axes with the data points.

Age (years)	Cumulative Percent Sexually Active
15	16.6
16	28.7
17	47.9
18	64.0
19	77.6
20	83.0

(Source: "National Longitudinal Survey of Youth," riskingthefuture.com)

16. *World Internet Usage* The number of worldwide users (in millions) of the Internet is given in the following table.

a. Find a logistic function that models the data, with *y* equal to the number of users in millions and *x* equal to the number of years after 1990.

b. Use the model to predict the number of users in 2015.

c. What will be the maximum number of worldwide users, according to the model?

d. Graph the aligned data and the model on the same axes. Is the model a good fit?

Year	Users (millions)	Year	Users (millions)
1995	16	2002	598
1996	36	2003	719
1997	70	2004	817
1998	147	2005	1018
1999	248	2006	1093
2000	361	2007	1215
2001	536		

(Source: www.internetworldstats.com)

17. *Internet Usage* The percent of the U.S. population that uses the Internet during selected years from 1997 is given in the table below.

a. Find the logistic function that models the percent as a function of the years, with *x* equal to the number of years after 1995.

b. Visually determine if the model is a good fit for the data.

c. Use the model to predict the percent of Internet users in the United States in 2010.

Year	Percent	Year	Percent
1997	22.2	2003	59.2
2000	44.1	2004	68.8
2001	50.0	2005	68.1
2002	58.0	2007	70.2

(Source: www.internetworldstats.com)

18. *Nonmarital Childbearing* The percents of live births to unmarried mothers for selected years 1970–2003 are shown in the table below.

Year	Percent	Year	Percent
1970	10.7	1990	28.0
1975	14.3	1995	32.2
1980	18.4	2000	33.2
1985	22.0	2003	34.6

(Source: *World Almanac and Book of Facts*)

a. Find a logistic function that models the data, with y the percent and x the number of years from 1960.

b. Graph the model and the aligned data on the same axes.

c. What percent does this model predict for 2010?

19. *Organizational Growth* A new community college predicts that its student body will grow rapidly at first and then begin to level off according to the Gompertz curve with equation $N = 10{,}000(0.4)^{0.2^t}$ students, where t is the number of years after the college opens.

a. What does this model predict the number of students to be when the college opens?

b. How many students are predicted to attend the college after 4 years?

c. Graph the equation for $0 \le t \le 10$ and estimate an upper limit on the number of students at the college.

20. *Organizational Growth* A new technology company started with 6 employees and predicted that its number of employees would grow rapidly at first and then begin to level off according to the Gompertz function

$$N = 150(0.04)^{0.5^t}$$

where t is the number of years after the company started.

a. What does this model predict the number of employees to be in 8 years?

b. Graph the function for $0 \le t \le 10$ and estimate the maximum predicted number of employees.

21. *Sales Growth* The president of a company predicts that sales will increase rapidly after a new product is brought to the market and that the number of units sold monthly can be modeled by $N = 40{,}000(0.2)^{0.4^t}$, where t represents the number of months after the product is introduced.

a. How many units will be sold by the end of the first month?

b. Graph the function for $0 \le t \le 10$.

c. What is the predicted upper limit on sales?

22. *Company Growth* Because of a new research grant, the number of employees in a firm is expected to grow, with the number of employees modeled by $N = 1600(0.6)^{0.2^t}$, where t is the number of years after the grant was received.

a. How many employees did the company have when the grant was received?

b. How many employees did the company have at the end of 3 years after the grant was received?

c. What is the expected upper limit on the number of employees?

d. Graph the function.

23. *Company Growth* Suppose that the number of employees in a new company is expected to grow, with the number of employees modeled by $N = 1000(0.01)^{0.5^t}$, where t is the number of years after the company was formed.

a. How many employees did the company have when it started?

b. How many employees did the company have at the end of 1 year?

c. What is the expected upper limit on the number of employees?

d. Use graphical or numerical methods to find the year in which 930 people are employed.

24. *Sales Growth* The president of a company predicts that sales will increase rapidly after a new advertising campaign, with the number of units sold weekly modeled by $N = 8000(0.1)^{0.3^t}$, where t represents the number of weeks after the advertising campaign begins.

a. How many units per week were sold at the beginning of the campaign?

b. How many units were sold at the end of the first week?

c. What is the expected upper limit on the number of units sold per week?

d. Use graphical or numerical methods to find the first week in which 6500 units were sold.

25. *Spread of Disease* An employee brings a contagious disease to an office with 150 employees. The number

of employees infected by the disease t days after the employees are first exposed to it is given by

$$N = \frac{100}{1 + 79e^{-0.9t}}$$

Use graphical or numerical methods to find the number of days until 99 employees have been infected.

26. *Advertisement* The number of people in a community of 15,360 who are reached by a particular advertisement t weeks after it begins is given by

$$N(t) = \frac{14,000}{1 + 100e^{-0.6t}}$$

Use graphical or numerical methods to find the number of weeks until at least half of the community is reached by this advertisement.

27. *Population Growth* A pair of deer are introduced on a small island, and the population grows until

the food supply and natural enemies of the deer on the island limit the population. If the number of deer is

$$N = \frac{180}{1 + 89e^{-0.5554t}}$$

where t is the number of years after the deer are introduced, how long does it take for the deer population to reach 150?

28. *Spread of Disease* A student brings a contagious disease to an elementary school of 1200 students. If the number of students infected by the disease t days after the students are first exposed to it is given by

$$N = \frac{800}{1 + 799e^{-0.9t}}$$

use numerical or graphical methods to find in how many days at least 500 students will be infected.

chapter

5 Summary

In this chapter, we discussed exponential and logarithmic functions and their applications. We explored the fact that logarithmic and exponential functions are inverses of each other. Many applications in today's world involve exponential and logarithmic functions, including the pH of substances, stellar magnitude, intensity of earthquakes, the growth of bacteria, the decay of radioactive isotopes, and compound interest. Logarithmic and exponential equations are solved by using both algebraic and graphical solution methods.

Key Concepts and Formulas

5.1 Exponential Functions

Exponential function	If a and b are real numbers, $b \neq 1$, then the function $f(x) = a(b^x)$ is an exponential function.
Exponential growth	Whenever the base b of $f(x) = a(b^x)$ is greater than 1 and $a > 0$, the exponential function is increasing and can be used to model exponential growth. The graph of an exponential growth model resembles the graph at the right. The x-axis (on the left) is a horizontal asymptote for the graph. 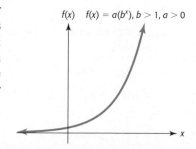

Exponential decay	Whenever the base b of $f(x) = a(b^{-x})$ is greater than 1 and $a > 0$, the exponential function is decreasing and can be used to model exponential decay. The graph of an exponential decay model resembles the graph at the right. The x-axis (on the right) is a horizontal asymptote for the graph. The function $f(x) = a(c^x)$, $a > 0, 0 < c < 1$, is also an exponential decay function.	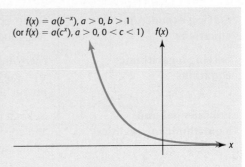
Number e	Many real applications involve exponential functions with the base e. The number e is an irrational number with decimal approximation 2.718281828 (to nine decimal places).	

5.2 Logarithmic Functions; Properties of Logarithms

Logarithmic function	For $x > 0, a > 0$, and $a \neq 1$, the logarithmic function to the base a is $y = \log_a x$, which is defined by $x = a^y$. The graph of $y = \log_a x$ for the base $a > 0$ is shown at the right. The y-axis is a vertical asymptote for the graph.	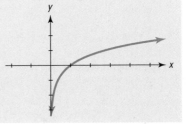
Logarithmic and exponential forms	These forms are equivalent: $$y = \log_a x \text{ and } x = a^y$$	
Common logarithms	Logarithms with a base of 10; $\log_{10} x$ is denoted $\log x$.	
Natural logarithms	Logarithms with a base of e; $\log_e x$ is denoted $\ln x$.	
Properties of logarithms	For $a > 0, a \neq 1, k$ a real number, and M and N positive real numbers, Property 1. $\log_a a = 1$ Property 2. $\log_a 1 = 0$ Property 3. $\log_a a^x = x$ Property 4. $a^{\log_a x} = x$ Property 5. If $M = N$, then $\log_a M = \log_a N$ Property 6. $\log_a(MN) = \log_a M + \log_a N$ Property 7. $\log_a\left(\dfrac{M}{N}\right) = \log_a M - \log_a N$ Property 8. $\log_a M^k = k \log_a M$	

5.3 Exponential and Logarithmic Equations

Solving exponential equations	Many exponential equations can be solved by converting to logarithmic form; that is, $x = a^y$ is equivalent to $y = \log_a x$.
Change of base	If $a > 0, a \neq 1, b > 0, b \neq 1$, and $x > 0$, then $\log_b x = \dfrac{\log_a x}{\log_a b}$.

Solving exponential equations	Some exponential equations can be solved by taking the logarithm of both sides and using logarithmic properties.
Solving logarithmic equations	Many logarithmic equations can be solved by converting to exponential form; that is, $y = \log_a x$ is equivalent to $x = a^y$. Sometimes properties of logarithms can be used to rewrite the equation in a form that can be more easily solved.
Exponential and logarithmic inequalities	Solving inequalities involving exponential and logarithmic functions involves rewriting and solving the related equation. Then graphical methods are used to find the values of the variable that satisfy the inequality.

5.4 Exponential and Logarithmic Models

Exponential models	Many sources provide data that can be modeled by exponential growth and decay functions according to the model $y = ab^x$ for real numbers a and b, $b \neq 1$.
Constant percent change	If the percent change of the outputs of a set of data is constant for equally spaced inputs, an exponential function will be a perfect fit for the data. If the percent change of the outputs is approximately constant for equally spaced inputs, an exponential function will be an approximate fit for the data.
Exponential model	If a set of data has initial value a and has a constant percent change r, the data can be modeled by the exponential function $$y = a(1 + r)^x$$
Comparison of models	To determine which type of model is the best fit for a set of data, we can look at the first differences and second differences of the outputs and the percent change of the outputs.
Logarithmic models	Models can be created involving $\ln x$. Data that exhibit an initial rapid increase and then have a slow rate of growth can often be described by the function $f(x) = a + b \ln x$ for $b > 0$.

5.5 Exponential Functions and Investing

Future value with annual compounding	If P dollars are invested for n years at an interest rate r compounded annually, then the future value of P is given by $S = P(1 + r)^n$ dollars.
Future value of an investment	If P dollars are invested for t years at the annual interest rate r, where the interest is compounded k times per year, then the interest rate per period is $\dfrac{r}{k}$, the number of compounding periods is kt, and the future value that results is given by $S = P\left(1 + \dfrac{r}{k}\right)^{kt}$ dollars.
Future value with continuous compounding	If P dollars are invested for t years at an annual interest rate r compounded continuously, then the future value S is given by $S = Pe^{rt}$ dollars.
Present value of an investment	The lump sum P that is necessary to have it grow to the future value S if invested for n periods at interest rate i per compounding period is $$P = \frac{S}{(1 + i)^n} \quad \text{or} \quad P = S(1 + i)^{-n}$$

5.6 Annuities; Loan Repayment

Future value of an ordinary annuity	If R dollars are contributed at the end of each period for n periods to an annuity that pays at rate i at the end of the period, the future value of the annuity is $$S = R\left[\frac{(1 + i)^n - 1}{i}\right]$$
Present value of an ordinary annuity	If a payment of R is to be made at the end of each period for n periods from an account that earns interest at a rate of i per period, then the account is an ordinary annuity, and the present value is $$A = R\left[\frac{1 - (1 + i)^{-n}}{i}\right]$$
Amortization formula	If a debt of A, with interest at a rate of i per period, is amortized by n equal periodic payments made at the end of each period, then the size of each payment is $$R = A\left[\frac{i}{1 - (1 + i)^{-n}}\right]$$

5.7 Logistic and Gompertz Functions

Logistic function	For real numbers a, b, and c, the function $$f(x) = \frac{c}{1 + ae^{-bx}}$$ is a logistic function. A logistic function increases when $a > 0$ and $b > 0$. The graph of a logistic function resembles the graph at right. 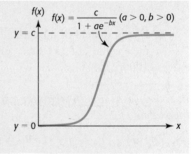
Logistic modeling	If data points indicate that growth begins slowly, then increases very rapidly and finally slows over time to a rate that is almost zero, the data may fit the logistic model $$f(x) = \frac{c}{1 + ae^{-bx}}$$
Gompertz function	For real numbers C, a, and R $(0 < R < 1)$, the function $N = Ca^{R^t}$ is a Gompertz function. A Gompertz function increases rapidly then levels off as it approaches a limiting value. The graph of a Gompertz function resembles the graph at right. 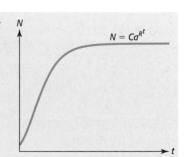

chapter

5

Skills Check

1. a. Graph the function $f(x) = 4e^{-0.3x}$.

b. Find $f(-10)$ and $f(10)$.

2. Is $f(x) = 4e^{-0.3x}$ an increasing or a decreasing function?

3. Graph $f(x) = 3^x$ on $[-10, 10]$ by $[-10, 20]$.

4. Graph $y = 3^{(x-1)} + 4$ on $[-10, 10]$ by $[0, 15]$.

5. How does the graph of Exercise 4 compare with the graph of Exercise 3?

6. Is $y = 3^{(x-1)} + 4$ an increasing or a decreasing function?

7. a. If $y = 1000(2^{-0.1x})$, find the value of y when $x = 10$.

b. Use graphical or numerical methods to determine the value of x that gives $y = 250$ if $y = 1000(2^{-0.1x})$.

In Exercises 8 and 9, write the exponential equations in logarithmic form.

8. $x = 6^y$
9. $y = 7^{3x}$

In Exercises 10–12, write the logarithmic equations in exponential form.

10. $y = \log_4 x$
11. $y = \log(x)$

12. $y = \ln x$

13. Write the inverse of $y = 4^x$ in logarithmic form.

In Exercises 14–16, evaluate the logarithms, if possible. Give approximate solutions to four decimal places.

14. $\log 22$
15. $\ln 56$

16. $\log 10$

In Exercises 17–19, find the value of the logarithms without using a calculator.

17. $\log_2 16$
18. $\ln(e^4)$

19. $\log 0.001$

In Exercises 20 and 21, use a change of base formula to evaluate each of the logarithms. Give approximate solutions to four decimal places.

20. $\log_3(54)$
21. $\log_8(56)$

In Exercises 22 and 23, graph the functions with technology.

22. $y = \ln(x - 3)$
23. $y = \log_3 x$

In Exercises 24–27, solve the exponential equations. Give approximate solutions to four decimal places.

24. $340 = e^x$

25. $1500 = 300e^{8x}$

26. $9200 = 23(2^{3x})$

27. $4(3^x) = 36$

28. Rewrite $\ln\dfrac{(2x - 5)^3}{x - 3}$ as the sum, difference, or product of logarithms, and simplify if possible.

29. Rewrite $6\log_4 x - 2\log_4 y$ as a single logarithm.

30. Determine whether a linear or an exponential function is the better fit for the data in the table. Then find the equation of the function with three-decimal-place accuracy.

x	1	2	3	4	5
y	2	5	12	30	75

31. Evaluate $P\left(1 + \dfrac{r}{k}\right)^{kn}$ for $P = 1000$, $r = 8\%$, $k = 12, n = 20$, to two decimal places.

32. Evaluate $1000\left[\dfrac{1 - 1.03^{-240+n}}{0.03}\right]$ for $n = 120$, to two decimal places.

33. a. Graph $f(t) = \dfrac{2000}{1 + 8e^{-0.8t}}$ for $t = 0$ to $t = 10$.

   b. Find $f(0)$ and $f(8)$, to two decimal places.

   c. What is the limiting value of this function?

34. a. Graph $y = 500(0.1^{0.2^x})$ for $x = 0$ to $x = 10$.

   b. What is the initial value (the value of y when $x = 0$) of this function?

   c. What is the limiting value of this function?

chapter

5 Review

35. *Prisoners* The total number of prisoners (in thousands) in state and federal prisons from 1970 to 1997 can be modeled by the function $y = 84.7518$ (1.0746^x), where x is the number of years after 1960. According to this model, how many prisoners were incarcerated in 1990?
(Source: Index of Leading Cultural Indicators)

36. *Sales Decay* At the end of an advertising campaign, the daily sales (in dollars) declined, with daily sales given by the equation $y = 2000(2^{-0.1x})$, where x is the number of weeks after the end of the campaign. According to this model, what will sales be 4 weeks after the end of the ad campaign?

37. *Corporate Revenue* Prior to the November 1994 $1.7 billion takeover proposal by Quaker Oats, Snapple Beverage Corporation's annual revenues were given by the function $B(t) = 1.337e^{0.718t}$ million dollars, where t is the number of years after 1985. When does this model indicate that the annual revenue is more than $10 million?
(Source: *U.S. News and World Report*, November 14, 1994)

38. *Earthquakes*
   a. If an earthquake has an intensity of 1000 times I_0, what is the magnitude of the earthquake?

   b. An earthquake that measured 6.5 on the Richter scale occurred in Pakistan in February 1991. Express this intensity in terms of I_0.

39. *Earthquakes* On January 23, 2001, an earthquake registering 7.9 hit western India, killing more than 15,000 people. On the same day, an earthquake registering 4.8 hit the United States, killing no one. How much more intense was the Indian earthquake than the American earthquake?

40. *Investments* The number of years needed for an investment of $10,000 to grow to $30,000, when money is invested at 12% compounded annually, is given by

$$t = \log_{1.12}\dfrac{30,000}{10,000} = \log_{1.12}3$$

In how many years will this occur?

41. *Investments* If $1000 is invested at 10% compounded quarterly, the future value of the investment is given by $S = 1000(2^{x/7})$, where x is the number of years after the investment is made.

   a. Use the logarithmic form of this function to solve the equation for x.

   b. Find when the future value of the investment is $19,504.

42. *Sales Decay* At the end of an advertising campaign, the weekly sales (in dollars) declined, with weekly sales given by the equation $y = 2000(2^{-0.1x})$, where x is the number of weeks after the end of the campaign. In how many weeks will sales be half of the sales at the end of the ad campaign?

43. *Prisoners* The rate (number per 100,000 people in the U.S. population) of prisoners in state and federal prisons from 1970 to 1997 can be modeled by the function $y = 45.6786(1.06386^x)$, where x is the number of years from 1960. In what year does this model indicate a rate of 275 per 100,000?

44. *Mobile Home Sales* A company that buys and sells used mobile homes estimates its cost to be given by $C(x) = 2x + 50$ thousand dollars when x mobile homes are purchased. The same company estimates that its revenue from the sale of x mobile homes is given by $R(x) = 10(1.26)^x$ thousand dollars.

a. Combine C and R into a single function that gives the profit for the company when x used mobile homes are bought and sold.

b. How many mobile homes must the company sell so that revenue is at least $30,000 more than cost?

45. *Sales Decay* At the end of an advertising campaign, the sales (in dollars) declined, with weekly sales given by the equation $y = 40,000(3^{-0.1x})$, where x is the number of weeks after the end of the campaign. In how many weeks will sales be less than half of the sales at the end of the ad campaign?

46. *Carbon-14 Dating* An exponential decay function can be used to model the number of atoms of a radioactive material that remain after a period of time. Carbon-14 decays over time, with the amount remaining after t years given by

$$y = y_0 e^{-0.00012097t}$$

where y_0 is the original amount.

a. If a sample of carbon-14 weighs 100 grams originally, how many grams will be present in 5000 years?

b. If a sample of wood at an archaeological site contains 36% as much carbon-14 as living wood, determine when the wood was cut.

47. *Purchasing Power* If a retired couple has a fixed income of $60,000 per year, the purchasing power (adjusted value of the money) after t years of 5% inflation is given by the equation $P = 60,000e^{-0.05t}$. In how many years will the purchasing power of their income fall below half of their current income?

48. *Investments* If $2000 is invested at 8% compounded continuously, the future value of the investment after t years is given by $S = 2000e^{0.08t}$ dollars. What is the future value of this investment in 10 years?

49. *Investments* If $3300 is invested for x years at 10% compounded annually, the future value that will result is $S = 3300(1.10)^x$ dollars. In how many years will the investment result in $13,784.92?

50. *Starbucks* The number of Starbucks stores increased rapidly during 1992–2007, as the following table shows. Find an exponential function that models these data, with x equal to the number of years from 1990 and y equal to the number of U.S. stores.

Year	Number of U.S. Stores	Year	Number of U.S. Stores
1992	113	2000	2119
1993	163	2001	2925
1994	264	2002	3756
1995	430	2003	4453
1996	663	2004	5452
1997	974	2005	6423
1998	1321	2006	7715
1999	1657	2007	9401

(Source: starbucks.com)

51. *Students per Computer* The following table gives the average number of students per computer in public schools for the school years that ended in 1985 through 2006.

a. Find an exponential model for these data. Let x be the number of years past 1980.

b. Is this model an exponential growth function or an exponential decay function?

c. How many students per computer in public schools does this model predict for 2010?

Year	Students per Computer	Year	Students per Computer
1985	75	1995	10.5
1986	50	1996	10
1987	37	1997	7.8
1988	32	1998	6.1
1989	25	1999	5.7
1990	22	2000	5.4
1991	20	2001	5.0
1992	18	2002	4.9
1993	16	2004	4.4
1994	14	2006	3.9

(Source: Quality Education Data, Inc., Denver, Colorado)

52 *Cell Phones* The table on the next page gives the number of thousands of U.S. cellular telephone sub-scriberships.

a. Find the exponential function that is the best fit for these data, with x equal to the number of years after 1980 and y equal to the number of subscriberships in thousands.

b. Use the model to estimate the number of subscriberships in 2007.

c. Why does this estimate seem overly optimistic? Is this a good model for the data?

Year	Subscriberships (thousands)	Year	Subscriberships (thousands)
1985	340	1996	44,043
1986	682	1997	55,312
1987	1,231	1998	69,209
1988	2,069	1999	86,047
1989	3,509	2000	109,478
1990	5,283	2001	128,375
1991	7,557	2002	140,767
1992	11,033	2003	158,722
1993	16,009	2004	182,140
1994	24,134	2005	207,896
1995	33,786	2006	233,000

(Source: Semiannual CTIA Wireless Industry Survey)

53. *World Tourism* The table below shows the receipts (in billions of dollars) for world tourism. Write an exponential function that models the data, with $x = 0$ in 1980, and predict the receipts in 2015 with the model.

Year	Receipts (in $ billions)	Year	Receipts (in $ billions)
1990	264	1998	445
1991	278	1999	455
1992	317	2000	473
1993	323	2001	459
1994	356	2002	474
1995	405	2003	524
1996	439	2004	623
1997	443	2007	735

54. *U.S. Internet Users* The percent of U.S. residents using the Internet is given in the table below.

a. Find a logarithmic function that models the data, with x equal to the number of years after 1995.

b. Use the model to predict the percent of users in 2015.

c. When will the model no longer be valid? Explain.

Year	Users (percent)	Year	Users (percent)
1997	22.2	2003	59.2
2000	44.1	2004	68.8
2001	50.0	2005	68.1
2002	58.0	2007	70.2

(Source: www.internetworldstats.com)

55. *Corvette Acceleration* The following table shows the times that it takes a 2008 Corvette to reach speeds from 0 mph to 100 mph, in increments of 10 mph after 30 mph.

Time (sec)	Speed (mph)	Time (sec)	Speed (mph)
1.7	30	5.3	70
2.4	40	6.5	80
3.3	50	7.9	90
4.1	60	9.5	100

Find a logarithmic function that gives the speed as a function of the time.

56. *Japan's Population* The table on the next page gives the population of Japan for the years 1984–2004.

a. Find the logistic function that models the population N, using an input x equal to the number of years from 1980.

b. Comment on the goodness of fit of the model to the data.

Year	Population (millions)	Year	Population (millions)	Year	Population (millions)
1984	120.235	1991	124.043	1998	126.486
1985	121.049	1992	124.452	1999	126.686
1986	121.672	1993	124.764	2000	126.926
1987	122.264	1994	125.034	2001	127.291
1988	122.783	1995	125.570	2002	127.435
1989	123.255	1996	125.864	2003	127.619
1990	123.611	1997	126.166	2004	127.687

(Source: www.jinjapan.org/stat/)

57. *Investment* Suppose $12,500 is invested in an account earning 5% annual interest compounded continuously. What is the future value in 10 years?

58. *Investments* Find the 7-year future value of an investment of $20,000 placed into an account that pays 6% compounded annually.

59. *Annuities* At the end of each quarter, $1000 is placed into an account that pays 12%, compounded quarterly. What is the future value of this annuity in 6 years?

60. *Annuities* Find the 10-year future value of an ordinary annuity with a contribution of $1500 at the end of each month, placed into an account that pays 8% compounded monthly.

61. *Present Value* Find the present value of an annuity that will pay $2000 at the end of each month for 15 years if the interest rate is 8% compounded monthly.

62. *Present Value* Find the present value of an annuity that will pay $500 at the end of each 6-month period for 12 years if the interest rate is 10% compounded semiannually.

63. *Loan Amortization* A debt of $2000 with interest at 12% compounded monthly is amortized by equal monthly payments for 36 months. What is the size of each payment?

64. *Loan Amortization* A debt of $120,000 with interest at 6% compounded monthly is amortized by equal monthly payments for 25 years. What is the size of each monthly payment?

65. *Out-of-Wedlock Births* The percent of all teen mothers who were unmarried during the years 1960–1996 can be modeled by the logistic function

$$y = \frac{96.3641}{1 + 12.3313e^{-0.0844x}}$$

where x is the number of years after 1950.

a. Use this model to estimate the percent in 1990 and in 1996.

b. What is the upper limit of the percent of teen mothers who were unmarried according to this model?

66. *Spread of Disease* A student brings a contagious disease to an elementary school of 2000 students. The number of students infected by the disease is given by

$$n = \frac{1400}{1 + 200e^{-0.5x}}$$

where x is the number of days after the student brings the disease.

a. How many students will be infected in 14 days?

b. How many days will it take for 1312 students to be infected?

67. *Organizational Growth* The president of a new campus of a university predicts that the student body will grow rapidly after the campus is open, with the number of students at the beginning of year t given by

$$N = 4000(0.06^{0.4^{t-1}})$$

a. How many students does this model predict for the beginning of the second year ($t = 2$)?

b. How many students are predicted for the beginning of the tenth year?

c. What is the limit on the number of students that can attend this campus, according to the model?

68. *Sales Growth* The number of units of a new product that were sold each month after the product was introduced is given by

$$N = 18,000(0.03^{0.4^t})$$

where t is the number of months.

a. How many units were sold 10 months after the product was introduced?

b. What is the limit on sales if this model is accurate?

69. *Endangered Species* The following table gives the numbers of species of plants that were endangered in various years from 1980 to 2006.

a. Find the logistic function that models these data. Use x as the number of years past 1975.

b. Use the model to predict the number of species of endangered plants in 2010.

c. When does the model predict that 627 plant species will be endangered?

Year	Endangered Plant Species
1980	50
1985	93
1990	179
1995	432
1998	567
1999	581
2000	593
2001	595
2002	596
2003	599
2006	598

(Source: U.S. Fish and Wildlife Service)

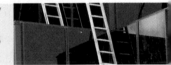

group activities / extended applications

1. Chain Letters

Suppose you receive a chain letter asking you to send $1 to the person at the top of the list of six names, then to add your name to the bottom of the list, and finally to send the revised letter to six people. The promise is that when the letters have been sent to each of the people above you on the list, you will be at the top of the list and you will receive a large amount of money. To investigate whether this is worth the dollar that you are asked to spend, create the requested models and answer the following questions.

1. Suppose each of the six persons on the original list sent six letters. How much money would the person on the top of the list receive?
2. Consider the person who is second on the original list. How much money would this person receive?
3. Complete the partial table of amounts that will be sent to the person on the top of the list during each "cycle" of the chain letter.

Cycle Number (after original six names)	Money Sent to Person on Top of List ($)
1	$6 \times 6 = 36$
2	$6 \times 36 = 216$
3	
4	
5	

4. If you were going to start such a chain letter (don't, it's illegal), would you put your name at the top of the first 6 names or as number 5?
5. Find the best quadratic, power, and exponential models for the data in the table. Which of these function types gives the best model for the data?

6. Use the exponential function found in part (5) to determine the amount of money the person who was at the bottom of the original list would receive if all people contacted send the $1 and mail the letter to 6 additional people.

7. How many people will have to respond to the chain letter for the sixth person on the original list to receive all the money that was promised?

8. If you receive the letter in its tenth cycle, how many other people have been contacted, along with you, assuming that everyone who receives the letter cooperates with its suggestions?

9. Who remains in the United States for you to send your letter to, if no one sends a letter to someone who has already received it?

10. Why do you think the federal government has made it illegal to send chain letters in the U.S. mail?

2. Modeling

Your mission is to find real-world data for company sales, a stock price, biological growth, or some sociological trend over a period of years that fits an exponential, logistic, or logarithmic function. To do this, you can look at a statistical graph called a histogram which describes the situation or at a graph or table describing it. Once you have found data consisting of numerical values, you are to find the equation that is the best exponential, logistic, or logarithmic fit for the data and then write the model that describes the relationship. Some helpful steps for this process are given in the instructions below.

1. Look in newspapers, in periodicals, on the Internet, or in statistical abstracts for a graph or table of data containing at least six data points. The data must have numerical values for the independent and dependent variables.

2. If you decide to use a relation determined by a graph that you have found, read the graph very carefully to determine the data points or (for full credit) contact the source of the data to obtain the data from which the graph was drawn.

3. If you have a table of values for different years, create a graph of the data to determine the type of function that will be the best fit for the data. To do this, first align the independent variable by letting the input represent the number of years from some convenient year and then enter the data into your graphing technology and draw a scatter plot. Check to see if the points on the scatter plot lie near some curve that could be described by an exponential, logistic, or logarithmic

function. If not, save the data for possible later use and renew your search. You can also test to see if the percent change of the outputs is nearly constant for equally spaced inputs. If so, the data can be modeled by an exponential function.

4. Use your calculator to create the equation of the function that is the best fit for the data. Graph this equation and the data points on the same axes to see if the equation is reasonable.

5. Write some statements about the data that you have been working with. For example, describe how the quantity is increasing or decreasing over periods of years, why it is changing as it is, or when this model is no longer appropriate, and why.

6. Your completed project should include the following:
   a. A proper bibliographical reference for the source of your data.
   b. An original copy or photocopy of the data being used.
   c. A scatter plot of the data and reasons why you chose the model that you did to fit it.
   d. The equation that you have created and labeled with appropriate units of measure and variable descriptions.
   e. A graph of the scatter plot and the model on the same axes.
   f. One or more statements about how you think the model you have created could actually be used. Include any restrictions that should be placed on the model.

Higher-Degree Polynomial and Rational Functions

The amount of money spent by domestic tourists can be modeled by the function

$$y = 0.0504x^3 - 1.4659x^2 + 28.2691x - 96.8781$$

where y represents the amount spent by domestic travelers (in billions of dollars) and x represents the number of years after 1980. If the total cost of producing x units of a product is $C(x) = 100 + 30x + 0.01x^2$, the function

$$\overline{C}(x) = \frac{100 + 30x + 0.01x^2}{x}$$

models the average cost per unit. These are examples of polynomial and rational functions, which we will discuss in this chapter.

Algebra Toolbox

Polynomials

Recall from Chapter 1 that an expression containing a finite number of additions, subtractions, and multiplications of constants and positive integer powers of variables is called a **polynomial**. The general form of a polynomial in x is

$$a_n x^n + a_{n-1} x^{n-1} + \cdots + a_1 x + a_0$$

where a_0 and each coefficient of x are real numbers and each power of x is a positive integer. If $a_n \neq 0$, n is the highest power of x, a_n is called the **leading coefficient**, and n is the **degree** of the polynomial. Thus, $5x^4 + 3x^2 - 6$ is a fourth-degree (quartic) polynomial with leading coefficient 5.

example 1

Polynomials

What are the degree and the leading coefficient of each of the following polynomials?

a. $4x^2 + 5x^3 - 17$ **b.** $30 - 6x^5 + 3x^2 - 7$

Solution

a. It is a third-degree polynomial. The leading coefficient is 5, the coefficient of the highest-degree term.

b. It is a fifth-degree polynomial. The leading coefficient is -6, the coefficient of the highest-degree term. ∎

Factoring Higher-Degree Polynomials

Some higher-degree polynomials can be factored by expressing them as the product of a monomial and another polynomial, which sometimes can be factored to complete the factorization.

example 2

Factoring

Factor: **a.** $3x^3 - 21x^2 + 36x$ **b.** $-6x^4 - 10x^3 + 4x^2$

Solution

a. $3x^3 - 21x^2 + 36x = 3x(x^2 - 7x + 12) = 3x(x - 3)(x - 4)$

b. $-6x^4 - 10x^3 + 4x^2 = -2x^2(3x^2 + 5x - 2) = -2x^2(3x - 1)(x + 2)$ ∎

Some polynomials can be written in *quadratic form* by using substitution. Then quadratic factoring methods can be used to begin the factorization.

example 3

Factoring Polynomials in Quadratic Form

Factor: **a.** $x^4 - 6x^2 + 8$ **b.** $x^4 - 18x^2 + 81$

Solution

a. The expression $x^4 - 6x^2 + 8$ is not a quadratic polynomial, but substituting u for x^2 converts it into the quadratic expression $u^2 - 6u + 8$, so we say that the original expression is in *quadratic form*. We factor it using quadratic methods, as follows:

$$u^2 - 6u + 8 = (u - 2)(u - 4)$$

Replacing u with x^2 gives

$$(x^2 - 2)(x^2 - 4)$$

Completing the factorization by factoring $x^2 - 4$ gives

$$x^4 - 6x^2 + 8 = (x^2 - 2)(x - 2)(x + 2)$$

b. Substituting u for x^2 converts $x^4 - 18x^2 + 81$ into the quadratic expression $u^2 - 18u + 81$. Factoring this expression gives

$$u^2 - 18u + 81 = (u - 9)(u - 9)$$

Replacing u with x^2 and completing the factorization gives

$$x^4 - 18x^2 + 81 = (x^2 - 9)(x^2 - 9)$$
$$= (x - 3)(x + 3)(x - 3)(x + 3) = (x - 3)^2(x + 3)^2 \qquad \blacksquare$$

Rational Expressions

An expression that is the quotient of two nonzero polynomials is called a **rational expression**. For example, $\dfrac{3x + 1}{x^2 - 2}$ is a rational expression. A rational expression is undefined when the denominator equals 0.

Fundamental Principle of Rational Expressions

The **Fundamental Principle of Rational Expressions** provides the means to simplify them.

$$\frac{ac}{bc} = \frac{a}{b} \quad \text{for} \quad b \neq 0, c \neq 0$$

We simplify a rational expression by factoring the numerator and denominator and dividing both the numerator and denominator by any common factors. We will assume that all rational expressions are defined for those values of the variables that do not make any denominator 0.

---

example 4 | **Rational Expressions**

Simplify the rational expression:

a. $\dfrac{2x^2 - 8}{x + 2}$ **b.** $\dfrac{3x}{3x + 6}$ **c.** $\dfrac{3x^2 - 14x + 8}{x^2 - 16}$

Solution

a. $\dfrac{2x^2 - 8}{x + 2} = \dfrac{2(x^2 - 4)}{x + 2} = \dfrac{2\overset{1}{\cancel{(x + 2)}}(x - 2)}{\underset{1}{\cancel{(x + 2)}}} = 2(x - 2)$

b. Note that we cannot divide both numerator and denominator by $3x$, because $3x$ is not a factor of $3x + 6$. Instead, we factor the denominator.

$$\dfrac{3x}{3x + 6} = \dfrac{\overset{1}{\cancel{3}}x}{\underset{1}{\cancel{3}}(x + 2)} = \dfrac{x}{x + 2}$$

c. $\dfrac{3x^2 - 14x + 8}{x^2 - 16} = \dfrac{(3x - 2)\overset{1}{\cancel{(x - 4)}}}{(x + 4)\underset{1}{\cancel{(x - 4)}}} = \dfrac{3x - 2}{x + 4}$ ∎

Multiplying and Dividing Rational Expressions

To multiply rational expressions, we write the product of the numerators divided by the product of the denominators and then simplify. We may also simplify prior to multiplying; this is usually easier than multiplying first.

example 5 | **Products of Rational Expressions**

Multiply and simplify:

a. $\dfrac{4x^2}{5y^2} \cdot \dfrac{25y}{12x}$

b. $\dfrac{6 - 3x}{2x + 4} \cdot \dfrac{4x - 20}{2 - x}$

Solution

a. $\dfrac{4x^2}{5y^2} \cdot \dfrac{25y}{12x} = \dfrac{4x^2 \cdot 25y}{5y^2 \cdot 12x} = \left(\dfrac{4 \cdot 25}{5 \cdot 12}\right)\left(\dfrac{x^2}{x}\right)\left(\dfrac{y}{y^2}\right) = \left(\dfrac{5}{3}\right)\left(\dfrac{x}{1}\right)\left(\dfrac{1}{y}\right) = \dfrac{5x}{3y}$

b. $\dfrac{6 - 3x}{2x + 4} \cdot \dfrac{4x - 20}{2 - x} = \dfrac{(6 - 3x)(4x - 20)}{(2x + 4)(2 - x)} = \dfrac{3\overset{1}{\cancel{(2 - x)}} \cdot 4\overset{2}{(x - 5)}}{2(x + 2) \cdot \underset{1}{\cancel{(2 - x)}}} = \dfrac{6x - 30}{x + 2}$ ∎

To divide rational expressions, we multiply by the reciprocal of the divisor.

example 6 | **Quotients of Rational Expressions**

Divide and simplify:

a. $\dfrac{a^2b}{c} \div \dfrac{ab^3}{c^2}$

b. $\dfrac{x^2 + 7x + 12}{2 - x} \div \dfrac{x^2 - 9}{x^2 - x - 2}$

Solution

a. $\dfrac{a^2b}{c} \div \dfrac{ab^3}{c^2} = \dfrac{a^2b}{c} \cdot \dfrac{c^2}{ab^3} = \left(\dfrac{a^2}{a}\right)\left(\dfrac{b}{b^3}\right)\left(\dfrac{c^2}{c}\right) = \dfrac{ac}{b^2}$

b. $\dfrac{x^2 + 7x + 12}{2 - x} \div \dfrac{x^2 - 9}{x^2 - x - 2} = \dfrac{x^2 + 7x + 12}{2 - x} \cdot \dfrac{x^2 - x - 2}{x^2 - 9}$

$\qquad = \dfrac{\cancel{(x + 3)}(x + 4)\cancel{(x - 2)}(x + 1)}{-\cancel{(x - 2)}(x - 3)\cancel{(x + 3)}} = \dfrac{(x + 4)(x + 1)}{-(x - 3)} = \dfrac{x^2 + 5x + 4}{3 - x}$ ∎

Adding and Subtracting Rational Expressions

We can add or subtract two fractions with the same denominator by adding or subtracting the numerators over their common denominator. If the denominators are not the same, we can write each fraction as an equivalent fraction with the common denominator. Our work is easier if we use the **least common denominator** (LCD). The LCD is the lowest-degree polynomial that all denominators will divide into. For example, if several denominators are $3x$, $6x^2y$, and $9y^3$, the lowest-degree polynomial that all three denominators will divide into is $18x^2y^3$. We can find the LCD as follows.

> ### Finding the Least Common Denominator of a Set of Fractions
>
> 1. Completely factor all the denominators.
> 2. The LCD is the product of each factor used the maximum number of times it occurs in any one denominator.

example 7	**Least Common Denominator**

Find the LCD of the fractions $\dfrac{3x}{x^2 - 4x - 5}$ and $\dfrac{5x - 1}{x^2 - 2x - 3}$.

Solution

The factored denominators are $(x - 5)(x + 1)$ and $(x - 3)(x + 1)$. The factors $(x - 5)$, $(x + 1)$, and $(x - 3)$ each occur a maximum of one time in any one denominator. Thus, the LCD is $(x - 5)(x + 1)(x - 3)$. ∎

The procedure for adding or subtracting rational expressions follows.

> ### Adding or Subtracting Rational Expressions
>
> 1. Find the LCD of all the rational expressions.
> 2. Write the equivalent of each rational expression with the LCD as its denominator.
> 3. Combine the like terms in the numerators and write this expression in the numerator with the LCD in the denominator.
> 4. Simplify the rational expression, if possible.

example 8

Adding Rational Expressions

Add $\dfrac{3}{x^2y} + \dfrac{5}{xy^2}$.

Solution

The factor x occurs a maximum of two times in any denominator, and the factor y occurs a maximum of two times in any denominator, so the LCD is x^2y^2. We rewrite the fractions with the common denominator by multiplying the numerator and denominator of the first fraction by y and multiplying the numerator and denominator of the second fraction by x, as follows:

$$\frac{3}{x^2y} + \frac{5}{xy^2} = \frac{3 \cdot y}{x^2y \cdot y} + \frac{5 \cdot x}{xy^2 \cdot x} = \frac{3y}{x^2y^2} + \frac{5x}{x^2y^2}$$

We can now add the fractions by adding the numerators over the common denominator:

$$\frac{3y}{x^2y^2} + \frac{5x}{x^2y^2} = \frac{3y + 5x}{x^2y^2}$$

This result cannot be simplified because there is no common factor of the numerator and denominator. ■

example 9

Subtracting Rational Expressions

Subtract $\dfrac{x + 1}{x^2 + x - 6} - \dfrac{2x - 1}{x - 2}$.

Solution

We factor the first denominator to get the LCD.

$$x^2 + x - 6 = (x + 3)(x - 2), \text{ so the LCD is } (x + 3)(x - 2)$$

We rewrite the second fraction so that it has the common denominator, subtract the second numerator, then combine the like terms of the numerators.

$$\frac{x + 1}{x^2 + x - 6} - \frac{2x - 1}{x - 2} = \frac{x + 1}{(x - 2)(x + 3)} - \frac{(2x - 1)(x + 3)}{(x - 2)(x + 3)}$$

$$= \frac{x + 1}{(x - 2)(x + 3)} - \frac{2x^2 + 5x - 3}{(x - 2)(x + 3)}$$

$$= \frac{x + 1 - (2x^2 + 5x - 3)}{(x - 2)(x + 3)}$$

$$= \frac{x + 1 - 2x^2 - 5x + 3}{(x - 2)(x + 3)}$$

$$= \frac{-2x^2 - 4x + 4}{(x - 2)(x + 3)}$$

This result cannot be simplified. ■

Division of Polynomials

When the divisor has a degree less than the dividend, the division of one polynomial by another is done in a manner similar to long division in arithmetic. The degree of the quotient polynomial will be less than the degree of the dividend polynomial, and the degree of the remainder will be less than the degree of the divisor. (If the remainder is 0, the divisor is a factor of the dividend.) The procedure follows.

Long Division of Polynomials

Divide $(4x^3 + 4x^2 + 5)$ by $(2x^2 + 1)$.

1. Write both the divisor and dividend in descending order of a variable. Include missing terms with 0 coefficient in the dividend.

$$2x^2 + 1 \overline{)4x^3 + 4x^2 + 0x + 5}$$

2. Divide the highest power of the divisor into the highest power of the dividend, and write this partial quotient above the dividend. Multiply the partial quotient times the divisor, write the product under the dividend, subtract, and bring down the next term, getting a new dividend.

$$
\begin{array}{r}
2x \\
2x^2 + 1 \overline{)4x^3 + 4x^2 + 0x + 5} \\
\underline{4x^3 + 2x } \\
4x^2 - 2x + 5
\end{array}
$$

3. Repeat until the degree of the new dividend is less than the degree of the divisor. If a nonzero expression remains, it is the remainder from the division.

$$
\begin{array}{r}
2x + 2 \\
2x^2 + 1 \overline{)4x^3 + 4x^2 + 0x + 5} \\
\underline{4x^3 + 2x } \\
4x^2 - 2x + 5 \\
\underline{4x^2 + 2} \\
-2x + 3
\end{array}
$$

Thus, the quotient is $2x + 2$, with remainder $-2x + 3$.

---

example 10   ## Division of Polynomials

Perform the indicated division: $\dfrac{x^4 + 6x^3 - 4x + 2}{x + 2}$.

Solution

Begin by placing $0x^2$ in the dividend.

$$
\begin{array}{r}
x^3 + 4x^2 - 8x + 12 \\
x + 2 \overline{)x^4 + 6x^3 + 0x^2 - 4x + 2} \\
\underline{x^4 + 2x^3 } \\
4x^3 + 0x^2 \\
\underline{4x^3 + 8x^2 } \\
-8x^2 - 4x \\
\underline{-8x^2 - 16x } \\
12x + 2 \\
\underline{12x + 24} \\
-22
\end{array}
$$

Thus, the quotient is $x^3 + 4x^2 - 8x + 12$, with remainder -22. We can write the result as

$$\frac{x^4 + 6x^3 - 4x + 2}{x + 2} = x^3 + 4x^2 - 8x + 12 - \frac{22}{x + 2} \qquad \blacksquare$$

toolbox exercises

For Exercises 1–4, (a) give the degree of the polynomial and (b) give the leading coefficient.

1. $3x^4 - 5x^2 + \dfrac{2}{3}$ **2.** $5x^3 - 4x + 7$

3. $7x^2 - 14x^5 + 16$ **4.** $2y^5 + 7y - 8y^6$

In Exercises 5–10, factor the polynomials completely.

5. $4x^3 - 8x^2 - 140x$

6. $4x^2 + 7x^3 - 2x^4$

7. $x^4 - 13x^2 + 36$

8. $x^4 - 21x^2 + 80$

9. $2x^4 - 8x^2 + 8$

10. $3x^5 - 24x^3 + 48x$

In Exercises 11–16, simplify each rational expression.

11. $\dfrac{x - 3y}{3x - 9y}$

12. $\dfrac{x^2 - 9}{4x + 12}$

13. $\dfrac{2y^3 - 2y}{y^2 - y}$

14. $\dfrac{4x^3 - 3x}{x^2 - x}$

15. $\dfrac{x^2 - 6x + 8}{x^2 - 16}$

16. $\dfrac{3x^2 - 7x - 6}{x^2 - 4x + 3}$

In Exercises 17–26, perform the indicated operations and simplify.

17. $\dfrac{x - 3}{x^3} \cdot \dfrac{x^2 - 4x}{x^2 - 7x + 12}$

18. $(x^2 - 4) \cdot \dfrac{2x - 3}{x + 2}$

19. $\dfrac{4x + 4}{x - 4} \div \dfrac{8x^2 + 8x}{x^2 - 6x + 8}$

20. $\dfrac{6x^2}{4x^2y - 12xy} \div \dfrac{3x^2 + 12x}{x^2 + x - 12}$

21. $3 + \dfrac{1}{x^2} - \dfrac{2}{x^3}$

22. $\dfrac{5}{x} - \dfrac{x - 2}{x^2} + \dfrac{4}{x^3}$

23. $\dfrac{a}{a^2 - 2a} - \dfrac{a - 2}{a^2}$

24. $\dfrac{5x}{x^4 - 16} + \dfrac{8x}{x + 2}$

25. $\dfrac{x - 1}{x + 1} - \dfrac{2}{x^2 + x}$

26. $\dfrac{2x + 1}{4x - 2} + \dfrac{5}{2x} - \dfrac{x + 1}{2x^2 - x}$

In Exercises 27–30, perform the long division.

27. $(x^5 + x^3 - 1) \div (x + 1)$

28. $(a^4 + 3a^3 + 2a^2) \div (a + 2)$

29. $(3x^5 - x^4 + 5x - 1) \div (x^2 - 2)$

30. $\dfrac{3x^4 + 2x^2 + 1}{3x^2 - 1}$

Higher-Degree Polynomial Functions

section preview ▪ U.S. Debt to United Nations

Although the United States is one of many countries that are members of the United Nations, the United States pays most of the operating expense of the United Nations. Table 6.1 shows the amount of U.S. debt and total debt (in millions of dollars) owed by all members to run the United Nations for the years 1990–2000 and the percent of the debt owed by the United States for each of these years.

The data can be used to find a **polynomial function**

$$y = -1.832x^3 + 22.111x^2 - 55.367x + 290.659$$

that models U.S. debt to the United Nations, in millions of dollars, as a function of the number of years, x, after 1990.

Table 6.1

Year	U.S. Debt ($ millions)	Total Debt ($ millions)	U.S. Share (%)
1990	296.2	403	73.5
1991	266.4	439.4	60.6
1992	239.5	500.6	47.8
1993	260.4	478	54.5
1994	248	480	51.6
1995	414.4	564	73.5
1996	376.8	510.7	73.8
1997	373.2	473.6	78.8
1998	315.7	417	75.7
1999	167.9	244.2	68.8
2000	164.6	222.4	74.0

(Source: United Nations, Institute for Global Communication)

In Chapters 2–4, we solved many real problems by modeling data with linear functions and with quadratic functions. (Recall that $ax + b$ is a first-degree polynomial and $ax^2 + bx + c$ is a second-degree polynomial.) Both of these functions are types of **polynomial functions**. **Higher-degree polynomial functions** are functions with degree higher than 2.

Examples of higher-degree polynomial functions are

$$y = x^3 - 16x^2 - 2x \qquad f(x) = 3x^4 - x^3 + 2 \qquad y = 2x^5 - x^3$$

In this section, we graph higher-degree polynomial functions and use technology to find local maxima and minima.

Cubic Functions

Using data from 1990 to 2000, the function

$$y = -1.832x^3 + 22.111x^2 - 55.367x + 290.659$$

can be used to model U.S. debt to the United Nations, in millions of dollars, with x equal to the number of years after 1990. This function is a third-degree function, or **cubic function**, because the highest-degree term in the function has degree 3. The graph of this function is shown in Figure 6.1.

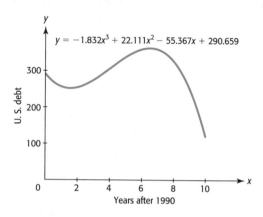

Figure 6.1

A **cubic function** in the variable x has the form

$$f(x) = ax^3 + bx^2 + cx + d \qquad (a \neq 0)$$

The basic cubic function $f(x) = x^3$ was discussed in Section 3.3. The graph of this function, shown in Figure 6.2(a), is one of the possible shapes of the graph of a cubic function. The graphs of two other cubic functions are shown in Figures 6.2(b) and (c).

The graph of a cubic function will have 0 or 2 turning points, as shown in Figures 6.1 and 6.2. Also, it is characteristic of a cubic polynomial function to have a graph that has one of the end behaviors shown in these figures. This end behavior can be described as "one end opening up and one end opening down." The specific end behavior is determined by the leading coefficient (the coefficient of the third-degree term); the curve opens up on the right if the leading coefficient is positive, and it opens down on the right if the leading coefficient is negative.

Notice that the graph of the cubic function shown in Figure 6.2(a) has one x-intercept, the graph of the cubic function shown in Figure 6.2(b) has three x-intercepts, and in Figure 6.2(c) the graph of the cubic function has two x-intercepts.

> In general, the graph of a polynomial function of degree n has at most n x-intercepts.

Recall that the graph of a quadratic (second-degree) function is a parabola with one turning point, called a vertex, and that the vertex is a maximum point or a minimum point. Notice that in each of the graphs of the cubic (third-degree) functions shown in Figures 6.2(b) and (c), there are two points where the function changes from increasing to decreasing or from decreasing to increasing. (Recall that a curve is increasing on an interval if the curve rises as it moves from left to right on that interval.) These *turning points* are called the **local extrema points** of the graph of the function. In particular,

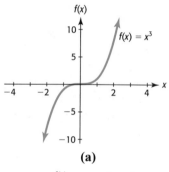

(a)

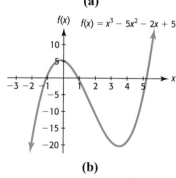

(b)

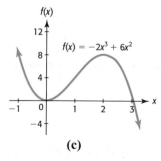

(c)

Figure 6.2

a point where the curve changes from decreasing to increasing is called a **local minimum point**, and a point where the curve changes from increasing to decreasing is called a **local maximum point**. For example, in Figure 6.2(c), the point (0, 0) is a local minimum point, and the point (2, 8) is a local maximum point. Notice that the graph in Figure 6.2(a) has no local extrema points. In general, a cubic equation will have 0 or 2 local extrema points.

The highest point on the graph over an interval is called the **absolute maximum point** on the interval, and the lowest point on the graph over an interval is called the **absolute minimum point** on that interval. The graph in Figure 6.2(c) is shown on the x-interval $[-1, 3.2]$; the absolute maximum point on this interval is $(-1, 8)$, and the absolute minimum point is $(3.2, -4.096)$ on this interval. Note also that at the point $(0, 0)$ the function changes from decreasing to increasing, and at the point $(2, 8)$ the function changes from increasing to decreasing.

| example 1 | **Cubic Graph** |

a. Using an appropriate window, graph $y = x^3 - 27x$.

b. Find the local maximum and local minimum, if possible.

Solution

a. This function is a third-degree (cubic) function, so it has 0 or 2 local extrema. We want to graph the function in a window that allows us to see any turning points that exist and recognize the end behavior of the function. Using the window $[-40, 30]$ by $[-100, 100]$, we get the graph shown in Figure 6.3(a). This graph has a shape like Figure 6.2(b), it has three x-intercepts, and it has two local extrema, so it is a complete graph. A different viewing window can be used to provide a more detailed view of the turning points of the graph (Figure 6.3(b)).

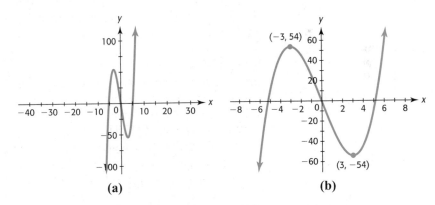

(a)   (b)

Figure 6.3

b. Using the "maximum" and "minimum" features of a graphing utility (found under 2nd TRACE) shows that there is a local maximum at $(-3, 54)$ and a local minimum at $(3, -54)$. ∎

| example 2 | **U.S. Foreign-Born Population** |

Table 6.2 gives the percents of U.S. population that were foreign born for selected years from 1900–2005.

Table 6.2

Years	Foreign Born (%)	Years	Foreign Born (%)
1900	13.6	1960	5.4
1910	14.7	1970	4.8
1920	13.2	1980	6.2
1930	11.6	1990	8.0
1940	8.8	2000	10.4
1950	6.9	2005	11.7

(Source: U.S. Census Bureau)

The data can be modeled by the cubic function

$$f(x) = 0.0000503x^3 - 0.00560x^2 + 0.0165x + 14.308$$

where y is the percent of foreign-born and x is the number of years after 1900.

a. Graph the data points and this function on the same axes, using the window [0, 110] by [0, 16].

b. Use the model and technology to find the approximate percent of the U.S. population that was foreign born in 1954.

c. Interpret the y-intercept of the graph of this function.

d. Find the local minimum and the local maximum. What do these two points on the graph indicate?

e. Over what interval is the function (and thus the percent) decreasing? What does this mean?

Solution

a. The graph is shown in Figure 6.4(a).

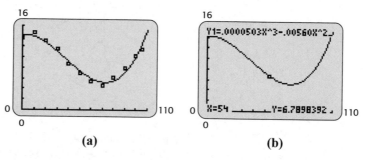

(a)   (b)

Figure 6.4

b. The approximate percent of the U.S. population that was foreign born in 1954 is $f(54) = 6.79$ percent (Figure 6.4(b)).

c. The y-intercept, 14.308, gives the percent of the U.S. population that was foreign born in 1900, according to the model.

d. Using "minimum" and "maximum" features on a graphing utility* gives the local minimum (72.72, 5.24) and the local maximum (1.50, 14.32) (Figure 6.5). This indicates that the minimum percent occurred in 1973 and was approximately 5.24%, and the maximum percent occurred in 1902 and was approximately 14.32%.

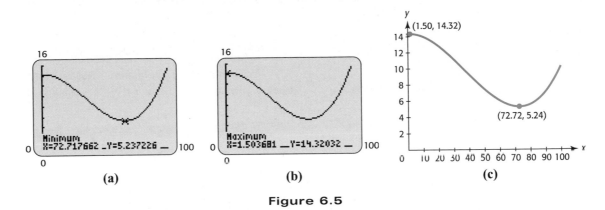

Figure 6.5

e. As x increases from 1.50 to 72.72, y decreases from the maximum at 14.32 to the minimum at 5.24 (Figure 6.5(c)). This means that the percent decreased from 14.32% to 5.24% in the years from 1902 to 1973. ■

Quartic Functions

A polynomial function of the form $f(x) = ax^4 + bx^3 + cx^2 + dx + e, a \neq 0$, is a fourth-degree, or **quartic**, function. The basic quartic function $f(x) = x^4$ has the graph shown in Figure 6.6(a). Notice that this graph resembles a parabola, although it is not actually a parabola (recall the shape of a parabola, which was discussed in Section 3.1). The graph in Figure 6.6(a) is one of the possible shapes of the graph of a quartic function. The graphs of two additional quartic functions are shown in Figure 6.6(b) and (c).

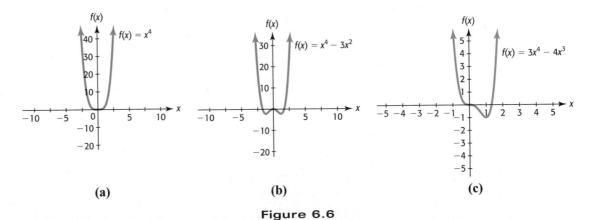

Figure 6.6

Notice that in the graph of the quartic function shown in Figure 6.6(b) there are 3 turning points and that in the graphs shown in Figures 6.6(a) and 6.6(c) there is 1 turning point each. The end behavior of a quartic function can be described as both ends

* See Graphing Calculator Guide, Appendix A, page 657.

opening up (if the leading coefficient is positive) or both ends opening down (if the leading coefficient is negative). Variations in the curvature may occur for different quartic functions, but these are the only possible end behaviors.

In Table 6.3, we summarize the information about turning points and end behavior for graphs of cubic and quartic functions as well as for linear and quadratic functions.

Table 6.3

Function	Possible Graphs	Degree	Number of Turning Points	End Behavior Positive leading coefficient	End Behavior Negative leading coefficient	Degree: Even or Odd
Linear	Positive slope Negative slope	1	0			odd
Quadratic	Positive leading coefficient Negative leading coefficient	2	1			even
Cubic Positive leading coefficient		3	2 or 0			odd
Negative leading coefficient						
Quartic Positive leading coefficient		4	3 or 1			even
Negative leading coefficient						

Table 6.3 helps us observe the following.

Polynomial Graphs

1. The graph of a polynomial function of degree n has at most $n - 1$ turning points.

2. The graph of a polynomial function of degree n has at most n x-intercepts.

3. The end behavior of the graph of a polynomial function with odd degree can be described as "one end opening up and one end opening down."

4. The end behavior of the graph of a polynomial function with even degree can be described as "both ends opening up" or "both ends opening down."

| example 3 | **Graphs of Polynomial Functions** |

Figures 6.7(a)–(d) show the complete graphs of several polynomial functions. For each function, determine

i. The number of x-intercepts.

ii. The number of turning points.

iii. Whether the leading coefficient is positive or negative.

iv. Whether the degree of the polynomial is even or odd.

v. The minimum possible degree.

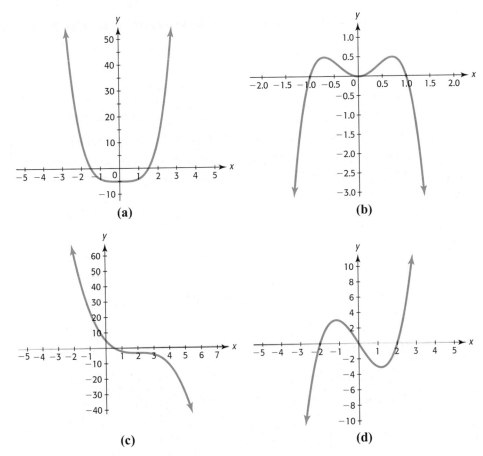

Figure 6.7

Solution

a. The graph clearly shows two x-intercepts and one turning point. Because the end behavior is "both ends opening up," the degree of the polynomial function is even and the leading coefficient is positive. Because there is only one turning point, the function could be quadratic, so the minimum degree of the polynomial is 2.

b. The graph shows 3 turning points and 3 x-intercepts. The end behavior is "both ends opening down," so the degree of the polynomial function is even and the leading coefficient is negative. The minimum possible degree is 4 because there is more than one turning point and thus the function cannot be quadratic.

c. The graph shows one x-intercept and no turning points. The degree of the function is odd, because the end behavior is "one end up and one end down." The leading coefficient is negative, because the left end is up. Because the graph is not a line, the minimum degree is 3.

d. The graph shows 3 x-intercepts and 2 turning points. The end behavior is "one end up and one end down," with the right end opening up, so the function has odd degree and the leading coefficient is positive. Because the graph has 3 x-intercepts, the minimum degree is 3. ∎

| example 4 | **Alcohol-Related Traffic Fatalities** |

Using data from 1982 to 2005, the total number of thousands of people killed in alcohol-related crashes can be modeled by the function

$$y = -0.000487x^4 + 0.0298x^3 - 0.602x^2 + 4.125x + 15.232$$

where x is the number of years from 1980.

a. Graph the function using the x-values $0 \le x \le 25$.

b. Use the graph and technology to approximate the year in which the number of alcohol-related traffic deaths is a maximum and in which year the number is a minimum during this period.

c. Is it likely that this model can be used to estimate the total number of people killed in alcohol-related crashes for long after 2007?

Solution

a. The graph is shown in Figure 6.8(a).

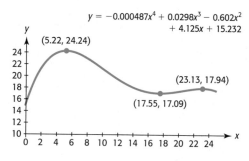

Figure 6.8(a)

b. The graph in Figure 6.8(a) shows two local maxima with the absolute maximum occurring at (5.22, 24.24). This indicates that the number of traffic deaths is a maximum in the year 1986. The minimum of this function occurs at (17.55, 17.09). This indicates that the number of traffic deaths was a minimum in the year 1998.

c. Observing the graph in Figure 6.8(b), we see that it drops very rapidly after $x = 27$ (which corresponds to 2007), with outputs becoming negative before $x = 32$ (which corresponds to the year 2012). Thus, the model ceases to be valid eventually, certainly before 2012.

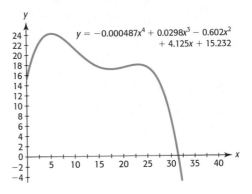

$$y = -0.000487x^4 + 0.0298x^3 - 0.602x^2 + 4.125x + 15.232$$

Figure 6.8(b)

skills check

6.1

1. Graph the function $h(x) = 3x^3 + 5x^2 - x - 10$ on the windows given in parts (a) and (b). Which window gives a complete graph?

   a. $[-5, 5]$ by $[-5, 5]$

   b. $[-5, 5]$ by $[-20, 20]$

2. Graph the function $f(x) = 2x^3 - 3x^2 - 6x$ on the windows given in parts (a) and (b). Which window gives a complete graph?

   a. $[-5, 5]$ by $[-5, 5]$

   b. $[-10, 10]$ by $[-10, 10]$

3. Graph the function $g(x) = 3x^4 - 12x^2$ on the windows given in parts (a) and (b). Which window gives a complete graph?

   a. $[-5, 5]$ by $[-5, 5]$

   b. $[-3, 3]$ by $[-12, 10]$

4. Graph the function $g(x) = 3x^4 - 4x^2 + 10$ on the windows given in parts (a) and (b). Which window gives a complete graph?

   a. $[-10, 10]$ by $[-10, 10]$

   b. $[-5, 5]$ by $[-20, 20]$

For Exercises 5–10, use the given graph of the polynomial function to (a) estimate the x-intercept(s), (b) state whether the leading coefficient is positive or negative, and (c) determine whether the polynomial function is cubic or quartic.

5.

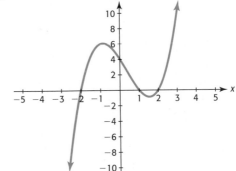

6.

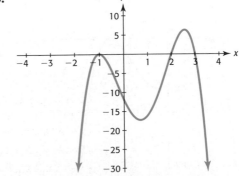

7.

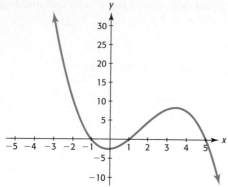

8.

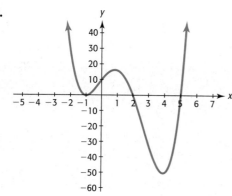

9.

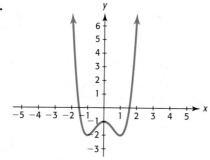

10.

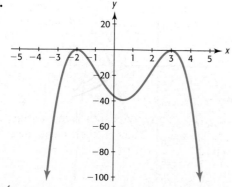

For Exercises 11–16, match the polynomial function with its graph.

11. $y = 2x^3 + 3x^2 - 23x - 12$

12. $y = -2x^3 - 8x^2 + 0.5x + 2$

13. $y = -6x^3 + 5x^2 + 17x - 6$

14. $y = x^4 - 0.5x^3 - 14.5x^2 + 17x + 12$

15. $y = x^4 + 3$

16. $y = -x^4 + 16x^2$

A.

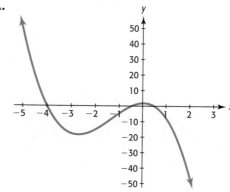

B.

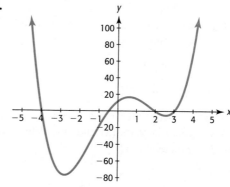

C.

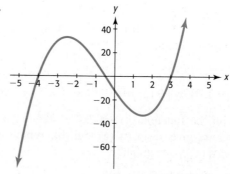

D.

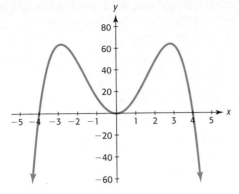

E.

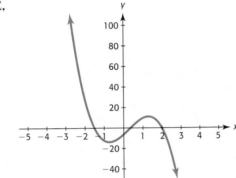

F.

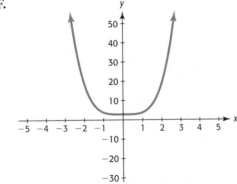

For Exercises 17–20, use the equation of the polynomial function to (a) state the degree and the leading coefficient and (b) describe the end behavior of the graph of the function. (c) Support your answer by graphing the function.

17. $f(x) = 2x^3 - x$

18. $g(x) = 0.3x^4 - 6x^2 + 17x$

19. $f(x) = -2(x - 1)(x^2 - 4)$

20. $g(x) = -3(x - 3)^2(x - 1)^2$

21. a. Graph the function $y = x^3 - 3x^2 - x + 3$ using the window $[-10, 10]$ by $[-10, 10]$.

   b. Is the graph complete?

22. a. Graph $y = x^3 + 6x^2 - 4x$ using the window $[-10, 10]$ by $[-10, 10]$.

   b. Is the graph complete?

23. Graph $y = 25x - x^3$ using

   a. the window $[-10, 10]$ by $[-10, 10]$.

   b. a window that shows two turning points.

24. Graph the function $y = x^3 - 16x$ using

   a. the window $[-10, 10]$ by $[-10, 10]$.

   b. a window that shows two turning points.

25. Graph the function $y = x^4 - 4x^3 + 4x^2$ using

   a. the window $[-10, 10]$ by $[-10, 10]$.

   b. the window $[-4, 4]$ by $[-4, 4]$.

   c. Which window gives a more detailed view of the graph near the turning points?

26. a. Graph $y = x^4 - 4x^2$ using the window $[-10, 10]$ by $[-10, 10]$.

   b. Is the graph complete?

27. a. Graph $y = x^4 - 4x^2 - 12$ on $[-8, 8]$ by $[-20, 10]$.

   b. How many turning points does the graph have?

   c. Could the graph of this function have more turning points?

28. a. Graph $y = x^4 + 6x^2$ on $[-10, 10]$ by $[-10, 50]$.

   b. Is it possible for the graph of this function to have only one turning point?

29. Sketch a graph of any cubic polynomial function that has a negative leading coefficient and one x-intercept.

30. Sketch a graph of any polynomial function that has degree 3, a positive leading coefficient, and three x-intercepts.

31. Sketch a graph of any polynomial function that has degree 4, a positive leading coefficient, and two x-intercepts.

32. Sketch a graph of any cubic polynomial function that has a negative leading coefficient and three x-intercepts.

33. **a.** Graph $y = x^3 + 4x^2 + 5$ on a window that shows a local maximum and a local minimum.

    b. A local maximum occurs at what point?

    c. A local minimum occurs at what point?

34. **a.** Graph $y = x^4 - 8x^2$ on a window that shows two local minima and one local maximum.

    b. A local maximum occurs at what point?

    c. The local minima occur at what points?

35. Use technology to find a local maximum and two local minima of the graph of the function $y = x^4 - 4x^3 + 4x^2$.

36. **a.** Graph the function $y = -x^3 - x^2 + 9x$ using the window $[-5, 5]$ by $[-15, 10]$.

    b. Graph the function on an interval with $x \geq 0$ and with $y \geq 0$.

    c. The graph of what other type of function resembles the *piece* of the graph of $y = -x^3 - x^2 + 9x$ shown in part (b)?

exercises

6.1

37. *Daily Revenue* The daily revenue from the sale of a product is given by $R = -0.1x^3 + 11x^2 - 100x$ dollars, where x is the number of units sold.

    a. Graph this function on the window $[-100, 100]$ by $[-5000, 25,000]$. How many turning points do you see?

    b. Because x represents the number of units sold, what restriction should be placed on x in the context of the problem? What restriction should be placed on R?

    c. Using your result from part (b), graph the function on a new window that makes sense for the problem.

    d. What does this function give as the revenue if 50 units are produced?

38. *Weekly Revenue* A firm has total weekly revenue for its product given by $R(x) = 2000x + 30x^2 - 0.3x^3$, where x is the number of units sold.

    a. Graph this function on the window $[-100, 150]$ by $[-30,000, 220,000]$.

    b. Because x represents the number of units sold, what restrictions should be placed on x in the context of the problem? What restrictions should be placed on y?

    c. Using your result from part (b), graph the function on a new window that makes sense for the problem.

    d. What does this function give as the revenue if 60 units are produced?

39. *Daily Revenue* The daily revenue from the sale of a product is given by $R = 600x - 0.1x^3 + 4x^2$ dollars, where x is the number of units sold.

    a. Graph this function on the window $[0, 100]$ by $[0, 30,000]$.

    b. Use the graph of this function to estimate the number of units that will give the maximum daily revenue and the maximum possible daily revenue.

    c. Find a window that will show a complete graph, that is, will show all of the turning points and intercepts. Graph the function using this window.

    d. Does the graph in part (a) or the graph in part (c) better represent the revenue from the sale of x units of a product?

    e. Over what interval of x-values is the revenue increasing, if $x \geq 0$?

40. *Weekly Revenue* A firm has total weekly revenue for its product given by $R(x) = 2800x - 8x^2 - x^3$, where x is the number of units sold.

a. Graph this function on the window [0, 50] by [0, 51,000].

b. Use technology to find the maximum possible revenue and the number of units that gives the maximum revenue.

c. Find a window that will show a complete graph, that is, will show all of the turning points and intercepts. Graph the function using this window.

d. Does the graph in part (a) or the graph in part (c) better represent the revenue from the sale of x units of a product?

e. Over what interval of x-values is the revenue increasing, if $x \geq 0$?

41. *Investment* If $2000 is invested for 3 years at rate r compounded annually, the future value of this investment is given by $S = 2000(1 + r)^3$, where r is the rate written as a decimal.

a. Complete the following table to see how increasing the interest rate affects the future value of this investment.

r (rate)	S (future value, $)
0%	
5%	
10%	
15%	
20%	

b. Graph this function for $0 \leq r \leq 0.24$.

c. Use the table and/or graph to compare the future value if $r = 10\%$ and if $r = 20\%$. How much more money is earned at 20%?

d. Which interest rate, 10% or 20%, is more realistic for an investment?

42. *Investment* The future value of $10,000 invested for 5 years at rate r compounded annually is given by $S = 10,000(1 + r)^5$, where r is the rate written as a decimal.

a. Complete the following table to see how increasing the interest rate affects the future value of this investment.

r (rate)	S (future value, $)
0%	
5%	
7%	
12%	
18%	

b. Graph this function for $0 \leq r \leq 0.24$.

c. Use the graph to compare the future value if $r = 10\%$ and if $r = 24\%$. How much more money is earned at 24%?

d. Is it more likely that you can get an investment paying 10% or 24%?

43. *Homicide Rate* The number of homicides per 100,000 people is given by the equation $y = -0.000710x^4 + 0.0265x^3 - 0.300x^2 + 0.684x + 9.337$ with x equal to the number of years after 1990.

a. Graph this function on a window with values for x representing the years 1990–2007.

b. What does this model give as the number of homicides per 100,000 people in 2007?

c. Use this model and technology to estimate the year after 1990 in which the number of homicides is at a maximum.

d. In what year can we be certain that the model is not valid? Why?
(Source: U.S. Department of Justice, Bureau of Justice Statistics)

44. *Drunk Driving Fatalities* Using data for 1982–2005, the total number of fatalities in drunk driving crashes in South Carolina can be modeled by the function $y = -0.0395x^4 + 2.101x^3 - 35.079x^2 + 194.109x + 100.148$, where x is the number of years after 1980.

a. Graph this function on a window with values for x representing the years 1980–2005.

b. How many such fatalities occurred in 2005, according to this model?

c. Use this model and technology to estimate the year in which the number of fatalities from drunk driving crashes was at a maximum.

d. Use this model and technology to estimate the year between 1981 and 2005 in which the number of fatalities from drunk driving crashes was at a minimum.
(Source: Anheuser-Busch, www.beeresponsible.com)

45. *Hybrid Vehicle Sales* The total hybrid electric passenger vehicle sales for the years from 1997 to 2006 are shown in the figure below. The sales can be modeled by the equation

$$S(x) = 0.744x^3 - 20.255x^2 + 184.729x - 553.098$$

where $S(x)$ is in thousands of vehicles and x is the number of years after 1990.

a. Graph this model for values of x representing 1990–2006 and positive values of $S(x)$.

b. Find the number of vehicles sold in 2000 and in 2006, according to this model.

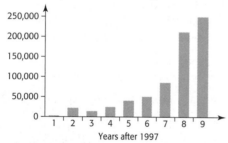

Total Hybrid Electric Passenger Vehicle Sales

Years after 1997

46. *Criminal Executions* Using data for 1982–2005, the number of criminal executions can be modeled by the function $y = -0.0453x^3 + 1.931x^2 - 20.714x + 79.591$, with x equal to the number of years after 1980.

a. Graph the function on an interval for the years 1982–2007.

b. In what year does this model estimate that the maximum number of executions occurred during 1982–2007?

c. What does the model estimate as the number of executions in 2006?
(Source: www.editonnine.deathrowbook.com)

47. *United Nations Debt* The cubic function

$$y = -1.832x^3 + 22.111x^2 - 55.367x + 290.659$$

models the U.S. debt to the United Nations, in millions of dollars, as a function of the number of years after 1990.

a. Graph this function using the window [0, 10] by [0, 400].

b. According to the model, what was the approximate U.S. debt in 1998?

c. Use technology to find the local maximum and the local minimum. What does this indicate about the maximum and minimum U.S. debt over this period?

48. *Salaries* The median salary for male workers for the years 1975–2006 can be modeled by the function

$$y = -0.0000929x^4 + 0.00784x^3$$
$$- 0.226x^2 + 2.508x + 34.347$$

thousand dollars, where x is the number of years from 1970.

a. Find the minimum on the graph of this function between $x = 10$ and $x = 30$ and interpret it.

b. Is the median salary found in (a) the lowest median salary in the period 1975–2006?

49. *Profit* The weekly revenue for a product is given by $R(x) = 120x - 0.015x^2$, and the weekly cost is $C(x) = 10,000 + 60x - 0.03x^2 + 0.00001x^3$, where x is the number of units produced and sold.

a. How many units will give maximum profit?

b. What is the maximum possible profit?

50. *Profit* The annual revenue for a product is given by $R(x) = 60,000x - 50x^2$, and the annual cost is $C(x) = 800 + 100x^2 + x^3$, where x is the number of thousands of units produced and sold.

a. How many units will give maximum profit?

b. What is the maximum possible profit?

6.2

Modeling with Cubic and Quartic Functions

section preview ▪ Poverty Level

One way to judge the success of an economy is to observe the number of citizens who earn less than a designated amount called the poverty level. The numbers of persons living below the poverty level in the United States for the years 1960–2005 are shown in Table 6.4. The scatter plot of the data, shown in Figure 6.9, shows that neither a line nor a parabola fits the data well, and it appears that a cubic function would be a better fit.

The use of graphing utilities and spreadsheets permits us to find cubic and quartic functions that model nonlinear data. In this section, we model and apply these functions.

Table 6.4

Year	Persons below Poverty Level (millions)	Year	Persons below Poverty Level (millions)
1960	39.9	1986	32.4
1965	33.2	1990	33.6
1970	25.4	1995	36.4
1975	25.9	2000	31.1
1980	29.3	2005	37.0

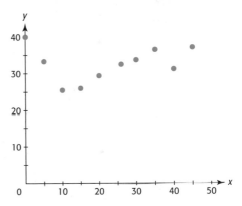

Figure 6.9

Modeling with Cubic Functions

The scatter plot of the data in Figure 6.9 indicates that the data may fit close to one of the possible shapes of the graph of a cubic function, so it is possible to find a cubic function that models the data.

example 1

Poverty Level

Use the data in Table 6.4 to do the following:

a. Find a cubic function that models the data in the table, with x equal to the number of years after 1960 and y equal to the number of millions of persons.

b. Graph the function reported in part (a) and the data on the same axes to see how well the model fits the data.

c. Use the model reported in part (a) to estimate the number of persons below the poverty level in 2008.

Solution

a. A cubic function that models the data is $y = -0.001x^3 + 0.095x^2 - 2.018x + 39.735$ where y is millions of persons and x is the number of years after 1960.

b. Figure 6.10 shows the graph of the function and the data on the same axes. The model is a good visual fit for the data.

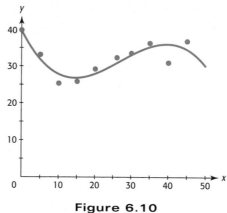

Figure 6.10

c. The number of persons below the poverty level in 2008 is estimated from the model with $y = f(48) = 32.4$ million. ∎

Sometimes a set of data points can be modeled reasonably well by both a linear function and a cubic function. Consider the following example.

| example 2 |

Tourism

The importance of the tourism industry in the United States is illustrated in Table 6.5, which shows the money spent by domestic travelers in the United States for each of the years 1987–2004.

Table 6.5

Year	Domestic Traveler Spending ($ billions)	Year	Domestic Traveler Spending ($ billions)
1987	235	1996	383
1988	258	1997	406
1989	273	1998	425
1990	291	1999	458
1991	296	2000	488
1992	306	2001	479
1993	323	2002	474
1994	339	2003	491
1995	360	2004	532

(Source: *2007 World Almanac*)

a. Find the linear function that is the best fit for the data with x in years after 1985 and y (in billions of dollars). Graph the function and the data points on the same axes.

b. Find the cubic function that is the best fit for the data, and graph the function and the data points on the same axes.

c. Use the graphs to determine which function appears to be the better model for the data.

Solution

a. The linear function that is the best fit for the data is

$$y = 17.287x + 197.210$$

The graphs of this function and the data points are shown in Figure 6.11(a).

b. The cubic function that is the best fit for the data is

$$y = -0.053x^3 + 1.738x^2 + 0.905x + 237.265$$

with x equal to the number of years after 1985. The graphs of this function and the data points are shown in Figure 6.11(b).

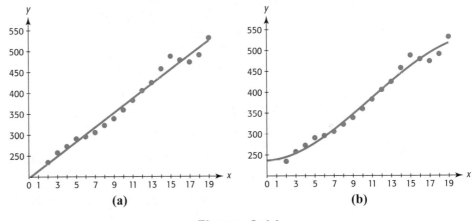

(a)   (b)

Figure 6.11

c. The graphs indicate that the cubic function is a slightly better fit than the linear function. ∎

---

spreadsheet solution

We can use graphing calculators, software programs, and spreadsheets to find the polynomial function that is the best fit for a set of data. Table 6.6 shows a partial Excel spreadsheet for the data of Example 2. Selecting the cells containing the data, using Chart Wizard to get the scatter plot of the data, selecting Add Trendline, and picking Polynomial with the order (degree) of the polynomial gives the equation of the polynomial function that is the best fit for the data, along with the scatter plot and the graph of the best-fitting curve. Figure 6.12 shows an Excel worksheet with the data from Example 2 and the graph of the cubic function that fits the data.

Table 6.6

	A	B
1	Year after 1985 x	Spending ($ billion) y
2	2	235
3	3	258
4	4	273
5	5	291
6	6	296
7	7	306
8	8	323
9	9	339
10	10	360
11	11	383

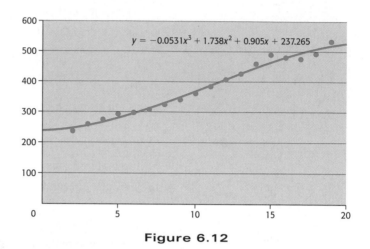

$y = -0.0531x^3 + 1.738x^2 + 0.905x + 237.265$

Figure 6.12

Modeling with Quartic Functions

If a scatter plot of data indicates that the data may fit close to one of the possible shapes that are graphs of quartic functions, it is possible to find a quartic function that models the data. Consider the following example.

example 3

Alcohol-Related Deaths

Table 6.7 gives the number of thousands of people killed in alcohol-related traffic accidents for the years 1983–2005.

Table 6.7

Years	Total Killed (thousands)	Years	Total Killed (thousands)	Years	Total Killed (thousands)
1983	23.646	1991	20.159	1999	16.572
1984	23.758	1992	18.290	2000	17.380
1985	22.716	1993	17.908	2001	17.400
1986	24.045	1994	17.307	2002	17.524
1987	23.641	1995	17.732	2003	17.105
1988	23.646	1996	17.749	2004	16.919
1989	22.404	1997	16.711	2005	16.885
1990	22.587	1998	16.673		

(Source: www.madd.org/stats/fatalities)

a. Make a scatter plot of the data with $x = 0$ in 1980 to determine if the data can be modeled by a quartic function.

b. Find the quartic function that models the data, with $x = 0$ in 1980.

c. Graph the unrounded model and the scatter plot of the data on the same axes.

d. Use the unrounded model to estimate the number of alcohol-related deaths in 2006.

Solution

a. The scatter plot of the data is shown in Figure 6.13(a).

b. The quartic function that models these data is

$$y = f(x) = -0.0005x^4 + 0.02977x^3 - 0.6022x^2 + 4.1245x + 15.2316$$

where x is the number of years after 1980.

c. The graphs of the model and the scatter plot of the data are shown in Figure 6.13(b).

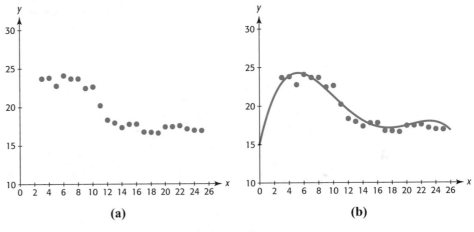

(a)   (b)

Figure 6.13

d. Because $f(26) \approx 15.891$, the estimated number of deaths for 2006 is 15,891. ■

Model Comparisons

Even when a scatter plot does not show all the characteristics of the graph of a quartic function, a quartic function may be a good fit for the data. Consider the following example, which shows a comparison of a cubic and a quartic model for a set of data. It also shows that many decimal places are sometimes required in reporting a model, so that the leading coefficient is not written as zero.

| example 4 |

Foreign-Born Population

The percents of U.S. population that were foreign born for selected years during 1900–2005 are shown in Figure 6.14 on the next page.

a. Create a scatter plot for the data, using the number of years after 1900 as the input x and the percent as the output y.

b. Find a cubic function to model the data and graph the function on the same axes with the data points. Round coefficients to seven decimal places.

c. Find a quartic function to model the data and graph the function on the same axes with the data points. Round coefficients to seven decimal places.

d. Use the graphs of the functions in parts (b) and (c) to determine which function is the better fit for the data.

Year	Percent
1900	13.6
1910	14.7
1920	13.2
1930	11.6
1940	8.8
1950	6.9
1960	5.4
1970	4.8
1980	6.2
1990	8.0
2000	10.4
2005	11.7

(Source: U.S. Census Bureau)

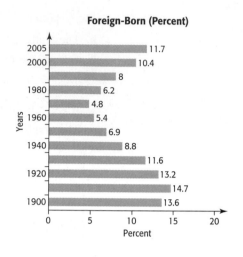

Figure 6.14

Solution

a. The scatter plot of the data is shown in Figure 6.15.

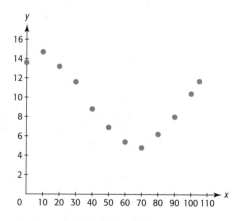

Figure 6.15

b. A cubic function that models the data is

$$y = 0.00005x^3 - 0.00560x^2 + 0.01652x + 14.30802$$

with $x = 0$ representing 1900. Figure 6.16(a) shows the graph of the unrounded model on the same axes with the data points.

c. A quartic function that models the data is

$$y = -0.0000008x^4 + 0.0002196x^3 - 0.0167097x^2 + 0.2503283x + 13.5809230$$

with $x = 0$ representing 1900. Figure 6.16(b) shows the graph of the unrounded model on the same axes with the data points. Note that the graph of this quartic function is not complete on this window because only two turning points are shown. Increasing the x-interval on the window would show another turning point.

d. Although both models are relatively good fits for the data, the graph of the quartic model appears to be a slightly better fit for the data than the cubic model.

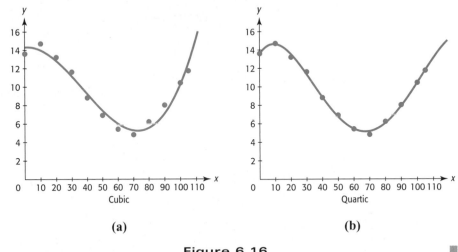

Figure 6.16

Third and Fourth Differences

Recall that for equally spaced inputs, the first differences of outputs are constant for data modeled by linear functions, and the second differences are constant for data modeled by quadratic functions. In a like manner, the third differences of outputs are constant for data modeled by cubic functions, and the fourth differences are constant for data modeled by quartic functions, if the inputs are equally spaced.

example 5

Model Comparisons

Table 6.8 has a set of equally spaced inputs x in column 1 and three sets of outputs in columns 2, 3, and 4. Use third and/or fourth differences with each set of outputs to determine if each set of data can be modeled by a cubic or quartic function. Then find the function that is the best fit for each set of outputs and the input, using:

a. Output Set I   **b.** Output Set II   **c.** Output Set III.

Table 6.8

Inputs, x	Output Set I	Output Set II	Output Set III
1	3	4	−16
2	14	4	−13
3	47	18	1
4	114	88	33
5	227	280	87
6	398	684	173

Solution

a. The inputs (x-values in column 1) are equally spaced. The third differences for Output Set I data in column 2 follow.

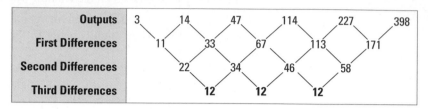

Outputs	3		14		47		114		227		398
First Differences		11		33		67		113		171		
Second Differences			22		34		46		58			
Third Differences				12		12		12				

The third differences are constant, so the data points fit exactly on the graph of a cubic function. The function that models the data is $f(x) = 2x^3 - x^2 + 2$. Figure 6.17 shows the graph of this function for $0 \le x \le 6$.

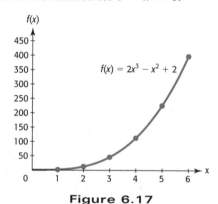

Figure 6.17

b. The third and fourth differences for Output Set II data in column 3 follow.

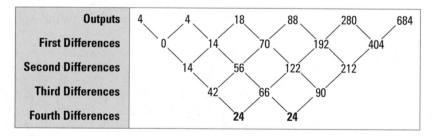

Outputs	4		4		18		88		280		684
First Differences		0		14		70		192		404		
Second Differences			14		56		122		212			
Third Differences				42		66		90				
Fourth Differences					24		24					

The fourth differences are constant, so the data points fit exactly on the graph of a quartic function. The function that models the data is $g(x) = x^4 - 3x^3 + 6x$ (Figure 6.18).

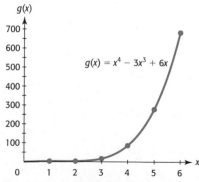

Figure 6.18

c. The third differences for Output Set III data in column 4 follow.

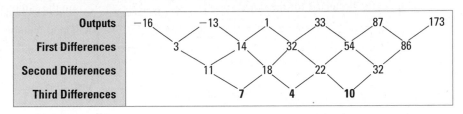

Outputs	−16		−13		1		33		87		173
First Differences		3		14		32		54		86	
Second Differences			11		18		22		32		
Third Differences				7		4		10			

The third differences are not constant, but are relatively close, so the data points can be approximately fitted by a cubic function. The cubic function that is the best fit for the data is $y = h(x) = 1.083x^3 - 1.107x^2 - 1.048x - 15$. Figure 6.19 shows that the graph is a good fit for the data points.

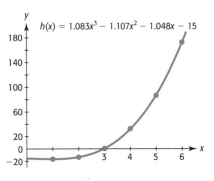

$h(x) = 1.083x^3 - 1.107x^2 - 1.048x - 15$

Figure 6.19

Note that the third (or fourth) differences will not be constant unless the data points fit on the graph of the cubic (or quartic) function exactly, and real data points will rarely fit on the graph of a function exactly. As with other models, statistical measures exist to measure the goodness of fit of the model to the data.

skills check

6.2

1. Find the cubic function that models the data in the table below.

x	−2	−1	0	1	2	3	4
y	−16	−3	0	−1	0	9	32

2. Find the cubic function that is the best fit for the data in the table below.

x	0	5	10	15
y	1	270	2600	9220

3. Find the quartic function that models the data in the table below.

x	−2	−1	0	1	2	3	4
y	0	−3	0	−3	0	45	192

4. Find the quartic function that is the best fit for the data in the table below.

x	−2	−1	0	1	4
y	17	−0.25	0	−0.25	356

5. a. Make a scatter plot of the data in the table on the next page.

b. Does it appear that a linear model or a cubic model is the better fit for the data?

x	1	2	3	4	5	6
y	−2	−1	0	4	8	16

6. a. Make a scatter plot of the data in the table below.

 b. Does it appear that a cubic model or a quartic model is the better fit for the data?

x	−2	−1	0	1	2	3
y	−8	7	1	2	9	57

7. a. Find a cubic function that models the data in the table in Exercise 5.

 b. Find a linear function that models the data.

 c. Visually determine which model is the better fit for the data.

8. a. Find a cubic function that models the data in the table in Exercise 6.

 b. Find a quartic function that models the data.

 c. Visually determine which model is the better fit for the data.

9. a. Find the cubic function that is the best fit for the data in the table below.

 b. Find the quartic function that is the best fit for the data in the table.

x	1	2	3	4	5	6	7
y	−2	0	4	16	54	192	1500

10. a. Graph each of the functions found in Exercise 9 on the same axes with the data in the table.

 b. Does it appear that a cubic model or a quartic model is the better fit for the data?

11. Find the quartic function that is the best fit for the data in the table below.

x	−3	−2	−1	0	1	2	3
y	55	7	1	1	−5	−5	37

12. Is the model found in Exercise 11 an exact fit to the data?

13. The following table has the inputs, x, and the outputs for two functions, f and g. Use third differences to determine whether a cubic function exactly fits the data with input x and output $f(x)$.

x	0	1	2	3	4	5
$f(x)$	0	1	5	24	60	110
$g(x)$	0	.5	4	13.5	32	62.5

14. Use third differences with the data in the table in Exercise 13 to determine whether a cubic function exactly fits the data with input x and output $g(x)$.

15. Find the cubic function that is the best fit for $f(x)$ defined by the table in Exercise 13.

16. Find the cubic function that is the best fit for $g(x)$ defined by the table in Exercise 13.

exercises

6.2

Use unrounded models for graphing and calculations unless otherwise stated. Report models to three decimal places unless otherwise instructed.

17. *Worldwide Internet* The table on the next page gives the number of worldwide users of the Internet, in millions.

 a. Find the cubic function that is the best fit for the data, with y equal to the number of millions of users and x equal to the number of years from 1990.

 b. Use the model to predict the number of users in 2013.

c. Compare the graphs of the data and the model to determine if the model is a good or poor fit for the data.

Year	Users (millions)	Year	Users (millions)
1995	16	2002	598
1996	36	2003	719
1997	70	2004	817
1998	147	2005	1018
1999	248	2006	1093
2000	361	2007	1215
2001	536		

18. *Japanese Economy* Japan experienced an economic setback, and some economists were fearful in 2002 that the Japanese economy was slipping into recession. The percent changes in gross domestic product (GDP) for each year since 1997, with forecasts for 2001 and 2002, are given in the figure.

a. Find a quartic function that models the data, where *x* equals the number of years after 1990 and *y* is the percent change.

b. Use the model to estimate the percent change in GDP for 2002.

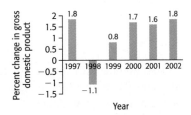

19. *Homicide Rates* The table at right above gives the U.S. homicide rates per 100,000 people for the years from 1990 through 2006.

a. Find the cubic and quartic functions that model these data, with *x* representing the number of years after 1990. Report models with four decimal places.

b. Graph each model on the same axes with the data. Which model appears to be the better fit?

Year	Homicide Rate	Year	Homicide Rate
1990	9.4	1999	5.7
1991	9.8	2000	5.5
1992	9.3	2001	5.6
1993	9.5	2002	5.6
1994	9.0	2003	5.7
1995	8.2	2004	5.5
1996	7.4	2005	5.6
1997	6.8	2006	5.7
1998	6.3		

(Source: www.ojp.usdog.gov/bjs)

20. *Social Security Beneficiaries* The table below gives the numbers of Social Security beneficiaries (in millions) for selected years from 1950 to 2000, with projections through 2030.

a. Find the cubic equation that models the data, with *x* equal to the number of years from 1950 and $B(x)$ equal to the number of millions of beneficiaries. Report the model with four decimal places.

b. Graph the data points and the model on the same axes.

c. Is the model a good fit for the data?

Year	Beneficiaries (millions)	Year	Beneficiaries (millions)
1950	2.9	2000	44.8
1960	14.3	2010	53.3*
1970	25.2	2020	68.8*
1980	35.1	2030	82.7*
1990	39.5		

(Source: 2000 Social Security Trustees Report)
*Projected.

21. *Teen Pregnancy* The table on the next page gives the number of pregnancies per 1000 U.S. females from 15 to 19 years of age for the years 1972–2002.

Year	Pregnancies/ 1000 Teens	Year	Pregnancies/ 1000 Teens	Year	Pregnancies/ 1000 Teens
1972	95	1983	109	1993	108
1973	96	1984	108	1994	105
1974	99	1985	109	1995	100
1975	101	1986	107	1996	96
1976	101	1987	107	1997	91
1977	105	1988	111	1998	89
1978	105	1989	115	1999	86
1979	109	1990	117	2000	84
1980	111	1991	115	2001	80
1981	110	1992	111	2002	75
1982	110				

(Source: U.S. Census Bureau)

a. Make a scatter plot of the data with $x = 0$ in 1970 to determine if the data can be modeled by a quartic function.

b. Find the quartic function that models the data, with $x = 0$ in 1970. Report the model with five decimal places.

c. Graph the model and the scatter plot of the data on the same axes.

d. Use the model to estimate the number of pregnancies per thousand in 2007.

22. *Methamphetamine Labs* The table below gives the numbers of methamphetamine labs seized in South Carolina during the years 2000–2004.

a. Find the quartic function that models the data, with x equal to the number of years from 2000.

b. Does this model fit the data points exactly or approximately?

Year	2000	2001	2002	2003	2004
Labs Seized	7	4	9	36	42

(Source: www.dea.gov/pubs/states/southcarolinap.html)

23. *Births* The numbers of births to females in the United States between 15 and 17 years of age for certain years are shown in the table below.

a. Find the cubic function that models the data, where y is the number of thousands of births and x is the number of years from 1960. Round the coefficients to four decimal places.

b. Use the rounded model reported in part (a) to estimate the number of births that occurred in 2004.

Year	Births (thousands)	Year	Births (thousands)
1960	182.408	1995	192.508
1970	223.590	1996	185.724
1980	198.222	1997	180.154
1986	168.572	1998	173.231
1990	183.327	1999	163.588
1991	188.226	2000	157.209
1992	187.549	2001	145.324
1993	190.535	2002	138.731
1994	195.169	2003	134.384

(Source: guttmacher.org)

24. *Births* The numbers of births to females in the United States under 15 years of age for certain years are shown in the following table.

a. Find the cubic function that models the data, where y is the number of thousands of births and x is the number of years from 1970. Round the coefficients to four decimal places.

b. Use the rounded model reported in part (a) to estimate the number of such births that occurred in 2005.

c. When do we know this model is no longer valid?

Year	Births (thousands)	Year	Births (thousands)
1970	11.752	1996	11.146
1980	10.169	1997	10.121
1986	10.176	1998	9.462
1990	11.657	1999	9.054
1991	12.014	2000	8.519
1992	12.220	2001	7.781
1993	12.554	2002	7.315
1994	12.901	2003	6.661
1995	12.242		

(Source: guttmacher.org)

25. *Domestic Leisure Travel* The graph below gives the number of millions of person-trips of 50 miles or more.

    a. Find the cubic function $y = f(x)$ that is the best fit for the data, with y equal to the number of millions of person-trips and x equal to the number years after 1990.

    b. Find and interpret $f(18)$.

(Source: *World Almanac and Book of Facts*)

26. *Congressional Grants* Congress earmarks millions of dollars each year for academic projects. These awards are often the result of congressional connections. The accompanying table gives the congressional money earmarked for academic projects, in millions of dollars, for the years 1990–2003.

    a. Use a scatter plot, with x equal to the number of years after 1990, to determine the degree of the function that is the best fit for the data.

    b. Use technology to find a function that models the millions of dollars in earmarked grants as a function of the number of years after 1990.

c. Graph the unrounded model on the same axes with the data. Does the model appear to be an exact fit, a good fit, or a poor fit for the data?

Year	$ Millions	Year	$ Millions
1990	230	1996	640
1991	450	1997	410
1992	640	1998	470
1993	680	1999	800
1994	610	2000	1040
1995	600	2003	2010

(Source: Associated Press)

27. *Alcohol-Related Traffic Fatalities* The table below shows the total number of people killed in alcohol-related traffic accidents for the years 1990–2005.

    a. Find a cubic function that models the data, where x is the number of years after 1990 and y is the number of thousands of fatalities.

    b. Find a quartic model that models the data, where x is the number of years after 1990 and y is the number of thousands of fatalities.

    c. Graph both models on $0 \le x \le 16$ to determine if one of these models is the better fit for the data, or if the fits are nearly equal.

Year	Total Killed	Year	Total Killed
1990	22,587	1998	16,673
1991	20,159	1999	16,572
1992	18,290	2000	17,380
1993	17,908	2001	17,480
1994	17,307	2002	17,524
1995	17,732	2003	17,185
1996	17,749	2004	16,919
1997	16,711	2005	16,885

(Source: madd.org/stats/fatalities)

28. *U.S. Consumer Price Index* The graph on the next page gives the consumer price index (the price of goods and services costing $100 in 1967) for selected years 1950–2006.

a. Find the cubic function that is the best model for the data, with *x* equal to the number of years from 1950.

b. What does this model predict the 2012 consumer price index to be?

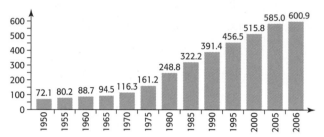

(Source: U.S. Dept of Labor)

29–30. *Age at First Marriage* The table below shows the U.S. median age at first marriage for men and women for selected years 1900–2004. Use the table in Problems 29–30.

Median Age at First Marriage

Year	Men	Women	Year	Men	Women
1900	25.9	21.9	1960	22.8	20.3
1910	25.1	21.6	1970	23.2	20.8
1920	24.6	21.2	1980	24.7	22.0
1930	24.3	21.3	1990	26.1	23.9
1940	24.3	21.5	2000	26.8	25.1
1950	22.8	20.3	2004	27.1	25.8

(Source: U.S. Census Bureau)

29. a. Find the cubic function $y = f(x)$ (to 5 decimal places) that is the best fit for the data, with *y* equal to the age at first marriage for women and *x* equal to the number of years after 1900.

b. What does the model predict to be the age at first marriage for women in 2012?

30. a. Find the cubic function $y = f(x)$ (to 6 decimal places) that is the best fit for the data, with *y* equal to the age at first marriage for men and *x* equal to the number of years after 1900.

b. What does the model predict to be the age at first marriage for men in 2012?

31. *Juvenile Crime* The table gives the annual numbers of arrests for violent crime per 100,000 juveniles

from 10 to 17 years of age (with some juveniles having multiple arrests).

a. Create a scatter plot and determine the degree of the function that is the best fit for the data, with *x* equal to the number of years after 1980.

b. Create a cubic function that gives the annual number of arrests for violent crime per 100,000 juveniles 10 to 17 years of age, with *x* = 0 in 1980.

c. Graph the data and the model reported in part (b) on the same axes.

d. Using the rounded model from part (b), what is the ratio of juvenile crimes in 2004 to those in 1986?

Year	Number of Arrests (per 100,000)	Year	Number of Arrests (per 100,000)
1980	334.09	1991	461.06
1981	322.64	1992	482.14
1982	314.48	1993	504.48
1983	295.98	1994	526.64
1984	297.46	1995	517.74
1985	302.97	1996	460.27
1986	316.70	1997	442.46
1987	310.55	1998	369.64
1988	326.48	1999	339.14
1989	381.56	2003	298.44
1990	428.48		

(Source: U.S. Department of Justice, Bureau of Justice Statistics)

32. *Inflation Rate* The annual changes in the Consumer Price Index (CPI) for 1990–2005 are shown in the following table.

Year	CPI Annual Change (%)	Year	CPI Annual Change (%)	Year	CPI Annual Change (%)
1990	5.4	1996	3.0	2002	1.6
1991	4.2	1997	2.3	2003	2.3
1992	3.0	1998	1.6	2004	2.7
1993	3.0	1999	2.2	2005	3.4
1994	2.6	2000	3.4		
1995	2.8	2001	2.8		

(Source: Bureau of Labor Statistics)

a. Find the quartic function that models the data, with $x = 0$ in 1990. Report the model to 4 decimal places.

b. Is the fit of this model to the data best described as fair, good, or exact?

33. *Teen Alcohol Use* The percent of U.S. 12th graders who used alcohol in the last 12 months during the years 1991–2006 is given in the table below.

a. Find the quartic function that is the best fit for the data, using $x = 0$ in 1990.

b. Use the model to find a local minimum on the graph for $0 \le x \le 15$.

c. Interpret the answer in (b).

d. Is there an absolute minimum percent between 1990 and 2010 according to the model? In what year?

Years	Percent	Years	Percent
1991	77.7	1999	73.8
1992	76.8	2000	73.2
1993	72.7	2001	73.3
1994	73.0	2002	71.5
1995	73.7	2003	70.1
1996	72.5	2004	70.6
1997	74.8	2005	68.6
1998	74.3	2006	66.5

34. *Gasoline Prices* The table at the right above gives the city average retail prices (in cents) for unleaded regular gasoline for the years 1992–2007.

a. Find the cubic function $y = f(x)$ that is the best fit for the data, with y equal to the retail price in cents and x equal to the number of years after 1990.

b. Find and interpret $f(18)$.

Year	Price (cents)	Year	Price (cents)
1992	112.7	2000	151.0
1993	110.8	2001	146.1
1994	111.2	2002	135.8
1995	114.7	2003	159.8
1996	123.1	2004	188.0
1997	123.4	2005	229.5
1998	105.9	2006	260.5
1999	116.5	2007	300.1

(Source: *Monthly Energy Review*)

35. *U.S. Budget Surpluses and Deficits* The table below gives the U.S. budget surpluses and deficits for the years 1990–2006, in billions of dollars.

a. Find a quartic function that models the data using $x = 0$ in 1990.

b. If this model applies beyond 2006, what does it estimate for 2008? Use the unrounded model.

Year	Surplus or Deficit	Year	Surplus or Deficit
1990	−221.036	1999	125.610
1991	−269.238	2000	236.241
1992	−290.321	2001	128.236
1993	−255.051	2002	−157.758
1994	−203.186	2003	−377.585
1995	−163.952	2004	−412.727
1996	−107.431	2005	−318.346
1997	−21.884	2006	−247.698
1998	69.270		

(Source: U.S. Department of the Treasury)

section

6.3

Solution of Polynomial Equations

section preview ▪ Worldcom Stock

The closing price of a share of Worldcom, Inc. stock for the years 1994–2000 can be modeled by the equation $y = -2.56x^3 + 53.73x^2 - 356.47x + 772.74$, where x represents the number of years after 1990.

The graph of this model, shown in Figure 6.20, shows that in 1995 the Worldcom stock value dropped to nearly \$10, then increased in value, and then experienced a sharp decline in its value after 1999. If this trend were to continue, in what year after 1999 will the closing price be worth \$15?

To answer this question, we would solve the equation

$$-2.56x^3 + 53.73x^2 - 356.47x + 772.74 = 15$$

The technology of graphing utilities allows us to find approximate solutions of equations such as this one.

Although many equations derived from real data frequently require technology to find or to approximate solutions, some higher-degree equations can be solved by factoring, and some can be solved by the root method. In this section, we discuss these methods for solving polynomial equations; in the next section, we discuss additional methods of solving higher-degree polynomial equations.

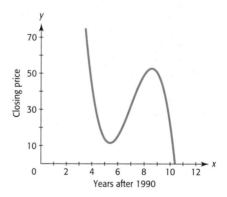

Figure 6.20

(Source: S&P Comstock)

Solving Polynomial Equations by Factoring

Some polynomial equations can be solved by factoring and using the zero-product property, much like the method used to solve quadratic equations described in Chapter 3. From our discussion in Chapter 3, we know that the following statements regarding a function *f* are equivalent:

- $(x - a)$ is a factor of $f(x)$.
- a is a real zero of the function f.
- a is a real solution to the equation $f(x) = 0$.
- a is an x-intercept of the graph of $y = f(x)$.
- The graph crosses the x-axis at the point $(a, 0)$.

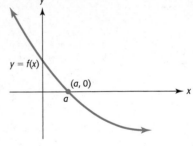

example 1   **Factors, Zeros, Intercepts, and Solutions**

If $P(x) = (x - 1)(x + 2)(x + 5)$, find

a. The zeros of $P(x)$.

b. The solutions of $P(x) = 0$.

c. The x-intercepts of the graph of $y = P(x)$.

Solution

a. The factors of $P(x)$ are $(x - 1)$, $(x + 2)$, and $(x + 5)$; thus, the zeros of $P(x)$ are $x = 1, x = -2$, and $x = -5$.

b. The solutions to $P(x) = 0$ are the zeros of $P(x)$: $x = 1, x = -2$, and $x = -5$.

c. The x-intercepts of the graph of $y = P(x)$ are $1, -2$, and -5 (Figure 6.21).

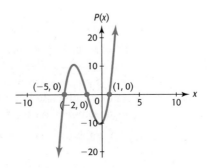

Figure 6.21

example 2   **Finding the Maximum Volume**

A box is to be formed by cutting squares x inches per side from each corner of a square piece of cardboard that is 24 inches on each side and folding up the sides. This will give a box whose height is x inches, with the length of each side of the bottom $2x$ less than the original length of the cardboard (Figure 6.22 on the next page). Using $V = lwh$, its volume is given by

$$V = (24 - 2x)(24 - 2x)x$$

where x is length of the side of the square that is cut out.

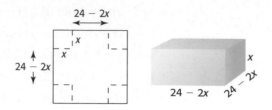

Figure 6.22

a. Use the factors of V to find the values of x that give volume 0.

b. Graph the function that gives the volume as a function of the side of the square that is cut out over an x-interval that includes the x-values found in part (a).

c. What input values make sense for this problem (that is, actually result in a box)?

d. Graph the function using the input values that make sense for the problem (found in part (c)).

e. Use technology to determine the size of the squares that should be cut out to give the maximum volume.

Solution

a. We use the zero-product property to solve the equation $0 = (24 - 2x)(24 - 2x)x$.

$$0 = (24 - 2x)(24 - 2x)x$$

$$\underline{x = 0 \quad \text{or} \quad 24 - 2x = 0}$$

$$x = 0 \quad \Big| \quad \begin{array}{c} -2x = -24 \\ x = 12 \end{array}$$

b. A graph of $V(x)$ from $x = 0$ to $x = 12$ is shown in Figure 6.23(a).

c. The box can exist only if each of the dimensions is positive, resulting in a volume that is positive. Thus, $24 - 2x > 0$, or $x < 12$, so the box can only exist if the value of x is greater than 0 and less than 12. Note that if squares 12 inches on a side are removed, no material remains to make a box, and no squares larger than 12 inches on a side can be removed from all four corners.

d. The graph of $V(x)$ over the x-interval $0 < x < 12$ is shown in Figure 6.23(b).

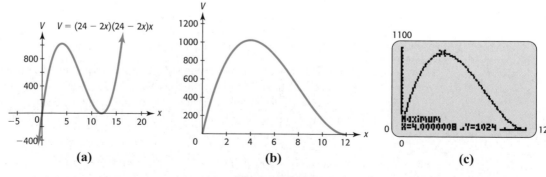

Figure 6.23

e. By observing the graph, we see that the function has a maximum value, which gives the maximum volume over the interval $0 < x < 12$. Using the "maximum" feature on a graphing utility, we see that the maximum occurs at or near $x = 4$ (Figure 6.23(c)). Cutting the squares of side 4 inches from each corner and folding up the sides gives a box that is 16 inches by 16 inches by 4 inches. The volume of this box is $16 \cdot 16 \cdot 4 = 1024$ cubic inches, and testing values near 4 shows that this is the maximum possible volume. ∎

| example 3 | **Photosynthesis** |

The amount y of photosynthesis that takes place in a certain plant depends on the intensity x of the light (in lumens) present, according to the function $y = 120x^2 - 20x^3$. The model is valid only for nonnegative x-values that produce a positive amount of photosynthesis.

a. Graph this function with a viewing window large enough to see two turning points.

b. Use factoring to find the x-intercepts of the graph.

c. What nonnegative values of x will give positive photosynthesis?

d. Graph the function with a viewing window that contains only the values of x found in part (c) and nonnegative values of y.

Solution

a. The graph in Figure 6.24 shows the function with two turning points.

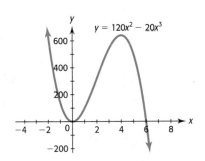

$$y = 120x^2 - 20x^3$$

Figure 6.24

b. We can find the x-intercepts of the graph by solving $120x^2 - 20x^3 = 0$ for x. Factoring gives

$$20x^2(6 - x) = 0$$
$$20x^2 = 0 \quad \text{or} \quad 6 - x = 0$$
$$x = 0 \quad \text{or} \quad x = 6$$

Thus, the x-intercepts are 0 and 6.

c. From the graph of $y = 120x^2 - 20x^3$, in Figure 6.24, we see that positive output values result when the input values are between 0 and 6. Thus, the nonnegative x-values that produce positive amounts of photosynthesis are $0 < x < 6$.

d. Graphing the function in the viewing window [0, 6] by [0, 700] will show the region that is appropriate for the model (see Figure 6.25 on the next page).

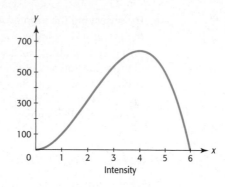

Figure 6.25

Solution Using Factoring by Grouping

Some higher-degree equations can be solved by using the method of factoring by grouping.* Consider the following examples.

| example 4 | **Cost** |

The total cost of producing a product is given by the function

$$C(x) = x^3 - 12x^2 + 3x + 9 \text{ thousand dollars}$$

where x is the number of hundreds of units produced. How many units must be produced to give a total cost of \$45,000?

Solution

Because $C(x)$ is the total cost in thousands of dollars, a total cost of \$45,000 is represented by $C(x) = 45$. So we seek to solve

$$45 = x^3 - 12x^2 + 3x + 9 \quad \text{or} \quad 0 = x^3 - 12x^2 + 3x - 36$$

This equation appears to have a form that permits factoring by grouping. The solution steps follow.

$0 = (x^3 - 12x^2) + (3x - 36)$ Separate the terms into two groups, each having a common factor.

$0 = x^2(x - 12) + 3(x - 12)$ Factor x^2 from the first group, then 3 from the second group.

$0 = (x - 12)(x^2 + 3)$ Factor the common factor $(x - 12)$ from each of the two terms.

$x - 12 = 0 \quad \text{or} \quad x^2 + 3 = 0$ Set each factor equal to 0 and solve for x.

$x = 12 \quad \text{or} \quad x^2 = -3$

$x = 12$

Because no real number x can satisfy $x^2 = -3$, the only solution is 12 hundred units. Thus, producing 1200 units gives a total cost of \$45,000.

---

* Factoring by grouping was discussed in the Chapter 3 Algebra Toolbox.

The Root Method

We solved quadratic equations of the form $x^2 = C$ in Chapter 3 by taking the square root of both sides and using a $\pm$ symbol to indicate that we get both the positive and the negative root: $x = \pm\sqrt{C}$. In the same manner, we can solve $x^3 = C$ by taking the cube root of both sides. There will not be a $\pm$ sign because there is only one real cube root of a number. In general, there is only one real nth root if n is odd, and there are zero or two real roots if n is even.

> ### Root Method
>
> The real solutions of the equation $x^n = C$ are found by taking the nth root of both sides:
>
> $$x = \sqrt[n]{C} \text{ if } n \text{ is odd}, \quad \text{and} \quad x = \pm\sqrt[n]{C} \text{ if } n \text{ is even and } C \geq 0$$

example 5

Root Method

Solve the following equations.

a. $x^3 = 125$ **b.** $5x^4 = 80$ **c.** $4x^2 = 18$

Solution

a. Taking the cube root of both sides gives

$$x^3 = 125$$
$$x = \sqrt[3]{125} = 5$$

No $\pm$ sign is needed because the root is odd.

b. We divide both sides by 5 to isolate the x^4 term

$$5x^4 = 80$$
$$x^4 = 16$$

and take the fourth root of both sides, using a $\pm$ sign (because the root is even).

$$x = \pm\sqrt[4]{16} = \pm 2$$

The solutions are 2 and -2.

c. We divide both sides by 4 to isolate the x^2 term:

$$4x^2 = 18$$
$$x^2 = \frac{9}{2}$$

Then we take the square root of both sides and use a $\pm$ sign (because the root is even).

$$x = \pm\sqrt{\frac{9}{2}} = \pm\frac{3}{\sqrt{2}}$$

The *exact* solutions are $\pm\dfrac{3}{\sqrt{2}}$, and these solutions are *approximately* 2.121 and -2.121. ■

| example 6 | Future Value of an Investment |

The future value of $10,000 invested for 4 years at interest rate r compounded annually is given by $S = 10,000(1 + r)^4$. Find the rate r, as a percent, for which the future value is $14,641.

Solution

We seek to solve $14,641 = 10,000(1 + r)^4$ for r. We can rewrite this equation with $(1 + r)^4$ on one side by dividing both sides by 10,000.

$$14,641 = 10,000(1 + r)^4$$
$$1.4641 = (1 + r)^4$$

We can now remove the exponent by taking the fourth root of both sides of the equation, and using a $\pm$ sign.

$$\pm \sqrt[4]{1.4641} = 1 + r$$
$$\pm 1.1 = 1 + r$$

Solving for r gives

$$
\begin{array}{c|c}
1.1 = 1 + r & -1.1 = 1 + r \\
0.1 = r & -2.1 = r
\end{array}
$$

The negative value cannot represent an interest rate, so the interest rate that gives this future value is

$$r = 0.1 = 10\%$$

The graphs of $y_1 = 10,000(1 + x)^4$ and $y_2 = 14,641$ shown in Figure 6.26 support the conclusion that $r = 0.10$ gives the future value $14,641.$*

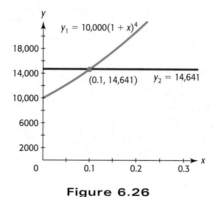

Figure 6.26

Estimating Solutions with Technology

Although we can solve some higher-degree equations by using factoring, many real problems do not have "nice" solutions, and factoring is not an appropriate solution method. Thus, many equations resulting from real data applications require graphical or numerical methods of solution.

---

* The graph of the quartic function $y = 10,000(1 + x)^4$ is actually curved, but within this small window it appears to be a line.

| example 7 | **Juvenile Crime Arrests** |

The annual number of arrests for crime per 100,000 juveniles from 10 to 17 years of age can be modeled by the function

$$f(x) = -0.357x^3 + 9.417x^2 - 51.852x + 361.208$$

where x is the number of years after 1980.* The number of arrests per 100,000 peaked in 1994 and then decreased. Use this model to estimate graphically the year after 1980 in which the number of arrests fell to 234 per 100,000 juveniles.

Solution

To find the year when the number of arrests fell to 234 per 100,000 juveniles, we solve

$$234 = -0.357x^3 + 9.417x^2 - 51.852x + 361.208$$

To solve this equation graphically, we can use either the intersection method or the x-intercept method. To use the intersection method, we graph

$$y_1 = -0.357x^3 + 9.417x^2 - 51.852x + 361.208 \quad \text{and} \quad y_2 = 234$$

for $x \geq 0$, and find the x-coordinate of the point of intersection (Figure 6.27).

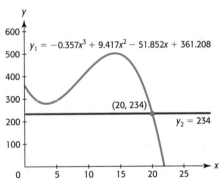

Figure 6.27

The graphs intersect at the point where $x \approx 20$. Thus, according to the model, the number of arrests for crime fell to 234 per 100,000 juveniles during the year 2000. ∎

| example 8 | **Stock Prices** |

The closing price of a share of Worldcom, Inc. stock for the years 1994–2000 can be modeled by the equation $y = -2.56x^3 + 53.73x^2 - 356.47x + 772.74$, where x represents the number of years after 1990.

Assuming that this trend continues, graphically determine the years after 1999 in which the closing price was $15 (roughly the cost of the stock in 1997).
(Source: S&P Comstock)

---

* Note that some juveniles have multiple arrests.

Solution

To answer this question, we solve the equation

$$-2.56x^3 + 53.73x^2 - 356.47x + 772.74 = 15 \quad \text{or}$$
$$-2.56x^3 + 53.73x^2 - 356.47x + 757.74 = 0$$

To solve this equation graphically with the *x*-intercept method, we graph

$$y = -2.56x^3 + 53.73x^2 - 356.47x + 757.74$$

and find the *x*-intercepts of the graph (Figure 6.28(a)).

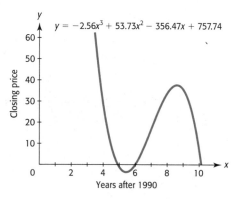

Figure 6.28(a)

It appears that this graph crosses the *x*-axis three times, so there are three solutions to the equation $-2.56x^3 + 53.73x^2 - 356.47x + 757.74 = 0$. The approximate solutions are shown in Figures 6.28(b), 6.28(c), and 6.28(d).

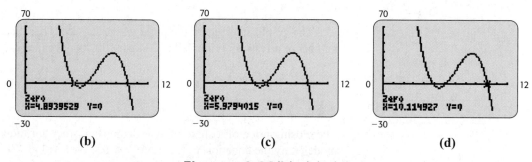

Figures 6.28(b), (c), (d)

We see that $y = 15$ when $x \approx 5, x \approx 6$, or $x \approx 10.1$. Thus, we might expect the closing price to be approximately $15 in 1995, in 1996, and in 2001. Worldcom declared bankruptcy in 2002, after accounting irregularities were discovered. ∎

**skills
check**

6.3

In Exercises 1–4, solve the polynomial equation.

1. $(2x - 3)(x + 1)(x - 6) = 0$

2. $(3x + 1)(2x - 1)(x + 5) = 0$

3. $(x + 1)^2(x - 4)(2x - 5) = 0$

4. $(2x + 3)^2(5 - x)^2 = 0$

In Exercises 5–10, solve the polynomial equations by factoring and check the solutions graphically.

5. $x^3 - 16x = 0$ **6.** $2x^3 - 8x = 0$

7. $x^4 - 4x^3 + 4x^2 = 0$

8. $x^4 - 6x^3 + 9x^2 = 0$

9. $4x^3 - 4x = 0$ **10.** $x^4 - 3x^3 + 2x^2 = 0$

In Exercises 11–14, use factoring by grouping to solve the equations.

11. $x^3 - 4x^2 - 9x + 36 = 0$

12. $x^3 + 5x^2 - 4x - 20 = 0$

13. $3x^3 - 4x^2 - 12x + 16 = 0$

14. $4x^3 + 8x^2 - 36x - 72 = 0$

In Exercises 15–18, solve the polynomial equations by using the root method, and check the solutions graphically.

15. $2x^3 - 16 = 0$ **16.** $3x^3 - 81 = 0$

17. $\frac{1}{2}x^4 - 8 = 0$ **18.** $2x^4 - 162 = 0$

In Exercises 19–24, use factoring and the root method to solve the polynomial equations.

19. $4x^4 - 8x^2 = 0$ **20.** $3x^4 - 24x^2 = 0$

21. $0.5x^3 - 12.5x = 0$ **22.** $0.2x^3 - 24x = 0$

23. $x^4 - 6x^2 + 9 = 0$ **24.** $x^4 - 10x^2 + 25 = 0$

In Exercises 25–30, use the graph of the polynomial function $f(x)$ (a) to solve $f(x) = 0$, and (b) find the factorization of $f(x)$.

25. $f(x) = x^3 - 2x^2 - 11x + 12$

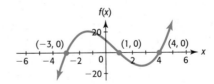

26. $f(x) = -x^3 + 6.5x^2 + 13x - 8$

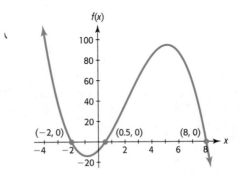

27. $y = -2x^4 + 6x^3 + 6x^2 - 14x - 12$

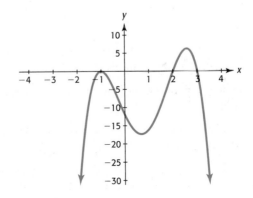

28. $y = x^4 - 5x^3 - 3x^2 + 13x + 10$

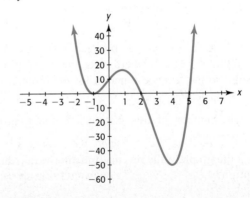

29. $y = -0.5x^3 + 2.5x^2 + 0.5x - 2.5$

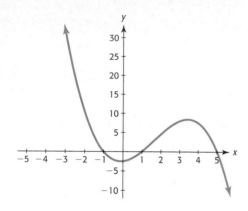

30. $y = x^3 - x^2 - 4x + 4$

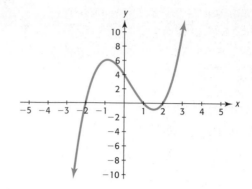

In Exercises 31 and 32, find the solutions graphically.

31. $4x^3 - 15x^2 - 31x + 30 = 0$

32. $2x^3 - 15x^2 - 62x + 120 = 0$

exercises

6.3

33. *Revenue* The revenue from the sale of a product is given by the function $R = 400x - x^3$.

   a. Use factoring to find the numbers of units that must be sold to give zero revenue.

   b. Does the graph of the revenue function verify this solution?

34. *Revenue* The revenue from the sale of a product is given by the function $R = 12{,}000x - 0.003x^3$.

   a. Use factoring and the root method to find the numbers of units that must be sold to give zero revenue.

   b. Does the graph of the revenue function verify this solution?

35. *Revenue* The price for a product is given by $p = 100{,}000 - 0.1x^2$, where x is the number of units sold, so the revenue function for the product is $R = px = (100{,}000 - 0.1x^2)x$.

   a. Find the numbers of units that must be sold to give zero revenue.

   b. Does the graph of the revenue function verify this solution?

36. *Revenue* If the price from the sale of x units of a product is given by the function $p = 100x - x^2$, the revenue function for this product is given by $R = px = (100x - x^2)x$.

   a. Find the numbers of units that must be sold to give zero revenue.

   b. Does the graph of the revenue function verify this solution?

37. *Future Value* The future value of $2000 invested for 3 years at rate r compounded annually is given by $S = 2000(1 + r)^3$.

   a. Complete the table below to determine the future value of $2000 at certain interest rates.

Rate	Future Value
4%	
5%	
7.25%	
10.5%	

   b. Graph this function on the window $[0, 0.24]$ by $[0, 5000]$.

   c. Use the root method to find the rate r, as a percent, for which the future value is $2662.

d. Use the root method to find the rate r, as a percent, for which the future value is $3456.

38. *Future Value* The future value of $5000 invested for 4 years at rate r compounded annually is given by $S = 5000(1 + r)^4$.

 a. Graph this function on the window [0, 0.24] by [0, 12,000].

 b. Use the root method to find the rate r, as a percent, for which the future value is $10,368.

 c. What rate, as a percent, gives $2320.50 in interest on this investment?

39. *Constructing a Box* A box can be formed by cutting squares out of each corner of a piece of tin and folding the "tabs" up. Suppose the piece of tin is 18 inches by 18 inches and each side of the square that is cut out has length x.

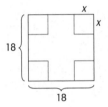

 a. Write an expression for the height of the box that is constructed.

 b. Write an expression for the dimensions of the base of the box that is constructed.

 c. Use the formula $V = lwh$ to find an equation that represents the volume of the box.

 d. Use the equation that you constructed to find the values of x that make $V = 0$.

 e. For which of these values of x does a box exist if squares of length x are cut out and the tabs are folded up?

40. *Constructing a Box* A box can be formed by cutting squares out of each corner of a piece of cardboard and folding the "tabs" up. If the piece of cardboard is 12 inches by 12 inches and each side of the square that is cut out has length x, the function that gives the volume of the box is $V = 144x - 48x^2 + 4x^3$.

 a. Find the values of x that make $V = 0$.

 b. For each of the values of x in part (a), discuss what happens to the box if squares of length x are cut out.

 c. For what values of x does a box exist?

d. Graph the function that gives the volume as a function of the side of the square that is cut out using an x-interval that makes sense for the problem (see part (c)).

41. *Profit* The profit function for a product is given by $P(x) = -x^3 + 2x^2 + 400x - 400$, where x is the number of units produced and sold and P is in hundreds of dollars. Use factoring by grouping to find the number of units that will give a profit of $40,000.

42. *Cost* The total cost function for a product is given by $C(x) = 3x^3 - 6x^2 - 300x + 1800$, where x is the number of units produced and C is the cost in hundreds of dollars. Use factoring by grouping to find the number of units that will give a total cost of $120,000.

43. *Ballistics* Ballistics experts are able to identify the weapon that fired a certain bullet by studying the markings on the bullet after it is fired. They test the rifling (the grooves in the bullet as it travels down the barrel of the gun) by comparing it to that of a second bullet fired into a bale of paper. The speed, s, in centimeters per second, that the bullet travels through the paper is given by $s = 30(3 - 10t)^3$, where t is the time after the bullet strikes the bale and $t \leq 0.3$ second.

 a. Complete the following table to find the speed for given values of t.

t	0	0.1	0.2	0.3
s (cm/sec)				

 b. Use the root method to find the number of seconds to give $s = 0$. Does this agree with the data in the table?

44. *Homicide Rate* The number of homicides per 100,000 people is given by the function $y = -0.000710x^4 + 0.0265x^3 - 0.300x^2 + 0.684x + 9.337$, with x equal to the number of years from 1990. Use technology to find the year or years in which the number of homicides per 100,000 people was 5.59. (Source: U.S. Dept. of Justice, Bureau of Justice Statistics)

45. *Arrests* The annual number of arrests by the New York State Police for driving infractions is given by $y = 1.646x^3 - 18.494x^2 + 39.403x + 838.950$, where y is in thousands and x is the number of years from 1990.

 a. What does the model estimate the number of arrests to be in 1995?

b. Use technology to find the year after 1995 when there will be 813,310 arrests, according to this model.

46. *Stock Price* The closing stock price of Priceline.com is given by $y = -0.104x^3 + 3.369x^2 - 36.268x + 184.703$, where x is the number of months after March 31, 1999 and y is in dollars.

    a. Use graphical or numerical methods to find the value of x that gives a closing price of $54.98.

    b. At the end of what month does this model say the Priceline.com stock is $54.98?

    (Source: Priceline.com Inc, PCLN Ticker)

47. *Japanese Economy* The percent change y in the Japanese gross domestic product (GDP) for years since 1997 is given by $y = 0.148x^4 - 5.829x^3 + $

$85.165x^2 - 545.851x + 1293.932$, where x is the number of years after 1990.

    a. Use graphical or numerical methods with this model to find the year before 1999 when the change in GDP was approximately 1.3 percentage points.

    b. According to this model, will the change in GDP ever be 1.3% again?

    (Source: *USA Today*, April 2, 2001)

48. *Foreign-Born Population* The percent of the U.S. population that was foreign born during the years 1900–2000 is given by the function $y = 0.0000503x^3 - 0.00560x^2 + 0.0165x + 14.308$, where x is the number of years from 1900. Use graphical methods to determine the year after 1900 when 14.7% of the population was foreign born.

6.4

Polynomial Equations Continued; Fundamental Theorem of Algebra

key concepts

- Polynomial division
- Synthetic division
- Using division to solve cubic equations
- Combining graphical and algebraic methods to solve polynomial equations
- Rational solutions test
- Solving quartic equations
- Fundamental Theorem of Algebra
- Complex solutions of polynomial equations

section preview ▪ Break-Even

Suppose the profit function for a product is given by

$$P(x) = -x^3 + 98x^2 - 700x - 1800 \text{ dollars}$$

where x is the number of units produced and sold. To find the number of units that gives break-even for the product, we must solve $P(x) = 0$. Solving this equation by factoring would be difficult. Finding all the zeros of $P(x)$ by using a graphical method may also be difficult without knowing an appropriate viewing window. In general, if we seek to find the exact solutions to a polynomial equation, a number of steps are involved if the polynomial is not easily factored. In this section, we investigate methods of finding additional solutions to polynomial equations after one or more solutions have been found. Specifically, if we know that a is one of the solutions to a polynomial equation, we know that $(x - a)$ is a factor of $P(x)$, and we can divide $P(x)$ by $(x - a)$ to find a second factor of $P(x)$ that may give additional solutions.

Division of Polynomials; Synthetic Division

Suppose $f(x)$ is a cubic function and we know that a is a solution of $f(x) = 0$. We can write $(x - a)$ as a factor of $f(x)$, and if we divide this factor into the cubic function, the quotient will be a quadratic factor of $f(x)$. If there are additional real solutions to $f(x) = 0$, this quadratic factor can be used to find the remaining solutions.

Division of a polynomial by a binomial was discussed in the Algebra Toolbox for this chapter. When we divide a polynomial by a binomial of the form

$$x - a$$

the division process can be simplified by using a technique called **synthetic division**. Note in the division of $x^4 + 6x^3 + 5x^2 - 4x + 2$ by $x + 2$ on the left below, many terms are recopied. Because the divisor is linear, the quotient has degree one less than the degree of the dividend and the degree decreases by one in each remaining term. This is true in general, so we can simplify the division process by omitting all variables and writing only the coefficients. If we rewrite the work without the terms that were recopied in the division and without the x's, the process looks like the one below on the right.

$$
\begin{array}{r}
x^3 + 4x^2 - 3x + 2 \\
x + 2 \overline{)\, x^4 + 6x^3 + 5x^2 - 4x + 2} \\
\underline{x^4 + 2x^3} \\
4x^3 + 5x^2 \\
\underline{4x^3 + 8x^2} \\
-3x^2 - 4x \\
\underline{-3x^2 - 6x} \\
2x + 2 \\
\underline{2x + 4} \\
-2
\end{array}
\qquad
\begin{array}{r}
1 + 4 - 3 + 2 \\
+2 \overline{)\, 1 + 6 + 5 - 4 + 2} \\
\underline{2} \\
4 \\
\underline{8} \\
-3 \\
\underline{-6} \\
2 \\
\underline{4} \\
-2
\end{array}
$$

Look at the process on the right above and observe that the coefficient of the first term of the dividend and the differences found in each step are identical to the coefficients of the quotient; the last difference, -2, is the remainder. Because of this, we can compress the division process as follows:

$$
\begin{array}{r}
+2 \overline{)\, 1 + 6 + 5 - 4 + 2} \\
\underline{2 + 8 - 6 + 4} \\
1 + 4 - 3 + 2 - 2
\end{array}
$$

Note that the last line in this process gives the coefficients of the quotient, with the last number equal to the remainder. Changing the sign in the divisor allows us to use addition instead of subtraction in the process, as follows:

$$
\begin{array}{r}
-2 \overline{)\, 1 + 6 + 5 - 4 + 2} \\
\underline{-2 - 8 + 6 - 4} \\
1 + 4 - 3 + 2 - 2
\end{array}
$$

The bottom line of this process gives the coefficients of the quotient, and the degree of the quotient is one less than the degree of the dividend, so the quotient is

$$x^3 + 4x^2 - 3x + 2, \quad \text{with remainder } -2$$

just as it was in the original division.

The synthetic division process uses the following steps.

Synthetic Division Steps to Divide a Polynomial By $x - a$

To divide a polynomial by $x - a$:

Divide $2x^3 - 9x - 27$
by $x - 3$.

1. Arrange the coefficients in descending powers of x, with a 0 for any missing power. Place a from $x - a$ to the left of the coefficients.

$$3\overline{)2 + 0 - 9 - 27}$$

2. Bring down the first coefficient to the third line. Multiply the last number in the third line by a and write the product in the second line under the next term.

$$3\overline{)2 + 0 - 9 - 27}$$
$$\underline{6}$$
$$2$$

3. Add the last number in the second line to the number above it in the first line. Continue this process until all numbers in the first line are used.

$$3\overline{)2 + 0 - 9 - 27}$$
$$\underline{6 + 18 + 27}$$
$$2 + 6 + 9 + 0$$

4. The third line represents the coefficients of the quotient, with the last number the remainder. The quotient is a polynomial of degree one less than the dividend.

The quotient is

$$2x^2 + 6x + 9$$

If the remainder is 0, $x - a$ is a factor of the polynomial, and the polynomial can be written as the product of the divisor $x - a$ and the quotient.

The remainder is 0, so $x - 3$ is a factor of $2x^3 - 9x - 27$.
$$2x^3 - 9x - 27$$
$$= (x - 3)(2x^2 + 6x + 9).$$

Using Synthetic Division to Solve Cubic Equations

If we know one solution of a cubic equation $P(x) = 0$, we can divide by the corresponding factor using synthetic division to obtain a second quadratic factor that can be used to find any additional solutions of the polynomial.

| example 1 | **Solving an Equation with Synthetic Division**

If $x = -2$ is a solution of $x^3 + 8x^2 + 21x + 18 = 0$, find the remaining solutions.

Solution

If -2 is a solution of the equation $x^3 + 8x^2 + 21x + 18 = 0$, then $x + 2$ is a factor of

$$f(x) = x^3 + 8x^2 + 21x + 18$$

Synthetically dividing $f(x)$ by $x + 2$ gives

$$-2\overline{)\,1 + 8 + 21 + 18}$$
$$\underline{-2 - 12 - 18}$$
$$1 + 6 + 9 + 0$$

Thus, the quotient is $x^2 + 6x + 9$, with remainder 0, so $x + 2$ is a factor of $x^3 + 8x^2 + 21x + 18$ and

$$x^3 + 8x^2 + 21x + 18 = (x + 2)(x^2 + 6x + 9)$$

Because $x^2 + 6x + 9 = (x + 3)(x + 3)$, we have

$$x^3 + 8x^2 + 21x + 18 = 0$$
$$(x + 2)(x^2 + 6x + 9) = 0$$
$$(x + 2)(x + 3)(x + 3) = 0$$

and the solutions to the equation are $x = -2$, $x = -3$, and $x = -3$.

Because one of the solutions of this equation, -3, occurs *twice*, we say that it is a double solution, or a solution of **multiplicity** 2.

To check the solutions, we can graph the function $f(x) = x^3 + 8x^2 + 21x + 18$ and observe x-intercepts at -2 and -3 (Figure 6.29). Observe that the graph touches but does not cross the x-axis at $x = -3$, where a *double* solution to the equation (and zero of the function) occurs.

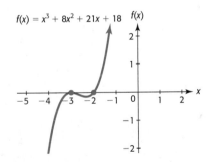

Figure 6.29

> A graph touches but does not cross the x-axis at a zero of even multiplicity, and it crosses the x-axis at a zero of odd multiplicity.

Graphs and Solutions

Sometimes a combination of graphical methods and factoring can be used to solve a polynomial equation. If we are seeking exact solutions to a cubic equation and one solution can be found graphically, we can use that solution and division to find the remaining quadratic factor and to find the remaining two solutions. Consider the following example.

example 2

Combining Graphical and Algebraic Methods

Solve the equation $x^3 + 9x^2 - 610x + 600 = 0$.

Solution

If we graph $P(x) = x^3 + 9x^2 - 610x + 600$ in a standard window, shown in Figure 6.30 on the next page, we see that the graph appears to cross the x-axis at $x = 1$, so one solution to $P(x) = 0$ may be $x = 1$. (Notice that this window does not give a complete graph of the cubic polynomial function.)

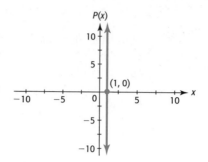

Figure 6.30

To verify that $x = 1$ is a solution to $P(x) = 0$, we use synthetic division to divide $x^3 + 9x^2 - 610x + 600$ by $x - 1$.

$$
\begin{array}{r}
1)\overline{1 + 9 - 610 + 600} \\
\underline{1 + 10 - 600} \\
1 + 10 - 600 + 0
\end{array}
$$

The fact that the division results in a 0 remainder verifies that 1 is a solutions to the equation $P(x) = 0$. The quotient is $x^2 + 10x - 600$, so the remaining solutions to $P(x) = 0$ can be found by factoring or by using the quadratic formula to solve $x^2 + 10x - 600 = 0$.

$$x^3 + 9x^2 - 610x + 600 = 0$$
$$(x - 1)(x^2 + 10x - 600) = 0$$
$$x - 1 = 0 \quad \text{or} \quad x^2 + 10x - 600 = 0$$
$$x = 1 \qquad\qquad (x + 30)(x - 20) = 0$$
$$x = 1 \qquad\qquad x = -30 \quad \text{or} \quad x = 20$$

Graphing the function $P(x) = x^3 + 9x^2 - 610x + 600$ using a viewing window that includes x-values of -30, 1, and 20 verifies that -30, 1, and 20 are zeros of the function and thus solutions to the equation $x^3 + 9x^2 - 610x + 600 = 0$ (Figure 6.31).

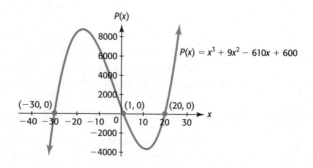

Figure 6.31

example 3	**Break-Even**

The weekly profit for a product is $P(x) = -0.1x^3 + 11x^2 - 80x - 2000$ thousand dollars, where x is the number of thousands of units produced and sold. To find the number of units that gives break-even,

a. Graph the function using a window representing up to 50 thousand units and find one x-intercept of the graph.

b. Use synthetic division to find a quadratic factor of $P(x)$.

c. Find all of the zeros of $P(x)$.

d. Determine the levels of production that give break-even.

Solution

a. Break-even occurs where the profit is 0, so we seek the solution to

$$0 = -0.1x^3 + 11x^2 - 80x - 2000$$

for $x \geq 0$ (because x represents the number of units). Because $x = 50$ represents 50 thousand units and we seek an x-value where $P(x) = 0$, we use the viewing window [0, 50] by [−2500, 8000]. The graph of

$$P(x) = -0.1x^3 + 11x^2 - 80x - 2000$$

is shown in Figure 6.32. (Note that this is not a complete graph of the polynomial function.)

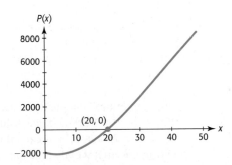

Figure 6.32

An x -intercept of this graph appears to be $x = 20$ (see Figure 6.32).

b. To verify that $P(x) = 0$ at $x = 20$ and to determine if any other x-values give break-even, we divide

$$P(x) = -0.1x^3 + 11x^2 - 80x - 2000$$

by the factor $x - 20$.

$$
\begin{array}{r}
20\overline{)\;-0.1 + 11 - \;\;\;80 - 2000} \\
-\;\;2 + 180 + 2000 \\
\hline
-0.1 + \;\;9 + 100 + \;\;\;\;\;0
\end{array}
$$

Thus, $P(x) = 0$ at $x = 20$ and the quotient is the quadratic factor $-0.1x^2 + 9x + 100$, so

$$-0.1x^3 + 11x^2 - 80x - 2000 = (x - 20)(-0.1x^2 + 9x + 100)$$

c. The solution to $P(x) = 0$ follows.

$$-0.1x^3 + 11x^2 - 80x - 2000 = 0$$
$$(x - 20)(-0.1x^2 + 9x + 100) = 0$$
$$(x - 20)[-0.1(x^2 - 90x - 1000)] = 0$$
$$-0.1(x - 20)(x - 100)(x + 10) = 0$$
$$x - 20 = 0 \quad \text{or} \quad x - 100 = 0 \quad \text{or} \quad x + 10 = 0$$
$$x = 20 \quad \text{or} \quad x = 100 \quad \text{or} \quad x = -10$$

Thus, the zeros are 20, 100 and -10. Figure 6.33 shows the graph of $P(x) = -0.1x^3 + 11x^2 - 80x - 2000$ on a window that shows all three x-intercepts.

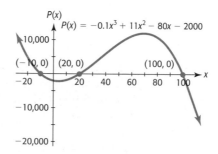

Figure 6.33

d. The solutions indicate that break-even occurs at $x = 20$ and at $x = 100$. The value $x = -10$ also makes $P(x) = 0$, but a negative number of units does not make sense in the context of this problem. Thus producing and selling 20,000 or 100,000 units gives break-even. ■

Combining graphical methods and synthetic division, as we did in Example 3, is especially useful in finding exact solutions to polynomial equations. If some of the solutions are irrational or nonreal solutions, the quadratic formula can be used rather than factoring. Remember that the quadratic formula can always be used to solve a quadratic equation, and factoring cannot always be used.

Rational Solutions Test

The Rational Solutions Test provides information about the rational solutions of a polynomial equation with integer coefficients.

Rational Solutions Test

The rational solutions of the polynomial equation

$$a_n x^n + a_{n-1} x^{n-1} + \cdots + a_1 x + a_0 = 0$$

with integer coefficients must be of the form $\dfrac{p}{q}$, where p is a factor of the constant term a_0 and q is a factor of a_n, the leading coefficient.

Recall that if a is a solution to $f(x) = 0$, then a is also a zero of the function $f(x)$. The Rational Solutions Test provides a list of *possible* rational zeros of a polynomial function. For a polynomial function with integer coefficients, if there is a rational zero, then it will be in this list of possible rational zeros. Keep in mind that there may be *no* rational zeros of a polynomial function. Graphing a polynomial function $f(x)$ will give us a better sense of the location of the zeros, and we can also use our list of possible zeros to find an appropriate viewing window for the function.

For example, the rational solutions of $x^3 + 8x^2 + 21x + 18 = 0$ must be numbers that are factors of 18 divided by factors of 1. Thus, the possible rational solutions are ± 1, ± 18, ± 2, ± 9, ± 3, and ± 6. A viewing window containing x-values from -18 to 18 will show all possible rational solutions. Note that the solutions to this equation, found in Example 1 to be -2 and -3, are contained in this list.

Many cubic and quartic equations can be solved by using the following steps.

Solving Cubic and Quartic Equations of the Form $f(x) = 0$

1. Determine the possible rational solutions of $f(x) = 0$.

2. Graph $y = f(x)$ to see if any of the values from Step 1 are x-intercepts. Those values that are x-intercepts are rational solutions to $f(x) = 0$.

3. Find the factors associated with the x-intercepts from Step 2.

4. Use synthetic division to divide $f(x)$ by the factors from Step 3 to confirm the graphical solutions and find additional factors. Continue until a quadratic factor remains.

5. Use factoring or the quadratic formula to find the solutions associated with the quadratic factor. These solutions are also solutions to $f(x) = 0$.

Example 4 shows how the Rational Solutions Test is used when the leading coefficient of the polynomial is not 1.

| example 4 | **Solving a Quartic Equation** |

Solve the equation $2x^4 + 10x^3 + 13x^2 - x - 6 = 0$.

Solution

The rational solutions of this equation must be factors of the constant -6 divided by factors of 2, the leading coefficient. Thus, the possible rational solutions are

$$\pm 1, \pm 2, \pm 3, \pm 6 \quad \text{and} \quad \pm\frac{1}{2}, \pm\frac{2}{2}, \pm\frac{3}{2}, \pm\frac{6}{2}^*$$

All of these possible rational solutions are between -6 and 6, so we graph

$$y = P(x) = 2x^4 + 10x^3 + 13x^2 - x - 6$$

---

* Some of these possible solutions are duplicates.

in a window with x-values from -4 to 4. Figure 6.34 shows that the graph appears to cross the x-axis at $x = -1$ and at $x = -2$, so two of the solutions to $y = 0$ appear to be $x = -1$ and $x = -2$. We will use synthetic division to confirm that these values are solutions to the equation and to find the two remaining solutions. (We know there are two additional solutions, which may be irrational, because the graph crosses the x-axis at two other points.)

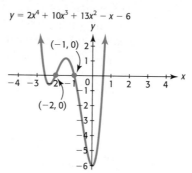

$y = 2x^4 + 10x^3 + 13x^2 - x - 6$

Figure 6.34

We can confirm that $x = -1$ is a solution and find the remaining factor of $P(x)$ by using synthetic division to divide $2x^4 + 10x^3 + 13x^2 - x - 6$ by $x + 1$.

$$
\begin{array}{r}
-1\overline{)2 + 10 + 13 - 1 - 6} \\
\underline{-\ 2 -\ 8 - 5 + 6} \\
2 +\ 8 +\ 5 - 6 + 0
\end{array}
$$

Thus, we have $2x^4 + 10x^3 + 13x^2 - x - 6 = (x + 1)(2x^3 + 8x^2 + 5x - 6)$.

We can also confirm that $x = -2$ is a solution of $P(x)$ by showing that $x + 2$ is a factor of $2x^3 + 8x^2 + 5x - 6$ and thus a factor of $P(x)$. Dividing $x + 2$ into $2x^3 + 8x^2 + 5x - 6$ using synthetic division gives a quadratic factor of $P(x)$.

$$
\begin{array}{r}
-2\overline{)2 + 8 + 5 - 6} \\
\underline{-\ 4 - 8 + 6} \\
2 + 4 - 3 + 0
\end{array}
$$

So the quotient is the quadratic factor $2x^2 + 4x - 3$, and $P(x)$ has the following factorization.

$$2x^4 + 10x^3 + 13x^2 - x - 6 = (x + 1)(x + 2)(2x^2 + 4x - 3)$$

Thus, $P(x) = 0$ is equivalent to $(x + 1)(x + 2)(2x^2 + 4x - 3) = 0$, so the solutions can be found by solving $x + 1 = 0$, $x + 2 = 0$, and $2x^2 + 4x - 3 = 0$.

The two remaining x-intercepts of the graph of $y = 2x^2 + 4x - 3$ do not appear to cross the x-axis at integer values, and $2x^2 + 4x - 3$ does not appear to be factorable. Using the quadratic formula is the obvious method to solve $2x^2 + 4x - 3 = 0$ to find the two remaining solutions.

$$x = \frac{-4 \pm \sqrt{16 - 4(2)(-3)}}{2(2)} = \frac{-4 \pm \sqrt{40}}{4} = \frac{-2 \pm \sqrt{10}}{2}$$

So the solutions are $-1, -2, \dfrac{-2 + \sqrt{10}}{2}$, and $\dfrac{-2 - \sqrt{10}}{2}$. Note that two of these solutions are rational and two are irrational, approximated by 0.58114 and -2.58114. The two irrational solutions can also be verified graphically (Figure 6.35).

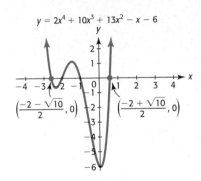

Figure 6.35

Note that some graphing utilities and spreadsheets have commands that can be used to find or to approximate solutions to polynomial equations.

Fundamental Theorem of Algebra

As we learned in Section 3.2, not all quadratic equations have real solutions. However, every quadratic equation has complex solutions. If the complex number a is a solution to $P(x) = 0$, then a is a complex zero of P. The **Fundamental Theorem of Algebra** states that every polynomial function has a complex zero.

> ### Fundamental Theorem of Algebra
> If $f(x)$ is a polynomial function of degree $n \geq 1$, then f has at least one complex zero.

As a result of this theorem, it is possible to prove the following.

> ### Complex Zeros
> Every polynomial function $f(x)$ of degree $n \geq 1$ has exactly n complex zeros. Some of these zeros may be imaginary, and some may be repeated. Nonreal zeros occur in conjugate pairs, $a + bi$ and $a - bi$.

example 5

Solution in the Complex Number System

Find the complex solutions to $x^3 + 2x^2 - 3 = 0$.

Solution

Because the function is cubic, there are three solutions, and because any nonreal solutions occur in conjugate pairs, at least one of the solutions must be real. To estimate that solution, we graph

$$y = x^3 + 2x^2 - 3$$

on a window that includes the possible rational solutions, which are ± 1 and ± 3 (see Figure 6.36 on the next page).

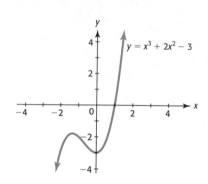

Figure 6.36

The graph appears to cross the x-axis at 1, so we (synthetically) divide $x^3 + 2x^2 - 3$ by $x - 1$.

$$
\begin{array}{r}
1\,)\,1 + 2 + 0 - 3 \\
\underline{1 + 3 + 3} \\
1 + 3 + 3 + 0
\end{array}
$$

The division confirms that 1 is a solution to $x^3 + 2x^2 - 3 = 0$ and shows that

$$x^2 + 3x + 3$$

is a factor of $x^3 + 2x^2 - 3$. Thus, any remaining solutions to the equation satisfy

$$x^2 + 3x + 3 = 0$$

Using the quadratic formula gives the remaining solutions.

$$x = \frac{-3 \pm \sqrt{3^2 - 4(1)(3)}}{2(1)} = \frac{-3 \pm \sqrt{-3}}{2} = \frac{-3 \pm i\sqrt{3}}{2}$$

Thus, the solutions are $1, -\dfrac{3}{2} + \dfrac{\sqrt{3}}{2}i$, and $-\dfrac{3}{2} - \dfrac{\sqrt{3}}{2}i$. Because there can only be three solutions to a cubic equation, we have all of the solutions.   ■

skills check

6.4

In Exercises 1–4, use synthetic division to find the quotient and the remainder.

1. $(x^4 - 4x^3 + 3x + 10) \div (x - 3)$

2. $(x^4 + 2x^3 - 3x^2 + 1) \div (x + 4)$

3. $(2x^4 - 3x^3 + x - 7) \div (x - 1)$

4. $(x^4 - 1) \div (x + 1)$

In Exercises 5 and 6, determine whether the given constant is a solution to the given polynomial equation.

5. $2x^4 - 4x^3 + 3x + 18 = 0;\ 3$

6. $x^4 + 3x^3 - 10x^2 + 8x + 40 = 0;\ -5$

In Exercises 7 and 8, determine whether the second polynomial is a factor of the first polynomial.

7. $(-x^4 - 9x^2 + 3x);\ (x + 3)$

8. $(2x^4 + 5x^3 - 6x - 4);\ (x + 2)$

In Exercises 9–12, one or more solutions of a polynomial equation are given. Use synthetic division to find any remaining solutions.

9. $-x^3 + x^2 + x - 1 = 0$; -1

10. $x^3 + 4x^2 - x - 4 = 0$; 1

11. $x^4 + 2x^3 - 21x^2 - 22x = -40$; $-5, 1$

12. $2x^4 - 17x^3 + 51x^2 - 63x + 27 = 0$; 3, 1

In Exercises 13–16, find one solution graphically and then find the remaining solutions using synthetic division.

13. $x^3 + 3x^2 - 18x - 40 = 0$

14. $x^3 - 3x^2 - 9x - 5 = 0$

15. $3x^3 + 2x^2 - 7x + 2 = 0$

16. $4x^3 + x^2 - 27x + 18 = 0$

In Exercises 17–20, determine all possible rational solutions of the polynomial equation.

17. $x^3 - 6x^2 + 5x + 12 = 0$

18. $4x^3 + 3x^2 - 9x + 2 = 0$

19. $9x^3 + 18x^2 + 5x - 4 = 0$

20. $6x^4 - x^3 - 42x^2 - 29x + 6 = 0$

In Exercises 21–24 find all rational zeros of the polynomial function.

21. $f(x) = x^3 - 6x^2 + 5x + 12$

22. $f(x) = 4x^3 + 3x^2 - 9x + 2$

23. $g(x) = 9x^3 + 18x^2 + 5x - 4$

24. $g(x) = 6x^3 + 19x^2 - 19x + 4$

Solve each of the equations in Exercises 25–30 exactly in the complex number system. (Find one solution graphically and then use the quadratic formula.)

25. $x^3 = 10x - 7x^2$

26. $t^3 - 2t^2 + 3t = 0$

27. $w^3 - 5w^2 + 6w - 2 = 0$

28. $2w^3 + 3w^2 + 3w + 2 = 0$

29. $z^3 - 8 = 0$

30. $x^3 + 1 = 0$

exercises

6.4

In Exercises 31–36, use synthetic division and factoring to solve the problems.

31. *Break-Even* The profit function for a product is given by $P(x) = -0.2x^3 + 66x^2 - 1600x - 60,000$ dollars, where x is the number of units produced and sold. If break-even occurs when 50 units are produced and sold,

   a. Use synthetic division to find a quadratic factor of $P(x)$.

   b. Use factoring to find a number of units other than 50 that gives break-even for the product, and verify your answer graphically.

32. *Break-Even* The profit function for a product is given by $P(x) = -x^3 + 98x^2 - 700x - 1800$ dollars, where x is the number of units produced and sold. If break-even occurs when 10 units are produced and sold,

   a. Use synthetic division to find a quadratic factor of $P(x)$.

   b. Find a number of units other than 10 that gives break-even for the product, and verify your answer graphically.

33. *Break-Even* The weekly profit for a product is $P(x) = -0.1x^3 + 50.7x^2 - 349.2x - 400$ thousand dollars, where x is the number of thousands of units produced and sold. To find the number of units that gives break-even,

a. Graph the function on the window $[-8, 10]$ by $[-60, 100]$.

b. Graphically find one x-intercept of the graph.

c. Use synthetic division to find a quadratic factor of $P(x)$.

d. Find all of the zeros of $P(x)$.

e. Determine the levels of production and sale that give break-even.

34. *Break-Even* The weekly profit for a product is $P(x) = -0.1x^3 + 10.9x^2 - 97.9x - 108.9$ thousand dollars, where x is the number of thousands of units produced and sold. To find the number of units that gives break-even,

a. Graph the function on the window $[-10, 20]$ by $[-100, 100]$.

b. Graphically find one x-intercept of the graph.

c. Use synthetic division to find a quadratic factor of $P(x)$.

d. Find all of the zeros of $P(x)$.

e. Determine the levels of production and sale that give break-even.

35. *Revenue* The revenue from the sale of a product is given by $R = 1810x - 81x^2 - x^3$. If the sale of 9 units gives a total revenue of $9000, use synthetic division to find another number of units that will give $9000 in revenue.

36. *Revenue* The revenue from the sale of a product is given by $R = 250x - 5x^2 - x^3$. If the sale of 5 units gives a total revenue of $1000, use synthetic division to find another number of units that will give $1000 in revenue.

37. *Drunk Driving Crashes* Suppose the total number of fatalities in drunk driving crashes in South Carolina is modeled by the function $y = 0.4566x^3 - 14.3085x^2 + 117.2978x + 107.8456$, where x is the number of years from 1980. This model indicates that the number of fatalities was 244 in 1992. To find any other years that the number was 244,

a. Set the function equal to 244 and rewrite the resulting equation with 0 on one side.

b. Graph the function from part (a) on the window $[-10, 25]$ by $[-75, 200]$ and use the graph to find an integer solution to this equation.

c. Use synthetic division to find a quadratic factor.

d. Use the quadratic formula to find any other solutions to this equation.

e. Use this information to determine the years in which the number of fatalities in drunk driving crashes was 244.
(Source: Anheuser-Busch, www.beeresponsible.com)

38. *College Enrollment* The number of thousands of college enrollments by recent high school graduates is given by $f(x) = 0.20x^3 - 13.71x^2 + 265.06x + 1612.56$, where x is the number of years from 1960. To find the year(s) after 1960 when the number of enrollments was 2862 thousand:

a. Set the function equal to 2862, and rewrite the resulting equation with 0 on one side.

b. Set the nonzero side of the equation in (a) equal to y and graph this function on the window $[25, 50]$ by $[-50, 50]$. Use the graph to find an integer solution to this equation.

c. Use synthetic division to find a quadratic factor.

d. Use the quadratic formula to find any other solutions to this equation.

e. Use this information to determine the years in which the number of enrollments was 2,862,000.
(Source: U.S. Census Bureau)

39. *Births* The number of births to females in the United States under 15 years of age can be modeled by the function $y = -0.0014x^3 + 0.1025x^2 - 2.2687x + 25.5704$, where y is the number of thousands of births and x is the number of years from 1970. If the model indicates that 10,808 births occurred in 1994, find two other years when the model indicates that 10,808 births occurred.
(Source: guttmacher.org)

40. *Births* The number of births to females in the United States between 15 and 17 years of age can be modeled by the cubic function $y = -0.0015x^3 + 0.0108x^2 + 1.1001x + 189.918$, where y is the number of thousands of births and x is the number of years from 1960. If the model indicates that 200,733 births occurred in 1985, find another year after 1960 when the model indicates that 200,733 births occurred.
(Source: Child Trends, Inc. MSNBC)

6.5

Rational Functions and Rational Equations

section preview ▪ Average Cost

The function that gives the daily average cost (in $hundred) for the production of Stanley golf carts is formed by dividing the total cost function for the production of the golf carts,

$$C(x) = 25 + 13x + x^2$$

by x, the number of golf carts produced. Thus, the average cost per golf cart is given by

$$\overline{C}(x) = \frac{25 + 13x + x^2}{x} \qquad (x > 0)$$

where x is the number of golf carts produced. The graph of this function with x restricted to positive values is shown in Figure 6.37(a). The graph of the function

$$\overline{C}(x) = \frac{25 + 13x + x^2}{x}$$

with x not restricted is shown in Figure 6.37(b). (See Example 3.)

This function, and any other function that is formed by taking the quotient of two polynomials, is called a **rational function**. In this section, we discuss the characteristics of the graphs of rational functions.

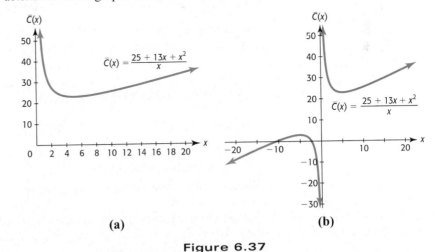

(a)   (b)

Figure 6.37

Graphs of Rational Functions

A rational function is defined as follows.

Rational Function

The function f is a rational function if

$$f(x) = \frac{P(x)}{Q(x)}$$

where $P(x)$ and $Q(x)$ are polynomials and $Q(x) \neq 0$.

In Section 3.3, we discussed a simple rational function, $y = \dfrac{1}{x}$, whose graph is shown in Figure 6.38. We now expand this discussion.

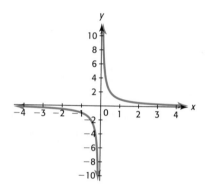

Figure 6.38

Because division by 0 is not possible, those real values of x for which $Q(x) = 0$ are not in the domain of the rational function $f(x) = \dfrac{P(x)}{Q(x)}$. Observe that the graph of $y = \dfrac{1}{x}$ approaches, but does not touch, the y-axis. Thus the y-axis is a **vertical asymptote**. The graph of a rational function frequently has vertical asymptotes.

Vertical Asymptote

A vertical asymptote occurs in the graph of $f(x) = \dfrac{P(x)}{Q(x)}$ at those values of x where $Q(x) = 0$ and $P(x) \neq 0$; that is, at values of x where the denominator equals 0 but the numerator does not equal 0.

To find where the vertical asymptote(s) occur, set the denominator equal to 0 and solve for x. If any of these values of x do not also make the numerator equal to 0, a vertical asymptote occurs at those values.

| example 1 | Rational Function |

Find the vertical asymptote and sketch the graph of $y = \dfrac{x + 1}{(x + 2)^2}$.

Solution

Setting the denominator equal to 0, taking the square root of both sides of the equation, and solving gives

$$(x + 2)^2 = 0$$
$$(x + 2) = 0$$
$$x = -2$$

The value $x = -2$ makes the denominator of the function equal to 0 and does not make the numerator equal to 0, so a vertical asymptote occurs at $x = -2$. The graph of the function $y = \dfrac{x + 1}{(x + 2)^2}$ is shown in Figure 6.39.

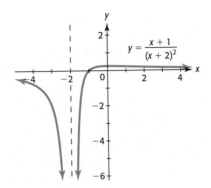

Figure 6.39

---

example 2

Cost-Benefit

Suppose that for specified values of p, the function

$$C(p) = \frac{800p}{100 - p}$$

can be used to model the cost of removing $p\%$ of the particulate pollution from the exhaust gases at an industrial site.

a. Graph this function on the window $[-100, 200]$ by $[-4000, 4000]$.

b. Does the graph of this function have a vertical asymptote on this window? Where?

c. For what values of p does this function serve as a model for the cost of removing particulate pollution?

d. Use the information determined in part (c) to graph the model.

e. What does the part of the graph near the vertical asymptote tell us about the cost of removing particulate pollution?

Solution

a. The graph is shown in Figure 6.40(a).

b. The denominator is equal to 0 at $p = 100$, and the numerator does not equal 0 at $p = 100$, so a vertical asymptote occurs at $p = 100$.

c. Because p represents the percent of pollution, p must be limited to values from 0 to 100. The domain cannot include 100, so the p-interval is [0, 100).

d. The graph of the function on the window [0, 100] by [−1000, 10,000] is shown in Figure 6.40(b).

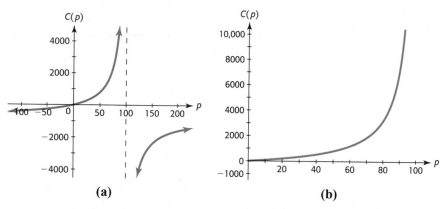

(a)   (b)

Figure 6.40

e. From Figure 6.40(b), we see that the graph is increasing rapidly as it approaches the vertical asymptote at $p = 100$, which tells us that the cost of removing pollution gets extremely high as the amount of pollution removed approaches 100%. It is impossible to remove 100% of the particulate pollution.   ■

We see in Figure 6.41 that the graph of

$$y = \frac{3}{x + 1}$$

approaches the x-axis asymptotically on the left and on the right. Because the x-axis is a horizontal line, we say that the x-axis is a **horizontal asymptote** for this graph.

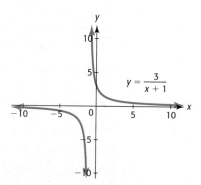

Figure 6.41

We can study the graph of a rational function to determine if the curve approaches a horizontal asymptote. If the graph approaches the horizontal line $y = a$ as $|x|$ gets very large, the graph has a horizontal asymptote at $y = a$. We can denote this as follows.

Horizontal Asymptote

If y approaches a as x approaches $+\infty$ or as x approaches $-\infty$, the graph of $y = f(x)$ has a horizontal asymptote at $y = a$.

We can also compare the degrees of the numerator and denominator of the rational function to determine if the curve approaches a horizontal asymptote, and to find the horizontal asymptote if it exists.

Determining the Horizontal Asymptotes of a Rational Function

Consider the rational function

$$f(x) = \frac{P(x)}{Q(x)} = \frac{a_n x^n + \cdots + a_1 x + a_0}{b_m x^m + \cdots + b_1 x + b_0}, a_n \neq 0 \quad \text{and} \quad b_m \neq 0$$

1. If $n < m$ (that is, if the degree of the numerator is less than the degree of the denominator), a horizontal asymptote occurs at $y = 0$ (the x-axis).

2. If $n = m$ (that is, if the degree of the numerator is equal to the degree of the denominator), a horizontal asymptote occurs at $y = \dfrac{a_n}{b_m}$. (This is the ratio of the leading coefficients.)

3. If $n > m$ (that is, if the degree of the numerator is greater than the degree of the denominator), there is no horizontal asymptote.

In the function $y = \dfrac{3}{x + 1}$, the degree of the numerator is less than the degree of the denominator, so a horizontal asymptote occurs at $y = 0$, the x-axis (Figure 6.41).

In the function $C(p) = \dfrac{800p}{100 - p}$ of Example 2, the degrees of the numerator and the denominator are equal, so the graph has a horizontal asymptote at $y = C(p) = \dfrac{800}{-1} = -800$. Observing the graph of the function, shown in Figure 6.40(a) on the previous page, we see that this horizontal asymptote is reasonable for the graph of $C(p) = \dfrac{800p}{100 - p}$.

Although it is impossible for a graph to cross a vertical asymptote (because the function is undefined at that value of x), the curve *may* cross a horizontal asymptote because the horizontal asymptote describes the end behavior (as x approaches ∞ or x approaches $-\infty$) of a graph. Figure 6.42 on the next page shows the graph of $y = \dfrac{6 - 5x}{x^2}$; it crosses the x-axis but approaches the line $y = 0$ (the x-axis) as x approaches ∞ and as x approaches $-\infty$.

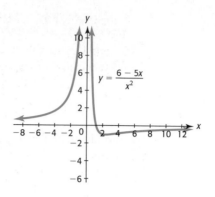

Figure 6.42

example 3   ## Average Cost

The function $\overline{C}(x) = \dfrac{25 + 13x + x^2}{x}$ represents the daily average cost (in $hundred) for the production of Stanley golf carts, with x equal to the number of golf carts produced.

a. Graph the function on the window $[-20, 20]$ by $[-30, 50]$.

b. Does the graph in (a) have a horizontal asymptote?

c. Graph the function on the window $[0, 20]$ by $[0, 50]$.

d. Does the graph of the function using the window in part (a) or part (b) better model the average cost function? Why?

e. Use technology to find the minimum daily average cost and the number of golf carts that gives the minimum daily average cost.

Solution

a. The graph on the window $[-20, 20]$ by $[-30, 50]$ is shown in Figure 6.43(a). Note that the graph has a vertical asymptote at $x = 0$ (the $\overline{C}(x)$-axis).

b. The graph in (a) does not have a horizontal asymptote, because the degree of the numerator is greater than the degree of the denominator.

c. The graph on the window $[0, 20]$ by $[0, 50]$ is shown in Figure 6.43(b).

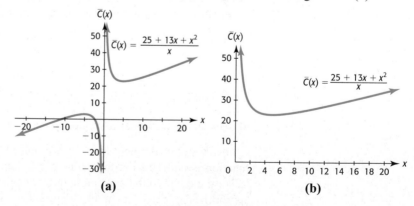

Figure 6.43

d. Because the number of golf carts produced cannot be negative, the window used in part (b) gives a better representation of the graph of the average cost function.

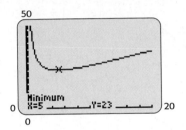

Figure 6.44

e. It appears that the graph reaches a low point somewhere between $x = 4$ and $x = 6$. By using technology, we can find the minimum point. Figure 6.44 shows that the minimum value of y is 23 at $x = 5$. Thus, the minimum daily average cost is $2300 per golf cart when 5 golf carts are produced. ∎

Slant Asymptotes and Missing Points

The average cost function from Example 3,

$$\overline{C}(x) = \frac{25 + 13x + x^2}{x}$$

does not have a horizontal asymptote because the degree of the numerator is greater than the degree of the denominator. However, if the degree of the numerator of a rational function is *one* more than the degree of the denominator, its graph approaches a **slant asymptote**, which we can find by using the following method:

1. Divide the numerator by the denominator, getting a linear function plus a rational expression with a constant numerator (the remainder).

2. Observe that the rational expression will approach 0 as x approaches ∞ or $-\infty$, so the graph of the original function will have a slant asymptote, which is the graph of the linear function.

Thus the average cost function

$$\overline{C}(x) = \frac{25 + 13x + x^2}{x} \quad \text{can be written as} \quad \overline{C}(x) = \frac{25}{x} + 13 + x$$

so its graph approaches the graph of the slant asymptote $y = 13 + x$ (Figure 6.45).

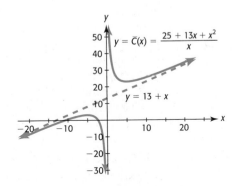

Figure 6.45

There is another type of rational function that is undefined at a value of x but does not have a vertical asymptote at that value. Note that if there is a value a such that $Q(a) = 0$ and $P(a) = 0$, then the function

$$f(x) = \frac{P(x)}{Q(x)}$$

is undefined at $x = a$, but its graph may have a "hole" in it at $x = a$ rather than an asymptote. Figure 6.46 on the next page shows the graph of such a function,

$$y = \frac{x^2 - 4}{x - 2}$$

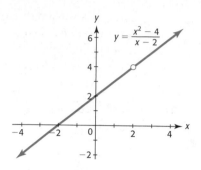

Figure 6.46

Observe that the function is undefined at $x = 2$, but the graph does not have an asymptote at this value of x. Notice that for all values of x except 2,

$$y = \frac{x^2 - 4}{x - 2} = \frac{(x + 2)(x - 2)}{x - 2} = x + 2$$

Thus, the graph of this function looks like the graph of $y = x + 2$, except that it has a missing point (hole) at $x = 2$ (Figure 6.46).

Algebraic and Graphical Solution of Rational Equations

To solve an equation involving rational expressions algebraically, we multiply both sides of the equation by the least common denominator (LCD) of the fractions in the equation. This will result in a linear or polynomial equation which can be solved by using the methods discussed earlier in the text. For these types of equations it is essential that all solutions be checked in the original equation, because some solutions to the resulting polynomial equation may not be solutions to the original equation. In particular, some solutions to the polynomial equation may result in a zero in the denominator of the original equation and thus cannot be solutions to the equation. Such solutions are called **extraneous solutions**.

The steps used to solve a rational equation follow.

Solving a Rational Equation Algebraically

To solve a rational equation algebraically, do the following:

1. Multiply both sides of the equation by the LCD of the fractions in the equation.

2. Solve the resulting polynomial equation for the variable.

3. Check each solution in the original equation. Some solutions to the polynomial equation may not be solutions to the original rational equation (these are called extraneous solutions).

| example 4 |

Solving a Rational Equation

Solve the equation $x^2 + \dfrac{x}{x - 1} = x + \dfrac{x^3}{x - 1}$ for x.

Solution

We first multiply both sides of the equation by $x - 1$, the LCD of the fractions in the equation, and then we solve the resulting polynomial equation.

$$x^2 + \frac{x}{x - 1} = x + \frac{x^3}{x - 1}$$

$$(x - 1)\left(x^2 + \frac{x}{x - 1}\right) = (x - 1)\left(x + \frac{x^3}{x - 1}\right)$$

$$(x - 1)x^2 + (x - 1)\frac{x}{x - 1} = (x - 1)x + (x - 1)\frac{x^3}{x - 1}$$

$$x^3 - x^2 + x = x^2 - x + x^3$$

$$0 = 2x^2 - 2x$$

$$0 = 2x(x - 1)$$

$$x = 0 \quad \text{or} \quad x = 1$$

We must check these answers in the original equation.

  Check $x = 0$:

$$0^2 + \frac{0}{0 - 1} = 0 + \frac{0^3}{0 - 1} \quad \text{checks}$$

  Check $x = 1$:

$$1^2 + \frac{1}{1 - 1} = 1 + \frac{1^3}{1 - 1} \quad \text{or} \quad 1 + \frac{1}{0} = 1 + \frac{1}{0}$$

But $1 + \dfrac{1}{0}$ cannot be evaluated, so $x = 1$ does not check in the original equation. This means that 1 is an extraneous solution of the equation. Thus, the only solution is $x = 0$.   ◼

  As with polynomial equations derived from real data, some equations involving rational expressions may require technology to find or to approximate their solutions. The following example shows how applied problems involving rational functions can be solved algebraically and graphically.

---

example 5   ## Advertising and Sales

Monthly sales y (in thousands of dollars) for Yang products are related to monthly advertising expenses x (in thousands of dollars) according to the function

$$y = \frac{300x}{15 + x}$$

Determine the amount of money that must be spent on advertising to generate $100,000 in sales

a. Algebraically.

b. Graphically.

Solution

a. To solve the equation $100 = \dfrac{300x}{15 + x}$ algebraically, we multiply both sides of the equation by $15 + x$, getting

$$100(15 + x) = 300x$$

Solving this equation gives

$$1500 + 100x = 300x$$
$$1500 = 200x$$
$$x = 7.5$$

Checking this solution in the original equation gives

$$100 = \frac{300(7.5)}{15 + 7.5} \quad \text{or} \quad 100 = \frac{2250}{22.5}$$

so the solution checks. Thus, spending 7.5 thousand dollars ($7500) per month on advertising results in monthly sales of $100,000.

b. To solve the equation graphically, we graph the rational functions $y_1 = \dfrac{300x}{15 + x}$ and $y_2 = 100$ on the same axes on an interval with $x \geq 0$ (Figure 6.47). We then find the point of intersection $(7.5, 100)$, which gives the same solution as in part (a).

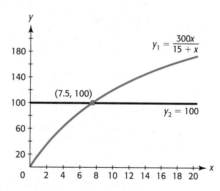

Figure 6.47

Inverse Variation

Recall from Section 2.1 that two variables are directly proportional (or vary directly) if their quotient is constant. When two variables x and y are **inversely proportional** (or vary inversely), an increase in one variable results in a decrease in the other. That is, y is inversely proportional to x (or y varies inversely as x) if there exists a nonzero number k such that

$$y = \frac{k}{x}$$

Also, y is inversely proportional to the nth power of x if there exists a nonzero number k such that

$$y = \frac{k}{x^n}$$

We can also say that y varies inversely as the nth power of x.

example 6	**Illumination**

The illumination produced by a light varies inversely as the square of the distance from the source of the light. If the illumination 30 feet from a light source is 60 candela, what is the illumination 20 feet from the source?

Solution

If L represents the illumination and d represents the distance, the relation is

$$L = \frac{k}{d^2}$$

Substituting for L and d and solving for k gives

$$60 = \frac{k}{30^2} \Rightarrow k = 54{,}000$$

Thus the relation is $L = \dfrac{54{,}000}{d^2}$ and when $d = 20$ feet

$$L = \frac{54{,}000}{20^2} = 135 \text{ candela}$$

So the illumination 20 feet from the source of the light is 135 candela.

skills check

6.5

Give the equations of any (a) vertical and (b) horizontal asymptotes for the graphs of the rational functions $y = f(x)$ in Exercises 1–6.

1. $f(x) = \dfrac{3}{x - 5}$

2. $f(x) = \dfrac{7}{x - 4}$

3. $f(x) = \dfrac{x - 4}{5 - 2x}$

4. $f(x) = \dfrac{2x - 5}{3 - x}$

5. $f(x) = \dfrac{x^3 + 4}{x^2 - 1}$

6. $f(x) = \dfrac{x^2 + 6}{x^2 + 3}$

7. Which of the following functions has a graph that does not have a vertical asymptote? Why?

a. $f(x) = \dfrac{x - 3}{x^2 - 4}$

b. $f(x) = \dfrac{3}{x^2 - 4x}$

c. $f(x) = \dfrac{3x^2 - 6}{x^2 + 6}$

d. $f(x) = \dfrac{x + 4}{(x - 6)(x + 2)}$

8. Which of the following functions has a graph that does not have a horizontal asymptote? Why?

a. $f(x) = \dfrac{2x + 3}{x - 4}$

b. $f(x) = \dfrac{5x}{x^2 - 16}$

c. $f(x) = \dfrac{x^3}{3x^2 + 2}$

d. $f(x) = \dfrac{5}{(x + 2)(x - 3)}$

In Exercises 9–14, match each function with its graph below and on the next page.

9. $y = \dfrac{3}{x - 1}$

10. $y = \dfrac{x + 1}{x - 2}$

11. $y = \dfrac{x - 3}{x^2 - 2x - 8}$

12. $y = \dfrac{3}{x^2 - 4}$

13. $y = \dfrac{-3x^2 + 2x}{x - 3}$

14. $y = \dfrac{2x^2 - 3}{x - 4}$

A.

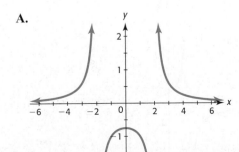

B.

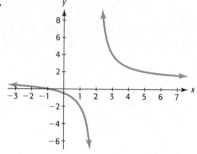

C.

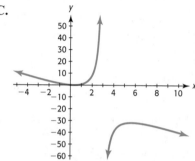

D.

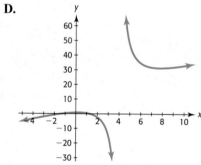

E.

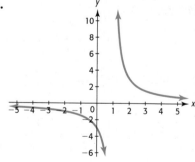

F.

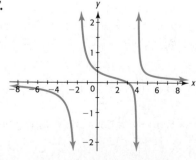

In Exercises 15–18, find (a) the horizontal asymptotes and (b) the vertical asymptotes and (c) sketch a graph of the function.

15. $f(x) = \dfrac{x + 1}{x - 2}$ **16.** $f(x) = \dfrac{5x}{x - 3}$

17. $f(x) = \dfrac{5 + 2x}{1 - x^2}$ **18.** $f(x) = \dfrac{2x^2 + 1}{2 - x}$

19. Graph the function $y = \dfrac{x^2 - 9}{x - 3}$. What happens at $x = 3$?

20. Graph the function $y = \dfrac{x^2 - 16}{x + 4}$. What happens at $x = -4$?

In Exercises 21–24, use graphical methods to find any turning points of the graph of the function.

21. $f(x) = \dfrac{x^2}{x - 1}$ **22.** $f(x) = \dfrac{(x - 1)^2}{x}$

23. $g(x) = \dfrac{x^2 + 4}{x}$ **24.** $g(x) = \dfrac{x}{x^2 + 1}$

25. a. Graph $y = \dfrac{1 - x^2}{x - 2}$ on the window $[-10, 10]$ by $[-10, 10]$.

   b. Use the graph to find y when $x = 1$ and when $x = 3$.

   c. Use the graph to find the value(s) of x that give $y = -7.5$.

   d. Use algebraic methods to solve $-7.5 = \dfrac{1 - x^2}{x - 2}$.

26. a. Graph $y = \dfrac{2 + 4x}{x^2 + 1}$ on the window $[-10, 10]$ by $[-5, 5]$.

   b. Use the graph to find y when $x = -3$ and when $x = 3$.

   c. Use the graph to find the value(s) of x that give $y = 2$.

   d. Use algebraic methods to solve $2 = \dfrac{2 + 4x}{x^2 + 1}$.

27. a. Graph $y = \dfrac{3 - 2x}{x}$ on the window $[-10, 10]$ by $[-10, 10]$.

b. Use the graph to find y when $x = -3$ and when $x = 3$.

c. Use the graph to find the value(s) of x that give $y = -5$.

d. Use algebraic methods to solve $-5 = \dfrac{3 - 2x}{x}$.

28. a. Graph $y = \dfrac{x^2}{(x + 1)^2}$ on the window $[-10, 10]$ by $[-6, 6]$.

b. Use the graph to find y when $x = 0$ and when $x = -2$.

c. Use the graph to find the value(s) of x that give $y = \dfrac{1}{4}$.

d. Use algebraic methods to solve $\dfrac{1}{4} = \dfrac{x^2}{(x + 1)^2}$.

29. Use algebraic methods to solve
$$\dfrac{x^2 + 1}{x - 1} + x = 2 + \dfrac{2}{x - 1}.$$

30. Use algebraic methods to solve
$$\dfrac{x}{x - 2} - x = 1 + \dfrac{2}{x - 2}.$$

31. If y varies inversely as the 4th power of x and $y = 5$ when $x = -1$, what is y when $x = 0.5$?

32. If S varies inversely as the square root of T and $S = 4$ when $T = 4$, what is S when $T = 16$?

exercises
6.5

33. *Average Cost* The average cost per unit for the production of a certain brand of DVD player is given by

$$\overline{C} = \dfrac{400 + 50x + 0.01x^2}{x}$$

where x is the number of units produced.

a. What is the average cost per unit when 500 units are produced?

b. What is the average cost per unit when 60 units are produced?

c. What is the average cost per unit when 100 units are produced?

d. Is it reasonable to say that the average cost continues to fall as the number of units produced rises?

34. *Average Cost* The average cost per set for the production of a certain brand of television sets is given by

$$\overline{C}(x) = \dfrac{1000 + 30x + 0.1x^2}{x}$$

where x is the number of units produced.

a. What is the average cost per set when 30 sets are produced?

b. What is the average cost per set when 300 sets are produced?

c. What happens to the function when $x = 0$? What does this tell you about the average cost when 0 units are produced?

35. *Advertising and Sales* The monthly sales volume y (in thousands of dollars) of a product is related to monthly advertising expenditures x (in thousands of dollars) according to the equation

$$y = \dfrac{400x}{x + 20}$$

a. What monthly sales will result if $5000 is spent monthly on advertising?

b. What value of x makes the denominator of this function 0? Will this ever happen in the context of this problem?

36. *Productivity* As an 8-hour shift progresses, the rate at which workers produce picture frames, in units per hour, changes during the day according to the equation

$$f(t) = \dfrac{100(t^2 + 3t)}{(t^2 + 3t + 12)^2} \qquad (0 \le t \le 8)$$

where t is the number of hours after the beginning of the shift.

a. Graph this function using the window $[-10, 10]$ by $[-5, 5]$.

b. Graph this function using the window $[0, 8]$ by $[0, 5]$.

c. Does the graph in part (a) or part (b) better represent the rate of change of productivity function? Why?

d. Is the rate of productivity higher near lunch $(t = 4)$ or near quitting time $(t = 8)$?

37. *Average Cost* The average cost per set for the production of television sets is given by

$$\overline{C}(x) = \frac{1000 + 30x + 0.1x^2}{x}$$

where x is the number of hundreds of units produced.

a. Graph this function using the window $[-10, 10]$ by $[-200, 300]$.

b. Graph this function using the window $[0, 50]$ by $[0, 300]$.

c. Which window makes sense for this application?

d. Use the graph with the window $[0, 200]$ by $[0, 300]$ to find the minimum average cost, and the number of units that gives the minimum average cost.

38. *Average Cost* The average cost per unit for the production of a certain brand of DVD players is given by

$$\overline{C}(x) = \frac{400 + 50x + 0.01x^2}{x}$$

where x is the number of hundreds of units produced.

a. Graph this function using the window $[-100, 100]$ by $[-100, 400]$.

b. Graph this function using the window $[0, 300]$ by $[0, 400]$.

c. Which window is more appropriate for this problem?

d. Use the graph to find the minimum average cost and the number of units that gives the minimum average cost.

39. *Population* Suppose the number of employees of a start-up company is given by

$$f(t) = \frac{30 + 40t}{5 + 2t}$$

where t is the number of months after the company is organized.

a. Graph this function using the window $[0, 20]$ by $[0, 20]$.

b. Use the graph to find $f(0)$. What does this represent?

c. Use the graph to find $f(12)$. What does this represent? If $f(t)$ represents a number of individuals, how should you report your answer?

40. *Drug Concentration* Suppose the concentration of a drug (as percent) in a patient's bloodstream t hours after injection is given by

$$C(t) = \frac{200t}{2t^2 + 32}$$

a. Graph the function using the window $[0, 20]$ by $[0, 20]$.

b. What is the drug concentration 1 hour after injection? 5 hours?

c. What is the highest percent concentration? In how many hours will it occur?

d. Describe how the end behavior of the graph of this function relates to the drug concentration.

41. *Cost-Benefit* Suppose the cost C of removing $p\%$ of the impurities from the waste water in a manufacturing process is given by

$$C(p) = \frac{3600p}{100 - p}$$

a. Where does the graph of this function have a vertical asymptote?

b. What does this tell us about removing the impurities from this process?

42. *Cost-Benefit* The percent p of particulate pollution that can be removed from the smokestacks of an industrial plant by spending C dollars is given by

$$p = \frac{100C}{8300 + C}$$

a. Find the percent of pollution that could be removed if spending were allowed to increase without bound.

b. Can 100% of the pollution be removed? Explain.

43. *Sales Volume* Suppose the weekly sales volume (in thousands of units) for a product is given by

$$V = \frac{640}{(p + 2)^2}$$

where p is the price in dollars per unit.

a. Graph this function on the p-interval $[-10, 10]$. Does the graph of this function have a vertical asymptote on this interval? Where?

b. Complete the table below.

c. What values of p give a weekly sales volume that makes sense? Does the graph of this function have a vertical asymptote on this interval?

d. What is the horizontal asymptote of the graph of the weekly sales volume? Explain what this means.

Price/Unit ($)	5	20	50	100	200	500
Weekly Sales Volume						

44. *Population* Suppose the number of employees of a start-up company is given by

$$N = \frac{30 + 40t}{5 + 2t}$$

where t is the number of months after the company is organized.

a. Does the graph of this function have a vertical asymptote on $[-10, 10]$? Where?

b. Does the graph of this function have a vertical asymptote for values of t in this application?

c. Does the graph of this function have a horizontal asymptote? Where?

d. What is the maximum number of employees predicted by this model?

45. *Demand* The quantity of a product demanded by consumers is defined by the function

$$p = \frac{100,000}{(q + 1)^2}$$

where p is the price and q is the quantity demanded.

a. Graph this function using the window $[0, 100]$ by $[0, 1000]$.

b. What is the horizontal asymptote that this graph approaches?

c. Use the graph to explain what happens to quantity demanded as price becomes lower.

46. *Advertising and Sales* Weekly sales y (in hundreds of dollars) are related to weekly advertising expenses x (in hundreds of dollars) according to the equation

$$y = \frac{800x}{20 + 5x}$$

a. Graph this function using the window $[0, 100]$ by $[0, 300]$.

b. Find the horizontal asymptote for this function.

c. Complete the table below.

d. Use your results from parts (a) and (b) to determine the maximum weekly sales even if an unlimited amount of money is spent on advertising.

Weekly Expenses ($)	0	50	100	200	300	500
Weekly Sales						

47. *Sales and Training* During the first 3 months of employment, the monthly sales S (in thousands of dollars) for an average new salesperson depends on the number of hours of training x, according to

$$S = \frac{40}{x} + \frac{x}{4} + 10 \quad \text{for} \quad x \ge 4$$

a. Combine the terms of this function over a common denominator to create a rational function.

b. How many hours of training should result in monthly sales of $21,000?

48. *Sales and Training* The average monthly sales volume (in thousands of dollars) for a company depends on the number of hours of training x of its sales staff, according to

$$S(x) = \frac{20}{x} + 40 + \frac{x}{2} \quad \text{for} \quad 4 \le x \le 120$$

a. Graph this function.

b. How many hours of training will give average monthly sales of $51,000?

49. *Worker Productivity* Suppose the average time (in hours) that a new production team takes to assemble 1 unit of a product is given by

$$H = \frac{5 + 3t}{2t + 1}$$

where t is the number of days of training for the team.

  a. Graph this function using the window [0, 20] by [0, 8].

  b. What is the horizontal asymptote? What does this mean in this application?

  c. Use a graphical or numerical solution method to find the number of days of training necessary to reduce the production time to 1.6 hours.

50. *Advertising and Sales* Weekly sales y (in hundreds of dollars) are related to weekly advertising expenses x (in hundreds of dollars) according to the equation

$$y = \frac{800x}{20 + 5x}$$

Use graphical or numerical methods to find the amount of weekly advertising expenses that will result in weekly sales of $14,000, according to this model.

51. *Age at First Marriage* The table below shows the U.S. median age at first marriage for men and women for selected years 1900–2004.

Median Age at First Marriage

Year	Men	Women
1900	25.9	21.9
1910	25.1	21.6
1920	24.6	21.2
1930	24.3	21.3
1940	24.3	21.5
1950	22.8	20.3
1960	22.8	20.3
1970	23.2	20.8
1980	24.7	22.0
1990	26.1	23.9
2000	26.8	25.1
2004	27.1	25.8

(Source: U.S. Census Bureau)

Using this data, the ratio of age at first marriage for women to the age at first marriage for men can be modeled by

$$y = \frac{0.00126x^2 - 0.101x + 22.5}{0.00127x^2 - 0.123x + 26.293}$$

where x is the number of years from 1900.

  a. Create a scatter plot with x representing the years from 1900 and y representing the ratio of age at first marriage for women to the age at first marriage for men for those years.

  b. Graph the model above on the same axis with the scatterplot in (a).

  c. Is the fit poor, good, or perfect?

52. *Fences* Suppose a rectangular field is to have an area of 51,200 square feet, and that it needs to be enclosed by fence on three of its four sides (see figure). What is the minimum length of fence needed? What dimensions should this field have to minimize the amount of fence needed? To solve this problem,

  a. Write an equation that describes the area of the proposed field, with x representing the length and y representing the width of the field.

  b. Write an equation that gives the length of fence L as a function of the dimensions of the field, remembering that only three sides are fenced in.

  c. Use the equation from part (a) to write the length L as a function of one of the dimensions.

  d. Graph the function L on a window that applies to the context of the problem: [0, 400] by [0, 1000].

  e. Use graphical or numerical methods to find the minimum length of fence needed and the dimensions required to give this length.

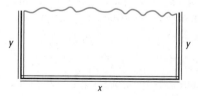

Polynomial and Rational Inequalities

section preview ▪ Average Cost

The average cost per set for the production of 42-inch plasma televisions is given by

$$\overline{C}(x) = \frac{5000 + 80x + x^2}{x}$$

where x is the number of hundreds of units produced. To find the number of televisions that must be produced to keep the average cost to at most \$590 per TV, we solve the **rational inequality**

$$\frac{5000 + 80x + x^2}{x} \le 590$$

(See Example 2.) In this section, we use algebraic and graphical methods similar to those used in Section 4.4 to solve inequalities involving polynomial functions and rational functions.

Polynomial Inequalities

The quadratic inequalities studied in Section 4.4 are examples of polynomial inequalities. Other examples of polynomial inequalities are

$$3x^3 + 3x^2 - 4x \ge 2, \qquad x - 3 < 8x^4, \qquad x \ge (x - 2)^3$$

We can solve the inequality $144x - 48x^2 + 4x^3 > 0$ with a graphing utility. The graph of the function $y = 144x - 48x^2 + 4x^3$ is shown in Figure 6.48.

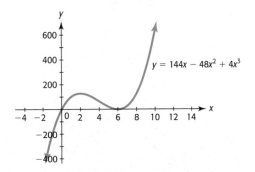

Figure 6.48

We can determine that the x-intercepts of the graph are $x = 0$ and $x = 6$. The graph touches but does not cross the x-axis at $x = 6$, and the graph of the function is above the x-axis for values of $x > 0$ and $x \ne 6$. The solution to $144x - 48x^2 + 4x^3 > 0$ is the set of all x that give positive values for this function: $0 < x < 6$ and $x > 6$; that is, for x in the intervals $(0, 6)$ and $(6, \infty)$.

499

The steps used to solve polynomial inequalities algebraically are similar to those used to solve quadratic inequalities.

> ## Solving Polynomial Inequalities
>
> To solve a polynomial inequality algebraically:
> 1. Write an equivalent inequality with 0 on one side and with the function $f(x)$ on the other side.
> 2. Solve $f(x) = 0$.
> 3. Create a sign diagram that uses the solutions from Step 2 to divide the number line into intervals. Pick a test value in each interval to determine whether $f(x)$ is positive or negative in that interval.*
> 4. Identify the intervals that satisfy the inequality in Step 1. The values of x that define these intervals are the solutions to the original inequality.

* The numerical feature of your graphing utility can be used to test the x-values.

We illustrate the use of this method in the following example.

| example 1 |

Constructing a Box

A box can be formed by cutting squares out of each corner of a piece of tin and folding the "tabs" up. If the piece of tin is 12 inches by 12 inches and each side of the square that is cut out has length x, the function that gives the volume of the box is $V(x) = 144x - 48x^2 + 4x^3$.

a. Use factoring and then find the values of x that give positive values for $V(x)$.

b. Which of the values of x that give positive values for $V(x)$ result in a box?

Solution

a. We seek those values of x that give positive values for $V(x)$—that is, for which

$$144x - 48x^2 + 4x^3 > 0$$

Writing the related equation $V(x) = 0$ and solving for x gives

$$144x - 48x^2 + 4x^3 = 0$$
$$4x(x^2 - 12x + 36) = 0$$
$$4x(x - 6)(x - 6) = 0$$
$$x = 0 \quad \text{or} \quad x = 6$$

These two values of x, 0 and 6, divide the number line into three intervals. Testing a value in each of the intervals (Figure 6.49(a)) determines the intervals where $V(x) > 0$ and thus determines the solutions to the inequality. The values of x that give positive values for V satisfy $0 < x < 6$ and $x > 6$. We found the same solutions to this inequality using graphical methods (see Figure 6.48) on the previous page.

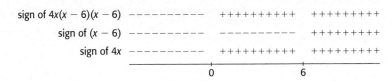

Figure 6.49(a)

The interval notation for this solution set is $(0, 6) \cup (6, \infty)$, and the graph of the solution set is shown in Figure 6.49(b).

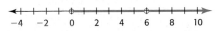

Figure 6.49(b)

b. The box is formed by cutting 4 squares of length x inches from a piece of tin that is 12 inches by 12 inches, so it is impossible to cut squares longer than 6 inches. Thus, the solution $x > 6$ does not apply to the physical building of the box, and so the values of x that result in a box satisfy $0 < x < 6$. ■

Rational Inequalities

To solve a **rational inequality** using an algebraic method, we first get 0 on the right side of the inequality; then, if necessary, we get a common denominator and combine the rational expressions on the left side. We find the numbers that make the numerator of the rational expression equal to zero and those that make the denominator equal to zero. These numbers are used to create a sign diagram much like we used when solving polynomial inequalities. Note that we should avoid multiplying both sides of an inequality by any term containing a variable, because we cannot easily determine when the term is positive or negative.

| example 2 |

Average Cost

The average cost per set for the production of 42-inch plasma televisions is given by

$$\overline{C}(x) = \frac{5000 + 80x + x^2}{x}$$

where x is the number of hundreds of units produced. Find the number of televisions that must be produced to keep the average cost to at most $590 per TV.

Solution

To find the number of televisions that must be produced to keep the average cost to at most $590 per TV, we solve the rational inequality

$$\frac{5000 + 80x + x^2}{x} \leq 590$$

To solve this inequality with algebraic methods, we rewrite the inequality with 0 on the right side of the inequality, getting

$$\frac{5000 + 80x + x^2}{x} - 590 \leq 0$$

Combining the terms over a common denominator and factoring give

$$\frac{5000 + 80x + x^2 - 590x}{x} \leq 0$$

$$\frac{x^2 - 510x + 5000}{x} \leq 0$$

$$\frac{(x - 10)(x - 500)}{x} \leq 0$$

The values of x that make the numerator equal to 0 are 10 and 500, and the value $x = 0$ makes the denominator equal to 0. We cannot have a negative number of units, so the average cost is defined only for values of $x > 0$. The values 0, 10, and 500 divide a number line into three intervals. Next we find the sign of $f(x) = \dfrac{(x - 10)(x - 500)}{x}$ in each interval. Testing a value in each interval determines which interval(s) satisfy the inequality. We do this by using the following sign diagram (Figure 6.50).

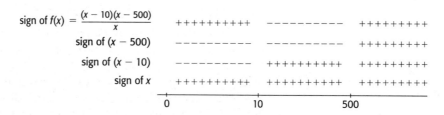

Figure 6.50

The function $f(x)$ is negative on $10 < x < 500$, so the solution to

$$\frac{5000 + 80x + x^2}{x} - 590 \leq 0,$$

and thus to the original inequality, is $10 \leq x \leq 500$.

Because x represents the number of hundreds, between 1000 and 50,000 televisions must be produced to keep the average cost to at most \$590 per TV. ■

We can also use graphical methods to solve rational inequalities. For example, we can solve the inequality

$$\frac{5000 + 80x + x^2}{x} \leq 590$$

for $x > 0$ by graphing

$$y_1 = \frac{5000 + 80x + x^2}{x} \quad \text{and} \quad y_2 = 590$$

on the same axes. We are only interested in the graph for positive values of x, so we restrict our graph to the first quadrant. The points of intersection can be found by the intersection method (Figure 6.51). The inequality is satisfied by the x-interval for which the graph of y_1 is below (or on) the graph of y_2, that is, for $10 \leq x \leq 500$.

Recall that this is the solution that we found using algebraic methods in Example 2.

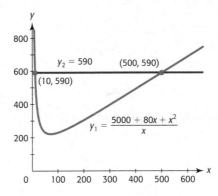

Figure 6.51

In Exercises 1–16, use algebraic and/or graphical methods to solve the inequalities.

1. $16x^2 - x^4 \geq 0$ **2.** $x^4 - 4x^2 \leq 0$

3. $2x^3 - x^4 < 0$ **4.** $3x^3 \geq x^4$

5. $(x - 1)(x - 3)(x + 1) \geq 0$

6. $(x - 3)^2(x + 1) < 0$

7. $\dfrac{4 - 2x}{x} > 2$ **8.** $\dfrac{x - 3}{x + 1} \geq 3$

9. $\dfrac{x}{2} + \dfrac{x - 2}{x + 1} \leq 1$ **10.** $\dfrac{x}{x - 1} \leq 2x + \dfrac{1}{x - 1}$

11. $(x - 1)^3 > 27$ **12.** $(2x + 3)^3 \leq 8$

13. $(x - 1)^3 < 64$ **14.** $(x + 4)^3 - 125 \geq 0$

15. $-x^3 - 10x^2 - 25x \leq 0$

16. $x^3 + 10x^2 + 25x < 0$

For Exercises 17–20, use the graph of $y = f(x)$ to solve the requested inequality.

17.

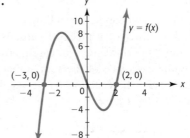

    a. Solve $f(x) < 0$.

    b. Solve $f(x) \geq 0$.

18.

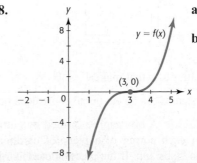

    a. Solve $f(x) < 0$.

    b. Solve $f(x) \geq 0$.

19. 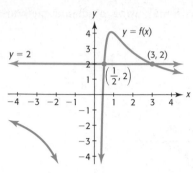 Solve $f(x) \geq 2$.

20. 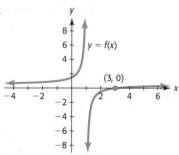 **a.** Solve $f(x) < 0$.

b. Solve $f(x) \geq 0$.

exercises

6.6

Combine factoring with graphical and/or numerical methods to solve Exercises 21–26.

21. *Revenue* The revenue from the sale of a product is given by the function $R = 400x - x^3$. Selling how many units will give positive revenue?

22. *Revenue* The price for a product is given by $p = 1000 - 0.1x^2$, where x is the number of units sold.

   a. Form the revenue function for the product.

   b. Selling how many units gives positive revenue?

23. *Constructing a Box* A box can be formed by cutting squares out of each corner of a piece of cardboard and folding the "tabs" up. If the piece of cardboard is 36 cm by 36 cm and each side of the square that is cut out has length x cm, the function that gives the volume of the box is $V = 1296x - 144x^2 + 4x^3$.

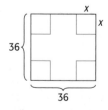

   a. Find the values of x that make $V > 0$.

   b. What size squares can be cut out to construct a box?

24. *Constructing a Box* A box can be formed by cutting squares out of each corner of a piece of cardboard and folding the "tabs" up. If the piece of cardboard is 12 inches by 16 inches and each side of the square that

is cut out has length x inches, the function that gives the volume of the box is $V = 192x - 56x^2 + 4x^3$. What size squares can be cut out to construct a box?

25. *Cost* The total cost function for a product is given by $C(x) = 3x^3 - 6x^2 - 300x + 1800$, where x is the number of units produced and C is the cost in hundreds of dollars. Use factoring by grouping and then find the number of units that will give a total cost of at least \$120,000. Verify your conclusion with a graphing utility.

26. *Profit* The profit function for a product is given by $P(x) = -x^3 + 2x^2 + 400x - 400$, where x is the number of units produced and sold and P is in hundreds of dollars. Use factoring by grouping to find the number of units that will give a profit of at least \$40,000. Verify your conclusion with a graphing utility.

27. *Advertising and Sales* The monthly sales volume y (in thousands of dollars) is related to monthly advertising expenditures x (in thousands of dollars) according to the equation

$$y = \frac{400x}{x + 20}$$

Spending how much money on advertising will result in sales of at least \$200,000 per month?

28. *Average Cost* The average cost per set for the production of a portable stereo system is given by

$$\overline{C} = \frac{100 + 30x + 0.1x^2}{x}$$

where x is the number of hundreds of units produced. What number of units can be produced while keeping the average cost to at most \$41?

29. *Future Value* The future value of $2000 invested for 3 years at rate r, compounded annually, is given by $S = 2000(1 + r)^3$. Find the rate r that gives a future value from $2662 to $3456, inclusive.

30. *Future Value* The future value of $5000 invested for 4 years at rate r, compounded annually, is given by $S = 5000(1 + r)^4$. Find the rate r that gives a future value of at least $6553.98.

31. *Revenue* The revenue from the sale of a product is given by the function $R = 4000x - 0.1x^3$ dollars. Use graphical or numerical methods to determine how many units should be sold to give a revenue of at least $39,990.

32. *Revenue* The price for a product is given by $p = 1000 - 0.1x^2$, where x is the number of units sold.

   a. Form the revenue function for the product.

   b. Use graphical or numerical methods to determine how many units should be sold to give a revenue of at most $37,500.

33. *Supply and Demand* Suppose the supply function for a product is given by $6p - q = 180$ and the demand is given by $(p + 20)q = 30,000$. Over

what meaningful price interval does supply exceed demand?

34. *Drug Concentration* A pharmaceutical company claims that the concentration of a drug in a patient's bloodstream will be at least 10% for 8 hours. Suppose clinical tests show that the concentration of a drug (as percent) t hours after injection is given by

$$C(t) = \frac{200t}{2t^2 + 32}$$

During what time period is the concentration at least 10%? Is the company's claim supported by the evidence?

35. *Population* Suppose the number of employees of a start-up company is given by

$$f(t) = \frac{30 + 40t}{5 + 2t}$$

where t is the number of months after the company is organized.

   a. For what values of t is $f(t) \leq 18$?

   b. During what months is the number of employees below 18?

chapter
6
Summary

In this chapter, we studied higher-degree polynomial functions, their graphs, and the solution of polynomial equations by factoring, by the root method, by graphical methods, and by synthetic division. We also studied rational functions, their graphs, vertical asymptotes, horizontal asymptotes, slant asymptotes, and the solution of equations and inequalities containing rational functions.

Key Concepts and Formulas

6.1 Higher-Degree Polynomial Functions

Polynomial function	A polynomial function of degree n with independent variable x is a function in the form $P(x) = a_n x^n + a_{n-1} x^{n-1} + \cdots + a_1 x + a_0$ with $a_n \neq 0$. Each a_i represents a real number and each n is a positive integer.
Cubic function	A polynomial function of the form $f(x) = ax^3 + bx^2 + cx + d$, with $a \neq 0$.
Quartic function	A polynomial function of the form $f(x) = ax^4 + bx^3 + cx^2 + dx + e$, with $a \neq 0$.
Extrema points	"Turning points" on the graph of a polynomial function: local minimum, local maximum, absolute minimum, absolute maximum.

Graphs of cubic functions	End behavior: one end opening up and one end opening down.

	Turning points: 0 or 2
Graphs of quartic functions	End behavior: Both ends opening up or both ends opening down.
	Turning points: 1 or 3

6.2 Modeling with Cubic and Quartic Functions

Cubic models	If a scatter plot of data indicates that the data may fit close to one of the possible shapes of the graph of a cubic function, it is possible to find a cubic function that is an approximate model for the data.
Quartic models	If a scatter plot of data indicates that the data may fit close to one of the possible shapes of the graph of a quartic function, it is possible to find a quartic function that is an approximate model for the data.
Third differences	For equally spaced inputs, the third differences are constant for data fitting cubic functions exactly.
Fourth differences	For equally spaced inputs, the fourth differences are constant for data fitting quartic functions exactly.

6.3 Solution of Polynomial Equations

Solving polynomial equations	
By factoring	Some polynomial equations can be solved by factoring and using the zero-product property.
By the root method	The real solutions of the equation $x^n = C$ are $x = \pm \sqrt[n]{C}$ if n is even and $C \geq 0$, or $x = \sqrt[n]{C}$ if n is odd.
Factors, zeros, intercepts, and solutions	If $(x - a)$ is a factor of $f(x)$, then a is an x-intercept of the graph of $y = f(x)$, a is a solution to the equation $f(x) = 0$, and a is a zero of the function f.
Approximating solutions	Solving equations derived from real data requires graphical or numerical methods using technology.

6.4 Polynomial Equations Continued; Fundamental Theorem of Algebra

Division of polynomials	If $f(x)$ is a function and we know that $x = a$ is a solution of $f(x) = 0$, then we can write $x - a$ as a factor of $f(x)$. If we divide this factor into the function, the quotient will be another factor of $f(x)$.
Synthetic division	To divide a polynomial by a binomial of the form $x - a$ (or $x + a$), we can use synthetic division to simplify the division process by omitting the variables and writing only the coefficients.

Solving cubic equations	If we know one exact solution of a cubic equation $P(x) = 0$, we can divide by the corresponding factor using synthetic division, obtaining a second quadratic factor that can be used to find any additional solutions of the equation.
Combining graphical and algebraic methods	To find exact solutions to a polynomial equation when one solution can be found graphically, we can use that solution and synthetic division to find a second factor.
Rational solutions test	The rational solutions of the polynomial equation $a_nx^n + a_{n-1}x^{n-1} + \cdots + a_1x + a_0 = 0$ with integer coefficients must be of the form $\dfrac{p}{q}$, where p is a factor of the constant term a_0 and q is a factor of the leading coefficient a_n.
Fundamental Theorem of Algebra	If $f(x)$ is a polynomial function of degree $n \geq 1$, then f has at least one complex zero.
Complex solutions of polynomial equations	Every polynomial function $f(x)$ of degree $n \geq 1$ has exactly n complex zeros. Some of these solutions may be imaginary, and some may be repeated.

6.5 Rational Functions and Rational Equations

Rational function	The function f is a rational function if $f(x) = \dfrac{P(x)}{Q(x)}$ where $P(x)$ and $Q(x)$ are polynomials and $Q(x) \neq 0$.
Graphs of rational functions	The graph of a rational function may have vertical asymptotes and/or a horizontal asymptote. The graph may have a "hole" at $x = a$ rather than a vertical asymptote.
Vertical asymptotes	A vertical asymptote occurs in the graph of $f(x) = \dfrac{P(x)}{Q(x)}$ at those values of x where $Q(x) = 0$ and $P(x) \neq 0$; that is, where the denominator equals 0 but the numerator does not equal 0.
Horizontal asymptotes	Consider the rational function $$f(x) = \frac{P(x)}{Q(x)} = \frac{a_nx^n + \cdots + a_1x + a_0}{b_mx^m + \cdots + b_1x + b_0}$$ 1. If $n < m$, a horizontal asymptote occurs at $y = 0$ (the x-axis). 2. If $n = m$, a horizontal asymptote occurs at $y = \dfrac{a_n}{b_m}$. 3. If $n > m$, there is no horizontal asymptote.
Slant asymptote	If the degree of the numerator of a rational function is *one* more than the degree of the denominator, its graph approaches a slant asymptote. The equation of the slant asymptote can be found by dividing the numerator by the denominator, getting a linear function plus a rational expression with a constant numerator. The graph of the original function will have a slant asymptote, which is the graph of the linear function.
Missing point (hole)	If there is a value a such that $Q(a) = 0$ and $P(a) = 0$, then the graph of the function is undefined at $x = a$ but may have a "hole" at $x = a$ rather than an asymptote.

Solution of rational equations	
Algebraically	Multiplying both sides of the equation by the LCD of any fractions in the equation converts the equation to a polynomial equation that can be solved. All solutions must be checked in the original equation because some may be extraneous.
Graphically	Solutions to rational equations can be found or approximated by graphical methods.

6.6 Polynomial and Rational Inequalities

Polynomial inequalities	The steps used to solve polynomial inequalities graphically are similar to those used to solve quadratic inequalities. To solve a polynomial inequality algebraically, solve the corresponding polynomial equation and use a sign diagram to complete the solution.
Rational inequalities	To solve a rational inequality, get 0 on the right side of the inequality; then, if necessary, get a common denominator and combine the rational expressions on the left side. Find the numbers that make the numerator of the rational expression equal to 0 and those that make the denominator equal to 0. Use these numbers to create a sign diagram and complete the solution. Graphical methods can be used to find or to verify the solutions.

chapter

6

Skills Check

1. What is the degree of the polynomial $5x^4 - 2x^3 + 4x^2 + 5$?

2. What is the type of the polynomial function $y = 5x^4 - 2x^3 + 4x^2 + 5$?

3. Graph $y = -4x^3 + 4x^2 + 1$ using the window $[-10, 10]$ by $[-10, 10]$. Is the graph complete on this window?

4. Graph the function $y = x^4 - 4x^2 - 20$ using

   a. The window $[-8, 8]$ by $[-8, 8]$.

   b. A window that gives a complete graph.

5. a. Graph $y = x^3 - 3x^2 - 4$ using a window that shows a local maximum and local minimum.

   b. A local maximum occurs at what point?

   c. A local minimum occurs at what point?

6. a. Graph $y = x^3 + 11x^2 - 16x + 40$ using the window $[-10, 10]$ by $[0, 100]$.

   b. Is the graph complete?

   c. Sketch a complete graph of this function.

   d. Find the points where a local maximum and a local minimum occur.

7. Solve $x^3 - 16x = 0$.

8. Solve $2x^4 - 8x^2 = 0$.

9. Solve $x^4 - x^3 - 20x^2 = 0$.

10. Solve $0 = x^3 - 15x^2 + 56x$ by factoring.

11. Use factoring by grouping to solve $4x^3 - 20x^2 - 4x + 20 = 0$.

12. Solve $12x^3 - 9x^2 - 48x + 36 = 0$ using factoring by grouping.

13. Use technology to find the solutions to $x^4 - 3x^3 - 3x^2 + 7x + 6 = 0$.

14. Use technology to solve $6x^3 - 59x^2 - 161x + 60 = 0$.

15. Use the root method to solve $(x - 4)^3 = 8$.

16. Use the root method to solve $5(x - 3)^4 = 80$.

17. Use synthetic division to divide $4x^4 - 3x^3 + 2x - 8$ by $x - 2$.

18. Use synthetic division to solve $2x^3 + 5x^2 - 11x + 4 = 0$, given that $x = 1$ is one solution.

19. Find one solution of $3x^3 - x^2 - 12x + 4 = 0$ graphically, and use synthetic division to find any additional solutions.

20. Find one solution of $2x^3 + 5x^2 - 4x - 3 = 0$ graphically, and use synthetic division to find any additional solutions.

21. For the function $y = \dfrac{1 - x^2}{x + 2}$,

    a. Find the x-intercepts and y-intercept if they exist.

    b. Find any horizontal asymptotes and vertical asymptotes that exist.

    c. Sketch the graph of the function.

22. For the function $y = \dfrac{3x - 2}{x - 3}$,

    a. Find the x-intercepts and y-intercept if they exist.

    b. Find any horizontal asymptotes and vertical asymptotes that exist.

    c. Sketch the graph of the function.

23. Use a graph and technology to find any local maxima and/or minima of the function $y = \dfrac{x^2}{x - 4}$.

24. Use a graph and technology to find any local maxima and/or minima of the function $y = \dfrac{x^2 + 3}{1 - x}$.

25. a. Graph $y = \dfrac{1 + 2x^2}{x + 2}$ using the window $[-10, 10]$ by $[-30, 10]$.

    b. Use the graph to find y when $x = 1$ and when $x = 3$.

    c. Use the graph to find the value(s) of x that give $y = \dfrac{9}{4}$.

    d. Use an algebraic method to solve $\dfrac{9}{4} = \dfrac{1 + 2x^2}{x + 2}$.

26. Solve $x^4 - 13x^2 + 36 = 0$.

In Exercises 27 and 28, find the exact solutions to $f(x) = 0$ in the complex number system.

27. $x^3 + x^2 + 2x - 4 = 0$

28. $4x^3 + 10x^2 + 5x + 2 = 0$

29. Solve $x^3 - 5x^2 \geq 0$.

30. Use a graphing utility to solve $x^3 - 5x^2 + 2x + 8 \geq 0$.

31. Solve $2 < \dfrac{4x - 6}{x}$.

32. Solve $\dfrac{5x - 10}{x + 1} \geq 20$.

chapter

6   Review

33. *Revenue* The monthly revenue for a product is given by $R = -0.1x^3 + 15x^2 - 25x$, where x is the number of thousands of units sold.

    a. Graph this function using the window $[-100, 200]$ by $[-2000, 60,000]$.

    b. Graph the function on a window that makes sense for the problem; that is, with nonnegative x and R.

    c. What is the revenue when 50,000 units are produced?

34. *Revenue* The monthly revenue for a product is given by $R = -0.1x^3 + 13.5x^2 - 150x$, where x is the number of thousands of units sold. Use technology to find the number of units that gives the maximum possible revenue.

35. *Investment* If $5000 is invested for 6 years at interest rate r (as a decimal) compounded annually, the future value of the investment is given by $S = 5000(1 + r)^6$ dollars.

   a. Find the future value of this investment for selected interest rates by completing the following table.

   b. Graph this function for $0 \le r \le 0.20$.

r (%)	S (future value, $)
1	
5	
10	
15	

   c. Compute the future value if $r = 10\%$ and $r = 20\%$. How much more money is earned at 20%?

36. *Immigration* The percent of the U.S. population that is foreign born is given by the graph below. The data can be modeled by the function $P = 0.0000506x^3 - 0.00565x^2 + 0.0178x + 14.304$, where x is the number of years from 1900.

   a. Graph this function on the window [0, 110] by [0, 16], along with the data.

   b. What does the model give as the percent in 2000? How does this compare with the data?

   c. Use graphical or numerical methods to find the years after 1900 during which the percent of the U.S. population that is foreign born is 7.6%, according to the model.

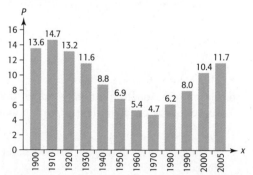

37. *United Nations Debt* The cubic function $y = -0.532x^3 - 1.003x^2 + 41.964x + 401.708$ models the total debt to the United Nations, in millions of dollars, as a function of the number of years from 1990.

   a. Graph this function using the window [0, 10] by [0, 600].

   b. According to the model, what was the approximate debt to the United Nations in 1998?

   c. Use graphical methods to estimate the maximum debt over the period 1990–2000.
     (Source: United Nations, Institute for Global Communication)

38. *United Nations Debt* The following table shows the amount of U.S. debt and total debt owed by all members to run the United Nations for the years 1990–2000 and the percent of the debt owed by the United States for each of these years.

Year	U.S. Debt ($ millions)	Total Debt ($ millions)	U.S. Share (%)
1990	296.2	403.0	73.5
1991	266.4	439.4	60.6
1992	239.5	500.6	47.8
1993	260.4	478.0	54.5
1994	248.0	480.0	51.6
1995	414.4	564.0	73.5
1996	376.8	510.7	73.8
1997	373.2	473.6	78.8
1998	315.7	417.0	75.7
1999	167.9	244.2	68.8
2000	164.6	222.4	74.0

(Source: United Nations, Institute for Global Communication)

   a. Create a scatter plot from the table that gives U.S. debt to the United Nations, in millions of dollars, as a function of the number of years from 1990.

   b. Find the cubic function that models U.S. debt to the United Nations, in millions of dollars, as a function of the number of years from 1990.

39. *Photosynthesis* The amount y of photosynthesis that takes place in a certain plant depends on the intensity x of the light present, according to the function

$y = 120x^2 - 20x^3$. The model is only valid for x-values that are nonnegative and that produce a positive amount of photosynthesis. Use graphical or numerical methods to find what intensity allows the maximum amount of photosynthesis.

40. *Investment* The future value of an investment of $8000 at interest rate r (as a decimal) compounded annually for 3 years is given by $S = 8000(1 + r)^3$. Use the root method to find the rate r that gives a future value of $9261.

41. *Constructing a Box* A box can be formed by cutting squares out of each corner of a piece of tin and folding the "tabs" up. If the piece of tin is 18 inches by 18 inches and each side of the square that is cut out has length x, the function that gives the volume of the box is $V = 324x - 72x^2 + 4x^3$.

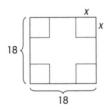

a. Use factoring to find the values of x that make $V = 0$.

b. For these values of x, discuss what happens to the box if squares of length x are cut out.

c. What values of x are reasonable for the squares that can be cut out to make a box?

42. *Break-Even* The daily profit for a product is given by $P(x) = -0.2x^3 + 20.5x^2 - 48.8x - 120$ dollars, where x is the number of hundreds of units produced. To find the number of units that gives break-even,

a. Graph the function on the window $[0, 40]$ by $[-15, 20]$, and graphically find an x-intercept of the graph.

b. Use synthetic division to find a quadratic factor of $P(x)$.

c. Find all of the zeros of $P(x)$.

d. Determine the levels of production that give break-even.

43. *Global Warming* Greenland's ice sheet melted at a record rate in 2007, giving further evidence of global climate change. The ice sheet mass (in billions of tons) lost during 2003–2007 is given in the following table.

a. Find the quartic function that is the best fit for these data, with x in years after 2000 and y in billions of tons.

b. What does the model predict Greenland's ice sheet loss will be in 2010?

Year	Ice sheet mass lost from melt seasons (billion tons)
2003	435
2004	410
2005	530
2006	405
2007	552

(Source: NASA; Steffen Research Group)

44. *Drugs in the Bloodstream* The concentration of a drug in a patient's bloodstream is given by

$$C = \frac{0.3t}{t^2 + 1}$$

where t is the number of hours after the drug was ingested and C is in mg/cm^3.

a. What is the horizontal asymptote for the graph of this function?

b. What does this say about the concentration of the drug?

c. Use technology to find the maximum concentration of the drug, and when it occurs, for $0 \le t \le 4$.

45. *Average Cost* The average cost for production of a product is given by

$$\overline{C}(x) = \frac{50x + 5600}{x}$$

where x is the number of units produced.

a. Find $\overline{C}(0)$, if it exists. What does this tell us about average cost?

b. Find the horizontal asymptote for the graph of this function. What does this tell us about the average cost of this product?

c. Does this function increase or decrease for $x > 0$?

46. *Average Cost* The average cost for production of a product is given by

$$\overline{C}(x) = \frac{30x^2 + 12{,}000}{x}$$

where x is the number of units produced.

a. Graph the function for $x > 0$.

b. Use technology to find the number of units that gives the minimum average cost. What is the minimum average cost?

47. *Demand* The quantity of a certain product that is demanded is related to the price per unit by the equation $p = \dfrac{30{,}000 - 20q}{q}$, where p is the price per unit. Graph the function on a window that fits the context of the problem, with $0 \le q \le 1000$, with q on the horizontal axis.

48. *Printing* A printer has a contract to run 20,000 posters for a state fair. He can use any number of plates from 1 to 8 to run the posters, so that each stamp of the press produces a number of posters equivalent to the number of plates that have been created. The cost of printing all the posters is given by $C(x) = 200 + 20x + \dfrac{180}{x}$, where x is the number of plates.

a. Combine the terms of this function over a common denominator to create a rational function.

b. Use numerical or graphical methods to find the number of plates that will make the cost $336.

c. How many plates should the printer create to produce the posters at the minimum cost?

49. *Computer Usage* The numbers of students per computer in U.S. public schools from the 1983–84 through 1998–99 school years are given in the table above at right. A function that can be used to model the data is $y = \dfrac{375.5 - 15x}{x + 0.03}$, where x is the number of years from the 1980–81 school year.

a. Graph the data from the table on the same axes as the model.

b. What is the number of students per computer in the 1993–94 school year, according to the model? How does this compare with the data in the table?

School Year	Students per Computer	School Year	Students per Computer
1983–84	125	1993–94	14
1985–86	50	1995–96	10
1987–88	32	1997–98	6.1
1989–90	22	1998–99	5.7
1991–92	18		

50. *Sales and Training* During the first month of employment, the monthly sales S (in thousands of dollars) for an average new salesperson depends on the number of hours of training x, according to

$$S = \frac{40}{x} + \frac{x}{4} + 10 \quad \text{for} \quad x \ge 4$$

How many hours of training should result in monthly sales greater than $23,300?

51. *Homicide Rate* The homicide rate per 100,000 people is given by the function $y = 0.00380x^3 - 0.0704x^2 - 0.0780x + 9.780$, with x equal to the number of years from 1990. Use technology to find the year or years after 1990 in which the homicide rate per 100,000 people is no more than 8.8.
(Source: U.S. Department of Justice, Bureau of Justice Statistics)

52. *Revenue* The revenue from the sale of x units of a product is given by the function $R = 1200x - 0.003x^3$. Selling how many units will give a revenue of at least $59,625?

53. *Average Cost* The average cost per set for the production of television sets is given by

$$\overline{C} = \frac{100 + 30x + 0.1x^2}{x}$$

where x is the number of hundreds of units produced. For how many units is the average cost at most $37?

54. *Cost-Benefit* The percent p of particulate pollution that can be removed from the smokestacks of an industrial plant by spending C dollars per week is given by $p = \dfrac{100C}{9600 + C}$. How much would it cost to remove at least 30.1% of the particulate pollution?

group activities / extended applications

1. Global Climate Change

1. Suppose the annual cost C (in hundreds of dollars) of removing $p\%$ of the particulate pollution from the smokestack of a power plant is given by

$$C(p) = \frac{242{,}000}{100 - p} - 2420$$

   a. What is the domain of the function described by this equation?

   b. Taking into account the meaning of the variables, what is the domain and range of the function?

2. Find $C(60)$ and $C(80)$ and interpret $C(80)$.

3. Find the asymptotes of the function. What does the vertical asymptote tell you about the cost of removing pollution?

4. Find and interpret the y-intercept.

5. Suppose the annual fine for removing less than 80% of the particulate pollution from the smokestack is $700,000. If you were the company's Chief Financial Officer, would you advise the company to remove the pollution or pay the fine? Why?

6. If the company has already paid to remove 60% of the particulate pollution, would you advise the company to reduce it by another 20% or pay the fine?

7. If you were a lawmaker, to what amount would you raise the fine in order to encourage power plants to comply with the pollution regulations?

8. Comparing the costs of removing 80% and 90% of the pollution from the smokestack, do you think the benefit of removing 90% of particulate pollution rather than 80% is worth the cost?

2. Printing

A printer has a contract to print 100,000 invitations for a political candidate. He can run the invitations by using any number of metal printing plates from 1 to 20 on his press. For example, if he prepares 10 plates, each impression of his press makes 10 invitations. Preparing each plate costs $8 and he can make 1000 impressions per plate per hour. If it costs $128 per hour to run the press, how many metal printing plates should the printer use to minimize the cost of printing the 100,000 invitations?

A. To understand the problem:
   1. Determine how many hours it would take to print the 100,000 invitations with 10 plates.
   2. How much would it cost to print the invitations with 10 plates?
   3. Determine how much it would cost to prepare 10 printing plates.
   4. What is the total cost to make the 10 plates and print 100,000 invitations?

B. To find the number of plates that will minimize the total cost of printing the invitations:
   1. Find the cost of preparing x plates.
   2. Find the number of invitations that can be made with one impression made by the x plates.
   3. Find the number of invitations that can be made per hour with x plates and the number of hours it would take to print 100,000 invitations.
   4. Write a function of x that gives the total cost of producing the invitations.
   5. Use technology to graph the cost function and to estimate the number of plates that will give the minimum cost of producing 100,000 plates.
   6. Use numerical methods to find the number of plates that will give the minimum cost and the minimum possible cost of producing 100,000 invitations.

Systems of Equations and Matrices

To improve the highway infrastructure, departments of transportation can use systems of equations to analyze traffic flow. Solving systems of equations is also useful in finding break-even and market equilibrium. We can represent systems with matrices and use matrix reduction to solve the systems and related problems.

    We can also use matrices to make comparisons of data, to solve pricing problems and advertising problems, and to encode and decode messages.

Algebra Toolbox

key concepts

- Proportion
- Proportional triples
- Linear equations in three variables

Proportion

Often in mathematics we need to express relationships between quantities. One relationship that is frequently used in applied mathematics occurs when two quantities are proportional. Two variables x and y are proportional to each other (or vary directly) if their quotient is constant. That is, y is **directly proportional** (or simply **proportional**) to x if y and x are related by the equation

$$\frac{y}{x} = k \quad \text{or equivalently by} \quad y = kx$$

where $k \neq 0$ is called the **constant of proportionality**.

example 1

Circles

Is the circumference of a circle proportional to the radius of the circle?

Solution

The circumference C of a circle is proportional to the radius r of the circle because $C = 2\pi r$. In this case, C and r are the variables, and 2π is the constant of proportionality. We could write an equivalent equation,

$$\frac{C}{r} = 2\pi$$

which says that the quotient of the circumference divided by the radius is constant, and that constant is 2π. ∎

Proportionality

One pair of numbers is said to be **proportional** to another pair of numbers if the ratio of one pair is equal to the ratio of the other pair. In general, the pair a, b is proportional to the pair c, d if and only if

$$\frac{a}{b} = \frac{c}{d} \quad (b \neq 0, d \neq 0)$$

This proportion can be written $a:b$ as $c:d$.

If $b \neq 0$, $d \neq 0$, we can verify that $\frac{a}{b} = \frac{c}{d}$ if $ad = bc$; that is, if the cross-products are equal.

example 2	**Proportional Pairs**

Is the pair 4, 3 proportional to the pair 12, 9?

Solution

It is true that $\dfrac{4}{3} = \dfrac{12}{9}$ because we can reduce $\dfrac{12}{9}$ to $\dfrac{4}{3}$; or we can show that $\dfrac{4}{3} = \dfrac{12}{9}$ is true by showing that the cross-products are equal: $4 \cdot 9 = 3 \cdot 12$. Thus, the pair 4, 3 is proportional to 12, 9. ■

Proportional Triples

For two sets of triples a, b, c, and d, e, f, if there exists some number k such that multiplying each number in the first set of triples by k results in the second set of triples, then the two sets of triples are proportional. That is, a, b, c is proportional to d, e, f if there exists a number k such that $ak = d$, $bk = e$, and $ck = f$.

example 3	**Proportional Triples**

Determine whether the sets of triples are proportional.

a. 2, 3, 5 and 4, 6, 10

b. 4, 6, 10 and 6, 9, 12

Solution

a. If the sets of triples are proportional, then there is a number k such that $2k = 4$, $3k = 6$, and $5k = 10$. Because $k = 2$ is the solution to all three equations, we conclude that the sets of triples are proportional.

b. If the sets of triples are proportional, then there is a number k such that $4k = 6$, $6k = 9$, and $10k = 12$. Because $k = 1.5$ is the solution to $4k = 6$ but not a solution to $10k = 12$, the sets of triples are not proportional. ■

Linear Equations in Three Variables

If a, b, c, and d represent nonzero constants, then

$$ax + by + cz = d$$

is a first-degree equation in three variables. We also call these equations linear equations in three variables, but unlike linear equations in two variables, their graphs are not lines. When equations of this form are graphed in a three-dimensional coordinate system, their graphs are planes. Similar to the graphs of linear equations in two variables, the graphs of two linear equations in three variables will be parallel planes if the respective coefficients of x, y, and z are proportional and the constants are not in the same proportion. If the respective coefficients of x, y, and z in two equations are not proportional, then the two planes will intersect along a line. And if the two equations are equivalent, the equations represent the same plane.

| example 4 | **Planes** |

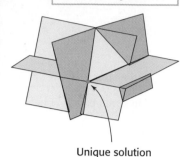

Unique solution

Figure 7.1

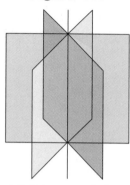

Infinitely many solutions

Figure 7.2

No solution

Figure 7.3

Determine if the planes represented by the following pairs of equations represent the same plane, parallel planes, or neither.

a. $3x + 4y + 5z = 1$ and $6x + 8y + 10z = 2$

b. $2x - 5y + 3z = 4$ and $-6x + 10y - 6z = -12$

c. $3x - 6y + 9z = 12$ and $x - 2y + 3z = 26$

Solution

a. If we multiply both sides of the first equation by 2, the second equation results, so the equations are equivalent, and thus the graphs of both equations are the same plane.

b. The respective coefficients of the variables are not in proportion, so the graphs are not the same plane and are not parallel planes, so the planes representing these equations must intersect in a line.

c. The respective coefficients of the variables are in proportion and the constants are not in the same proportion, so the graphs are parallel planes. ∎

   Three different planes may intersect in a single point (Figure 7.1), may intersect in a line (as in a paddle wheel; See Figure 7.2), or may not have a common intersection (Figure 7.3). Thus, three linear equations in three variables may have a unique solution, infinitely many solutions, or no solution. For example, the solution of the system

$$\begin{cases} 3x + 2y + z = 6 \\ x - y - z = 0 \\ x + y - z = 4 \end{cases}$$

is the single point with coordinates $x = 1$, $y = 2$, and $z = -1$, because these three values satisfy all three equations, and these are the only values that satisfy them.

| example 5 | **Solutions of Systems of Equations** |

Determine whether the given values of x, y, and z satisfy the system of equations.

a. $\begin{cases} 2x + y + z = -3 \\ x - y - 2z = -4; \quad x = -2, y = 0, z = 1 \\ x + y + 3z = 1 \end{cases}$

b. $\begin{cases} x + y + z = 4 \\ x - y - z = 2; \quad x = 3, y = -1, z = 2 \\ 2x + y - 2z = 3 \end{cases}$

Solution

a. Substituting the values of the variables into each equation gives

$$2(-2) + 0 + 1 = -4 + 1 = -3 \quad \checkmark$$
$$-2 - 0 - 2(1) = -2 - 2 = -4 \quad \checkmark$$
$$-2 + 0 + 3(1) = -2 + 3 = 1 \quad \checkmark$$

The values of x, y, and z satisfy all three equations, so we conclude that $x = -2, y = 0, z = 1$ is a solution to the system.

b. Substituting the values of the variables into each equation gives

$$3 + (-1) + 2 = 2 + 2 = 4 \qquad \checkmark$$
$$3 - (-1) - 2 = 4 - 2 = 2 \qquad \checkmark$$
$$2(3) + (-1) - 2(2) = 5 - 4 = 1 \neq 3$$

Although the values of x, y, and z satisfy the first two equations, they do not satisfy the third equation, so we conclude that $x = 3, y = -1, z = 2$ is not a solution to the system. ∎

toolbox exercises

1. If y is proportional to x, and $y = 12x$, what is the constant of proportionality?

2. If x is proportional to y, and $y = 10x$, what is the constant of proportionality?

3. Is the ratio 5:6 proportional to 10:18?

4. Is the ratio 8:3 proportional to 24:9?

5. Is the pair of numbers 4, 6 proportional to the pair 2, 3?

6. Is the pair of numbers 5.1, 2.3 proportional to the pair 51, 23?

7. Suppose y is proportional to x, and when $x = 2$, $y = 18$. Find y when $x = 11$.

8. Suppose x is proportional to y, and when $x = 3$, $y = 18$. Find y when $x = 13$.

In Exercises 9–12, determine if the pairs of triples are proportional.

9. $8, 6, 2$ and $20, 12.5, 5$

10. $4.1, 6.8, 9.3$ and $8.2, 13.6, 18.6$

11. $-1, 6, 4$ and $-3.4, 10.2, 13.6$

12. $5, -12, 16$ and $2, -3, 4$

In Exercises 13–16, determine whether the given values of x, y, and z satisfy the system of equations.

13. $\begin{cases} x - 2y + z = 8 \\ 2x - y + 2z = 10; \quad x = 1, y = -2, z = 3 \\ 3x - 2y + z = 5 \end{cases}$

14. $\begin{cases} x - 2y + 2z = -9 \\ 2x - 3y + z = -14; \quad x = -1, y = 4, z = 0 \\ x + 2y - 3z = 7 \end{cases}$

15. $\begin{cases} x + y - z = 4 \\ 2x + 3y - z = 5; \quad x = 5, y = -2, z = -1 \\ 3x - 2y + 5z = 14 \end{cases}$

16. $\begin{cases} x - 2y - 3z = 2 \\ 2x + 3y + 3z = 19; \quad x = 2, y = 3, z = -2 \\ -x - 2y + 2z = -12 \end{cases}$

17. Determine if the planes represented by the following pair of equations represent the same plane, parallel planes, or neither:

$$x + 3y - 7z = 12 \quad \text{and} \quad 2x + 6y - 14z = 18$$

18. Determine if the planes represented by the following pair of equations represent the same plane, parallel planes, or neither:

$$x - 3y - 2z = 15 \quad \text{and} \quad 2x - 6y - 4z = 30$$

19. Determine if the planes represented by the following pair of equations represent the same plane, parallel planes, or neither:

$$2x - y - z = 2 \quad \text{and} \quad 2x + y - 4z = 8$$

20. Determine if the planes represented by the following pair of equations represent the same plane, parallel planes, or neither:

$$3x - y + 2z = 9 \quad \text{and} \quad 9x - 3y + 6z = 27$$

Systems of Linear Equations in Three Variables

section preview ▪ Manufacturing

A manufacturer of furniture has three models of chairs, Anderson, Blake, and Colonial. The numbers of hours required for framing, upholstery, and finishing for each type of chair are given in Table 7.1. The company has 1500 hours per week for framing, 2100 hours for upholstery, and 850 hours for finishing.

Finding how many of each type of chair can be produced under these conditions involves solving a system with three linear equations in three variables. It is not possible to solve this equation graphically, and new algebraic methods are needed. In this section, we discuss how to solve a system of equations by using the left-to-right elimination method.

Table 7.1

	Anderson	Blake	Colonial
Framing	2	3	1
Upholstery	1	2	3
Finishing	1	2	1/2

Systems in Three Variables

In Section 2.3, we used the elimination method to reduce a system of two linear equations in two variables to one equation in one variable. If a system of three linear equations in three variables has a unique solution, we can reduce the system to an equivalent system where at least one of the equations contains only one variable. We can then easily solve that equation and use the resulting value to find the values of the remaining variables. For example, we will show later that the system

$$\begin{cases} x + y + z = 2 \\ x - y - z = 0 \\ x + y - z = 4 \end{cases}$$

can be reduced to the equivalent system

$$\begin{cases} x + y + z = 2 \quad (1) \\ \phantom{x + {}} y + z = 1 \quad (2) \\ \phantom{x + y + {}} z = -1 \quad (3) \end{cases}$$

Clearly, Equation (3) in the reduced system has solution $z = -1$. Using this value of z in Equation (2) gives

$$y + (-1) = 1 \quad \text{or} \quad y = 2$$

Using $z = -1$ and $y = 2$ in Equation (1) gives

$$x + 2 + (-1) = 2 \quad \text{or} \quad x = 1$$

Thus, the solution to the reduced system is $x = 1, y = 2, z = -1$. The process of using the value of the third variable to work back to the values of the other variables is called **back substitution**.

We say that the reduced system is equivalent to the original system because both systems have the same solutions. We verify that by showing that these values of x, y, and z satisfy the equations of the original system.

$$\begin{cases} 1 + 2 - 1 = 2 \\ 1 - 2 + 1 = 0 \\ 1 + 2 + 1 = 4 \end{cases}$$

Left-to-Right Elimination

The reduced system of equations that we discussed above is in a special form, called the **echelon form**, which, in general, looks like:

$$\begin{cases} x + ay + bz = d \\ \quad\quad y + cz = e \\ \quad\quad\quad\quad z = f \end{cases}$$

where a, b, c, d, e, and f are constants. Notice that in this form, each equation begins with a new (leading) variable with coefficient 1. Any system of linear equations with a unique solution can be reduced to this form. By using the following operations, we can be assured that the new system will be equivalent to the original system.

Operations on a System of Equations

The following operations result in an equivalent system.

1. Interchange any two equations.

2. Multiply both sides of any equation by the same nonzero number.

3. Multiply any equation by a number, add the result to a second equation, and then replace the second equation with the sum.

The following process, which uses these operations to reduce a system of linear equations, is called the **left-to-right elimination method**.

Left-to-Right Elimination Method of Solving Systems of Linear Equations in Three Variables x, y, and z

1. If necessary, interchange two equations or use multiplication to make the coefficient of x in the first equation a 1.

2. Add a multiple of the first equation to each of the following equations so that the coefficients of x in the second and third equations become 0.

3. Multiply (or divide) both sides of the second equation by a number that makes the coefficient of y in the second equation equal to 1.

(continued)

Left-to-Right Elimination Method of Solving Systems of Linear Equations in Three Variables x, y and z (continued)

4. Add a multiple of the (new) second equation to the (new) third equation so that the coefficient of y in the newest third equation becomes 0.

5. Multiply (or divide) both sides of the third equation by a number that makes the coefficient of z in the third equation equal to 1. This gives the solution for z in the system of equations.

6. Use the solution for z to solve for y in the second equation. Then substitute values for y and z to solve for x in the first equation. (This is called **back substitution.**)

---

example 1

Solution by Elimination

Solve the system

$$\begin{cases} 2x - 3y + z = -1 & (1) \\ x - y + 2z = -3 & (2) \\ 3x + y - z = 9 & (3) \end{cases}$$

Solution

1. To write the system with x-coefficient 1 in the first equation, interchange equations (1) and (2).

$$\begin{cases} x - y + 2z = -3 \\ 2x - 3y + z = -1 \\ 3x + y - z = 9 \end{cases}$$

2. To eliminate x in the second equation, multiply the (new) first equation by -2 and add it to the second equation. To eliminate x in the third equation, multiply the (new) first equation by -3 and add it to the third equation. This gives the equivalent system

$$\begin{cases} x - y + 2z = -3 \\ -y - 3z = 5 \\ 4y - 7z = 18 \end{cases}$$

3. To get 1 as the coefficient of y in the second equation, multiply the (new) second equation by -1.

$$\begin{cases} x - y + 2z = -3 \\ y + 3z = -5 \\ 4y - 7z = 18 \end{cases}$$

4. To eliminate y from the (new) third equation, multiply the (new) second equation by -4 and add it to the third equation.

$$\begin{cases} x - y + 2z = -3 \\ y + 3z = -5 \\ -19z = 38 \end{cases}$$

5. To solve for z, divide both sides of the (new) third equation by -19.

$$\begin{cases} x - y + 2z = -3 \\ \quad\quad y + 3z = -5 \\ \quad\quad\quad\quad z = -2 \end{cases}$$

This gives the solution $z = -2$.

6. Substituting -2 for z in the second equation and solving for y gives $y + 3(-2) = -5$, so $y = 1$. Substituting -2 for z and 1 for y in the first equation and solving for x gives $x - 1 + 2(-2) = -3$, so $x = 2$, and the solution is $(2, 1, -2)$.

The process used in step 6 is called *back substitution* because we use the solution for the third variable to work back to the solutions of the first and second variables. Checking in the original system verifies that the solution satisfies all the equations.

$$\begin{cases} 2(2) - 3(1) + (-2) = -1 \\ 2 - 1 + 2(-2) = -3 \\ 3(2) + 1 - (-2) = 9 \end{cases}$$

Modeling Systems of Equations

Solution of real problems sometimes requires us to create three or more equations whose simultaneous solution is the solution to the problem.

---

example 2

Manufacturing

A manufacturer of furniture has three models of chairs, Anderson, Blake, and Colonial. The numbers of hours required for framing, upholstery, and finishing for each type of chair are given in Table 7.2. The company has 1500 hours per week for framing, 2100 hours for upholstery, and 850 hours for finishing. How many of each type of chair can be produced under these conditions?

Table 7.2

	Anderson	Blake	Colonial
Framing	2	3	1
Upholstery	1	2	3
Finishing	1	2	$\dfrac{1}{2}$

Solution

If we represent the number of units of the Anderson model by x, the number of units of the Blake model by y, and the number of units of Colonial by z, then we have the following equations:

$$\begin{aligned} \text{Framing:} \quad & 2x + 3y + z = 1500 \\ \text{Upholstery:} \quad & x + 2y + 3z = 2100 \\ \text{Finishing:} \quad & x + 2y + \tfrac{1}{2}z = 850 \end{aligned}$$

We seek the values of x, y, and z that satisfy all three equations above. That is, we seek the solution to the system

$$\begin{cases} 2x + 3y + z = 1500 & (1) \\ x + 2y + 3z = 2100 & (2) \\ x + 2y + \dfrac{1}{2}z = 850 & (3) \end{cases}$$

We use the following steps to solve the system by left-to-right elimination. The notation that we use to represent each step is shown beside the system. We use this notation in subsequent examples.

To write the system with x-coefficient 1 in the first equation, interchange Equations (1) and (2).

$$\begin{cases} x + 2y + 3z = 2100 \\ 2x + 3y + z = 1500 \qquad \text{Eq1} \leftrightarrow \text{Eq2} \\ x + 2y + 0.5z = 850 \end{cases}$$

To eliminate x in the second equation, multiply the first equation by -2 and add it to the second equation. To eliminate x in the third equation, multiply the first equation by -1 and add it to the third equation.

$$\begin{cases} x + 2y + 3z = 2100 \\ -y - 5z = -2700 \qquad (-2)\text{Eq1} + \text{Eq2} \rightarrow \text{Eq2} \\ - 2.5z = -1250 \qquad (-1)\text{Eq1} + \text{Eq3} \rightarrow \text{Eq3} \end{cases}$$

Observe that we can solve Equation (3) for z, getting $z = -1250/(-2.5) = 500$. Substituting $z = 500$ in Equation (2) gives $-y - 5(500) = -2700$ or $y = 200$. Substituting $z = 500$ and $y = 200$ in Equation (1) gives $x + 2(200) + 3(500) = 2100$, or $x = 200$.

Thus, 200 Anderson, 200 Blake, and 500 Colonial chairs should be made under the given conditions.   ∎

Nonunique Solutions

It is not always possible to reduce a system of three equations in three variables to a system in which the third equation contains one variable. For example, the system

$$\begin{cases} x + 2y + z = 2 & (1) \\ x - y - 2z = -4 & (2) \\ 2x + y - z = 4 & (3) \end{cases}$$

reduces to

$$\begin{cases} x + 2y + z = 2 \\ y + z = 2 \\ 0 + 0 = 6 \end{cases}$$

The third equation in the reduced system is $0 = 6$, which is impossible. Thus, this system and the original system have no solution. Such a system is called an **inconsistent system**.

When a system of three linear equations in three variables has no solution, the graphs of the equations will not have a common intersection, as shown in Figure 7.4.

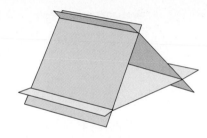

No solution

Figure 7.4

The system

$$\begin{cases} x + 2y + z = 2 & (1) \\ x - y - 2z = -4 & (2) \\ 2x + y - z = -2 & (3) \end{cases}$$

looks similar but is slightly different from the inconsistent system just discussed. Reducing this system with left-to-right elimination results in the system

$$\begin{cases} x + 2y + z = 2 \\ y + z = 2 \\ 0 + 0 = 0 \end{cases}$$

The third equation in this system is $0 = 0$, and we will see that the system has infinitely many solutions. Notice that the z is not a leading variable in any of the equations, so we can solve for the other variables in terms of z. Back substitution gives

$$y = 2 - z$$
$$x = 2 - 2(2 - z) - z = -2 + z$$

Thus, the solutions have the form $x = -2 + z$, $y = 2 - z$, and $z =$ any number.

By selecting different values of z, we can find some particular solutions of this system.

For example,

$$z = 1 \quad \text{gives} \quad x = -1, \quad y = 1, \quad z = 1$$
$$z = 2 \quad \text{gives} \quad x = 0, \quad y = 0, \quad z = 2$$
$$z = 5 \quad \text{gives} \quad x = 3, \quad y = -3, \quad z = 5$$

A system of this type, with infinitely many solutions, is called a **dependent system**. When a system of three equations in three variables has infinitely many solutions, the graphs of the equations may be planes that intersect in a line, as in the paddle wheel shown in Figure 7.5.

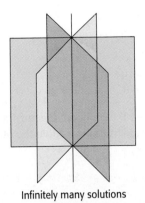

Infinitely many solutions

Figure 7.5

| example 3 |

Find the solution to each of the following systems, if it exists.

a. $\begin{cases} x - y + 4z = -5 \\ 3x + z = 0 \\ -x + y - 4z = 20 \end{cases}$ **b.** $\begin{cases} -x + y + 2z = 0 \\ x + 2y + z = 6 \\ -2x - y + z = -6 \end{cases}$

Solution

a. The system

$$\begin{cases} x - y + 4z = -5 & (1) \\ 3x \quad\quad + z = 0 & (2) \\ -x + y - 4z = 20 & (3) \end{cases}$$

reduces as follows:

$$\begin{cases} x - y + 4z = -5 \\ \quad 3y - 11z = 15 & (-3)\text{Eq1} + \text{Eq2} \rightarrow \text{Eq2} \\ \quad\quad\quad 0 = 15 & \text{Eq1} + \text{Eq3} \rightarrow \text{Eq3} \end{cases}$$

The third equation in the reduced system is $0 = 15$, which is impossible. Thus, the system is inconsistent and has no solution.

b. The system

$$\begin{cases} -x + y + 2z = 0 & (1) \\ x + 2y + z = 6 & (2) \\ -2x - y + z = -6 & (3) \end{cases}$$

reduces as follows:

$$\begin{cases} x - y - 2z = 0 & (-1)\text{Eq1} \rightarrow \text{Eq1} \\ x + 2y + z = 6 \\ -2x - y + z = -6 \end{cases}$$

$$\begin{cases} x - y - 2z = 0 \\ \quad 3y + 3z = 6 & (-1)\text{Eq1} + \text{Eq2} \rightarrow \text{Eq2} \\ \quad -3y - 3z = -6 & (2)\text{Eq1} + \text{Eq3} \rightarrow \text{Eq3} \end{cases}$$

$$\begin{cases} x - y - 2z = 0 \\ \quad y + z = 2 & (1/3)\text{Eq2} \rightarrow \text{Eq2} \\ \quad -3y - 3z = -6 \end{cases}$$

$$\begin{cases} x - y - 2z = 0 \\ \quad y + z = 2 \\ \quad\quad\quad 0 = 0 & (3)\text{Eq2} + \text{Eq3} \rightarrow \text{Eq3} \end{cases}$$

Because the third equation in the reduced system is $0 = 0$, the system has infinitely many solutions. We let z be any number and use back substitution to solve the other variables in terms of z.

$$y = 2 - z$$
$$x = 2 - z + 2z = 2 + z$$

Thus, the solution has the form

$$x = 2 + z$$
$$y = 2 - z$$
$$z = \text{any number}$$

    ■

If a system of equations has fewer equations than variables, the system must be dependent. Consider the following example.

| example 4 | **Purchasing** |

A young man wins $200,000 and (foolishly) decides to buy a new car for each of the seven days of the week. He wants to choose from cars that are priced at $40,000, $30,000, and $25,000. How many cars of each price can he buy with the $200,000?

Solution

If he buys x cars costing $40,000 each, y cars costing $30,000 each, and z cars costing $25,000 each, then

$$x + y + z = 7 \quad \text{and} \quad 40{,}000x + 30{,}000y + 25{,}000z = 200{,}000$$

Writing these two equations as a system of equations gives

$$\begin{cases} x + y + z = 7 & (1) \\ 40{,}000x + 30{,}000y + 25{,}000z = 200{,}000 & (2) \end{cases}$$

Because there are three variables and only two equations, the solution cannot be unique. The solution follows.

$$\begin{cases} x + y + z = 7 \\ -10{,}000y - 15{,}000z = -80{,}000 \end{cases} \quad -40{,}000\text{Eq1} + \text{Eq2} \rightarrow \text{Eq2}$$

$$\begin{cases} x + y + z = 7 \\ y + 1.5z = 8 \end{cases} \quad -0.0001\text{Eq2} \rightarrow \text{Eq2}$$

In the solution to this system, we see that z can be any value and that x and y depend on z.

$$y = 8 - 1.5z$$
$$x = -1 + 0.5z$$

In the context of this application, the solutions must be nonnegative integers (none of x, y, or z can be negative), so z must be at least 2 (or x will be negative), and z cannot be more than 5 (or y will be negative). In addition, z must be an even number (to get integer solutions). Thus, the only possible selections are

$$z = 2, x = 0, y = 5 \quad \text{or} \quad z = 4, x = 1, y = 2$$

Thus, the man can buy two $25,000 cars, zero $40,000 cars, and five $30,000 cars, or he can buy four $25,000 cars, one $40,000 car, and two $30,000 cars. So even though there are an infinite number of solutions to this system of equations, there are only two solutions in the context of this application. ■

skills check

7.1

In Exercises 1–4, solve the systems of equations.

1. $\begin{cases} x + 2y - z = 3 \\ y + 3z = 11 \\ z = 3 \end{cases}$
2. $\begin{cases} x - 4y - 3z = 3 \\ y - 2z = 11 \\ z = 4 \end{cases}$

3. $\begin{cases} x + 2y - z = 6 \\ y + 3z = 3 \\ z = -2 \end{cases}$

4. $\begin{cases} x + 2y - z = 22 \\ y + 3z = 21 \\ z = -5 \end{cases}$

In Exercises 5–16, use left-to-right elimination to solve the systems of equations.

5. $\begin{cases} x - y - 4z = 0 \\ y + 2z = 4 \\ 3y + 7z = 22 \end{cases}$ 6. $\begin{cases} x + 4y - 11z = 33 \\ y - 3z = 11 \\ 2y + 7z = -4 \end{cases}$

7. $\begin{cases} x - 2y + 3z = 0 \\ y - 2z = -1 \\ y + 5z = 6 \end{cases}$ 8. $\begin{cases} x - 2y + 3z = -10 \\ y - 2z = 7 \\ y - 3z = 6 \end{cases}$

9. $\begin{cases} x + 2y - 2z = 0 \\ x - y + 4z = 3 \\ x + 2y + 2z = 3 \end{cases}$ 10. $\begin{cases} x + 4y - 14z = 22 \\ x + 5y + z = -2 \\ x + 4y - z = 9 \end{cases}$

11. $\begin{cases} x + 3y - z = 0 \\ x - 2y + z = 8 \\ x - 6y + 2z = 6 \end{cases}$ 12. $\begin{cases} x - 5y - 2z = 7 \\ x - 3y + 4z = 21 \\ x - 5y + 2z = 19 \end{cases}$

13. $\begin{cases} 2x + 4y - 14z = 0 \\ 3x + 5y + z = 19 \\ x + 4y - z = 12 \end{cases}$

14. $\begin{cases} 3x - 6y + 6z = 24 \\ 2x - 6y + 2z = 6 \\ x - 3y + 2z = 4 \end{cases}$

15. $\begin{cases} 3x - 4y + 6z = 10 \\ 2x - 4y - 5z = -14 \\ x + 2y - 3z = 0 \end{cases}$

16. $\begin{cases} x - 3y + 4z = 7 \\ 2x + 2y - 3z = -3 \\ x - 3y + z = -2 \end{cases}$

The systems in Exercises 17–26 do not have unique solutions. Solve each system, if possible.

17. $\begin{cases} x - 3y + z = -2 \\ y - 2z = -4 \end{cases}$ 18. $\begin{cases} x + 2y + 2z = 4 \\ y + 3z = 6 \end{cases}$

19. $\begin{cases} x - y + 2z = 4 \\ y + 8z = 16 \end{cases}$ 20. $\begin{cases} x - 3y - 2z = 3 \\ y + 4z = 5 \\ 0 = 4 \end{cases}$

21. $\begin{cases} x - y + 2z = -1 \\ 2x + 2y + 3z = -3 \\ x + 3y + z = -2 \end{cases}$

22. $\begin{cases} x + 5y - 2z = 5 \\ x + 3y - 2z = 23 \\ x + 4y - 2z = 14 \end{cases}$

23. $\begin{cases} 3x - 4y - 6z = 10 \\ 2x - 4y - 5z = -14 \\ x - z = 0 \end{cases}$

24. $\begin{cases} 3x - 9y + 4z = 24 \\ 2x - 6y + 2z = 6 \\ x - 3y + 2z = 20 \end{cases}$

25. $\begin{cases} 2x + 3y - 14z = 16 \\ 4x + 5y - 30z = 34 \\ x + y - 8z = 9 \end{cases}$

26. $\begin{cases} 2x + 4y - 4z = 6 \\ x + 5y - 3z = 3 \\ 3x + 3y - 5z = 9 \end{cases}$

exercises

7.1

27. *Rental Cars* A car rental agency rents compact, midsize, and luxury cars. Its goal is to purchase 60 cars with a total of $1,400,000 and to earn a daily rental of $2200 from all the cars. The compact cars cost $15,000 and earn $30 per day in rental, the midsize cars cost $25,000 and earn $40 per day, and the luxury cars cost $45,000 and earn $50 per day. To find the number of each type of car the agency should purchase, solve the system of equations

$$\begin{cases} x + y + z = 60 \\ 15{,}000x + 25{,}000y + 45{,}000z = 1{,}400{,}000 \\ 30x + 40y + 50z = 2200 \end{cases}$$

where x, y, and z are the numbers of compact cars, midsize cars, and luxury cars, respectively.

28. *Ticket Pricing* A theater owner wants to divide an 1800 seat theater into three sections, with tickets costing $20, $35, and $50, depending on the section. He wants to have twice as many $20 tickets as the sum of the other tickets, and he wants to earn $48,000 from a full house. To find how many seats he should have in each section, solve the system of equations

$$\begin{cases} x + y + z = 1800 \\ x = 2(y + z) \\ 20x + 35y + 50z = 48{,}000 \end{cases}$$

where x, y, and z are the numbers of $20 tickets, $35 tickets, and $50 tickets, respectively.

29. *Pricing* A concert promoter needs to make $120,000 from the sale of 2600 tickets. The promoter charges $40 for some tickets and $60 for the others.

a. If there are x of the $40 tickets and y of the $60 tickets, write an equation that states that the total number of the tickets sold is 2600.

b. How much money is made from the sale of x tickets for $40 each?

c. How much money is made from the sale of y tickets for $60 each?

d. Write an equation that states that the total amount made from the sale is $120,000.

e. Solve the equations simultaneously to find how many tickets of each type must be sold to yield the $120,000.

30. *Investment* A trust account manager has $500,000 to be invested in three different accounts. The accounts pay 8%, 10%, and 14%, respectively, and the goal is to earn $49,000 with the amount invested at 8% equal to the sum of the other two investments. To accomplish this, assume that x dollars are invested at 8%, y dollars are invested at 10%, and z dollars are invested at 14%.

a. Write an equation that describes the sum of money in the three investments.

b. Write an equation that describes the total amount of money earned by the three investments.

c. Write an equation that describes the relationship among the three investments.

d. Solve the system of equations to find how much should be invested in each account to satisfy the conditions.

31. *Investment* A man has $400,000 invested in three rental properties. One property earns 7.5% per year on the investment, the second earns 8%, and the third earns 9%. The total annual earnings from the three properties are $33,700 and the amount invested at 9% equals the sum of the first two investments. Let x equal the investment at 7.5%, y equal the investment at 8%, and z represent the investment at 9%.

a. Write an equation that represents the sum of the three investments.

b. Write an equation that states the sum of the returns from all three investments is $33,700.

c. Write an equation that states that the amount invested at 9% equals the sum of the other two investments.

d. Solve the system of equations to find how much is invested in each property.

32. *Loans* A bank loans $285,000 to a development company to purchase three business properties. If one of the properties costs $45,000 more than the other and the third costs twice the sum of these two properties,

a. Write an equation that represents the total loaned as the sum of the cost of the three properties, if the costs are x, y, and z, respectively.

b. Write an equation that states that one cost is $45,000 more than another.

c. Write an equation that states that the third cost is equal to twice the sum of the other two.

d. Solve the system of equations to find the cost of each property.

33. *Transportation* Ace Trucking Company has an order for three products, A, B, and C, for delivery. The table below gives the volume in cubic feet, the weight in pounds, and the value for insurance in dollars for a unit of each of the products. If the carrier can carry 8000 cubic feet and 12,400 pounds and is insured for $52,600, how many units of each product can be carried?

	Product A	Product B	Product C
Unit volume (cu ft)	24	20	40
Weight (lb)	40	30	60
Value ($)	150	180	200

34. *Nutrition* The following table gives the calories, fat, and carbohydrates per ounce for three brands of cereal, and the total amount of calories, fat, and carbohydrates required for a special diet.

Cereal	Calories	Fat	Carbohydrates
All Bran	50	0	22.0
Sugar Frosted Flakes	108	0.1	25.5
Natural Mixed Grain	127	5.5	18.0
Total required	393	5.7	91.0

To find the number of ounces of each brand that should be combined to give the required totals of calories, fat, and carbohydrates, let x represent the number of ounces of All Bran, y represent the number of ounces of Sugar Frosted Flakes, and z represent the number of ounces of Natural Mixed Grain.

a. Use the information in the table to write a system of three equations.

b. Solve the system to find the amount of each brand of cereal that will give the necessary nutrition.

Exercises 35–40 have nonunique solutions.

35. *Investment* A trust account manager has $500,000 to be invested in three different accounts. The accounts pay 8%, 10%, and 14%, and the goal is to earn $49,000. To solve this problem, assume that x dollars are invested at 8%, y dollars are invested at 10%, and z dollars are invested at 14%. Then x, y, and z must satisfy the equations $x + y + z = 500{,}000$ and $0.08x + 0.10y + 0.14z = 49{,}000$.

a. Explain the meaning of each of these equations.

b. Solve the systems containing these two equations. Find x and y in terms of z.

c. What limits must there be on z so that all investment values are nonnegative?

36. *Loans* A bank gives three loans totaling $200,000 to a development company for the purchase of three business properties. The largest loan is $45,000 more than the sum of the other two. Represent the amount of money in each loan as x, y, and z, respectively.

a. Write a system of two equations in three variables to represent the problem.

b. Does the system have a unique solution?

c. Find the amount of the largest loan.

d. What is the relationship between the remaining two loans?

37. *Manufacturing* A company manufactures three types of air conditioners, the Acclaim, the Bestfrig, and the Cool King. The hours needed for the assembly, testing, and packing for each product are given in the table below. The daily number of hours that are available are 300 for assembly, 120 for testing, and 210 for packing. Can all of the available hours be used to produce these three types of air conditioners, and if so, how many of each type can be produced?

38. *Transportation* Hohman Trucking Company has an

	Assembly	Testing	Packing
Acclaim	5	2	1.4
Bestfrig	4	1.4	1.2
Cool King	4.5	1.7	1.3

order for three products, A, B, and C, for delivery. The following table gives the volume in cubic feet, the weight in pounds, and the value for insurance in dollars for a unit of each of the products. If one of the company's trucks can carry 4000 cubic feet and 6000 pounds and is insured to carry $24,450, how many units of each product can be carried on the truck?

	Product A	Product B	Product C
Unit Volume (cu ft)	24	20	50
Weight (lb)	36	30	75
Value ($)	150	180	120

39. *Nutrition* A psychologist studying the effects of good nutrition on the behavior of rabbits feeds one group a combination of three foods, I, II, and III. Each of these foods contains three additives, A, B, and C, that are used in the study. The following table gives the percent of each additive that is present in each food. If the diet being used requires 6.88 g per day of A, 6.72 g of B, and 6.8 g of C, find the number of grams of each food that should be used each day.

	Food I	Food II	Food III
Additive A (%)	12	14	8
Additive B (%)	8	6	16
Additive C (%)	10	10	12

40. *Investment* A woman has $1,400,000 to purchase three investment plans. One investment earns 8%, a second earns 7.5%, and the third earns 10%.

a. What investment strategies will give annual earnings of $120,000 from the these investments?

b. What one investment strategy minimizes her risk and returns $120,000 if the 10% investment has the most risk?

section 7.2

Matrix Solution of Systems of Linear Equations

key concepts

- Matrix representation of systems of equations
  - *Matrix*
  - *Dimension*
  - *Square matrix*
  - *Identity matrix*
  - *Augmented matrix*
  - *Coefficient matrix*
- Row-echelon form
- Matrix row operations
- Reduced row-echelon form
- Gauss–Jordan elimination method
- Solution with technology
- Nonunique solutions

section preview ▪ Transportation

Ace Trucking Company has an order for three products, A, B, and C, for delivery. The table below gives the volume in cubic feet, the weight in pounds, and the value for insurance in dollars for a unit of each of the products. If the carrier can carry 30,000 cubic feet and 62,000 pounds and is insured for $276,000, how many units of each product can be carried?

	Product A	Product B	Product C
Unit Volume (cubic feet)	25	22	30
Weight (pounds)	25	38	70
Value (dollars)	150	180	300

If the number of units of products A, B, and C are x, y, and z, respectively, the system of linear equations that satisfies these conditions is

$$\begin{cases} 25x + 22y + 30z = 30,000 & \text{Volume} \\ 25x + 38y + 70z = 62,000 & \text{Weight} \\ 150x + 180y + 300z = 276,000 & \text{Value} \end{cases}$$

As we will see in Example 5, there is a solution, but not a unique solution. In this section, we use matrices to solve systems with unique solutions and with infinitely many solutions, and we see how to determine when a system has no solution.

Matrix Representation of Systems of Equations

Recall from Section 7.1 that in solving a system of three linear equations in three variables, our goal in each step was to use elimination to remove each variable, in turn, from the equations below it in the system. This process can be simplified if we represent the system in a new way without including the symbols for the variables. We can do this by creating a rectangular array containing the coefficients and constants from the system. This array, called a **matrix** (the plural is **matrices**) uses each row (horizontal) to represent an equation and each column (vertical) to represent a variable. For example, the system of equations that was solved in Example 1 of Section 7.1,

$$\begin{cases} 2x - 3y + z = -1 \\ x - y + 2z = -3 \\ 3x + y - z = 9 \end{cases}$$

can be represented in a matrix as follows:

$$\left[\begin{array}{ccc|c} 2 & -3 & 1 & -1 \\ 1 & -1 & 2 & -3 \\ 3 & 1 & -1 & 9 \end{array}\right]$$

In this matrix, the first column contains the coefficients of x for each of the equations, the second column contains the coefficients of y, and the third column contains the coefficients of z. The column to the right of the vertical line, containing the constants of the equations, is called the **augment** of the matrix, and a matrix containing an augment is called an **augmented matrix**. The matrix that contains only the coefficients of the variables is called a **coefficient matrix**. The coefficient matrix for this system is

$$\left[\begin{array}{ccc} 2 & -3 & 1 \\ 1 & -1 & 2 \\ 3 & 1 & -1 \end{array}\right]$$

The augmented matrix above has 3 rows and 4 columns, so its **dimension** is 3×4 (three by four). The coefficient matrix above is a 3×3 matrix. Matrices like this one are called **square matrices**. A square matrix that has 1's down its diagonal and 0's everywhere else, as in matrix I below, is called an **identity matrix**. Matrix I is a 3×3 identity matrix.

$$I = \left[\begin{array}{ccc} 1 & 0 & 0 \\ 0 & 1 & 0 \\ 0 & 0 & 1 \end{array}\right]$$

Echelon Forms of Matrices; Solving Systems with Matrices

The matrix that represents the system of equations

$$\begin{cases} x - y + 2z = -3 \\ y + 3z = -5 \\ z = -2 \end{cases} \text{ is } \left[\begin{array}{ccc|c} 1 & -1 & 2 & -3 \\ 0 & 1 & 3 & -5 \\ 0 & 0 & 1 & -2 \end{array}\right]$$

The coefficient part of this augmented matrix has all 1's on its diagonal and all 0's below its diagonal.

Any augmented matrix that has 1's or 0's on the diagonal of its coefficient part and 0's below the diagonal is said to be in **row-echelon form**.

If the rows of an augmented matrix represent the equations of a system of linear equations, we can perform **row operations** on the matrices that are equivalent to operations that we performed on equations to solve a system of equations. These operations are as follows.

Matrix Row Operations

Row Operation	**Corresponding Equation Operation**
1. Interchange two rows of the matrix.	1. Interchange two equations.
2. Multiply a row by any nonzero constant.	2. Multiply an equation by any nonzero constant.
3. Add a multiple of one row to another row.	3. Add a multiple of one equation to another.

When a new matrix results from one or more of these row operations performed on a matrix, the new matrix is **equivalent** to the original matrix.

example 1

Matrix Solution of Linear Systems

Use matrix row operations to solve the system of equations

$$\begin{cases} 2x - 3y + z = -1 & (1) \\ x - y + 2z = -3 & (2) \\ 3x + y - z = 9 & (3) \end{cases}$$

Solution

We begin by writing the augmented matrix that represents the system.

$$\left[\begin{array}{ccc|c} 2 & -3 & 1 & -1 \\ 1 & -1 & 2 & -3 \\ 3 & 1 & -1 & 9 \end{array}\right]$$

To get a first row that has 1 in the first column, we interchange row 1 and row 2. (Compare this to the equivalent system of equations, which is to the right.)

$$\xrightarrow{R_1 \leftrightarrow R_2} \left[\begin{array}{ccc|c} 1 & -1 & 2 & -3 \\ 2 & -3 & 1 & -1 \\ 3 & 1 & -1 & 9 \end{array}\right] \qquad \begin{cases} x - y + 2z = -3 \\ 2x - 3y + z = -1 \\ 3x + y - z = 9 \end{cases}$$

To get a 0 as the first entry in the second row, we multiply the first row by -2 and add it to the second row. To get a 0 as the first entry in the third row, we multiply the first row by -3 and add it to the third row. (Compare this to the equivalent system of equations, which is to the right.)

$$\begin{array}{c} (-2)R_1 + R_2 \rightarrow R_2 \\ (-3)R_1 + R_3 \rightarrow R_3 \\ \xrightarrow{\hspace{2cm}} \end{array} \left[\begin{array}{ccc|c} 1 & -1 & 2 & -3 \\ 0 & -1 & -3 & 5 \\ 0 & 4 & -7 & 18 \end{array}\right] \qquad \begin{cases} x - y + 2z = -3 \\ -y - 3z = 5 \\ 4y - 7z = 18 \end{cases}$$

To get 1 as the second entry in the second row, we multiply the second row by -1. (Compare this to the equivalent system of equations, which is to the right.)

$$\xrightarrow{(-1)R_2 \to R_2} \begin{bmatrix} 1 & -1 & 2 & -3 \\ 0 & 1 & 3 & -5 \\ 0 & 4 & -7 & 18 \end{bmatrix} \qquad \begin{cases} x - y + 2z = -3 \\ y + 3z = -5 \\ 4y - 7z = 18 \end{cases}$$

To get a 0 as the second entry in the third row, we multiply the new second row by -4 and add it to the third row. (Compare this to the equivalent system of equations, which is to the right.)

$$\xrightarrow{(-4)R_2 + R_3 \to R_3} \begin{bmatrix} 1 & -1 & 2 & -3 \\ 0 & 1 & 3 & -5 \\ 0 & 0 & -19 & 38 \end{bmatrix} \qquad \begin{cases} x - y + 2z = -3 \\ y + 3z = -5 \\ -19z = 38 \end{cases}$$

To get 1 as the third entry in the third row, we multiply the third row by $-\dfrac{1}{19}$. (Compare this to the equivalent system of equations, which is to the right.)

$$\xrightarrow{(-1/19)R_3 \to R_3} \begin{bmatrix} 1 & -1 & 2 & -3 \\ 0 & 1 & 3 & -5 \\ 0 & 0 & 1 & -2 \end{bmatrix} \qquad \begin{cases} x - y + 2z = -3 \\ y + 3z = -5 \\ z = -2 \end{cases}$$

The matrix is now in row-echelon form. The equivalent system can be solved by back substitution, giving the solution $(2, 1, -2)$, the same solution that was found in Example 1 of Section 7.1. ∎

Gauss–Jordan Elimination

One method that can be used to solve a system of n equations in n variables is called the **Gauss–Jordan elimination method**. To use this method, we attempt to reduce the $n \times n$ coefficient matrix to one that contains 1's on its diagonal and 0's everywhere else.

> The augmented matrix representing n equations in n variables is said to be in **reduced row-echelon form** if it has 1's or 0's on the diagonal of its coefficient part and 0's everywhere else.

When the original augmented matrix represents a system of n equations in n variables with a unique solution, this method will reduce the coefficient matrix of the augmented matrix to an identity matrix. When this happens, we can easily find the solution to the system, as we will see below.

Rather than use back substitution to complete the solution to the system of Example 1, we can apply several additional row operations to reduce the matrix to reduced row-echelon form. That is, we can reduce the matrix

$$\begin{bmatrix} 1 & -1 & 2 & -3 \\ 0 & 1 & 3 & -5 \\ 0 & 0 & 1 & -2 \end{bmatrix}$$

to reduced row-echelon form with the following steps:

1. Add row 2 to row 1:

$$\xrightarrow{R_2 + R_1 \to R_1} \begin{bmatrix} 1 & 0 & 5 & -8 \\ 0 & 1 & 3 & -5 \\ 0 & 0 & 1 & -2 \end{bmatrix}$$

2. Add (-5) times row 3 to row 1 and (-3) times row 3 to row 2:

$$\begin{array}{c}(-5)R_3 + R_1 \rightarrow R_1 \\ (-3)R_3 + R_2 \rightarrow R_2 \\ \hline \end{array} \left[\begin{array}{ccc|c} 1 & 0 & 0 & 2 \\ 0 & 1 & 0 & 1 \\ 0 & 0 & 1 & -2 \end{array}\right]$$

3. The coefficient matrix part of this reduced row-echelon matrix is the identity matrix, so the solutions to the system can be easily "read" from the reduced augmented matrix. The rows of this reduced row-echelon matrix translate into the equations

$$\begin{cases} x + 0y + 0z = & 2 \\ 0x + y + 0z = & 1 \\ 0x + 0y + z = -2 \end{cases} \text{ or } \begin{cases} x = & 2 \\ y = & 1 \\ z = -2 \end{cases}$$

This solution agrees with the solution to the system found in Example 1.

---

example 2

Investment

The Trust Department of Century Bank divided a $150,000 investment among three mutual funds with different levels of risk and return. The Potus Fund returns 10% per year, the Stong Fund returns 8% per year, and the Franklin Fund returns 7%. If the annual return from the combined investments is $12,900 and if the investment in the Potus Fund has $20,000 less than the sum of the investments in the other two funds, how much is invested in each fund?

Solution

If we represent the amount invested in the Potus Fund by x, the amount invested in the Stong Fund by y, and the amount invested in the Franklin Fund by z, the equations that represent this situation are

$$x + y + z = 150,000$$

$$0.10x + 0.08y + 0.07z = 12,900$$

$$x = y + z - 20,000$$

We can write these equations in a system, create an augmented matrix, and reduce the matrix representing the system to reduced row-echelon form.

$$\begin{cases} x + y + z = 150,000 \\ 0.10x + 0.08y + 0.07z = 12,900 \\ x - y - z = -20,000 \end{cases}$$

$$\left[\begin{array}{ccc|c} 1 & 1 & 1 & 150,000 \\ 0.10 & 0.08 & 0.07 & 12,900 \\ 1 & -1 & -1 & -20,000 \end{array}\right] \begin{array}{c} -0.10R_1 + R_2 \rightarrow R_2 \\ -R_1 + R_3 \rightarrow R_3 \\ \hline \end{array} \left[\begin{array}{ccc|c} 1 & 1 & 1 & 150,000 \\ 0 & -0.02 & -0.03 & -2,100 \\ 0 & -2 & -2 & -170,000 \end{array}\right]$$

$$\begin{array}{c} -50R_2 \rightarrow R_2 \\ \hline \end{array} \left[\begin{array}{ccc|c} 1 & 1 & 1 & 150,000 \\ 0 & 1 & 1.5 & 105,000 \\ 0 & -2 & -2 & -170,000 \end{array}\right] \begin{array}{c} -R_2 + R_1 \rightarrow R_1 \\ 2R_2 + R_3 \rightarrow R_3 \\ \hline \end{array} \left[\begin{array}{ccc|c} 1 & 0 & -0.5 & 45,000 \\ 0 & 1 & 1.5 & 105,000 \\ 0 & 0 & 1 & 40,000 \end{array}\right]$$

$$\begin{array}{c} 0.5R_3 + R_1 \rightarrow R_1 \\ -1.5R_3 + R_2 \rightarrow R_2 \\ \hline \end{array} \left[\begin{array}{ccc|c} 1 & 0 & 0 & 65,000 \\ 0 & 1 & 0 & 45,000 \\ 0 & 0 & 1 & 40,000 \end{array}\right]$$

From the reduced row-echelon matrix, we see that

$$x = 65,000, y = 45,000, \text{ and } z = 40,000$$

Thus, the investments are \$65,000 in the Potus Fund, \$45,000 in the Stong Fund, and \$40,000 in the Franklin Fund.

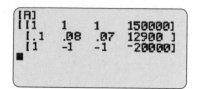

Figure 7.6

Solution with Technology

Spreadsheets, computer programs, and graphing calculators are very useful in performing matrix row operations used to reduce an augmented matrix and solve a system of linear equations. Some types of technology have a direct command that finds the equivalent row-echelon form of an augmented matrix. Figure 7.6 shows the augmented matrix for the system of equations in Example 2. We enter this matrix in a graphing utility as a 3 × 4 matrix because the matrix has 3 rows and 4 columns.

The reduced row-echelon form of this augmented matrix that results from the use of technology is shown in Figure 7.7. This matrix is the same as the one found in Example 2, and yields the same solution.

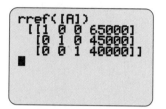

Figure 7.7

Solving systems of linear equations in more than three variables can also be done using Gauss–Jordan elimination. Because of the increased number of steps involved, technology is especially useful in solving larger systems.*

example 3

Four Equations in Four Variables

Solve the system

$$\begin{cases} x + y + z + w = 3 \\ x - 2y + z - 4w = -5 \\ x \quad\quad - z + w = 0 \\ y + z + w = 2 \end{cases}$$

Solution

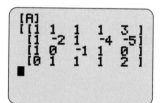

This system can be represented by the augmented matrix

$$\begin{bmatrix} 1 & 1 & 1 & 1 & 3 \\ 1 & -2 & 1 & -4 & -5 \\ 1 & 0 & -1 & 1 & 0 \\ 0 & 1 & 1 & 1 & 2 \end{bmatrix}$$

We can enter this augmented matrix into a graphing calculator and reduce the matrix to its reduced row-echelon form (Figure 7.8). The solution to the system can be "read" from the reduced matrix:

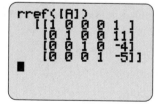

$$x = 1 \quad y = 11 \quad z = -4, \text{ and } w = -5 \quad \text{or} \quad (1, 11, -4, -5)$$

Figure 7.8

Nonunique Solution

If we represent a system of three equations in three variables by an augmented matrix and the coefficient matrix reduces to a 3 × 3 identity matrix, then the system has a unique solution. If the coefficient matrix representing a system of three equations in three variables does not reduce to an identity matrix, then either there is no solution to the system, or there are an infinite number of solutions.

---

* For details of this solution method, see Appendix A, page 668.

If a row of the reduced row-echelon coefficient matrix associated with a system contains all 0's and the augment of that row contains a nonzero number, the system has no solution and is an **inconsistent system**.

If a row of the reduced 3×3 row-echelon coefficient matrix associated with a system contains all 0's and the augment of that row also contains 0, then there are infinitely many solutions. If any column of the reduced matrix does not contain a leading 1, that variable can assume any value, and the variables corresponding to columns containing leading 1's will have values dependent on that variable. (There may be more than one variable that can assume any value, with other variables depending on them.) This is a **dependent system**, and it has infinitely many solutions.

Of course, any system with two equations in three variables cannot have a unique solution.

> A system with fewer equations than variables has either infinitely many solutions or no solutions.

Dependent Systems

The following example illustrates the solution of a dependent system.

example 4

A Dependent System

Solve the system

$$\begin{cases} 5x + 10y + 12z = 6{,}140 \\ 10x + 18y + 30z = 13{,}400 \\ 300x + 480y + 1080z = 435{,}600 \end{cases}$$

Solution

The augmented matrix that represents the system is

$$\left[\begin{array}{ccc|c} 5 & 10 & 12 & 6{,}140 \\ 10 & 18 & 30 & 13{,}400 \\ 300 & 480 & 1080 & 435{,}600 \end{array} \right]$$

The procedure follows:

$$\left[\begin{array}{ccc|c} 5 & 10 & 12 & 6{,}140 \\ 10 & 18 & 30 & 13{,}400 \\ 300 & 480 & 1080 & 435{,}600 \end{array} \right] \xrightarrow{(1/5)R_1 \to R_1} \left[\begin{array}{ccc|c} 1 & 2 & 2.4 & 1{,}228 \\ 10 & 18 & 30 & 13{,}400 \\ 300 & 480 & 1080 & 435{,}600 \end{array} \right]$$

$$\xrightarrow[\substack{-10R_1 + R_2 \to R_2 \\ -300R_1 + R_3 \to R_3}]{} \left[\begin{array}{ccc|c} 1 & 2 & 2.4 & 1{,}228 \\ 0 & -2 & 6 & 1{,}120 \\ 0 & -120 & 360 & 67{,}200 \end{array} \right]$$

$$\xrightarrow{(-1/2)R_2 \to R_2} \left[\begin{array}{ccc|c} 1 & 2 & 2.4 & 1{,}228 \\ 0 & 1 & -3 & -560 \\ 0 & -120 & 360 & 67{,}200 \end{array} \right]$$

$$\xrightarrow[\substack{120R_2 + R_3 \to R_3 \\ -2R_2 + R_1 \to R_1}]{} \left[\begin{array}{ccc|c} 1 & 0 & 8.4 & 2348 \\ 0 & 1 & -3 & -560 \\ 0 & 0 & 0 & 0 \end{array} \right]$$

Note that the third row contains only 0's and there is no leading 1 in the "z" column of the reduced matrix, and that z occurs in the equations which correspond to the first two rows of the matrix. Thus, we can solve for x in terms of z in the first equation and for y in terms of z in the second equation. These solutions are

$$x = -8.4z + 2348$$
$$y = 3z - 560$$
$$z = \text{any real number}$$

Different values of z give different solutions to the system. Two sample solutions are $z = 0, y = -560, x = 2348$ and $z = 10, y = -530, x = 2264$. If $z = a$, the solution is $(-8.4a + 2348, 3a - 560, a)$. ■

Notice that the row-reduction technique is the same for systems that have nonunique solutions as it is for systems that have unique solutions. To solve a system of n linear equations in n variables, we attempt to reduce the coefficient matrix to the identity matrix. If we succeed, the system has a unique solution. If we are unable to reduce the coefficient matrix to the identity matrix, either there is no solution or the system is dependent and we can solve it in terms of one or more of the variables.

Figure 7.9(a) shows the original augmented matrix for the system of Example 4 on a calculator screen, and Figure 7.9(b) shows the reduced row-echelon form that is equivalent to the original matrix. This is the same as the reduced matrix found in Example 4 and yields the same solution.

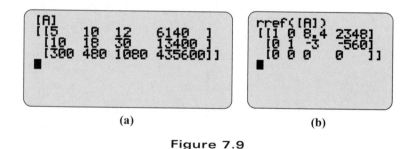

(a)   (b)

Figure 7.9

example 5   Transportation

Ace Trucking Company has an order for three products, A, B, and C, for delivery. The table below gives the volume in cubic feet, the weight in pounds, and the value for insurance in dollars for a unit of each of the products. If the company can carry 30,000 cubic feet and 62,000 pounds and is insured for $276,000, how many units of each product can be carried?

	Product A	Product B	Product C
Unit Volume (cubic feet)	25	22	30
Weight (pounds)	25	38	70
Value (dollars)	150	180	300

Solution

If we represent the number of units of product A by x, the number of units of product B by y, and the number of units of product C by z, then we can write a system of equations to represent the problem.

$$\begin{cases} 25x + 22y + 30z = 30{,}000 & \text{Volume} \\ 25x + 38y + 70z = 62{,}000 & \text{Weight} \\ 150x + 180y + 300z = 276{,}000 & \text{Value} \end{cases}$$

The Gauss–Jordan elimination method gives

$$\begin{bmatrix} 25 & 22 & 30 & 30{,}000 \\ 25 & 38 & 70 & 62{,}000 \\ 150 & 180 & 300 & 276{,}000 \end{bmatrix} \xrightarrow{\left(\frac{1}{25}\right)R_1 \to R_1} \begin{bmatrix} 1 & \frac{22}{25} & \frac{6}{5} & 1{,}200 \\ 25 & 38 & 70 & 62{,}000 \\ 150 & 180 & 300 & 276{,}000 \end{bmatrix}$$

$$\xrightarrow[\substack{(-25)R_1 + R_2 \to R_2 \\ (-150)R_1 + R_3 \to R_3}]{} \begin{bmatrix} 1 & \frac{22}{25} & \frac{6}{5} & 1{,}200 \\ 0 & 16 & 40 & 32{,}000 \\ 0 & 48 & 120 & 96{,}000 \end{bmatrix}$$

$$\xrightarrow{\left(\frac{1}{16}\right)R_2 \to R_2} \begin{bmatrix} 1 & \frac{22}{25} & \frac{6}{5} & 1{,}200 \\ 0 & 1 & \frac{5}{2} & 2{,}000 \\ 0 & 48 & 120 & 96{,}000 \end{bmatrix}$$

$$\xrightarrow[\substack{\left(-\frac{22}{25}\right)R_2 + R_1 \to R_1 \\ (-48)R_2 + R_3 \to R_3}]{} \begin{bmatrix} 1 & 0 & -1 & -560 \\ 0 & 1 & \frac{5}{2} & 2000 \\ 0 & 0 & 0 & 0 \end{bmatrix}$$

Reducing this augmented matrix by algebraic methods (above) is quite time consuming. Much time and effort can be saved if we use technology to obtain the reduced row-echelon form (Figure 7.10).

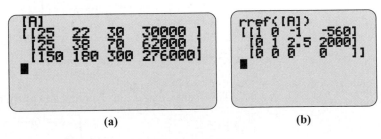

(a)   (b)

Figure 7.10

The solution to this system is $x = -560 + z$, $y = 2000 - 2.5z$, with the values of z limited so that all values are nonnegative integers. Because x must be nonnegative, z

must be at least 560, and because y must be a nonnegative integer, z must be an even number that is no more than 800. This also means that the size of the shipments of products A and B is limited by the size of the shipments of product C, because the limits on z put limits on x and y. We can write the solution as follows:

Product C: $560 \leq z \leq 800$ (z is an even integer)

Product B: $y = 2000 - 2.5z$

Product A: $x = -560 + z$

Note that the number of solutions is large, but limited.   ▪

Inconsistent Systems

The following example shows how we can use matrices to determine that a system has no solution.

example 6   ## An Inconsistent System

Solve the following system of equations if a solution exists.

$$\begin{cases} 5x + 10y + 12z = 6{,}140 \\ 10x + 18y + 30z = 13{,}400 \\ 300x + 480y + 1080z = 214{,}800 \end{cases}$$

Solution

The augmented matrix for this system is $\begin{bmatrix} 5 & 10 & 12 & 6{,}140 \\ 10 & 18 & 30 & 13{,}400 \\ 300 & 480 & 1080 & 214{,}800 \end{bmatrix}$

We reduce this matrix as follows.

$\begin{bmatrix} 5 & 10 & 12 & 6{,}140 \\ 10 & 18 & 30 & 13{,}400 \\ 300 & 480 & 1080 & 214{,}800 \end{bmatrix}$ $\xrightarrow{(1/5)R_1 \rightarrow R_1}$ $\begin{bmatrix} 1 & 2 & 2.4 & 1{,}228 \\ 10 & 18 & 30 & 13{,}400 \\ 300 & 480 & 1080 & 214{,}800 \end{bmatrix}$

$\xrightarrow[-300R_1 + R_3 \rightarrow R_3]{-10R_1 + R_2 \rightarrow R_2}$ $\begin{bmatrix} 1 & 2 & 2.4 & 1{,}128 \\ 0 & -2 & 6 & 1{,}120 \\ 0 & -120 & 360 & -153{,}600 \end{bmatrix}$

$\xrightarrow{(-1/2)R_2 \rightarrow R_2}$ $\begin{bmatrix} 1 & 2 & 2.4 & 1{,}228 \\ 0 & 1 & -3 & -560 \\ 0 & -120 & 360 & -153{,}600 \end{bmatrix}$

$\xrightarrow[120R_2 + R_3 \rightarrow R_3]{-2R_2 + R_1 \rightarrow R_1}$ $\begin{bmatrix} 1 & 0 & 8.4 & 2{,}348 \\ 0 & 1 & -3 & -560 \\ 0 & 0 & 0 & -220{,}800 \end{bmatrix}$

This matrix represents the reduced system that is equivalent to the original system.

$$\begin{cases} x + 0y + 8.4z = 2{,}348 \\ 0x + y - 3z = -560 \\ 0x + 0y + 0z = -220{,}800 \end{cases}$$

The third equation is $0 = -220{,}800$, which is impossible. Thus, the system has no solution.   ▪

Figure 7.11 shows the augmented matrix and resulting reduced row-echelon form for the system of equations in Example 6.

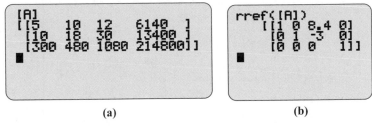

(a)   (b)

Figure 7.11

Finally, we consider systems with more equations than variables. Such systems of equations may have zero, one, or many solutions, and they are solved in the same manner as other systems.

1. If the reduced augmented matrix contains a row of 0's in the coefficient matrix with a nonzero number in the augment, the system has no solution.

2. If the coefficient matrix in the reduced augmented matrix contains an identity matrix and all remaining rows of the reduced augmented matrix contain all 0's, there is a unique solution to the system.

3. Otherwise, the system has many solutions.

skills check

7.2

In Exercises 1–4, write the augmented matrix associated with the given system.

1. $\begin{cases} x + y - z = 4 \\ x - 2y - z = -2 \\ 2x + 2y + z = 11 \end{cases}$ **2.** $\begin{cases} 3x - 4y + 6z = 10 \\ 2x - 4y - 5z = -14 \\ x + 2y - 3z = 0 \end{cases}$

3. $\begin{cases} 5x - 3y + 2z = 12 \\ 3x + 6y - 9z = 4 \\ 2x + 3y - 4z = 9 \end{cases}$ **4.** $\begin{cases} x - 3y + 4z = 7 \\ 2x + 2y - 3z = -3 \\ x - 3y + z = -2 \end{cases}$

In Exercises 5–14, the matrix associated with the solution to a system of linear equations in x, y, and z is given. Write the solution to the system, if it exists.

5. $\left[\begin{array}{ccc|c} 1 & 0 & 0 & -1 \\ 0 & 1 & 0 & 4 \\ 0 & 0 & 1 & -2 \end{array}\right]$ **6.** $\left[\begin{array}{ccc|c} 1 & 0 & 0 & -2 \\ 0 & 1 & 0 & 4 \\ 0 & 0 & 1 & 8 \end{array}\right]$

7. $\left[\begin{array}{ccc|c} 1 & 1 & -1 & 4 \\ 1 & -2 & -1 & -2 \\ 2 & 2 & 1 & 11 \end{array}\right]$

8. $\left[\begin{array}{ccc|c} 1 & -1 & 3 & 4 \\ 2 & 3 & -1 & 11 \\ 4 & 2 & -4 & 12 \end{array}\right]$

9. $\left[\begin{array}{ccc|c} 2 & -3 & 4 & 13 \\ 1 & -2 & 1 & 3 \\ 2 & -3 & 1 & 4 \end{array}\right]$ **10.** $\left[\begin{array}{ccc|c} 3 & 1 & 2 & 1 \\ 2 & 3 & -4 & -20 \\ 2 & 4 & 8 & 14 \end{array}\right]$

11. $\left[\begin{array}{ccc|c} 1 & 0 & 4 & 3 \\ 0 & 1 & 2 & 2 \\ 0 & 0 & 0 & 1 \end{array}\right]$ **12.** $\left[\begin{array}{ccc|c} 1 & 0 & 2 & 1 \\ 0 & 1 & 3 & 5 \\ 0 & 0 & 0 & 0 \end{array}\right]$

13. $\left[\begin{array}{ccc|c} 1 & 0 & 3 & 2 \\ 0 & 1 & -5 & 5 \\ 0 & 0 & 0 & 0 \end{array}\right]$ **14.** $\left[\begin{array}{ccc|c} 1 & 0 & -1 & 3 \\ 0 & 1 & 2 & -2 \\ 0 & 0 & 0 & 0 \end{array}\right]$

Solve the systems in Exercises 15–22.

15. $\begin{cases} x + y - z = 0 \\ x - 2y - z = 6 \\ 2x + 2y + z = 3 \end{cases}$

16. $\begin{cases} x - 2y + z = -5 \\ 2x - y + 2z = 6 \\ 3x + 2y - z = 1 \end{cases}$

17. $\begin{cases} 3x - 2y + 5z = 15 \\ x - 2y - 2z = -1 \\ 2x - 2y = 0 \end{cases}$

18. $\begin{cases} x + 3y + 5z = 5 \\ 2x + 4y + 3z = 9 \\ 2x + 3y + z = 1 \end{cases}$

19. $\begin{cases} 2x + 3y + 4z = 5 \\ 6x + 7y + 8z = 9 \\ 2x + y + z = 1 \end{cases}$

20. $\begin{cases} 4x + 3y + 8z = 1 \\ 2x + 3y + 8z = 5 \\ 2x + 5y + 5z = 6 \end{cases}$

21. $\begin{cases} x - y + z - w = -2 \\ 2x + 4z + w = 5 \\ 2x - 3y + z = -5 \\ y + 2z + 20w = 4 \end{cases}$

22. $\begin{cases} x - 2y + z - 3w = 10 \\ 2x - 3y + 4z + w = 12 \\ 2x - 3y + z - 4w = 7 \\ x - y + z + w = 4 \end{cases}$

In Exercises 23–32, find the solutions, if any exist, to the systems.

23. $\begin{cases} -2x + 3y + 2z = 13 \\ -2x - 2y + 3z = 0 \\ 4x + y + 4z = 11 \end{cases}$

24. $\begin{cases} 2x + 3y + 4z = 5 \\ x + y + z = 1 \\ 6x + 7y + 8z = 9 \end{cases}$ 25. $\begin{cases} 2x + 5y + 6z = 6 \\ 3x - 2y + 2z = 4 \\ 5x + 3y + 8z = 10 \end{cases}$

26. $\begin{cases} -x + 5y - 3z = 10 \\ 3x + 7y + 2z = 5 \\ 4x + 12y - z = 15 \end{cases}$

27. $\begin{cases} -x - 5y + 3z = -2 \\ 3x + 7y + 2z = 5 \\ 4x + 12y - z = 7 \end{cases}$

28. $\begin{cases} x - 3y + 2z = 12 \\ 2x - 6y + z = 7 \end{cases}$

29. $\begin{cases} 2x - 3y + 2z = 5 \\ 4x + y - 3z = 6 \end{cases}$ 30. $\begin{cases} 3x + 2y - z = 4 \\ 2x - 3y + z = 3 \end{cases}$

31. $\begin{cases} x - 3z - 3w = -2 \\ x + y + z + 3w = 2 \\ 2x + y - 2z - 2w = 0 \\ 3x + 2y - z + w = 2 \end{cases}$

32. $\begin{cases} x + y + z + 5w = 10 \\ x + 2y + z + 6w = 16 \\ x + y + 2z + 7w = 11 \\ 2x + 3y + 3z + 13w = 27 \end{cases}$

exercises

7.2

33. *Ticket Pricing* A theater owner wants to divide a 3600 seat theater into three sections, with tickets costing $40, $70, and $100, depending on the section. He wants to have twice as many $40 tickets as the sum of the other tickets, and he wants to earn $192,000 from a full house. Find how many seats he should have in each section.

34. *Rental Cars* A car rental agency rents compact, mid-size, and luxury cars. Its goal is to purchase 90 cars with a total of $2,270,000 and to earn a daily rental of $3150 from all the cars. The compact cars cost $18,000 each and earn $25 per day in rental, the mid-size cars cost $25,000 each and earn $35 per day, and the luxury cars cost $40,000 each and earn $55 per day. Find the number of each type of car the agency should purchase to meet its goal.

35. *Testing* A professor wants to create a test that has 15 true-false questions, 10 multiple-choice questions, and 5 essay questions. She wants the test to be worth 100 points, with each multiple-choice question worth twice as many points as a true-false question, and with each essay question equal to three times the number of points of a true-false question.

a. Write a system of equations to represent this problem, with x, y, and z equal to the number of points for a true-false, multiple-choice, and essay question, respectively.

b. How many points should be assigned to each type of problem? Note that only positive integer answers are useful.

36. *Testing* A professor wants to create a test that has true-false questions, multiple-choice questions, and essay questions. She wants the test to have 35 questions, with twice as many multiple-choice questions as essay questions, and with twice as many true-false questions as multiple-choice questions.

a. Write a system of equations to represent this problem.

b. How many of each type question should the professor put on the test? Note that only positive integer answers are useful.

37. *Investment* A company offers three mutual fund plans for its employees. Plan I consists of 4 blocks of common stocks and 2 municipal bonds. Plan II consists of 8 blocks of common stock, 4 municipal bonds, and 6 blocks of preferred stock. Plan III consists of 14 blocks of common stocks, 6 municipal bonds, and 6 blocks of preferred stock. If an employee wants to combine these plans so that she has 42 blocks of common stock, 20 municipal bonds, and 18 blocks of preferred stock, how many units of each plan does she need?

38. *Manufacturing* A manufacturer of swing sets has three models, Deluxe, Premium, and Ultimate, which must be painted, assembled, and packaged for shipping. The following table gives the number of hours required for each of these operations for each type of swing set. If the manufacturer has 55 hours available per day for painting, 75 hours for assembly, and 40 hours for packaging, how many of each type swing set can be produced each day?

	Deluxe	Premium	Ultimate
Painting	0.8	1	1.4
Assembly	1	1.5	2
Packaging	0.6	0.75	1

39. *Nutrition* A psychologist studying the effects of good nutrition on the behavior of rabbits feeds one group a combination of three foods, I, II, and III. Each of these foods contains three additives, A, B, and C, that are used in the study. The table below gives the percent of each additive that is present in each food. If the diet being used requires 3.74 grams per day of A, 2.04 grams of B, and 1.35 grams of C, find the number of grams of each food that should be used each day.

	Food I	Food II	Food III
Additive A	12%	15%	28%
Additive B	8%	6%	16%
Additive C	15%	2%	6%

40. *Manufacturing* To expand its manufacturing capacity, Krug Industries borrowed $440,000, part at 6%, part at 8%, and part at 10%. The sum of the money borrowed at 6% and 8% was three times that borrowed at 10%. The loan was repaid in full at the end of 5 years and the annual interest paid was $34,400. How much was borrowed at each rate?

Some of the following exercises have nonunique solutions.

41. *Investment* A trust account manager has $400,000 to be invested in three different accounts. The accounts pay 8%, 10%, and 12%, and the goal is to earn $42,400 with a minimum risk. To solve this problem, assume that x dollars are invested at 8%, y dollars are invested at 10%, and z dollars are invested at 12%.

a. Write a system of two equations in three variables to represent the problem.

b. How much can be invested in each account with the largest possible amount invested at 8%?

42. *Investment* A man has $235,000 invested in three rental properties. One property earns 7.5% per year on the investment, a second earns 10%, and the third earns 8%. The annual earnings from the properties total $18,000.

  a. Write a system of two equations to represent the problem with x, y, and z representing the 7.5%, 10%, and 8% investments, respectively.

  b. Solve this system.

  c. If $60,000 is invested at 8%, how much is invested in each of the other properties?

43. *Investment* A brokerage house offers three stock portfolios for its clients. Portfolio I consists of 10 blocks of common stock, 2 municipal bonds, and 3 blocks of preferred stock. Portfolio II consists of 12 blocks of common stock, 8 municipal bonds, and 5 blocks of preferred stock. Portfolio III consists of 10 blocks of common stock, 6 municipal bonds, and 4 blocks of preferred stock. A client wants to combine these portfolios so that she has 180 blocks of common stock, 140 municipal bonds, and 110 blocks of preferred stock. Can she do this? To answer this question, let x equal the number of units of portfolio I, y equal the number of units of portfolio II, and z equal the number of units of portfolio III, so that the equation $10x + 12y + 10z = 180$ represents the total number of blocks of common stock.

  a. Write the remaining two equations to create a system of three equations.

  b. Solve the system of equations, if possible.

44. *Investment* A company offers three mutual fund plans for its employees. Plan I consists of 14 blocks of common stock, 4 municipal bonds, and 6 blocks of preferred stock. Plan II consists of 4 blocks of common stock and 2 municipal bonds. Plan III consists of 18 blocks of common stock, 6 municipal bonds, and 6 blocks of preferred stock. If an employee wants to combine these plans so that she has 58 blocks of common stock, 20 municipal bonds, and 18 blocks of preferred stock, how many units of each plan does she need?

45. *Purchasing* A young man wins $100,000 and decides to buy four new cars. He wants to choose from cars that are priced at $40,000, $30,000, and $20,000 and spend all of the money.

  a. Write a system of two equations in three variables to represent the problem.

  b. Can this system have a unique solution?

  c. Solve the system.

  d. Use the context of the problem to find how many cars of each price he can buy with the $100,000.

46. *Social Services* A social agency is charged with providing services to three types of clients, A, B, and C. A total of 500 clients are to be served, with $300,000 available for counseling and $200,000 available for emergency food and shelter. Type A clients require an average of $400 for counseling and $600 for emergencies. Type B clients require an average of $1000 for counseling and $400 for emergencies. Type C clients require an average of $600 for counseling and $200 for emergencies. How many of each type of client can be served?

47. *Traffic Flow* In an analysis of traffic, a certain city estimates the traffic flow as illustrated in the following figure, where the arrows indicate the flow of the traffic. If x_1 represents the number of cars traveling from intersection A to intersection B, x_2 represents the number of cars traveling from intersection B to intersection C, and so on, we can formulate equations based on the principle that the number of vehicles entering the intersection equals the number leaving it.

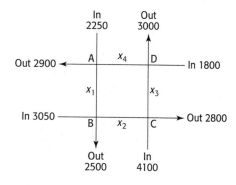

The arrows indicate that $3050 + x_1$ cars are entering intersection B and that $2500 + x_2$ cars are leaving intersection B. Thus, the equation that describes the number of cars entering and leaving intersection B is $3050 + x_1 = 2500 + x_2$. The equations that describe the number of cars entering and leaving all intersections are

$$
\begin{aligned}
\text{B:} &\quad 3050 + x_1 = 2500 + x_2 \\
\text{C:} &\quad 4100 + x_2 = x_3 + 2800 \\
\text{D:} &\quad x_3 + 1800 = 3000 + x_4 \\
\text{A:} &\quad x_4 + 2250 = x_1 + 2900
\end{aligned}
$$

Solve the system of these four equations to find how traffic between the other intersections is related to the traffic from intersection D to intersection A.

48. *Traffic Flow* In the analysis of traffic, a retirement community estimates the traffic flow on their "town square" at 6 p.m. to be as illustrated in the figure. If x_1 illustrates the number of cars moving from intersection A to intersection B, x_2 represents the number of cars traveling from intersection B to intersection C, and so on, we can formulate equations based on the principle that the number of vehicles entering the intersection equals the number leaving it. For example, the equation that represents the traffic through A is $x_4 + 470 = x_1 + 340$.

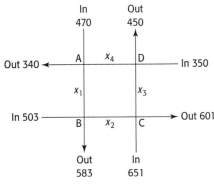

a. Formulate an equation for the traffic at each of the four intersections.

b. Solve the system of these four equations, to find how traffic between the other intersections is related to the traffic from intersection D to intersection A.

49. *Irrigation* An irrigation system allows water to flow in the pattern shown in the figure below. Water flows into the system at A and exits at B, C, and D with amounts shown. If x_1 represents the number of gallons of water moving from A to B, x_2 represents the number of gallons moving from A to C, and so on, we can formulate equations using the fact that at each point the amount of water entering the system equals the amount exiting. For example, the equation that represents the water flow through C is $x_2 = x_3 + 200{,}000$.

a. Formulate an equation for the water flow at each of the other three points.

b. Solve the system of these four equations.

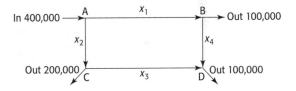

section

7.3

Matrix Operations

key concepts

- Matrix operations

  Addition

  Subtraction

  Multiplication by a constant

  Matrix multiplication

section preview ▪ Expected Life Span

Table 7.3 on the next page gives the years of life expected at birth for male and female blacks and whites born in the United States in the years 1920, 1940, 1960, 1980, 2000, 2002, and 2004. To determine what these data tell us about the relationships among race, sex, and life expectancy, we can make a matrix W containing the information for whites and a matrix B for blacks, and use these matrices to find additional information. For example, we can find a matrix D that shows how many more years whites in each category are expected to live than blacks.

We used matrices to solve systems of equations in the last section. In this section, we consider the operations of addition, subtraction, and multiplication of matrices and we use these operations in applications like the one above.

Table 7.3 Life Expectancy

YEARS	WHITES		BLACKS	
	Males	Females	Males	Females
1920	54.4	55.6	45.5	54.9
1940	62.1	66.6	51.1	45.2
1960	67.4	74.1	61.1	67.4
1980	70.7	78.1	63.8	72.5
2000	74.9	80.1	68.3	75.2
2002	75.1	80.3	68.8	75.6
2004	75.7	80.8	69.5	76.3

(Source: National Center for Health Statistics)

Addition and Subtraction of Matrices

If two matrices have the same numbers of rows and columns (the same dimensions) we can add them by adding the corresponding entries of the two matrices.

> ### Matrix Addition
>
> The sum of two matrices with the same dimensions is the matrix that is formed by adding the corresponding entries of the two matrices. Addition is not defined if the matrices do not have the same number of rows and the same number of columns.

example 1 ### Matrix Addition

Find the following sums of matrices.

a. $\begin{bmatrix} a & b \\ c & d \end{bmatrix} + \begin{bmatrix} w & y \\ x & z \end{bmatrix}$

b. $\begin{bmatrix} 1 & 3 & -8 & 0 \\ 3 & 6 & 1 & -3 \\ -4 & 5 & 3 & 2 \end{bmatrix} + \begin{bmatrix} -1 & 4 & 2 & 4 \\ 5 & -2 & 4 & 2 \\ -5 & 1 & 4 & -1 \end{bmatrix}$

c. $A + B$ if $A = \begin{bmatrix} 2 & -3 \\ -1 & 4 \end{bmatrix}$ and $B = \begin{bmatrix} -2 & 3 \\ 1 & -4 \end{bmatrix}$

Solution

a. $\begin{bmatrix} a & b \\ c & d \end{bmatrix} + \begin{bmatrix} w & y \\ x & z \end{bmatrix} = \begin{bmatrix} a+w & b+y \\ c+x & d+z \end{bmatrix}$

b. $\begin{bmatrix} 1 & 3 & -8 & 0 \\ 3 & 6 & 1 & -3 \\ -4 & 5 & 3 & 2 \end{bmatrix} + \begin{bmatrix} -1 & 4 & 2 & 4 \\ 5 & -2 & 4 & 2 \\ -5 & 1 & 4 & -1 \end{bmatrix}$

$= \begin{bmatrix} 1+(-1) & 3+4 & -8+2 & 0+4 \\ 3+5 & 6+(-2) & 1+4 & -3+2 \\ -4+(-5) & 5+1 & 3+4 & 2+(-1) \end{bmatrix} = \begin{bmatrix} 0 & 7 & -6 & 4 \\ 8 & 4 & 5 & -1 \\ -9 & 6 & 7 & 1 \end{bmatrix}$

c. $A + B = \begin{bmatrix} 2 & -3 \\ -1 & 4 \end{bmatrix} + \begin{bmatrix} -2 & 3 \\ 1 & -4 \end{bmatrix} = \begin{bmatrix} 0 & 0 \\ 0 & 0 \end{bmatrix}$ ■

The matrix that is the sum in Example 1(c) is called a **zero matrix** because each of its elements is 0. Matrix B in Example 1(c) is called the **negative of matrix** A, denoted $-A$, because the sum of matrices A and B is a zero matrix. Similarly, matrix A is the negative of matrix B, and can be denoted by $-B$.

> ### Matrix Subtraction
>
> If matrices M and N have the same dimension, the difference $M - N$ is found by subtracting the elements of N from the corresponding elements of M. This difference can also be defined as
>
> $$M - N = M + (-N)$$

example 2 | ## Matrix Subtraction

Complete the following matrix operations.

a. $\begin{bmatrix} 2 & -1 & 3 \\ 5 & -4 & 2 \\ 1 & 4 & -2 \end{bmatrix} - \begin{bmatrix} 1 & 2 & 4 \\ 5 & -2 & 3 \\ 6 & 2 & -3 \end{bmatrix}$

b. $\begin{bmatrix} 1 & 4 \\ 3 & 6 \\ -2 & 1 \end{bmatrix} - \begin{bmatrix} -5 & 2 \\ 3 & 5 \\ 12 & 3 \end{bmatrix} + \begin{bmatrix} 2 & 3 \\ -4 & -8 \\ 2 & 0 \end{bmatrix}$

Solution

a. $\begin{bmatrix} 2 & -1 & 3 \\ 5 & -4 & 2 \\ 1 & 4 & -2 \end{bmatrix} - \begin{bmatrix} 1 & 2 & 4 \\ 5 & -2 & 3 \\ 6 & 2 & -3 \end{bmatrix}$

$= \begin{bmatrix} 2-1 & -1-2 & 3-4 \\ 5-5 & -4-(-2) & 2-3 \\ 1-6 & 4-2 & -2-(-3) \end{bmatrix} = \begin{bmatrix} 1 & -3 & -1 \\ 0 & -2 & -1 \\ -5 & 2 & 1 \end{bmatrix}$

b. $\begin{bmatrix} 1 & 4 \\ 3 & 6 \\ -2 & 1 \end{bmatrix} - \begin{bmatrix} -5 & 2 \\ 3 & 5 \\ 12 & 3 \end{bmatrix} + \begin{bmatrix} 2 & 3 \\ -4 & -8 \\ 2 & 0 \end{bmatrix}$

$= \begin{bmatrix} 6 & 2 \\ 0 & 1 \\ -14 & -2 \end{bmatrix} + \begin{bmatrix} 2 & 3 \\ -4 & -8 \\ 2 & 0 \end{bmatrix} = \begin{bmatrix} 8 & 5 \\ -4 & -7 \\ -12 & -2 \end{bmatrix}$ ■

We can use technology to perform the operations of addition and subtraction of matrices, such as those in Examples 1 and 2. Figure 7.12 on the next page shows the computations for Example 2(b).*

———

* For more details, see Appendix A, page 666.

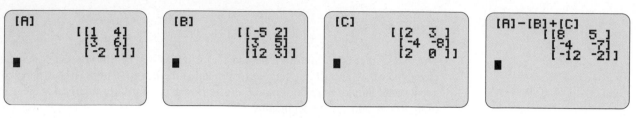

Figure 7.12

example 3

Life Expectancy

Table 7.4 gives the years of life expected at birth for male and female blacks and whites born in the United States in the years 1920, 1940, 1960, 1980, 2000, 2002, and 2004.

Table 7.4

YEARS	WHITES		BLACKS	
	Males	Females	Males	Females
1920	54.4	55.6	45.5	54.9
1940	62.1	66.6	51.1	45.2
1960	67.4	74.1	61.1	67.4
1980	70.7	78.1	63.8	72.5
2000	74.9	80.1	68.3	75.2
2002	75.1	80.3	68.8	75.6
2004	75.7	80.8	69.5	76.3

(Source: National Center for Health Statistics)

a. Make a matrix W containing the life expectancy data for whites and a matrix B for blacks.

b. Use these matrices to find matrix $D = W - B$ that represents the difference between the white and black life expectancy.

c. What does this tell us about race and life expectancy?

Solution

$$\textbf{a. } W = \begin{bmatrix} 54.4 & 55.6 \\ 62.1 & 66.6 \\ 67.4 & 74.1 \\ 70.7 & 78.1 \\ 74.9 & 80.1 \\ 75.1 & 80.3 \\ 75.7 & 80.8 \end{bmatrix} \quad B = \begin{bmatrix} 45.5 & 54.9 \\ 51.1 & 45.2 \\ 61.1 & 67.4 \\ 63.8 & 72.5 \\ 68.3 & 75.2 \\ 68.8 & 75.6 \\ 69.5 & 76.3 \end{bmatrix} \quad \textbf{b. } D = W - B = \begin{bmatrix} 8.9 & 0.7 \\ 11.0 & 21.4 \\ 6.3 & 6.7 \\ 6.9 & 5.6 \\ 6.6 & 4.9 \\ 6.3 & 4.7 \\ 6.2 & 4.5 \end{bmatrix}$$

c. Because each element in the difference matrix D is positive, we conclude that the life expectancy for whites is longer than that for blacks for all birth years and for both sexes.

spreadsheet solution

We can use spreadsheets as well as calculators to perform operations with matrices. Table 7.5 shows spreadsheets of matrices W, B, and D from Example 3.*

Table 7.5

	A	B	C
1	W	54.4	55.6
2		62.1	66.6
3		67.4	74.1
4		70.7	78.1
5		74.9	80.1
6		75.1	80.3
7		75.7	80.8
8			
9	B	45.5	54.9
10		51.1	45.2
11		61.1	67.4
12		63.8	72.5
13		68.3	75.2
14		68.8	75.6
15		69.5	76.3
16			
17	D = W − B	8.9	0.7
18		11	21.4
19		6.3	6.7
20		6.9	5.6
21		6.6	4.9
22		6.3	4.7
23		6.2	4.5

Multiplication of a Matrix by a Number

As in operations with real numbers, we can multiply a matrix by a positive integer to find the sum of repeated additions. For example, if

$$A = \begin{bmatrix} 1 & -2 & 4 \\ 6 & 3 & -3 \end{bmatrix}$$

---

* For more details, see Appendix B, page 689.

then we can find $2A$ in two ways:

$$2A = A + A = \begin{bmatrix} 1 & -2 & 4 \\ 6 & 3 & -3 \end{bmatrix} + \begin{bmatrix} 1 & -2 & 4 \\ 6 & 3 & -3 \end{bmatrix} = \begin{bmatrix} 2 & -4 & 8 \\ 12 & 6 & -6 \end{bmatrix}$$

and

$$2A = \begin{bmatrix} 2 \cdot 1 & 2(-2) & 2 \cdot 4 \\ 2 \cdot 6 & 2 \cdot 3 & 2(-3) \end{bmatrix} = \begin{bmatrix} 2 & -4 & 8 \\ 12 & 6 & -6 \end{bmatrix}$$

In general, we define multiplication of a matrix by a real number as follows.

Product of a Number and a Matrix

Multiplying a matrix A by a real number c results in a matrix in which each entry of matrix A is multiplied by the number c.

For example, if $A = \begin{bmatrix} a & b & c & d \\ e & f & g & h \end{bmatrix}$, then $nA = \begin{bmatrix} na & nb & nc & nd \\ ne & nf & ng & nh \end{bmatrix}$.

Note that $-A = (-1)A = \begin{bmatrix} -a & -b & -c & -d \\ -e & -f & -g & -h \end{bmatrix}$.

example 4 ## Price Increases

Table 7.6 contains the purchase prices and delivery costs (per unit) for plywood, siding, and 2×4 lumber. If the supplier announces a 5% increase in all of these prices and delivery costs, find the matrix that gives the new prices and costs.

Table 7.6

	Plywood	Siding	2×4's
Purchase Price ($)	32	23	2.60
Delivery Costs ($)	3	1	0.60

Solution

The matrix that represents the original prices and costs is

$$\begin{bmatrix} 32 & 23 & 2.60 \\ 3 & 1 & 0.60 \end{bmatrix}$$

To find the prices and costs after a 5% increase, we multiply the matrix by 1.05 (100% of the old prices plus the 5% increase).

$$1.05\begin{bmatrix} 32 & 23 & 2.60 \\ 3 & 1 & 0.60 \end{bmatrix} = \begin{bmatrix} 33.60 & 24.15 & 2.73 \\ 3.15 & 1.05 & 0.63 \end{bmatrix} \qquad \blacksquare$$

Matrix Multiplication

Suppose Circuitown has made a special purchase for one of its stores, consisting of 22 televisions, 15 washers, and 12 dryers. If the value of each television is $550, each washer is $435, and each dryer is $325, then the value of this purchase is

$$550 \cdot 22 + 435 \cdot 15 + 325 \cdot 12 = 22{,}525 \text{ dollars}$$

If we write the value of each item in a 1×3 *row matrix*

$$A = [550 \quad 435 \quad 325]$$

and the number of each of the items in the special purchase in a 3×1 *column matrix*

$$B = \begin{bmatrix} 22 \\ 15 \\ 12 \end{bmatrix}$$

then the value of the special purchase can be represented by the **matrix product**

$$AB = [550 \quad 435 \quad 325]\begin{bmatrix} 22 \\ 15 \\ 12 \end{bmatrix}$$

$$= [550 \cdot 22 + 435 \cdot 15 + 325 \cdot 12] = [22{,}525]$$

In general, we have the following.

Product of a Row Matrix and a Column Matrix

The product of a $1 \times n$ row matrix and an $n \times 1$ column matrix is a 1×1 matrix given by

$$[a_1 \quad a_2 \cdots a_n]\begin{bmatrix} b_1 \\ b_2 \\ \vdots \\ b_n \end{bmatrix} = [a_1b_1 + a_2b_2 + \cdots + a_nb_n]$$

We can expand the multiplication to larger matrices. For example, suppose Circuitown has a second store and purchases 28 televisions, 21 washers, and 26 dryers for it. Rather than writing two matrices to represent the two stores, we can use a two-column matrix C to represent the purchases for the two stores.

$$
\begin{array}{c}
\text{Store I} \quad \text{Store II} \\
C = \begin{bmatrix} 22 & 28 \\ 15 & 21 \\ 12 & 26 \end{bmatrix} \begin{array}{l} \text{TVs} \\ \text{Washers} \\ \text{Dryers} \end{array}
\end{array}
$$

If these products have the same values as given above, we can find the value of the purchases for each store by multiplying the row matrix times each of the column matrices. The value of the store I purchases is found by multiplying the row matrix A times the first column of matrix C, and the value of the store II purchases is found by multiplying matrix A times the second column of matrix C. The result is

$$AC = [550 \quad 435 \quad 325]\begin{bmatrix} 22 & 28 \\ 15 & 21 \\ 12 & 26 \end{bmatrix}$$

$$= [550 \cdot 22 + 435 \cdot 15 + 325 \cdot 12 \quad 550 \cdot 28 + 435 \cdot 21 + 325 \cdot 26]$$

$$= [22{,}525 \quad 32{,}985]$$

This indicates that the value of the store I purchase is \$22,525 (the value found before) and the value of the store II purchase is \$32,985.

The matrix AC on the previous page is the product of the 1×3 matrix A and the 3×2 matrix C. This product is a 1×2 matrix. In general, the product of an $m \times n$ matrix and an $n \times k$ matrix is an $m \times k$ matrix, and the product is undefined if the number of columns in the first matrix does not equal the number of rows in the second matrix.

In general, we can define the product of two matrices by defining how each element of the product is formed.

Product of Two Matrices

The product of an $m \times n$ matrix A and an $n \times k$ matrix B is the $m \times k$ matrix $C = AB$. The element in the ith row and jth column of matrix C has the form

$$c_{ij} = [a_{i1} \quad a_{i2} \cdots a_{in}]\begin{bmatrix} b_{1j} \\ b_{2j} \\ \vdots \\ b_{nj} \end{bmatrix} = [a_{i1}b_{1j} + a_{i2}b_{2j} + \cdots + a_{in}b_{nj}]$$

We illustrate below the product AB, with each of the c_{ij} elements found as shown in the box above:

$$C = AB = \begin{bmatrix} a_{11} & a_{12} \ldots a_{1n} \\ a_{21} & a_{22} \ldots a_{2n} \\ \vdots & \vdots \quad \vdots \\ a_{i1} & a_{i2} \ldots a_{in} \\ \vdots & \vdots \quad \vdots \\ a_{m1} & a_{m2} \ldots a_{mn} \end{bmatrix} \begin{bmatrix} b_{11} & b_{12} \ldots b_{1j} \ldots b_{1k} \\ b_{21} & b_{22} \ldots b_{2j} \ldots b_{2k} \\ \vdots & \vdots \quad \vdots \quad \vdots \\ b_{n1} & b_{n2} \ldots b_{nj} \ldots b_{nk} \end{bmatrix} = \begin{bmatrix} c_{11} & c_{12} \ldots c_{1j} \ldots c_{1k} \\ c_{21} & c_{22} \ldots c_{2j} \ldots c_{2k} \\ \vdots & \vdots \quad \vdots \quad \vdots \\ c_{i1} & c_{i2} \ldots c_{ij} \ldots c_{ik} \\ \vdots & \vdots \quad \vdots \quad \vdots \\ c_{m1} & c_{m2} \ldots c_{mj} \ldots c_{mk} \end{bmatrix}$$

$$m \times n \qquad\qquad n \times k \qquad\qquad m \times k$$

---

example 5 ## Matrix Product

Compute the products AB and BA for the matrices $A = \begin{bmatrix} 1 & 2 \\ 3 & 4 \end{bmatrix}$ and $B = \begin{bmatrix} a & b \\ c & d \end{bmatrix}$.

Solution

$$AB = \begin{bmatrix} 1 & 2 \\ 3 & 4 \end{bmatrix}\begin{bmatrix} a & b \\ c & d \end{bmatrix} = \begin{bmatrix} 1a + 2c & 1b + 2d \\ 3a + 4c & 3b + 4d \end{bmatrix}$$

$$BA = \begin{bmatrix} a & b \\ c & d \end{bmatrix}\begin{bmatrix} 1 & 2 \\ 3 & 4 \end{bmatrix} = \begin{bmatrix} 1a + 3b & 2a + 4b \\ 1c + 3d & 2c + 4d \end{bmatrix}$$

Note that in Example 5 the product AB is quite different from the product BA. That is, $AB \neq BA$. For some matrices, but not all, $AB \neq BA$. We indicate this by saying that *matrix multiplication is not commutative.*

---

example 6 ## Advertising

A business plans to use three methods of advertising: newspapers, radio, and cable TV in each of its two markets, I and II. The cost per ad type in each market (in thousands of dollars) is given by matrix A. The business has three target groups, teenagers, single women, and men aged 35 to 50. Matrix B gives the number of ads per week directed at each of these groups.

$$
A = \begin{array}{c} \\ \\ \\ \end{array}
\begin{array}{cc} \text{Mkt I} & \text{Mkt II} \\ \left[\begin{array}{cc} 12 & 10 \\ 10 & 8 \\ 5 & 9 \end{array} \right] & \begin{array}{l} \text{Paper} \\ \text{Radio} \\ \text{TV} \end{array} \end{array}
\qquad
B = \begin{array}{cc} \text{Paper} \quad \text{Radio} \quad \text{TV} \\ \left[\begin{array}{ccc} 3 & 12 & 15 \\ 5 & 16 & 10 \\ 10 & 11 & 6 \end{array} \right] \begin{array}{l} \text{Teens} \\ \text{Single females} \\ \text{Men 35–50} \end{array} \end{array}
$$

a. Does AB or BA give the matrix that represents the cost of ads for each target group of people in each market?

b. Find this matrix.

c. For what group of people is the most money spent on advertising?

Solution

a. Multiplying matrix B times matrix A gives the total cost of ads for each target group in each market. Note that the product AB is undefined, because multiplying a 3×2 matrix times a 3×3 matrix is not possible.

b. $BA = \begin{bmatrix} 3 & 12 & 15 \\ 5 & 16 & 10 \\ 10 & 11 & 6 \end{bmatrix} \begin{bmatrix} 12 & 10 \\ 10 & 8 \\ 5 & 9 \end{bmatrix}$

$= \begin{bmatrix} 36 + 120 + 75 & 30 + 96 + 135 \\ 60 + 160 + 50 & 50 + 128 + 90 \\ 120 + 110 + 30 & 100 + 88 + 54 \end{bmatrix} = \begin{bmatrix} 231 & 261 \\ 270 & 268 \\ 260 & 242 \end{bmatrix}$

The columns of this matrix represent the markets and the rows represent the target groups.

$$
\begin{array}{cc} \text{Mkt I} \quad\ \text{Mkt II} \\ \left[\begin{array}{cc} 231 & 261 \\ 270 & 268 \\ 260 & 242 \end{array} \right] \begin{array}{l} \text{Teens} \\ \text{Single females} \\ \text{Men 35–50} \end{array} \end{array}
$$

c. The largest amount is spent on single females, \$270,000 in market I and \$268,000 in market II. ∎

Multiplication with Technology

We can multiply two matrices by using technology. We can find the product of matrix B times matrix A on a graphing utility by entering $[B]*[A]$ (or $[B][A]$) and pressing $\boxed{\text{ENTER}}$.*
Figure 7.13 shows the product BA from Example 6.

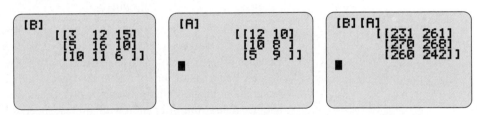

Figure 7.13

---

* For more details, see Appendix A, page 667.

| example 7 | **Multiplying with Technology** |

Use technology to compute BA and AB if $A = \begin{bmatrix} 1 & 2 \\ 0 & -1 \\ 3 & -2 \end{bmatrix}$ and $B = \begin{bmatrix} 2 & 4 & -1 \\ 3 & -2 & 1 \\ 2 & 0 & 2 \\ 1 & -3 & 0 \end{bmatrix}$.

Solution

Figure 7.14 shows displays with matrix A, matrix B, and the product BA.

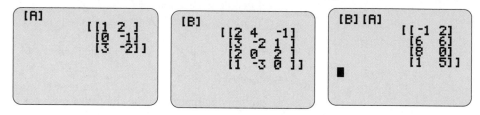

Figure 7.14

Figure 7.15 shows that the matrix product AB does not exist because the dimensions do not match (that is, the number of columns of A does not equal the number of rows of B).

Figure 7.15

┌─ **Spreadsheet Solution**

Like graphing calculators and software programs, computer spreadsheets can be used to perform operations with matrices.* Consider the following example.

| example 8 | **Manufacturing** |

A furniture company manufactures two products, A and B, which are constructed using steel, plastic, and fabric. The numbers of units of each raw material that are required for each product are given by the table below.

	Steel	Plastic	Fabric
Product A	2	3	8
Product B	3	1	10

Because of transportation costs to the company's two plants, the unit costs for some of the raw materials differ. The table on page 555 gives the unit costs for each of the raw materials at the two plants. Create two matrices from the information in the two tables and use matrix multiplication with a spreadsheet to find the cost of manufacturing each product at each plant.

---

* For more details, see Appendix B, page 691.

	Plant I	Plant II
Steel	$15	$16
Plastic	$11	$10
Fabric	$ 6	$ 7

Solution

We represent the number of units of raw materials for each product by the matrix P and the costs of the raw materials at each plant by matrix C. The cost of manufacturing each product at each plant is given by the matrix product PC. The spreadsheet shown in Table 7.7 gives matrix P, matrix C, and the matrix product PC.

Table 7.7

	A	B	C	D
1	Matrix P	2	3	8
2		3	1	10
3				
4	Matrix C	15	16	
5		11	10	
6		6	7	
7				
8	Product PC	111	118	
9		116	128	

The rows of matrix PC represent Product A and Product B, respectively, and the columns represent Plant I and Plant II. The entries give the costs of each product at each plant.

$$\begin{array}{cc} \text{Plant I} & \text{Plant II} \end{array}$$
$$\begin{bmatrix} 111 & 118 \\ 116 & 128 \end{bmatrix} \begin{array}{l} \text{Product A} \\ \text{Product B} \end{array}$$

Recall that a square ($n \times n$) matrix with 1's on the diagonal and 0's elsewhere is an identity matrix. For any $n \times k$ matrix A and the $n \times n$ matrix I, $IA = A$, and for any $m \times n$ matrix B and the $n \times n$ matrix I, $BI = B$. If matrix C is an $n \times n$ matrix, then for the $n \times n$ matrix I, $IC = C$ and $CI = C$, so the product of an identity matrix and another square matrix is *commutative*. (Recall that multiplication of matrices, in general, is not commutative.)

example 9

The Identity Matrix

a. Write the 2×2 identity matrix I.

b. For the matrix $A = \begin{bmatrix} 3 & -2 \\ -1 & 5 \end{bmatrix}$, find IA and AI.

Solution

a. $\begin{bmatrix} 1 & 0 \\ 0 & 1 \end{bmatrix}$

b. $\begin{bmatrix} 1 & 0 \\ 0 & 1 \end{bmatrix}\begin{bmatrix} 3 & -2 \\ -1 & 5 \end{bmatrix} = \begin{bmatrix} 3 & -2 \\ -1 & 5 \end{bmatrix}$ and $\begin{bmatrix} 3 & -2 \\ -1 & 5 \end{bmatrix}\begin{bmatrix} 1 & 0 \\ 0 & 1 \end{bmatrix} = \begin{bmatrix} 3 & -2 \\ -1 & 5 \end{bmatrix}$

Thus, $IA = A$ and $AI = A$. ■

skills check

7.3

Use the following matrices for Exercises 1–12.

$$A = \begin{bmatrix} 1 & 3 & -2 \\ 3 & 1 & 4 \\ -5 & 3 & 6 \end{bmatrix} \qquad B = \begin{bmatrix} 2 & 1 & -1 \\ 3 & 2 & 4 \end{bmatrix}$$

$$C = \begin{bmatrix} 1 & 3 \\ 2 & 1 \\ 3 & -1 \end{bmatrix} \qquad D = \begin{bmatrix} 2 & 3 & 1 \\ 3 & 4 & -1 \\ 2 & 5 & 1 \end{bmatrix}$$

$$E = \begin{bmatrix} 9 & 2 & -7 \\ -5 & 0 & 5 \\ 7 & -4 & -1 \end{bmatrix} \qquad F = \begin{bmatrix} 2 & 1 & 3 \\ 4 & 0 & 1 \end{bmatrix}$$

1. Which pairs of the matrices can be added?

2. Use letters to represent the matrix products that are defined.

3. Find the sum of A and D if it is defined.

4. **a.** Find $D + E$ and $E + D$.

   b. Are the sums equal?

5. Find $3A$.

6. Find $-4F$.

7. Find $2D - 4A$.

8. Find $2B - 4F$.

9. **a.** Find AD and DA if these products exist.

   b. Are these products equal?

   c. Do these products have the same dimension?

10. **a.** Find BC and CB if these products exist.

    b. Are these products equal?

    c. Do these products have the same dimension?

11. **a.** Compute DE and ED.

    b. What is the name of $\dfrac{1}{10} DE$?

12. If I is a 3×3 identity matrix, find ID and DI.

13. Compute the sum of $A = \begin{bmatrix} 1 & 5 \\ 3 & 2 \end{bmatrix}$ and $B = \begin{bmatrix} 2a & 3b \\ -c & -2d \end{bmatrix}$.

14. Compute the difference $A - B$ if $A = \begin{bmatrix} a & b \\ c & d \\ f & g \end{bmatrix}$ and $B = \begin{bmatrix} 1 & 2 \\ 3 & 4 \\ 5 & 6 \end{bmatrix}$.

15. Compute $3A - 2B$ if $A = \begin{bmatrix} a & b \\ c & d \end{bmatrix}$ and $B = \begin{bmatrix} 1 & 2 \\ 3 & 4 \end{bmatrix}$.

16. Compute $2A - 3B$ if $A = \begin{bmatrix} 1 & -3 & 2 \\ 2 & 2 & -1 \\ 3 & 4 & 2 \end{bmatrix}$ and $B = \begin{bmatrix} 2 & 2 & 2 \\ 3 & -2 & -1 \\ 1 & 1 & 2 \end{bmatrix}$.

17. If an $m \times n$ matrix is multiplied by an $n \times k$ matrix, what is the dimension of the matrix that is the product?

18. If A and B are any two matrices, does $AB = BA$ always, sometimes, or never?

19. If A is a 2×3 matrix and B is a 4×2 matrix:

    a. Which product is defined, AB or BA?

    b. What is the dimension of the product that is defined?

20. If C is a 3×4 matrix and D is a 4×3 matrix, what are the dimensions of CD and DC?

21. Find AB and BA if $A = \begin{bmatrix} a & b & c \\ d & e & f \end{bmatrix}$ and

$B = \begin{bmatrix} 1 & 2 \\ 3 & 4 \\ 5 & 6 \end{bmatrix}$.

22. Find EF and FE if $E = \begin{bmatrix} a & b \\ c & d \end{bmatrix}$ and $F = \begin{bmatrix} e & f \\ g & h \end{bmatrix}$.

23. If $A = \begin{bmatrix} 1 & 5 \\ 3 & 2 \end{bmatrix}$ and $B = \begin{bmatrix} 2 & 3 \\ -1 & -2 \end{bmatrix}$, compute AB and BA, if possible.

24. If $A = \begin{bmatrix} 1 & 4 \\ 3 & -1 \\ -2 & 2 \end{bmatrix}$ and $B = \begin{bmatrix} 4 & 2 & 2 \\ -1 & 3 & 1 \end{bmatrix}$, compute AB and BA, if possible.

25. If $A = \begin{bmatrix} 1 & -1 & 2 \\ 3 & 4 & 4 \end{bmatrix}$ and $B = \begin{bmatrix} 3 & 1 \\ 1 & 3 \\ -2 & 1 \end{bmatrix}$, compute AB and BA, if possible.

26. Suppose $A = \begin{bmatrix} 1 & -\dfrac{1}{2} & -\dfrac{1}{4} \\ -\dfrac{1}{2} & 0 & \dfrac{1}{2} \\ 0 & \dfrac{1}{2} & -\dfrac{1}{4} \end{bmatrix}$ and

$B = \begin{bmatrix} 2 & 2 & 2 \\ 1 & 2 & 3 \\ 2 & 4 & 2 \end{bmatrix}$.

a. Compute AB and BA, if possible.

b. Are the products equal?

exercises

7.3

27. *Endangered Species* The tables give the number of some species of threatened and endangered wildlife in the United States and in foreign countries in 2007.

a. Form the matrix A that contains the number of each of these species in the United States in 2007 and matrix B that contains the number of each of these species outside the United States in 2007.

b. Write a matrix containing the total number of each of these species, assuming that the U.S. and foreign species are different.

United States	Mammals	Birds	Reptiles	Amphibians	Fishes
Endangered	70	76	13	13	74
Threatened	12	15	24	10	64

Foreign	Mammals	Birds	Reptiles	Amphibians	Fishes
Endangered	255	175	65	8	11
Threatened	20	6	16	1	1

(Source: U.S. Census Bureau; *The 2008 Statistical Abstract*)

28. *Endangered Species*

   a. Use the information and matrices in Exercise 27 to find the matrix $C = B - A$ and tell what the elements of this matrix signify.

   b. What do the negative elements of matrix C mean?

29. *Trade Balances* The table below gives the U.S. exports and imports in three categories for the years 2000 and 2006.

   a. Form the matrix A that contains the number of millions of dollars of U.S. exports in 2000 and 2006.

   b. Form the matrix B that contains the number of millions of dollars of U.S. imports in 2000 and in 2006.

   c. Write the matrix $C = A - B$, which gives the *balance of trade* for these U.S. product categories.

   d. In what categories and years does the U.S. have a positive trade balance?

   e. In which category and year is the trade deficit the greatest? Why?

U.S. Exports ($ millions)	2000	2006
Agricultural	51,296	70,912
Manufactured goods	625,894	785,599
Mineral fuels	13,179	34,711

U.S. Imports ($ millions)	2000	2006
Agricultural	39,186	65,459
Manufactured goods	1,012,855	1,416,302
Mineral fuels	135,367	332,500

(Source: U.S. Dept. of Commerce)

30. *Exports*

   a. Use the tables in Exercise 29 to create four 3×1 matrices, one for each of the columns in the tables.

   b. Use two of these matrices to find a matrix E that gives the balance of trade for 2000.

   c. Use two of these matrices to find a matrix F that gives the balance of trade for 2006.

   d. Find the matrix $E - F$ that compares the balance of trade for the two years.

   e. Did the balance of trade improve in any categories from 2000–2006?

31. *Income* The table below gives the median annual income for different sexes and races in 2005. Create a matrix containing the data and use a matrix operation to create a matrix that contains the median income if the 2005 median incomes are increased by 12% in all categories.

Median Annual Income, 2005($)		
	Male	Female
White	32,179	18,669
Black	22,653	17,631
Hispanic	22,089	15,036

(Source: U.S. Census Bureau)

32. *Exchange Rates* The table below gives the national currency units per U.S. dollar for Germany (mark) and for Japan (yen) for the years 1999 and 2006. If the fee for exchanging dollars to these units reduces the national currency returned per U.S. dollar by 5%, place this data in a matrix and use a matrix operation to find the currency units actually returned per dollar.

National Currency Units per U.S. Dollar		
	Germany	Japan
1999	1.8361	113.907
2006	1.5613	116.299

(Source: International Monetary Fund)

33. *Advertising* A political candidate plans to use three methods of advertising: newspapers, radio, and cable TV. The cost per ad (in thousands of dollars) for each type of media is given by matrix A. Matrix B shows the number of ads per month in these three media that are targeted to single people, to married males aged 35 to 55, and to married females over 65 years of age. Find the matrix that gives the cost of ads for each target group.

$$A = \begin{bmatrix} 12 \\ 15 \\ 5 \end{bmatrix} \begin{matrix} \text{TV} \\ \text{Radio} \\ \text{Papers} \end{matrix}$$

$$B = \begin{bmatrix} 30 & 45 & 35 \\ 25 & 32 & 40 \\ 22 & 12 & 30 \end{bmatrix} \begin{matrix} \text{Singles} \\ \text{Males 35–55} \\ \text{Females 65+} \end{matrix}$$

with column headings TV, Radio, Papers and Cost above matrix A.

34. *Cost* Men and women in a church choir wear choir robes in the sizes shown in matrix A. Matrix B contains the prices (in dollars) of new robes and hoods according to size.

   a. Find the product BA and label the rows and columns to show what each row represents.

   b. What is the cost of the robes for all the men? For all the women?

$$A = \begin{bmatrix} 10 & 24 \\ 22 & 10 \\ 33 & 3 \end{bmatrix} \begin{matrix} \text{Small} \\ \text{Medium} \\ \text{Large} \end{matrix}$$

with column headings Men, Women.

$$B = \begin{bmatrix} 45 & 50 & 55 \\ 20 & 20 & 20 \end{bmatrix} \begin{matrix} \text{Robes} \\ \text{Hoods} \end{matrix}$$

with column headings S, M, L.

35. *Manufacturing* Two departments of a firm, A and B, need differing amounts of steel, wood, and plastic. The following tables give the amounts of the products needed by the departments.

	Steel	Wood	Plastic
Department A	60	40	20
Department B	40	20	40

These three products are supplied by two suppliers, DeTuris and Marriott, with the unit prices given in the following table.

	DeTuris	Marriott
Steel	600	560
Wood	300	200
Plastic	300	400

   a. Use matrix multiplication to determine how much these orders will cost each department at each of the two suppliers.

   b. From which supplier should each department make its purchase?

36. *Manufacturing* A furniture manufacturer produces three styles of chairs, with the number of units of each type of raw material needed for each style given in the table below.

	Wood	Nylon	Velvet	Springs
Style A	5	20	0	10
Style B	10	9	0	0
Style C	5	10	10	10

The cost in dollars per unit for each of the raw materials is given in the table below.

Wood	15
Nylon	12
Velvet	14
Springs	30

Create two matrices to represent these data and use matrix multiplication to find the price of manufacturing each style of chair.

37. *Politics* In a midwestern state, it is determined that 90% of all Republicans vote for Republican candidates and the remainder for Democratic candidates, while 80% of all Democrats vote for Democratic candidates and the remainder vote for Republican candidates. The percent of each party predicted to win the next election is given by

$$\begin{bmatrix} R \\ D \end{bmatrix} = \begin{bmatrix} 0.90 & 0.20 \\ 0.10 & 0.80 \end{bmatrix} \begin{bmatrix} a \\ b \end{bmatrix}$$

where a is the percent of Republicans and b is the percent of Democrats that won the last election. If 50% of those winning the election last time were Republicans and 50% were Democrats, what are the percents of each party that are predicted to win the next election?

38. *Competition* Two phone companies compete for customers in the southeastern region of a state. Company X retains $\frac{3}{5}$ of its customers and loses $\frac{2}{5}$ of its customers to company Y, and company Y retains $\frac{2}{3}$ of its customers and loses $\frac{1}{3}$ to company X. If we represent the fraction of the market held last year by

$$\begin{bmatrix} a \\ b \end{bmatrix}$$

where a is the fraction that company X had and b is the fraction that company Y had, then the fraction that each company will have this year can be found by

$$\begin{bmatrix} x \\ y \end{bmatrix} = \begin{bmatrix} 3/5 & 1/3 \\ 2/5 & 2/3 \end{bmatrix} \begin{bmatrix} a \\ b \end{bmatrix}$$

If company X had 120,000 customers and company Y had 90,000 customers last year, how many customers did each have this year?

39. *Wages* The graph below gives the median weekly earnings for nonunion men and women for different age groups in 2005.

a. Make a 2 × 6 matrix containing the data in this graph.

b. Suppose the median weekly earnings of men union workers are found to be 10% more than the earnings given in the graph, and the median weekly earnings of women union workers are found to be 25% more than the earnings given in the graph. Use matrix multiplication by a 2 × 2 matrix to find the new median weekly earnings by age for the male and female union members.

Median Usual Weekly Earnings of Full-time Wage and Salary Workers by Age and Sex, 2006 Annual Averages

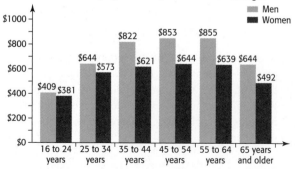

(Source: U.S. Dept. of Labor, Bureau of Labor Statistics)

40. *Libraries* Suppose the numbers of libraries and the operating incomes of public libraries in selected states for 2000 and for 2006 are shown in the tables below.

a. Form the matrix A that contains the numbers of libraries and operating incomes for 2000.

b. Form the matrix B that contains the numbers of libraries and operating incomes for 2006.

c. Use these two matrices to find the increase in the numbers of libraries and operating incomes from 2000 through 2006.

d. For which state is the increase in the number of libraries largest?

e. For which state is the increase in operating income the largest?

2000	Number of Libraries	Operating Income ($ million)
Illinois	606	358
Iowa	518	49
Kansas	324	50
Alabama	207	48
California	170	566

2006	Number of Libraries	Operating Income ($ million)
Illinois	799	407
Iowa	556	58
Kansas	374	60
Alabama	274	56
California	1039	655

Inverse Matrices; Matrix Equations

section preview ▪ Encryption

Security with credit card numbers on the Internet depends on encryption of the data. Encryption involves providing a way for the sender to encode a message so that the message is not apparent and a way for the receiver to decode the message so it can be read. Throughout history, different military units have used encoding and decoding systems of varying sophistication. In this section, we learn how to find the inverse of a matrix, how to use it to decode messages, and how to use it to solve matrix equations that have unique solutions.

Inverse Matrices

If the product of matrices A and B is an identity matrix, I, we say that B is the inverse of A (and A is the inverse of B). B is called the **inverse matrix** of A and is denoted A^{-1}.

> **Inverse Matrices**
>
> Two square matrices, A and B, are called **inverses** of each other if
>
> $$AB = I \quad \text{and} \quad BA = I$$
>
> where I is the identity matrix. We denote this by $B = A^{-1}$ and $A = B^{-1}$.

example 1   **Inverse Matrices**

Show that A and B are inverse matrices if

$$A = \begin{bmatrix} 1 & -0.6 & -0.2 \\ 0 & 0.4 & -0.2 \\ -1 & 0.4 & 0.8 \end{bmatrix} \quad \text{and} \quad B = \begin{bmatrix} 2 & 2 & 1 \\ 1 & 3 & 1 \\ 2 & 1 & 2 \end{bmatrix}$$

Solution

$$AB = \begin{bmatrix} 1 & -0.6 & -0.2 \\ 0 & 0.4 & -0.2 \\ -1 & 0.4 & 0.8 \end{bmatrix}\begin{bmatrix} 2 & 2 & 1 \\ 1 & 3 & 1 \\ 2 & 1 & 2 \end{bmatrix}$$

$$= \begin{bmatrix} 2 - 0.6 - 0.4 & 2 - 1.8 - 0.2 & 1 - 0.6 - 0.4 \\ 0 + 0.4 - 0.4 & 0 + 1.2 - 0.2 & 0 + 0.4 - 0.4 \\ -2 + 0.4 + 1.6 & -2 + 1.2 + 0.8 & -1 + 0.4 + 1.6 \end{bmatrix}$$

$$= \begin{bmatrix} 1 & 0 & 0 \\ 0 & 1 & 0 \\ 0 & 0 & 1 \end{bmatrix}$$

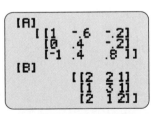

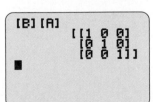

Figure 7.16

The product AB is the 3×3 identity matrix; we can use technology (Figure 7.16) to see that the product BA is also the identity matrix. Thus, A and B are inverse matrices. ■

We have used elementary row operations on augmented matrices to solve systems of equations. We can also find the inverse of a matrix A, if it exists, by using elementary row operations. Note that if a matrix is not square, then it does not have an inverse, and that not all square matrices have inverses.

Finding the Inverse of a Square Matrix

1. Write the matrix with the same dimension identity matrix in its augment, getting a matrix of the form $[A|I]$.

2. Use elementary row operations on $[A|I]$ to attempt to transform A into an identity matrix, giving a new matrix of the form $[I|B]$. The matrix B is the inverse of A.

3. If A does not have an inverse, the reduction process will yield a row of zeros in the left half (representing the original matrix) of the augmented matrix.

---

example 2 **Finding an Inverse Matrix**

Find the inverse of $A = \begin{bmatrix} 2 & 2 \\ 2 & 1 \end{bmatrix}$.

Solution

Creating the matrix $[A|I]$ and performing the operations to convert A to I gives

$$\left[\begin{array}{cc|cc} 2 & 2 & 1 & 0 \\ 2 & 1 & 0 & 1 \end{array}\right] \xrightarrow{\left(\frac{1}{2}\right)R_1 \rightarrow R_1} \left[\begin{array}{cc|cc} 1 & 1 & \frac{1}{2} & 0 \\ 2 & 1 & 0 & 1 \end{array}\right]$$

$$\xrightarrow{-2R_1 + R_2 \rightarrow R_2} \left[\begin{array}{cc|cc} 1 & 1 & \frac{1}{2} & 0 \\ 0 & -1 & -1 & 1 \end{array}\right]$$

$$\xrightarrow{-R_2 \rightarrow R_2} \left[\begin{array}{cc|cc} 1 & 1 & \frac{1}{2} & 0 \\ 0 & 1 & 1 & -1 \end{array}\right] \xrightarrow{-R_2 + R_1 \rightarrow R_1} \left[\begin{array}{cc|cc} 1 & 0 & -\frac{1}{2} & 1 \\ 0 & 1 & 1 & -1 \end{array}\right]$$

Thus, the inverse matrix of A is $A^{-1} = \begin{bmatrix} -\frac{1}{2} & 1 \\ 1 & -1 \end{bmatrix}$. We can verify this by observing

that $\begin{bmatrix} 2 & 2 \\ 2 & 1 \end{bmatrix}\begin{bmatrix} -\frac{1}{2} & 1 \\ 1 & -1 \end{bmatrix} = \begin{bmatrix} 1 & 0 \\ 0 & 1 \end{bmatrix}$.

---

example 3 **Inverse of a 3 × 3 Matrix**

Find the inverse of $A = \begin{bmatrix} -2 & 1 & 2 \\ 1 & 0 & -1 \\ 4 & -2 & -3 \end{bmatrix}$.

Solution

Creating the matrix $[A|I]$ and performing the operations to convert A to I gives

$$\begin{bmatrix} -2 & 1 & 2 & | & 1 & 0 & 0 \\ 1 & 0 & -1 & | & 0 & 1 & 0 \\ 4 & -2 & -3 & | & 0 & 0 & 1 \end{bmatrix} \xrightarrow{R_1 \leftrightarrow R_2} \begin{bmatrix} 1 & 0 & -1 & | & 0 & 1 & 0 \\ -2 & 1 & 2 & | & 1 & 0 & 0 \\ 4 & -2 & -3 & | & 0 & 0 & 1 \end{bmatrix}$$

$$\xrightarrow[\substack{2R_1 + R_2 \rightarrow R_2 \\ -4R_1 + R_3 \rightarrow R_3}]{} \begin{bmatrix} 1 & 0 & -1 & | & 0 & 1 & 0 \\ 0 & 1 & 0 & | & 1 & 2 & 0 \\ 0 & -2 & 1 & | & 0 & -4 & 1 \end{bmatrix}$$

$$\xrightarrow{2R_2 + R_3 \rightarrow R_3} \begin{bmatrix} 1 & 0 & -1 & | & 0 & 1 & 0 \\ 0 & 1 & 0 & | & 1 & 2 & 0 \\ 0 & 0 & 1 & | & 2 & 0 & 1 \end{bmatrix}$$

$$\xrightarrow{R_3 + R_1 \rightarrow R_1} \begin{bmatrix} 1 & 0 & 0 & | & 2 & 1 & 1 \\ 0 & 1 & 0 & | & 1 & 2 & 0 \\ 0 & 0 & 1 & | & 2 & 0 & 1 \end{bmatrix}$$

Thus we have the inverse of A.

$$A^{-1} = \begin{bmatrix} 2 & 1 & 1 \\ 1 & 2 & 0 \\ 2 & 0 & 1 \end{bmatrix}$$

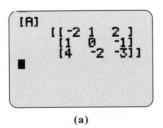

(a)

(b)

Figure 7.17

Figure 7.18

Inverses and Technology

Computer software, spreadsheets, and calculators can be used to find the inverse of a matrix. We will see that if the inverse of a square matrix exists, it can be found easily with technology.* Figure 7.17(a) shows the matrix A from Example 3 and Figure 7.17(b) shows the inverse of A found with a graphing calculator. We can confirm that these matrices are inverses by computing AA^{-1} (Figure 7.18).

---

spreadsheet solution

Table 7.8 shows a spreadsheet that gives the inverse of the matrix A.†

Table 7.8

		A	B	C	D
1	Matrix A	−2	1	2	
2			1	0	−1
3			4	−2	−3
4					
5	A inverse	2	1	1	
6			1	2	0
7			2	0	1

---

* For more details, see Appendix A, page 667.
† For more details, see Appendix B, page 692.

| example 4 | **Does the Inverse Exist?** |

Find the inverse of $A = \begin{bmatrix} 1 & 2 & -2 \\ 2 & 0 & 2 \\ 6 & 4 & 0 \end{bmatrix}$, if it exists.

Solution

Attempting to find the inverse of matrix A with technology results in an error statement, indicating that the inverse does not exist (Figure 7.19).

Recall that if A does not have an inverse, the reduction process using elementary row operations will yield a row of zeros in the left half of the augmented matrix.*

$$\begin{bmatrix} 1 & 2 & -2 & 1 & 0 & 0 \\ 2 & 0 & 2 & 0 & 1 & 0 \\ 6 & 4 & 0 & 0 & 0 & 1 \end{bmatrix} \xrightarrow[-6R_1 + R_3 \rightarrow R_3]{-2R_1 + R_2 \rightarrow R_2} \begin{bmatrix} 1 & 2 & -2 & 1 & 0 & 0 \\ 0 & -4 & 6 & -2 & 1 & 0 \\ 0 & -8 & 12 & -6 & 0 & 1 \end{bmatrix}$$

$$\xrightarrow{(-1/4)R_2 \rightarrow R_2} \begin{bmatrix} 1 & 2 & -2 & 1 & 0 & 0 \\ 0 & 1 & -\dfrac{3}{2} & \dfrac{1}{2} & -\dfrac{1}{4} & 0 \\ 0 & -8 & 12 & -6 & 0 & 1 \end{bmatrix}$$

$$\xrightarrow{8R_2 + R_3 \rightarrow R_3} \begin{bmatrix} 1 & 2 & -2 & 1 & 0 & 0 \\ 0 & 1 & -\dfrac{3}{2} & \dfrac{1}{2} & -\dfrac{1}{4} & 0 \\ 0 & 0 & 0 & -2 & -2 & 1 \end{bmatrix}$$

We see that the left half of the bottom row of the reduced matrix contains all 0's, so it is not possible to reduce the original matrix to the identity matrix, and thus the matrix A does not have an inverse. ∎

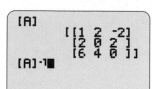

Figure 7.19

Encoding and Decoding Messages

In sending messages during military maneuvers, in business transactions, and in sending secure data on the Internet, encoding (or encryption) of messages is important. Suppose we want to encode the message "Cheer up." The following simple code could be used to change letters of the alphabet to the numbers 1 to 26, respectively, with the number 27 representing a blank space.

$a \quad b \quad c \quad d \quad e \quad f \quad g \quad h \quad i \quad j \quad k \quad l \quad m \quad n \quad o \quad p \quad q \quad r \quad s \quad t \quad u \quad v \quad w \quad x \quad y \quad z$
1 2 3 4 5 6 7 8 9 10 11 12 13 14 15 16 17 18 19 20 21 22 23 24 25 26 27

Then "Cheer up" can be represented by the numbers

3 8 5 5 18 27 21 16

To further encode the message, we put these numbers in pairs and then create a 2×1 matrix for each pair of numbers. Next we choose an *encoding matrix*, like

$$A = \begin{bmatrix} 3 & -2 \\ -1 & 1 \end{bmatrix}$$

---

* The methods used by technology will sometimes yield approximations for numbers that should be zeros.

and multiply each pair of numbers by the encoding matrix, as follows.

$$\begin{bmatrix} 3 & -2 \\ -1 & 1 \end{bmatrix}\begin{bmatrix} 3 \\ 8 \end{bmatrix} = \begin{bmatrix} -7 \\ 5 \end{bmatrix} \qquad \begin{bmatrix} 3 & -2 \\ -1 & 1 \end{bmatrix}\begin{bmatrix} 5 \\ 5 \end{bmatrix} = \begin{bmatrix} 5 \\ 0 \end{bmatrix}$$

$$\begin{bmatrix} 3 & -2 \\ -1 & 1 \end{bmatrix}\begin{bmatrix} 18 \\ 27 \end{bmatrix} = \begin{bmatrix} 0 \\ 9 \end{bmatrix} \qquad \begin{bmatrix} 3 & -2 \\ -1 & 1 \end{bmatrix}\begin{bmatrix} 21 \\ 16 \end{bmatrix} = \begin{bmatrix} 31 \\ -5 \end{bmatrix}$$

Because multiplying a 2×2 matrix by a 2×1 matrix gives a 2×1 matrix, the resulting products are pairs of numbers. Combining the pairs of numbers gives the encoded numerical message

$$-7\ 5\ 5\ 0\ 0\ 9\ 31\ -5$$

Note that another encoding matrix could be used rather than the one used above. We can also encode a message by putting triples of numbers in 3×1 matrices and multiplying each 3×1 matrix by a 3×3 encoding matrix.

example 5

Encoding Messages

Use the encoding matrix $A = \begin{bmatrix} 1 & -3 & 2 \\ 2 & -2 & 2 \\ 3 & -1 & 1 \end{bmatrix}$ to encode the message "Meet me for lunch."

Solution

Converting the letters of the message to triples of numbers gives:

13 5 5   20 27 13   5 27 6   15 18 27   12 21 14   3 8 27

with 27 used to complete the last triple.

Placing these triples of numbers in 3×1 matrices and multiplying by matrix A gives

$$\begin{bmatrix} 1 & -3 & 2 \\ 2 & -2 & 2 \\ 3 & -1 & 1 \end{bmatrix}\begin{bmatrix} 13 \\ 5 \\ 5 \end{bmatrix} = \begin{bmatrix} 8 \\ 26 \\ 39 \end{bmatrix} \qquad \begin{bmatrix} 1 & -3 & 2 \\ 2 & -2 & 2 \\ 3 & -1 & 1 \end{bmatrix}\begin{bmatrix} 20 \\ 27 \\ 13 \end{bmatrix} = \begin{bmatrix} -35 \\ 12 \\ 46 \end{bmatrix}$$

$$\begin{bmatrix} 1 & -3 & 2 \\ 2 & -2 & 2 \\ 3 & -1 & 1 \end{bmatrix}\begin{bmatrix} 5 \\ 27 \\ 6 \end{bmatrix} = \begin{bmatrix} -64 \\ -32 \\ -6 \end{bmatrix} \qquad \begin{bmatrix} 1 & -3 & 2 \\ 2 & -2 & 2 \\ 3 & -1 & 1 \end{bmatrix}\begin{bmatrix} 15 \\ 18 \\ 27 \end{bmatrix} = \begin{bmatrix} 15 \\ 48 \\ 54 \end{bmatrix}$$

$$\begin{bmatrix} 1 & -3 & 2 \\ 2 & -2 & 2 \\ 3 & -1 & 1 \end{bmatrix}\begin{bmatrix} 12 \\ 21 \\ 14 \end{bmatrix} = \begin{bmatrix} -23 \\ 10 \\ 29 \end{bmatrix} \qquad \begin{bmatrix} 1 & -3 & 2 \\ 2 & -2 & 2 \\ 3 & -1 & 1 \end{bmatrix}\begin{bmatrix} 3 \\ 8 \\ 27 \end{bmatrix} = \begin{bmatrix} 33 \\ 44 \\ 28 \end{bmatrix}$$

The products are triples of numbers. Combining the triples of numbers gives the encoded message

8 26 39 −35 12 46 −64 −32 −6 15 48 54 −23 10 29 33 44 28   ∎

We can also find the encoded triples of numbers for the message of Example 5 by writing the original triples of numbers as columns in one matrix and then multiplying this matrix by the encoding matrix. The columns of the product will be the encoded triples of numbers.

$$\begin{bmatrix} 1 & -3 & 2 \\ 2 & -2 & 2 \\ 3 & -1 & 1 \end{bmatrix} \begin{bmatrix} 13 & 20 & 5 & 15 & 12 & 3 \\ 5 & 27 & 27 & 18 & 21 & 8 \\ 5 & 13 & 6 & 27 & 14 & 27 \end{bmatrix} = \begin{bmatrix} 8 & -35 & -64 & 15 & -23 & 33 \\ 26 & 12 & -32 & 48 & 10 & 44 \\ 39 & 46 & -6 & 54 & 29 & 28 \end{bmatrix}$$

Observe that the columns of this product have the same triples, respectively, as the individual products in Example 5. Figure 7.20 shows a calculator display (in two screens with some columns repeated) of the product.

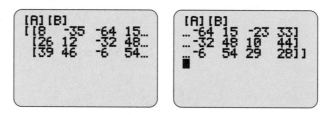

Figure 7.20

Sending an encoded message is of little value if the recipient is not able to decode it. If the message is encoded with a matrix, then the message can be decoded with the inverse of the matrix.

Recall that the message "Cheer up" was represented by the numbers 3 8 5 5 18 27 21 16 and encoded to -7 5 5 0 0 9 31 -5 with the encoding matrix

$$A = \begin{bmatrix} 3 & -2 \\ -1 & 1 \end{bmatrix}$$

To decode the encoded message, we multiply each pair of numbers in the encoded message on the left by the inverse of matrix A.

$$A^{-1} = \begin{bmatrix} 1 & 2 \\ 1 & 3 \end{bmatrix}$$

Multiplying the pairs of the coded message by A^{-1} gives us back the original message.

$$\begin{bmatrix} 1 & 2 \\ 1 & 3 \end{bmatrix}\begin{bmatrix} -7 \\ 5 \end{bmatrix} = \begin{bmatrix} 3 \\ 8 \end{bmatrix} \qquad \begin{bmatrix} 1 & 2 \\ 1 & 3 \end{bmatrix}\begin{bmatrix} 5 \\ 0 \end{bmatrix} = \begin{bmatrix} 5 \\ 5 \end{bmatrix}$$

$$\begin{bmatrix} 1 & 2 \\ 1 & 3 \end{bmatrix}\begin{bmatrix} 0 \\ 9 \end{bmatrix} = \begin{bmatrix} 18 \\ 27 \end{bmatrix} \qquad \begin{bmatrix} 1 & 2 \\ 1 & 3 \end{bmatrix}\begin{bmatrix} 31 \\ -5 \end{bmatrix} = \begin{bmatrix} 21 \\ 16 \end{bmatrix}$$

Thus, the decoded numbers are 3 8 5 5 18 27 21 16, which correspond to the letters in the message "Cheer up." ∎

example 6 | Decoding Messages

Suppose we encoded messages using a simple code that changed letters of the alphabet to numbers, with the number 27 representing a blank space, and the encoding matrix

$$A = \begin{bmatrix} 1 & 2 & -1 \\ 2 & 1 & 2 \\ 3 & 2 & -3 \end{bmatrix}$$

Decode the following coded message:

$$18 \ \ 41 \ \ 34 \ \ 61 \ \ 93 \ \ 75 \ \ 39 \ \ 77 \ \ 9 \ \ 34 \ \ 62 \ \ 46 \ \ 3 \ \ 62 \ \ -23$$

Solution

To find the numbers representing the message, we place triples of numbers from the coded message into columns of a 3×5 matrix C, and multiply that matrix by A^{-1}. The inverse of A is

$$A^{-1} = \begin{bmatrix} \dfrac{-7}{16} & \dfrac{1}{4} & \dfrac{5}{16} \\ \dfrac{3}{4} & 0 & \dfrac{-1}{4} \\ \dfrac{1}{16} & \dfrac{1}{4} & \dfrac{-3}{16} \end{bmatrix}$$

Multiplying by A^{-1} on the left gives

$$A^{-1}\begin{bmatrix} 18 & 61 & 39 & 34 & 3 \\ 41 & 93 & 77 & 62 & 62 \\ 34 & 75 & 9 & 46 & -23 \end{bmatrix} = \begin{bmatrix} \dfrac{-7}{16} & \dfrac{1}{4} & \dfrac{5}{16} \\ \dfrac{3}{4} & 0 & \dfrac{-1}{4} \\ \dfrac{1}{16} & \dfrac{1}{4} & \dfrac{-3}{16} \end{bmatrix}\begin{bmatrix} 18 & 61 & 39 & 34 & 3 \\ 41 & 93 & 77 & 62 & 62 \\ 34 & 75 & 9 & 46 & -23 \end{bmatrix}$$

$$= \begin{bmatrix} 13 & 20 & 5 & 15 & 7 \\ 5 & 27 & 27 & 14 & 8 \\ 5 & 13 & 20 & 9 & 20 \end{bmatrix}$$

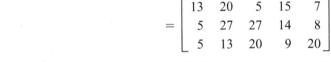

Figure 7.21

If we use technology to perform this multiplication, we do not need to display the elements of A^{-1} in the computation. We enter $[A]$ and $[C]$ and compute $[A]^{-1}[C]$ (Figure 7.21).

Reading down the columns of the product gives the numbers representing the message:

$$13 \ \ 5 \ \ 5 \ \ 20 \ \ 27 \ \ 13 \ \ 5 \ \ 27 \ \ 20 \ \ 15 \ \ 14 \ \ 9 \ \ 7 \ \ 8 \ \ 20$$

The message is "Meet me tonight." ■

Matrix Equations

The system of equations

$$\begin{cases} -2x + \ y + 2z = 5 \\ x - \ z = 2 \\ 4x - 2y - 3z = 4 \end{cases}$$

can be written as the matrix equation

$$\begin{bmatrix} -2x + \ y + 2z \\ x - \ z \\ 4x - 2y - 3z \end{bmatrix} = \begin{bmatrix} 5 \\ 2 \\ 4 \end{bmatrix}$$

Because

$$\begin{bmatrix} -2 & 1 & 2 \\ 1 & 0 & -1 \\ 4 & -2 & -3 \end{bmatrix}\begin{bmatrix} x \\ y \\ z \end{bmatrix} = \begin{bmatrix} -2x + y + 2z \\ x - z \\ 4x - 2y - 3z \end{bmatrix}$$

we can write the system of equations as the matrix equation in the form

$$\begin{bmatrix} -2 & 1 & 2 \\ 1 & 0 & -1 \\ 4 & -2 & -3 \end{bmatrix}\begin{bmatrix} x \\ y \\ z \end{bmatrix} = \begin{bmatrix} 5 \\ 2 \\ 4 \end{bmatrix}$$

Note that the 3×3 matrix on the left side of the matrix equation is the coefficient matrix for the original system. We will call the coefficient matrix A, the matrix containing the variables X, and the matrix containing the constants C, giving the form

$$AX = C$$

If A^{-1} exists and we multiply both sides of this equation on the left by A^{-1}, the product is as follows:*

$$A^{-1}AX = A^{-1}C$$

Because $A^{-1}A = I$ and because multiplication of matrix X by an identity matrix gives the matrix X, we have

$$IX = A^{-1}C \quad \text{or} \quad X = A^{-1}C$$

In general, multiplying both sides of the matrix equation $AX = C$ on the left by A^{-1} gives the solution to the system that the matrix equation represents, if the solution is unique.

| example 7 | ## Solution of Matrix Equations with Inverses

Solve the system

$$\begin{cases} -2x + y + 2z = 5 \\ x - z = 2 \\ 4x - 2y - 3z = 4 \end{cases}$$

by writing a matrix equation and using an inverse matrix.

Solution

We can write this system of equations as $AX = C$ where

$$A = \begin{bmatrix} -2 & 1 & 2 \\ 1 & 0 & -1 \\ 4 & -2 & -3 \end{bmatrix}, \quad X = \begin{bmatrix} x \\ y \\ z \end{bmatrix} \quad \text{and} \quad C = \begin{bmatrix} 5 \\ 2 \\ 4 \end{bmatrix}$$

That is,

$$\begin{bmatrix} -2 & 1 & 2 \\ 1 & 0 & -1 \\ 4 & -2 & -3 \end{bmatrix}\begin{bmatrix} x \\ y \\ z \end{bmatrix} = \begin{bmatrix} 5 \\ 2 \\ 4 \end{bmatrix}$$

---

* Recall that multiplication on the right may give a different product than multiplication on the left.

As we saw in Example 3, the inverse of the matrix A is

$$A^{-1} = \begin{bmatrix} 2 & 1 & 1 \\ 1 & 2 & 0 \\ 2 & 0 & 1 \end{bmatrix}$$

Thus, we can solve the system by multiplying both sides of the matrix equation on the left by A^{-1} as follows:

$$A^{-1}A \begin{bmatrix} x \\ y \\ z \end{bmatrix} = A^{-1} \begin{bmatrix} 5 \\ 2 \\ 4 \end{bmatrix}$$

$$I \begin{bmatrix} x \\ y \\ z \end{bmatrix} = \begin{bmatrix} 2 & 1 & 1 \\ 1 & 2 & 0 \\ 2 & 0 & 1 \end{bmatrix} \begin{bmatrix} 5 \\ 2 \\ 4 \end{bmatrix}$$

$$\begin{bmatrix} x \\ y \\ z \end{bmatrix} = \begin{bmatrix} 16 \\ 9 \\ 14 \end{bmatrix}$$

Thus, we have the solution to the system, $x = 16$, $y = 9$, and $z = 14$, or $(16, 9, 14)$. ∎

Matrix Equations and Technology

We can use technology to multiply both sides of the matrix equation $AX = C$ by the inverse of A and get the solution to the system that the matrix equation represents. If we use technology to solve a system of linear equations, it is not necessary to display the inverse of the coefficient matrix.

| example 8 |

Manufacturing

Sharper Technology Company manufactures three types of calculators, a business calculator, a scientific calculator, and a graphing calculator. The production requirements are given in Table 7.9. If during each month the company has 134,000 circuit components, 56,000 hours for assembly, and 14,000 cases, how many of each type of calculator can it produce each month?

Table 7.9

	Business Calculator	Scientific Calculator	Graphing Calculator
Circuit Components	5	7	12
Assembly Time (hours)	2	3	5
Cases	1	1	1

Solution

If the company produces x business calculators, y scientific calculators, and z graphing calculators, then the problem can be represented using the following system of equations.

$$\begin{cases} 5x + 7y + 12z = 134{,}000 \\ 2x + 3y + 5z = 56{,}000 \\ x + y + z = 14{,}000 \end{cases}$$

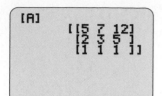

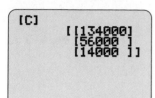

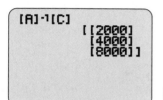

Figure 7.22

This system can be represented by the matrix equation $AX = C$, where A is the coefficient matrix and C is the constant matrix.

$$\begin{bmatrix} 5 & 7 & 12 \\ 2 & 3 & 5 \\ 1 & 1 & 1 \end{bmatrix} \begin{bmatrix} x \\ y \\ z \end{bmatrix} = \begin{bmatrix} 134{,}000 \\ 56{,}000 \\ 14{,}000 \end{bmatrix}$$

Multiplying both sides of this equation by the inverse of the coefficient matrix gives the solution (Figure 7.22):

$$\begin{bmatrix} x \\ y \\ z \end{bmatrix} = A^{-1} \begin{bmatrix} 134{,}000 \\ 56{,}000 \\ 14{,}000 \end{bmatrix} = \begin{bmatrix} 2000 \\ 4000 \\ 8000 \end{bmatrix}$$

This shows that the company can produce 2000 business calculators, 4000 scientific calculators, and 8000 graphing calculators each month. ■

Recall that we could also solve this system of linear equations by using elementary row operations with an augmented matrix, as discussed in Section 7.2.

spreadsheet solution

We can also use inverse matrices with Excel to solve a system of linear equations if a unique solution exists. We can solve such a system by finding the inverse of the coefficient matrix and multiplying this inverse times the matrix containing the constants. The spreadsheet (Table 7.10) shows the solution of Example 8 found with Excel.

Table 7.10

	A	B	C	D
1	Matrix A	5	7	12
2		2	3	5
3		1	1	1
4				
5	Inverse of A	2	−5	1
6		−3	7	1
7		1	−2	−1
8				
9	Matrix C	134,000		
10		56,000		
11		14,000		
12				
13	Solution X	2000		
14		4000		
15		8000		

We can use the inverse of the coefficient matrix to solve a system of linear equations only if the system has a unique solution. Otherwise, the inverse of the coefficient matrix will not exist, and an attempt to use technology to solve the system with an inverse will give an error message.

skills check

7.4

1. Suppose $A = \begin{bmatrix} 3 & 1 \\ 4 & 2 \end{bmatrix}$ and $B = \begin{bmatrix} 1 & -0.5 \\ -2 & 1.5 \end{bmatrix}$.

   a. Compute AB and BA, if possible.

   b. What is the relationship between A and B?

2. Suppose $A = \begin{bmatrix} 1 & -\dfrac{1}{2} & -\dfrac{1}{4} \\ -\dfrac{1}{2} & 0 & \dfrac{1}{2} \\ 0 & \dfrac{1}{2} & -\dfrac{1}{4} \end{bmatrix}$ and

   $B = \begin{bmatrix} 2 & 2 & 2 \\ 1 & 2 & 3 \\ 2 & 4 & 2 \end{bmatrix}$.

   a. Compute AB and BA, if possible.

   b. What is the relationship between A and B?

3. Show that A and B are inverse matrices if
$A = \begin{bmatrix} 1 & -1 & 1 \\ 2 & -1 & 0 \\ -2 & 2 & -1 \end{bmatrix}$ and $B = \begin{bmatrix} 1 & 1 & 1 \\ 2 & 1 & 2 \\ 2 & 0 & 1 \end{bmatrix}$.

4. Show that C and D are inverse matrices if
$C = \begin{bmatrix} -\dfrac{1}{2} & \dfrac{1}{2} & 1 \\ \dfrac{1}{2} & \dfrac{1}{2} & 0 \\ -\dfrac{1}{2} & \dfrac{3}{2} & 1 \end{bmatrix}$ and $D = \begin{bmatrix} 1 & 2 & -1 \\ -1 & 0 & 1 \\ 2 & 1 & -1 \end{bmatrix}$.

In Exercises 5–14, find the inverse of the given matrix.

5. $A = \begin{bmatrix} 1 & 3 \\ 2 & 7 \end{bmatrix}$ **6.** $A = \begin{bmatrix} 2 & 4 \\ 2 & 5 \end{bmatrix}$

7. $A = \begin{bmatrix} 2 & -2 & 2 \\ 2 & 1 & 2 \\ 2 & 0 & 1 \end{bmatrix}$

8. $A = \begin{bmatrix} \dfrac{1}{2} & -1 & 1 \\ 1 & -2 & 0 \\ 1 & 2 & -1 \end{bmatrix}$ **9.** $A = \begin{bmatrix} 1 & 1 & -1 \\ -1 & 0 & 1 \\ 1 & 1 & 2 \end{bmatrix}$

10. $C = \begin{bmatrix} 0 & 1 & -1 \\ 1 & 0 & 0 \\ 0 & 0 & 1 \end{bmatrix}$ **11.** $A = \begin{bmatrix} 2 & 3 & 1 \\ 3 & 4 & -1 \\ 2 & 5 & 1 \end{bmatrix}$

12. $B = \begin{bmatrix} 9 & 2 & -7 \\ -5 & 0 & 5 \\ 7 & -4 & -1 \end{bmatrix}$

13. $C = \begin{bmatrix} 1 & 0 & -1 & -1 \\ 0 & 1 & -1 & -1 \\ 0 & 0 & 1 & 0 \\ 0 & 0 & 0 & 1 \end{bmatrix}$

14. $A = \begin{bmatrix} -1 & 0 & 1 & 1 \\ 0 & 1 & 1 & 1 \\ 1 & -1 & 0 & 0 \\ 0 & 0 & 0 & 1 \end{bmatrix}$

15. If $A^{-1} = \begin{bmatrix} 1 & 2 \\ 4 & 3 \end{bmatrix}$, solve $AX = \begin{bmatrix} 2 \\ 4 \end{bmatrix}$ for X.

16. If $A^{-1} = \begin{bmatrix} 1 & 2 & -1 \\ 0 & 2 & 1 \\ 2 & 0 & -2 \end{bmatrix}$, solve $AX = \begin{bmatrix} 1 \\ 2 \\ -2 \end{bmatrix}$ for X.

17. Solve the matrix equation
$\begin{bmatrix} -1 & 1 & 0 \\ -2 & 3 & -2 \\ 2 & -2 & 1 \end{bmatrix} \begin{bmatrix} x \\ y \\ z \end{bmatrix} = \begin{bmatrix} 3 \\ 5 \\ 8 \end{bmatrix}$

18. Solve the matrix equation

$$\begin{bmatrix} -2 & 1 & 0 \\ -1 & 3 & -2 \\ 2 & -2 & 1 \end{bmatrix} \begin{bmatrix} x \\ y \\ z \end{bmatrix} = \begin{bmatrix} 2 \\ 4 \\ 8 \end{bmatrix}$$

In Exercises 19–24, use an inverse matrix to find the solution to the systems.

19. $\begin{cases} 4x - 3y + z = 2 \\ -6x + 5y - 2z = -3 \\ x - y + z = 1 \end{cases}$

20. $\begin{cases} 5x + 3y + z = 12 \\ 4x + 3y + 2z = 9 \\ x + y + 2z = 2 \end{cases}$

21. $\begin{cases} 2x + y + z = 4 \\ x + 4y + 2z = 4 \\ 2x + y + 2z = 3 \end{cases}$ **22.** $\begin{cases} x + 2y - z = 2 \\ -x + z = 4 \\ 2x + y - z = 6 \end{cases}$

23. $\begin{cases} x_1 + x_3 + x_4 = 90 \\ x_2 + x_3 + x_4 = 72 \\ 2x_1 + 5x_2 + x_3 = 108 \\ 3x_2 + x_4 = 144 \end{cases}$

24. $\begin{cases} 2x_1 + x_2 + 3x_3 + 4x_4 = 1 \\ x_1 + x_2 + x_3 - x_4 = 2 \\ x_1 + x_2 + x_3 + x_4 = 10 \\ 2x_1 - x_2 + 3x_3 - x_4 = 5 \end{cases}$

exercises

7.4

25. *Competition* Two phone companies compete for customers in the southeastern region of a state. Each year company X retains $\frac{3}{5}$ of its customers and loses $\frac{2}{5}$ of its customers to company Y, while company Y retains $\frac{2}{3}$ of its customers and loses $\frac{1}{3}$ to company X. If we represent the fraction of the market held last year by

$$\begin{bmatrix} a \\ b \end{bmatrix}$$

where a is the number that company X had last year and b is the number that company Y had last year, then the number that each company will have this year can be found by

$$\begin{bmatrix} A \\ B \end{bmatrix} = \begin{bmatrix} \frac{3}{5} & \frac{1}{3} \\ \frac{2}{5} & \frac{2}{3} \end{bmatrix} \begin{bmatrix} a \\ b \end{bmatrix}$$

If company X has $A = 150,000$ customers and company Y has $B = 120,000$ customers this year, how many customers did each have last year?

26. *Competition* A satellite company and a cable company compete for customers in a city. The satellite company retains $\frac{2}{3}$ of its customers and loses $\frac{1}{3}$ of its customers to the cable company, while the cable company retains $\frac{2}{5}$ of its customers and loses $\frac{3}{5}$ to the satellite company. If we represent the fraction of the market held last year by

$$\begin{bmatrix} a \\ b \end{bmatrix}$$

where a is the number of customers that the satellite company had last year and b is the number that the cable company had last year, then the number that each company will have this year can be found by

$$\begin{bmatrix} A \\ B \end{bmatrix} = \begin{bmatrix} \frac{2}{3} & \frac{2}{5} \\ \frac{1}{3} & \frac{3}{5} \end{bmatrix} \begin{bmatrix} a \\ b \end{bmatrix}$$

If the satellite company has $A = 90,000$ customers and the cable company has $B = 85,000$ customers this year, how many customers did each have last year?

27. *Politics* In a midwestern state, it is observed that 90% of all Republicans vote for the Republican candidate for governor and the remainder for the Democratic candidate, while 80% of all Democrats vote for the Democratic candidate and the remainder vote for the Republican candidate. The percent voting for each party in the next election is given by

$$\begin{bmatrix} R \\ D \end{bmatrix} = \begin{bmatrix} 0.90 & 0.20 \\ 0.10 & 0.80 \end{bmatrix} \begin{bmatrix} r \\ d \end{bmatrix}$$

where r and d are the respective percents voting for each party in the last election. If 55% of the votes in this election were for a Republican and 45% were for a Democrat, what are the percents for each party candidate in the last election?

28. *Politics* Suppose that in a certain city, the Democratic, Republican, and Consumer parties always nominate candidates for mayor. The percent of people voting for each party candidate in the next election depends on the percent voting for that party in the last election. The percent voting for each party in the next election is given by

$$\begin{bmatrix} D \\ R \\ C \end{bmatrix} = \begin{bmatrix} 50 & 40 & 30 \\ 40 & 50 & 30 \\ 10 & 10 & 40 \end{bmatrix} \begin{bmatrix} d \\ r \\ c \end{bmatrix}$$

where d, r, and c are the respective percents (in decimals) voting for each party in the last election. If the percent voting for a Democrat in this election is $D = 42\%$, the percent voting for a Republican in this election is $R = 42\%$, and the percent voting for the Consumer party candidate in this election is $C = 16\%$, what were the respective percents of people voting for these parties in the last election?

29. *Loans* A bank gives three loans totaling $400,000 to a development company for the purchase of three business properties. The largest loan is $100,000 more than the sum of the other two, and the smallest loan is one-half of the next larger loan. Represent the amount of money in each loan as x, y, and z, respectively.

   a. Write a system of three equations in three variables to represent the problem.

   b. Use a matrix equation to solve the system and find the amount of each loan.

30. *Transportation* Ross Freightline has an order for two products, A and B, to be delivered to a store. The table in the next column gives the volume and weight for each unit of the two products.

	Product A	Product B
Unit Volume (cubic ft)	40	65
Unit Weight (pounds)	320	360

If the truck that can deliver the products can carry 5200 cubic feet and 31,360 pounds, how many units of each product can the truck carry?

31. *Investment* An investor has $400,000 in three accounts, paying 6%, 8%, and 10%, respectively. If she has twice as much invested at 8% as she has at 6%, how much does she have invested in each account if she earns $36,000 in interest?

32. *Pricing* A theater has 1000 seats divided into orchestra, main, and balcony. The orchestra seats cost $80, the main seats cost $50, and the balcony seats cost $40. If all the seats are sold, the revenue is $50,000. If all the orchestra and balcony seats are sold and $\frac{3}{4}$ of the main seats are sold, the revenue is $42,500. How many of each type of seat does the theater have?

33. *Venture Capital* Suppose a bank draws its venture-capital funds annually from three sources of income: business loans, auto loans, and home mortgages. One spreadsheet shows the income from each of these sources for the years 2001, 2002, and 2003, and the second spreadsheet shows the venture capital for these years. If the bank uses a fixed percent of its income from each of the business loans, find the percent of income from each of these loans.

	A	B	C	D
1		Income	From Loans	(Millions)
2	Years	Business	Auto	Home
3	2001	532	58	682
4	2002	562	62	695
5	2003	578	69	722

	A	B
1	Venture	Capital (Millions)
2	2001	483.94
3	2002	503.28
4	2003	521.33

34. *Bookcases* A company produces three types of bookcases, 4-shelf metal, 6-shelf metal, and 4-shelf wooden. The 4-shelf metal bookcase requires 2 hours for fabrication, 1 hour to paint, and one-half hour to package for shipping. The 6-shelf bookcase requires 3 hours for fabrication, 1.5 hours to paint, and $\frac{1}{2}$ hour to package. The wooden bookcase requires 3 hours to fabricate, 2 hours to paint, and $\frac{1}{2}$ hour to package. The daily amount of time available is 124 hours for fabrication, 68 hours for painting, and 24 hours for packaging. How many of each type of bookcase can be produced each day?

Encoding Messages We have encoded messages by assigning the numbers 1 to 26 to the letters a to z of the alphabet, respectively, and assigning 27 to a blank space.

$$
\begin{array}{cccccccccccccccc}
a & b & c & d & e & f & g & h & i & j & k & l & m & n & o & p \\
1 & 2 & 3 & 4 & 5 & 6 & 7 & 8 & 9 & 10 & 11 & 12 & 13 & 14 & 15 & 16 \\
q & r & s & t & u & v & w & x & y & z & blank \\
17 & 18 & 19 & 20 & 21 & 22 & 23 & 24 & 25 & 26 & 27
\end{array}
$$

To further encode the messages, we can use an encoding matrix A to convert these numbers into new pairs or triples of numbers. In Exercises 35–38, use the given matrix A to encode the given message.

35. *Encoding Messages*
  a. Convert "Just do it" from letters to numbers.

  b. Multiply $A = \begin{bmatrix} 4 & 4 \\ 1 & 2 \end{bmatrix}$ times 2×1 matrices created with columns containing pairs of numbers from part (a) to encode the message "Just do it" into pairs of coded numbers.

36. *Encoding Messages*
  a. Convert "Call home" from letters to numbers.

  b. Multiply $A = \begin{bmatrix} 2 & 3 \\ 2 & 2 \end{bmatrix}$ times 2×1 matrices created with columns containing pairs of numbers from part (a) to encode the message "Call home" into pairs of coded numbers.

37. *Encoding Messages*

Use the matrix $\begin{bmatrix} 4 & 4 & 4 \\ 1 & 2 & 3 \\ 2 & 4 & 2 \end{bmatrix}$ to encode the message "Neatness counts" into triples of numbers.

38. *Encoding Messages*

Use the matrix $\begin{bmatrix} 4 & 4 & 4 \\ 1 & 2 & 3 \\ 2 & 4 & 2 \end{bmatrix}$ to encode the message "Meet for lunch" into triples of numbers.

Decoding Messages We have encoded messages by assigning the numbers 1 to 26 to the letters a to z of the alphabet, respectively, and assigning 27 to a blank space. We can decode messages of this type by finding the inverse of the encoding matrix and multiplying it times the coded message. Use A^{-1} and the conversion table below to decode the messages in Exercises 39–44.

$$
\begin{array}{cccccccccccccccc}
a & b & c & d & e & f & g & h & i & j & k & l & m & n & o & p \\
1 & 2 & 3 & 4 & 5 & 6 & 7 & 8 & 9 & 10 & 11 & 12 & 13 & 14 & 15 & 16 \\
q & r & s & t & u & v & w & x & y & z & blank \\
17 & 18 & 19 & 20 & 21 & 22 & 23 & 24 & 25 & 26 & 27
\end{array}
$$

39. *Decoding Messages* The encoding matrix is $A = \begin{bmatrix} 3 & -1 \\ -2 & 1 \end{bmatrix}$ and the encoded message is 51, −29, 55, −35, 76, −49, −15 16, 11, 1.

40. *Decoding Messages* The encoding matrix is $A = \begin{bmatrix} 2 & 3 \\ 2 & 4 \end{bmatrix}$ and the encoded message is 59, 74, 72, 78, 84, 102, 81, 90, 121, 148.

41. *Decoding Messages* The encoding matrix is $A = \begin{bmatrix} -1 & 0 & 1 \\ -1 & 1 & 0 \\ 3 & -1 & -1 \end{bmatrix}$ and the encoded message is 1, −4, 16, 21, 23, −40, 3, 6, 6, −26, −14, 67, −9, 0, 23, 9, 1, 8.

42. *Decoding Messages* The encoding matrix is $A = \begin{bmatrix} 1 & 1 & 0 \\ 1 & 3 & 0 \\ 5 & 3 & 3 \end{bmatrix}$ and the encoded message is 16, 34, 128, 32, 86, 145, 32, 86, 109, 29, 33, 195, 6, 8, 61.

43. *Decoding Messages* Use the encoding matrix $A = \begin{bmatrix} 1 & 2 & -1 \\ 2 & 1 & -3 \\ -1 & 4 & 2 \end{bmatrix}$ to decode the message 29, −1, 75, −19, −66, 50, 46, 41, 47, 3, −38, 65.

44. *Decoding Messages* Use the encoding matrix $A = \begin{bmatrix} 2 & 3 & -2 \\ 3 & 2 & -4 \\ -2 & 5 & 3 \end{bmatrix}$ to decode the message 20, −33, 142, 74, 51, 107, 67, 87, −26, 22, −22, 99.

45. *Coded Message*

a. Choose a partner and create a message to send to the partner. Change the message from letters to numbers with the code

$$
\begin{array}{cccccccccccccccc}
a & b & c & d & e & f & g & h & i & j & k & l & m & n & o & p \\
1 & 2 & 3 & 4 & 5 & 6 & 7 & 8 & 9 & 10 & 11 & 12 & 13 & 14 & 15 & 16
\end{array}
$$

$$
\begin{array}{ccccccccccc}
q & r & s & t & u & v & w & x & y & z & blank \\
17 & 18 & 19 & 20 & 21 & 22 & 23 & 24 & 25 & 26 & 27
\end{array}
$$

b. Encode your message with encoding matrix

$$
A = \begin{bmatrix} 1 & 2 & 1 \\ 2 & 1 & -3 \\ -1 & 4 & 2 \end{bmatrix}
$$

c. Exchange coded messages.

d. Decode the message that you receive from your partner, and report your original message and your partner's message as you have decoded it.

section

7.5

Systems of Nonlinear Equations

key concepts

- Solving systems of nonlinear equations

  Algebraically

  Graphically

section preview ▪ Market Equilibrium

The supply function for a product is given by $p = q^2 + 804$, and the demand function for this product is $p = 1200 - 60q$, where p is the price in dollars and q is the number of hundreds of units. Suppose we are interested in finding the price that gives market equilibrium.

To answer this question, we create a system of equations, one of which is nonlinear, and solve the system. In this section, we use algebraic and graphical methods to solve systems of nonlinear equations.

We have solved systems of linear equations in two variables by substitution, by elimination, and by graphing. To solve systems of nonlinear equations, we can use these same methods. A system of nonlinear equations can have any number of solutions or no solution.

Algebraic Solution of Nonlinear Systems

example 1

Solving a Nonlinear System Algebraically

Use substitution to solve the system

$$
\begin{cases} x^2 - x - 3y = 0 \\ x - y = -4 \end{cases}
$$

and check the solution.

Solution

To solve the system

$$
\begin{cases} x^2 - x - 3y = 0 & (1) \\ x - y = -4 & (2) \end{cases}
$$

by substitution, notice that Equation (2) can easily be solved for y:

$$x - y = -4$$
$$y = x + 4$$

Substituting this expression for y in Equation (1) gives

$$x^2 - x - 3y = 0$$
$$x^2 - x - 3(x + 4) = 0$$
$$x^2 - x - 3x - 12 = 0$$
$$x^2 - 4x - 12 = 0$$
$$(x - 6)(x + 2) = 0$$
$$x = 6 \quad \text{or} \quad x = -2$$

Using these values for x in $y = x + 4$ gives the solutions $(6, 10)$ and $(-2, 2)$.

To check the solution, we substitute each ordered pair into both of the original equations:

Check $(6, 10)$:
$$x^2 - x - 3y = 0 \qquad x - y = -4$$
$$(6)^2 - (6) - 3(10) = 0 \qquad 6 - 10 = -4$$
$$36 - 6 - 30 = 0 \qquad -4 = -4$$
$$0 = 0$$

Check $(-2, 2)$:
$$x^2 - x - 3y = 0 \qquad x - y = -4$$
$$(-2)^2 - (-2) - 3(2) = 0 \qquad -2 - 2 = -4$$
$$4 + 2 - 6 = 0 \qquad -4 = -4$$
$$0 = 0$$

Both solutions check in each equation of the original system. ∎

example 2

Solving a Nonlinear System Algebraically

Solve the system algebraically, if $y = 0$:

$$\begin{cases} x + \dfrac{x}{y} = 6 \\ xy + y = 10 \end{cases}$$

Solution

To solve the system

$$\begin{cases} x + \dfrac{x}{y} = 6 & (1) \\ xy + y = 10 & (2) \end{cases}$$

algebraically, we begin by multiplying Equation (1) by y:

$$xy + x = 6y \qquad (3)$$

Next we solve Equation (2) for x:

$$xy + y = 10$$
$$xy = 10 - y$$
$$x = \frac{10 - y}{y}$$

Substituting this expression for x in Equation (3) gives

$$\frac{10 - y}{y}(y) + \frac{10 - y}{y} = 6y$$

$$10 - y + \frac{10 - y}{y} = 6y$$

Now we multiply both sides of this equation by y and solve the quadratic equation that results:

$$10 - y + \frac{10 - y}{y} = 6y$$

$$10y - y^2 + 10 - y = 6y^2$$

$$0 = 7y^2 - 9y - 10$$

$$0 = (y - 2)(7y + 5)$$

$$y = 2 \quad \text{or} \quad y = -\frac{5}{7}$$

If $y = 2$, then

$$x = \frac{10 - 2}{2} = \frac{8}{2} = 4$$

and if $y = -\frac{5}{7}$, then

$$x = \frac{10 + \dfrac{5}{7}}{-\dfrac{5}{7}} = -\frac{75}{5} = -15$$

Thus, the solutions are $(4, 2)$ and $\left(-15, -\dfrac{5}{7}\right)$.

Both solutions check in the original system of equations. ■

---

example 3

Market Equilibrium

The supply function for a product is given by $p = q^2 + 804$, and the demand function for this product is $p = 1200 - 60q$, where p is the price in dollars and q is the number of hundreds of units. Find the price that gives market equilibrium.

Solution

Market equilibrium occurs when the supply price (and quantity) equals the demand price (and quantity). Thus, the solution to the system

$$\begin{cases} p = q^2 + 804 & (1) \\ p = 1200 - 60q & (2) \end{cases}$$

gives the equilibrium quantity and price.

Substituting for p (from Equation (1)) into Equation (2) and solving gives

$$q^2 + 804 = 1200 - 60q$$

$$q^2 + 60q - 396 = 0$$

$$(q - 6)(q + 66) = 0$$

$$q = 6 \quad \text{or} \quad q = -66$$

The negative value of q has no meaning in this application, so we have $q = 6$ and $p = 6^2 + 804 = 840$. Thus, market equilibrium occurs when the price is $840, and the quantity is 600 units.

Graphical Solution of Nonlinear Systems

As with systems of linear equations in two variables, we can use graphical methods to check and to solve systems of nonlinear equations.

example 4 | ### Solving a Nonlinear System Graphically

Solve the system

$$\begin{cases} y = -x^2 - 2x + 8 \\ 2x + y = 4 \end{cases}$$

graphically.

Solution

To solve the system

$$\begin{cases} y = -x^2 - 2x + 8 & (1) \\ 2x + y = 4 & (2) \end{cases}$$

graphically, we first solve Equation (2) for y, getting $y = -2x + 4$. Graphing the two equations on the same axes gives the graph in Figure 7.23.

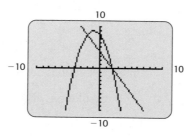

Figure 7.23

The points of intersection are $(2, 0)$ and $(-2, 8)$, as shown in Figures 7.24(a) and (b).

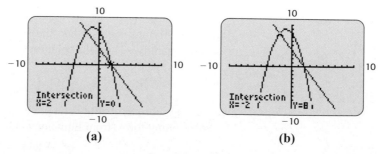

(a)   (b)

Figure 7.24

Thus, the solutions to the system are $(2, 0)$ and $(-2, 8)$.

| **example 5** | ### Constructing a Box |

An open box with square base is to be constructed so that the area of the material used is 300 cm^2 and the volume of the box is 500 cm^2. What are the dimensions of the box?

Solution

Let x represent the length of a side of the base of the box and let y represent the height of the box (Figure 7.25). The area of the base of the box is x^2, and the area of each of the four sides is xy, so the area of the material is

$$x^2 + 4xy = 300$$

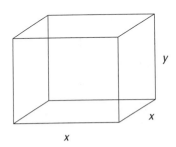

Figure 7.25

The volume of the box is $x^2y = 500$. So the system whose solution we seek is

$$\begin{cases} x^2 + 4xy = 300 & (1) \\ x^2y = 500 & (2) \end{cases}$$

Solving Equation (2) for y and substituting in Equation (1) gives

$$y = \frac{500}{x^2}$$

$$x^2 + 4x\left(\frac{500}{x^2}\right) = 300$$

$$x^2 + \left(\frac{2000}{x}\right) = 300$$

The resulting equation is difficult to solve algebraically, so we solve the original system by graphing. Solving Equation (1) and Equation (2) for y gives

$$y = \frac{300 - x^2}{4x} \quad \text{and} \quad y = \frac{500}{x^2}$$

Figure 7.26 shows the graphs of these equations on the same axes and the point of intersection.

Thus, $x = 10$ cm gives the length of each side of the base of the box, and $y = 5$ cm is the height of the box. ∎

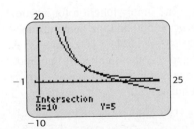

Figure 7.26

skills check

7.5

In Exercises 1–8, solve the system algebraically.

1. $\begin{cases} x^2 - y = 0 \\ 3x + y = 0 \end{cases}$
 2. $\begin{cases} x^2 - 2y = 0 \\ 2y - x = 0 \end{cases}$

3. $\begin{cases} x^2 - 3y = 4 \\ 2x + 3y = 4 \end{cases}$
 4. $\begin{cases} x^2 - 5y = 2 \\ 3x + 5y = 8 \end{cases}$

5. $\begin{cases} x^2 + y^2 = 80 \\ y = 2x \end{cases}$
 6. $\begin{cases} x^2 + y^2 = 72 \\ x + y = 0 \end{cases}$

7. $\begin{cases} x + y = 8 \\ xy = 12 \end{cases}$
 8. $\begin{cases} 2xy + y = 36 \\ 2x - 3y = 2 \end{cases}$

In Exercises 9–16, solve the system algebraically or graphically.

9. $\begin{cases} x^2 + 5x - y = 6 \\ 2x - y + 4 = 0 \end{cases}$
 10. $\begin{cases} x^2 - y - 8x = 6 \\ y + 10x = 18 \end{cases}$

11. $\begin{cases} 2x^2 - 2y + 7x = 19 \\ 2y - x = 61 \end{cases}$

12. $\begin{cases} 2x^2 + 4x - 2y = 24 \\ 2x - y + 37 = 0 \end{cases}$

13. $\begin{cases} x^2 + 4y = 28 \\ \sqrt{x} - y = -1 \end{cases}$
 14. $\begin{cases} x^2 - y = 0 \\ 2x + \sqrt{y} = 4 \end{cases}$

15. $\begin{cases} 4xy + x = 10 \\ x - 4y = -8 \end{cases}$
 16. $\begin{cases} xy + y = 4 \\ 2x - 3y = -4 \end{cases}$

In Exercises 17–20, solve the system graphically.

17. $\begin{cases} y = x^3 - 7x^2 + 10x \\ y = 70x - 10x^2 - 100 \end{cases}$

18. $\begin{cases} y = x^3 - 14x^2 + 40x \\ y = 70x - 5x^2 - 200 \end{cases}$

19. $\begin{cases} x^3 + x^2 + y = 15 \\ 2x^2 + y = 11 \end{cases}$
 20. $\begin{cases} x^3 - 3x + 2y = 2 \\ x^2 + x + y = 4 \end{cases}$

21. a. Algebraically eliminate y from the system in Exercise 19.

b. Use synthetic division to confirm that there is only one real solution to the system.

22. a. Algebraically eliminate y from the system in Exercise 20.

b. Use synthetic division to confirm that the system has three real solutions.

23. Use graphical methods to find two solutions to the system

$$\begin{cases} x^2 + 4y = 17 \\ x^2 y - 5x = 3 \end{cases}$$

24. Use graphical or algebraic methods to find two solutions to the system

$$\begin{cases} xy + 4x = 10 \\ \dfrac{y}{x} + y = \dfrac{3}{2} \end{cases}$$

exercises

7.5

25. *Supply and Demand* The supply function for a product is given by $p = q^2 + 2q + 122$ and the demand function for this product is $p = 650 - 30q$, where p is the price in dollars and q is the number of hundreds of units. Find the price that gives market equilibrium and the equilibrium quantity.

26. *Supply and Demand* The supply function for a product is given by $p = q^2 + 500$ and the demand function for this product is $p = 1124 - 40q$, where p is the price in dollars and q is the number of hundreds of units. Find the price that gives market equilibrium and the equilibrium quantity.

27. *Supply and Demand* The supply function for a product is given by $p = 0.1q^2 + 50q + 1027.50$ and the demand function for this product is $p = 6000 - 20q$, where p is the price in dollars and q is the number of hundreds of units. Find the price that gives market equilibrium and the equilibrium quantity.

28. *Supply and Demand* The supply function for a product is given by $p = q^3 + 800q + 6000$ and the demand function for this product is $p = 12,200 - 10q - q^2$, where p is the price in dollars and q is the number of hundreds of units. Find the price that gives market equilibrium and the equilibrium quantity.

29. *Break-Even* The weekly total cost function for a piece of equipment is $C(x) = 2000x + 18,000 + 60x^2$ dollars, and the weekly revenue is $R(x) = 4620x - 12x^2 - x^3$, where x is the number of thousands of units. The x-values that are the solutions to the system

$$\begin{cases} y = 2000x + 18,000 + 60x^2 \\ y = 4620x - 12x^2 - x^3 \end{cases}$$

are the quantities that give break-even. Find these quantities.

30. *Break-Even* The daily total cost function for a product is $C(x) - 400x + 4800$ dollars, and the daily revenue is $R(x) = 600x + 4x^2 - 0.1x^3$, where x is the number of units. Use graphical methods to find the numbers of units that give break-even.

31. *Break-Even* The monthly total cost function for a product is $C(x) - 400x + 0.1x^2 + 17,901$ dollars, and the monthly revenue is $R(x) = 1000x - 0.01x^2$, where x is the number of units. Find the numbers of units that give break-even for the product. (*Hint:* Use graphical methods to find one solution less than 100 units and one greater than 5000 units.)

32. *Break-Even* The weekly total cost function for a product is $C(x) = 3x^2 + 1228$ dollars, and the weekly revenue is $R(x) = 177.50x$, where x is the number of hundreds of units. Use graphical methods to find the number of units that gives break-even.

33. *Constructing a Box* A rectangular piece of cardboard with area 180 square inches is made into an open box by cutting squares 2 inches on a side from each corner and turning up the sides. If the box is to have a volume of 176 cubic inches, what are the dimensions of the original piece of cardboard?

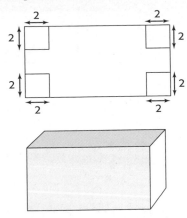

34. *Constructing a Box* An open box is constructed from a rectangular piece of tin by cutting 4-centimeter squares from each corner and turning up the sides. If the box has a volume of 768 cubic centimeters and the area of the original piece of tin was 512 square centimeters, what are the dimensions of the original piece of tin?

35. *Dimensions of a Box* An open box has a rectangular base and its height is equal to the length of the shortest side of the base. What dimensions give a volume of 2000 cubic centimeters using 800 square centimeters of material?

36. *Constructing a Box* A box with a top is to be constructed with the width of its base equal to half of the length of its base. If it is to have volume 6000 cubic centimeters and uses 2200 square centimeters of material, what are the dimensions of the box?

37. *Trust Fund* A father gives a trust fund to his daughter by depositing $50,000 into an account paying 10% compounded annually. This account grows according to $y = 50,000(1.10^t)$. He will give her $12,968.72 per year from another account until she has received in total the value to which the trust fund has grown. For how many years will she receive the $12,968.72?

38. *Constructing a Box* An open box with a square base is to be constructed so that the area of the material used is 500 cm and the volume of the box is 500 cm^3. What are the dimensions of the box?

chapter

7

Summary

In this chapter, we studied systems of equations and their applications. We first solved systems of linear equations in three variables by using left-to-right elimination. Matrices were introduced to help solve systems of equations, and other applications of matrices were investigated. We also solved systems of nonlinear equations by algebraic and by graphical methods.

Key Concepts and Formulas

7.1 Systems of Linear Equations in Three Variables

Left-to-right elimination method	An extension of the elimination method used to solve systems of two equations in two variables.
Possible solutions to a system of linear equations in three variables	
• **Unique solution**	Planes intersect in a single point.
• **No solution**	Planes have no common intersection.
• **Many solutions**	Planes intersect in a line.
Modeling systems of equations	Solution of real problems sometimes requires us to create three or more equations whose simultaneous solution is the solution to the problem.

7.2 Matrix Solution of Systems of Linear Equations

Matrix representation of systems of equations	Systems of linear equations can be represented using matrices, and the solution can be found easily.
Matrices	
• **Dimension**	The numbers of rows and columns of a matrix.
• **Square matrix**	Equal number of rows and columns.
• **Coefficient matrix**	Coefficients of the variables in a linear system placed in a matrix.
• **Augmented matrix**	A constant column is added to a coefficient matrix.
• **Identity matrix**	A square matrix that has 1's down its diagonal and 0's everywhere else.
• **Row-echelon form**	The coefficient part has all 1's or 0's on its diagonal and all 0's below its diagonal.
• **Reduced row-echelon form**	The coefficient part has 1's or 0's on its diagonal and 0's elsewhere.
Matrix row operations	1. Interchange two rows of the matrix.
	2. Multiply a row by a nonzero constant.
	3. Add a multiple of one row to another row.

Gauss–Jordan elimination method	Row operations are used to reduce a matrix to reduced row-echelon form. If the coefficient part reduces to the identity matrix, the system has a unique solution.
Using technology to find reduced row-echelon form	Graphing utilities and spreadsheets can be used to find the reduced row-echelon form.
Unique/nonunique solutions	Systems of equations can have a unique solution, no solution, or infinitely many solutions.
• Unique solution	The reduced coefficient matrix contains the identity matrix.
• No solution	A row of the reduced coefficient matrix contains all 0's and the augment of that row contains a nonzero number.
• Infinite number of solutions	If neither of the above occurs.

7.3 Matrix Operations

Addition of matrices	Matrices of the same dimension are added by adding the corresponding entries of the two matrices.
Zero matrix	Each element is zero.
Negative of a matrix	The sum of a matrix and its negative is the zero matrix.
Subtraction of matrices	If the matrices M and N are the same dimension, the difference $M - N$ is found by subtracting the elements of N from the corresponding elements of M.
Multiplication of matrices by a number	Multiplying a matrix A by a real number c results in a matrix in which each entry in matrix A is multiplied by the number c.
Matrix multiplication	The product of an $m \times n$ matrix A and an $n \times k$ matrix B is an $m \times k$ matrix $C = AB$ with the element in the ith row and jth column of matrix C given by the product of the ith row of A and the jth column of B.

7.4 Inverse Matrices; Matrix Equations

Inverse matrices	Two square matrices, A and B, are called inverses of each other if $AB = I$ and $BA = I$.
Finding the inverse of a square matrix	Placing the same dimension identity matrix in its augment and reducing the matrix to its reduced row-echelon form will give the inverse matrix in the augment if the inverse exists.
Inverses and technology	Computer software, spreadsheets, and calculators can be used to find the inverse of a matrix, if it exists.
Matrix equations	Multiplying both sides of the matrix equation $AX = C$ on the left by A^{-1}, if it exists, gives the solution to the system that the matrix equation represents.
Encoding messages and decoding messages	Matrix multiplication can be used to encode messages. The inverse of an encoding matrix can be used to decode messages.

7.5 Systems of Nonlinear Equations

Systems of nonlinear equations	A system of nonlinear equations contains at least one equation that is nonlinear. These systems can have any number of solutions or no solution.
Algebraic solution	Systems of nonlinear equations can be solved algebraically by substitution or elimination.

Graphical solution	Solutions to systems of nonlinear equations can be checked with graphical methods, or, if solving the system algebraically is difficult, graphical methods may be used to solve the system.

chapter

7

Skills Check

In Exercises 1–10, solve the systems of linear equations, if possible, by (a) left-to-right elimination, and (b) using technology.

1. $\begin{cases} 2x - 3y + z = 2 \\ 3x + 2y - z = 6 \\ x - 4y + 2z = 2 \end{cases}$ 2. $\begin{cases} 3x - 2y - 4z = 9 \\ x + 3y + 2z = -1 \\ 2x + 4y + 4z = 2 \end{cases}$

3. $\begin{cases} 3x + 2y - z = 6 \\ 2x - 4y - 2z = 0 \\ 5x + 3y + 6z = 2 \end{cases}$

4. $\begin{cases} 3x + 6y + 9z = 27 \\ 2x + 3y - z = -2 \\ 4x + 5y + z = 6 \end{cases}$

5. $\begin{cases} x + 2y - 2z = 1 \\ 2x - y + 5z = 15 \\ 3x - 4y + z = 7 \end{cases}$

6. $\begin{cases} -6x + 4y - 2z = 4 \\ 3x - 2y + 5z = -6 \\ x - 4y + z = -8 \end{cases}$

7. $\begin{cases} 2x + 5y + 8z = 30 \\ 18x + 42y + 18z = 60 \end{cases}$

8. $\begin{cases} 9x + 21y + 15z = 60 \\ 2x + 5y + 8z = 30 \\ x + 2y - 3z = -10 \end{cases}$

9. $\begin{cases} x + 3y + 2z = 5 \\ 9x + 12y + 15z = 6 \\ 2x + y + 3z = -10 \end{cases}$

10. $\begin{cases} 5x + 7y + 10z = 6 \\ 2x + 5y + 6z = 1 \\ 3x + 2y + 4z = 6 \end{cases}$

In Exercises 11 and 12, solve the systems of linear equations using technology.

11. $\begin{cases} 3x + 2y - z + w = 12 \\ x - 4y + 3z - w = -18 \\ x + y + 3z + 2w = 0 \\ 2x - y + 3z - 3w = -10 \end{cases}$

12. $\begin{cases} 2x + y - 3z + 4w = 7 \\ x - 2y + z - 2w = 0 \\ 3x + y + 4z + w = -2 \\ x + 3y + 2z + 2w = -1 \end{cases}$

Perform the matrix operations, if possible, in Exercises 13–21, with

$$A = \begin{bmatrix} 1 & 3 & -3 \\ 2 & 4 & 1 \\ -1 & 3 & 2 \end{bmatrix} \qquad B = \begin{bmatrix} 1 & 2 & 1 \\ 2 & -1 & 3 \end{bmatrix}$$

$$C = \begin{bmatrix} 2 & 3 \\ -1 & 2 \\ 3 & -2 \end{bmatrix} \qquad D = \begin{bmatrix} -2 & 3 & 1 \\ -3 & 2 & 2 \end{bmatrix}$$

13. $B + D$ 14. $D - B$ 15. $5C$

16. AB 17. BA 18. CD

19. DC 20. $A^2 = A \cdot A$ 21. A^{-1}

Find the inverse of each matrix in Exercises 22–24.

22. $\begin{bmatrix} 2 & 1 & 1 \\ 1 & 2 & -1 \\ 2 & 2 & 1 \end{bmatrix}$ 23. $\begin{bmatrix} 1 & 2 & 1 \\ 1 & 1 & 0 \\ 0 & 2 & 1 \end{bmatrix}$

24. $\begin{bmatrix} -1 & 0 & 1 & 1 \\ 1 & 1 & 2 & 1 \\ 2 & -1 & 0 & 1 \\ 0 & 0 & 0 & 1 \end{bmatrix}$

In Exercises 25–28, solve the system of equations by using the inverse of the coefficient matrix.

25. $\begin{cases} x + y - 3z = 8 \\ 2x + 4y + z = 15 \\ -x + 3y + 2z = 5 \end{cases}$

26. $\begin{cases} 2x + 3y + z = 20 \\ 3x + 4y - z = 40 \\ 2x + 5y + z = 60 \end{cases}$

27. $\begin{cases} -x_1 + + x_3 + x_4 = 6 \\ x_2 + x_3 + x_4 = 12 \\ 2x_1 + 5x_2 + x_3 = 20 \\ 3x_2 + x_4 = 24 \end{cases}$

28. $\begin{cases} 2x_1 + 2x_2 + x_4 = 4 \\ 2x_1 + x_2 + 2x_4 = 12 \\ x_1 + x_2 + x_3 + x_4 = 4 \\ x_2 + x_4 = 8 \end{cases}$

29. Solve the nonlinear system
$$\begin{cases} x^2 - y = x \\ 4x - y = 4 \end{cases}$$

30. Find two solutions to the system
$$\begin{cases} x^2 y = 2000 \\ x^2 + 2y = 140 \end{cases}$$

Review

31. *Pricing* A concert promoter needs to make $200,000 from the sale of 4000 tickets. The promoter charges $40 for some tickets, $60 for some tickets, and $100 for the others. If the number of $60 tickets is $\dfrac{1}{4}$ the sum of the remaining tickets, how many tickets of each type should be sold?

32. *Medication* Medication A is given six times per day, and medication B is given twice per day. Medication C is given once every two days, and its dosage is $\dfrac{1}{2}$ the sum of the dosages of the other two medications. For a certain patient, the total intake of the three medications is limited to 28.7 milligrams on days when all three medications are given. If the ratio of dosage of medication A to the dosage of medication B is 2 to 3, how many milligrams are in each dosage?

33. *Rental Income* A woman has $750,000 invested in three rental properties. One yields an annual return of 12% on her investment, and the other returns 15% per year on her investment. The third investment returns 10% and is limited to $\dfrac{1}{2}$ the sum of the other

two investments. Her total annual return from the three investments is $89,500. How much is invested in each property?

34. *Investment* A woman invests $360,000 in three different mutual funds, one that averages 12% per year and another that averages 16% per year. A third fund averages 8% per year, and because it has much less risk, she invests twice as much in it as in the sum of the other two funds. To realize an annual return of $35,200, how much should she invest in each fund?

35. *Loans* A bank loans $1,180,000 to a development company to purchase three business properties. If one of the properties costs $75,000 more than the other and the third costs three times the sum of these two properties, find the cost of each property.

36. *Investment* A brokerage house offers three stock portfolios for its clients. Portfolio I consists of 10 blocks of common stock, 2 municipal bonds, and 3 blocks of preferred stock. Portfolio II consists of 12 blocks of common stock, 8 municipal bonds, and 5 blocks of preferred stock. Portfolio III consists of 10 blocks of common stock, 4 municipal bonds, and 8 blocks of preferred stock. If a client wants to combine these portfolios so that she has 290 blocks of common stock, 138 municipal bonds, and 161 blocks of preferred stock, how many units of each portfolio does she need? To answer this question, let x equal the number of units of portfolio I, y equal the number of units of portfolio II, and z equal the number of units of portfolio III.

37. *Property* A realty partnership purchases three business properties with a total cost of $375,000. One property costs $50,000 more than the second, and the third property costs half the sum of these two properties. Find the cost of each property.

38. *Nutrition* A nutritionist wants to create a diet that uses a combination of three foods—I, II, and III. Each of these foods contains three additives, A, B, and C, that are used in the study. The following table gives the percent of each additive that is present in each food. If the diet being used requires 12.5 grams per day of A, 9.1 grams of B, and 9.6 grams of C, find the number of grams of each food that should be used each day for this diet.

	Food I	Food II	Food III
Additive A (%)	10	11	18
Additive B (%)	12	9	10
Additive C (%)	14	12	8

39. *Transportation* A delivery service has three types of aircraft, each of which carries three types of cargo. The payload of each type is summarized in the table below. Suppose on a given day the airline must move 2200 next-day delivery letters, 3860 two-day delivery letters, and 920 units of air freight. How many aircraft of each type should be scheduled?

	Aircraft Type		
Units Carried	Passenger	Transport	Jumbo
Next-Day Letters	200	200	200
Two-Day Letters	300	40	700
Air Freight	40	130	70

40. *Finance* A financial planner promises a long-term return of 12% from a combination of three mutual funds. The cost per share and average annual return

per share for each mutual fund are given in the table below. If the annual returns are accurate, how many shares of each mutual fund will return an average of 12% on a total investment of $210,000?

	Tech Fund	Balanced Fund	Utility Fund
Cost per Share ($)	180	210	120
Annual Return per Share ($)	18	42	18

41. *Transportation* Marshall Trucking Company has an order for three products, A, B, and C, for delivery. The following table gives the volume in cubic feet, the weight in pounds, and the value for insurance in dollars for a unit of each of the products. If one of the company's trucks can carry 9260 cubic feet and 12,000 pounds and is insured to carry $52,600, how many units of each product can be carried on the truck?

	Product A	Product B	Product C
Unit Volume (cubic feet)	25	30	40
Weight (pounds)	30	36	60
Value (dollars)	150	180	200

42. *Nutrition* A biologist is growing three types of slugs (A, B, and C) in the same laboratory environment. Each day, the slugs are given different nutrients (I, II, and III). Each type A slug requires 2 units of I, 6 units of II, and 2 units of III per day. Each type B slug requires 2 units of I, 8 units of II, and 4 units of III per day. Each type C slug requires 4 units of I, 20 units of II, and 12 units of III per day. If the daily mixture contains 4000 units of I, 16,000 units of II, and 8000 units of III, find the number of slugs of each type that can be supported.

43. *Trade Balances* Table A and Table B on the next page give the values of U.S. exports and imports with Mexico and with Canada for selected years 2002–2006.

  a. Form the matrix A that contains the values of exports to Mexico and to Canada for these years.

  b. Form the matrix B that contains the values of imports from Mexico and from Canada for these years.

c. Use these two matrices to find the trade balances for the United States for these years.

d. Are the trade balances with these countries improving for the United States?

e. With which country is the trade balance worse in 2006?

Table A

Mexico ($ millions)		
Year	Exports	Imports
2002	97,470	134,616
2003	97,412	138,060
2004	110,835	155,902
2005	120,365	170,109
2006	133,979	198,253

Table B

Canada ($ millions)		
Year	Exports	Imports
2002	160,923	209,088
2003	169,924	221,595
2004	189,880	256,360
2005	211,899	290,384
2006	230,656	302,438

44. *Traffic Study* In the analysis of traffic, a city engineer estimates the traffic flow at noon on the town square to be as illustrated in the figure in the next column. If x_1 illustrates the number of cars moving from intersection A to intersection B, x_2 represents the number of cars traveling from intersection B to intersection C, and so on, we can formulate equations based on the principle that the number of vehicles entering the intersection equals the number leaving it. For example, the equation that represents the traffic through A is $x_4 + 2250 = x_1 + 2900$.

a. Formulate equations for the traffic at the other intersections.

b. Solve the system of these four equations, to find how traffic between the other intersections is related to the traffic from intersection D to intersection A.

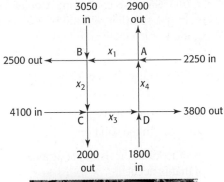

45. *Income and SAT Scores* The following table gives the average verbal and math SAT scores for certain family income levels in 2005.

a. Use linear regression to find a linear equation that gives the verbal SAT score as a function of family income.

b. Use linear regression to find a linear equation that gives the math SAT score as a function of family income.

c. At what income level do these models predict that the verbal SAT score will reach the math SAT?

Family Income ($ thousands)	Verbal SAT	Math SAT
5	426	438
15	443	463
25	463	474
35	480	487
45	496	500
55	505	509
65	511	515
75	517	522
90	529	534
110	554	565

(Source: College Board.com)

46. *Medication* Suppose combining x cubic centimeters (cc) of a 20% solution of a medication and y cc of a 5% solution of the medication gives 10 cc of a 15.5% solution.

   a. Write the equation that gives the total amount of solution in terms of x and y.

   b. Write the equation that gives the total amount of medication in the solution.

   c. Use substitution to find the amount of each solution that is needed.

47. *Supply and Demand* The supply function for a product is given by $p = q + 578$, and the demand function for this product is $p = 396 + q^2$, where p is the price in dollars and q is the number of units. Find the price that gives market equilibrium and the equilibrium quantity.

48. *Break-Even* The daily total cost function for a product is $C(x) = 2500x + x^2 + 27{,}540$ dollars, and the daily revenue is $R(x) = 3899x - 0.1x^2$, where x is the number of units. Find the numbers of units that give break-even.

group activities / extended applications

1. Salaries

Average Salaries of College Professors, 2006–2007

Academic Rank	Public	Private-Independent	Church-Related	Public	Private-Independent	Church-Related
		MEN			WOMEN	
(Doctoral)						
Professor	108,481	138,921	121,312	98,552	127,542	109,720
Associate	76,030	89,936	82,941	70,764	83,148	76,814
Assistant	65,498	78,079	70,227	60,155	71,207	65,882
Instructor	43,352	50,418	55,513	42,012	49,246	53,549
Lecturer	51,182	59,443	52,257	46,340	52,047	46,476
(Master's)						
Professor	82,834	93,020	84,493	79,493	86,476	78,355
Associate	66,102	71,130	66,448	63,630	66,881	63,338
Assistant	56,225	57,610	55,107	53,907	55,065	52,467
Instructor	41,495	46,444	45,997	40,820	44,315	44,174
Lecturer	46,030	51,037	49,064	44,018	43,907	45,631
(Baccalaureate)						
Professor	78,390	91,563	69,969	73,372	87,570	66,427
Associate	63,658	67,210	56,752	61,363	66,444	55,315
Assistant	52,960	55,180	47,853	50,714	54,011	47,085
Instructor	40,844	45,139	40,966	41,183	40,705	40,181
Lecturer	47,743	58,064	42,095	45,565	54,343	39,233

(Source: American Association of University Professors)

The table above gives the average salaries of college professors for men and women who teach at three types of institutions at three different levels of instruction, and at five job designations. To investigate the relationship between salaries of males and females:

1. Create matrix *A* and matrix *B*, which contain the salaries of male and female professors, respectively, in each type of school and at each level of instruction.

2. Subtract the matrix of male professor salaries from the matrix of female professor salaries.

3. What does this show about female salaries versus male salaries in every category? Based on this data, discuss gender bias in educational institutions.

4. Create a scatter plot with male professor salaries as the input and female professor salaries as the output.

5. Find a function that gives a relationship between the salaries of male and female professors from part 4.

6. Graph the scatter plot from part 4 and the function from part 5 on the same axes.

7. The graph of the function $y = x$ would represent the relationship between male and female professors' salaries if they were equal. Graph the function from part 5 and the function $y = x$ on the same axes.

8. Use the two graphs in part 7 to determine how female professors' salaries compare to male salaries as male salaries rise.

9. Solve the two equations whose graphs are discussed in part 8 to determine at what salary level male salaries become greater than female salaries.

2. Parts-Listing

Wingo Playgrounds must maintain numerous parts in its inventory for its Fun-in-the-Sun swing set. The swing set has 4 legs and a top connecting the legs. Each of the legs is made from 1 nine-foot pipe connected with 1 brace and 2 bolts, and the top is made of 1 nine-foot pipe, 2 clamps, and 6 bolts. The parts-listing for these swing sets can be described by the following matrix.

$$P = \begin{bmatrix} SS & L & T & P & C & Bc & B \\ 0 & 0 & 0 & 0 & 0 & 0 & 0 \\ 4 & 0 & 0 & 0 & 0 & 0 & 0 \\ 1 & 0 & 0 & 0 & 0 & 0 & 0 \\ 0 & 1 & 1 & 0 & 0 & 0 & 0 \\ 0 & 0 & 2 & 0 & 0 & 0 & 0 \\ 0 & 1 & 0 & 0 & 0 & 0 & 0 \\ 0 & 2 & 6 & 0 & 0 & 0 & 0 \end{bmatrix} \begin{matrix} \text{Swing sets} \\ \text{Legs} \\ \text{Top} \\ \text{Pipes} \\ \text{Clamps} \\ \text{Braces} \\ \text{Bolts} \end{matrix}$$

Each column indicates how many of each part are required to produce the part shown at the top of the column. For example, column 1 indicates that to produce a swing set requires 4 legs and 1 top, and the second column indicates that each leg (L) is constructed from 1 pipe, 1 brace, and 2 bolts.

Suppose an order is received for 8 complete swing sets plus the following spare parts: 2 legs, 1 top, 2 pipes, 4 clamps, 4 braces, and 8 bolts. We indicate the number of each part type that must be supplied to fill this order by the matrix

$$X = \begin{bmatrix} x_1 \\ x_2 \\ x_3 \\ x_4 \\ x_5 \\ x_6 \\ x_7 \end{bmatrix}$$

To find the numbers of each part that are needed, perform the following:

1. Form the 7×1 matrix D that contains the numbers of items to be delivered to fill the order for the 8 swing sets and the spare parts. Some parts of each type are needed to produce other parts of the swing set. The matrix PX gives the number of parts of each type used to produce other parts, and the matrix

$$X - PX$$

gives the numbers of parts of each type that are available to be delivered. That is,

$$X - PX = D,$$

where D is the number of parts of each type that are available for delivery.

2. The matrix equation $X - PX = D$ can be rewritten in the form $(I - P)X = D$, where I is the 7×7 identity matrix. Find the matrix $I - P$ that satisfies this equation.

3. Find the inverse of the matrix $I - P$.

4. Multiply both sides of the matrix equation $(I - P)X = D$ by the inverse of $I - P$. This gives the matrix X that contains the numbers of parts of each type that are needed.

5. The primary assembly parts are the pipes, clamps, braces, and bolts. What is the total number of primary parts of each type needed to fill the order?

Special Topics: Systems of Inequalities and Linear Programming; Sequences and Series; Preparing for Calculus

To determine how many cars of different types that it must buy and rent to maximize its profit, a car rental agency must consider the constraints on its purchases. The constraints include how much money it has to invest in cars and how many cars it needs. These constraints can be written as inequalities and "pictured" by graphs of systems of inequalities. The process of finding the optimal solution to a problem subject to the constraint inequalities is called linear programming. In addition, sequences and series can be used to solve several types of financial problems, including future values of lump sum investments and annuities. We also review numerous algebra skills in the context of calculus applications in this chapter.

section preview ▪ Car Rental

A rental agency has a maximum of $1,260,000 to invest in the purchase of at most 71 new cars of two different types, compact and midsize. The cost per compact car is $15,000 and the cost per midsize car is $28,000. The number of cars of each type is limited (constrained) by the budget available and the number of cars needed. To get a "picture" of the constraints on this rental agency, we can graph the inequalities determined by the maximum amount of money and the limit on the number of cars purchased. Because these inequalities deal with two types of cars, they will contain two variables. For example, we can denote the number of compact cars by x and the number of midsize cars by y, so the statement that the agency wants to purchase at most 71 cars can be written as the inequality

$$x + y \leq 71, \quad \text{where } x \text{ and } y \text{ are integers}$$

The collection of all **constraint** inequalities can be expressed by a **system of inequalities** in two variables. Finding the values that satisfy all these constraints at the same time is called solving the system of inequalities. In this section, we will graph solutions of linear and nonlinear inequalities in two variables.

Linear Inequalities in Two Variables

We solved linear inequalities in one variable in Section 2.4. We now consider linear inequalities in two variables. For example, consider the inequality $y > x$. The solutions to this inequality are the ordered pairs (x, y) that satisfy the statement $y > x$. The graph of $y > x$ consists of all points above the line $y = x$ (the shaded area in Figure 8.1). The line $y = x$ is dashed in Figure 8.1 because the inequality does not include $y = x$. This (dashed) line divides the plane into two **half-planes**, $y > x$ and $y < x$. We can determine which half-plane is the solution to the inequality by selecting any point not on the line. If the coordinates of this **test point** satisfy the inequality, then the half-plane containing that point is the graph of the solution. If the test point does not satisfy the inequality, then the other half-plane is the graph of the solution. For example, the point $(2, 4)$ is above the line $y = x$ and $x = 2, y = 4$ satisfies the inequality $y > x$, so the region above the line is the graph of the solution of the inequality. Note that any other point on the half-plane above the line also satisfies the inequality.

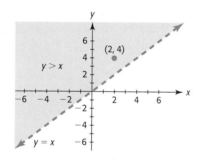

Figure 8.1

---

example 1 | **Graph of an Inequality**

Graph the solution of the inequality $6x - 3y \leq 15$.

Solution

First we graph the line $6x - 3y = 15$ (which we can also write as $y = 2x - 5$) as a solid line, because the inequality symbol "$\leq$" indicates that points on the line are part of the solution (Figure 8.2(a)). Next we pick a test point that is not on the line. If we use $(0, 0)$, we get $6(0) - 3(0) \leq 15$, or $0 \leq 15$, which is true, so the inequality is satisfied. Thus, the half-plane that contains the point $(0, 0)$ and the line $6x - 3y = 15$ is the graph of the solution, which we call the **solution region** of the inequality.

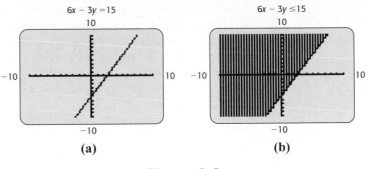

Figure 8.2

Technology Note

Graphing utilities can also be used to shade the solution region of an inequality. Figure 8.2(b) shows a graphing calculator screen with the solution to $6x - 3y \leq 15$ shaded.

example 2

Car Rental Agency

One of the constraint inequalities for the car rental company application given in the Section Preview is that the sum of x and y (the number of compact and midsize cars, respectively) is at most 71. This constraint can be described by the inequality

$$x + y \leq 71$$

where x and y are integers. Find the graphical solution of this inequality.

Solution

The inequality $x + y \leq 71$, where x and y are integers, has many integer solutions. The inequality is satisfied if x is any integer from 0 to 71 and y is any integer less than or equal to $71 - x$. For example, some of the solutions are $(0, 71)$, $(1, 70)$, $(60, 5)$, and $(43, 20)$. As we stated in Chapter 2, problems are frequently easier to solve if we treat the data as continuous rather than discrete. With this in mind, we graph this inequality as a region on a coordinate plane rather than attempting to write all of the possible integer solutions.

We create the graph of the inequality by first graphing the equation

$$x + y = 71, \quad \text{or equivalently,} \quad y = 71 - x$$

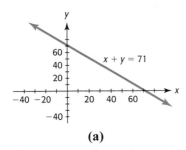

(a)

(Figure 8.3(a)). The graph of this line divides the coordinate plane into two half-planes, $y > 71 - x$ and $y < 71 - x$. We can find the half-plane that represents the graph of $y < 71 - x$ by testing the coordinates of a point and determining if it satisfies $y < 71 - x$. For example, the point $(0, 0)$ is in the half-plane below the line, and $x = 0, y = 0$ satisfies the inequality $y < 71 - x$ because $0 < 71 - 0$. Thus the half-plane below the line satisfies the inequality. If we shade the half-plane that satisfies $y < 71 - x$, the line and the shaded region constitute the graph of $x + y \leq 71$ (Figure 8.3(b)). This is the solution region for the inequality.

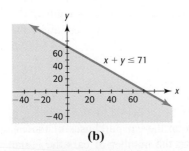

(b)

Figure 8.3

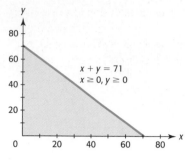

Figure 8.4

Note that the number of cars satisfying this constraint has other limitations. In the context of this application, the number of cars cannot be negative, so the values of x and y must be nonnegative. We can show the solution to the inequality for nonnegative x and y by using a window with $x \geq 0$ and $y \geq 0$. Figure 8.4 shows the solution of $x + y \leq 71$ for $x \geq 0$ and $y \geq 0$.

All points in this region with integer coordinates satisfy the stated conditions of this example. ■

We now investigate the second inequality in the car rental problem.

| example 3 | **Rental Cars** |

The car rental agency in the Section Preview has a maximum of \$1,260,000 to invest, with x compact cars costing \$15,000 each and y midsized cars costing \$28,000 each.

a. Write the inequality representing this information.

b. Graph this inequality on a coordinate plane.

Solution

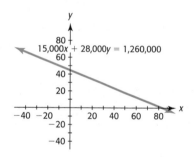

Figure 8.5

a. If x represents the number of compact cars, the cost of all the compact cars purchased is $15,000x$ dollars. Similarly, the cost of all the midsized cars purchased is $28,000y$ dollars. Because the total amount available to spend is 1,260,000 dollars, x and y satisfy

$$15,000x + 28,000y \leq 1,260,000$$

b. To graph the solutions to this inequality, we first graph the equation

$$15,000x + 28,000y = 1,260,000, \quad \text{or equivalently,}$$

$$y = \frac{1,260,000 - 15,000x}{28,000} = \frac{1260 - 15x}{28}$$

The graph is shown in Figure 8.5.

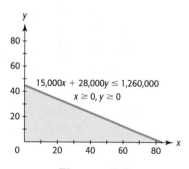

Figure 8.6

To decide which of the two half-planes determined by the line satisfies the inequality, we test points in one half-plane. The point $(0, 0)$ is in the half-plane below the line, and it satisfies the inequality because $15,000(0) + 28,000(0) < 1,260,000$. Thus, the half-plane containing $(0, 0)$, along with the line, forms the solution region. Because x and y must be nonnegative in this application, the graph of the inequality is shown only in the first quadrant (Figure 8.6). All points in this region with integer coordinates satisfy the stated conditions of this example. ■

Systems of Inequalities in Two Variables

We now have pictures of the limitations (constraints) on the car purchases of the car rental agency, but how are they related? We know that the number of cars the rental agency can buy is limited by *both* inequalities in Examples 2 and 3, so we seek the solution to both inequalities *simultaneously*. That is, we seek the solution to a **system of inequalities**. In general, if we have two or more inequalities in two variables and seek the values of the variables that satisfy all inequalities, we are solving a system of inequalities. The solution to the system can be found by finding the intersection of the solution sets of the inequalities.

---

example 4 | **System of Inequalities**

Solve the system of inequalities $\begin{cases} 2x - 4y \geq 12 \\ x + 3y > -4 \end{cases}$.

Solution

To find the solution, we first solve the inequalities for y.

$$y \leq \frac{1}{2}x - 3 \qquad (1)$$

$$y > -\frac{1}{3}x - \frac{4}{3} \qquad (2)$$

The *borders* of the inequality region are graphed as the solid line with equation $y = \frac{1}{2}x - 3$ and the dashed line with equation $y = -\frac{1}{3}x - \frac{4}{3}$. One of these borders is a solid line because $y = \frac{1}{2}x - 3$ is part of Inequality (1), and one is a dashed line because $y = -\frac{1}{3}x - \frac{4}{3}$ is not part of Inequality (2) (Figure 8.7(a)). We find a *corner* of the solution region by finding the intersection of the two lines. We can find this point of intersection by algebraically or graphically solving the equations simultaneously. Using substitution and solving gives the point of intersection:

$$\frac{1}{2}x - 3 = -\frac{1}{3}x - \frac{4}{3}$$

$$3x - 18 = -2x - 8$$

$$5x = 10$$

$$x = 2$$

$$y = \frac{1}{2}(2) - 3 = -2$$

The two border lines divide the plane into four regions, one of which is the solution region. We can determine what region satisfies both inequalities by choosing a test point in each of the four regions. For example:

- $(0, 0)$ and $(1, -2)$ do not satisfy Inequality (1).

- $(2, -4)$ does not satisfy Inequality (2).

- $(6, -1)$ satisfies both inequalities, so the region that contains this point is the solution region.

Figure 8.7(b) shows the solution region for the system of inequalities, with its corner at $(2, -2)$. Note that the graph of $y = -\frac{1}{3}x - \frac{4}{3}$ is not part of the solution. ■

We now turn our attention to finding the values of x and y that satisfy both the constraints on the car rental agency problem in the Section Preview.

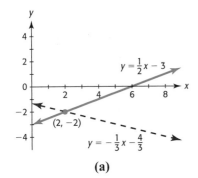

(a)

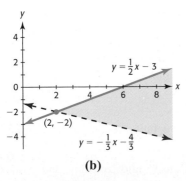

(b)

Figure 8.7

| example 5 | **Car Rental Agency** |

The inequalities that satisfy the conditions given in the car rental problem in the Section Preview are called the **constraint inequalities** for the problem. They form the system

$$\begin{cases} y \leq 71 - x \\ y \leq \dfrac{1260 - 15x}{28} \\ x \geq 0 \\ y \geq 0 \end{cases}$$

a. Graph the solution to the system.

b. Find the coordinates of the points where the borders intersect and the points where the region intersects the x- and y-axes.

Solution

a. We graphed the solutions to the first two inequalities discussed for the car rental agency application in Examples 2 and 3. If we graph them both in the first quadrant, the region where the two solution regions overlap is where both inequalities are true simultaneously (the coordinates of the points satisfy both inequalities) (Figure 8.8). Any point in the solution region with integer coordinates is a solution to the application. For example, $x = 50$, $y = 8$ is a solution, as is $x = 6$, $y = 40$.

b. The region is bordered on one side by the *line*

$$y = 71 - x$$

and on the other side by the *line*

$$y = \dfrac{1260 - 15x}{28}$$

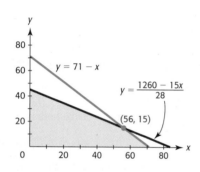

Figure 8.8

A **corner** of the graph of the solution set occurs at the *point* where these two lines intersect. We find this point of intersection graphically as shown in Figure 8.8 or by solving the two equations simultaneously.

$$\begin{cases} y = 71 - x \\ y = \dfrac{1260 - 15x}{28} \end{cases}$$

Substitution gives

$$71 - x = \dfrac{1260 - 15x}{28}$$

$$1988 - 28x = 1260 - 15x$$

$$728 = 13x$$

$$x = 56 \quad \text{and} \quad y = 71 - 56 = 15$$

Thus, this point of intersection of the borders of the solution region, or the *corner*, occurs at $x = 56$, $y = 15$.

This corresponds to purchasing 56 compact cars and 15 midsized cars, and is one of many possible solutions to the rental application.

Other corners occur where the region intersects the axes:

- At the origin ($x = 0$, $y = 0$, representing no cars purchased).
- Where $y = 71 - x$ intersects the x-axis ($x = 71$, $y = 0$, representing 71 compact and no midsize cars).
- Where $y = \dfrac{1260 - 15x}{28}$ intersects the y-axis ($x = 0$, $y = 45$, representing 0 compact and 45 midsize cars).

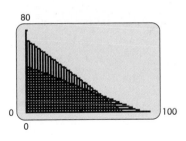

Figure 8.9

Computer and calculator programs can be used to save considerable time and energy in determining regions that satisfy systems of inequalities. We can also use technology to graph the two lines shown in Figure 8.8 and to find their point of intersection, and we can use $\boxed{\text{SHADE}}$ to show the region that is the intersection of the inequalities* (Figure 8.9).

---

example 6

Advertising

A candidate for mayor of a city wishes to use a combination of radio and television advertisements in her campaign. Research has shown that each 1-minute spot on television reaches 0.09 million people and each 1-minute spot on radio reaches 0.006 million. The candidate feels that she must reach at least 2.16 million people, and she can buy a total of no more than 80 minutes of advertisement time.

a. Write the inequalities that describe her advertising needs.

b. Graph the region determined by these constraint inequalities.

c. Interpret the solution region in the context of this problem.

Solution

a. If we represent by x the number of minutes of television time, the total number of people reached by television is $0.09x$ million, and if we represent by y the number of minutes of radio time, the total number reached by radio is $0.006y$ million. Thus, one condition that must be satisfied is given by the inequality

$$0.09x + 0.006y \geq 2.16$$

In addition, the statement that the total number of minutes of advertising can be no more than 80 minutes can be written as the inequality

$$x + y \leq 80$$

Neither x nor y can be negative in this application, so we also have

$$x \geq 0$$
$$y \geq 0$$

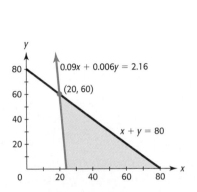

Figure 8.10

b. Graphing the equations corresponding to these inequalities and shading the regions satisfying the inequalities give the graph of the solution (Figure 8.10).

One corner of the region occurs where the boundary lines intersect. We can write the equations of the boundary lines as

$$0.09x + 0.006y = 2.16 \quad \text{and} \quad y = 80 - x$$

---

* See Appendix A, page 669.

Their y-values are equal where the lines intersect, so we substitute for y in the first equation and solve.

$$0.09x + 0.006(80 - x) = 2.16$$
$$0.09x + 0.48 - 0.006x = 2.16$$
$$0.084x = 1.68$$
$$x = 20$$
$$y = 80 - 20 = 60$$

The other corners occur where the region intersects the axes:

- $0.09x + 0.006y = 2.16$ intersects the x-axis at $x = 24$, $y = 0$.
- $x + y = 80$ intersects the x-axis at $x = 80$, $y = 0$.

c. Each point in this region represents a combination of television and radio time that the candidate for mayor can use in her campaign. For example, she could use 20 minutes of television time and 60 minutes of radio time, or 40 minutes of television time and 20 minutes of radio time. No point outside this region can represent a number of minutes used in television and radio advertising for this candidate because of the constraints given in the problem. ∎

Systems of Nonlinear Inequalities

As with systems of linear inequalities in two variables, the solutions of systems of nonlinear inequalities in two variables are regions in a plane. For example, the solution of the system

$$\begin{cases} y \le 8 - 2x - x^2 \\ y \ge 4 - 2x \end{cases}$$

is the area below the graph of $y = 8 - 2x - x^2$ and above the graph of $y = 4 - 2x$, as well as any points on the graphs of $y = 4 - 2x$ and $y = 8 - 2x - x^2$ between $(-2, 8)$ and $(2, 0)$ (Figure 8.11). The points of intersection of the two border curves are $(2, 0)$ and $(-2, 8)$.

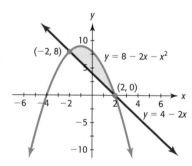

Figure 8.11

---

example 7 | ## Nonlinear System of Inequalities

Graphically solve the system

$$\begin{cases} x^2 - 10x + y \le 0 \\ -x^2 + 8x + y \ge 28 \end{cases}$$

Solution

To solve the system

$$\begin{cases} x^2 - 10x + y \le 0 \\ -x^2 + 8x + y \ge 28 \end{cases}$$

graphically, we first solve the inequalities for y:

$$\begin{cases} y \le 10x - x^2 \\ y \ge x^2 - 8x + 28 \end{cases}$$

Graphing the associated equations on the same axes gives the graphs in Figure 8.12(a).

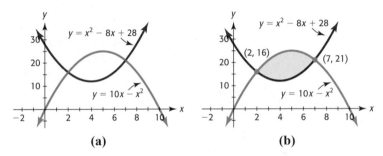

(a)   **(b)**

Figure 8.12

The points of intersection of the two border curves are (2, 16) and (7, 21). We can show this algebraically as follows:

$$x^2 - 8x + 28 = 10x - x^2$$
$$2x^2 - 18x + 28 = 0$$
$$2(x - 2)(x - 7) = 0$$
$$x = 2 \quad | \quad x = 7$$
$$y = 16 \quad | \quad y = 21$$

The region that we seek is the portion of the graph that is above the graph of $y = x^2 - 8x + 28$ and below the graph of $y = 10x - x^2$, as well as any points on the graphs of these two functions between (2, 16) and (7, 21), as shown in Figure 8.12(b). ■

| example 8 | ## Nonlinear System of Inequalities |

Graphically solve the system

$$\begin{cases} y > x^3 - 3x \\ y < x \\ x > -2.5 \\ x < 2.5 \end{cases}$$

Solution

To solve the system

$$\begin{cases} y > x^3 - 3x \\ y < x \\ x > -2.5 \\ x < 2.5 \end{cases}$$

we first graph the related equations (Figure 8.13(a)).

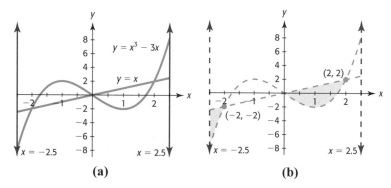

Figure 8.13

One region is bounded on the left by $x = -2.5$ and the two curves, with a corner at $(-2, -2)$. The second region is bounded by the two curves with corners at $(0, 0)$ and $(2, 2)$. These corner points can also be found algebraically as follows.

$$x^3 - 3x = x$$
$$x^3 - 4x = 0$$
$$x(x - 2)(x + 2) = 0$$
$$x = 0 \ \big| \ x = 2 \ \big| \ x = -2$$
$$y = 0 \ \big| \ y = 2 \ \big| \ y = -2$$

The solution to the system of inequalities consists of the two regions that are above the graph of $y = x^3 - 3x$ and below the graph of $y = x$ in the interval $[-2.5, 2.5]$. This solution is shown in Figure 8.13(b). ■

skills check

8.1

In Exercises 1–6, graph each inequality.

1. $y \le 5x - 4$

2. $y > 3x + 2$

3. $6x - 3y \ge 12$

4. $\dfrac{x}{2} + \dfrac{y}{3} \le 6$

5. $4x + 5y \le 20$

6. $3x - 5y \ge 30$

In Exercises 7–10, match the solution of the system of inequalities with the graph.

7. $\begin{cases} y > 3x - 4 \\ y < 2x + 3 \end{cases}$

8. $\begin{cases} y \ge x + 1 \\ y \le 2x + 1 \\ x \ge 0 \\ y \ge 0 \end{cases}$

9. $\begin{cases} 3x + y \geq 5 \\ x + 3y \geq 6 \\ x \geq 0 \\ y \geq 0 \end{cases}$

10. $\begin{cases} x + 2y \leq 20 \\ 3x + 2y \leq 42 \\ x \geq 0 \\ y \geq 0 \end{cases}$

13. $\begin{cases} 4x + 2y > 8 \\ 3x + y > 5 \\ x \geq 0, y \geq 0 \end{cases}$

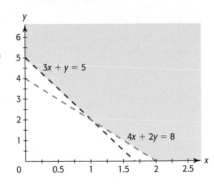

A.

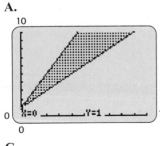

B.

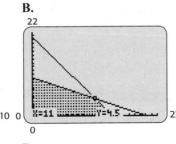

14. $\begin{cases} 2x + 4y \geq 12 \\ x + 3y \geq 8 \\ x \geq 0, y \geq 0 \end{cases}$

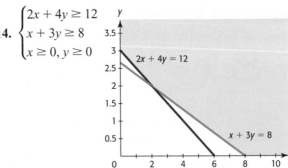

C.

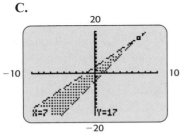

D.

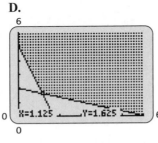

15. $\begin{cases} 2x + 6y \geq 12 \\ 3x + y \geq 5 \\ x + 2y \geq 5 \\ x \geq 0, y \geq 0 \end{cases}$

In Exercises 11–16, the graph of the boundary equations for each system of inequalities is shown with the system. Locate the solution region, and identify it by finding the corners.

11. $\begin{cases} x + y \leq 5 \\ 2x + y \leq 8 \\ x \geq 0, y \geq 0 \end{cases}$

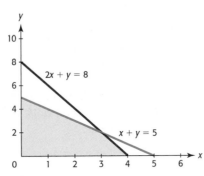

16. $\begin{cases} 2x + y \geq 12 \\ x + y \leq 8 \\ 2x + y \leq 14 \\ x \geq 0, y \geq 0 \end{cases}$

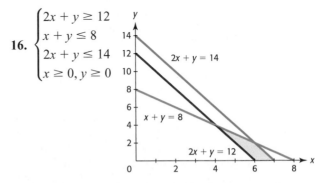

12. $\begin{cases} 2x + y \leq 12 \\ x + 5y \leq 15 \\ x \geq 0, y \geq 0 \end{cases}$

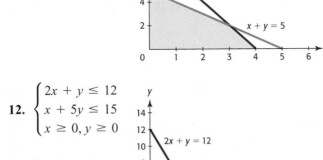

For each system of inequalities in Exercises 17–22, graph the solution region and identify the corners of the region.

17. $\begin{cases} y \leq 8 - 3x \\ y \leq 2x + 3 \\ y > 3 \end{cases}$

18. $\begin{cases} 2x + y \geq 10 \\ 3x + 2y \geq 17 \\ x + 2y \geq 7 \end{cases}$

19. $\begin{cases} 2x + y < 5 \\ 2x - y > -1 \\ x \geq 0, y \geq 0 \end{cases}$ 20. $\begin{cases} y < 2x \\ y > x + 2 \\ x \geq 0, y \geq 0 \end{cases}$

21. $\begin{cases} x + 2y \geq 4 \\ x + y \leq 5 \\ 2x + y \leq 8 \\ x \geq 0, y \geq 0 \end{cases}$ 22. $\begin{cases} x + y < 4 \\ x + 2y < 6 \\ 2x + y < 7 \\ x \geq 0, y \geq 0 \end{cases}$

In Exercises 23–26, graph the solution of the system of nonlinear inequalities.

23. $\begin{cases} x^2 - 3y < 4 \\ 2x + 3y < 4 \end{cases}$ 24. $\begin{cases} x + y \leq 8 \\ xy \geq 12 \end{cases}$

25. $\begin{cases} x^2 - y - 8x \leq -6 \\ y + 9x \leq 18 \end{cases}$ 26. $\begin{cases} x^2 - y > 0 \\ 2x + \sqrt{y} > 4 \end{cases}$

exercises

8.1

27. *Manufacturing* A company manufactures two types of leaf blowers, an electric Turbo model and a gas-powered Tornado model. The company's production plan calls for the production of at least 780 blowers per month.

    a. Write the inequality that describes the production plan, if x represents the number of Turbo blowers and y represents the number of Tornado blowers.

    b. Graph the region determined by this inequality in the context of the application.

28. *Manufacturing* A company manufactures two types of leaf blowers, an electric Turbo model and a gas-powered Tornado model. It costs $78 to produce each of the x Turbo models and $117 to produce each of the y Tornado models, and the company has at most $76,050 per month to use for production.

    a. Write the inequality that describes this constraint on production.

    b. Graph the region determined by this inequality in the context of the application.

29. *Sales* Trix Auto Sales sells used cars. To advertise its cars to target groups, they advertise with x 1-minute spots on cable television, at a cost of $240 per minute, and y 1-minute spots on radio, at a cost of $150 per minute. Suppose the company has at most $36,000 to spend on advertising.

    a. Write the inequality that describes this constraint on advertising.

    b. Graph the region determined by this inequality in the context of the application.

30. *Sales* Trix Auto Sales sells used cars. To advertise its cars to target groups, they advertise with x 1-minute spots on cable television and y 1-minute spots on radio. Research shows that they sell one vehicle for every 4 minutes of cable television advertising and they sell one vehicle for each 10 minutes of radio advertising.

    a. Write the inequality that describes the requirement that at least 33 cars be sold from this advertising.

    b. Graph the region determined by this inequality in the context of the application.

31. *Politics* A political candidate wishes to use a combination of x television and y radio advertisements in his campaign. Each 1-minute ad on television reaches 0.12 million eligible voters, and each 1-minute ad on radio reaches 0.009 million eligible voters. The candidate feels that he must reach at least 7.56 million eligible voters and that he must buy at least 100 minutes of advertisements.

    a. Write the inequalities that describes these advertising requirements.

    b. Graph the region determined by these inequalities in the context of the application.

32. *Housing* A contractor builds two models of homes, the Van Buren and the Jefferson. The Van Buren requires 200 worker-days of labor and $240,000 in capital, and the Jefferson requires 500 worker-days of labor and $300,000 in capital. The contractor has a total of 5000 worker-days and $3,600,000 in capital available per month. Let x represent the number of Van Buren models, and y represent the number of Jefferson models, and graph the region that satisfies these inequalities.

33. *Manufacturing* A company manufactures two types of leaf blowers, an electric Turbo model and a gas-powered Tornado model. The company's production plan calls for the production of at least 780 blowers per month. It costs $78 to produce each Turbo model and $117 to manufacture each Tornado model, and the company has at most $76,050 per month to use for production. Let x represent the number of Turbo models and y represent the number of Tornado models, and graph the region that satisfies these constraints.

34. *Sales* Trix Auto Sales sells used cars. To advertise its cars to target groups, they advertise with x 1-minute spots on cable television at a cost of $240 per minute and y 1-minute spots on radio, at a cost of $150 per minute. Research shows that they sell one vehicle for every 4 minutes of cable television advertising (that is, $\frac{1}{4}$ car per minute) and they sell one vehicle for each 10 minutes of radio advertising. Suppose the company has at most $36,000 to spend on advertising and that they sell at least 33 cars per month from this advertising. Let x represent the number of cable TV minutes and y represent the number of radio minutes, and graph the region that satisfies these constraints. That is, graph the solution to this system of inequalities.

35. *Production* A firm produces three different-size television sets on two assembly lines. The following table summarizes the production capacity of each assembly line and the minimum number of each size TV needed to fill orders.

    a. Write the inequalities that are the constraints.

    b. Graph the region that satisfies the inequalities, and identify the corners.

	Assembly Line 1	Assembly Line 2	Number Ordered
19-in. TV	80 per day	40 per day	3200
25-in. TV	20 per day	20 per day	1000
35-in. TV	100 per day	40 per day	3400

36. *Advertising* Tire Town is developing an advertising campaign. The following table indicates the cost per ad package in newspapers and the cost per ad package on radio, and the number of ads in each type of ad package.

	Newspaper	Radio
Cost per Ad Package ($)	1000	3000
Ads per Package	18	36

The owner of the company can spend no more than $18,000 per month, and he wants to have at least 252 ads per month.

a. Write the inequalities that are the constraints.

b. Graph the region that satisfies the inequalities, and identify the corners.

37. *Nutrition* A privately owned lake contains two types of fish, bass and trout. The owner provides two types of food, A and B, for these fish. Trout require 4 units of food A and 6 units of food B, and bass require 10 units of food A and 7 units of food B. If the owner has 1600 units of each food, graph the region that satisfies the constraints.

38. *Manufacturing* Evergreen Company produces two types of printers, the inkjet and the laserjet. It takes 2 hours to make the inkjet and 6 hours to make the laserjet. The company can make at most 120 printers per day and has 400 labor-hours available per day.

a. Write the inequalities that describe this application.

b. Graph the region that satisfies the inequalities, and identify the corners of the region.

39. *Manufacturing* Easyboy manufactures two types of chairs, Standard and Deluxe. Each Standard chair requires 4 hours to construct and finish, and each Deluxe chair requires 6 hours to construct and finish. Upholstering takes 2 hours for a Standard chair and 6 hours for a Deluxe chair. There are 480 hours available each day for construction and finishing and there are 300 hours available per day for upholstering.

a. Write the inequalities that describe the application.

b. Graph the solution of the system of inequalities, and identify the corners of the region.

40. *Manufacturing* Safeco Company produces two types of chainsaws, the Safecut and the Safecut Deluxe. The Safecut model requires 2 hours to assemble and 1 hour to paint, and the Deluxe model requires 3 hours to assemble and $\frac{1}{2}$ hour to paint. The daily maximum number of hours available for assembly is 36, and the daily maximum number of hours available for painting is 12.

a. Write the inequalities that describe the application.

b. Graph the solution of the system of inequalities, and identify the corners of the region.

41. *Manufacturing* A company manufactures commercial and domestic heating systems at two plant sites. It can produce no more than 1400 units per month, and it needs to fill orders of at least 500 commercial units and 750 domestic units.

a. Write the system of inequalities that describe the constraints on production for these orders.

b. Graph the solution set of this system of inequalities.

42. *Constructing a Box* An open box with square base is to be constructed so that the area of the material used is at most 500 cm^2. What dimensions of the base will yield a box which gives a volume of at least 500 cm^3?

section

8.2

Linear Programming: Graphical Methods

key concepts

- Linear programming
- Constraints
- Objective function
- Feasible solutions
- Feasible region
- Optimal values
  Maximum
  Minimum

section preview ▪ Maximizing Profit

A rental agency has a maximum of $1,260,000 to invest in the purchase of at most 71 new cars of two different types, compact and midsize. The cost per compact car is $15,000, and the cost per midsize car is $28,000. The number of cars of each type is limited (constrained) by the budget available and the number of cars needed. We saw in the previous section that purchases of the rental agency were constrained by the number of cars needed and the amount of money available for purchasing the cars. The constraints were given by a system of inequalities, and we found the solution region determined by the constraints.

If the average yearly profit is $6000 for each compact car and $9000 for each midsize car, buying a different number of each type will affect the profit. How do we find the correct number of each type so that the profit is maximized under these conditions? To answer questions of this type without knowledge of special techniques, people must use the "guess-and-check" method repeatedly to arrive at a solution. And when the number of conditions that enter into a problem of this type increases, guessing is a very unsatisfactory solution method. We can answer these questions with a mathematical technique called **linear programming**. Linear programming is a valuable tool in management; in fact, 85% of all Fortune 500 firms use linear programming. In this section, we see how to use graphical methods to find the **optimal solutions** to linear programming problems such as the one above.

Linear Programming

In any linear programming problem, we seek the point or points in the region determined by the constraint inequalities that give optimal (*maximum* or *minimum*) values for the **objective function** (the function we seek to optimize).* Any point in the region determined by the constraints is called a **feasible solution** to the linear programming problem, and the region itself is called the **feasible region**.

Note that the constraints in our car rental application above can be written as the system of inequalities

$$\begin{cases} x + y \leq 71 \\ 15x + 28y \leq 1260 \\ x \geq 0 \\ y \geq 0 \end{cases}$$

The solution set of the inequalities for the problem, the feasible region, is shown as the shaded region in Figure 8.14. Any point inside the shaded region or on its boundary satisfies the constraints, so it is a feasible solution. In the context of this application, only feasible solutions with integer coordinates can be solutions to the problem.

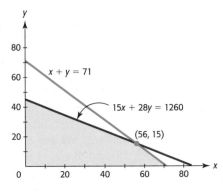

Figure 8.14

Because the profit P is 6 thousand dollars for each of the x compact cars and 9 thousand dollars for each of the y midsize cars, we can model the profit function as

$$P = 6x + 9y, \quad \text{where } P \text{ is in thousands of dollars}$$

In this application, we seek to find the maximum profit subject to the constraints listed above. Because we seek to maximize the profit subject to the constraints, we evaluate the **objective function**

$$P = 6x + 9y$$

for points (x, y) in the feasible region. For example, at $(20, 19)$, $P = 291$, at $(50, 10)$, $P = 390$, and at $(50, 15)$, $P = 435$. But there are a large number of points to test, so we look at a different way to find the maximum value of P. We can graph $P = 6x + 9y$ for different values of P by first writing the equation in the form

$$y = -\frac{2}{3}x + \frac{P}{9}$$

Then, for each possible value of P, the graph is a line with slope $-\dfrac{2}{3}$ and y-intercept $\dfrac{P}{9}$. Figure 8.15 shows the graphs of the feasible region and objective functions for

---

* The region determined by the constraints must be convex for the optimal solution to exist. A convex region is one such that for any two points in the region, the segment joining those points lies entirely within the region. We restrict our discussion to convex regions.

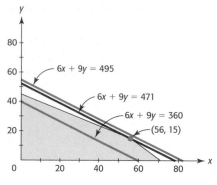

Figure 8.15

$P = 360, 471,$ and 495. Note that the graphs are parallel $\left(\text{with slope } -\dfrac{2}{3}\right)$, and that larger values of P coincide with larger y-intercepts $\left(\dfrac{P}{9}\right)$. Observe that if $P = 471$, the graph of $y = -\dfrac{2}{3}x + \dfrac{P}{9}$ intersects the feasible region and that values of P less than 471 will give lines that pass through the feasible region, but each of these values of P represent smaller profit. Note also that the value $P = 495$ (or any P-value greater than 471) results in a line that "misses" the feasible region, so this value of P is not a solution to the problem. Thus, the maximum value for P, subject to the constraints, is 471 (thousand dollars). We can verify that the graph of $471 = 6x + 9y$ intersects the region at $(56, 15)$ and that $P = 6x + 9y$ will not intersect the feasible region if $P > 471$. Thus, the profit is maximized at \$471,000, when $x = 56$ and $y = 15$; that is, when 56 compact and 15 midsize cars are purchased.

Note that the values of x and y that correspond to the maximum value of P occur at $(56, 15)$, which is a corner of the feasible region. In fact, the maximum value of an objective function always occurs at a corner of the feasible region if the feasible region is closed and bounded. (A feasible region is closed and bounded if it is enclosed by and includes the lines associated with the constraints.) This gives us the basis for finding the solutions to linear programming problems using the graphical method.

Solutions to Linear Programming Problems

1. If a linear programming problem has an optimal solution, then the optimum value (maximum or minimum) of an objective function occurs at a corner of the feasible region determined by the constraints.

2. When a feasible region for a linear programming problem is closed and bounded, the objective function has a maximum and a minimum value.

3. If the objective function has the same optimum value at two corners, then it also has that optimum value at any point on the boundary line segment connecting the two corners.

4. When the feasible region is not closed and bounded, the objective function may have a maximum only, a minimum only, or neither.

Thus, for a closed and bounded region, we can find the maximum and minimum values of an objective function by evaluating the function at each of the corners of the feasible region formed by the graphical solution of the constraint inequalities.

| example 1 | # Finding the Optimal Values of a Function Subject to Constraints |

Find (a) the maximum value and (b) the minimum value of $C = 5x + 3y$ subject to the constraints

$$\begin{cases} x + y \le 5 \\ 2x + y \le 8 \\ x \ge 0, y \ge 0 \end{cases}$$

Solution

The feasible region is in the first quadrant because it is bounded by $x \ge 0$ and $y \ge 0$. We can graph the remaining boundary lines with technology if we solve the first two inequalities for y.

$$\begin{cases} y \le 5 - x \\ y \le -2x + 8 \\ x \ge 0, y \ge 0 \end{cases}$$

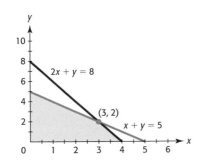

Figure 8.16

The solution set for the system of inequalities (that is, the feasible region) is bounded by the lines $y = 5 - x$, $y = -2x + 8$, and the x- and y-axes. The corners of the feasible region occur at the origin $(0, 0)$, at the x-intercept of one of the two boundary lines, at the y-intercept of one of the two boundary lines, and at the point of intersection of the lines $y = 5 - x$ and $y = -2x + 8$ (Figure 8.16).

There are two y-intercept points, $(0, 5)$ and $(0, 8)$, and the one that satisfies both inequalities is $(0, 5)$. The x-intercept points are $(4, 0)$ and $(5, 0)$; the one that satisfies both inequalities is $(4, 0)$. We can find the point of intersection of the two boundary lines by solving the system of equations

$$\begin{cases} y = 5 - x \\ y = -2x + 8 \end{cases}$$

Substitution gives

$$5 - x = -2x + 8$$

Solving gives $x = 3$, and substituting gives $y = 2$. Thus, the point of intersection of these two boundary lines is $(3, 2)$. This point of intersection can also be found by using a graphing utility.

a. Any point inside the shaded region or on the boundary is a feasible solution to the problem, but the maximum possible value of

$$C = 5x + 3y$$

occurs at a corner of the feasible region. We evaluate the objective function at each of the corner points:

$$\text{At } (0, 0): \quad C = 5x + 3y = 5(0) + 3(0) = 0$$
$$\text{At } (0, 5): \quad C = 5x + 3y = 5(0) + 3(5) = 15$$
$$\text{At } (4, 0): \quad C = 5x + 3y = 5(4) + 3(0) = 20$$
$$\text{At } (3, 2): \quad C = 5x + 3y = 5(3) + 3(2) = 21$$

Thus, the maximum value of the objective function is 21 when $x = 3, y = 2$.

b. Observing the values of $C = 5x + 3y$ at each corner point of the feasible region, we see that the minimum value of the objective function is 0 at $(0, 0)$.

example 2	Cost Minimization

Star Manufacturing Company produces two types of DVD players, which are assembled at two different locations. Plant 1 can assemble 60 units of the Star model and 80 units of the Prostar model per hour, and plant 2 can assemble 300 units of the Star model and 80 units of the Prostar model per hour. The company needs to produce at least 5400 units of the Star model and 4000 units of the Prostar model to fill an order. If it costs \$2000 per hour to run plant 1 and \$3000 per hour to run plant 2, how many hours should each plant spend on manufacturing DVD players to minimize its cost for this order? What is the minimum cost for this order?

Solution

If we let x equal the number of hours of assembly time at plant 1 and we let y equal the number of hours of assembly time at plant 2, then the cost function that we seek to minimize is

$$C = 2000x + 3000y$$

The constraints on the assembly hours follow:

Star model units:	$60x + 300y \geq 5400$	
Prostar model units:	$80x + 80y \geq 4000$	
Nonnegative hours:	$x \geq 0, y \geq 0$	

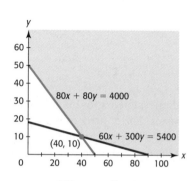

Figure 8.17

We can solve the first two inequalities for y, write these constraints as a system of inequalities, and graph the feasible region (Figure 8.17):

$$\begin{cases} y \geq 18 - 0.2x \\ y \geq 50 - x \\ x \geq 0, y \geq 0 \end{cases}$$

The two boundary lines intersect where $18 - 0.2x = 50 - x$, at $x = 40$, which then gives $y = 10$. The boundary lines of the feasible region intersect the axes at $(0, 50)$ and at $(90, 0)$.

Thus, the corners of the feasible region are

- $(0, 50)$, the y-intercept that satisfies all the inequalities.
- $(90, 0)$, the x-intercept that satisfies all the inequalities.
- $(40, 10)$, the intersection of the two boundary lines.

To find the hours of assembly time at each plant that will minimize the cost, we test the corner points:

At $(0, 50)$:	$C = 2000(0) + 3000(50) = 150{,}000$
At $(90, 0)$:	$C = 2000(90) + 3000(0) = 180{,}000$
At $(40, 10)$:	$C = 2000(40) + 3000(10) = 110{,}000$

Thus, the cost is minimized when assembly time is 40 hours at plant 1 and 10 hours at plant 2. The minimum cost is \$110,000. ∎

Note that the feasible region in Example 2 is not closed and bounded, and that there is no maximum value for the objective function even though there is a minimum value. It should also be noted that some applied linear programming problems require discrete solutions; for example, the optimal solution to the earlier car rental problem required a number of cars, so only nonnegative integer solutions were possible.

| example 3 | **Maximizing Profit** |

Smoker Meat Packing Company makes two different types of hot dogs, regular and all-beef. Each pound of all-beef hot dogs requires 0.8 pound of beef and 0.2 pound of spices, and each pound of regular hot dogs requires 0.3 pound of beef and 0.2 pound of spices, with the remainder nonbeef meat products. The company has at most 1020 pounds of beef and has at most 500 pounds of spices available for hot dogs. If the profit is $0.90 on each pound of all-beef hot dogs and $1.20 on each pound of regular hot dogs, how many of each type should be produced to maximize the profit?

Solution

Letting x equal the number of pounds of all-beef hot dogs and y equal the number of pounds of regular hot dogs, we seek to maximize the function $P = 0.90x + 1.20y$ subject to the constraints

$$\begin{cases} 0.8x + 0.3y \le 1020 \\ 0.2x + 0.2y \le 500 \\ x \ge 0, y \ge 0 \end{cases}$$

The graph of this system of inequalities is more easily found if the first two inequalities are solved for y.

$$\begin{cases} y \le 3400 - \dfrac{8}{3}x \\ y \le 2500 - x \\ x \ge 0, y \ge 0 \end{cases}$$

Graphing the boundary lines and testing the four regions determine the feasible region (Figure 8.18). The boundary lines intersect where

$$3400 - \frac{8}{3}x = 2500 - x \quad \text{or} \quad \text{at} \quad x = 540. \text{ This gives } y = 1960.$$

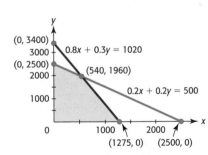

Figure 8.18

The x-intercept (1275, 0) and the y-intercept (0, 2500) satisfy the inequalities. Thus, the corners of this region are (0, 0), (0, 2500), (1275, 0), and (540, 1960).

$P = 0$ at (0, 0). Testing the objective function at the other corner points gives:

At (0, 2500): $P = 0.90x + 1.20y = 0.90(0) + 1.20(2500) = 3000$

At (1275, 0): $P = 0.90x + 1.20y = 0.90(1275) + 1.20(0) = 1147.50$

At (540, 1960): $P = 0.90x + 1.20y = 0.90(540) + 1.20(1960) = 2838$

This indicates that profit will be maximized if 2500 pounds of regular hot dogs and no all-beef hot dogs are produced. ∎

| example 4 | Manufacturing |

A company manufactures air conditioning units and heat pumps at its factories in Atlanta, Georgia, and Newark, New Jersey. The Atlanta plant can produce no more than 1000 items per day, and the number of air conditioning units cannot exceed 100 more than half the number of heat pumps. The Newark plant can produce no more than 850 units per day. The profit on each air conditioning unit is $400 at the Atlanta plant and $390 at the Newark plant. The profit on each heat pump is $200 at the Atlanta plant and $215 at the Newark plant. If there is an order for 500 air conditioning units and 750 heat pumps,

a. Graph the feasible region and identify the corners.

b. Find the maximum profit that can be made on this order and what production distribution will give the maximum profit.

Solution

a. Because the total number of the air conditioning units needed is 500, we can represent the number produced at Atlanta by x and the number produced at Newark by $(500 - x)$. Similarly, we can represent the number of heat pumps produced at Atlanta by y and the number produced at Newark by $(750 - y)$. Table 8.1 summarizes the constraints.

Table 8.1

	Air Conditioning Units	Heat Pumps	Total
Atlanta	x	$+ y$	≤ 1000
Newark	$500 - x$	$+ (750 - y)$	≤ 850
Other Constraints	x		$\leq 0.5y + 100$
	x		≥ 0
		y	≥ 0
Profit	$400x + 390(500 - x)$	$+ 200y + 215(750 - y)$	

Using this information gives the following system:

$$\begin{cases} x + y \leq 1000 & (1) \\ (500 - x) + (750 - y) \leq 850 & (2) \\ x \leq 0.5y + 100 & (3) \\ x \geq 0, y \geq 0 \end{cases}$$

We solve each of these inequalities for y and graph the solution set (Figure 8.19.):

$$\begin{cases} y \leq 1000 - x & (1) \\ y \geq 400 - x & (2) \\ y \geq 2x - 200 & (3) \\ x \geq 0, y \geq 0 \end{cases}$$

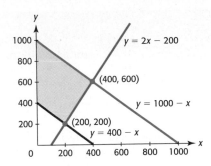

Figure 8.19

The intersections of the boundary lines associated with the inequalities follow.

- Inequalities (1) and (3): where $1000 - x = 2x - 200$, at $x = 400$, $y = 600$.
- Inequalities (2) and (3): where $400 - x = 2x - 200$, at $x = 200$, $y = 200$.
- Inequalities (1) and (2): where $1000 - x = 400 - x$, which has no solution, so they do not intersect.

The other corners are on the y-axis, at $(0, 400)$ and $(0, 1000)$.

b. The objective function gives the profit for the products.

$$P = 400x + 390(500 - x) + 200y + 215(750 - y) \quad \text{or} \quad P = 10x - 15y + 356{,}250$$

Testing the profit function at the corners of the feasible region determines where the profit is maximized.

$$
\begin{aligned}
\text{At } (0, 400)\text{:} \quad & P = 350{,}250 \\
\text{At } (200, 200)\text{:} \quad & P = 355{,}250 \\
\text{At } (0, 1000)\text{:} \quad & P = 341{,}250 \\
\text{At } (400, 600)\text{:} \quad & P = 351{,}250
\end{aligned}
$$

Thus, the profit is maximized at $355{,}250 when 200 air conditioning units and 200 heat pumps are manufactured at Atlanta. The remainder are manufactured at Newark, so 300 air conditioning units and 550 heat pumps are manufactured at Newark.   ■

Solution with Technology

As we have seen, graphing utilities can be used to graph the feasible region and to find the corners of the region satisfying the constraint inequalities. The graphical method that we have been using cannot be used if the problem involves more than two variables, and most real linear programming applications involve more than two variables. In fact, some real applications involve as many as 40 variables. Problems involving more than two variables can be solved with spreadsheet programs such as Lotus 1-2-3, Microsoft Excel, and Quatro Pro. Use of these programs to solve linear programming problems is beyond the scope of this text.

skills check

8.2

In Exercises 1–4, use the given feasible region determined by the constraint inequalities to find the maximum possible value and the minimum possible value of the objective function.

1. $f = 4x + 9y$ subject to the constraints

$$\begin{cases} 2x + 3y \le 134 \\ x + 5y \le 200 \\ x \ge 0, y \ge 0 \end{cases}$$

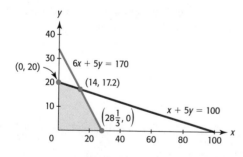

2. $f = 2x + 3y$ subject to the constraints

$$\begin{cases} 6x + 5y \le 170 \\ x + 5y \le 100 \\ x \ge 0, y \ge 0 \end{cases}$$

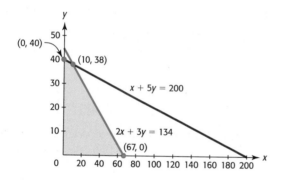

3. $f = 4x + 2y$ subject to the constraints

$$\begin{cases} 3x + y \le 15 \\ x + 2y \le 10 \\ -x + y \le 2 \\ x \ge 0, y \ge 0 \end{cases}$$

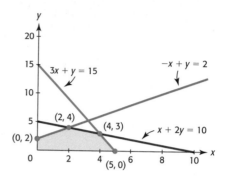

4. $f = 3x + 9y$ subject to the constraints

$$\begin{cases} 2x + y \le 12 \\ x + 3y \le 15 \\ x + y \le 7 \\ x \ge 0, y \ge 0 \end{cases}$$

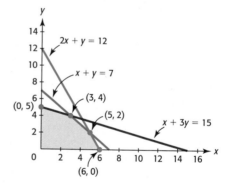

5. Perform the following steps to maximize $f = 3x + 5y$ subject to the constraints

$$\begin{cases} y \le \dfrac{54 - 2x}{3} \\ y \le 22 - x \\ x \ge 0, y \ge 0 \end{cases}$$

a. Graph the region that satisfies the system of inequalities, and identify the corners of the region.

b. Test the objective function $f = 3x + 5y$ at each of the corners of the feasible region to determine which corner gives the maximum value. Give the maximum possible value of f and the values of x and y that give the value.

6. Perform the following steps to maximize $f = 3x + 4y$ subject to the constraints

$$\begin{cases} x + 2y \le 16 \\ x + y \le 10 \\ x \ge 0, y \ge 0 \end{cases}$$

   a. Graph the region that satisfies the system of inequalities, and identify the corners of the region.

   b. Test the objective function $f = 3x + 4y$ at each of the corners of the feasible region to determine which corner gives the maximum value. Give the maximum possible value of f and the values of x and y that give that value.

7. Perform the following steps to minimize $g = 4x + 2y$ subject to the constraints

$$\begin{cases} x + 2y \ge 15 \\ x + y \ge 10 \\ x \ge 0, y \ge 0 \end{cases}$$

   a. Graph the region that satisfies the system of inequalities, and identify the corners of the region.

   b. Test the objective function $g = 4x + 2y$ at each of the corners of the feasible region to determine which corner gives the minimum value. Give the minimum possible value of g and the values of x and y that give that value.

8. Find the maximum possible value of $f = 400x + 300y$ and the values of x and y that give that value, subject to the constraints

$$\begin{cases} 3x + 2y \ge 6 \\ 2x + y \le 7 \\ x \ge 0, y \ge 0 \end{cases}$$

9. Find the maximum possible value of $f = 20x + 30y$ and the values of x and y that give that value, subject to the constraints

$$\begin{cases} y \le -\dfrac{1}{2}x + 4 \\ y \le -x + 6 \\ y \le -\dfrac{1}{3}x + 4 \\ x \ge 0, y \ge 0 \end{cases}$$

10. Find the maximum possible value of $f = 100x + 100y$ and the values of x and y that give that value, subject to the constraints

$$\begin{cases} x + 2y \le 6 \\ x + 4y \le 10 \\ 2x + y \le 9 \\ x \ge 0, y \ge 0 \end{cases}$$

11. Find the maximum possible value of $f = 80x + 160y$ and the values of x and y that give that value, subject to the constraints

$$\begin{cases} 3x + 4y \le 32 \\ x + 2y \le 15 \\ 2x + y \le 18 \\ x \ge 0, y \ge 0 \end{cases}$$

12. Find the minimum possible value of $g = 30x + 40y$ and the values of x and y that give that value, subject to the constraints

$$\begin{cases} x + 2y \ge 16 \\ x + y \ge 10 \\ x \ge 0, y \ge 0 \end{cases}$$

13. Find the minimum possible value of $g = 40x + 30y$ and the values of x and y that give that value, subject to the constraints

$$\begin{cases} 2x + y \ge 6 \\ 4x + y \ge 8 \\ x + 2y \ge 6 \\ x \ge 0, y \ge 0 \end{cases}$$

14. Find the minimum possible value of $g = 30x + 40y$ and the values of x and y that give that value, subject to the constraints

$$\begin{cases} x + 2y \ge 5 \\ x + y \ge 4 \\ 2x + y \ge 6 \\ x \ge 0, y \ge 0 \end{cases}$$

15. Find the minimum possible value of $g = 46x + 23y$ and the values of x and y that give that value, subject to the constraints

$$\begin{cases} 3x + y \ge 6 \\ x + y \ge 4 \\ x + 5y \le 8 \\ x \ge 0, y \ge 0 \end{cases}$$

16. Consider the constraints

$$\begin{cases} 2x + y \le 9 \\ 3x + y \ge 11 \\ x + y \ge 5 \\ x \ge 0, y \ge 0 \end{cases}$$

a. Find the maximum possible value of $f = 4x + 5y$ and the values of x and y that give that value.

b. Find the minimum possible value of $g = 3x + 2y$ and the values of x and y that give that value.

17. Find the minimum possible value of $g = 60x + 10y$ and the values of x and y that give that value, subject to the constraints

$$\begin{cases} 3x + 2y \ge 12 \\ 2x + y \ge 7 \\ x \ge 0, y \ge 0 \end{cases}$$

18. Find the maximum possible value of $f = 10x + 10y$ and the values of x and y that give that value subject to the constraints

$$\begin{cases} x + 2y \le 10 \\ x + y \le 6 \\ 2x + y \le 10 \\ x \ge 0, y \ge 0 \end{cases}$$

exercises
8.2

19. *Manufacturing* A company manufactures two types of leaf blowers, an electric Turbo model and a gas-powered Tornado model. The company's production plan calls for the production of at least 780 blowers per month. It costs $78 to produce each Turbo model and $117 to manufacture each Tornado model, and the company has at most $76,050 per month to use for production. Find the number of units that should be produced to maximize the profit for the company if the profit on each Turbo model is $32 and the profit on each Tornado model is $45.

20. *Manufacturing* Evergreen Company produces two types of printers, the Inkjet and the Laserjet. The company can make at most 120 printers per day and has 400 labor-hours available per day. It takes 2 hours to make the Inkjet and 6 hours to make the Laserjet. If the profit on the Inkjet is $80 and the profit on the Laserjet is $120, find the maximum possible daily profit and the number of each type of printer that gives it.

21. *Manufacturing* Safeco Company produces two types of chainsaws, the Safecut and the Safecut Deluxe. The Safecut model requires 2 hours to assemble and 1 hour to paint, and the Deluxe model requires 3 hours to assemble and $\frac{1}{2}$ hour to paint. The daily maximum number of hours available for assembly is 36, and the daily maximum number of hours available for painting is 12. If the profit is $24 per unit on the Safecut model and $30 per unit on the Deluxe model, how many units of each type will maximize the daily profit?

22. *Production* Two models of riding mowers, the Lawn King and the Lawn Master, are produced on two assembly lines. Producing the Lawn King requires 2 hours on line I and 1 hour on line II. Producing the Lawn Master requires 1 hour on line I and 3 hours on line II. The number of hours available for production is limited to 60 on line I and 40 on line II. If there is $150 profit per mower on the Lawn King and $200 profit per mower on the Lawn Master, producing how many of each model maximizes the profit? What is the maximum possible profit?

23. *Sales* Trix Auto Sales sells used cars. To advertise its cars to target groups, they advertise with x 1-minute spots per month on cable television at a cost of $240 per minute and y 1-minute spots per month on radio at a cost of $150 per minute. Research shows that they sell one vehicle for every 4 minutes of cable television advertising (that is, $\frac{1}{4}$ car per minute) and they sell one vehicle for each 10 minutes of radio advertising. Suppose the company has at most $36,000 per month to spend on advertising and they sell at least 33 cars per month from this advertising. Graph the region that satisfies these two inequalities; that is, graph the solution to this system of inequalities.

a. If the profit on cars advertised on television averages $500 and the profit on cars advertised on radio averages $550, how many minutes per month of advertising should be spent on television advertising and how many should be spent on radio advertising?

b. What is the maximum possible profit?

24. *Advertising* Tire Town is developing an advertising campaign. The table below indicates the cost per ad package in newspapers and the cost per ad package on radio, and the number of ads in each type of ad package.

	Newspaper	Radio
Cost per Ad Package ($)	1000	3000
Ads per Package	18	36

The owner of the company can spend no more than $18,000 per month, and he wants to have at least 252 ads per month. If each newspaper ad package reaches 6000 people over 20 years of age and each radio ad reaches 8000 of these people, how many newspaper ad packages and radio ad packages should he buy to maximize the number of people over age 20 reached?

25. *Production* A firm produces three different-size television sets on two assembly lines. The following table summarizes the production capacity of each assembly line, the number of each size TV ordered by a retailer, and the daily operating costs for each assembly line. How many days should each assembly line run to fill this order with minimum cost? What is the minimum cost?

	Assembly Line 1	Assembly Line 2	Number Ordered
19-in. TV	80 per day	40 per day	3200
27-in. TV	20 per day	20 per day	1000
35-in. TV	100 per day	40 per day	3400
Daily Cost ($)	20,000	40,000	

26. *Nutrition* A privately owned lake contains two types of fish, bass and trout. The owner provides two types of food, A and B, for these fish. Trout require 4 units of food A and 5 units of food B, and bass require 10 units

of food A and 4 units of food B. If the owner has 1600 units of food A and 1000 units of food B, find the maximum number of fish that the lake can support.

27. *Housing* A contractor builds two models of homes, the Van Buren and the Jefferson. The Van Buren requires 200 worker-days of labor and $240,000 in capital, and the Jefferson requires 500 worker-days of labor and $300,000 in capital. The contractor has a total of 5000 worker-days and $3,600,000 in capital available per month. The profit is $60,000 on the Van Buren and $75,000 on the Jefferson. Building how many of each model will maximize the monthly profit? What is the maximum possible profit?

28. *Manufacturing* Easyboy manufactures two types of chairs, Standard and Deluxe. Each Standard chair requires 4 hours to construct and finish, and each Deluxe chair requires 6 hours to construct and finish. Upholstering takes 2 hours for a Standard chair and 6 hours for a Deluxe chair. There are 480 hours available each day for construction and finishing, and there are 300 hours available per day for upholstering. If the revenue is $178 for each Standard chair and $267 for each Deluxe chair,

a. What is the maximum possible daily revenue?

b. How many of each type should be produced each day to maximize the daily revenue?

29. *Production* Ace Company produces three models of VCRs, the Ace, the Ace Plus, and the Ace Deluxe models, at two facilities, A and B. The company has orders for at least 4000 of the Ace models, at least 1800 of the Ace Plus, and at least 2400 of the Ace Deluxe models. The weekly production capacity of each model at each facility and the cost per week to operate each facility are given in the table below. For how many weeks should each facility operate to minimize the cost of filling the orders?

	A	B
Ace	400	400
Ace Plus	300	100
Ace Deluxe	200	400
Cost per Week ($)	15,000	20,000

30. *Politics* A political candidate wishes to use a combination of television and radio advertisements in his campaign. Each 1-minute ad on television reaches

0.12 million eligible voters and each 1-minute ad on radio reaches 0.009 million eligible voters. The candidate feels that he must reach at least 7.56 million eligible voters and that he must buy at least 100 minutes of advertisements. If television ads cost $1000 per minute and radio ads cost $200 per minute, how many minutes of radio and television advertisement does he need to minimize costs?

31. *Nutrition* In a hospital ward, patients are grouped into two general nutritional categories depending on the amount of solid food in their diet, and are provided two different diets with different amounts of the solid foods and with some detrimental substances. The following table gives the patient groups, the weekly diet requirements for each group, and the amount of detrimental substance in each diet. How many servings of each diet will satisfy the nutritional requirements and minimize the detrimental substance? What is the minimum amount of detrimental substance?

	Diet A (oz/serving)	Diet B (oz/serving)	Minimum Daily Requirements (oz)
Group 1	2	1	18
Group 2	4	1	26
Detrimental Substance	0.09	0.035	

32. *Manufacturing* Kitchen Pride manufactures toasters and can openers, which are manufactured on two different assembly lines. Line 1 can assemble 15 toasters and 20 can openers per hour, and line 2 can assemble 75 toasters and 20 can openers per hour. The company needs to produce at least 540 toasters and 400 can openers. If it costs $300 per hour to run line 1 and $600 per hour to run line 2, how many hours should each line be run to fill all the orders at the minimum cost? What is the minimum cost?

section 8.3

Sequences and Discrete Functions

key concepts
- Sequences
- Arithmetic sequences
- Geometric sequences

section preview ■ Football Contracts

Suppose a football player is offered a chance to sign a contract for 18 games with one of the following salary plans:

  Plan A: $10,000 for the first game and a $10,000 raise for each game thereafter

  Plan B: $2 for the first game with his salary doubled for each game thereafter

Which salary plan should he accept if he wants to make the most money for his last game?

  The payments for the games form the sequences

  Plan A: 10,000, 20,000, 30,000, . . .

  Plan B: 2, 4, 8, 16, . . .

In this section, we answer this question after we have developed formulas that apply to sequences, and we use sequences to solve applied problems.

Sequences

In Chapter 5 we found the future values of investments at simple and compound interest. Consider a $5000 investment that pays 1% simple interest for each of 6 months, shown in Table 8.2.

Table 8.2

Month	Interest, $I = Prt$ ($)	Future Value of the Investment ($)
1	$(5000)(0.01)(1) = 50$	$5000 + 50 = 5050$
2	$(5000)(0.01)(1) = 50$	$5050 + 50 = 5100$
3	$(5000)(0.01)(1) = 50$	$5100 + 50 = 5150$
4	$(5000)(0.01)(1) = 50$	$5150 + 50 = 5200$
5	$(5000)(0.01)(1) = 50$	$5200 + 50 = 5250$
6	$(5000)(0.01)(1) = 50$	$5250 + 50 = 5300$

These future values are outputs that result when the inputs are positive integers that correspond to the number of months of the investment. Outputs (such as these future values) that result uniquely from the positive integer inputs define a function whose domain is a set of positive integers. Such a function is called a **sequence**, and the ordered outputs corresponding to the integer inputs are called the **terms of the sequence**.

Sequences have the same properties as other functions that we have studied, *except* that the domains of sequences are positive integers, and so sequences are **discrete functions**. Rather than denoting a functional output with y, we denote the output of the sequence f that corresponds to input n with $f(n) = a_n$, where n is a positive integer.

> ## Sequence
>
> The functional values $a_1, a_2, a_3, \ldots$ of a sequence are called the terms of the sequence, with a_1 the first term, a_2 the second term, and so on. The general term (or nth term) is denoted by a_n.

If the domain of a sequence is the set of all positive integers, the outputs form an **infinite sequence**. If the domain is the set of positive integers from 1 to n, the outputs form a **finite sequence**. Sequences are important because they permit us to apply discrete functions to real problems without having to approximate them with continuous functions.

By looking at the future values in Table 8.2, we can see that the future value of $5000 at the end of each month is given by

$$a_1 = f(1) = 5000 + 50(1) = 5050$$
$$a_2 = f(2) = 5000 + 50(2) = 5100$$
$$a_3 = f(3) = 5000 + 50(3) = 5150$$
$$a_4 = f(4) = 5000 + 50(4) = 5200$$
$$a_5 = f(5) = 5000 + 50(5) = 5250$$
$$a_6 = f(6) = 5000 + 50(6) = 5300$$

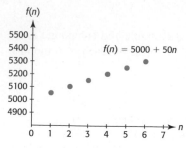

$f(n) = 5000 + 50n$

Figure 8.20

These six terms can be written in the form 5050, 5100, 5150, 5200, 5250, 5300. By observing this pattern, we write the general term as $a_n = 5000 + 50n$, which represents

$$f(n) = 5000 + 50n$$

As with other functions, we can represent sequences in tables and on graphs. Figure 8.20 shows the graph of the function for these values of n.

example 1   Finding Terms of Sequences

Find the first four terms of the sequence (for $n = 1, 2, 3, 4$) defined by:

a. $a_n = \dfrac{8}{n}$

b. $b_n = (-1)^n n(n + 1)$

Solution

a. $a_1 = \dfrac{8}{1} = 8, a_2 = \dfrac{8}{2} = 4, a_3 = \dfrac{8}{3}$, and $a_4 = \dfrac{8}{4} = 2$.

We write these terms in the form $8, 4, \dfrac{8}{3}, 2$.

b. $b_1 = (-1)^1(1)(1 + 1) = -2, b_2 = (-1)^2(2)(2 + 1) = 6,$
$b_3 = (-1)^3(3)(3 + 1) = -12, b_4 = (-1)^4(4)(4 + 1) = 20.$

Notice that $(-1)^n$ causes the signs of the terms to alternate. The terms are $-2, 6, -12, 20$. ■

Technology Note

We can find n terms of a sequence with a calculator if we set the calculator to sequence mode. Figure 8.21(a) shows the first four terms defined in Example 2(a), in decimal and fractional form. The graph of the sequence and the table of values for the sequence are shown in Figure 8.21(b) and Figure 8.21(c), respectively.*

* For more details, see Appendix A, pages 669–670.

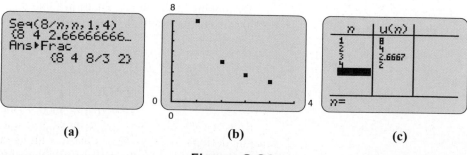

(a)   (b)   (c)

Figure 8.21

example 2

Depreciation

For tax purposes, a firm depreciates its \$900,000 building over 30 years by the straight-line method, which depreciates the value of the building by $\dfrac{900,000}{30} = 30,000$ dollars each year. Write a sequence that gives the value of the building at the end of each of the first 5 years.

Solution

The value would be reduced each year by 30,000 dollars, so the first five terms of the sequence are

$$870,000, \ 840,000, \ 810,000, \ 780,000, \ 750,000$$

Arithmetic Sequences

The future values of the investment described in Table 8.2 are the first six terms of the sequence

$$5050, \ 5100, \ 5150, \ 5200, \ 5250, \ 5300, \ldots$$

We can define this sequence **recursively**, with each term after the first defined from the previous term, as

$$a_1 = 5050, \ a_n = a_{n-1} + 50, \ \text{for } n > 1$$

This sequence is an example of a special sequence called an **arithmetic sequence**. In such a sequence, each term after the first term is found by adding a constant to the preceding term. Thus we have the following definition.

> ### Arithmetic Sequence
> A sequence is called an arithmetic sequence if there exists a number d, called the **common difference**, such that
> $$a_n = a_{n-1} + d, \qquad \text{for } n > 1$$

example 3

Arithmetic Sequences

Write the next three terms of the arithmetic sequences:

a. $1, 4, 7, 10, \ldots$ **b.** $11, 9, 7, \ldots$ **c.** $\dfrac{1}{2}, \dfrac{2}{3}, \dfrac{5}{6}, \ldots$

Solution

a. The common difference that gives each term from the previous one is 3, so the next three terms are 13, 16, 19.

b. The common difference is -2, so the next three terms are 5, 3, 1.

c. The common difference is $\dfrac{1}{6}$, so the next three terms are $1, \dfrac{7}{6}, \dfrac{4}{3}$.

Note that the differences of terms of an arithmetic sequence are constant, so the function defining the sequence is linear, with its rate of change equal to the common difference. That is, an arithmetic sequence is really a linear function whose domain is restricted to a subset of the positive integers. Because each term after the first in an arithmetic sequence is obtained by adding d to the preceding term, the second term is $a_1 + d$, the third term is $(a_1 + d) + d = a_1 + 2d$, the fourth term is $a_1 + 3d, \ldots,$ and the nth term is $a_1 + (n - 1)d$. Thus, we have a formula for the nth term of an arithmetic sequence.

> ### *n*th Term of an Arithmetic Sequence
>
> The nth term of an arithmetic sequence is given by
>
> $$a_n = a_1 + (n - 1)d$$
>
> where a_1 is the first term of the sequence, n is the number of the term, and d is the common difference between consecutive terms.

Thus, the 25th term of the arithmetic sequence $1, 4, 7, 10, \ldots$ (see Example 3) is

$$a_{25} = 1 + (25 - 1)3 = 73$$

---

example 4

Depreciation

For tax purposes, a firm depreciates its $900,000 building over 30 years by the straight-line method, which depreciates the value of the building by $\dfrac{900,000}{30} = 30,000$ dollars each year. What is the value of the building after 12 years?

Solution

The description indicates that the value of the building is 900,000 dollars at the beginning, that its value is $900,000 - 30,000 = 870,000$ dollars after 1 year, and that it decreases by $30,000 each year for 29 additional years. Because the value of the building is reduced by the same amount each year, after n years the value is given by the arithmetic sequence with first term 870,000, common difference $-30,000$, and nth term

$$a_n = 870,000 + (n - 1)(-30,000), \qquad \text{for } n = 1, 2, 3, \ldots, 30$$

Thus, the value of the building at the end of 12 years is

$$a_{12} = 870,000 + (12 - 1)(-30,000) = 540,000 \text{ dollars.} \qquad \blacksquare$$

---

example 5

Football Contract Plan A

Recall from the Section Preview that the first payment plan offered to the football player was

   Plan A: 10,000, 20,000, 30,000, ...

a. Do the payments form an arithmetic sequence?

b. What is the 18th payment under this payment plan?

c. If the contract is extended into the postseason, what is the 20th payment under this payment plan?

Solution

a. The payments form a finite sequence with 18 terms. The common difference between terms is 10,000, so the sequence is an arithmetic sequence.

b. Because this is an arithmetic sequence with first term 10,000 and common difference $d = 10,000$, the 18th payment is

$$a_{18} = 10,000 + (18 - 1)10,000 = 180,000 \text{ dollars}$$

c. The 20th payment is

$$a_{20} = 10,000 + (20 - 1)10,000 = 200,000 \text{ dollars}$$ ■

Geometric Sequences

Notice in the Section Preview that the payment plan B for the football player pays \$2, \$4, \$8, ... It is not an arithmetic sequence, because there is no constant difference between the terms. Each successive term of this sequence is doubled (that is, multiplied by 2). This is an example of another special sequence, where each term is found by multiplying the previous term by the same number. This sequence is called a **geometric sequence**.

> ### Geometric Sequence
>
> A sequence is called a geometric sequence if there exists a number r, called the **common ratio**, such that
>
> $$a_n = ra_{n-1} \qquad \text{for } n > 1$$

| example 6 | ## Geometric Sequences |

Write the next three terms of the geometric sequences:

a. $1, 3, 9, \ldots$ **b.** $64, 16, 4, \ldots$ **c.** $2, -4, 8, \ldots$

Solution

a. The common ratio that gives each term from the previous one is 3, so the next three terms are 27, 81, 243.

b. The common ratio is $\dfrac{1}{4}$, so the next three terms are $1, \dfrac{1}{4}, \dfrac{1}{16}$.

c. The common ratio is -2, so the next three terms are $-16, 32, -64$. ■

Note that there is a **constant percent change** in the terms of a geometric sequence, so the function defining the sequence is exponential, and a geometric sequence is really an exponential function with its domain restricted to the positive integers. Because each term after the first in a geometric sequence is obtained by multiplying r times the preceding term, the second term is $a_1 r$, the third term is $a_1 r \cdot r = a_1 r^2$, the fourth term is $a_1 r^3, \ldots$, and the nth term is $a_1 r^{n-1}$. Thus, we have a formula for the nth term of a geometric sequence.

> ## nth Term of a Geometric Sequence
>
> The nth term of a geometric sequence is given by
>
> $$a_n = a_1 r^{n-1}$$
>
> where a_1 is the first term of the sequence, n is the number of the term, and r is the common ratio of consecutive terms.

Thus, the 25th term of the geometric sequence 2, −4, 8, . . . (see Example 6) is

$$a_{25} = 2(-2)^{25-1} = 33{,}554{,}432$$

example 7	### Rebounding

A ball is dropped from a height of 100 feet and rebounds $\frac{2}{5}$ of the height from which it falls every time it hits the ground. How high will it bounce after it hits the ground the fourth time?

Solution

The first rebound is $\frac{2}{5}$ of 100 = 40 feet, the second is $\frac{2}{5}$ of 40 = 16 feet, and the rebounds form a geometric sequence with first term 40 and common ratio $\frac{2}{5}$.
   Thus, the fourth rebound is

$$40\left(\frac{2}{5}\right)^{4-1} = 2.56 \text{ feet}$$

   We found the future value of money invested at compound interest in Chapter 5 by using exponential functions. Because geometric sequences are really exponential functions with domains restricted to the positive integers, the future value of an investment with interest compounded over a number of discrete periods can be found using a geometric sequence.

example 8	### Compound Interest

The future value of $1000 invested for 3 years at 6% compounded annually can be found using the simple interest formula $S = P + Prt$, as follows.

$$\begin{aligned}
\text{Year 1:} \quad S &= 1000 + 1000(0.06)(1) \\
&= 1000(1 + 0.06) \\
&= 1000(1.06) \text{ dollars} \\
\text{Year 2:} \quad S &= 1000(1.06) + [1000(1.06)](0.06)(1) \\
&= 1000(1.06)(1 + 0.06) \\
&= 1000(1.06)^2 \text{ dollars} \\
\text{Year 3:} \quad S &= 1000(1.06)^2 + [1000(1.06)^2](0.06)(1) \\
&= 1000(1.06)^2(1 + 0.06) \\
&= 1000(1.06)^3 \text{ dollars}
\end{aligned}$$

a. Do the future values of this investment form a geometric sequence?

b. What is the value of this investment in 25 years?

Solution

a. The future values of the investment form a sequence with first term $1000(1.06) = 1060$, and common ratio 1.06, so the sequence is a geometric sequence.

b. The future value of the investment is the 25th term of the geometric sequence:

$$1060(1.06)^{25-1} = 4291.87 \text{ dollars}$$

■

---

example 9   **Football Contract Plan B**

Recall from the Section Preview that the second payment plan offered to the football player was

    Plan B: $2, 4, 8, \ldots$.

a. Do the payments form a geometric sequence?

b. What is the 18th payment under this payment plan?

Solution

a. The payments form a finite sequence with 18 terms. The common ratio between terms is 2 because each payment is doubled. Thus, the sequence is a geometric sequence.

b. Because this is a geometric sequence with first term 2 and common ratio $r = 2$, the 18th payment is

$$a_{18} = 2 \cdot 2^{18-1} = 262{,}144 \text{ dollars}$$

Thus, the 18th payment ($262,144) from Plan B is significantly larger than the 18th payment ($180,000) from Plan A. ■

    But does the fact that the 18th payment from Plan B is much larger than the payment from Plan A mean that the total payment from Plan B is larger than Plan A? We answer this question in the next section.

---

skills check
8.3

1. Find the first 6 terms of the sequence defined by $f(n) = 2n + 3$.

2. Find the first 4 terms of the sequence defined by $f(n) = \dfrac{1}{2n} + n$.

3. Find the first 5 terms of the sequence defined by $a_n = \dfrac{10}{n}$.

4. Find the first 5 terms of the sequence defined by $a_n = (-1)^n(2n)$.

5. Find the next 3 terms of the arithmetic sequence $1, 3, 5, 7, \ldots$.

6. Find the next 3 terms of the arithmetic sequence 2, 5, 8, ...

7. Find the eighth term of the arithmetic sequence with first term -3 and common difference 4.

8. Find the 40th term of the arithmetic sequence with first term 5 and common difference 15.

9. Write 4 additional terms of the geometric sequence 3, 6, 12, ...

10. Write 4 additional terms of the geometric sequence 8, 20, 50, ...

11. Find the sixth term of the geometric sequence with first term 10 and common ratio 3.

12. Find the tenth term of the geometric sequence with first term 48 and common ratio $-\dfrac{1}{2}$.

13. Find the first 4 terms of the sequence with first term 5 and nth term $a_n = a_{n-1} - 2$.

14. Find the first 6 terms of the sequence with first term 8 and nth term $a_n = a_{n-1} + 3$.

15. Find the first 4 terms of the sequence with first term 2 and nth term $a_n = 2a_{n-1} + 3$.

16. Find the first 5 terms of the sequence with first term 26 and nth term $a_n = \dfrac{a_{n-1} + 4}{2}$.

exercises

8.3

17. *Salaries* Suppose you are offered a job with a relatively low starting salary but with a $1500 raise for each of the next 7 years. How much more than your starting salary would you be making in the eighth year?

18. *Depreciation* A new car costing $35,000 is purchased for business and is depreciated with the straight-line depreciation method over a 5-year period, which means that it is depreciated by the same amount each year. Write a sequence that gives the value after depreciation for each of the 5 years.

19. *Landscaping* Grading equipment used for landscaping costs $300 plus $60 per hour or part of an hour thereafter.

   a. Write an expression for the cost of a job lasting n hours.

   b. If the answer to part (a) is the nth term of a sequence, write the first 6 terms of this sequence.

20. *Profit* A new firm loses $2000 in its first month, but its profit increases by $400 in each succeeding month for the rest of the year. What is its profit in the 12th month?

21. *Salaries* Suppose you are offered two identical jobs, one paying a starting salary of $40,000 with yearly raises of $2000 and a second one paying a starting salary of $36,000 with yearly raises of $2400.

   a. Which job will be paying more in 5 years? By how much?

   b. Which job will be paying more in 10 years? By how much?

   c. What factor is important in deciding which job to take?

22. *Phone Calls* Suppose a long-distance call costs 99¢ for the first minute plus 25¢ for each additional minute.

   a. Write the cost of a call lasting n minutes.

   b. If the answer to part (a) is the nth term of a sequence, write the first 6 terms of this sequence.

23. *Interest* If $1000 is invested at 5% interest compounded annually, write a sequence that gives the amount in the account at the end of each of the first 4 years.

24. *Interest* If $10,000 is invested at 6% interest compounded annually, write a sequence that gives the amount in the account at the end of each of the first 3 years.

25. *Depreciation* A new car costing $50,000 depreciates by 20% of its original value each year.

    a. What is the value of the car at the end of the third year?

    b. Write an expression that gives the value at the end of the *n*th year.

    c. Write the first 5 terms of the sequence of the values after depreciation.

26. *Salaries* If you accept a job paying $32,000 for the first year, with a guaranteed 8% raise each year, how much will you earn in your fifth year?

27. *Bacteria* The size of a certain bacteria culture doubles each hour. If the number of bacteria present initially is 5000, how many will be present at the end of 6 hours?

28. *Bacteria* If a bacteria culture increases by 20% every hour and 2000 are present initially, how many will be present at the end of 10 hours?

29. *Bouncing Ball* A ball is dropped from 128 feet. It rebounds $\frac{1}{4}$ of the height from which it falls every time it hits the ground. How high will it bounce after it hits the ground for the fourth time?

30. *Bouncing Ball* A ball is dropped from 64 feet. If it rebounds $\frac{3}{4}$ of the height from which it falls every time it hits the ground, how high will it bounce after it hits the ground for the fourth time?

31. *Profit* If changing market conditions cause a company earning a profit of $8,000,000 this year to project decreases of 2% of its profit in each of the next 5 years, what profit does it project 5 years from now?

32. *Pumps* A pump removes $\frac{1}{3}$ of the water in a container with every stroke. What amount of water is removed on the fifth stroke if the container originally had 81 cm^3 of water?

33. *Salaries* If you begin a job making $54,000 and are given a $3600 raise each year, use numerical methods to find in how many years your salary will double.

34. *Interest* If $2500 is invested at 8% interest compounded annually, how long will it take the account to reach $5829.10?

35. *Compound Interest* Find the future value of $10,000 invested for 10 years at 8% compounded daily.

36. *Compound Interest* Find the future value of $10,000 invested for 10 years at 8% compounded annually.

37. *Rabbit Breeding* The number of pairs of rabbits in the population during the first 6 months can be written in the Fibonacci sequence

$$1, 1, 2, 3, 5, 8, \ldots$$

in which each term after the second is the sum of the two previous terms. Find the number of pairs of rabbits for the next 4 months.

38. *Bee Ancestry* A female bee hatches from a fertilized egg whereas a male bee hatches from an unfertilized egg. Thus, a female bee has a male parent and a female parent whereas a male bee has only a female parent. Therefore, the number of ancestors of a male bee follow the Fibonacci sequence

$$1, 2, 3, 5, 8, 13, \ldots$$

    a. Observe the pattern and write three more terms of the sequence.

    b. What do the 1, the 2, and the 3, respectively, represent for a given male bee?

39. *Credit Card Debt* Kirsten has a $10,000 credit card debt. Each month she pays 1% interest plus a payment of 10% of the monthly balance. If she makes no new purchases, write a sequence that gives the payments for the first 4 months.

section preview ▪ Football Contracts

In Section 8.3, we discussed a football player's opportunity to sign a contract for 18 games for either

Plan A: $10,000 for the first game and a $10,000 raise for each game thereafter

Plan B: $2 for the first game with his salary doubled for each game thereafter

Which salary plan should he accept if he wants to make the most money?
The payments for the games form the sequences:

Plan A: 10,000, 20,000, 30,000, . . .

Plan B: 2, 4, 8, 16, . . .

We found that Plan B paid more than Plan A for the 18th game, but we did not determine which plan pays the most for all 18 games. To answer this question, we need to find the *sum* of the first 18 terms of these sequences. The sum of the terms of a sequence is called a series. Series are useful in computing finite and infinite sums, and in finding depreciation and future values of annuities.

Finite and Infinite Series

Suppose a firm loses $2000 in its first month of operation, but its profit increases by $500 in each succeeding month for the remainder of its first year. What is its profit for the year? The description indicates that the profit is −$2000 for the first month and that it increases by $500 each month for 11 months. The monthly profits form the sequence that begins

$$-2000, -1500, -1000, \ldots$$

The profit for the 12 months is the sum of these terms:

$$(-2000) + (-1500) + (-1000) + (-500) + 0 + 500 + 1000 + 1500 + 2000$$
$$+ 2500 + 3000 + 3500 = 9000$$

The total profit for the year is this sum, $9000.

The sum of the 12 terms of this sequence can be referred to as a **finite series**. In general, a **series** is defined as the sum of the terms of a sequence. If there are infinitely many terms, the series is an **infinite series**.

> ### Series
>
> A finite series is defined by
>
> $$a_1 + a_2 + a_3 + \cdots + a_n$$
>
> where $a_1, a_2, \ldots a_n$ are terms of a sequence.
> An infinite series is defined by
>
> $$a_1 + a_2 + a_3 + \cdots + a_n + \cdots$$

We can use the Greek letter Σ (sigma) to express the sum of numbers or expressions. For example, we can write

$$\sum_{i=1}^{3} i \quad \text{to denote} \quad 1 + 2 + 3$$

and we can write the general finite series in the form

$$\sum_{i=1}^{n} a_i = a_1 + a_2 + a_3 + \cdots + a_n$$

The symbol $\sum_{i=1}^{n} a_i$ may be read as "The sum of a_i as i goes from 1 to n one unit at a time."

The general infinite series can be written in the form

$$\sum_{i=1}^{\infty} a_i = a_1 + a_2 + a_3 + \cdots + a_n + \cdots$$

where ∞ indicates that there is no end to the number of terms.

example 1

Finite Series

Find the sum of the first six terms of the sequence $6, 3, \dfrac{3}{2}, \ldots$.

Solution

Each succeeding term of this sequence is found by multiplying the preceding term by $\dfrac{1}{2}$, so the sequence is

$$6, 3, \frac{3}{2}, \frac{3}{4}, \frac{3}{8}, \frac{3}{16}, \cdots$$

and the sum of the first six terms is $6 + 3 + \dfrac{3}{2} + \dfrac{3}{4} + \dfrac{3}{8} + \dfrac{3}{16} = \dfrac{189}{16}$. ∎

Arithmetic Series

The sum of the payments to the football player in the Section Preview for the first three games under payment Plan A is

$$10{,}000 + 20{,}000 + 30{,}000 = 60{,}000 \text{ dollars}$$

and the sum of the payments for the first three games under Plan B is

$$2 + 4 + 8 = 14 \text{ dollars}$$

so the player would be well advised to take Plan A if there were only three games. To find the sum of the payments for 18 games and to answer other questions, it would be more efficient to have formulas to use in finding the sums of terms of sequences. We begin by finding a formula for the sum of n terms of an arithmetic sequence.

Given the first term a_1 in an arithmetic sequence with common difference d, we use s_n to represent the sum of the first n terms and write it as follows:

$$s_n = a_1 + (a_1 + d) + (a_1 + 2d) + \cdots + (a_n - 2d) + (a_n - d) + a_n$$

Writing this sum in reverse order gives

$$s_n = a_n + (a_n - d) + (a_n - 2d) + \cdots + (a_1 + 2d) + (a_1 + d) + a_1$$

Adding these two equations term by term gives

$$s_n + s_n = (a_1 + a_n) + (a_1 + a_n) + (a_1 + a_n) + \cdots + (a_1 + a_n)$$
$$+ (a_1 + a_n) + (a_1 + a_n)$$

so $2s_n = n(a_1 + a_n)$ because there are n terms, and we have $s_n = \dfrac{n(a_1 + a_n)}{2}$.

Sum of *n* Terms of an Arithmetic Sequence

The sum of the first n terms of an arithmetic sequence is given by

$$s_n = \frac{n(a_1 + a_n)}{2}$$

where a_1 is the first term of the sequence and a_n is the nth term.

Recall that Plan A of the football player's contract began with $a_1 = 10{,}000$ and had 18th payment $a_{18} = 180{,}000$. Thus, the sum of the 18 payments is

$$s_{18} = \frac{18(10{,}000 + 180{,}000)}{2} = 1{,}710{,}000 \text{ dollars}$$

example 2

Depreciation

An automobile valued at \$24,000 is depreciated over 5 years with the *sum-of-the-years-digits* depreciation method. Under this method, annual depreciation is found by multiplying the value of the property by a fraction whose denominator is the sum of the 5 years' digits and whose numerator is 5 for the first year's depreciation, 4 for the second year's, 3 for the third, and so on.

a. Write a sequence that represents the parts of the automobile value that are depreciated in each year, and show that the sum of the terms is 100%.

b. Find the yearly depreciation and show that the sum of the depreciations is \$24,000.

Solution

a. The sum of the 5 years' digits is $1 + 2 + 3 + 4 + 5 = 15$, so the parts of the value that are depreciated for each of the 5 years are the numbers

$$\frac{5}{15}, \frac{4}{15}, \frac{3}{15}, \frac{2}{15}, \frac{1}{15}$$

This is an arithmetic sequence with first term $\dfrac{5}{15}$, common difference $-\dfrac{1}{15}$, and fifth term $\dfrac{1}{15}$. Thus, the sum of the terms is

$$s_5 = \frac{5\left(\dfrac{5}{15} + \dfrac{1}{15}\right)}{2} = \frac{5}{2} \cdot \frac{6}{15} = 1 = 100\%$$

b. The respective annual depreciations for the automobile are

$$\frac{5}{15} \cdot 24{,}000 = 8000 \qquad \frac{4}{15} \cdot 24{,}000 = 6400 \qquad \frac{3}{15} \cdot 24{,}000 = 4800$$

$$\frac{2}{15} \cdot 24{,}000 = 3200 \qquad \frac{1}{15} \cdot 24{,}000 = 1600$$

The depreciations form an arithmetic sequence with difference -1600, first term 8000, and fifth term 1600, so the sum of these depreciations is

$$s_n = \frac{5(8000 + 1600)}{2} = 24{,}000 \text{ dollars}$$

Note that the sum is the original value of the automobile, so it is totally depreciated in 5 years. ■

Geometric Series

In Section 5.6, we found the future value of the annuity with payments of $1000 at the end of each of 5 years, with interest at 10% compounded annually by finding the sum

$$S = 1000 + 1000(1.10)^1 + 1000(1.10)^2 + 1000(1.10)^3 + 1000(1.10)^4$$

$$= 6105.10 \text{ dollars}$$

We can see that this is a finite geometric series with first term 1000 and common ratio 1.10. We can develop a formula to find the sum of this and any other finite geometric series.

The sum of the first n terms of a geometric sequence with first term a_1 and common ratio r is

$$s_n = a_1 + a_1 r + a_1 r^2 + \cdots + a_1 r^{n-1}$$

To find a formula for this sum, we multiply it by r to get a new expression, and then find the difference of the two expressions.

$$s_n = a_1 + a_1 r + a_1 r^2 + a_1 r^3 + \cdots + a_1 r^{n-1}$$

$$r s_n = a_1 r + a_1 r^2 + a_1 r^3 + \cdots + a_1 r^{n-1} + a_1 r^n$$

$$s_n - r s_n = a_1 - a_1 r^n$$

Rewriting this difference and solving for s_n gives a formula for the sum.

$$s_n(1 - r) = a_1(1 - r^n)$$

$$s_n = \frac{a_1(1 - r^n)}{(1 - r)}$$

Sum of *n* Terms of a Geometric Sequence

The sum of the first n terms of a geometric sequence is

$$s_n = \frac{a_1(1 - r^n)}{1 - r}, \quad r \neq 1$$

where a_1 is the first term of the sequence and r is the common ratio.

---

example 3	### Sums of Terms of Geometric Sequences

a. Find the sum of the first 10 terms of the geometric sequence with first term 4 and common ratio -3.

b. Find the sum of the first 8 terms of the sequence $2, 1, \dfrac{1}{2}, \ldots$.

Solution

a. The sum of the first 10 terms of the geometric sequence with first term 4 and common ratio -3 is given by

$$s_{10} = \frac{a_1(1 - r^{10})}{1 - r} = \frac{4(1 - (-3)^{10})}{1 - (-3)} = -59{,}048$$

b. This sequence is geometric, with first term 2 and common ratio $\dfrac{1}{2}$, so the sum of the first 8 terms is

$$s_8 = \frac{a_1(1 - r^8)}{1 - r} = \frac{2\left(1 - \left(\dfrac{1}{2}\right)^8\right)}{1 - \left(\dfrac{1}{2}\right)} = 4\left(1 - \frac{1}{256}\right) = \frac{255}{64}$$ ■

---

example 4	### Annuities

Show that if regular payments of $1000 are made at the end of each year for 5 years into an account that pays interest at 10% per year compounded annually, the future value of this annuity is $6105.10.

Solution

As we saw on page 629, the future values of the individual payments into this annuity form a geometric sequence with first term 1000 and common ratio 1.10, so the sum of the first five terms is

$$s_5 = \frac{1000(1 - 1.10^5)}{1 - 1.10} = 6105.10$$ ■

We found that the total payment to the football player for 18 games under Plan A is $1,710,000. The payments under Plan B form a geometric sequence with first term 2 and common ratio 2, so the sum of the payments under Plan B is

$$s_{18} = \frac{2(1 - 2^{18})}{1 - 2} = 524{,}286$$

Thus, Plan A gives total earnings for the 18 games that are more than three times that for Plan B. However, if the season has 20 games, the sum of the payments for Plan A is

$$s_{20} = \frac{20(10{,}000 + 200{,}000)}{2} = 2{,}100{,}000 \text{ dollars}$$

and the sum of the payments for Plan B is

$$s_{20} = \frac{2(1 - 2^{20})}{1 - 2} = 2{,}097{,}150 \text{ dollars}$$

In this case, the plans are nearly equal.

Infinite Geometric Series

If we add the terms of an infinite geometric sequence, we have an **infinite geometric series**. It is perhaps surprising that we can add an infinite number of positive numbers and get a finite sum. To see that this is possible, consider a football team that has only 5 yards remaining to score a touchdown. How many penalties on the opposing team will it take for the sum of the penalties to be 5 yards (giving them a touchdown), if each penalty is "half the distance to the goal"? The answer is that the team can never reach the goal line because every penalty is half the distance to the goal, which always leaves some distance remaining to travel. Thus, the sum of any number of penalties is less than 5 yards. To see this, let's find the sum of the first 10 such penalties. The first few penalties are $\frac{5}{2}, \frac{5}{4},$ and $\frac{5}{8},$ so the sequence of penalties has first term $\frac{5}{2}$ and common ratio $\frac{1}{2}$ and the sum of the first n penalties is

$$s_n = \frac{a_1(1 - r^n)}{1 - r} = \frac{\left(\frac{5}{2}\right)\left[1 - \left(\frac{1}{2}\right)^n\right]}{1 - \frac{1}{2}}$$

The sum of the first 10 penalties is found by letting $n = 10$, getting

$$\frac{\frac{5}{2}\left[1 - \left(\frac{1}{2}\right)^{10}\right]}{1 - \frac{1}{2}} \approx 4.9951$$

We can see that this sum is less than 5 yards, but close to 5 yards. Notice that as n gets larger, $\left(\frac{1}{2}\right)^n$ gets smaller, approaching 0, and the sum approaches $\frac{5/2(1 - 0)}{1 - (1/2)} = 5$. Thus, the sum of an *infinite number* of penalties is the finite number 5 yards.

In general, if $|r| < 1$, it can be shown that r^n approaches 0 as n gets large without bound. Thus, $1 - r^n$ approaches 1, and $\frac{a_1(1 - r^n)}{1 - r}$ approaches $\frac{a_1}{1 - r}$.

Sum of an Infinite Geometric Series

The sum of an infinite geometric series with common ratio r, $|r| < 1$, is

$$S = \frac{a_1}{1 - r}$$

where a_1 is the first term.
  If $|r| \geq 1$, the sum does not exist (is not a finite number).

We can use this formula to compute the sum of an infinite number of "half the distance to the goal" penalties from the 5-yard line.

$$S = \frac{\frac{5}{2}}{1 - \frac{1}{2}} = 5$$

| example 5 | **Infinite Geometric Series** |

Find the sum represented by the following series, if possible.

a. $81 + 9 + 1 + \cdots$

b. $\sum_{i=1}^{\infty} \left(\dfrac{2}{3}\right)^i$

Solution

a. This is an infinite geometric series with first term $a_1 = 81$ and common ratio $r = \dfrac{1}{9} < 1$, so the sum is

$$S = \frac{81}{1 - \dfrac{1}{9}} = \frac{729}{8}$$

b. This is the sigma notation for an infinite series. Each new term results in one higher power of $\dfrac{2}{3}$ (because i increases by one), so the common ratio is $\dfrac{2}{3} < 1$. The first term is $\dfrac{2}{3}$ (the value when $i = 1$), so the sum is

$$S = \frac{\dfrac{2}{3}}{1 - \dfrac{2}{3}} = 2$$ ■

skills check

8.4

1. Find the sum of the first 6 terms of the geometric sequence $9, 3, 1, \ldots$.

2. Find the sum of the first 7 terms of the arithmetic sequence $1, 3, 5, \ldots$.

3. Find the sum of the first 10 terms of the arithmetic sequence $7, 10, 13, \ldots$.

4. Find the sum of the first 20 terms of the arithmetic sequence $5, 12, 19, \ldots$.

5. Find the sum of the first 15 terms of the arithmetic sequence with first term -4 and common difference 2.

6. Find the sum of the first 10 terms of the arithmetic sequence with first term 50 and common difference 3.

7. Find the sum of the first 15 terms of the geometric sequence with first term 3 and common ratio 2.

8. Find the sum of the first 12 terms of the geometric sequence with first term 48 and common ratio $\dfrac{1}{2}$.

9. Find the sum of the first 10 terms of the geometric sequence $5, 10, 20, \ldots$.

10. Find the sum of the first 15 terms of the geometric sequence $100, 50, 25, \ldots$.

11. Find the sum represented by the series $81 + 9 + 1 + \cdots$, if possible.

12. Find the sum represented by the series $64 + 32 + 16 + \cdots$, if possible.

In Exercises 13–16, evaluate the sums.

13. $\sum_{i=1}^{6} 2^i$ **14.** $\sum_{i=1}^{5} 4^i$

15. $\sum_{i=1}^{4} \dfrac{1+i}{i}$ **16.** $\sum_{i=1}^{5} \left(\dfrac{1}{2}\right)^i$

17. Find the sum of the series $\sum_{i=1}^{\infty} \left(\dfrac{3}{4}\right)^i$ if possible.

18. Find the sum $\sum_{i=1}^{\infty} \left(\dfrac{5}{6}\right)^i$ if possible.

19. Find the sum $\sum_{i=1}^{\infty} \left(\dfrac{4}{3}\right)^i$ if possible.

20. Find the following sum, if possible.

$$800 + 600 + 450 + \cdots$$

exercises

8.4

21. *Profit* A new firm loses $2000 in its first month, but its profit increases by $400 in each succeeding month for the rest of the year. What is its profit for the first year?

22. *Salaries* Suppose you are offered a job with a relatively low starting salary but with a $1500 raise for each of the next 7 years. What is the total amount of your raises over the next 7 years?

23. *Bee Ancestry* A female bee hatches from a fertilized egg whereas a male bee hatches from an unfertilized egg. Thus, a female bee has a male parent and a female parent whereas a male bee has only a female parent. Therefore, the numbers of ancestors of a male bee follow the Fibonacci sequence

$$1, 2, 3, 5, 8, 13, \ldots$$

Use this sequence to determine the total number of ancestors of a male bee back four generations.

24. *Salaries* Suppose you are offered two identical jobs, one paying a starting salary of $40,000 with yearly raises of $2000 and one paying a starting salary of $36,000 with yearly raises of $3000.

  a. Which job will pay more over the first 3 years? How much?

  b. Which job will pay more over the first 12 years? How much?

  c. What factor is important in deciding which job to take?

25. *Clocks* A grandfather clock strikes a chime indicating each hour of the day, so that it chimes 3 times at 3:00, 10 times at 10:00, and so on.

  a. How many times will it chime in a 12-hour period?

  b. How many times will it chime in a 24-hour day?

26. *Depreciation* Under the sum-of-the-years depreciation method, annual depreciation for 4 years is found by multiplying the value of the property by a fraction whose denominator is the sum of the 4 years' digits and whose numerator is 4 for the first year's depreciation, 3 for the second year's, 2 for the third, and so on.

  a. Write a sequence that gives the depreciation for each year on a truck with value $36,000.

  b. Show that the sum of the depreciations for the 4 years adds to 100% of the original value.

27. *Profit* Suppose a new business makes a profit of $2000 in its first month, and its profit increases by 10% in each of the next 11 months. How much profit did it earn in its first year?

28. *Profit* If changing market conditions cause a company earning a profit of $8,000,000 this year to project decreases of 2% of its profit in each of the next 5 years, what is the sum of the yearly profits that it projects for the next 5 years?

29. *Pumps* A pump removes $\dfrac{1}{3}$ of the water in a container with every stroke. After 4 strokes, what amount of water is still in a container that originally had 81 cm^3 of water?

30. *Salaries* If you accept a job paying $32,000 for the first year with a guaranteed 8% raise each year, how much will you earn in your first 5 years?

31. *Chain Letters* Suppose you receive a chain letter with 5 names on it, and to keep the chain unbroken, you mail a dollar to the person whose name is at the top of the list, cross out the top name, add your name to the bottom, and mail it to 5 friends.

    a. How many friends will receive your letter?

    b. If your friends cross the second person from the original list off the list and mail out 5 letters each, how many people will receive letters from your friends?

    c. How many "descendants" of your letters will receive letters in the next two levels if no one breaks the chain?

    d. Describe the type of sequence that gives the number of people receiving the letter at each level.

32. *Chain Letters* Mailing chain letters that involve money in the U.S. mail is illegal because many people would receive nothing while a comparative few would profit. Suppose the chain letter in Exercise 31 were to go through 12 unbroken levels.

    a. How many people would receive a letter asking for money?

    b. What is the significance of this number?

33. *Credit Card Debt* Kirsten has a $10,000 credit card debt. She makes a minimum payment of 10% of the balance at the end of each month. If she makes no new purchases, how much does she owe at the end of 1 year?

34. *Credit Card Debt* Kirsten has a $10,000 credit card debt. She makes a minimum payment of 10% of the balance at the end of each month. If she makes no new purchases, in how many months will she have half of her debt repaid?

35. *Bouncing Ball* A ball is dropped from 128 feet. It rebounds $\frac{1}{4}$ of the height from which it falls every time it hits the ground. How far will it have traveled up and down when it hits the ground for the fifth time?

36. *Bouncing Ball* A ball is dropped from 64 feet. If it rebounds $\frac{3}{4}$ of the height from which it falls every time it hits the ground, how far will it have traveled up and down when it hits the ground for the fourth time?

37. *Depreciation* A new car costing $35,000 depreciates by 16% of its current value each year.

    a. Write an expression that gives the sum of the depreciations for the first n years.

    b. What is the value of the car at the end of the nth year?

38. *Loans* An interest-free loan of $15,000 requires monthly payments of 8% of the unpaid balance. What is the unpaid balance after

    a. 1 year?

    b. 2 years?

39. *Annuities* Find the 8-year future value of an annuity with a contribution of $100 at the end of each month into an account that pays 12% per year compounded monthly.

40. *Annuities* Find the 10-year future value of an ordinary annuity with a contribution of $300 per quarter into an account that pays 8% per year compounded quarterly.

41. *Present Value of Annuities* The present value of an annuity with n payments of $$R$, at interest rate i per period is

$$A = R(1 + i)^{-1} + R(1 + i)^{-2}$$
$$+ R(1 + i)^{-3} + \cdots + R(1 + i)^{-(n-1)}$$
$$+ R(1 + i)^{-n}$$

Treat this as the sum of a geometric series with n terms, $a_1 = R(1 + i)^{-n}$, and $r = (1 + i)$, and show that the sum is

$$A = R\left[\frac{1 - (1 + i)^{-n}}{i}\right]$$

8.5

Preparing for Calculus

section preview ▪ Calculus

In this section, we show how algebra is used in the context of a calculus course; that is, we show how algebra skills that we have already acquired are applied in the solution of calculus problems. Three fundamental concepts in calculus are limit, derivative, and integral. Although we do not define nor use these concepts here, we do illustrate that algebra skills are necessary for computing, simplifying, and applying these concepts.

example 1

Simplifying Expressions

The following expressions represent the derivatives of functions. Simplify these expressions by multiplying to remove parentheses and combining like terms.

a. $(x^3 - 3x)(2x) + (x^2 - 1)(3x^2 - 3)$ **b.** $\dfrac{x^3(20x^4) - (4x^5 - 2)(3x^2)}{(x^3)^2}$

Solution

a. $(x^3 - 3x)(2x) + (x^2 - 1)(3x^2 - 3) = 2x^4 - 6x^2 + 3x^4 - 3x^2 - 3x^2 + 3$

$$= 5x^4 - 12x^2 + 3$$

b. $\dfrac{x^3(20x^4) - (4x^5 - 2)(3x^2)}{(x^3)^2} = \dfrac{20x^7 - (12x^7 - 6x^2)}{x^6}$

$$= \dfrac{8x^7 + 6x^2}{x^6} = \dfrac{8x^5 + 6}{x^4}$$ ■

example 2

Evaluating Functions

In calculus, the derivative of $f(x)$ is denoted by $f'(x)$, and the value of the derivative at $x = a$ is denoted $f'(a)$. If $f'(x) = 3x^2 - 4x + 2$, find $f'(-1)$ and $f'(3)$.

Solution

$$f'(-1) = 3(-1)^2 - 4(-1) + 2 = 9$$
$$f'(3) = 3(3)^2 - 4(3) + 2 = 17$$ ■

example 3

Slope of a Tangent Line

Calculus can be used to find the slope of the tangent to a curve. The slope of the tangent to the curve $y = f(x)$ at $x = a$ is $f'(a)$. If $f'(x) = 4x^3 - 3x^2 + 1$, find the slope of the tangent at $x = 2$.

Solution

The slope of the tangent to the curve at $x = 2$ is $f'(2) = 4(2)^3 - 3(2)^2 + 1 = 21$. ■

635

| example 4 | **Equation of a Tangent Line** |

Calculus can be used to find the slope of the tangent to a curve, and then we can write the equation of the tangent line. If the slope of the line tangent to the graph of $f(x) = x^2$ at $(1, 1)$ is 2, write the equation of this line (Figure 8.22).

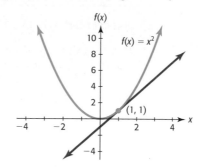

Figure 8.22

Solution

The point-slope form of the equation of a line is

$$y - y_1 = m(x - x_1)$$

Using the given point and the slope of the tangent line gives the equation

$$y - 1 = 2(x - 1) \quad \text{or} \quad y = 2x - 1$$

| example 5 | **Difference Quotient** |

a. An important step in finding the derivative of a function involves reducing fractions like

$$\frac{(2(x + h) + 1) - (2x + 1)}{h}$$

which is called the difference quotient. Reduce this fraction, if $h \neq 0$.

b. An important calculation in calculus is the evaluation of the expression $\frac{f(x + h) - f(x)}{h}$, where $h \neq 0$. Find $\frac{f(x + h) - f(x)}{h}$ if $f(x) = 3x + 2$.

Solution

a. $\dfrac{(2(x + h) + 1) - (2x + 1)}{h} = \dfrac{2x + 2h + 1 - 2x - 1}{h} = \dfrac{2h}{h} = 2$

b. $\dfrac{f(x + h) - f(x)}{h} = \dfrac{(3(x + h) + 2) - (3x + 2)}{h}$

$$= \dfrac{3x + 3h + 2 - 3x - 2}{h} = \dfrac{3h}{h} = 3$$

| example 6 | **Solving Equations** |

Finding where the derivative is zero is an important skill in calculus.

a. Find where $f'(x) = 0$ if $f'(x) = 3x - 6$.

b. Find where $f'(x) = 0$ if $f'(x) = 8x - 4$.

Solution

a. $f'(x) = 0$ where $0 = 3x - 6$, or where $6 = 3x$, at $x = 2$.

b. $f'(x) = 0$ where $0 = 8x - 4$, or where $4 = 8x$, at $x = \dfrac{1}{2}$. ■

---

example 7 **Rewriting Radicals**

To perform calculus operations on functions, it is frequently desirable or necessary to write a radical function in the form $y = ax^b$.

a. Convert the function $y = 3\sqrt[3]{x^4}$ to this form.

b. Convert the function $y = 2x\sqrt{x}$ to this form.

Solution

a. $3\sqrt[3]{x^4} = 3x^{4/3}$, so the form is $y = 3x^{4/3}$.

b. $2x\sqrt{x} = 2x \cdot x^{1/2} = 2x^{1+1/2} = 2x^{3/2}$, so the form is $y = 2x^{3/2}$. ■

---

example 8 **Rewriting Fractional Exponents**

Sometimes a derivative contains fractional exponents, and simplification requires that all fractional exponents be converted to radicals. Write the expression $f'(x) = -3x^{1/3} + 2x^{1/2} + 1$ without fractional exponents.

Solution

$$f'(x) = -3x^{1/3} + 2x^{1/2} + 1 = -3\sqrt[3]{x} + 2\sqrt{x} + 1$$ ■

---

example 9 **Evaluating the Difference Quotient**

An important calculation in calculus is the evaluation of the difference quotient $\dfrac{f(x + h) - f(x)}{h}$, where $h \neq 0$. Find each of the following, if $f(x) = 3x^2 - 2x$.

a. $f(x + h)$ b. $f(x + h) - f(x)$ c. $\dfrac{f(x + h) - f(x)}{h}$

Solution

a. $f(x + h) = 3(x + h)^2 - 2(x + h) = 3(x^2 + 2xh + h^2) - 2(x + h)$

$\qquad = 3x^2 + 6xh + 3h^2 - 2x - 2h$

b. $f(x + h) - f(x) = (3x^2 + 6xh + 3h^2 - 2x - 2h) - (3x^2 - 2x)$

$\qquad\qquad = 6xh + 3h^2 - 2h$

c. $\dfrac{f(x + h) - f(x)}{h} = \dfrac{6xh + 3h^2 - 2h}{h} = 6x + 3h - 2$ ■

---

example 10 **Equation Solution**

Finding where the derivative is zero is an important skill in calculus.

a. Find where $f'(x) = 0$ if $f'(x) = 3x^2 + x - 2$.

b. Find where $f'(x) = 0$ if $f'(x) = 3x^2 + 5x - 2$.

Solution

a. To find where $f'(x) = 0$, we solve $3x^2 + x - 2 = 0$.

$$3x^2 + x - 2 = 0$$
$$(3x - 2)(x + 1) = 0$$
$$3x - 2 = 0 \quad \text{or} \quad x + 1 = 0$$
$$x = \frac{2}{3} \quad \text{or} \quad x = -1$$

b. To find where $f'(x) = 0$, we solve $3x^2 + 5x - 2 = 0$.

$$3x^2 + 5x - 2 = 0$$
$$(3x - 1)(x + 2) = 0$$
$$3x - 1 = 0 \quad \text{or} \quad x + 2 = 0$$
$$x = \frac{1}{3} \quad \text{or} \quad x = -2 \qquad \blacksquare$$

| example 11 | **Simplifying Expressions by Factoring** |

After a derivative is found, the expression frequently must be simplified. Simplify the following expression by factoring.

$$x[2(6x - 1)6] + (6x - 1)^2$$

Solution

Factoring $(6x - 1)$ from both terms gives

$$x[2(6x - 1)6] + (6x - 1)^2 = (6x - 1)[x \cdot 2 \cdot 6 + (6x - 1)]$$
$$= (6x - 1)(18x - 1) \qquad \blacksquare$$

| example 12 | **Simplifying Expressions by Factoring** |

Simplify by factoring.

$$x^4[6(3x^2 - 2x)^5 (6x - 2)] + (3x^2 - 2x)^6(4x^3)$$

Solution

Each term contains the factor $(3x^2 - 2x)$ to some power, and it would be difficult to raise this factor to the fifth and sixth power, so we factor $(3x^2 - 2x)^5$ from each term.

$$x^4[6(3x^2 - 2x)^5(6x - 2)] + (3x^2 - 2x)^6(4x^3)$$
$$= (3x^2 - 2x)^5\{x^4[6(6x - 2)] + (3x^2 - 2x)(4x^3)\}$$
$$= (3x^2 - 2x)^5(36x^5 - 12x^4 + 12x^5 - 8x^4) = (3x^2 - 2x)^5 (48x^5 - 20x^4)$$

Because $4x^4$ is a factor of $(48x^5 - 20x^4)$, we can write the expression as $4x^4(12x - 5)(3x^2 - 2x)^5 = 4x^4(12x - 5) \cdot [x(3x - 2)]^5 = 4x^9(12x - 5)(3x - 2)^5$. $\blacksquare$

In calculus, it is sometimes useful to find two functions whose composition gives a function. For example, if $y = (3x - 2)^2$, we can write this function as $y = u^2$, where $u = 3x - 2$. Going backwards from a function to two functions whose composition gives the function is called finding the *decomposition* of the function. Another example of decomposition is writing $y = \dfrac{1}{x^3 - 1}$ as $y = \dfrac{1}{u}$, where $u = x^3 - 1$. It is possible to write several decompositions for a given function. For example, $y = \dfrac{1}{x^3 - 1}$ could also be written as $y = \dfrac{1}{u - 1}$ where $u = x^3$.

example 13	**Composition of Functions**

An equation of the form $y = (ax^k + b)^n$ can be thought of as the composition of functions and can be written as $y = u^n$, where $u = ax^k + b$. Use this idea to write two functions whose composition gives $y = (x^4 + 5)^6$.

Solution

We can let $y = u^6$, where $u = x^4 + 5$. Then $y = (x^4 + 5)^6$. ∎

example 14	**Function Decomposition**

Find two functions, f and g, such that their composition is $(f \circ g)(x) = \sqrt{x^2 + 1}$.

Solution

One decomposition is found by letting $f(u) = \sqrt{u}$ and $g(x) = x^2 + 1$; then $(f \circ g)(x) = f(g(x)) = \sqrt{x^2 + 1}$. ∎

example 15	**Rewriting Expressions Using Negative Exponents**

Many functions must be written in a form with no variables in the denominator in order to take the derivative. Use negative exponents where necessary to write the following function with all variables in the numerator of a term. That is, write each term in the form cx^n.

$$f(x) = \frac{4}{x^2} + \frac{3}{x} - 2x$$

Solution

Using the definition $a^{-n} = \dfrac{1}{a^n}, a \neq 0$, we get

$$f(x) = 4\frac{1}{x^2} + 3\frac{1}{x} - 2x = 4x^{-2} + 3x^{-1} - 2x$$ ∎

example 16	**Rewriting Expressions Using Positive Exponents**

Sometimes a derivative contains negative exponents, and simplification requires that all exponents be positive. Write the expression $f'(x) = -3x^{-4} + 2x^{-3} + 1$ without negative exponents.

Solution

$$f'(x) = -3x^{-4} + 2x^{-3} + 1$$

$$= -3\frac{1}{x^4} + 2\frac{1}{x^3} + 1 = -\frac{3}{x^4} + \frac{2}{x^3} + 1 \qquad \blacksquare$$

example 17 | Using Logarithmic Properties

Taking derivatives involving logarithms frequently requires simplifying by converting the expressions using logarithmic properties. Rewrite the following expressions as the sum, difference, or product of logarithms without exponents.

a. $\log[x^2(2x + 3)^2]$

b. $\ln\dfrac{\sqrt{2x - 4}}{3x + 1}$

Solution

a. $\log[x^2(2x + 3)^2] = \log x^2 + \log(2x + 3)^2$

$$= 2\log x + 2\log(2x + 3)$$

b. $\ln\dfrac{\sqrt{2x - 4}}{3x + 1} = \ln\sqrt{2x - 4} - \ln(3x + 1)$

$$= \ln(2x - 4)^{1/2} - \ln(3x + 1)$$

$$= \frac{1}{2}\ln(2x - 4) - \ln(3x + 1) \qquad \blacksquare$$

example 18 | Reducing Fractions

Evaluating limits is a calculus skill that frequently requires reducing fractions, as in the following cases. Reducing these fractions may require factoring.

a. Simplify $\dfrac{x^2 - 4}{x - 2}$, if $x \neq 2$.

b. Simplify $\dfrac{x^2 - 7x + 12}{x - 4}$, if $x \neq 4$.

c. Simplify $\dfrac{x^3 - 4x}{2x^2 - x^3}$, if $x \neq 0, x \neq 2$.

Solution

a. $\dfrac{x^2 - 4}{x - 2} = \dfrac{(x - 2)(x + 2)}{x - 2} = x + 2$, if $x \neq 2$

b. $\dfrac{x^2 - 7x + 12}{x - 4} = \dfrac{(x - 3)(x - 4)}{x - 4} = x - 3$, if $x \neq 4$

c. $\dfrac{x^3 - 4x}{2x^2 - x^3} = \dfrac{x(x^2 - 4)}{x^2(2 - x)} = \dfrac{x(x - 2)(x + 2)}{x^2(2 - x)} = \dfrac{x(x - 2)(x + 2)}{-x^2(x - 2)} = -\dfrac{x + 2}{x}$,

if $x \neq 0, x \neq 2$ $\qquad \blacksquare$

| example 19 | **Simplifying Expressions** |

After a derivative is taken, the expression frequently must be simplified. Simplify the following expression by multiplying and combining like terms.

$$\frac{(4x - 3)(2x - 3) - (x^2 - 3x)(4)}{(4x - 3)^2}$$

Solution

$$\frac{(4x - 3)(2x - 3) - (x^2 - 3x)(4)}{(4x - 3)^2} = \frac{8x^2 - 12x - 6x + 9 - 4x^2 + 12x}{(4x - 3)^2}$$

$$= \frac{4x^2 - 6x + 9}{(4x - 3)^2}$$

∎

| example 20 | **Solving Equations** |

Finding where the derivative is zero is an important skill in calculus.

a. Find where $f'(x) = 0$, if $f'(x) = 2x^3 - 2x$.

b. Find where $f'(x) = 0$, if $f'(x) = 4x^3 - 10x^2 - 6x$.

c. Find where $f'(x) = 0$, if $f'(x) = (x^2 - 1)3(x - 5)^2 + (x - 5)^3(2x)$.

d. Find where $f'(x) = 0$, if $f'(x) = \dfrac{x[3(x - 2)^2] - (x - 2)^3}{x^2}$.

Solution

a. Solving $0 = 2x^3 - 2x$ gives

$$0 = 2x^3 - 2x$$
$$0 = 2x(x^2 - 1)$$
$$0 = 2x(x - 1)(x + 1)$$
$$x = 0, x = 1, \text{ and } x = -1 \text{ are the solutions.}$$

Thus, $f'(x) = 0$ at $x = 0$, $x = 1$, and $x = -1$.

b. Solving $0 = 4x^3 - 10x^2 - 6x$ gives

$$0 = 4x^3 - 10x^2 - 6x$$
$$0 = 2x(2x^2 - 5x - 3)$$
$$0 = 2x(2x + 1)(x - 3)$$
$$x = 0, x = -\frac{1}{2}, \text{ and } x = 3 \text{ are the solutions.}$$

Thus, $f'(x) = 0$ at $x = 0$, $x = -\frac{1}{2}$, and $x = 3$.

c. Each term contains the factor $(x - 5)$ to some power. Rather than raise this factor to the second and third power, we use factoring to simplify this expression. Factoring $(x - 5)^2$ from each term gives the following:

$$0 = (x^2 - 1)3(x - 5)^2 + (x - 5)^3(2x)$$
$$0 = (x - 5)^2[1(3)(x^2 - 1) + (x - 5)(2x)]$$
$$0 = (x - 5)^2(3x^2 - 3 + 2x^2 - 10x)$$
$$0 = (x - 5)^2(5x^2 - 10x - 3)$$

One of the solutions is $x = 5$. The remaining solutions can be found with the quadratic formula.

$$x = \frac{10 \pm \sqrt{100 - 4(5)(-3)}}{2(5)} = \frac{10 \pm 4\sqrt{10}}{10} = \frac{5 \pm 2\sqrt{10}}{5}$$

Thus, $f'(x) = 0$ at $x = 5$, $x = \dfrac{5 + 2\sqrt{10}}{5}$, and $x = \dfrac{5 - 2\sqrt{10}}{5}$.

d. To find where the fraction equals 0, we set the numerator equal to 0 and solve.

$$0 = x[3(x - 2)^2] - (x - 2)^3$$
$$0 = (x - 2)^2[3x - (x - 2)]$$
$$0 = (x - 2)^2(2x + 2)$$
$$0 = (x - 2)^2 \quad \text{or} \quad 0 = 2x + 2$$

$$\begin{array}{c|c} 0 = x - 2 & -2x = 2 \\ x = 2 & x = -1 \end{array}$$

Checking these values in the rational function $f'(x) = \dfrac{x[3(x - 2)^2] - (x - 2)^3}{x^2}$ verifies that $f'(x) = 0$ at $x = 2$ and at $x = -1$.   ■

skills check

8.5

The expressions in Exercises 1–4 represent the derivatives of functions. Simplify the expressions.

1. $f'(x) = (2x^3 + 3x + 1)(2x) + (x^2 + 4)(6x^2 + 3)$

2. $f'(x) = (x^2 - 3x)(3x^2) + (x^3 - 4)(2x - 3)$

3. $f'(x) = \dfrac{x^2(3x^2) - (x^3 - 3)(2x)}{(x^2)^2}$

4. Simplify the numerator of

$$f'(x) = \frac{(x^2 - 4)(3x^2 - 6x) - (x^3 - 3x^2 + 2)(2x)}{(x^2 - 4)^2}$$

In calculus, the derivative of $f(x)$ is denoted by $f'(x)$, and the value of the derivative at $x = a$ is $f'(a)$. Evaluate the derivatives in Exercises 5 and 6.

5. Find $f'(2)$ if $f'(x) = 4x^3 - 3x^2 + 4x - 2$.

6. Find $f'(-1)$ if

$$f'(x) = (x^2 - 1)(6x) + (3x^2 + 1)(2x).$$

7. Find $f'(3)$ if $f'(x) = \dfrac{(x^2 - 1)(2) + (2x)(2x)}{(x^2 - 1)^2}$.

8. Calculus can be used to find the slope of the line tangent to a curve. The slope of the tangent to the curve $y = f(x)$ at $x = a$ is $f'(a)$. Find the slope of the tangent to $y = f(x)$ at $x = 2$ if $f'(x) = 8x^3 + 6x^2 - 5x - 10$.

9. The derivative of a function can be used to find the slope of the line tangent to its graph. If the slope of the line tangent to the graph of $f(x) = x^3$ at $(2, 8)$ (see the following figure) is 12, write the equation of this tangent line.

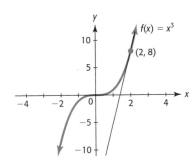

An important calculation in calculus is the simplification of the difference quotient $\dfrac{f(x + h) - f(x)}{h}$, where $h \neq 0$. In Exercises 10–12, find each of the following.

a. $f(x + h)$

b. $f(x + h) - f(x)$

c. $\dfrac{f(x + h) - f(x)}{h}$

10. $f(x) = 3x - 2$ 11. $f(x) = 4x + 5$

12. $f(x) = 12 - 2x$

Writing the equation defining the function in the form $y = f(x)$ usually makes it easier to find the derivative. Write each of the equations in Exercises 13–16 in this form by solving for y.

13. $8x^2 + 4y = 12$ 14. $3x^2 - 2y = 6$

15. $9x^3 + 5y = 18$ 16. $12x^3 = 6y - 24$

The vertex of a parabola occurs where the derivative of its equation is 0, so an equation associated with the equation of a parabola can be solved to find the x-coordinate of the vertex. For each parabola whose equation and "associated equation" are given in Exercises 17–20:

a. Find the solution of the associated equation.

b. Compare the solution with the x-coordinate of the vertex found by the methods of Section 3.1.

c. Find the vertex of the parabola.

Equation of Parabola	Associated Equation
17. $y = x^2 - 2x + 5$	$0 = 2x - 2$
18. $y = 5 + x - x^2$	$0 = 1 - 2x$
19. $y = 6x^2 - 24x + 15$	$0 = 12x - 24$
20. $y = -(4 + 3x + x^2)$	$0 = -3 - 2x$

21. Between $x = -3$ and $x = 10$, the largest value of the expression $y = x^3 - 9x^2 - 48x + 15$ occurs where $3x^2 - 18x - 48 = 0$.

a. Find the values of x that satisfy

$$3x^2 - 18x - 48 = 0$$

b. Find the value of $y = x^3 - 9x^2 - 48x + 15$ at each of the solutions found in part (a).

c. Which of the solutions from part (a) gives the maximum value of $y = x^3 - 9x^2 - 48x + 15$ on $[-3, 10]$? Test the expression at other values of x to confirm your conclusion.

22. Between $x = -2$ and $x = 4$, the largest value of the expression $y = x^3 - 3x^2 - 9x + 5$ occurs where $3x^2 - 6x - 9 = 0$.

a. Find the values of x that satisfy

$$3x^2 - 6x - 9 = 0$$

b. Find the value of $y = x^3 - 3x^2 - 9x + 5$ at each of the solutions found in part (a).

c. Which of the solutions from part (a) gives the maximum value of $y = x^3 - 3x^2 - 9x + 5$ on $[-2, 4]$? Test the expression at other values of x to confirm your conclusion.

To perform calculus operations on functions, it is frequently desirable or necessary to write the function in the form $y = cx^b$ where b is a constant. Write each of the functions in Exercises 23–26 in this form.

23. $y = 3\sqrt{x}$ 24. $y = 6\sqrt[3]{x}$

25. $y = 2\sqrt[3]{x^2}$ 26. $y = 4\sqrt{x^3}$

Write each of the functions in Exercises 27–32 without radicals by using fractional exponents.

27. $y = \sqrt[3]{x^2 + 1}$ 28. $y = \sqrt[4]{x^3 - 2}$

29. $y = \sqrt[3]{(x^3 - 2)^2}$ 30. $y = \sqrt{(5x - 3)^3}$

31. $y = \sqrt{x} + \sqrt[3]{2x}$ 32. $y = \sqrt[3]{(2x)^2} + \sqrt{(4x)^3}$

Equations containing y^2 occasionally must be solved for y. Sometimes this can be accomplished with the root method, and sometimes a method such as the quadratic formula is necessary. Solve the equations in Exercises 33–36 for y.

33. $y^2 + 4x - 3 = 0$ **34.** $5x - y^2 = 12$

35. $y^2 - 6x + y = 0$ **36.** $2x - 2y^2 + y + 4 = 0$

An equation of the form $y = (ax^k + b)^n$ can be thought of as the composition of the functions $f(u) = u^n$ and $u(x) = ax^k + b$ and can be written as $y = u^n$ where $u = ax^k + b$. This form is very useful in calculus. In Exercises 37–41, use this idea to fill in the blanks.

37. $y = (4x^3 + 5)^2$ can be written as $y = u^-$ where $u(x) = \underline{\hspace{1cm}}$.

38. $y = \underline{\hspace{1cm}}$ can be written as $y = u^7$ where $u(x) = 5x^2 - 4x$.

39. $y = \sqrt{\underline{\hspace{1cm}}}$ can be written as $y = u^-$ where $u(x) = x^3 + x$.

40. If $f(u) = u^6$ and $u = x^4 + 5$, then express $6u^5 \cdot 4x^3$ in terms of x alone.

41. If $f(u) = u^{3/2}$ and $u = x^2 - 1$, then express $\frac{3}{2}u^{1/2} \cdot 2x$ in terms of x alone.

In calculus it is often necessary to write expressions in the form cx^n, where n is a constant. Use negative exponents where necessary to write the expressions in Exercises 42–44 with all variables in the numerator of a term. That is, write each term in the form cx^n if the term contains a variable.

42. $y = 3x + \frac{1}{x} - \frac{2}{x^2}$ **43.** $y = \frac{3}{x} - \frac{4}{x^2} - 6$

44. $y = 5 + \frac{1}{x^3} - \frac{2}{\sqrt{x}} + \frac{3}{\sqrt[3]{x^2}}$

45. Write $y = \dfrac{1}{(x^2 - 3)^3}$ with no denominator.

Sometimes a derivative contains negative exponents, and simplification requires that all exponents be positive. Write each of the expressions in Exercises 46–49 without negative exponents.

46. $f'(x) = 8x^{-4} + 5x^{-2} + x$

47. $f'(x) = -6x^{-4} + 4x^{-2} + x^{-1}$

48. $f'(x) = -3(3x - 2)^{-3}$

49. $f'(x) = (4x^2 - 3)^{-1/2}(8x)$

Rewrite the expressions in Exercises 50–52 as the sum, difference, or multiple of logarithms.

50. $\ln \dfrac{3x - 2}{x + 1}$ **51.** $\log x^3(3x - 4)^5$

52. $\ln \dfrac{\sqrt[4]{4x + 1}}{4x^2}$

Find where the derivative is zero in Exercises 53–56.

53. Find where $f'(x) = 0$ if $f'(x) = 3x^2 - 3x$.

54. Find where $f'(x) = 0$ if $f'(x) = 3x^3 - 14x^2 + 8x$.

55. Find where $f'(x) = 0$ if

$$f'(x) = (x^2 - 4)[3(x - 3)^2] + (x - 3)^3(2x)$$

56. Find where $f'(x) = 0$ if

$$f'(x) = \frac{(x + 1)[3(2x - 3)^2 2] - (2x - 3)^3}{(x + 1)^2}$$

chapter

8

Summary

In this chapter we graphed linear inequalities in two variables and systems of linear inequalities in two variables as regions in the two-dimensional plane. We discussed an important application of linear inequalities, linear programming, where we seek a point or points in a region determined by constraints that give an optimum value for an objective function.

We also discussed sequences, which are special discrete functions whose domains are sets of positive integers. Sequences can be used to solve numerous problems involving discrete inputs, including investment problems. We also discussed series, which are the sums of terms of a sequence. Series are useful in finding the future value of annuities if the interest is compounded over discrete time periods.

Finally, we showed how important algebra is in the development and application of calculus.

Key Concepts and Formulas

8.1 Systems of Linear Inequalities

Linear inequalities in two variables	The solution to a linear inequality in two variables is the set of ordered pairs that satisfy the inequality.
System of linear inequalities	The solution to a system of linear inequalities in two or more variables consists of the values of the variables that satisfy all the inequalities.
Border	The lines that bound the solution region are called the borders of the system of linear inequalities.
Solution region	The region that includes the half-plane containing the ordered pairs that satisfy an inequality is called the solution region. The border line is included in the solution region if the inequality includes an equal sign.
Corner	A corner of the graph of a solution to a system of inequalities occurs at a point where boundary lines intersect.
Systems of nonlinear inequalities	As with systems of linear inequalities in two variables, the solutions of systems of nonlinear inequalities in two variables are regions in a plane.

8.2 Linear Programming: Graphical Methods

Linear programming	Linear programming is a technique that can be used to solve problems when the constraints on the variables can be expressed as linear inequalities and a linear objective function is to be maximized or minimized.
Constraints	The inequalities that limit the values of the variables in a linear programming application are called the constraints.
Feasible region	The constraints of a linear programming application form a feasible region in which the solution lies.
Feasible solution	Any point in the region determined by the constraints is called a feasible solution.
Optimum value	A maximum or minimum value of an objective function is called an optimum value.
Solving a linear programming problem	1. If a linear programming problem has a solution, then the optimum value (maximum or minimum) of an objective function occurs at a corner of the feasible region determined by the constraints. 2. When a feasible region for a linear programming problem is closed and bounded, the objective function has a maximum and a minimum value. 3. When the feasible region is not closed and bounded, the objective function may have a maximum only, a minimum only, or neither.

8.3 Sequences and Discrete Functions

Sequence	A sequence is a function whose outputs result uniquely from positive integer inputs. Because the domains of sequences are positive integers, sequences are discrete functions.
Terms of the sequence	The ordered outputs corresponding to the integer inputs of a sequence are called the terms of the sequence.
Infinite sequence	If the domain of a sequence is the set of all positive integers, the outputs form an infinite sequence.
Finite sequence	If the domain of a sequence is a set of positive integers from 1 to n, the outputs form a finite sequence.
Arithmetic sequence	A sequence is called an arithmetic sequence if there exists a number d, called the common difference, such that $a_n = a_{n-1} + d$ for $n > 1$.
nth term of an arithmetic sequence	The nth term of an arithmetic sequence is given by $a_n = a_1 + (n - 1)d$, where a_1 is the first term of the sequence, n is the number of the term, and d is the common difference between the terms.
Geometric sequence	A sequence is called a geometric sequence if there exists a number r, the common ratio, such that $a_n = ra_{n-1}$ for $n > 1$.
nth term of a geometric sequence	The nth term of a geometric sequence is given by $a_n = a_1 r^{n-1}$, where a_1 is the first term of the sequence, n is the number of the term, and r is the common ratio of the terms.

8.4 Series

Series	A series is the sum of the terms of a sequence.
Finite series	A finite series is defined by $a_1 + a_2 + a_3 + \cdots + a_n$, where $a_1, a_2, \ldots a_n$ are the terms of a sequence.
Infinite series	An infinite series is defined by $a_1 + a_2 + a_3 + \cdots + a_n + \cdots$.
Sum of n terms of an arithmetic sequence	The sum of the first n terms of an arithmetic sequence is given by the formula $s_n = \dfrac{n(a_1 + a_n)}{2}$, where a_1 is the first term of the sequence and a_n is the nth term.
Sum of n terms of a geometric sequence	The sum of the first n terms of a geometric sequence is $s_n = \dfrac{a_1(1 - r^n)}{1 - r}$, where a_1 is the first term of the sequence and r is the common ratio.
Infinite geometric series	An infinite geometric series sums to $S = \dfrac{a_1}{1 - r}$, where a_1 is the first term and r is the common ratio, if $|r| < 1$. If $|r| \geq 1$, the sum does not exist (is not a finite number).

8.5 Preparing for Calculus

Algebra review	Algebra topics from Chapters 1–6 are presented in calculus contexts.

chapter

8

Skills Check

Graph the following inequalities.

1. $5x + 2y \le 10$

2. $5x - 4y > 12$

Graph the solutions to the systems of inequalities in Exercises 3–6. Identify the corners of the solution regions.

3. $\begin{cases} 2x + y \le 3 \\ x + y \le 2 \\ x \ge 0, y \ge 0 \end{cases}$

4. $\begin{cases} 3x + 2y \le 6 \\ 3x + 6y \le 12 \\ x \ge 0, y \ge 0 \end{cases}$

5. $\begin{cases} 2x + y \le 30 \\ x + y \le 19 \\ x + 2y \le 30 \\ x \ge 0, y \ge 0 \end{cases}$

6. $\begin{cases} 2x + y \le 10 \\ x + 2y \le 11 \\ x \ge 0, y \ge 0 \end{cases}$

Graph the solutions to the systems of nonlinear inequalities in Exercises 7–8

7. $\begin{cases} 15x - x^2 - y \ge 0 \\ y - \dfrac{44x + 60}{x} \ge 0 \\ x \ge 0, y \ge 0 \end{cases}$

8. $\begin{cases} x^3 - 26x + 100 - y \ge 0 \\ 19x^2 - 20 - y \le 0 \\ x \ge 0, y \ge 0 \end{cases}$

9. Minimize $g = 3x + 4y$ subject to the constraints
$$\begin{cases} 3x + y \ge 8 \\ 2x + 5y \ge 14 \\ x \ge 0, y \ge 0 \end{cases}$$

10. Maximize the function $f = 7x + 12y$ subject to the constraints
$$\begin{cases} 7x + 3y \le 105 \\ 2x + 5y \le 59 \\ x + 7y \le 70 \\ x \ge 0, y \ge 0 \end{cases}$$

11. Find the maximum value of $f = 3x + 5y$ and the values of x and y that give that value, subject to the constraints
$$\begin{cases} 2x + y \le 30 \\ x + y \le 19 \\ x + 2y \le 30 \\ x \ge 0, y \ge 0 \end{cases}$$

Determine whether the sequences in Exercises 12–14 are arithmetic or geometric.

12. $\dfrac{1}{9}, \dfrac{2}{3}, 4, 24, \ldots$

13. $4, 16, 28, \ldots$

14. $16, -12, 9, \dfrac{-27}{4}$

15. Find the fifth term of the geometric sequence with first term 64 and eighth term $\dfrac{1}{2}$.

16. Find the sixth term of the geometric sequence $\dfrac{1}{9}, \dfrac{1}{3}, 1, \ldots$.

17. Find the sum of the first 10 terms of the geometric sequence with first term 5 and common ratio -2.

18. Find the sum of the first 12 terms of the sequence $3, 6, 9, \ldots$.

19. Find $\displaystyle\sum_{i=1}^{\infty} \left(\dfrac{4}{5}\right)^i$ if possible.

20. Simplify the expression
$$(x^3 + 2x + 3)(2x) + (x^2 - 5)(3x^2 + 2)$$

21. Between $x = 0$ and $x = 10$, the smallest value of the expression $y = 3x^4 - 32x^3 - 54x^2$ occurs where $x^3 - 8x^2 - 9x = 0$. Find the values of x that satisfy $x^3 - 8x^2 - 9x = 0$.

22. Write $y = x + \dfrac{1}{x^2} - \dfrac{2}{\sqrt{x^3}} + \dfrac{3}{\sqrt[3]{x}}$ with each term in the form cx^n.

23. Write $f'(x) = 8x^{-3} + 5x^{-1} + 4x$ with no negative exponents.

24. Find where $f'(x) = 0$ if $f'(x) = 2x^3 + x^2 - x$.

chapter

8

Review

25. *Manufacturing* Ace Manufacturing produces two types of DVD players, the Deluxe model and the Superior model, which also plays videotapes. Each Deluxe model requires 3 hours to manufacture and $40 for assembly parts, and each Superior model requires 2 hours to assemble and $60 for assembly parts. The company is limited to 1800 assembly hours and $36,000 for assembly parts each month. If the profit on the Deluxe model is $30 and the profit on the Superior model is $40, how many of each model should be produced to maximize profit?

26. *Manufacturing* A company manufactures two grades of steel, cold steel and stainless steel, at the Pottstown and Ethica factories. The table gives the daily cost of operation and the daily production capabilities of each of the factories, and the number of units of each type of steel that is required to fill an order.

  a. How many days should each factory operate to fill the order at minimum cost?

  b. What is the minimum cost?

	Units/ Day at Pottstown	Units/ Day at Ethica	Required for Order
Units of Cold Steel	20	40	At least 1600
Units of Stainless Steel	60	40	At least 2400
Cost/Day for Operation ($)	20,000	24,000	

27. *Nutrition* A laboratory wishes to purchase two different feeds, A and B, for its animals. Feed A has 2 units of carbohydrates per pound and 4 units of protein per pound, and feed B has 8 units of carbohydrates per pound and 2 units of protein per pound. Feed A costs $1.40 per pound, and feed B costs $1.60 per pound. If at least 80 units of carbohydrates and 132 units of protein are required, how many pounds of each feed are required to minimize the cost?

28. *Manufacturing* A company manufactures leaf blowers and weed wackers. One line can produce 260 leaf blowers per day, and another line can produce 240 weed wackers per day. The combined number of leaf blowers and weed wackers that shipping can handle is 460 per day. How many of each can be produced daily to maximize company profit if the profit is $5 on the leaf blowers and $10 on the weed wackers?

29. *Profits* A woman has a building with 60 two-bedroom and 40 three-bedroom apartments available to rent to students. She has set the rent at $800 per month for the two-bedroom units and $1150 for the three-bedroom units. She must rent to one student per bedroom and zoning laws limit her to at most 180 students in this building. How many of each type of apartment should she rent to maximize her revenue?

30. *Loans* A finance company has at most $30 million available for auto and home equity loans. The auto loans have an annual return rate of 8% to the company and the home equity loans have an annual rate of return of 7% to the company. If there must be at least twice as many auto loans as home equity loans, how much should be loaned in each type of loan to maximize the profit to the finance company?

31. *Salaries* Suppose you are offered two identical jobs, job 1 paying a starting salary of $20,000 with yearly raises of $1000 at the end of each year and job 2 paying a starting salary of $18,000 with yearly increases of $1600 at the end of each year.

  a. Which job will pay more in the fifth year on the job?

  b. Which job will pay more over the first 5 years?

32. *Drug in the Bloodstream* Suppose a 400-milligram dose of heart medicine is taken daily. During each 24-hour period, the body eliminates 40% of the drug (so that 60% remains in the body). Thus, the amount of drug in the body just after the 21 doses, over 21 days, is given by

$$400 + 400(0.6) + 400(0.6)^2 + 400(0.6)^3 + \cdots + 400(0.6)^{19} + 400(0.6)^{20}$$

Find the level of the drug in the bloodstream at this time.

33. *Chess Legend* Legend has it that when the king of Persia offered to reward the inventor of chess any prize he wanted, the inventor asked for one grain of wheat on the first square of the chessboard, with the number of grains doubled on each square thereafter for the remaining 63 squares.

    a. How many grains of wheat are there for the 64th square?

    b. What is the total number of grains of wheat for all 64 squares? (This is enough wheat to cover Alaska more than 3 inches deep in wheat.)

34. *Compound Interest* Find the future value of $20,000 invested for 5 years at 6% compounded annually.

35. *Annuities* Find the 5-year future value of an ordinary annuity with a contribution of $300 per month into an account that pays 12% per year compounded monthly.

36. *Profit* The weekly total cost function for a product is $C(x) = 3x^2 + 1228$ dollars and the weekly revenue is $R(x) = 177.50x$, where x is the number of hundreds of units. Use graphical methods to find the number of units that gives profit for the product.

group activity / extended application

Salaries

In the 1970s, a local teachers' organization in Pennsylvania asked for a $1000 raise at the end of each year for the next 3 years. When the school board said it could not afford to do this, the teachers' organization then said that the teachers would accept a $300 raise at the end of each 6-month period. The school board initially responded favorably because the members thought it would save them $400 per year per teacher.

    If you are an employee, would you rather be given a raise of $1000 at the end of each year (Plan I) or be given a $300 raise at the end of each 6-month period (Plan II)? To answer the question, complete the table for an employee whose base salary is $40,000 (or $20,000 for each 6-month period), and then answer the questions below.

1. Find the total additional money earned for the first 3 years from the raises of Plan I.
2. Find the total additional money earned for the first 3 years from the raises of Plan II.
3. Which plan gives more money from the raises, and how much more? Which plan is better for the employee?
4. Find the total additional money earned (per employee) from raises for each of Plan I and Plan II if they are extended to 4 years. Which plan gives more money in

raises, and how much more? Which plan is better for the employee?

5. Did the school board make a mistake in thinking that the $300 raises every 6 months instead of $1000 every year would save money? If so, what was the mistake?
6. If there were 200 teachers in the school district, how much extra would Plan II cost over Plan I for the 4 years?

Year	Period (in months)	Salary Received per 6-Month Period ($)	
		Plan I	Plan II
1	0–6	20,000	20,000
	6–12	20,000	20,300
2	12–18	20,500	20,600
	18–24	20,500	20,900
3	24–30		
	30–36		
Total for 3 Years			

Basic Calculator Guide

Operating the TI-84 Plus

Turning the Calculator On and Off

| ON | | Turns the calculator on |
| 2nd | ON | Turns the calculator off |

Adjusting the Display Contrast

| 2nd | ▲ | Increases the display (darkens the screen) |
| 2nd | ▼ | Decreases the contrast (lightens the screen) |

Note: If the display begins to dim (especially during calculations), and you must adjust the contrast to 8 or 9 in order to see the screen, then batteries are low and you should replace them soon.

The TI-84 Plus keyboard is divided into four zones: graphing keys, editing keys, advanced function keys, and scientific calculator keys (Figure 1).

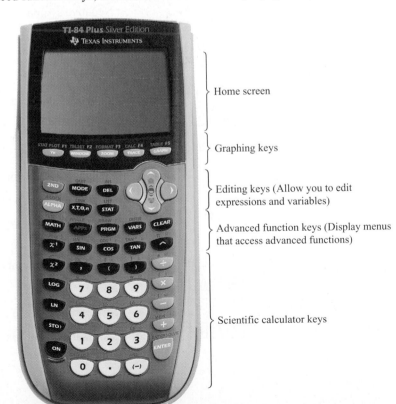

Home screen

Graphing keys

Editing keys (Allow you to edit expressions and variables)

Advanced function keys (Display menus that access advanced functions)

Scientific calculator keys

Figure 1

Keystrokes on the TI-83, TI-83 Plus, TI-84 Plus, and TI-Nspire

ENTER	Executes commands or performs a calculation
2nd	Pressing the 2nd key *before* another key accesses the character located above the key and printed in blue (yellow on the TI-83)
ALPHA	Pressing the ALPHA key *before* another key accesses the character located above the key and printed in green
2nd ALPHA	Locks in the ALPHA keyboard
CLEAR	Pressing CLEAR once clears the line Pressing CLEAR twice clears the screen
2nd MODE	Returns to the homescreen
DEL	Deletes the character at the cursor
2nd DEL	Inserts characters at the underline cursor
X,T,θ,n	Enters an X in Function Mode, a T in Parametric Mode, a θ in Polar Mode, or an n in Sequence Mode
STO▶	Stores a value to a variable
^	Raises to an exponent
2nd ^	the number π
(–)	Negative symbol
MATH 1	Converts a rational number to a fraction
MATH ▶ [1]	Computes the absolute value of a number or an expression in parentheses
2nd ENTER	Recalls the last entry
2nd ·	Used to enter more than one expression on a line
2nd (-)	Recalls the most recent answer to a calculation
x^2	Squares a number or an expression
x^{-1}	Inverse; can be used with a real number or a matrix
2nd x^2	Computes the square root of a number or an expression in parentheses
2nd LN	Returns the constant e raised to a power
ALPHA 0	Space
2nd ◄	Moves the cursor to the beginning of an expression
2nd ►	Moves the cursor to the end of an expression

I. Creating Scatter Plots

Clear Lists

Press [STAT] and under EDIT press 4:ClrList (Figure 2). Press [2nd] [1] (L_1) [ENTER] to clear List 1. Repeat the line by pressing [2nd] [ENTER], move the cursor to L_1. Press [2nd] [2] (L_2) and [ENTER] to clear List 2 (Figure 3).

Another way to clear a list in the STAT menu is, after accessing the list, press the up arrow until the name of the list is highlighted, then press [CLEAR] and [ENTER]. CAUTION: Do not press the [DEL] key. This will not actually delete the list, but hide it from view. To view the list, press [2nd] [DEL] (INS) and type the list name.

Entering Data into Lists

Press [STAT] and under EDIT press 1:Edit. This brings you to the screen where you enter data into lists.

Enter the *x*-values (input) in the column headed L_1 and the corresponding *y*-values (output) in the column headed L_2 (Figure 4).

Create a Scatter Plot of Data Points

Go to the Y= menu and turn off or clear any functions entered there. To turn off a function, move the cursor over the = sign and press [ENTER].

Press [2nd] [Y=], 1:Plot 1. Highlight On, and then highlight the first graph type (Scatter Plot), Enter Xlist:L1, Ylist:L2, and pick the point plot mark you want (Figure 5).

Choose an appropriate WINDOW for the graph and press [GRAPH], or press [ZOOM], 9:ZoomStat to plot the data points (Figure 6).

Trace Along the Plot

Press [TRACE] and the right arrow to move from point to point. The *x* and *y* coordinates are displayed at the bottom of the screen.

In the upper left-hand corner of the screen, P1:L1,L2 is displayed. This tells you that you are tracing along the scatter plot (Figure 7).

Turning Off Stat Plot

After creating a scatter plot, turn Stat Plot 1 off by pressing [Y=], moving up to Plot 1 with the cursor, and pressing [ENTER]. Stat Plot 1 can be turned on by pressing [ENTER] with the cursor on Plot 1. It will be highlighted when it is on.

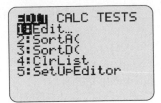

Figure 2

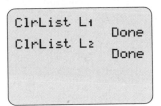

Figure 3

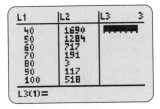

Figure 4

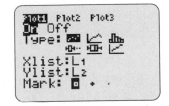

Figure 5

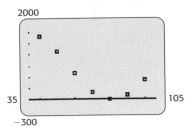

Figure 6

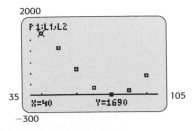

Figure 7

II. Graphing Equations

Setting Windows

The window defines the highest and lowest values of x and y on the graph of the function that will be shown on the screen. To set the window manually, press the WINDOW key and enter the values that you want (Figure 8).

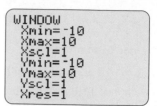

Figure 8

The values that define the viewing window can also be set by using ZOOM keys (Figure 9). Frequently the standard window (ZOOM 6) is appropriate. The standard window sets values Xmin=−10, Xmax=10, Ymin=−10, Ymax=10. Often a decimal or integer viewing window (ZOOM 4 or ZOOM 8) gives a better representation of the graph.

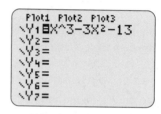

Figure 9

The window should be set so that the important parts of the graph are shown and the unseen parts are suggested. Such a graph is called complete.

Graphing Equations

To graph an equation in the variables x and y, first solve the equation for y in terms of x. If the equation has variables other than x and y, solve for the dependent variable and replace the independent variable with x.

Press the Y= key to access the function entry screen. You can input up to ten functions in the Y= menu (Figure 10). Enter the equation, using the X,T,Θ,n key to input x. Use the subtraction key − between terms. Use the negation key (-) when the first term is negative. Use parentheses as needed so that what is entered agrees with the order of operations.

Figure 10 The function $y = x^3 - 3x^2 - 13$ is entered in Y1.

The = sign is highlighted to show that Y1 is *on* and ready to be graphed. If the = sign is not highlighted, the equation will remain, but its graph is "turned off" and will not appear when GRAPH is pressed. (The graph is "turned on" by repeating the process.)

To erase an equation, press CLEAR. To return to the homescreen, press 2nd MODE (QUIT).

Determine an appropriate viewing window (Figures 11 and 12).

Pressing GRAPH or a ZOOM key will activate the graph.

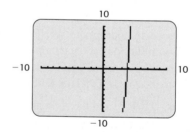

Figure 11 The graph of the function using the standard window.

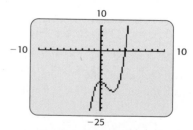

Figure 12 A better graph using the window [−10, 10] by [−25, 10].

III. Finding Function Values

Using TRACE, VALUE Directly on the Graph

Enter the function to be evaluated in Y1. Choose a window so that it contains the x-value whose y-value we seek (Figure 13).

Turn all Stat Plots off.

Press [TRACE] and then enter the selected x-value followed by [ENTER.] The cursor will move to the selected value and give the resulting y-value if the selected x-value is in the window.

If the selected x-value is not in the window, ERR: INVALID occurs.

If the x-value is in the window, the y-value will occur even if it is not in the window (Figures 14 and 15).

Using the TABLE ASK Feature

Enter the function with the [Y=] key. (Note: The = sign must be highlighted.) Press [2nd] [WINDOW] (TBLSET), move the cursor to Ask opposite Indpnt:, and press [ENTER] (Figure 16). This allows you to input specific values for x. Pressing [DEL] will clear entries in the table.

Then press [2nd] [GRAPH] (TABLE) and enter the specific values (Figure 17).

Making a Table of Values with Uniform Inputs

Press [2nd] [WINDOW] (TBLSET), enter an initial x-value in the table (Tblmin), and enter the desired change (ΔTbl) in the x-value in the table (Figure 18).

Enter [2nd] [GRAPH] (TABLE) to get the list of x-values and the corresponding y-values. The value of the function at the given value of x can be read from the table (Figure 19).

Use the up or down arrow to find the x-values where the function is to be evaluated (Figure 20).

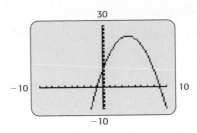

Figure 13 To evaluate $y = -x^2 + 8x + 9$ when $x = 3$ and when $x = -5$, graph the function using the window $[-10, 10]$ by $[-10, 30]$.

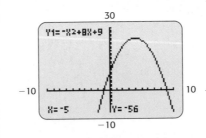

Figure 14

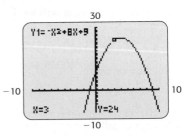

Figure 15

Figure 16

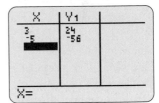

Figure 17

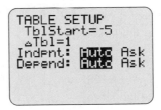

Figure 18

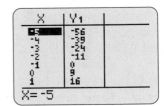

Figure 19

Figure 20

Using Y-VARS

Use the $\boxed{Y=}$ key to store Y1 = $f(x)$. Press $\boxed{2nd}$ (QUIT).

Press $\boxed{VARS}$, Y-VARS 1,1 to display Y1 on the homescreen. Then press $\boxed{(}$, the x-value, $\boxed{)}$ and $\boxed{ENTER}$ (Figure 21).

OR

Enter the x-values needed as follows:
Y$_1$ ({value 1, value 2, etc.}) $\boxed{ENTER}$ (Figure 22).
Values of the function will be displayed.

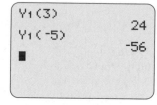

Figure 21

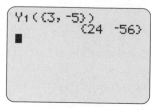

Figure 22

IV. Finding Intercepts of Graphs

Solve the equation for y. Enter the equation with the Y= key.

Finding the y-Intercept

Press $\boxed{TRACE}$ and enter the value 0. The resulting y value is the y-intercept of the graph (Figure 23).

Finding the x-Intercept(s)

Set the window so that the intercepts to be located can be seen. The graph of a linear equation will cross the x-axis at most one time; the graph of a quadratic equation will cross the x-axis at most two times, etc.

To find the point(s) where the graph crosses the x-axis, press $\boxed{2nd}$ $\boxed{TRACE}$ to access the CALC menu. Select 2:zero (Figure 24).

Answer the question "Left bound?" with $\boxed{ENTER}$ after moving the cursor close to and to the left of an x-intercept (Figure 25).

Answer the question "Right bound?" with $\boxed{ENTER}$ after moving the cursor close to and to the right of this x-intercept (Figure 26).

To the question "Guess?" press $\boxed{ENTER}$.

The coordinates of the x-intercept will be displayed (Figure 27). Repeat to get all x-intercepts (Figure 28). (Some of these values may be approximate.)

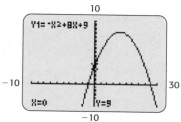

Figure 23 The y-intercept of the graph of the function $y = -x^2 + 8x + 9$ is 9.

Figure 24

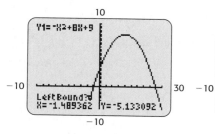

Figure 25

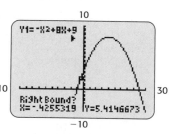

Figure 26

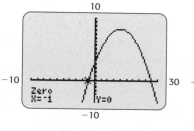

Figure 27

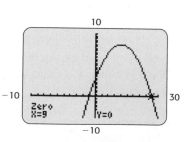

Figure 28

V. Finding Maxima and Minima

To locate a local maximum on the graph of an equation, graph the function using a window that shows all the turning points (Figure 29).

Press $\boxed{\text{2nd}}$ $\boxed{\text{TRACE}}$ to access the CALC menu (Figure 30). Select 4:maximum. The calculator asks the question "Left Bound?" (Figure 31). Move the cursor close to and to the left of the maximum and press $\boxed{\text{ENTER}}$. Next the calculator asks the question "Right Bound?" (Figure 32). Move the cursor close to and to the right of the maximum and press $\boxed{\text{ENTER}}$. When the calculator asks "Guess?" press $\boxed{\text{ENTER}}$ (Figure 33). The coordinates of the maximum will be displayed (Figure 34).

Note: When selecting a left bound or a right bound, you may also type in a value of x if you wish.

To locate a local minimum on the graph of an equation, press $\boxed{\text{2nd}}$ $\boxed{\text{TRACE}}$ to access the CALC menu. Select 3:minimum, and follow the same steps as above (Figure 35).

The graph of the function $y = x^3 - 27x$ is shown below on the window $[-10, 10]$ by $[-70, 70]$.

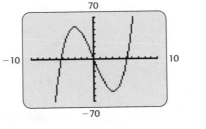

Figure 29

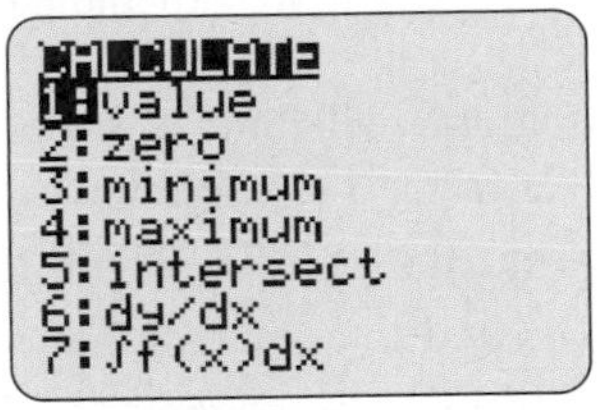

Figure 30

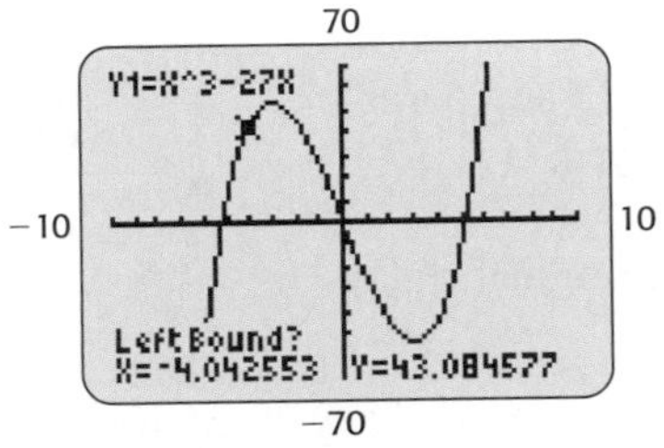

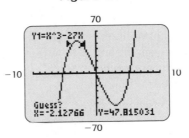

Figure 31

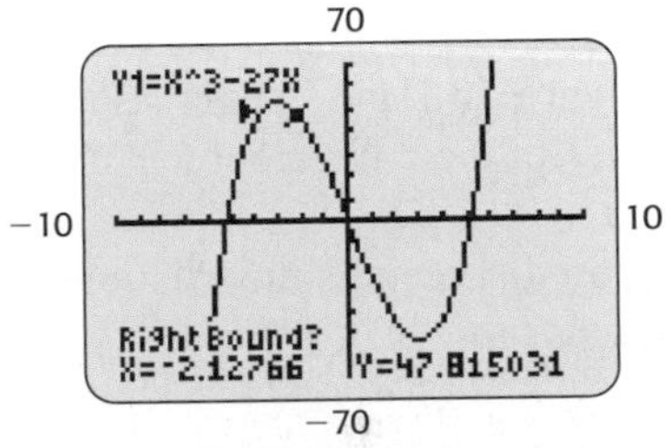

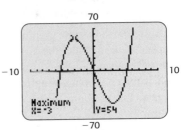

Figure 32

Figure 33

Figure 34

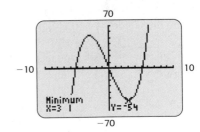

Figure 35

VI. Solving Equations Graphically

The x-Intercept Method

To solve an equation graphically using the x-intercept method, first rewrite the equation with 0 on one side of the equation. Using the $\boxed{\text{Y=}}$ key, enter the nonzero side of the equation equal to Y_1. Then graph the equation with a window that shows all points where the graph crosses the x-axis.

For example, to solve the equation $\frac{2x - 3}{4} = \frac{x}{3} + 1$, we enter $Y_1 = (2x - 3)/4 - (x/3) - 1$ and graph using the window $[-10, 15]$ by $[-5, 5]$ (Figures 36 and 37).

The graph will intersect the x-axis where $y = 0$; that is, when x is a solution to the original equation. To find the point(s) where the graph crosses the x-axis and the equation has solutions, press 2nd TRACE to access the CALC menu (Figure 38). Select 2:zero.

Select a "left bound" (Figure 39), "right bound" (Figure 40), and "guess" (Figure 41). The coordinates of the x-intercept will be displayed.

The x-intercept is the solution to the original equation (Figure 42).

Repeat to get all x-intercepts (and solutions).

The Intersection Method

To solve an equation graphically using the intersection method, under the Y= menu, assign the left side of the equation to Y1 and the right side of the equation to Y2. Then graph the equations using a window that contains the point(s) of intersection of the graphs.

To solve the equation $\frac{2x - 3}{4} = \frac{x}{3} + 1$ enter the left and right sides as shown in Figure 43 and graph using the window $[-10, 20]$ by $[-3, 10]$ (Figure 44).

Press 2nd TRACE to access the CALC menu (Figure 45). Select 5:intersect.

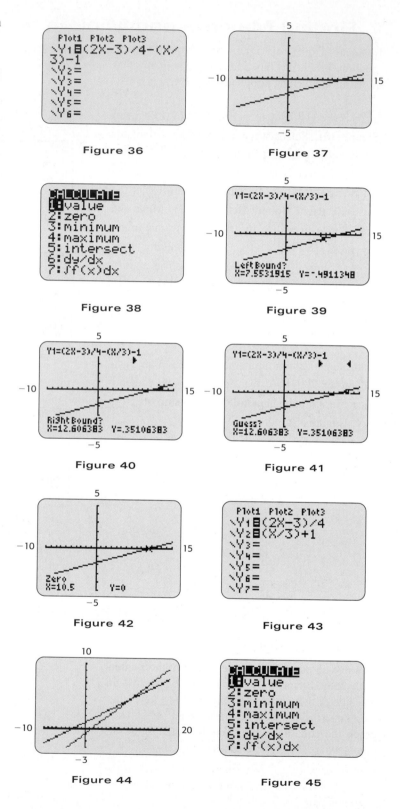

Figure 36

Figure 37

Figure 38

Figure 39

Figure 40

Figure 41

Figure 42

Figure 43

Figure 44

Figure 45

Answer the question "First curve?" by pressing ENTER (Figure 46) and "Second curve?" by pressing ENTER (Figure 47). (Or press the down arrow to move to one of the two curves.)

To the question "Guess?" move the cursor close to the desired point of intersection and press ENTER (Figure 48). The coordinates of the point of intersection will be displayed. Repeat to get all points of intersection.

The solution(s) to the equation will be the values of x from the points of intersection (Figure 49).

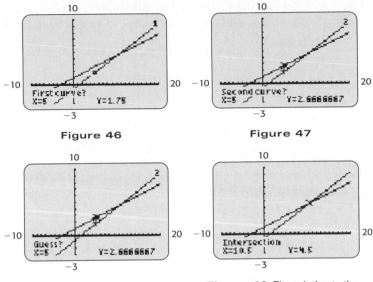

Figure 46

Figure 47

Figure 48

Figure 49 The solution to the equation is $x = 10.5$.

VII. Solving Systems of Equations

To find the solution of a system of equations graphically, first solve each equation for y and use the Y= key with Y_1 and Y_2 to enter the equations. Graph the equation with a friendly window.

For example, to solve the system $\begin{cases} 4x + 3y = 11 \\ 2x - 5y = -1 \end{cases}$

graphically, we solve for y:

$y_1 = \dfrac{11}{3} - \left(\dfrac{4}{3}\right)x$

$y_2 = \left(\dfrac{2}{5}\right)x + \dfrac{1}{5}$. We graph using ZOOM 4 and

then Intersect under the CALC menu to find the point of intersection. If the two lines intersect in one point, the coordinates give the x- and y-values of the solution (Figure 50). The solution of the system above is $x = 2, y = 1$. If the two lines are parallel, there is no solution; the system of equations is inconsistent. For example, to solve the system

$\begin{cases} 4x + 3y = 4 \\ 8x + 6y = 25 \end{cases}$ we solve for y (Figure 51) and

obtain the graph at right (Figure 52). This system has no solution. Note that if the lines are parallel, then when solving for y, the equations will show that the lines have the same slope and different y-intercepts.

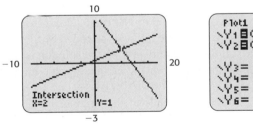

Figure 50

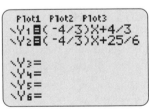

Figure 51

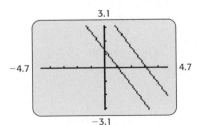

Figure 52

If the two graphs of the equations give only one line, every point on the line gives a solution to the system, and the system is dependent. For example, to solve the system $\begin{cases} 2x + 3y = 6 \\ 4x + 6y = 12 \end{cases}$, we solve for y (Figure 53) and obtain the graph in Figure 54. This system has many solutions and is dependent. Note that the two graphs will be the same graph if, when solving for y to use the graphing calculator, the equations are equivalent.

Figure 53

VIII. Solving Inequalities Graphically

To solve a linear inequality graphically, first rewrite the inequality with 0 on the right side and simplify. Then, under the Y= menu, assign the left side of the inequality to Y_1, so that $Y_1 = f(x)$. For example, to solve $3x > 6 + 5x$, rewrite the inequality as $3x - 5x - 6 > 0$, or $-2x - 6 > 0$. Set $Y_1 = -2x - 6$.

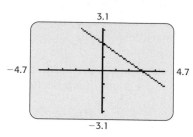

Figure 54

Graph this equation so that the point where the graph crosses the x-axis is visible. Note that the graph will cross the axis in at most one point because the function is of degree 1. (Using ZOOM OUT can help find this point.)

Using ZOOM 4, the graph of $Y_1 = -2x - 6$ is shown in Figure 55.

Use the ZERO command under the CALC menu to find the x-value where the graph crosses the x-axis (the x-intercept). This value can also be found by finding the solution to $0 = f(x)$ algebraically.

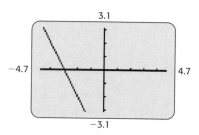

Figure 55

The x-intercept of the graph of $Y_1 = -2x - 6$ is $x = -3$ (Figure 56).

Observe the inequality as written with 0 on the right side. If the inequality is "<", the solution to the original inequality is the interval (bounded by the x-intercept) where the graph is below the x-axis. If the inequality is ">", the solution to the original inequality is the interval (bounded by the x-intercept) where the graph is above the x-axis.

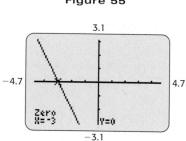

Figure 56

The solution to $-2x - 6 > 0$, and thus to the original inequality $3x > 6 + 5x$, is $x < -3$.

The region above the x-axis and under the graph can be shaded with 2nd PRGM (DRAW), 7 (Shade), and entering $(0, Y_1)$ to shade on the homescreen (Figure 57).

The x-interval where the shading occurs is the solution.

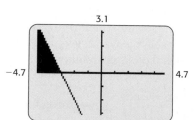

Figure 57

IX. Piecewise-Defined Functions

A piecewise-defined function is defined differently over two or more intervals.

To graph a piecewise-defined function

$$y = \begin{cases} f(x) \text{ if } x \le a \\ g(x) \text{ if } x > a \end{cases}$$

press [Y=] and enter
$Y_1 = (f(x))/(x \le a)$

and

$Y_2 = (g(x))/(x > a)$ (Figure 59)

Note that the inequality symbols are found by pressing [2nd] [MATH] to access the TEST menu (Figure 58).

Graph the functions using an appropriate window (Figure 60).

Evaluating a piecewise-defined function at a given value of x requires that the correct equation ("piece") be selected (Figures 61 and 62). For example, $f(-6)$ and $f(3)$ are shown in Figures 61 and 62, respectively.

Figure 58

Example To graph
$$f(x) = \begin{cases} x + 7 \text{ if } x \le -5 \\ -x + 2 \text{ if } x > -5 \end{cases}$$

Figure 59

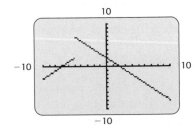

Figure 60

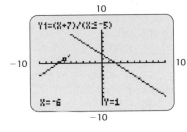

Figure 61

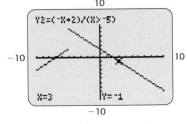

Figure 62

X. Modeling—Regression Equations

Finding an Equation That Models a Set of Data Points

Press [STAT] and under EDIT press 1:Edit. Enter the x-values (inputs) in the column headed L1 and the corresponding y-values (outputs) in the column headed L2 (Figure 63).

Press [2nd] [Y=] (STAT PLOT), 1:Plot 1. Highlight On, and then highlight the first graph Type. Enter Xlist:L1, Ylist:L2, and pick the point plot mark you want (Figure 64).

Example The average daily number of inmates in the Beaufort County, South Carolina, Detention Center for the years 1993–1998 are given by the data points (1993, 96), (1994, 109), (1995, 119), (1996, 116), (1997, 137), (1998, 143). Write the equation of the line that is the best fit for these data.

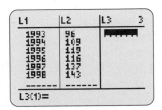

Figure 63

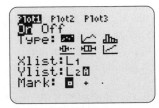

Figure 64

Press GRAPH with an appropriate window or ZOOM, 9:ZoomStat to plot the data points (Figure 65).

The graph looks like a line, so use the linear model, with LinReg.

Observe the point plots to determine what type of function would best model the data.

Press STAT, move to CALC, and select the function type to be used to model the data (Figure 66). Press the VARS key, over to Y-VARS, select 1:Function and 1:Y_1 (Figure 67). Press ENTER. The coefficients of the equation will appear on the screen (Figure 68), and the linear regression equation will appear as Y_1 on the Y= screen (see Figure 69).

To see how well the equation models the data, press GRAPH. If the graph does not fit the points well, another type function may be used to model the data.

For some models the "r" is a diagnostic value that gives the **correlation coefficient**, which determines the strength of the relationship between the independent and dependent variables. To display this value, select DiagnosticOn, after pressing 2nd 0 (CATALOG) (Figure 68).

Report the equation in a way that makes sense in the context of the problem, with the appropriate units, and the variables identified.

The equation that models the average daily number of inmates in the Beaufort County, South Carolina, Detention Center (see page 661) is $y = 9.0286x - 17896.5143$ (rounded to 4 decimal places), where x is the year and y is the number of inmates (Figures 69 and 70).

Using a Model to Find an Output

To use the model to find output values inside the data range (*interpolation*) or outside the data range (*extrapolation*), evaluate the function at the desired input value. This may be done using TABLE (Figure 71), Y-VARS (Figure 72), or TRACE (Figure 73). (Note that the WINDOW may have to be changed to see the x-value to which you are tracing.)

To predict the average daily inmate population in the Beaufort County jail in 2008, we compute Y_1 (2008).

According to the model, approximately 233 inmates were, on average, in Beaufort County jail in 2008.

When using TRACE, the WINDOW must be changed so that $x = 2008$ is visible (Figure 73).

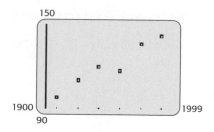

Figure 65

Figure 66

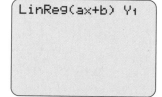

Figure 67

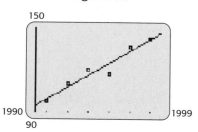

Figure 68

Figure 69

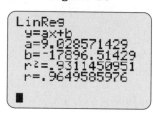

Figure 70

TABLE

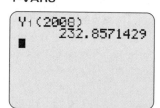

Figure 71

Y-VARS

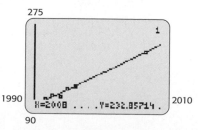

Figure 72

Figure 73

Using a Model to Find an Input

To use the model to estimate an input for a given output, set the model function equal to Y_1 and the desired output equal to Y_2 and solve the resulting equations for the input variable. An approximate solution can be found using TABLE or 2nd TRACE (CALC) 5:Intersect.

To use the model to determine in what year the population will be 260, we must solve the equations $Y_1 = f(x)$, $Y_2 = 260$. This may be approximated using the TABLE to find the x-value when y is approximately 260 (Figure 74).

We can also solve the equation $Y_1 = Y_2$ using the intersect method (Figures 75 and 76):

According to the model, the population of the jail will reach 260 in the year 2011.

XI. Complex Numbers

Calculations that involve complex numbers of the form $a + bi$ will be displayed as complex numbers only if the MODE is set to $a + bi$ (Figure 77).

Complex numbers can be added (Figure 78), subtracted (Figure 79), multiplied (Figure 80), divided (Figure 81), and raised to powers (Figure 82). The number i is accessed by pressing 2nd · .

Care must be taken when writing the quotient of two complex numbers. The answer in Figure 81 appears to have "i" in the denominator; however, the form of the number is $a + bi$, so the answer is actually $-\dfrac{14}{17} - \dfrac{5}{17}i$.

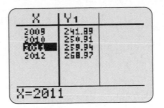

Figure 74

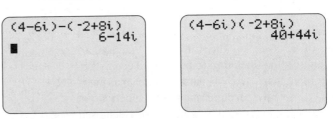

Figure 75

Figure 76

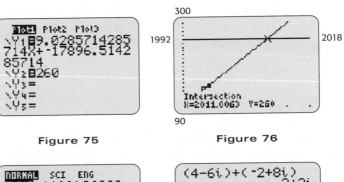

Figure 77

Figure 78

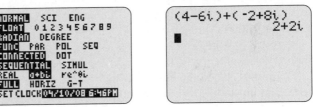

Figure 79

Figure 80

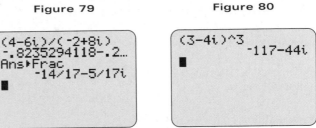

Figure 81

Figure 82

XII. Operations with Functions

Combinations of Functions

To find the graphs of combinations of two functions $f(x)$ and $g(x)$, enter $f(x)$ as Y_1 and $g(x)$ as Y_2 under the $Y=$ menu. Consider the example $f(x) = 4x - 8$ and $g(x) = x^2$. To graph $(f + g)(x)$, enter $Y_1 + Y_2$ as Y_3 under the $Y=$ menu (Figure 83). Place the cursor on the $=$ sign beside Y_1 and press $\boxed{\text{ENTER}}$ to turn off the graph of Y_1. Repeat with Y_2. Press $\boxed{\text{GRAPH}}$ with an appropriate window (Figure 84).

To graph $(f - g)(x)$, enter $Y_1 - Y_2$ as Y_3 under the $Y=$ menu (Figure 85). Place the cursor on the $=$ sign beside Y_1 and press $\boxed{\text{ENTER}}$ to turn off the graph of Y_1. Repeat with Y_2. Press $\boxed{\text{GRAPH}}$ (Figure 86).

To graph $(f*g)(x)$, enter Y_1*Y_2 as Y_3 under the $Y=$ menu (Figure 87). Place the cursor on the $=$ sign beside Y_1 and press $\boxed{\text{ENTER}}$ to turn off the graph of Y_1. Repeat with Y_2. Press $\boxed{\text{GRAPH}}$ (Figure 88).

To graph $(f/g)(x)$, enter Y_1/Y_2 as Y_3 under the $Y=$ menu (Figure 89). Place the cursor on the $=$ sign beside Y_1 and press $\boxed{\text{ENTER}}$ to turn off the graph of Y_1. Repeat with Y_2. Press $\boxed{\text{GRAPH}}$ with an appropriate window (Figure 90).

To evaluate $f + g, f - g, f*g$, or f/g at a specified value of x, enter Y_1, the value of x enclosed in parentheses, the operation to be performed, Y_2, the value of x enclosed in parentheses, and press $\boxed{\text{ENTER}}$. Or, if the combination of functions is entered as Y_3, enter Y_3 and the value of x enclosed in parentheses (Figure 91), or use $\boxed{\text{TABLE}}$.

The value of $(f + g)(3)$ is shown in Figure 92.

Note: Entering $(Y_1 + Y_2)(3)$ does not produce the correct result (Figure 93).

Composition of Functions

To graph the composition of two functions $f(x)$ and $g(x)$, enter $f(x)$ as Y_1 and $g(x)$ as Y_2 under the $Y=$ menu. Consider the example $f(x) = 4x - 8$ and $g(x) = x^2$.

To graph $(f \circ g)(x) = f(g(x))$, enter $Y_1(Y_2)$ as Y_3 under the $Y=$ menu (Figure 94). Place the cursor on the $=$ sign beside Y_1 and press $\boxed{\text{ENTER}}$ to turn off

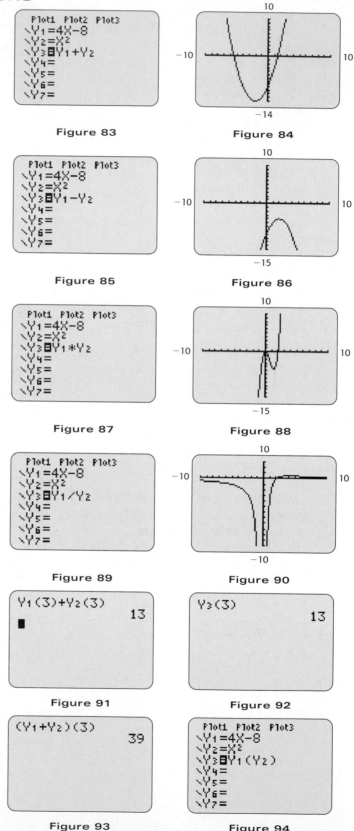

Figure 83

Figure 84

Figure 85

Figure 86

Figure 87

Figure 88

Figure 89

Figure 90

Figure 91

Figure 92

Figure 93

Figure 94

the graph of Y_1. Repeat with Y_2. Press GRAPH with an appropriate window (Figure 95).

To graph $(g \circ f)(x)$, enter $Y_2(Y_1)$ as Y_4 under the Y= menu (Figure 96). Turn off the graphs of Y_1, Y_2, and Y_3. Press GRAPH with an appropriate window (Figure 97).

To evaluate $(f \circ g)(x)$ at a specified value of x, enter $Y_1(Y_2)$ and the value of x enclosed in parentheses (Figure 98), and press ENTER. Or, if the combination of functions is entered as Y_3, enter Y_3 and the value of x enclosed in parentheses. The value of $(f \circ g)(-5) = f(g(-5))$ is shown in Figure 99.

XIII. Inverse Functions

To show that the graphs of two inverse functions are symmetrical about the line $y = x$, graph the functions on a square window. Enter $f(x)$ as Y_1 and the inverse function $g(x)$ as Y_2, and press ENTER with the cursor to the left of Y_2 (to make the graph **dark**). Press GRAPH with an appropriate window. Enter $Y_3 = x$ under the Y= menu and graph using ZOOM 5:Zsquare. For example, the graphs of $f(x) = x^3 - 3$ and its inverse $g(x) = \sqrt[3]{x + 3}$ are symmetric about the line $y = x$, as shown in Figures 100 and 101.

To graph a function $f(x)$ and its inverse, under the Y= menu, enter $f(x)$ as Y_1 (Figure 102). Choose a square window. Press 2nd PRGM (DRAW) (Figure 103), 8:DrawInv (Figure 104), press VARS, move to Y-VARS, and press ENTER three times. The graph of $f(x)$ and its inverse will be displayed.

To clear the graph of the inverse, press 2nd PRGM (DRAW), 1:ClrDraw. The graphs of $f(x) = 2x - 5$ and its inverse are shown in Figure 105.

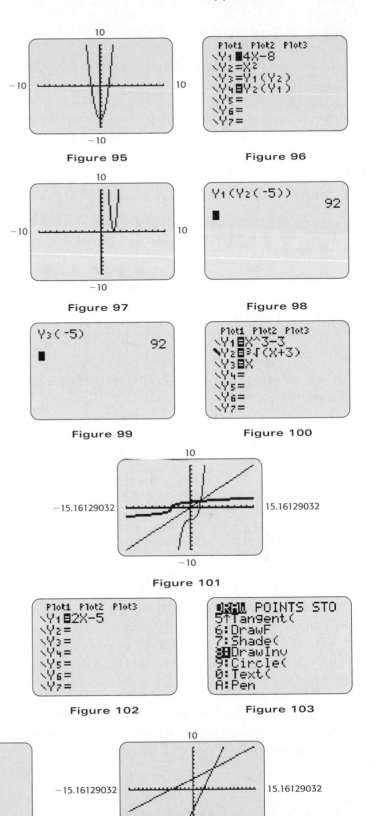

Figure 95

Figure 96

Figure 97

Figure 98

Figure 99

Figure 100

Figure 101

Figure 102

Figure 103

Figure 104

Figure 105

XIV. Matrices

Entering Data into Matrices

To enter data into matrices, press 2nd x^{-1} [MATRIX] to get the MATRIX menu, (Figure 106). Move the cursor to EDIT. Enter the number of the matrix into which the data are to be entered. Enter the dimensions of the matrix, and enter the value for each entry of the matrix (Figure 107). Press ENTER after each entry.

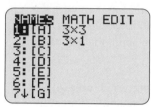

Figure 106

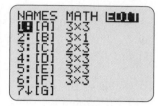

Figure 107

For example, to enter the matrix below as [A],

$$\begin{bmatrix} 1 & 2 & 3 \\ 2 & -2 & 1 \\ 3 & 1 & -2 \end{bmatrix}$$

enter 3's to set the dimension, and enter the elements of the matrix (Figure 108).

To perform operations with the matrix or leave the editor, first press 2nd MODE (QUIT).

To view the matrix, press MATRIX, the number of the matrix (Figure 109), and ENTER (Figure 110).

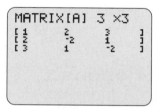

Figure 108

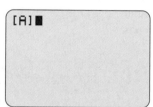

Figure 109

The Identity Matrix

To display an identity matrix of order n (an $n \times n$ matrix consisting of 1's on the main diagonal and 0's elsewhere), press MATRIX, move to MATH, enter 5:identity(, and the order of the identity matrix desired (Figure 111). The identity matrix of order 2 is shown in Figure 112.

An identity matrix can also be created by entering the numbers directly with MATRIX, EDIT.

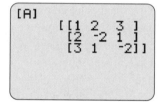

Figure 110

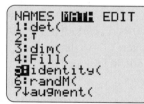

Figure 111

Operations with Matrices

To find the sum of two matrices, [A] and [D], enter the values of the elements of [A] using MATRIX and EDIT (Figure 113). Press 2nd MODE (QUIT). Enter the values of the elements of [D] using MATRIX and EDIT (Figure 114). Press 2nd QUIT.

Use MATRIX and NAMES to enter [A] + [D], and press ENTER. If the matrices have the same dimensions, they can be added (or subtracted). If they do not have the same dimensions, an error message will occur.

For example, the sum

$$\begin{bmatrix} 1 & 2 & 3 \\ 2 & -2 & 1 \\ 3 & 1 & -2 \end{bmatrix} + \begin{bmatrix} 7 & -3 & 2 \\ 4 & -5 & 3 \\ 0 & 2 & 1 \end{bmatrix}$$ is shown in Figure 115.

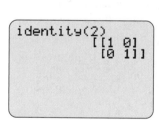

Figure 112

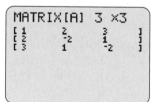

Figure 113

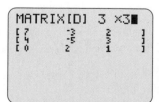

Figure 114

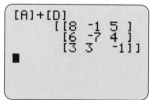

Figure 115

To find the difference of the matrices [A] and [D] enter [A] – [D] and press ENTER (Figure 116).

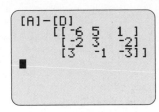

Figure 116

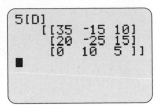

Figure 117

We can multiply a matrix [D] by a real number (scalar) k by pressing k [D]. (Or $k*$[D].) In Figures 117 and 118, we multiply the matrix [D] by 5.

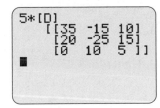

Figure 118

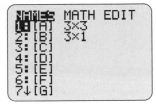

Figure 119

Multiplying Two Matrices

To find the product of matrices, [C][A], press MATRIX, move to EDIT, enter 1:[A], enter the dimensions of [A] (Figure 119), and enter the elements of [A] (Figure 120). Press 2nd MODE (QUIT). Enter the elements in matrix [C] (Figure 121). Press 2nd MODE (QUIT). Press MATRIX, 3 [C], *, MATRIX [A], and ENTER (Figure 122). (Or press MATRIX [C], MATRIX [A], and ENTER [Figure 123].)

For example, the product

$$\begin{bmatrix} 1 & 2 & 4 \\ -3 & 2 & -1 \end{bmatrix} \begin{bmatrix} 1 & 2 & 3 \\ 2 & -2 & 1 \\ 3 & 1 & -2 \end{bmatrix}$$

is shown in Figures 122 and 123.

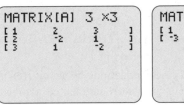

Figure 120

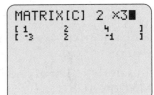

Figure 121

Note that [A][C] does not always equal [C][A]. The product [A][C] may be the same as [C][A], may be different from [C][A], or may not exist.

In Figures 124 and 125, [A][C] cannot be computed because their dimensions do not match.

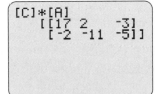

Figure 122

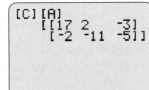

Figure 123

Figure 124

Figure 125

Finding the Inverse of a Matrix

To find the inverse of a matrix, enter the elements of the matrix using MATRIX and EDIT (Figure 126). Press 2nd MODE (QUIT). Press MATRIX, the number of the matrix, then press the x^{-1} key and ENTER (Figure 127).

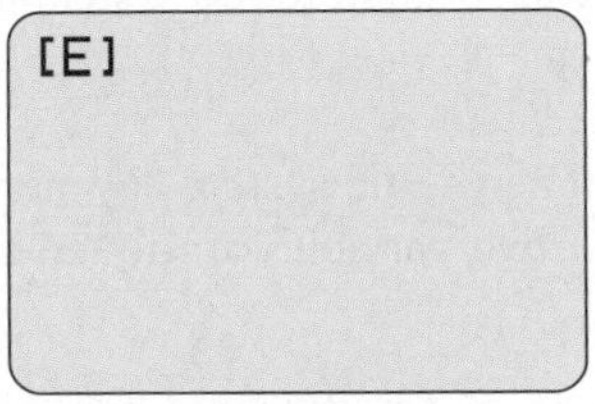

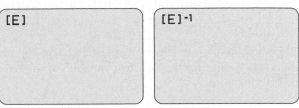

Figure 126

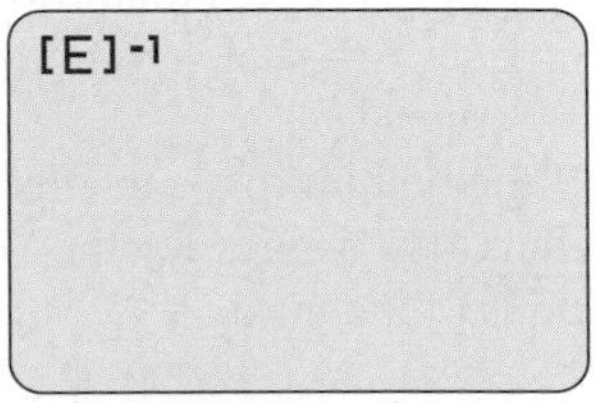

Figure 127

For example, the inverse of $E = \begin{bmatrix} 2 & 0 & 2 \\ -1 & 0 & 1 \\ 4 & 2 & 0 \end{bmatrix}$ is shown in Figure 128.

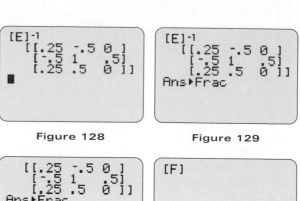

Figure 128   **Figure 129**

To see the entries as fractions, press $\boxed{\text{MATH}}$ and press 1:Frac, and press $\boxed{\text{ENTER}}$ (Figures 129 and 130).

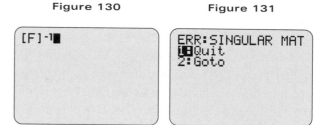

Figure 130   **Figure 131**

Not all matrices have inverses. Matrices that do not have inverses are called singular matrices (Figures 131–133).

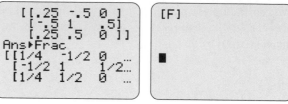

Figure 132   **Figure 133**

Solution of Systems—Reduced Echelon Form

To solve a system of linear equations by using rref under the MATRIX MATH menu, create an augmented matrix [A] with the coefficient matrix augmented by the constants (Figure 134). Use the MATRIX menu to produce a reduced row echelon form of Matrix A, as follows:

1. Press $\boxed{\text{MATRIX}}$, move to the right to MATH (Figure 135).

2. Scroll down to B:rref(, and press $\boxed{\text{ENTER}}$, or press $\boxed{\text{ALPHA}}$ B (Figure 136). Press $\boxed{\text{MATRIX}}$, 1:[A] to get rref([A]). Press $\boxed{\text{ENTER}}$. This gives the reduced echelon form.

Figure 134   **Figure 135**

For example, the system $\begin{cases} 2x - y + z = 6 \\ x + 2y - 3z = 9 \\ 3x - 3z = 15 \end{cases}$ is solved in Figure 137.

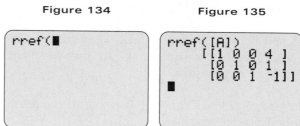

Figure 136   **Figure 137**

If each row in the coefficient matrix (first 3 columns) contains a 1 with the other elements 0's, the solution is unique and the number in column 4 of a row is the value of the variable corresponding to a 1 in that row. The solution to the system above is unique: $x = 4$, $y = 1$, and $z = -1$.

If the bottom row contains all zeros, the system has many solutions. The values for the first two variables are found as functions of the third.

If there is a nonzero element in the augment of row 3 and zeros elsewhere in row 3, there is no solution to the system.

XV. Linear Programming

To solve a linear programming problem involving two constraints graphically, write the inequalities as equations, solved for y. Graph the equations. The inequalities $x \geq 0$, $y \geq 0$ limit the graph to Quadrant I, so choose a window with $x$min $= 0$ and $y$min $= 0$. Use $\boxed{\text{TRACE}}$ or $\boxed{\text{INTERSECT}}$ to find the corners of the region, where the borders intersect.

For example, to find the region defined by the inequalities

$$5x + 2y \leq 54$$
$$2x + 4y \leq 60$$
$$x \geq 0, y \geq 0$$

write $y = 27 - 5x/2$ and $y = 15 - x/2$, graph, and find the intersection points as shown in Figures 138 to 140.

The corners of the region determined by the inequalities are $(0, 15)$, $(6, 12)$, and $(10.8, 0)$.

Press $\boxed{\text{2nd}}$ $\boxed{\text{PRGM}}$ (DRAW) and select 7: SHADE to shade the region determined by the inequalities (Figure 141). Shade under the border from $x = 0$ to a corner and shade under the second border from the corner to the x-intercept (Figure 142).

Evaluating the objective function at the coordinates of each of the corners determines where the objective function is maximized or minimized.

$$\text{At } (0, 15), f = 165$$
$$\text{At } (6, 12), f = 162$$
$$\text{At } (10.8, 0), f = 54$$

The maximum value of f is 165 at $x = 0$, $y = 15$.

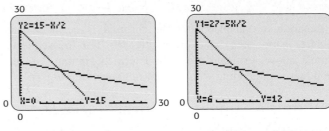

Figure 138   **Figure 139**

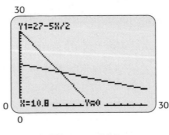

Figure 140   **Figure 141**

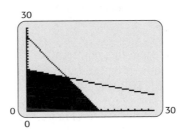

Figure 142

XVI. Sequences and Series

Evaluating a Sequence

To evaluate a sequence for different values of n, press $\boxed{\text{MODE}}$ and highlight Seq (Figure 143). Press $\boxed{\text{ENTER}}$ and $\boxed{\text{2nd}}$ $\boxed{\text{MODE}}$ (QUIT). Store the formula for the sequence (in quotes) in u, using $\boxed{\text{STO}}$ u. (Press the $\boxed{\text{X,T,}\Theta\text{,}n}$ n for the formula, and $\boxed{\text{2nd}}$ 7 to get u.) Press $\boxed{\text{ENTER}}$.

For example, to evaluate the sequence with nth term $n^2 + 1$ at $n = 1, 3, 5,$ and 9, we store the formula as shown in Figure 144.

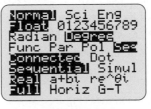

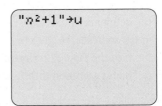

Figure 143   **Figure 144**

Enter u({a,b,c,..}) to evaluate the sequence at a,b,c,…, and press ENTER (Figure 145).

To generate a sequence after the formula is defined, enter u(*n*start, *n*stop, step), and press ENTER. For the sequence formula above, evaluate every third term beginning with the second term and ending with the eleventh term (Figure 146).

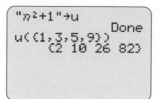
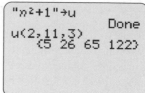

Figure 145   **Figure 146**

Arithmetic Sequences; *n*th Terms

To find the *n*th term of an arithmetic sequence with first term a and common difference d, press MODE and highlight Seq (Figure 147). Press ENTER and press 2nd MODE (QUIT).

Press Y=. At u(*n*) =, enter the formula for the *n*th term of an arithmetic sequence, using the x,T,Θ,*n* key to enter *n*. The formula is $a + (n - 1) * d$, where a is the first term and d is the common difference.

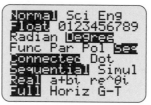
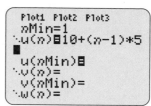

Figure 147   **Figure 148**

For example, to find the 12th term of the arithmetic sequence with first term 10 and common difference 5, substitute 10 for a and 5 for d, as shown in Figure 148, to get u(*n*) = $10 + (n - 1) * 5$.

Press 2nd MODE (QUIT). To find the *n*th term of the sequence, press 2nd u (above 7) followed by the value of *n*, in parentheses, to get u(*n*), then press ENTER (Figure 149).

Additional terms can be found in the same manner (Figure 150).

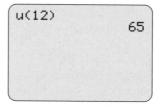

Figure 149 The 12th term.   **Figure 150** The 8th term.

Sums of Arithmetic Sequences

To find the sum of the first *n* terms of an arithmetic sequence, press MODE and highlight Seq (Figure 151). Press ENTER and press 2nd MODE (QUIT).

Press Y=. At v(*n*) =, enter the formula for the sum of the first *n* terms of an arithmetic sequence, using the x,T,Θ,*n* key to enter *n*. The formula is $(n/2)(a + (a + (n - 1)d))$, where a is the first term and d is the common difference (Figure 152).

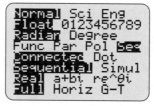
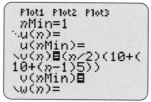

Figure 151   **Figure 152**

Press 2nd MODE (QUIT). To find the sum of the first *n* terms of the sequence, press 2nd v (above 8) followed by the value of *n*, in parentheses, to get v(*n*), then press ENTER. For example, the sum of the first 12 terms of the arithmetic sequence with first term 10 and common difference 5 is shown in Figure 153.

Other sums can be found in the same manner (Figure 154).

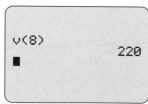

Figure 153 The sum of the first 12 terms.   **Figure 154** The sum of the first 8 terms.

Geometric Sequences; *n*th Terms and Sums

To find the *n*th term of a geometric sequence with first term *a* and common ratio *r*, press MODE and highlight Seq (Figure 155). Press ENTER and press 2nd MODE (QUIT).

Press Y= . At u(*n*) =, enter the formula for the *n*th term of a geometric sequence, using the x,T,Θ,*n* key to enter *n*. The formula is ar^{n-1}, where *a* is the first term and *r* is the common ratio.

For example, to find the geometric sequence with first term 40 and common ratio 1/2, we substitute 40 for *a* and (1/2) for *r* (Figure 156).

Press 2nd MODE (QUIT). To find the *n*th term of the sequence, press 2nd u (above 7) followed by the value of *n*, in parentheses, to get u(*n*); then press ENTER . The 8th and 12th terms of the geometric sequence above are shown in Figures 157 and 158, respectively. To get a fractional answer, press MATH , 1:Frac. Additional terms can be found in the same manner.

To find the sum of the first *n* terms of a geometric sequence, press Y= . At v(*n*) =, enter the formula for the sum of the first *n* terms of a geometric sequence, using the x,T,Θ,*n* key to enter *n*. The formula is $a(1 - r^n)/(1 - r)$, where *a* is the first term and *r* is the common ratio. Press 2nd MODE (QUIT). To find the sum of the first *n* terms of the sequence, press 2nd v (above 8) followed by the value of *n*, in parentheses, to get v(*n*), then press ENTER .

For example, to find the sum of the first 12 terms of the geometric sequence with first term 40 and common ratio 1/2, substitute 40 for *a* and (1/2) for *r* (Figure 159).

To get a fractional answer, press MATH , 1:Frac (Figure 160).

Other sums can be found in the same manner (Figure 161).

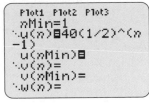

Figure 155

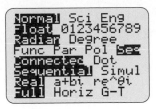

Figure 156

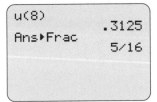

Figure 157

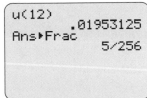

Figure 158

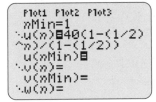

Figure 159

Figure 160 The sum of the first 12 terms.

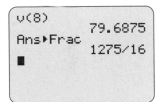

Figure 161 The sum of the first 8 terms.

Basic Guide to Excel

This appendix contains the basic features and operation of Excel. More details of specific features can be found in the **Graphing Calculator and Excel Manual**.

Excel Worksheet

When you start up Excel by using the instructions for your software and computer, the following screen will appear (Figure 1). The components of the **spreadsheet** are shown, and the grid shown is called a **worksheet**. By clicking on the tabs at the bottom, you can move to other worksheets.

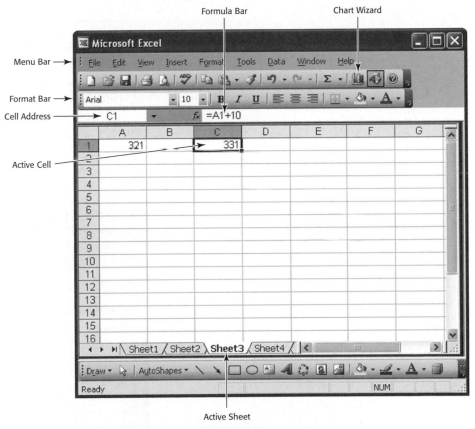

Figure 1

Addresses and Operations

Notice the letters at the top of the columns and the numbers identifying the rows. The cell addresses are given by the column and row; for example, the first cell has address A1. You can move from one cell to another with arrow keys, or you can select a cell with a mouse click. After you enter an entry in a cell, press enter to accept the entry. To edit the contents of a cell, press the F2 key and edit the contents in the formula bar at the top. To delete the contents, press the delete key.

The file operations such as "open a new file," "saving a file," and "printing a file" are similar to those in WORD. For example, <CTRL>S saves a file. You can also format a cell entry by selecting it and using menus similar to those in WORD.

Working with Cells

Cell entries, rows containing entries, and columns containing entries can be copied and pasted with the same commands as in WORD. For example, a highlighted cell can be copied with <CTRL>C. Sometimes entries exceed the width of the cell containing it, especially if they are text. To widen the cells in a column, place the mouse at the right side of the column heading, until you see the symbol ↔, then hold down the left button and move the mouse to the right (moving to the left makes it more narrow). If entering a number results in #####, the number is too long for the cell, and the cell should be widened.

If you don't want to widen a cell when an entry exceeds the width of the cell, select it, go to Format and then Cells, click the alignment tab, and check the **wrap text** box. The text will wrap within the selected cell.

You can work with a cell or with a range of cells. To select a range of cells:

1. Click on the beginning of the range of cells, hold down the left mouse button, and drag to the end of the desired range.

2. Release the mouse button. The range of cells will be highlighted. If the range of cells is from A2 through B6, the range would be indicated by a2:b6.

Creating Tables of Numbers in Excel

To Create Tables of Numbers in Excel with Independent Values

Type the column heading, such as x in cell A1, and enter each data value in a cell of the column (Figure 2).

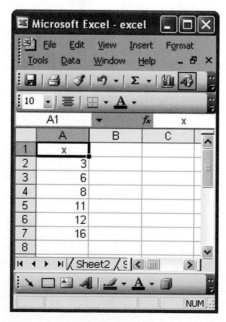

Figure 2

Fill Down Method for Values That Change by a Constant Increment

1. Type in the first two numbers (in C2 and C3, for example).

2. Select the cells C2 and C3 (C2:C3) (Figure 3).

Figure 3

3. Move the mouse to the lower right corner until the mouse becomes a thin "+" sign.

4. Drag the mouse down to the last cell where data are required (Figure 4). Press the right arrow to remove the highlight.

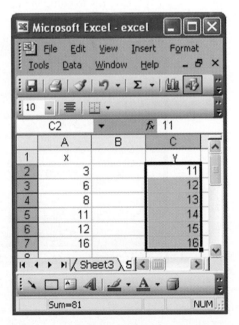

Figure 4

Evaluating a Function

To Evaluate a Function

1. Put headings on the two columns (x and $f(x)$, for example).

2. Begin a formula by entering =, then enter the function performing operations with data in cells by entering the operation side of the formula with appropriate cell address(es) representing the variable(s) (Figure 5). Use * to indicate multiplication and ^ to indicate power.

3. Enter the value(s) of the variable(s) for which the function is to be evaluated and press ENTER (Figure 6).

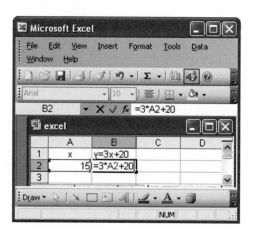

Figure 5

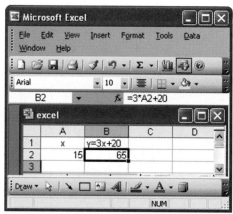

Figure 6

4. Changing the input values will give different function values (Figure 7). Press ENTER to complete the process (Figure 8).

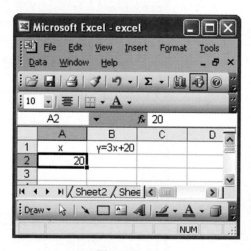

Figure 7

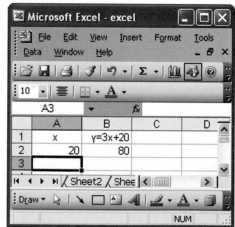

Figure 8

Creating a Table for a Function

To Create a Table for a Function Using Fill Down

1. Put headings on the two columns (x and $f(x)$, for example).

2. Fill the inputs (x-values) as previously described (page 674).

3. Enter the formula for the function as previously described (page 676).

4. Select the cell containing the formula for the function (B2, for example) (Figure 9).

5. Move the mouse to the lower right corner until there is a thin "+" sign.

6. Drag the mouse down to the last cell where the formula is required, and the values will be displayed (Figure 10). Using an arrow to move to a cell on the right will remove the highlight from the outputs.

Figure 9

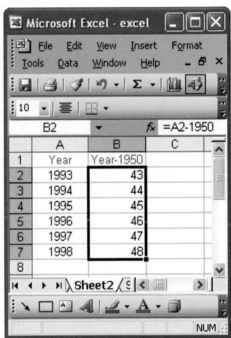

Figure 10

The worksheet below (Figure 11) has headings typed in, numbers entered with Fill Down in Column A, inputs entered in Column B, the numbers in Column D created with Fill Down using the formula A5 − 1950, and Column B copied in Column E.

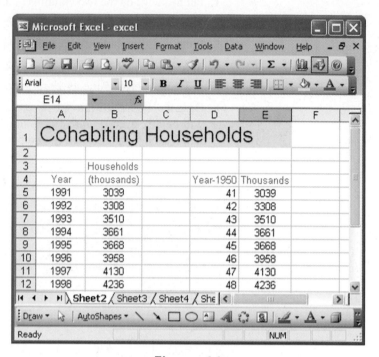

Figure 11

Graphing a Function of a Single Variable

To Graph a Function of a Single Variable ($y = f(x)$, for Example)

1. Create a table containing values for x and $f(x)$ as previously described (page 677).

2. Highlight the two columns containing the values of x and $f(x)$ (Figure 12).

Figure 12

3. Click the Chart Wizard icon and then select XY(Scatter) chart type with the smooth curve option (Figure 13).

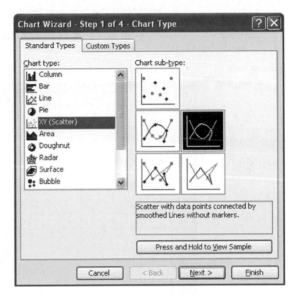

Figure 13

4. Click the Next button to get the Chart Source Data box (Figure 14). Then click Next to get the Chart Options box, and enter your chart title and labels for the *x*- and *y*-axes.

5. Click Next, select whether the graph should be within the current worksheet or on another, and click Finish.

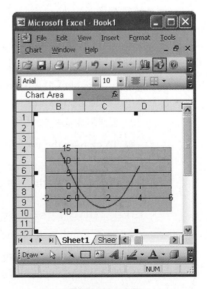

Figure 14

Graphing More Than One Function

1. Create a table with an input column and one column for each function, with multiple function headings.

2. Highlight the columns containing the variable and function values (Figure 15).

3. Proceed with the same steps as those used to graph a single function (Figure 16).

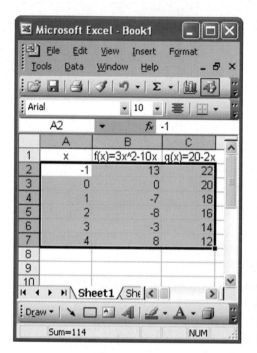

Figure 15

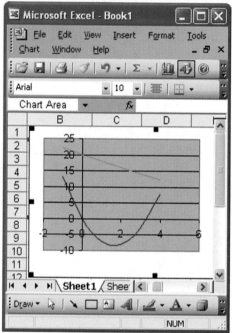

Figure 16

Changing Graphing Windows

1. To change the x-scale:

    a. Double click on the x-axis, then click on the scale tab.

    b. Uncheck the Auto boxes, and change the minimum and maximum values to the desired values.

    c. Click OK.

2. To change the y-scale, double click on the y-axis and proceed as with the x-axis.

3. To delete shading on graphs, double click on the plot area and check "None" for the format of the plot area.

Graphing Discontinuous Functions

An Excel graph will connect all points corresponding to values in the table, so:

1. If the function you are graphing is discontinuous for some x-value a, enter x-values near this value and leave (or make) the corresponding $f(a)$ cell blank.

2. Graph the function using the steps previously described on page 678 (Figure 17).

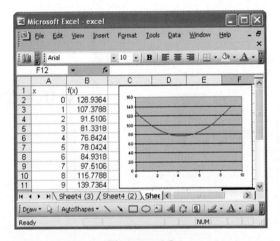

Figure 17

Finding a Minimum or Maximum of a Function with Solver

To Find the Minimum of a Function (if it has a minimum)

1. Enter values for x and for $y = f(x)$, and use them to graph the function over an interval that shows a minimum (Figure 18). (If a minimum exists but does not show, add more points.)

2. Use any value for x in A2 and the same function formula as in Step 1 (Figure 19).

Figure 18   **Figure 19**

3. Invoke Solver by choosing Tools>Solver.

4. Click on Min in the Solver dialog box.

5. Set the Target Cell to B2 by choosing the box and clicking on B2, then set the Changing Cells to A2 with a similar process (Figure 20).

6. Click Solve in the dialog box, and click OK to accept the solution (Figure 21).

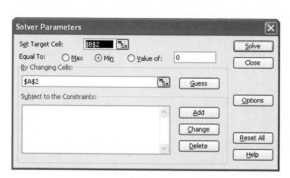

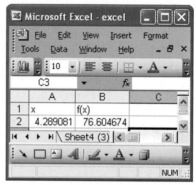

Figure 20   **Figure 21**

To Find the Relative Maximum of a Function (if it has a maximum)

Use the process above except Step 4, which is "Click on Max in the Solver dialog box."

Scatter Plots of Data

To Create a Scatter Plot of Data

1. Enter the inputs (*x*-values) in Column A and the outputs (*y*-values) in Column B.

2. Highlight the two columns and use Chart Wizard, XY(Scatter) and click Next (Figure 22). Click Next to move to Step 3.

3. In Step 3 of the Chart Wizard, enter the title and the *x*- and *y*-axis labels (Figure 23) and click Next.

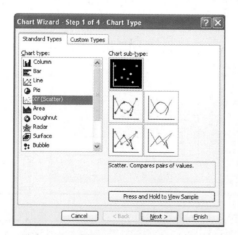

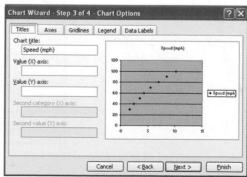

Figure 22   **Figure 23**

4. In Step 4, indicate that the scatter plot should be placed in the current worksheet, and click Finish (Figure 24).

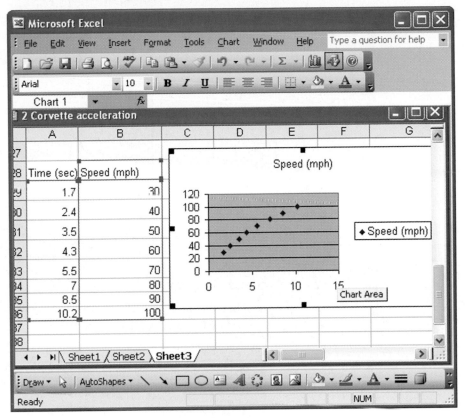

Figure 24

Curve Fitting

To Find the Equations of Lines or Curves That Best Fit a Given Set of Data Points

1. Place the scatter plot of the data in the worksheet, as previously described (page 682) (Figure 25).

2. Single click on the scatter plot in the workbook.

3. From the Chart menu choose Add Trendline (Figure 25).

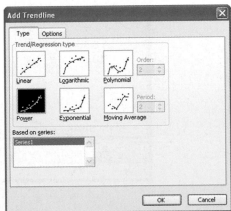

Figure 25

4. Click the regression type that appears to be the best function fit for the scatter plot. If Polynomial is selected, choose the appropriate Order (degree).

5. Click the Options tab and check "Display equation" on chart box (Figure 26).

Figure 26

6. Click OK and you will see the graph of the selected function that is the best fit, along with its equation (Figure 27).

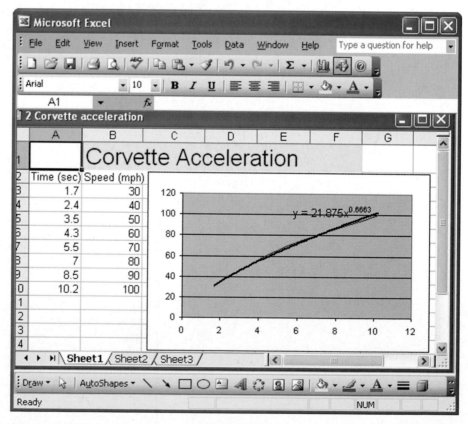

Figure 27

Solving Equations

Solving Linear Equations of the Form $f(x) = 0$ Using the Intercept Method

This solution is also the x-intercept of the graph and the zero of the function.

1. Enter an *x*-value in A2 and the function formula in B2, as previously described on page 678 (Figure 28).

2. Highlight B2. From the Tools menu, select Goal Seek.

3. In the dialog box:

   a. Click the Set cell box and click on the B2 cell.

   b. Enter 0 in the To Value box.

   c. Click the By changing cell box and click on the A2 cell.

4. Click OK to see the Goal Seek Status box (Figure 29).

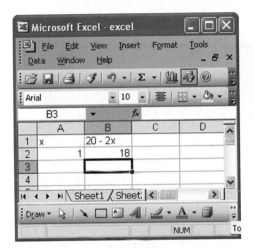

Figure 28

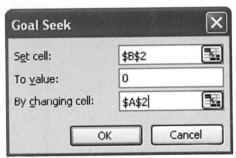

Figure 29

5. Click OK again. The solution is in A2 and 0 is in B2 (Figure 30).

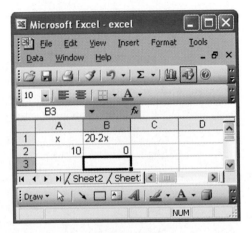

Figure 30

Solving Quadratic Equations of the Form $f(x) = 0$ Using the x-Intercept Method

The solutions are also the x-intercepts of the graph and the zeros of the function. Because the graph of a quadratic function $f(x) = ax^2 + bx + c$ can have two intercepts, it is wise to graph it using an x-interval with the x-coordinate $\left(h = \dfrac{-b}{2a} \right)$ near the center.

To Solve a Quadratic Equation

1. Enter x-values centered around h in Column A and use the function formula to find the values of $f(x)$ in Column B, as previously described on page 677 (Figure 31).

2. Graph the function as previously described (page 678) and observe where the function values are at or near 0.

3. Use Tools> Goal Seek entering a cell address with a function value in Column B at or near 0, and complete the process as previously described on page 685 (Figure 32). The solution may be approximate.

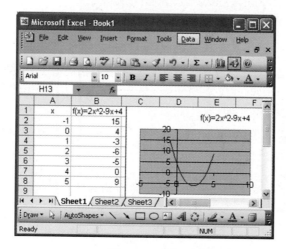

Figure 31

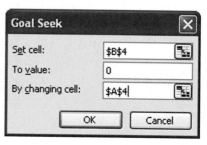

Figure 32

4. One solution is $x = 0.5$ (Figure 33).*

5. After finding the first solution, repeat the process using a second function value at or near 0.

* **Note 1:** The solution $x = 0.50001$, giving $f(x) = -7.287\mathrm{E}{-05}$, is an approximation of the exact solution $x = 0.5$, which gives $f(x) = 0$.

   Note 2: Solutions of other equations (logarithmic, etc) can be found with similar solution methods.

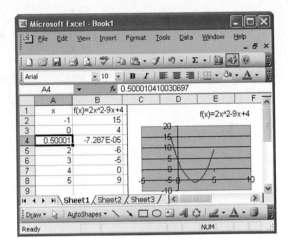

Figure 33

Solving an Equation of the Form $f(x) = g(x)$ Using the Intersection Method

If One Solution Is Sought

1. Enter a value for the input variable (x, for example) in cell A2 and the formula for each of the two sides of the equation in cells B2 and C2, respectively.

2. Enter "=B2−C2" in cell D2 (Figure 34).

3. Select Tools> Goal Seek to find x when B2−C2=0.

4. In the dialog box:

   a. Click the Set cell box and click on the D2 cell.

   b. Enter 0 in the To Value box.

   c. Click the By changing cell box and click on the A2 cell (Figure 35).

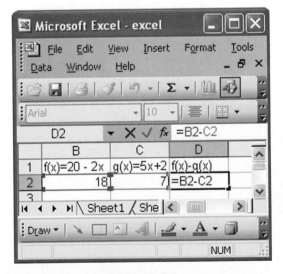

Figure 34

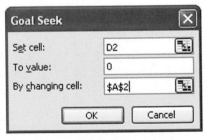

Figure 35

5. Click OK to find the *x*-value of the solution in cell A2 (Figure 36). The solution may be approximate.

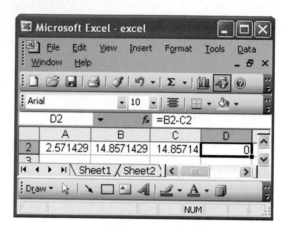

Figure 36

If Multiple Solutions Are Sought (as in the case of a quadratic equation)

1. Proceed as in Steps 1 and 2 above.

2. Enter values for $f(x)$ and for $g(x)$, and use them to graph the function over an interval that shows points of intersection. Input values near those of these points will give Column D values that are at or near 0. Entering *x*-values with $\left(h = \dfrac{-b}{2a}\right)$ near the center is useful when solving quadratic equations.

3. Use Tools> Goal Seek, entering a cell address with a function value in D2 at or near 0, and complete the process as above.

4. After finding the first solution, repeat the process using a second function value at or near 0.

Solving a System of Two Linear Equations in Two Variables

To Solve a System of Two Linear Equations in Two Variables

1. Write the two equations as linear functions in the form $y = mx + b$.

2. Enter a value for the input variable (*x*) in cell A2 and the formula for each of the two equations in cells B2 and C2, respectively (Figure 37).

3. Enter "=B2−C2" in cell D2.

4. Select Tools< Goal Seek to find *x* when B2−C2=0.

5. In the dialog box:

   a. Click the Set cell box and click on the D2 cell.

   b. Enter 0 in the To Value box.

   c. Click the By changing cell box and click on the A2 cell.

6. Click OK in the Goal Seek dialog box, getting 0.

7. The *x*-value of the solution is in cell A2, and the *y*-value is in both B2 and C2 (Figure 38).

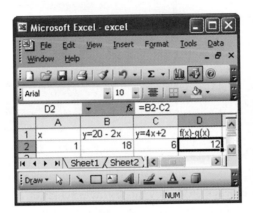

Figure 37

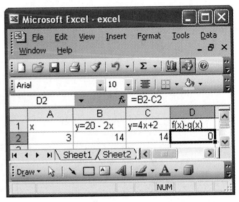

Figure 38

Matrices

Matrices are useful to add, subtract, multiply, and find inverses of matrices. To perform operations with a matrix, enter each of its elements in a cell of a worksheet. Addition and subtraction of matrices are intuitive with Excel, but finding products and inverses requires special commands.

To Add and Subtract Matrices (steps for two 3 × 3 matrices)

1. Type a name A in A1 to identify the first matrix.

2. Enter the matrix elements of matrix A in the cells B1:D3.

3. Type a name B in A5 to identify the second matrix.

4. Enter the matrix elements of matrix B in the cells B5:D7.

5. Type a name A+B in A9 to indicate the matrix sum.

6. Type the formula "=B1+B5" in B9 and press ENTER (Figure 39).

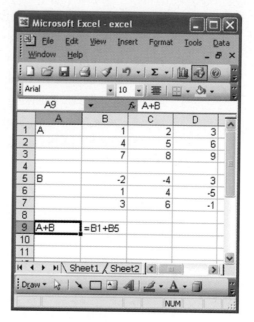

Figure 39

7. Use Fill Across to copy this formula across the row to C9 and D9.

8. Use Fill Down to copy the row B9:D9 to B11:D11, which gives the sum (Figure 40).

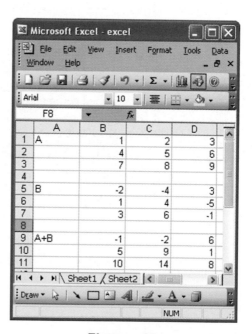

Figure 40

9. To subtract the matrices, change the formula in B9 to "=B1−B5" (Figure 41) and proceed as with addition (Figure 42).

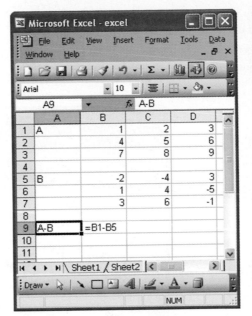

Figure 41

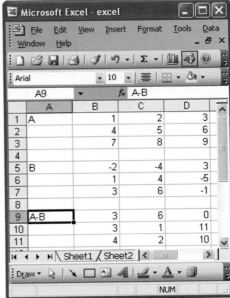

Figure 42

To Find the Product of Two Matrices (steps for two 3 × 3 matrices)

1. Enter the names and elements of the matrices as previously described (page 689).

2. Enter the name A×B in A9 to indicate the matrix product (Figure 43).

3. Select a range of cells that is the correct size to contain the product (B9:D11 in this case).

4. Type "=mmult(" in the formula bar, and then select the cells containing the elements of matrix A (B1:D3) (Figure 44).

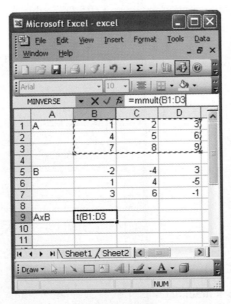

Figure 43

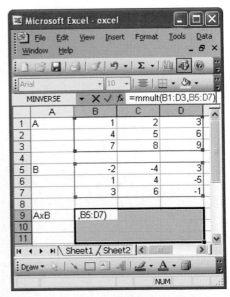

Figure 44

5. Stay in the formula bar, type a comma and select the matrix B elements (B5:D7), and close the parentheses.

6. Press the CTRL, SHIFT, and ENTER keys *all at the same time*, giving the product (Figure 45).

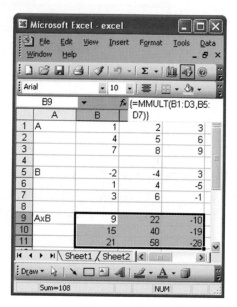

Figure 45

To Find the Inverse of a Matrix (steps for a 3 × 3 matrix)

1. Enter the name A in A1 and the elements of the matrix in B1:D3 as previously described (page 689).

2. Enter the name "Inverse(A)" in A5 and select a range of cells that is the correct size to contain the inverse (B5:D7 in this case) (Figure 46).

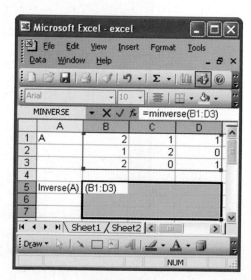

Figure 46

3. In the formula bar, enter "=minverse(", then select matrix A (B1:D3), and close the parentheses.

4. Press the CTRL, SHIFT, and ENTER keys *all at the same time*, getting the inverse (Figure 47).

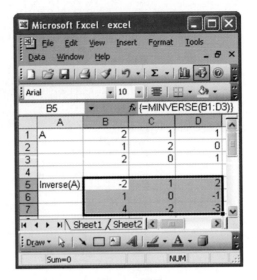

Figure 47

Solving Systems of Linear Equations; Matrix Inverses

A system of linear equations can be solved by multiplying the matrix containing the augment by the inverse of the coefficient matrix. The steps used to solve

$$\begin{cases} 2x + y + z = 8 \\ x + 2y \quad\quad = 6 \text{ follow.} \\ 2x \quad\quad + z = 5 \end{cases}$$

1. Enter the coefficient matrix A in B1:D3.

2. Compute the inverse of A in B5:D7, as previously described on page 692 (Figure 48).

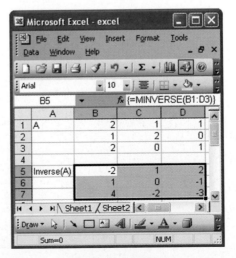

Figure 48

3. Enter B in cell A9 and enter the augment matrix in B9:B11.

4. Enter X in A13 and select the cells B13:B15.

5. In the formula bar, type "=mmult(", then select matrix inverse(A) in B5:D7, type a comma, select matrix B in B9:B11, and close the parentheses (Figure 49).

6. Press the CTRL, SHIFT, and ENTER keys *all at the same time*, getting the solution.

7. Matrix X gives the solution $x = 0, y = 3, z = 5$ (Figure 50).

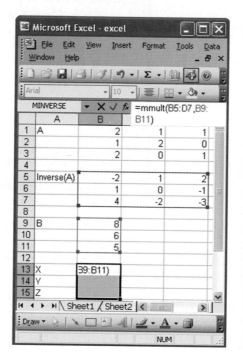

Figure 49

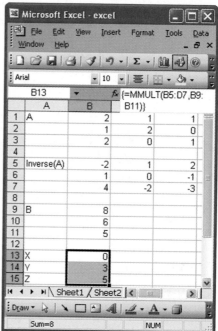

Figure 50

answers to selected exercises

Chapter 1 Functions, Graphs, and Models; Linear Functions

Toolbox Exercises

1. $\{1, 2, 3, 4, 5, 6, 7, 8\}$; $\{x : x$ is a natural number, $x < 9\}$ **2.** Yes
3. No   **4.** No   **5.** Yes   **6.** Yes   **7.** Integers, rational numbers
8. Rational   **9.** Irrational   **10.** $x > -3$ **11.** $-3 \le x \le 3$
12. $x \le 3$ **13.** $(-\infty, 7]$ **14.** $(3, 7]$,   **15.** $(-\infty, 4)$
16.   **17.**

18.   **19.**

20.   **21.**

22.   **23.**

24. $-3x^2$, -3; $-4x$, -4; constant 8
25. $5x^4$, 5; $7x^3$, 7; constant -3
26. $3z^4 + 4z^3 - 27z^2 + 20z - 11$
27. $7y^4 - 2x^3y^4 - 3y^2 - 2x - 9$ **28.** $4p + 4d$
29. $-6x + 14y$ **30.** $-ab - 8ac$ **31.** $x - 6y$
32. $6x + 9xy - 5y$ **33.** $3xyz - 5x$

1.1 Skills Check

1. a. Input   **b.** Output   **c.** D: $\{-9, -5, -7, 6, 12, 17, 20\}$;
$R = \{5, 6, 7, 4, 9, 10\}$ **d.** Each input gives exactly one output.
3. $5; 9$ **5.** No; input 9 for y gives two outputs for x.
7. a. $f(2) = -1$ **b.** $f(2) = 10 - 3(2^2) = -2$
c. $f(2) = -3$

9.

x	y
0	2
-2	-4

11. a. -7 **b.** 3   **c.** 18

13. Yes; $D = \{-1, 0, 1, 2, 3\}$; $R = \{5, 7, 2, -1, -8\}$
15. No   **17.** Yes   **19.** No; one value of x, 3, gives two
values of y.   **21.** Set b.   **23.** b
25. D: $\{-3, -2, -1, 1, 3, 4\}$; R: $\{-8, -4, 2, 4, 6\}$
27. D:$[-10, 8]$; R:$[-12, 2]$ **29.** $x \ge 2$ **31.** All real numbers
except -4 **33.** No   **35.** $C = 2\pi r$

1.1 Exercises

37. a. No.   **b.** Yes; x, p; there is one closing price p on any given day x.
39. a. Yes; a, p; there is one premium p for each age a.   **b.** No
41. Yes; y, I; there is one average income I for each input y
43. a. Yes; there is one price for each barcode.   **b.** No; numerous
items with the same price have different barcodes.   **45.** Yes; there
is exactly one output for each input.   **47. a.** Yes   **b.** The days
1–14 of May   **c.** $\{171, 172, 173, 174, 175, 176, 177, 178\}$
d. May 1, May 3   **e.** May 14   **f.** 3 days   **49. a.** 1096.78;
if the car is financed over 3 years, the payment is $1096.78
b. $42,580.80; $C(5) = 42,580.80$ **c.** 4   **d.** $3096.72
51. a. 3 to 1   **b.** 4; 4 is the projected ratio of the working-age
population (25- to 64-year-olds) to the elderly in 2005.
c. $\{1995, 2000, 2005, 2010, 2015, 2020, 2025, 2030\}$
d. Decreasing   **53. a.** 419,640   **b.** There were 419,640 nonfatal
firearm incidents in 2005.   **c.** 1,060,800; in 1994   **55. a.** 41.5 million
b. 52.4; 52.4 million U.S. homes used the Internet in 2003.   **c.** 1998
d. Increasing, it has increased rapidly.   **57. a.** 3.4; in 1990, there were
3.4 workers for each retiree.   **b.** 2030   **c.** Funding is in jeopardy.
59. a. 6400, the revenue from the sale of 200 hats is $6400.
b. $80,000; $R(2500) = 80,000$ **61. a.** 783.37; the charge for
1000 KWh is $783.37   **b.** $1166.87   **63. a.** 1200; the daily
profit for producing 100 bicycles is $1200.   **b.** $1560
65. a. Yes   **b.** $s \ge -1/4$ **c.** $s \ge 0$
67. a. $0 \le p < 100$ **b.** 355,500; 2,133,000

69. a. 8640; 11,664 **b.** $0 < x < 27$, so volume is positive and box exists **c.** Testing values in the table shows a maximum volume of 11,664 cubic inches when $x = 18$ inches.

x	y
10	6,800
15	10,800
20	11,200
21	10,584
19	11,552
18	11,664
17	11,560

The dimensions of the box are 18 in. by 18 in. by 36 in.

1.2 Skills Check

1. a.

x	-3	-2	-1	0	1	2	3
$y = x^3$	-27	-8	-1	0	1	8	27

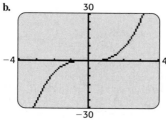

b.

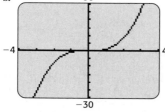

c. Graphs are the same.

3. Answers vary.

x	$f(x)$
-2	-7
-1	-4
0	-1
1	2
2	5

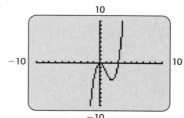

5. Answers vary.

x	$f(x)$
-4	8
-2	2
0	0
2	2
4	8

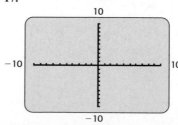

7. Answers vary.

x	$f(x)$
-2	$-1/4$
-1	$-1/3$
0	$-1/2$
1	-1
2	undefined
3	1
4	1/2

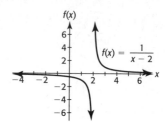

9.

11.

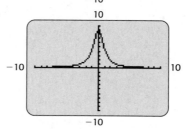

13.

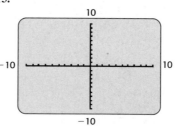

15.

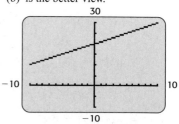

(b) is the better view.

17.

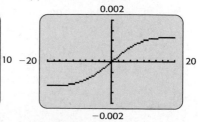

(b) is the better view.

19. Letting y vary from 0 to 180 gives one view.

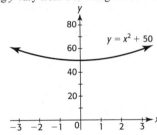

21. $y = x^3 + 3x^2 - 45x$ for values of x between -10 and 10.

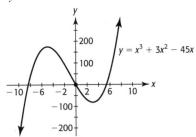

23.

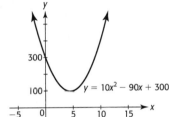

25.

t	S
12	51.9
16	72.7
28	135.1
43	213.1

27.

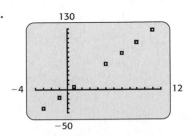

29. a.

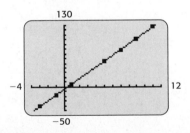

b.

c. Yes; yes

31. a. 300 **b.** 2020

1.2 Exercises

33. a. 10,963.87; in 1944 there were 10,963.87 thousand (10,963,870) women in the workforce **b.** 68,180,590 **35. a.** 1; 19 **b.** 2005; 65.8 **c.** 0; 20

37. a.

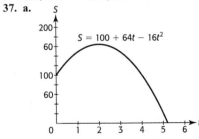

b. 148 ft and 148 ft; ball rising and falling **c.** 164 ft, in 2 seconds

39. a.

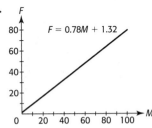

b. 50,460

41. a.

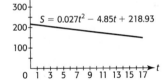

b. 152.255 **c.** 114,555

43. a.

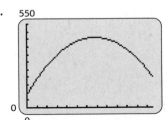

b. 1980 through 2005 **c.** decrease **d.** 461 in 1994 and 202 in 2005; decreasing **e.** yes

45.

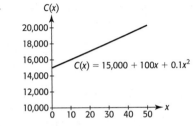

47.

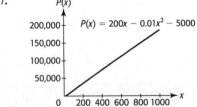

49. a. 0 through 15   **b.** 32,903.77; 47,634.67
c.

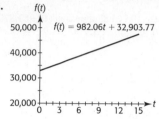

51. a. 299.9 million or 299,900,000
b.

Years After 2000	Population (millions)
0	275.3
10	299.9
20	324.9
30	351.1
40	377.4
50	403.7
60	432.0

c.

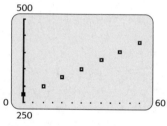

53. a.

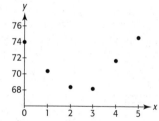

b.

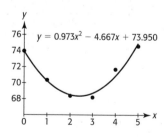

55. a. 3.5%   **b.**

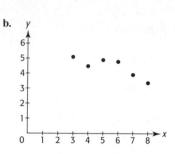

c.

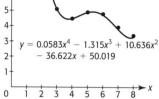

1.3 Skills Check

1. b   **3.** $-\dfrac{1}{2}$ **5.** 2

7. a. x: 3; y: -5 **b.**

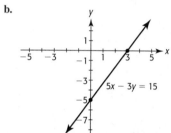

9. a. x: $\dfrac{3}{2}$; y: 3   **b.**

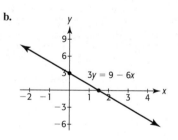

11. 0; undefined   **13. a.** Positive   **b.** Undefined

15. a. slope 4; y-intercept 8   **b.**

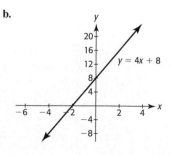

17. a. slope 0; y-intercept $\dfrac{2}{5}$ **b.**

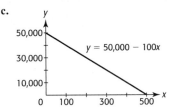

19. a. $4, 5$ **b.** Rising   **c.**

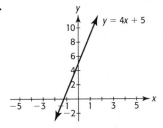

21. a. $-100, 50{,}000$ **b.** Falling

c.

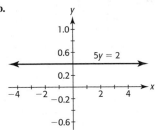

23. 4 **25.** -15 **27.** -2 **29. a.** ii   **b.** i

31. a. 0 **b.** 0

1.3 Exercises

33. Yes; it is written in the form $y = ax + b$.   **35. a.** It is written in the form $y = ax + b$ **b.** -0.762; the percent of unmarried women who get married has decreased at a rate of 0.762 per year since 1950.

37. a. $x = -\dfrac{30}{19}$ **b.** $p = 1$; 1% of high school seniors were using marijuana daily in 1990.   **c.** Integer values of $x \geq 0$ on the graph represent years 1990 and after.

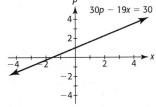

39. a. Constant function   **b.** $y = 11.81$ **c.** 0 **d.** 0
41. a. Negative   **b.** The percent is decreasing at a rate of 0.743 percent per year during this period.   **43. a.** Positive   **b.** As temperature increases, the number of chirps increases.   **45. a.** Yes   **b.** 0.959
c. Minority median annual salaries will increase by 0.959 for each $1 increase in median annual salaries for whites.
47. a. $\dfrac{19}{30}$ per year   **b.** $\dfrac{19}{30} \approx 0.6333\%$ per year

49. x: 50; R: 3500

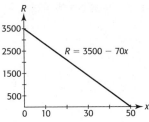

51. a. $m = 6.9; b = -3.18$ **b.** Percent with Internet access increased by 6.9 each year.   **53. a.** $-61{,}000$
b. $-\$61{,}000$ per year   **55.** $58 per unit   **57. a.** 0.56
b. $0.56 per ball   **c.** The cost will increase by $0.56 for each additional ball produced in a month.   **59. a.** 1.60 **b.** $1.60 per ball
c. The revenue will increase by $1.60 for each additional ball sold in a month.   **61.** $19 per unit

1.4 Skills Check

1. $y = 4x + \dfrac{1}{2}$ **3.** $y = \dfrac{1}{3}x + 3$ **5.** $y = \dfrac{-3}{4}x - 3$

7. $x = 9$ **9.** $y = x + 3$ **11.** $y = 2$ **13.** $y = \dfrac{4}{5}x + 4$

15. $y = -3x + 6$ **17.** $y = \dfrac{3}{2}x + \dfrac{23}{2}$ **19.** $y = 3x + 1$

21. $y = -15x + 12$ **23.** 1 **25.** -3 **27.** -15
29. $4x + 2h$ **31. a.** Yes; a scatterplot shows that data fit along a line.   **b.** $y = 3x + 555$

1.4 Exercises

33. $y = 8.95 + 0.0935x$ (dollars)   **35.** $y = -3600t + 36{,}000$
37. a. $P = 705x + 198$ **b.** 4428 **39. a.** $25{,}000$ **b.** $5000
c. $s = 26{,}000 - 5000t$ **41.** $y = x; x = $ deputies; $y = $ cars
43. $P(x) = 58x - 12{,}750; x = $ number of units
45. $V = -61{,}000\,x + 1{,}920{,}000$ **47.** $P = -1.64t + 73.34$
49. a. 0.05 **b.** $f(x) = 0.05x$ **51. a.** $y = 0.504x - 934.46$
b. Yes   **c.** They are the same.   **53. a.** 3.42 **b.** 3.42
c. No   **55. a.** $\dfrac{61}{36} \approx 1.69$ **b.** 1.69 **c.** The percent is increasing at a rate of approximately 1.69 per year.   **d.** $p = 1.69x - 1.74$
57. a. 28,886 per year   **b.** $28{,}886$ **c.** $y = 28{,}886x - 56{,}403{,}607$
d. No   **e.** 1997 and 2005   **59. a.** No   **b.** Yes, they appear to lie on a line.   **c.** -0.085 **d.** $y = -0.085x + 174.75$
61. a. Increasing at a rate of approximately 2621 thousand people, or approximately 2.621 million people per year   **b.** $y = 2621x + 152{,}271$
c. 217,796 thousand, or 217,796,000; it is close   **d.** The line does not model the data exactly.

Chapter 1 Skills Check

1. Each value of x is assigned exactly one value of y.
2. Domain: $\{-3, -1, 1, 3, 5, 7, 9, 11, 13\}$
Range: $\{9, 6, 3, 0, -3, -6, -9, -12, -15\}$
3. $f(3) = 0$ **4.** Yes; $y = -\dfrac{3}{2}x + \dfrac{9}{2}$ **5. a.** -2 **b.** 8
c. 14 **6. a.** 1 **b.** -10

7.

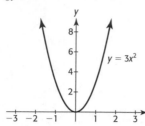

$f(x) = -2x^3 + 5x$

8.

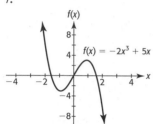

$y = 3x^2$

27.

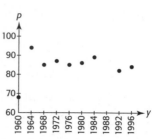

28. a. −0.0357
b. −0.0357
c. No   **d.** No

9. standard window

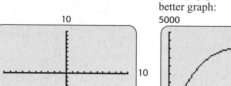

[0, 40] by [0, 5000] gives a better graph:

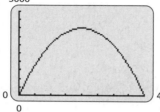

10.

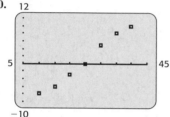

11. a. $x \geq 4$ **b.** All real numbers except 6   **12.** Perpendicular

13. $-\dfrac{7}{6}; \dfrac{6}{7}$ **14.** $-\dfrac{11}{6}$

15. a. $(0, -4)$ and $(6, 0)$ **b.**

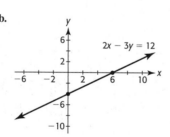

$2x - 3y = 12$

16. $\dfrac{2}{3}$ **17.** Slope −6, y-intercept 3   **18.** −6

19. $y = \dfrac{1}{3}x + 3$ **20.** $y = -\dfrac{3}{4}x - 3$ **21.** $y = x + 4$ **22.** 3

23. a. $5 - 4x - 4h$ **b.** $-4h$ **c.** −4
24. a. $10x + 10h - 50$ **b.** $10h$ **c.** 10

Chapter 1 Review

25. a. Yes   **b.** $f(1992) = 82$; in 1992, 82% of African American voters supported a Democratic candidate for president.   **c.** 1964; in 1964, 94% of African American voters supported a Democratic candidate for president.   **26. a.** {1960, 1964, 1968, 1972, 1976, 1980, 1984, 1992, 1996}   **b.** No; 1982 was not a presidential election year.   **c.** Discrete

29. a. As each input value changes by 5000, the output value changes by 89.62.   **b.** 448.11; the monthly payment to borrow $25,000 is $448.11.   **c.** $A = 20,000$ **30. a.** D: {10,000, 15,000, 20,000, 25,000, 30,000}; R: {179.25, 268.87, 358.49, 448.11, 537.73}
b. No   **c.** Discrete   **31. a.** $f(28,000) = 501.882$; If $28,000 is borrowed, the payment is $501.88.   **b.** Yes
32. a. $f(1960) = 15.9$; in 1960 the average woman was expected to live 15.9 years past the age of 65, or 80.9 years.
b. $65 + 19.4 = 84.4$ years   **c.** 1990   **33. a.** 16.9; a 65-year-old man in 2020 is expected to live 16.9 more years, or 81.9 years
b. 77.8 years   **c.** $f(1990) = 15$ **34. a.** $42,724.37;
$f(10) = 42,724.37$ **b.** $f(15) = 47,634.67$; the average is $47,634.67 in 2005.   **c.** Increasing

35. a.

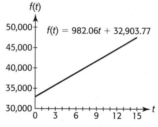

$f(t) = 982.06t + 32,903.77$

b. Years from 1990 to 2005

36. a. $m = 0.357$ **b.** Average rate of change = 0.357 million users per year   **37.** $f(x) = 4500$ **38. a.** $f(x) = 33.8$
b. It is a constant function.   **39. a.** $67,680   **b.** $47,680
c. $\overline{MR}$: 564; $\overline{MC}$: 64   **d.** $m = 64$
e.

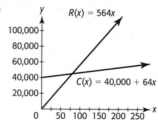

$R(x) = 564x$

$C(x) = 40,000 + 64x$

40. a. $P(x) = 500x - 40,000$ **b.** $20,000   **c.** 80   **d.** 500
e. Marginal revenue minus marginal cost   **41. a.** $y = 300,000$; the initial value of the property is $300,000.   **b.** $x = 100$; the value of the property after 100 years is zero dollars
42. a. Average rate of change = $4.4 per unit   **b.** Slope = 4.4
c. $P(x) = 4.4x - 205$ **d.** $4.4 per unit   **e.** Approximately 47

Chapter 2 Linear Models, Equations, and Inequalities

Toolbox Exercises

1. Division property; $x = 2$ **2.** Addition property; $x = 18$
3. Subtraction property; $x = 5$ **4.** Addition property; $x = 3$
5. Multiplication property; $x = 18$ **6.** Division property; $x = -2$

7. Subtraction property and division property; $x = -10$
8. Addition property and multiplication property; $x = 32$ **9.** $x = 3$
10. $x = \dfrac{3}{5}$ **11.** $x = 16$ **12.** $x = -4$ **13.** $x = 2$
14. $x = 1$ **15.** $x = \dfrac{1}{2}$ **16.** $x = 1$ **17.** $x = 4$
18. $x = 5$ **19.** $x = 3$ **20.** $x = 5$ **21.** Conditional
22. Contradiction   **23.** Identity   **24.** Conditional   **25.** $x > -\dfrac{6}{5}$
26. $x \le -2$ **27.** $x > -12$ **28.** $x > -12$ **29.** $x < \dfrac{-8}{19}$
30. $x < \dfrac{6}{11}$ **31.** $x > \dfrac{19}{3}$ **32.** $x > -16$

2.1 Skills Check

1. $x = -\dfrac{37}{2}$ **3.** $x = 10$ **5.** $x = -\dfrac{13}{24}$ **7.** $x = -63$
9. $t = -1$ **11.** $x = \dfrac{25}{36}$ **13. a.** -20 **b.** -20 **c.** -20
15. a. 4   **b.** 4   **c.** 4   **17. a.** 2   **b.** -34 **c.** 2
19. a. 40   **b.** 40   **21. a.** 25   **b.** 25   **c.** 25
23. a. -8.25 **b.** -8.25 **c.** -8.25 **25.** $x = 3$
27. $s = -5$ **29.** $t = -4$ **31.** $t = \dfrac{17}{4}$ **33. a.** $t = \dfrac{A - P}{Pr}$
b. $P = \dfrac{A}{1 + rt}$ **35.** $F = \dfrac{9}{5}C + 32$ **37.** $n = \dfrac{5m}{2} - \dfrac{P}{4} - \dfrac{A}{2}$
39. $y = \dfrac{5x - 5}{3}$

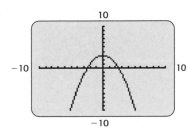

41. $y = \dfrac{6 - x^2}{2}$

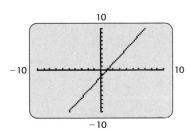

2.1 Exercises

43. 60 months, or 5 years   **45.** \$6000   **47.** \$54,000
49. $-40°$ **51.** 26 years from 1970, or 1996   **53. a.** 600
b. 200   **55.** $x = 50$, so 2010   **57.** $x = 22$, so 2012
59. a. 2004   **b.** No, gives 50.2%. Changing conditions can make models inaccurate.   **61.** 1997   **63.** 97   **65.** \$78.723 billion
67. 25%   **69.** \$1698   **71.** $t = \dfrac{A - P}{Pr}$ **73.** 10%   **75.** \$613.33

2.2 Skills Check

1. Should not be modeled by a linear function; data points do not lie close to a line.   **3.** Approximately, because all the data points do not lie on the line, but are close to the line.

5.

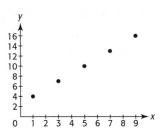

7. Exactly; first differences of the outputs are equal.
9. $y = 1.5x + 2.5$
11.

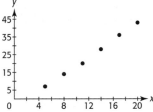

13. $y = 2.419x - 5.571$

15.

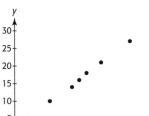

17. $y = 1.577x + 1.892$
19. $y = -1.5x + 8$
21. a. Exactly
b. Nonlinear
c. Approximately

2.2 Exercises

23. a. Discrete   **b.** No; the graph of the data has some curvature.
c. Yes   **25. a.** Yes; the first differences are constant for uniform inputs.   **b.** $S = 60t + 1000$ **c.** \$1420; extrapolation
d. Discretely, because the payments occur only at the end of the years.
27. a. Yes; the first differences are constant for uniform inputs.
b. $T = 0.15x - 782.50$ **c.** They agree   **d.** \$3736.25; interpolation   **e.** No, just for income in the 15% tax bracket.
29. a. 17,000

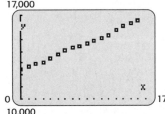

b. $y = 253.267x + 12,409.686$
c. Yes

31. a. $y = 17.906x + 857.886$ **b.** 1019.04 million (1.01904 billion)
c. 2006   **33. a.** $y = 0.316x + 25.860$ **b.** Not close to an exact fit
35. a.

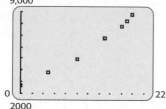

c. $y = 299.635x + 2409.236$ **d.** $9300.84 billion

37. a. $f(x) = 2.920x + 265.864$ **b.** 455.7, projected population
is 455.7 million in 2065.   **c.** 499.5 million; fairly close

39. a.   **b.** $y = 37.7x + 221.333$

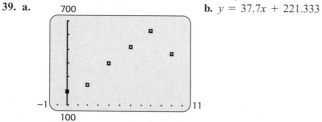

c. A line will not fit these data well.

41. a. $P = 17W + 25$ **b.** Fits exactly   **c.** 127 cents, or $1.27.
d. Discretely   **43. a.** $y = -0.720x + 37.644$ **b.** 22.5
c. 2038
45. a.   **b.** $y = 6.888x + 3.180$

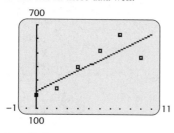

c. Good fit

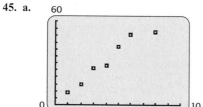

2.3 Skills Check

1. $(1, 1)$ **3.** $x = -14, y = -54$ **5.** $x = \dfrac{-4}{7}, y = \dfrac{4}{7}$

7. Many solutions; the two equations have the same graph.
9. $x = 2, y = -2$ **11.** $x = 1, y = -1$

13. $x = 2, y = 1$ **15.** $x = \dfrac{4}{7}, y = \dfrac{12}{7}$ **17.** $x = 4; y = 3$

19. Dependent; many solutions   **21.** No solution

23. $x = 2, y = 4$ **25.** $x = \dfrac{32}{11}, y = \dfrac{-14}{11}$ **27.** $x = 3, y = -2$

29. $x = 2, y = 1$ **31.** No Solution

2.3 Exercises

33. 60 units   **35.** 104   **37. a.** 8 supplied, 17 demanded
b. $80; 12 units   **39.** $3   **41.** Solution is (18.715, 53.362);
in 1979   **43.** $44,701   **45.** $699.8 million in 2008, $805.3 million
in 2009   **47. a.** $x + y = 2400$ **b.** $30x$ **c.** $45y$
d. $30x + 45y = 84,000$ **e.** 1600 $30 tickets and 800 $45 tickets
49. a. $75,000 at 8% and $25,000 at 12%   **b.** 12% account is
probably more risky; might lose money.   **51.** $200,000 at 6.6%,
$50,000 at 8.6%   **53.** 60 cc of 10% solution, 40 cc of 5% solution
55. 3 glasses of milk and 2 servings of meat   **57.** 8 cc of the 5% solution

and 12 cc of the 10% solution   **59. a.** $p = -\dfrac{1}{2}q + 155$

b. $p = \dfrac{1}{4}q + 50$ **c.** $85   **61.** No solution

63. a. $300x + 200y = 100,000$ **b.** 250 in the first group and
125 in the second group   **65.** 700 units at $30. When the price is
$30, the amount demanded equals the amount supplied equals 700.

2.4 Skills Check

1. $x \le 3$;

3. $x \le \dfrac{-1}{7}$;

5. $x < \dfrac{20}{23}$;

7. $x < \dfrac{61}{5}$;

9. $x \ge \dfrac{-30}{11}$;

11. $x \ge 2.8\overline{6}$;

13. $(-\infty, -2)$ **15.** $[-8, \infty)$ **17. a.** $x = -1$ **b.** $(-\infty, -1)$

19. $\dfrac{22}{3} \le x < 12$ **21.** $\dfrac{5}{2} \le x \le 10$ **23.** no solution

25. $x \ge \dfrac{32}{11}$ or $x \le \dfrac{3}{4}$ **27.** $72 \le x \le 146$

2.4 Exercises

29. a. $V = 12,000 - 2000t$ **b.** $12,000 - 2000t < 8000$
c. $12,000 - 2000t \geq 6000$ **31.** $C \leq 0$ (degrees)
33. More than $22,000   **35.** 80 through 100   **37.** $3.83 \leq p \leq 6.37$
39. a. $0.97x + 128.3829 \geq 1000; x \geq 899$
b.

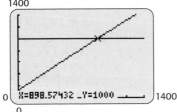

41. $1350 \leq x \leq 1650$ **43. a.** Before 1996   **b.** After 2002
45. a. $y = 20,000x + 190,000$ **b.** $11 \leq x \leq 14$
47. $x > 2000$ **49.** $x \geq 1500$ **51.** $364x - 345,000 > 0$
53. $245 < 0.155x + 244.37 < 248$; between 1974 and 1993
55. a. About 62.4%   **b.** $x \leq 22$ **c.** Before 2012

Chapter 2 Skills Check

1. $x = \dfrac{34}{5}$ **2.** $x = -14.5$ **3.** $x = -138$ **4.** $x = \dfrac{5}{28}$

5. $x = \dfrac{52}{51}$ **6.** $x = 5$ **7. a.** 15   **b.** 15   **c.** 15

8. $y = \dfrac{3 + m - 3Pa}{-3P}$

9. $y = \dfrac{4x - 6}{3}$

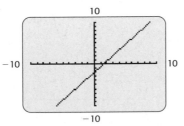

10.

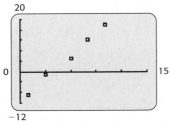

11. $y = 2.8947x - 11.211$

12.

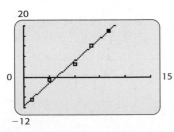

13. No   **14.** $x = 2, y = -3$ **15.** $x = \dfrac{-3}{13}, y = \dfrac{-15}{13}$

16. Many solutions   **17.** No solution   **18.** $x = -3, y = 5$

19. $x = \dfrac{1}{2}, y = -4$ **20.** $x < -\dfrac{4}{5}$ **21.** $x \leq \dfrac{25}{28}$

22. $6 \leq x < 18$

Chapter 2 Review

23. a. Yes   **b.** Yes   **24. a.** $P = f(A) = 0.018A + 0.010$
b. 501.882; the predicted monthly payment on a car loan of $28,000
is $501.88.   **c.** Yes   **d.** $27,895   **25.** 1998
26. $f(x) = 4500$ **27. a.** $f(x) = 34.6$ **b.** Yes, it is a constant
function.   **28. a.** $22,000   **b.** More than $22,000 per month
29. $25,440 each   **30.** $300,000 in the safe account and $120,000
in the risky account   **31.** 1994   **32.** More than 120
33. a. $P(x) = 500x - 40,000$ **b.** $x > 80$ **c.** More than 80
34. a. $x > 10$ **b.** 10 years after its purchase
35. a. $P(x) = 4.4x - 205$ **b.** At least 47 units
36. a. $y = 0.0638x + 15.702$
b.

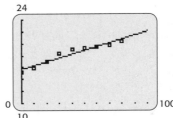

c. 22; the projected life span of a woman is 87 in 2049.
d. 2002 and after
37. a. $y = 0.0655x + 12.324$
b.

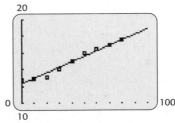

c. 20.8; the average man is expected to live to age 86 in 2080.
d. 2144   **e.** Before 2007   **38. a.** A linear equation is reasonable.
b. $y = 3.317x + 3.254$ **c.** $43.1 billion   **39. a.** A linear
equation is a reasonable model.   **b.** $y = 285.269x + 9875.17$,
where x is the number of years past 1980 and y is population in
thousands   **c.** 16,151 thousand
40. a. A linear equation is a reasonable model.   **b.** $y = 0.301x + 0.477$ **c.** The line seems to fit the data points very well.

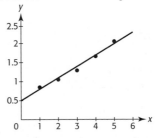

41. a. Before 1998   **b.** 2008 and after   **42.** 1995 through 2001

43. Between 17 and 31 months   **44.** $100,000 at 8% and $100,000 at 12%   **45.** 25   **46.** Each dosage of medication A contains 2.8 mg, and each dosage of medication B contains 4.2 mg.
47. $p = \$100$; $q = 80$ pairs   **48.** $p = \$79$; $q = 710$ units
49. a. $x + y = 2600$ **b.** $40x$ **c.** $60y$ **d.** $40x + 60y = 120{,}000$ **e.** 1800 $40 tickets and 800 $60 tickets.
50. a. $x + y = 500{,}000$ **b.** $0.12x$ **c.** $0.15y$
d. $0.12x + 0.15y = 64{,}500$ **e.** $350,000 in the 12% property and $150,000 in the 15% property

Chapter 3 Quadratic and Other Nonlinear Functions

Toolbox Exercises

1. $\dfrac{9}{4}$ **2.** $\dfrac{8}{27}$ **3.** $\dfrac{1}{100}$ **4.** $\dfrac{1}{64}$ **5.** $\dfrac{1}{8}$ **6.** $\dfrac{1}{256}$

7. 6   **8.** 4   **9. a.** $x^{3/2}$ **b.** $x^{3/4}$ **c.** $x^{3/5}$

d. $3^{1/2}y^{3/2}$ **e.** $27y^{3/2}$ **10. a.** $\sqrt[4]{a^3}$

b. $-15\sqrt[8]{x^5}$ **c.** $\sqrt[8]{(-15x)^5}$ **11.** $-12a^2x^5y^3$

12. $4x^3y^4 + 8x^2y^3z - 6xy^3z^2$ **13.** $2x^2 - 11x - 21$

14. $k^2 - 6k + 9$ **15.** $16x^2 - 49y^2$ **16.** $3x(x - 4)$

17. $12x^3(x^2 - 2)$ **18.** $(3x - 5m)(3x + 5m)$

19. $(x - 3)(x - 5)$ **20.** $(x - 7)(x + 5)$

21. $(x - 2)(3x + 1)$ **22.** $(2x - 5)(4x - 1)$

23. $3(2n + 1)(n + 6)$ **24.** $3(y - 2)(y + 2)(y^2 + 3)$

25. $(3p + 2)(6p - 1)$ **26.** $(5x - 3)(x - 2y)$

27. a. Imaginary   **b.** Pure imaginary   **c.** Real   **d.** Real
28. a. Imaginary   **b.** Real   **c.** Pure imaginary   **d.** Imaginary
29. $a = 4, b = 0$ **30.** $a = 15, b = -3$ **31.** $a = 2, b = 4$

3.1 Skills Check

1. a. Quadratic   **b.** Up   **c.** Minimum   **3.** Not quadratic
5. a. Quadratic   **b.** Down   **c.** Maximum
7. a.

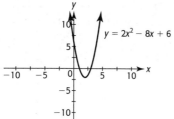

b. Yes

9. a.

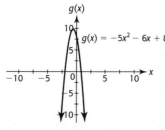

b. Yes

11. a.

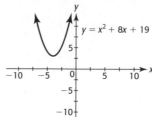

b. Yes

13. a.

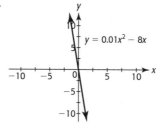

b. No; the complete graph will be a parabola

15. $y = (x - 2)^2 - 4$ **17.** y_1 **19.** $y = -5x^2 + 10x + 8$

21. a. $(1, 3)$ **b.**

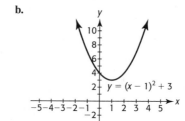

23. a. $(-8, 8)$ **b.**

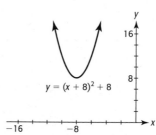

25. a. $(4, -6)$ **b.**

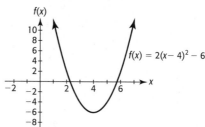

27. a. $(2, 12)$ **b.**

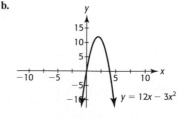

29. a. $(-3, -30)$ **b.**

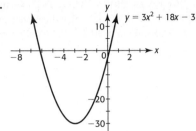

$y = 3x^2 + 18x - 3$

31. a. $x = 10$ **b.**

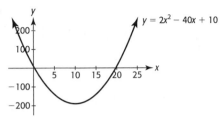

$y = 2x^2 - 40x + 10$

33. a. $x = -80$ **b.**

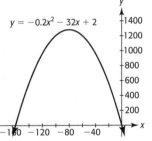

$y = -0.2x^2 - 32x + 2$

35.

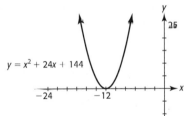

$y = x^2 + 24x + 144$

37.

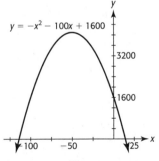

$y = -x^2 - 100x + 1600$

39.

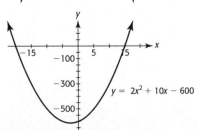

$y = 2x^2 + 10x - 600$

41. $x = 1, x = 3$ **43.** $x = -10, x = 11$ **45.** $x = -2, x = 0.8$

3.1 Exercises

47. a.

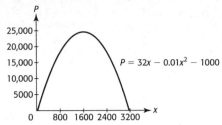

$P = 32x - 0.01x^2 - 1000$

b. It increases.   **c.** Profit decreases.

49. a.

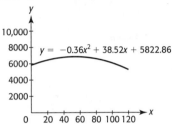

$y = -0.36x^2 + 38.52x + 5822.86$

b. 6449.3 million, or 6.4493 billion

51. a.

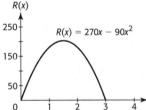

$y = 6.75x^2 - 122.94x + 1009.44$

b. \$1,251 billion   **c.** Extrapolation

53. a. A parabola opening down   **b.** $t = 3, S = 244$
c. The ball reaches its maximum height of 244 feet in 3 seconds.

55. a. $R(x)$

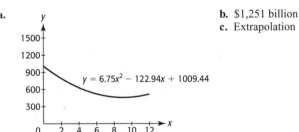

$R(x) = 270x - 90x^2$

b. 1.5 lumens

57. a. 2000 units   **b.** \$37,000   **59. a.** 37,500   **b.** \$28,125,000
61. a. Yes   **b.** $A = 100x - x^2$; maximum area of 2500 sq ft
when x is 50 ft   **63. a.** $(9.38, 29.51)$ **b.** 2000   **c.** 29.5%
65. a. Minimum   **b.** $(63.6, 6.5)$ **c.** The minimum percent of
foreign-born residents was 6.5% in 1964.
d.

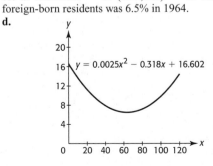

$y = 0.0025x^2 - 0.318x + 16.602$

67. a.

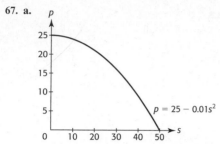

$p = 25 - 0.01s^2$

b. Decreasing
c. 25   **d.** When wind speed is 0 mph, amount of pollution is 25 oz per cubic yard.

69. The t-intercepts are 3.5 and -3.75. This means that the ball will strike the pool in 3.5 seconds; -3.75 is meaningless.

71. a. $y = -16t^2 + 32t + 3$ **b.** 19 feet

73. a.

Rent ($)	Number of Apartments	Revenue ($)
1200	100	120,000
1240	98	121,520
1280	96	122,880
1320	94	124,080

b. Yes
c. $1600

75. a. $x = 53.5, y = 6853.3$ **b.** World population will be maximized at 6,853,300,000 in 2044.   **c.** Until 2044

3.2 Skills Check

1. $x = 5, x = -2$ **3.** $x = 8, x = 3$ **5.** $x = -3, x = 2$

7. $\frac{3}{2}, 4$ **9.** $x = -2, x = \frac{1}{3}$ **11.** $x = -2, x = 5$

13. $x = \frac{2}{3}, x = 2$ **15.** $x = \frac{1}{2}, x = -4$ **17.** $x = -\frac{1}{2}, x = 3$

19. $x = 8, x = 32$ **21.** $s = 50, s = -15$ **23.** $x = \pm\frac{3}{2}$

25. $x = \pm 4\sqrt{2}$ **27.** $x = 2 \pm \sqrt{13}$ **29.** $x = 1, x = 2$

31. $x = \frac{5 \pm \sqrt{17}}{2}$ **33.** $x = 1, x = -\frac{8}{3}$ **35.** $x = -3, x = 2$

37. $x = \frac{2}{3} \approx 0.667, x = -\frac{3}{2} = -1.5$ **39.** $x = \frac{-1}{2} = -0.5,$

$x = \frac{2}{3} \approx 0.667$ **41.** $x = \pm 5i$ **43.** $x = 1 \pm 2i$

45. $x = -2 \pm 2i$ **47. a.** Positive   **b.** 2   **c.** $x = 3, x = -2$

3.2 Exercises

49. $t = 2$ sec and 4 sec   **51.** 20 units and 90 units
53. a. $P = 520x - 10,000 - x^2$ **b.** $-$964; loss of $964
c. $5616; profit of $5616   **d.** 20 units or 500 units
55. a. $s = 50, s = -50$ **b.** There is no particulate pollution.
c. $s = 50$; speed is positive.   **57. a.** $r = 0$ **b.** $r = 0.05$
c. $r = 0.1$; against the wall of the artery   **59.** $100 is the price that gives demand = supply = 97 trees

61. a.

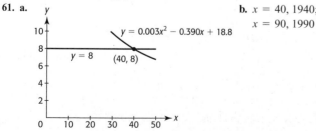

$y = 0.003x^2 - 0.390x + 18.8$
$y = 8$ $(40, 8)$

b. $x = 40, 1940;$ $x = 90, 1990$

63. 2016   **65.** 2011   **67. a.** Declined by 262,000
b. 1994   **c.** 37,851,047; that the model continues to apply
69. a. 9258 million in 2000; 13,526 million in 2008   **b.** 2009
71. a.

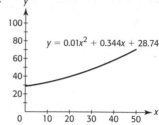

$y = 0.01x^2 + 0.344x + 28.74$

b. 70.94%   **c.** 1981

73. 2023

3.3 Skills Check

1.

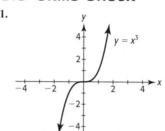

$y = x^3$

3.

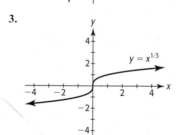

$y = x^{1/3}$

5.

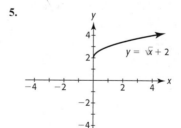

$y = \sqrt{x} + 2$

7.

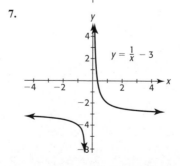

$y = \frac{1}{x} - 3$

9.

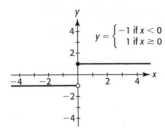

$$y = \begin{cases} -1 & \text{if } x < 0 \\ 1 & \text{if } x \geq 0 \end{cases}$$

11. a. $f(x)$

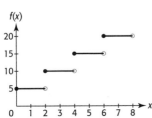

b. Piecewise, step function

13. a.

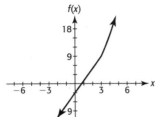

b. 5, 16
c. All real numbers

15. a.

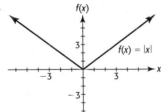

$f(x) = |x|$

b. 2, 5
c. All real numbers

17. a. 5   **b.** 6   **19. a.** 0   **b.** 29   **21. a.** Increasing
b. Increasing   **23.** Concave down   **25.** Concave up
27.

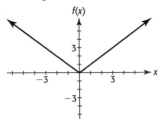

29.

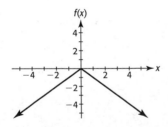

31. They are the same.   **33.** $x = \dfrac{7}{2}, x = -\dfrac{5}{2}$ **35.** $x = \dfrac{1}{7}$ **37.** 16

3.3 Exercises

39. a. $f(x) = \begin{cases} 7.10 + 0.06747x & \text{if } 0 \leq x \leq 1200 \\ 88.06 + 0.05788(x - 1200) & \text{if } x > 1200 \end{cases}$

b. \$71.87   **c.** \$110.05

41. a. $P(x) = \begin{cases} 42 & \text{if } 0 < x \leq 1 \\ 59 & \text{if } 1 < x \leq 2 \\ 76 & \text{if } 2 < x \leq 3 \\ 93 & \text{if } 3 < x \leq 4 \end{cases}$ **b.** 59; the postage on a 1.2 oz first class letter is 59¢.

c. $0 < x \leq 4$ **d.** 59; 76   **e.** 59¢; 76¢

43. a. $\begin{cases} 0.10x & \text{if } 0 < x \leq 15{,}650 \\ 1565 + 0.15(x - 15{,}650) & \text{if } 15{,}650 < x \leq 63{,}700 \\ 8772.50 + 0.25(x - 63{,}700) & \text{if } 63{,}700 < x \leq 128{,}500 \end{cases}$

b. \$5517.50   **c.** \$9097.50   **d.** The 25% rate applies to only the \$1 above \$63,700.   **45 a.** Power   **b.** 23,874; 23,874 suicides occurred in 1965.   **c.** 28,099

47. a.

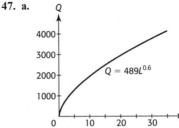

$Q = 489L^{0.6}$

b. 3912
c. Increases; yes

49. a. Increasing   **b.** Concave up   **c.** $x = 10.78$; 2001
51. a.

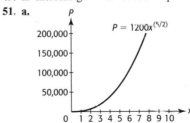

$P = 1200x^{(5/2)}$

b. Concave up

53. a.

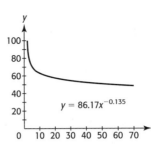

$y = 86.17x^{-0.135}$

b. Decreasing
c. 50.8   **d.** 50.3%
e. No

55. \$130   **57.** 6%

3.4 Skills Check

1. $y = 2x^2 - 3x + 1$ **3.** $f(x) = 3x^2 - 2x$ **5.** The x-values are not equally spaced.   **7.** $f(x) = 99.9x^2 + 0.64x - 0.75$
9. a. $y = 3.545x^{1.323}$ **b.** $y = 8.114x - 8.067$ **c.** Power function   **11. a.** $y = 1.292x^{1.178}$ **b.** $y = 2.065x - 1.565$
c. They are both good fits.   **13.** $y = 2.98x^{0.614}$

3.4 Exercises

Numerical results are computed with the unrounded models, unless otherwise stated.

15. a. $y = 6.829x^2 + 329.198x + 3550.236$ **b.** \$25,080; \$37,083 (from the unrounded model) **c.** Yes, but extrapolating will not always give accurate predictions.

17. a.

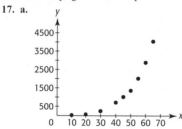

b. $y = 0.0324x^{2.737}$; $y = 2.0254x^2 - 86.722x + 853.890$

c. Quadratic **d.** 4707.8 billion

19. a.

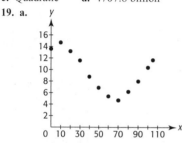

b. Yes **c.** $y = 0.0024x^2 - 0.3066x + 16.4920$ **d.** 11.6%

21. a. $y = 35.279x^{0.0409}$ **b.** 39.2 per 100,000 residents

23. a. $y = 514.143x^{-0.0526}$ **b.** 449.2 crimes per 100,000 residents

25. a. ; yes

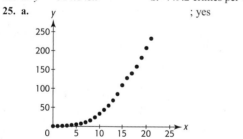

b. $y = 0.708x^2 - 4.031x + 4.865$ **c.** 347 million

d. Approximately 100% **27. a.** $y = -0.011x^2 + 0.313x + 7.176$

b. (14.2, 9.4); in 1975 the savings rate was 9.4%, the maximum during 1960–2005. **c.** 43.5 **d.** 2004 or after

29. a. ii, the quadratic model **b.** i, $y = 17\sqrt[3]{x}$

31. a. $y = 3.357x^{1.969}$ **b.** \$10.6654 trillion

c. $4.875x^2 - 94.825x + 830.814$ **d.** Quadratic

33. a. $y = 0.013x^2 - 0.054x + 1.564$

b. $y = 0.013x^2 + .080x + 1.630$ **c.** 1988

d. The results would be equivalent. **35. a.** $y = 252.576x^{0.296}$

b. \$613 billion **c.** 1992

37. a. $y = 3.993x^2 - 432.497x + 12,862.212$

b. $y = 0.046x^{2.628}$ **c.** Quadratic

39. a.

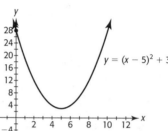

b. $y = -5.457x^2 + 684.592x - 2889.155$ **c.** 14,521

d. In 1963

Chapter 3 Skills Check

1. a. (5, 3) **b.**

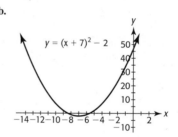
$y = (x - 5)^2 + 3$

2. a. $(-7, -2)$ **b.**

$y = (x + 7)^2 - 2$

3. a. $(1, -27)$ **b.**

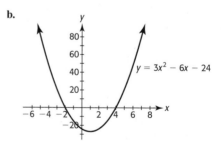
$y = 3x^2 - 6x - 24$

4. a. $(-2, -18)$ **b.**

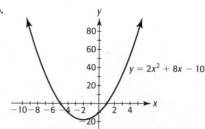
$y = 2x^2 + 8x - 10$

5. a. $(15, 80)$ **b.**

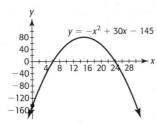

24.

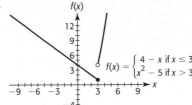

$$f(x) = \begin{cases} 4 - x & \text{if } x \le 3 \\ x^2 - 5 & \text{if } x > 3 \end{cases}$$

6. a. $(30, -400)$ **b.**

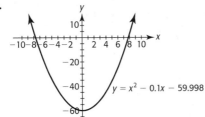

25.

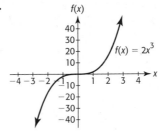

$f(x) = 2x^3$

7. a. $(0.05, -60.0005)$
b.

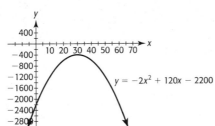

26.

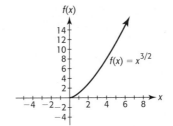

$f(x) = x^{3/2}$

8. a. $(-0.2, -100)$
b.

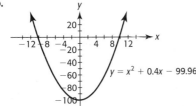

27.

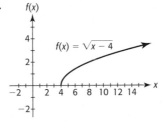

$f(x) = \sqrt{x - 4}$

9. $x = 4, x = 1$ **10.** $x = \dfrac{1}{2}, x = -\dfrac{2}{3}$ **11.** $x = -4/5, x = 1$

12. $x = -2, x = \dfrac{2}{3}$ **13.** $x = 3, x = 1$ **14.** $x = \dfrac{1}{2}, x = -\dfrac{3}{2}$

15. a. $4, -2$ **b.** $4, -2$ **16. a.** $-5, 1$ **b.** $-5, 1$

17. $2, -2$ **18.** $9, -1$ **19.** $z = 2 \pm i\sqrt{2}$ **20.** $w = 2 \pm i$

21. $x = \dfrac{5 \pm i\sqrt{23}}{8}$ **22.** $x = \dfrac{-1 \pm i\sqrt{3}}{4}$

28.

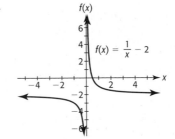

$f(x) = \dfrac{1}{x} - 2$

23.

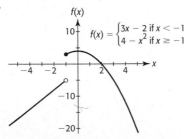

$$f(x) = \begin{cases} 3x - 2 & \text{if } x < -1 \\ 4 - x^2 & \text{if } x \ge -1 \end{cases}$$

29.

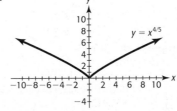

$y = x^{4/5}$

30.

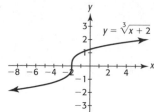

$y = \sqrt[3]{x + 2}$

31. a. Increasing   **b.** Decreasing   **32. a.** Concave up
b. Concave down   **33.** $x = 10, x = -6$ **34.** $x = 5, x = -8$
35. $y = 3.545x^{1.323}$ **36.** $y = 1.043x^2 - 0.513x + 0.977$
37. 128   **38.** $-26, -4, -5$

Chapter 3 Review

39. a. 3100   **b.** \$84,100   **40. a.** 4050   **b.** \$312,050
41. a. 2 sec   **b.** 256 ft   **42. a.** 1.5 sec   **b.** 82.05 m
43. a. 2004   **b.** 137 thousand   **c.** 2008
44. 30 units and 120 units   **45.** 5   **46.** 400 or 3700 units
47. a.

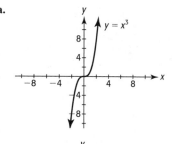

b. \$70.593 million
c. \$77.486 million
48. 2004

49. a. 92   **b.**

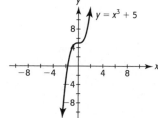

c. 1989

50. a. 5,544,000   **b.** 1999; model is valid after 1997
51. a. $y = 5.126x^2 + 0.370x - 221.289$
b.   **c.** Yes   **d.** 2011

52. a. $y = -3.423x^2 + 227.826x + 2298.951$ **b.** 1992
c. 2005   **53.** $y = 18.624x^2 - 440.198x + 20823.439$ thousand
people in the U.S. in year x
54. a. $y = 0.05143x^2 + 3.18286x + 65.40000$ **b.** 16 yr

55. a. $y = 750.487t^{-1.619}$; yes   **b.** Yes, but far in the
future, at $x \approx 59.8$,
which represents the
2040–2041 school year.

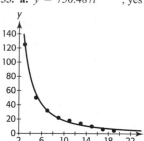

Yes, good fit
56. a. $y = 956.097x^{0.092}$ **b.** 2014
57. a. $y = 0.013x^2 - 0.223x + 7.297$ **b.** $y = 15.592x^{-0.381}$

c. $f(x) = \begin{cases} 0.013x^2 - 0.223x + 7.297 & \text{if } 0 \le x \le 11 \\ 15.592x^{-0.381} & \text{if } 12 \le x \le 24 \end{cases}$

d. i. 6.4; average stay in 1987 was 6.4 days.   **ii.** 1992   **iii.** 4.8 days
58. a. $y = 13.000x + 324.180$; excellent fit
b. $y = -0.017x^2 + 13.331x + 323.351$; excellent fit
c. 2014 with both models

Chapter 4 Additional Topics with Functions

Toolbox Exercises

1. y-axis   **2.** x-axis and y-axis   **3.** x-axis   **4.** Neither
5. Yes   **6.** Yes   **7.** No   **8.** Yes   **9.** No   **10.** Yes
11. Yes   **12.** Yes   **13.** Yes   **14.** Yes   **15.** No   **16.** Yes
17. Yes   **18.** No

4.1 Skills Check

1. a.

$y = x^3$

$y = x^3 + 5$

b. Vertical shift 5 units up

3. a.

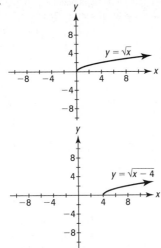

b. Horizontal shift 4 units right

5. a.

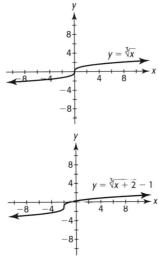

b. Horizontal shift 2 units left, vertical shift 1 unit down

7. a.

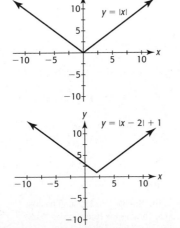

b. Horizontal shift 2 units right, vertical shift 1 unit up

9. a.

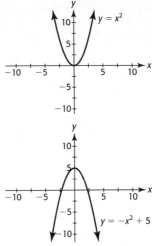

b. Reflection across x-axis, vertical shift 5 units up

11. a.

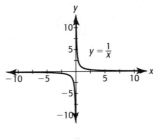

b. Vertical shift 3 units down

13. a.

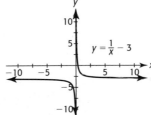

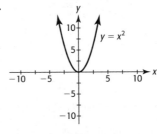

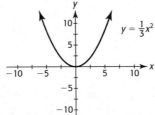

b. Vertical compression by a factor of 1/3

15. a.

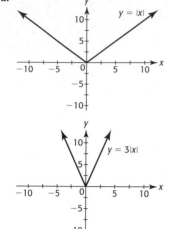

b. Vertical stretch by a factor of 3
17. Shifted to the right 2 units and up 3 units   **19.** $y = (x + 4)^{3/2}$
21. $y = 3x^{3/2} + 5$ **23.** $g(x) = -x^2 + 2$
25. $g(x) = |x + 3| - 2$ **27.** y-axis   **29.** Origin
31. Origin   **33.** x-axis, y-axis, and origin
35. Even   **37.** Even   **39.** Odd   **41.** Even
43.

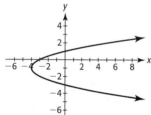

4.1 Exercises
45. a. $s = t^3$ **b.**   **c.** 27 in.

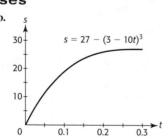

47. a. Linear   **b.** Reciprocal; vertically stretched by a factor of 30,000 and shifted down 20 units
49. a. Reciprocal function, shifted 1 unit to the left
b.   **c.** Decrease

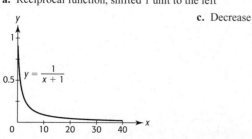

51. a. The graph is shifted to the left 10 units, is reflected about the x-axis, is stretched vertically by the factor 1000, and is shifted down 1 unit.
b.

$$P = -1000\left(\frac{1}{t+10} - 1\right)$$

53. $f(x) = 105.095(x + 5)^{1.5307}$
55. $y = 0.084(x + 4)^2 + 1.124(x + 4) + 4.028$
57. a. 240.5 million   **b.** $C(t) = 0.000442(t + 5)^{4.103}$
c. $t = 20$; yes
59. a.

$$\bar{C}(x) = \frac{100,000}{x} + 150$$

b. Decrease
c. Stretched by factor of 100,000 and shifted up 150 units

61. a. \$0   **b.** \$1200   **c.** Stretched by a factor of 120,000 and shifted down 1200 units

4.2 Skills Check
1. a. $2x - 1$ **b.** $4x - 9$ **c.** $-3x^2 + 17x - 20$
d. $\dfrac{3x - 5}{4 - x}$ **e.** All real numbers except 4

3. a. $x^2 - x + 1$ **b.** $x^2 - 3x - 1$ **c.** $x^3 - x^2 - 2x$
d. $\dfrac{x^2 - 2x}{1 + x}$ **e.** All real numbers except -1

5. a. $\dfrac{x^2 + x + 5}{5x}$ **b.** $\dfrac{-x^2 - x + 5}{5x}$ **c.** $\dfrac{x + 1}{5x}$
d. $\dfrac{5}{x(x + 1)}$ **e.** All real numbers except 0 and -1

7. a. $\sqrt{x} + 1 - x^2$ **b.** $\sqrt{x} - 1 + x^2$ **c.** $\sqrt{x}(1 - x^2)$
d. $\dfrac{\sqrt{x}}{1 - x^2}$ **e.** All real numbers $x \geq 0$ except 1

9. a. -8 **b.** 1   **c.** 196   **d.** 3.5   **11. a.** $6x - 8$
b. $6x - 19$ **13. a.** $\dfrac{1}{x^2}$ **b.** $\dfrac{1}{x^2}$ **15. a.** $\sqrt{2x - 8}$
b. $2\sqrt{x - 1} - 7$ **17. a.** $|4x - 3|$ **b.** $4|x - 3|$
19. a. x **b.** x **21. a.** 2   **b.** 1   **23. a.** -2 **b.** -1
c. -3 **d.** -3 **e.** 2

4.2 Exercises
25. a. $P(x) = 66x - 3420$ **b.** \$6480   **27. a.** The cost function is quadratic; the revenue function is linear.
b. $P = 1020x - x^2 - 10,000$ **c.** Quadratic

29. a. $P = 520x - x^2 - 10,000$ **b.** 260 units **c.** $57,600

31. a. $\overline{C}(x)$ is $C(x) = 50,000 + 105x$ divided by $f(x) = x$; that is,

$\overline{C}(x) = \dfrac{C(x)}{x}.$ **b.** $121.67 **33. a.** $\overline{C}(x) = \dfrac{3000 + 72x}{x}$

b. $102 per printer **35. a.** $S(p) + N(p) = 0.5p^2 + 78p + 12,900$

b. $0 \le p \le 100$ **c.** 23,970 **37. a.** $B(8) = 162$

b. $P(8) = $7.54 **c.** $(B \cdot P)(8) = $1221.48

d. $W(x) = (B \cdot P)(x) = 6(x + 1)^{3/2}(8.5 - 0.12x)$

39. a. $P(x) = 160x - 32,000$ **b.** $64,000 **c.** $160 per unit

41. a. $B(t) + G(t) = 0.014t^2 - 0.32t + 20.825$ **b.** 19.03 million

43. a. Meat in container **b.** Ground meat **c.** Meat ground and ground again **d.** Ground meat in the container **e.** Meat in container, then both ground **f.** d **45.** $p(s(x)) = x - 18.5$

47. 49.68 Russian rubles **49.** $100\dfrac{f(x)}{g(x)}$ **51.** $f(x) + g(x)$

53. 40%

4.3 Skills Check

1. Yes;

Domain   Range
f^{-1}

3. No inverse **5. a.** x **b.** Yes
7. Yes

9.

x	$f(x)$	x	$f^{-1}(x)$
-1	-7	-7	-1
0	-4	-4	0
1	-1	-1	1
2	2	2	2
3	5	5	3

11. a. $f^{-1}(x) = \dfrac{x + 4}{3}$ **b.** Yes

13. (b, a) **15.** $f^{-1}(x) = \dfrac{1}{x}$

17. $f^{-1}(x) = \dfrac{\sqrt{x}}{2}$

19.

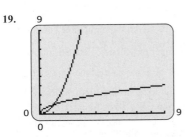

21. $x \ge 2$ **23.** Yes; yes **25.** Yes **27.** No

29.

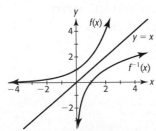

4.3 Exercises

31. a. $d^{-1}(x) = x - 0.5$ **b.** Size 8

33. $f^{-1}(t) = \dfrac{-t + 75.451}{0.707}$; 1990 **35. a.** $f^{-1}(x) = \dfrac{x - 493.432}{225.304}$;

the year when Ritalin consumption is at a given level x

b. 3 yr after 1990; 1993 **37. a.** Domain: all real $x \ge -\dfrac{1}{4}$;

range: real $f(x) \ge 0$ **b.** $f^{-1}(x) = \dfrac{x^2 - 16}{64}$ **c.** Domain: all real

$x \ge 0$; range: real $f^{-1}(x) \ge -\dfrac{1}{4}$ **d.** Domain: all real $x \ge 4$; range:

real $f^{-1}(x) \ge 0$ **39.** $C^{-1}(x) = x - 3$; THE_REAL_THING
41. Yes; each number identifies exactly one person.

43. a. Yes **b.** $f^{-1}(x) = \sqrt[3]{x}$ **c.** Domain: $x > 0$; range:

$f^{-1}(x) > 0$ **d.** To find the edge length of a cube from its volume
45. a. $f^{-1}(x) = 1.0071x$; dividing Canadian dollars by 0.99295 gives

U.S. dollars **b.** $500 **47. a.** No **b.** $x \ge 0$; it is a distance,

$x \ne 0.$ **c.** Yes **d.** $I^{-1}(x) = \sqrt{\dfrac{300,000}{x}}$; 2 ft

49. a. $f^{-1}(x) = 0.506765x$; converts U.S. dollars to U.K. pounds

b. $1000 **51. a.** 8 sec **b.** No **c.** $0 \le x \le 3$ or

$3 \le x \le 8$ **d.** $f^{-1}(x) = 3 - \dfrac{\sqrt{400 - x}}{4}$; the number of seconds,

from 0 sec to 3 sec, to attain height x

4.4 Skills Check

1. $x = 1$ **3.** $x = -7$ **5.** $x = 6$ **7.** $x = -2, x = 6$
9. $x = 9$ **11.** $x = 23, x = -31$ **13.** $-4 < x < 0$
15. $-3 \le x \le 3$ **17.** $4 < x < 5$ **19.** $x \le -2$ or $x \ge 6$
21. $-1 < x < 7$ **23.** $x \le 0.314$ or $x \ge 3.186$
25. $x \le -0.914$ or $x \ge 1.314$ **27.** $x < 0.587$
29. $x < 5$ **31.** $-1 < x < 2$ **33.** $x \le 4$ or $x \ge 8$
35. a. $x \le -2$ or $x \ge 3$ **b.** $-2 < x < 3$
37. a. No solution **b.** All real numbers **39.** $-3 \le x \le 2.5$

4.4 Exercises

41. $100 < x < 4000$ units **43.** $100 < x < 8000$ units
45. $2 < t < 8.83$ sec **47.** $1988 \le x \le 1997$
49. $0 < x < 130$, so 1900 to 2030 **51.** 1971 through 1994
53. $4.7 < x < 16$; 1995 through 2006 **55.** $2.56 < x < 32$;
1983 through 2012 **57. a.** $|x - 68| > 8$ **b.** $x < 60$ in. or
$x > 76$ in.

Chapter 4 Skills Check

1. Horizontal shift 8 units to right, vertical shift 7 units up
2. Shifted to left 1 unit, vertically stretched by a factor of 2 units,
reflected about the x-axis

3. a.

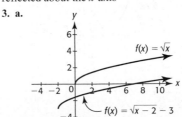

b. The graph of $g(x)$ can be obtained from the graph of $f(x)$ by shifting 2 units to the left and 3 units down.

4. $[-2, \infty)$; real numbers ≥ -2 **5.** $y = (x - 6)^{1/3} + 4$

6. $y = 3x^{1/3} - 5$ **7.** f   **8.** c   **9.** e   **10.** Origin

11. y-axis   **12.** Odd   **13.** $3x^2 + x - 4$ **14.** $9 - x^3 - 6x$

15. $18x^3 - 42x^2 + 20x$ **16.** $\dfrac{5 - x^3}{6x - 4}, \; x \neq \dfrac{2}{3}$ **17.** 38

18. $108x^2 - 174x + 68$ **19.** $18x^2 - 30x - 4$ **20.** 188

21. a. $x; x$ **b.** They are inverse functions.

22. $f^{-1}(x) = \dfrac{x + 2}{3}$ **23.** $g^{-1}(x) = x^3 + 1$

24.

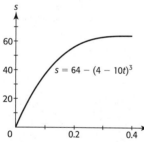

25. Yes   **26.** No   **27.** $x = -\dfrac{3}{8}$ **28.** $x = -2$

29. $[-2, 9]$ **30.** $x \leq -3$ or $x \geq \dfrac{1}{2}$ **31.** $-2 \leq x \leq 6$

32. $x \geq 4.5$ or $x \leq -3$ **33.** $x < 20$

34. $x \geq 16\sqrt{2} - 2$ or $x \leq -16\sqrt{2} - 2$

Chapter 4 Review

35. a.   **b.** 64 in.

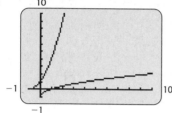

36. a. 4 sec   **b.** 380 ft   **c.** Stretched vertically by a factor of 16, and reflected across the x-axis, then shifted 4 units right and 380 units up.
37. a. 2004   **b.** Between 2001 and 2007   **38.** $20.5 \leq x \leq 44.6$, so 1971 through 1994   **39.** 2000 through 2010

40. a. $\overline{C}(x) = \dfrac{30x + 3150}{x}$ **b.** $\overline{C}(x)$

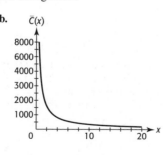

41. a. Demand function   **b.**

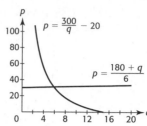

42. a. Demand function   **b.**

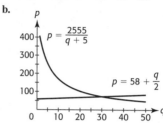

43. a. Reflected across the x-axis and compressed vertically by a factor of 0.2, shifted right 10.3 units, then shifted up 48.968 units
b. 31.67; 35.19; 41.03; 45.27
c. $M(x)$ **d.** $\{1, 2, 3, 4, 5, 6\}$

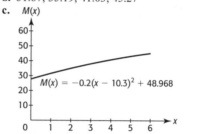

44. a. $R(E(t)) = 0.002805t^2 + 0.35706t + 1.104065$; the function calculates the revenue for Southwest Airlines given the number of years past 1990.   **b.** $R(E(3)) \approx 2.2$; in 1993 Southwest Airlines had revenue of \$2.2 billion.   **c.** $E(7) = 24.042$; in 1997 Southwest Airlines had 24,042 employees.   **d.** \$3.7 billion

45. a. $f^{-1}(x) = \dfrac{x + 2.886}{0.554}$ **b.** The mean sentence length can be found as a function of the mean time in prison.

46. a. $C^{-1}(x) = \dfrac{9.929 - x}{0.093}$ **b.** Given the number of cows used for milk production, the inverse function will calculate the number of years past 1990.   **c.** 8   **47. a.** $y = \dfrac{1}{t}$; stretched by a factor of 380, shifted horizontally 0.3 unit to the left, shifted vertically 15 units down.   **b.** The function $y = f(t)$; its function values will never be negative.   **c.** $y = C(t)$; no   **48.** Between 200 and 6000 units   **49.** Between 1960 and 1996

Chapter 5   Exponential and Logarithmic Functions

Toolbox Exercises

1. a. x^7 **b.** x^5 **c.** $256a^4y^4$ **d.** $\dfrac{81}{z^4}$ **e.** $2^5 = 32$

f. x^8 **2. a.** y^6 **b.** w^6 **c.** $216b^3x^3$ **d.** $\dfrac{125z^3}{8}$

e. 3^5 or 243   **f.** $16y^{12}$ **3.** 10   **4.** 256   **5.** $\dfrac{1}{x^7}$ **6.** $\dfrac{1}{y^8}$

7. $\dfrac{1}{c^{18}}$ **8.** $\dfrac{1}{x^8}$ **9.** a **10.** b^2 **11.** $x^{1/6}$ **12.** $y^{1/15}$

13. $\dfrac{6}{ab^2}$ **14.** $\dfrac{-8a^2}{b^2}$ **15.** $\dfrac{x^{10}}{4}$ **16.** $\dfrac{8y^{18}}{27}$ **17.** $\dfrac{-7}{a^2b}$

18. $\dfrac{-6y^2}{x}$ **19.** 4.6×10^7 **20.** 8.62×10^{11}

21. 9.4×10^{-5} **22.** 2.78×10^{-6} **23.** $437{,}200$

24. $7{,}910{,}000$ **25.** 0.00056294 **26.** 0.0063478

27. 3.708125×10^6 **28.** 6.460833515×10^1 **29.** $x^{4/3}$

30. $y^{13/20}$ **31.** $c^{5/3}$ **32.** $x^{9/8}$ **33.** $x^{1/4}$ **34.** $y^{1/8}$

5.1 Skills Check

1. c, e

3. a.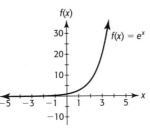

b. 2.718; 0.368; 54.598

c. x-axis **d.** 1

5.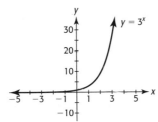

7.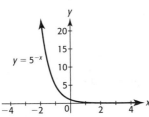

9.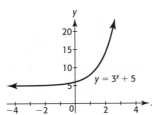

11.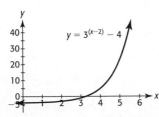

13. B **15.** A **17.** E

19. Vertical shift, 2 units up

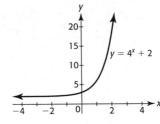

21. Reflection across the y-axis

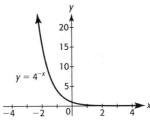

23. Vertical stretch by a factor of 3

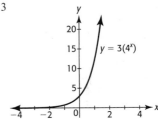

25. The graph of $y = 3 \cdot 4^{(x-2)} - 3$ is the graph of $y = 3(4^x)$ shifted to the right 2 units and shifted down 3 units.

27. a.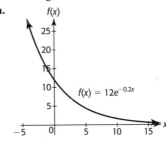

b. $f(10) = 1.624$; $f(-10) = 88.669$

c. Decay

5.1 Exercises

29. a. $12,000 **b.** $8603.73 **c.** No

31. a.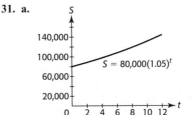

b. 130,311.57

33. a.

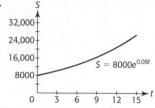

b. $x \approx 11.45$ yr

c.

t(Year)	S ($)
10	17,804.33
20	39,624.26
22	46,499.50

35. a. 376.84 g   **b.** $A(t)$
c. About 24.5 yr

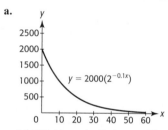

37. a.

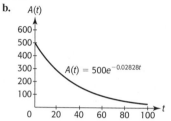

b. $1000   **c.** Sales declined drastically after the end of the ad campaign. (Answers may vary.)

39. a. 14,339.44   **b.** Because of future inflation, they should plan to save money. (Answers may vary.)   **41. a.** $122,140.28
b. In about 14 yr   **43. a.** Increasing   **b.** 57,128   **c.** 61,577
d. Approximately 858 people per year   **45. a.** About 88.6 g
b. About 19,034 yr
47. a.

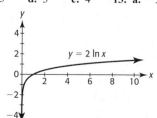

b. 50

5.2 Skills Check

1. $3^y = x$ **3.** $e^y = 2x$ **5.** $\log_4 x = y$ **7.** $\log_2 32 = 5$
9. a. 0.845   **b.** 4.454   **c.** 4.806   **11. a.** 5   **b.** 2
c. 3   **d.** 3   **e.** 4   **13. a.** -3 **b.** 0   **c.** 1   **d.** -4
15.

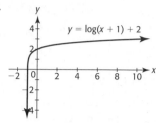

17.

19. a. $y = 4^x$, $x = 4^y$, $\log_4 x = y$; therefore, the inverse function is $y = \log_4 x$.
b. The graphs are symmetric with respect to the line $y = x$.

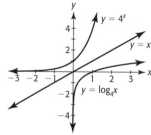

21. $a^x = a$; $x = 1$ **23.** 14   **25.** 12   **27.** 2.2146
29. 2.322   **31.** $\ln(3x - 2) - \ln(x + 1)$
33. $\dfrac{1}{4} \log_3 (4x + 1) - \log_3 (4) - 2 \log_3 (x)$

35. $\log_2 x^3 y$ **37.** $\ln \dfrac{(2a)^4}{b}$

5.2 Exercises

39. a. 57; 78   **b.** Improved health care and better diet. (Answers may vary.)   **41. a.** $7622; $10,699   **b.** Increasing   **c.** Increased cost of living, raise in minimum wage. (Answers may vary.)
43. $75.50   **45. a.** 42.9%, 43.5%   **b.** Increasing
47. ≈ 6.9 yr   **49. b.** 35 quarters, or $8\dfrac{3}{4}$ yr   **51.** 9 years
53. 4.4   **55.** $2{,}511{,}886 I_0$ **57.** $I = 10^{7.1} I_0 \approx 12{,}589{,}254 I_0$
59. About 14.1 times as intense   **61.** 43   **63.** $I = 10{,}000 I_0$
65. The decibel reading of a busy street is approximately 49 more than the decibel reading for a whisper.   **67.** 4.2
69. $10^{-14} \leq [H^+] \leq 10^{-1}$

5.3 Skills Check

1. 3.204   **3.** 7.824   **5.** $x \approx 1.819$ **7.** 0.374
9. $x \approx 1.204$ **11.** 1.6131   **13.** 0.1667   **15.** $x = \dfrac{10}{3}$
17. $x = 2$ **19.** $x \approx 3.5$ **21.** $x \approx 0.769$ **23.** $x = 8$
25. $x = e^{1.5} \approx 4.482$ **27.** $x = \dfrac{e^6}{2} \approx 201.71$ **29.** $x = 50$
31. $x = \dfrac{1}{3}$ **33.** $x = 40$ **35.** $x = 5$ **37.** $x < 5$
39. $x \geq 9$

5.3 Exercises

41. 5   **43. a.** $\ln\left(\dfrac{S}{25{,}000}\right) = -0.072x$ **b.** 6   **45. a.** 3200
b. 9 days   **47. a.** $1755.358 million   **b.** 1996

49. 13.87 yr   **51.** 23.11 yr   **53. a.** 500 g   **b.** About 24.51 yr

55. 3 hours   **57.** about 24,101 yr   **59.** 7 units   **61.** $t = \dfrac{\ln 2}{\ln (1.07)}$

63. 13 years   **65.** 5 years   **67. a.** 2008   **b.** 2008; yes
69. 1500 units   **71.** 1996   **73. a.** ≈ 35 **b.** 8.75 yr
75. $t \approx 14.2$; in 15 yr   **77.** 1994 and after   **79.** Months after 46
81. About 2075 yr
83. a. 16   **b.**

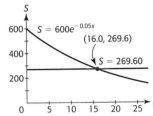

c. The exponential function appears to be the better fit.

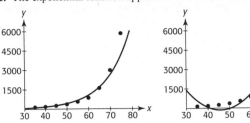

25. a. $y = 2.708(1.042^x)$ **b.** 286.7   **c.** 2015
27. a. $y = 10.963 + 14.321 \ln x$

b. $y = -0.00201x^2 + 0.514x + 45.607$ **c.** Logarithmic model

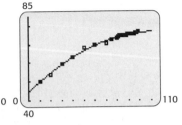

d. Logarithmic: 78.3; quadratic: 77.9
29. a. $y = 33.266 + 7.297 \ln x$ **b.** 47.5%
31. a. $y = 85.533(1.031^x)$ **b.** 14,043
33. a. $y = -681.976 + 251.829 \ln x$ **b.** 31.5%
c. $y = 0.627x^2 - 7.400x - 26.675$ **d.** Quadratic

5.4 Skills Check

1. $y = 2(3^x)$ **3.** not exponential   **5.** $f(x) = 4^x$
7. a.

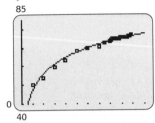

b. Linear

9. Exponential   **11.** $y = 0.876(2.494^x)$
13. a.

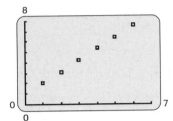

b. Logarithmic

15. a.

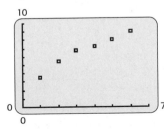

b. $y = 3.671x^{0.505}$

c. $y = -0.125x^2 + 1.886x + 1.960$ **d.** $y = 3.468 + 2.917 \ln x$

5.4 Exercises

17. a. $y = 30,000(1.04^t)$ **b.** $54,028
19. a. $y = 20,000(.98^x)$ **b.** $18,078
21. a. $y = 445.172(1.076^x)$ **b.** $17,644.700 billion   **c.** 2013
23. a. $y = 4.304(1.096^x)$
b. $y = 6.182x^2 - 565.948x + 12,810.482$

5.5 Skills Check

Approximate answers to two decimal places.
1. 49,801.75   **3.** 17,230.47   **5.** 26,445.08   **7.** 12,311.80
9. 1,723,331.03   **11.** 1123.60; 1191.00; 1191.00

13. $P = \dfrac{S}{\left(1 + \dfrac{r}{k}\right)^{kn}} = S\left(1 + \dfrac{r}{k}\right)^{-kn}$

5.5 Exercises

15. a. $16,288.19   **b.** $88,551.38
17. a.

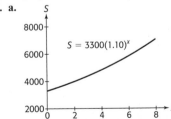

b. $x \approx 7.27$; 8 yr because compounded annually

19. $32,620.38   **21. a.** $33,194.62   **b.** More; compounded more frequently   **23.** $49,958.02   **25. a.** $20,544.33
b. $29,446.80   **27. a.** $28,543.39   **b.** $903.41 more with continuous compounding   **29. a.** ≈ 7.27, 8 yr   **b.** ≈ 6.93 yr
31. a. $2954.91   **b.** $4813.24   **33.** $6152.25
35. a.

Years	0	7	14	21	28
Future Value ($)	1000	2000	4000	8000	16,000

b. $y = 1000(1.104)^x$ **c.** $1640.01, $2826.02

37. $29,303.36   **39.** $5583.95   **41.** $4996.09
43. $24,215.65   **45.** $t \approx 17.39$; 17 yr, 4 mo   **47.** 12 yr and after

49.
$$2 = \left(1 + \frac{r}{m}\right)^{mt}$$
$$\ln 2 = mt \ln \left(1 + \frac{r}{m}\right)$$
$$\frac{\ln 2}{m \ln (1 + r/m)} = t$$

5.6 Skills Check

1. $P = \dfrac{S}{(1 + i)^n}$ **3.** $A = R\left[\dfrac{1 - (1 + i)^{-n}}{i}\right]$

5. $R = A\left[\dfrac{i}{1 - (1 + i)^{-n}}\right]$

5.6 Exercises

7. $52,723.18   **9.** $21,824.53   **11.** $486,043.02
13. $9549.11   **15.** $7023.58   **17.** $530,179.96
19. $372,845.60   **21. a.** $702,600.91   **b.** $121,548.38
c. The $100,000 plus the annuity gives approximately $2600 more in present value   **23. a.** $198,850.99   **b.** $576,000
c. $377,149.01   **25. a.** 2%   **b.** 16   **c.** $736.50
27. a. $1498.88   **b.** $639,596.80   **c.** $289,596.80

5.7 Skills Check

1. 73.83   **3.** 995.51; 999.82
5. a.
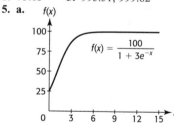
b. 25; 99.99
c. Increasing
d. 100

7. a.
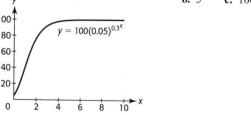
b. 5   **c.** 100

5.7 Exercises

9. a.
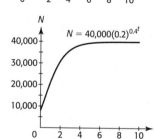
b. 5   **c.** 5000

11. a.
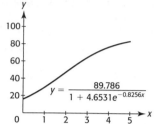
b. 29.56%
c. 86.93%
d. The limiting value is 89.786.

13. a. Approximately 218 people   **b.** 1970   **c.** The seventh day

15. a. $y = \dfrac{89.786}{1 + 4.6531e^{-0.8256x}}$ **b.** Yes

c. $y = 14.137x + 17.624$ **d.** The logistic model is the better fit.

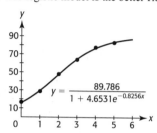

17. a. $y = \dfrac{73.921}{1 + 5.441e^{-0.415x}}$ **b.** a good fit   **c.** 73.1%

19. a. 4000   **b.** Approximately 9985
c. Upper limit 10,000

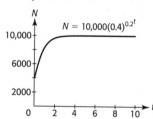

21. a. 21,012 units   **b.**
c. 40,000 units

23. a. 10   **b.** 100   **c.** 1000   **d.** 6 yr after the company was formed.   **25.** 10 days   **27.** 11 yr

Chapter 5 Skills Check

1. a. $f(x) = 4e^{-0.3x}$ **b.** ≈ 80.342; ≈ 0.19915

2. Decreasing

3.

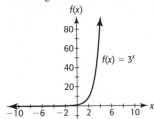

4.

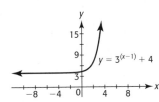

5. The graph of $y = 3^{(x-1)} + 4$ is the graph of $f(x) = 3^x$ shifted to the right 1 unit and up 4 units. **6.** Increasing. **7. a.** 500 **b.** $x = 20$
8. $y = \log_6 x$ **9.** $3x = \log_7 y$ **10.** $x = 4^y$ **11.** $x = 10^y$
12. $x = e^y$ **13.** $y = \log_4 x$ **14.** 1.3424 **15.** 4.0254
16. 1 **17.** 4 **18.** 4 **19.** -3 **20.** 3.6309 **21.** 1.9358
22.

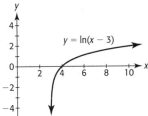

23.

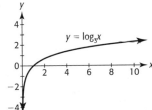

24. 5.8289 **25.** 0.2012 **26.** 2.8813 **27.** 2
28. $\ln(2x - 5)^3 - \ln(x - 3) = 3\ln(2x - 5) - \ln(x - 3)$

29. $\log_4 \dfrac{x^6}{y^2}$ **30.** Exponential function; $y = 0.810(2.470)^x$

31. 4926.80 **32.** 32,373.02

33. a.

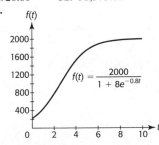

b. 222.22; 1973.76 **c.** 2000

34. a. **b.** 50 **c.** 500

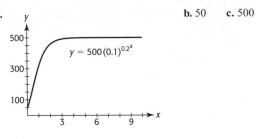

Chapter 5 Review

35. 733,760 **36.** \$1515.72 **37.** 1988 and after
38. a. 3 on Richter scale **b.** 3,162,278 I_0
39. Approximately 1259 times **40.** 10 yr
41. a. $x = 7 \log_2 \dfrac{S}{1000}$ **b.** 30 yr

42. 10 weeks **43.** $x = 29$; 1989
44. a. $P(x) = R(x) - C(x)$
$P(x) = 10(1.26)^x - (2x + 50)$
$= 10(1.26)^x - 2x - 50$
b. Applying the intersection of graphs method, selling at least 10 mobile homes produces a profit of at least \$30,000.

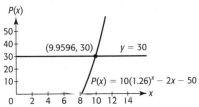

45. $x = 6.3$; 7 wk **46. a.** $y = 54.62$ grams **b.** 8445 yr ago
47. $x \approx 13.9$; 14 yr **48.** ≈ 4451.08 **49.** 15 yr
50. $y = 96.510(1.336^x)$ **51. a.** $y = 98.221(0.870^x)$
b. Decay **c.** 1.5 **52. a.** $y = 220.936(1.347^x)$
b. 686,377 thousand **c.** Is twice U.S. population; not a good model **53.** $y = 165.893(1.055^x)$; \$1065 billion
54. a. $y = 0.940 + 28.672 \ln x$ **b.** 86.8%
c. 2027; predicted percent >100 **55.** $y = 4.337 + 40.890 \ln x$

56. a. $N = \dfrac{130.439}{1 + 0.11e^{-0.0693x}}$ **b.** Excellent fit

57. \$20,609.02 **58.** \$30,072.61 **59.** \$34,426.47
60. \$274,419.05 **61.** \$209,281.18 **62.** \$6899.32
63. \$66.43 **64.** \$773.16 **65. a.** 67.79%; 76.84%
b. 96.3641% **66. a.** Approximately 1184 students
b. 16 days **67. a.** 1298 **b.** 3997 **c.** 4000
68. a. Approximately 17,993 Units **b.** 18,000

69. a. $y = \dfrac{627.044}{1 + 268.609e^{-0.324x}}$ **b.** 628 **c.** 2022

Chapter 6 Higher-Degree Polynomial and Rational Functions

Toolbox Exercises

1. a. Fourth **b.** 3 **2. a.** Third **b.** 5 **3. a.** Fifth
b. -14 **4. a.** Sixth **b.** -8 **5.** $4x(x - 7)(x + 5)$

6. $-x^2(2x + 1)(x - 4)$ **7.** $(x - 3)(x + 3)(x - 2)(x + 2)$

8. $(x - 4)(x + 4)(x^2 - 5)$ **9.** $2(x^2 - 2)^2$

10. $3x(x - 2)^2(x + 2)^2$ **11.** $\dfrac{1}{3}$ **12.** $\dfrac{x - 3}{4}$

13. $2y + 2$ **14.** $\dfrac{4x^2 - 3}{x - 1}$ **15.** $\dfrac{x - 2}{x + 4}$ **16.** $\dfrac{3x + 2}{x - 1}$

17. $\dfrac{1}{x^2}$ **18.** $2x^2 - 7x + 6$ **19.** $\dfrac{x - 2}{2x}$ **20.** $\dfrac{1}{2y}$

21. $\dfrac{3x^3 + x - 2}{x^3}$ **22.** $\dfrac{4x^2 + 2x + 4}{x^3}$ **23.** $\dfrac{4a - 4}{a^3 - 2a^2}$

24. $\dfrac{8x^4 - 16x^3 + 32x^2 - 59x}{x^4 - 16}$ **25.** $\dfrac{x - 2}{x}$ **26.** $\dfrac{2x^2 + 9x - 7}{4x^2 - 2x}$

27. $x^4 - x^3 + 2x^2 - 2x + 2, R(-3)$ **28.** $a^3 + a^2$

29. $3x^3 - x^2 + 6x - 2, R(17x - 5)$ **30.** $x^2 + 1, R2$

6.1 Skills Check

1. a.   **b.** Window (b) gives a complete graph

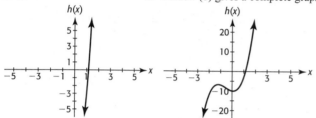

3. a.   **b.** Window (b) gives a complete graph.

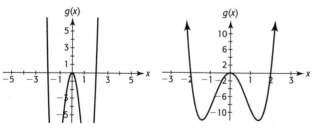

5. a. $-2, 1, 2$ **b.** Positive   **c.** Cubic   **7. a.** $-1, 1, 5$
b. Negative   **c.** Cubic   **9. a.** $-1.5, 1.5$ **b.** Positive
c. Quartic   **11.** C   **13.** E   **15.** F
17. a. Degree 3, leading coefficient 2   **b.** Opening up to right because of positive leading coefficient, down to left because cubic graphs have end behaviors in opposite directions

c.

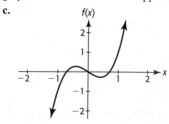

19. a. Expanding shows that this is a third-degree polynomial function with leading coefficient -2.   **b.** Opening down to right because of negative leading coefficient, up to left because cubic graphs have end behaviors in opposite directions   **c.**

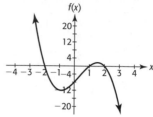

21. a.   **b.** Three x-intercepts are shown, along with the y-intercept, so the graph is complete.

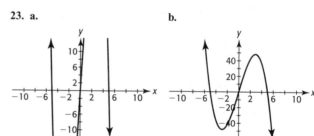

23. a.   **b.**

25. a.   **b.**

c. The graph shown in part (b).
27. a.   **b.** 3   **c.** No

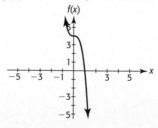

29. Answers will vary. One such graph is the graph of $f(x) = -4x^3 + 4$, shown below.

31. Answers will vary. One such graph is the graph of $f(x) = x^4 - 3x^2 - 4$, shown below.

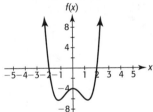

33. a.

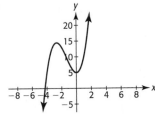

b. $(-2.67, 14.48)$
c. $(0, 5)$

35. Maximum: $(1, 1)$; minima: $(0, 0)$, $(2, 0)$

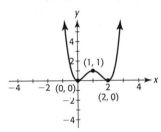

6.1 Exercises

37. a. Two turning points   **b.** Nonnegative; nonnegative

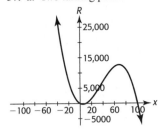

c.   **d.** $10,000

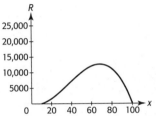

39. a.   **b.** 60 units gives $28,800.

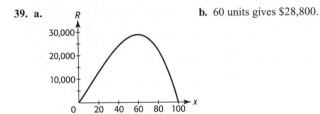

c.   **d.** Part (a)   **e.** $0 < x < 60$

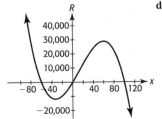

41. a.   **b.**

$r\,(\%)$	S (future value)
0	2000.0
5	2315.25
10	2662.00
15	3041.75
20	3456.00

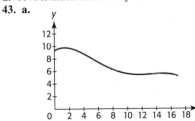

c. $2662.00 at 10% and $3456.00 at 20%; $794
d. 10% is much more likely than 20%.

43. a.   **b.** 5.2
c. 1992
d. 2011; rate is negative.

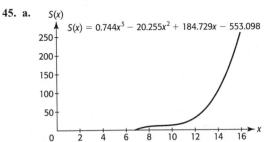

45. a.

$$S(x) = 0.744x^3 - 20.255x^2 + 184.729x - 553.098$$

b. $S(10) = 12.692$, so there were 12,692 vehicles sold in 2000; $S(16) = 264.71$, so the number sold in 2006 was 264,710 according to the model.

47. a.

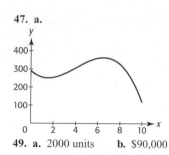

b. The value $x = 8$ represents 1998, so the approximate debt is the corresponding value of y, $324.84 million.   **c.** The local minimum is (1.55, 251.14) and the local maximum is (6.495, 361.85). The minimum debt is approximately $251 million, and the maximum debt is approximately $362 million.

49. a. 2000 units   **b.** $90,000

6.2 Skills Check

1. $y = x^3 - 2x^2$ **3.** $y = x^4 - 4x^2$

5. a.

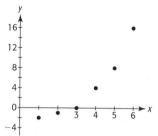

b. Cubic

7. a.   **b.**

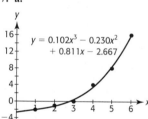

$y = 0.102x^3 - 0.230x^2 + 0.811x - 2.667$

$y = 3.457x - 7.933$

c. The cubic model is better.

9. a. $y = 35x^3 - 333.667x^2 + 920.762x - 677.714$

b. $y = 12.515x^4 - 165.242x^3 + 748x^2 - 1324.814x + 738.286$

11. $y = x^4 - 4x^2 - 3x + 1$ **13.** f is not exactly cubic.

15. $y = 0.565x^3 + 2.425x^2 - 4.251x + 0.556$

6.2 Exercises

17. a. $y = -0.613x^3 + 23.835x^2 - 179.586x + 385.670$
b. 1405 million   **c.** Good fit

19. a. $y = 0.0038x^3 - 0.0704x^2 - 0.0781x + 9.7800$;

$y = -0.0007x^4 + 0.0265x^3 - 0.2999x^2 + 0.6836x + 9.3366$

b.

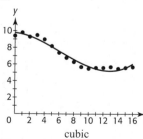

cubic   quartic

The quartic model is a better fit.

21. a.

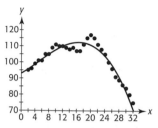

b. $y = 0.00005x^4 - 0.00608x^3 + 0.08688x^2 + 1.16438x + 93.0013$

c.   **d.** 43

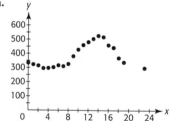

23. a. $y = -0.0015x^3 + 0.0108x^2 + 1.1001x + 189.9179$

b. 131.455 thousand

25. a. $y = 0.197x^3 - 4.910x^2 + 50.488x + 666.037$
b. 1131; the number of millions of person-trips is estimated to be 1131 in 2008.

27. a. $y = -0.008x^3 + 0.221x^2 - 1.903x + 22.052$

b. $y = 0.001x^4 - 0.025x^3 + 0.386x^2 - 2.416x + 22.326$

c. Nearly equal   **29. a.** $y = 0.00002x^3 - 0.00116x^2 - 0.00465x + 21.855$ **b.** 28.3 years old

31. a.

b. $y = -0.146x^3 + 3.878x^2 - 15.219x + 319.025$

c.   **d.** 0.504

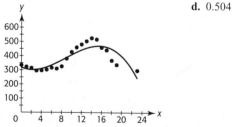

33. a. $y = 0.002x^4 - 0.075x^3 + 1.002x^2 - 5.177x + 82.325$
b. (4.7, 73.2)   **c.** Model says percent is lower in 1995 than in 1994 and 1996.   **d.** Yes; 2008

35. a. $y = 0.258x^4 - 8.676x^3 + 87.467x^2 - 244.413x - 150.928$
b. $286.228 billion

6.3 Skills Check

1. $\dfrac{3}{2}, -1, 6$ **3.** $-1, 4, \dfrac{5}{2}$ **5.** $0, 4, -4$ **7.** $0, 2$

9. $0, 1, -1$ **11.** $4, 3, -3$ **13.** $2, -2, \dfrac{4}{3}$ **15.** 2

17. $2, -2$ **19.** $0, \pm\sqrt{2}$ **21.** $0, 5, -5$ **23.** $\pm\sqrt{3}$

25. a. $-3, 1, 4$ **b.** $(x + 3)(x - 1)(x - 4)$

27. a. $-1, 2, 3$ **b.** $(x + 1)^2(x - 2)(x - 3)$

29. a. $-1, 1, 5$ **b.** $(x + 1)(x - 1)(x - 5)$

31. $-2, 5, \dfrac{3}{4}$

6.3 Exercises

33. a. $0, 20$
b. Yes

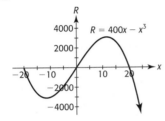

35. a. $0, 1000$ **b.** Yes

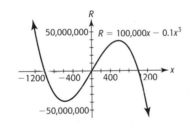

37. a. \$2249.73; 2315.25; 2467.30; 2698.47
b. **c.** 10% **d.** 20%

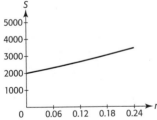

39. a. x in. **b.** $18 - 2x$ in. by $18 - 2x$ in.

c. $V = (18 - 2x)(18 - 2x)x = 324x - 72x^2 + 4x^3$

d. $x = 0, x = 9$ **e.** Neither

41. 2 units or 20 units

43. a.

t	0	0.1	0.2	0.3
s (cm/sec)	810	240	30	0

b. 0.3 sec; yes

45. a. $y = 779{,}365$ **b.** 1998 **47. a.** $x = 7.11$; 1998
b. Yes, in 1999

6.4 Skills Check

1. $x^3 - x^2 - 3x - 6 - \dfrac{8}{x - 3}$ **3.** $2x^3 - x^2 - x - \dfrac{7}{x - 1}$

5. No **7.** No **9.** 1 (a double solution) **11.** $-2, 4$

13. $4, -2, -5$ **15.** $-2, 1, \dfrac{1}{3}$ **17.** $\pm 1, \pm 2, \pm 3, \pm 4, \pm 6, \pm 12$

19. $\pm 1, \pm 2, \pm 4, \pm\dfrac{1}{3}, \pm\dfrac{2}{3}, \pm\dfrac{4}{3}, \pm\dfrac{1}{9}, \pm\dfrac{2}{9}, \pm\dfrac{4}{9}$ **21.** $-1, 3, 4$

23. $\dfrac{1}{3}, -1, -\dfrac{4}{3}$ **25.** $x = 0, x = \dfrac{-7 \pm \sqrt{89}}{2}$

27. $w = 1, w = 2 \pm \sqrt{2}$ **29.** $z = 2, w = -1 \pm i\sqrt{3}$

6.4 Exercises

31. a. $-0.2x^2 + 56x + 1200$ **b.** $x = 300, x = -20$, so 300 units
gives break-even.
33. a.

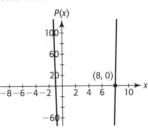

b. 8
c. $-0.1x^2 + 49.9x + 50$

d. $x = 8, x = 500, x = -1$ **e.** 8 units or 500 units
35. $x = 9, x = 10, x = -100$; 9 units or 10 units
37. a. $y = 0.4566x^3 - 14.3085x^2 + 117.2978x - 136.1544 = 0$

b. $x = 12$ **c.** $0.4566x^2 - 8.8293x + 11.3462$

d. $x = 17.953, x = 1.384$

e. 1982, 1992, 1998

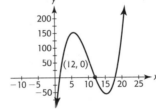

39. $x = 11.7, x = 37.5$; 1982 and 2008

6.5 Skills Check

1. a. $x = 5$ **b.** $y = 0$ **3. a.** $x = \dfrac{5}{2}$ **b.** $y = -\dfrac{1}{2}$

5. a. $x = 1, x = -1$ **b.** None **7.** c; no value of x makes the
denominator 0. **9.** E **11.** F **13.** C

15. a. $y = 1$ **b.** $x = 2$ **c.**

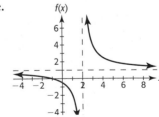

17. a. $y = 0$ **b.** $x = -1, x = 1$
c.

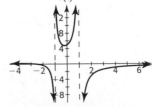

19. A hole in the graph

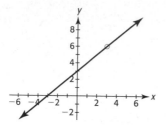

21. (0, 0) and (2, 4)

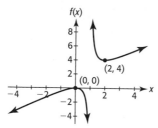

23. (−2, −4) and (2, 4)

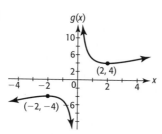

25. a.

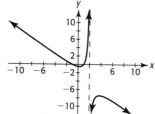

b. 0; −8
c. $x = 4, x = 3.5$
d. $x = 4, x = 3.5$

27. a.

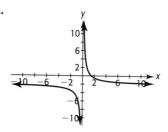

b. −3, −1
c. $x = -1$
d. $x = -1$

29. $x = \dfrac{1}{2}$ **31.** 80

6.5 Exercises

33. a. $55.80 per unit   **b.** $57.27 per unit   **c.** $55 per unit
d. No; for example, at 600 units the average cost is $56.67 per unit.
35. a. $80,000   **b.** −20; no

37. a.

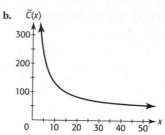

b.

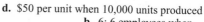

c. [0, 50] by [0, 300]   **d.** $50 per unit when 10,000 units produced

39. a.

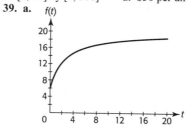

b. 6; 6 employees when the company starts
c. 17.586; the number of employees after 12 months; approximately 18 employees

41. a. At $p = 100$ **b.** It is impossible to remove 100% of the impurities.

43. a. Yes; $p = -2$

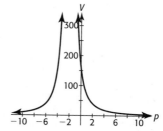

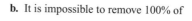

b.

Price per Unit ($)	5	20	50	100	200	500
Weekly Sales Volume	13,061	1322	237	62	16	3

c. Nonnegative values of p; no
d. $y = 0$; weekly sales approach 0 as price increases.

45. a.

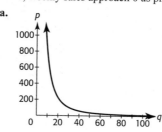

b. $p = 0$
c. As the price drops, the quantity demanded increases.

47. a. $S = \dfrac{x^2 + 40x + 160}{4x}$ **b.** 4 hours or 40 hours

49. a.

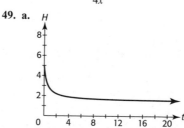

b. $H = \dfrac{3}{2}$; as the hours of training increase, the time a production team takes to assemble 1 unit approaches 1.5 hours.
c. 17 days

51. a.

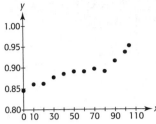

b.

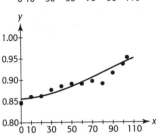

c. Good

6.6 Skills Check

1. $-4 \le x \le 4$ **3.** $x < 0$ or $x > 2$ **5.** $-1 \le x \le 1$ or $x \ge 3$

7. $0 < x < 1$ **9.** $x \le -3$ or $-1 < x \le 2$ **11.** $x > 4$

13. $x < 5$ **15.** $x = -5$ or $x \ge 0$ **17. a.** $x < -3$ or $0 < x < 2$

b. $-3 \le x \le 0$ or $x \ge 2$ **19.** $\dfrac{1}{2} \le x \le 3$

6.6 Exercises

21. $0 < x < 20$; more than 0 and less than 20 units

23. a. $0 < x < 18$ or $x > 18$ **b.** More than 0 cm and less than 18 cm   **25.** $0 \le x \le 2$ or $x \ge 10$; between 0 and 2 units or at least 10 units   **27.** $x \ge 20$; at least \$20,000   **29.** $0.10 \le r \le 0.20$; between 10% and 20%   **31.** $10 \le x \le 194$; between 10 and 194 units   **33.** $p > 80$; if the price is above \$80 per unit

35. a. $0 \le t \le 15$ **b.** The first 15 months

Chapter 6 Skills Check

1. 4   **2.** Quartic   **3.** Yes

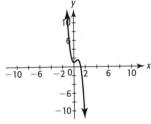

4. a.

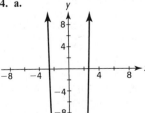

b.

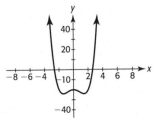

5. a.

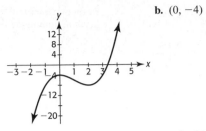

b. $(0, -4)$ **c.** $(2, -8)$

6. a.

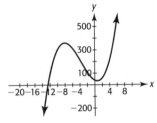

b. No

c.

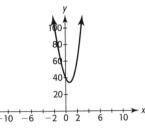

d. Maximum: $(-8, 360)$; minimum: $(0.667, 34.5)$

7. $0, -4, 4$ **8.** $0, 2, -2$ **9.** $0, 5, -4$ **10.** $0, 7, 8$

11. $-1, 1, 5$ **12.** $\dfrac{3}{4}, 2, -2$ **13.** $-1, 2, 3$ **14.** $12, -2.5, \dfrac{1}{3}$

15. 6   **16.** 5, 1   **17.** $4x^3 + 5x^2 + 10x + 22\ R36$

18. $1, -4, \dfrac{1}{2}$ **19.** $2, \dfrac{1}{3}, -2$ **20.** $1, -3, -\dfrac{1}{2}$

21. a. y-intercept: $\left(0, \dfrac{1}{2}\right)$; x-intercepts: $(-1, 0), (1, 0)$

b. Vertical asymptote: $x = -2$; no horizontal asymptote

c.

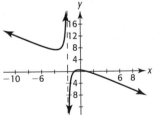

22. a. x-intercept $\left(\dfrac{2}{3}, 0\right)$; y-intercept $\left(0, \dfrac{2}{3}\right)$

b. Horizontal asymptote: $y = 3$; Vertical asymptote: $x = 3$

c.

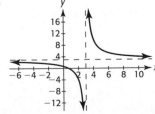

23.

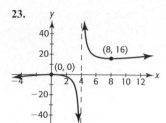

Local maximum: (0, 0); local minimum: (8, 16)

24.

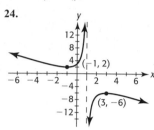

Local minimum: (−1, 2); local maximum: (3, −6)

25. a.

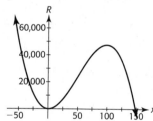

b. $x = 1, y = 1; x = 3, y = 3.8$

c. 2, −0.875   **d.** 2, $-\dfrac{7}{8}$

26. −3, 3, −2, 2   **27.** 1, −1 − $i\sqrt{3}$, −1 + $i\sqrt{3}$

28. −2, $-\dfrac{1}{4} \pm \dfrac{\sqrt{3}}{4}i$ **29.** $x = 0, x \geq 5$ **30.** $x \geq 4$ or

$-1 \leq x \leq 2$ **31.** $x > 3$ or $x < 0$ **32.** $-2 \leq x < -1$

Chapter 6 Review

33. a.

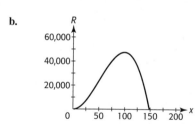

b.

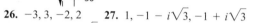

c. $23,750

34. When 84,051 units are sold, the maximum revenue of $23,385.63 is generated.

35. a.

r (rate)	S (future value, $)
1%	5307.60
5%	6700.48
10%	8857.81
15%	11,565.30

c. $6072.11

b.

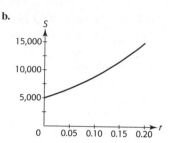

36. a.

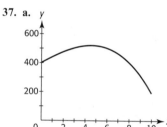

b. In 2000 the model predicts the percent to be 10.2. The prediction is close to the actual value of 10.4%. **c.** In 1949 and again in 1993, the percent is 7.6.

37. a.

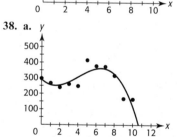

b. $400.844 million
c. $521.768 million

38. a.

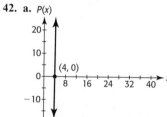

b. $y = -1.832x^3 + 22.111x^2 - 55.367x + 290.659$
39. The local maximum is (4, 640). Thus, an intensity of 4 allows the maximum amount of photosynthesis.   **40.** 5%   **41. a.** 0, 9
b. No sides if $x = 0$; no box material left if $x = 9$.   **c.** $0 < x < 9$
42. a. $P(x)$

b. The remaining quadratic factor is $-0.2x^2 + 19.7x + 30$.
c. 4, −1.5, 100   **d.** Because negative solutions do not make sense in the context of the question, break-even occurs when 400 units are produced or when 10,000 units are produced.

43. a. $y = 37.792x^4 - 745.250x^3 + 5349.708x^2 - 16,512.250x + 18,885$ **b.** 21,400 billion tons   **44. a.** $C = 0$ **b.** As t increases without bound, concentration of drug approaches 0.
c. 0.15 when $t = 1$ **45. a.** $\overline{C}(0)$ does not exist. If no units are produced, an average cost per unit cannot be calculated.

b. Because the degree of the numerator equals the degree of the denominator, the horizontal asymptote is $\overline{C}(x) = \dfrac{50}{1} = 50$. As the number of units produced increases without bound, the average cost per unit approaches \$50.   **c.** The function decreases as x increases.

46. a.   **b.** 20 units; \$1200

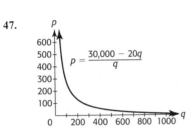

47.

$$p = \frac{30,000 - 20q}{q}$$

48. a. $C = \dfrac{200x + 20x^2 + 180}{x}$ **b.** Five plates   **c.** Three plates

49. a.   **b.** 13.85 students per computer; the prediction is close to the actual value in the table of 14 students per computer.

50. 50 or more hours   **51.** $y \le 8.8$ when $3.6 \le x \le 18.8$. Between 1994 and 2009 inclusive, the homicide rate is no more than 8.8 per 100,000 people.   **52.** At least 50 units, but no more than 605 units   **53.** $\overline{C} \le 37$ when $20 \le x \le 50$. The average cost is at most \$37 when between 20 and 50 units inclusive are produced.
54. \$4134 or more

Chapter 7   Systems of Equations and Matrices

Toolbox Exercises

1. 12   **2.** $\dfrac{1}{10}$ **3.** No   **4.** Yes   **5.** Yes   **6.** Yes
7. 99   **8.** 78   **9.** No   **10.** Yes   **11.** No   **12.** No
13. No   **14.** Yes   **15.** Yes   **16.** No   **17.** Parallel planes
18. Same plane   **19.** Neither   **20.** Same plane

7.1 Skills Check

1. $x = 2, y = 2, z = 3$ **3.** $x = -14, y = 9, z = -2$
5. $x = 24, y = -16, z = 10$ **7.** $x = -1, y = 1, z = 1$
9. $x = \dfrac{1}{2}, y = \dfrac{1}{2}, z = \dfrac{3}{4}$ **11.** $x = 2, y = 4, z = 14$

13. $x = 1, y = 3, z = 1$ **15.** $x = 2, y = 2, z = 2$
17. $x = 5z - 14, y = 2z - 4, z = z$ (any number)
19. $x = 20 - 10z, y = 16 - 8z, z = z$ (any number)
21. $x = -\dfrac{7}{4}z - \dfrac{5}{4}, y = \dfrac{1}{4}z - \dfrac{1}{4}, z = z$
23. No solution   **25.** $x = 10z + 11, y = -2z - 2, z = z$

7.1 Exercises

27. 30 compact, 20 midsize, 10 luxury   **29. a.** $x + y = 2600$
b. $40x$ **c.** $60y$ **d.** $40x + 60y = 120,000$
e. 1800 at \$40, 800 at \$60   **31. a.** $x + y + z = 400,000$
b. $0.075x + 0.08y + 0.09z = 33,700$ **c.** $z = x + y$
d. \$60,000 at 7.5%, \$140,000 at 8%, \$200,000 at 9%
33. A: 100 units; B: 120 units; C: 80 units
35. a. First equation: sum of investments is \$500,000; 2nd equation: sum of interest earned is \$49,000   **b.** $x = 50,000 + 2z$, $y = 450,000 - 3z$ **c.** $0 \le z \le 150,000$
37. Inconsistent system; can't use all hours
39. Food I $= 132 - (4.4)$ food III, food II $= (3.2)$ food III $- 64$, where $20 \le$ food III ≤ 30 g

7.2 Skills Check

1. $\begin{bmatrix} 1 & 1 & -1 & | & 4 \\ 1 & -2 & -1 & | & -2 \\ 2 & 2 & 1 & | & 11 \end{bmatrix}$ **3.** $\begin{bmatrix} 5 & -3 & 2 & | & 12 \\ 3 & 6 & -9 & | & 4 \\ 2 & 3 & -4 & | & 9 \end{bmatrix}$

5. $x = -1, y = 4, z = -2$ **7.** $x = 3, y = 2, z = 1$
9. $x = 2, y = 1, z = 3$ **11.** No solution   **13.** $x = 2 - 3z$, $y - 5z + 5, z =$ any number   **15.** $x = 3, y = -2, z = 1$
17. $x = -\dfrac{25}{3}, y = -\dfrac{25}{3}, z = \dfrac{14}{3}$ **19.** $x = 0, y = -1, z = 2$
21. $x = 40, y = 22, z = -19, w = 1$ **23.** $x = 0, y = 3, z = 2$
25. $x = \dfrac{32}{19} - \dfrac{22z}{19}, y = \dfrac{10}{19} - \dfrac{14z}{19}, z = z$
27. $x = \dfrac{11}{8} - \dfrac{31z}{8}, y = \dfrac{1}{8} + \dfrac{11z}{8}, z = z$
29. $x = \dfrac{23}{14} + \dfrac{z}{2}, y = -\dfrac{4}{7} + z, z = z$
31. $x = -2 + 3z, y = 4 - 4z, z = z, w = 0$

7.2 Exercises

33. 2400 at \$40, 800 at \$70, 400 at \$100
35. a. $\begin{cases} 15x + 10y + 5z = 100 \\ y = 2x \\ z = 3x \end{cases}$
b. 2 points for T-F, 4 points for MC, 6 points for essay
37. 3 units of plan I, 2 units of plan II, 1 unit of plan III
39. 5 g of food I, 6 g of food II, 8 g of food III
41. a. $\begin{cases} x + y + z = 400,000 \\ 0.08x + 0.10y + 0.12z = 42,400 \end{cases}$
b. \$260,000 at 12%, \$0 at 10%, and \$140,000 at 8%
43. a. $\begin{cases} 2x + 8y + 6z = 140 \\ 3x + 5y + 4z = 110 \end{cases}$ **b.** Not possible

45. a. $\begin{cases} x + y + z = 4 \\ 40{,}000x + 30{,}000y + 20{,}000z = 100{,}000 \end{cases}$ **b.** No

c. $x = z - 2, y = 6 - 2z, 2 \le z \le 3$ **d.** 2 at \$30,000, 2 at \$20,000, and 0 at \$40,000 or 1 at \$40,000, 0 at \$30,000, and 3 at \$20,000   **47.** Traffic from intersection A to intersection B is 650 less than the traffic from intersection D to intersection A. Traffic from intersection B to intersection C is 100 less than the traffic from intersection D to intersection A. Traffic from intersection C to intersection D is 1200 plus the traffic from intersection D to intersection A.
49. a. At A, $400{,}000 = x_1 + x_2$; at B, $x_1 = x_4 + 100{,}000$; at D, $x_3 + x_4 = 100{,}000$ **b.** $x_1 = 100{,}000 + x_4, x_2 = 300{,}000 - x_4$, $x_3 = 100{,}000 - x_4$, where x_4 is the number of gallons flowing from B to D; $x_4 \le 100{,}000$

7.3 Skills Check

1. A and D, A and E, D and E, B and F

3. $\begin{bmatrix} 3 & 6 & -1 \\ 6 & 5 & 3 \\ -3 & 8 & 7 \end{bmatrix}$ **5.** $\begin{bmatrix} 3 & 9 & -6 \\ 9 & 3 & 12 \\ -15 & 9 & 18 \end{bmatrix}$

7. $\begin{bmatrix} 0 & -6 & 10 \\ -6 & 4 & -18 \\ 24 & -2 & -22 \end{bmatrix}$

9. a. $AD = \begin{bmatrix} 7 & 5 & -4 \\ 17 & 33 & 6 \\ 11 & 27 & -2 \end{bmatrix}, DA = \begin{bmatrix} 6 & 12 & 14 \\ 20 & 10 & 4 \\ 12 & 14 & 22 \end{bmatrix}$

b. Not equal   **c.** Yes

11. a. $DE = ED = \begin{bmatrix} 10 & 0 & 0 \\ 0 & 10 & 0 \\ 0 & 0 & 10 \end{bmatrix}$ **b.** 3×3 identity matrix

13. $\begin{bmatrix} 1 + 2a & 5 + 3b \\ 3 - c & 2 - 2d \end{bmatrix}$ **15.** $\begin{bmatrix} 3a - 2 & 3b - 4 \\ 3c - 6 & 3d - 8 \end{bmatrix}$

17. $m \times k$ **19. a.** BA **b.** 4×3

21. $AB = \begin{bmatrix} a + 3b + 5c & 2a + 4b + 6c \\ d + 3e + 5f & 2d + 4e + 6f \end{bmatrix}$;

$BA = \begin{bmatrix} a + 2d & b + 2e & c + 2f \\ 3a + 4d & 3b + 4e & 3c + 4f \\ 5a + 6d & 5b + 6e & 5c + 6f \end{bmatrix}$

23. $AB = \begin{bmatrix} -3 & -7 \\ 4 & 5 \end{bmatrix}, BA = \begin{bmatrix} 11 & 16 \\ -7 & -9 \end{bmatrix}$

25. $AB = \begin{bmatrix} -2 & 0 \\ 5 & 19 \end{bmatrix}, BA = \begin{bmatrix} 6 & 1 & 10 \\ 10 & 11 & 14 \\ 1 & 6 & 0 \end{bmatrix}$

7.3 Exercises

27. a. $A = \begin{bmatrix} 70 & 76 & 13 & 13 & 74 \\ 12 & 15 & 24 & 10 & 64 \end{bmatrix}$;

$B = \begin{bmatrix} 255 & 175 & 65 & 8 & 11 \\ 20 & 6 & 16 & 1 & 1 \end{bmatrix}$ **b.** $\begin{bmatrix} 325 & 251 & 78 & 21 & 85 \\ 32 & 21 & 40 & 11 & 65 \end{bmatrix}$

29. a. $A = \begin{bmatrix} 51{,}296 & 70{,}912 \\ 625{,}894 & 785{,}599 \\ 13{,}179 & 34{,}711 \end{bmatrix}$

b. $B = \begin{bmatrix} 39{,}186 & 65{,}459 \\ 1{,}012{,}855 & 1{,}416{,}302 \\ 135{,}367 & 332{,}500 \end{bmatrix}$

c. $C = \begin{bmatrix} 12{,}110 & 5453 \\ -386{,}961 & -630{,}703 \\ -122{,}188 & -297{,}789 \end{bmatrix}$

d. 2000: agricultural; 2006: agricultural   **e.** 2006: manufactured goods; this is the largest negative entry in the matrix.

31. $\begin{bmatrix} 36{,}040.48 & 20{,}909.28 \\ 25{,}371.36 & 19{,}746.72 \\ 24{,}739.68 & 16{,}840.32 \end{bmatrix}$

33. $BA = \begin{matrix} \text{Cost} \\ \begin{bmatrix} 1210 \\ 980 \\ 594 \end{bmatrix} \end{matrix} \begin{matrix} \\ \text{Singles} \\ \text{Males } 35-55 \\ \text{Females } 65+ \end{matrix}$

35. a.
$$\begin{matrix} & \text{DeTuris} \quad \text{Marriott} \\ \text{Dept. A} \\ \text{Dept. B} \end{matrix} \begin{bmatrix} 54{,}000 & 49{,}600 \\ 42{,}000 & 42{,}400 \end{bmatrix}$$

b. Dept. A should purchase from Marriott; dept. B should purchase from DeTuris.
37. 55% Republican; 45% Democrat
39. a.

	16–24	25–34	35–44	45–54	55–64	65 and older
Men	409	644	822	853	855	644
Women	381	573	621	644	639	492

b. The 2×2 matrix is $\begin{bmatrix} 1.1 & 0 \\ 0 & 1.25 \end{bmatrix}$.

	16–24	25–34	35–44	45–54	55–64	65 and older
Men	449.90	708.40	904.20	938.30	940.5	708.40
Women	476.25	716.25	776.25	805	798.75	615

7.4 Skills Check

1. a. $AB = BA = \begin{bmatrix} 1 & 0 \\ 0 & 1 \end{bmatrix}$ **b.** They are inverses.

3. $AB = BA = \begin{bmatrix} 1 & 0 & 0 \\ 0 & 1 & 0 \\ 0 & 0 & 1 \end{bmatrix}$ **5.** $A^{-1} = \begin{bmatrix} 7 & -3 \\ -2 & 1 \end{bmatrix}$

7. $A^{-1} = \begin{bmatrix} -1/6 & -1/3 & 1 \\ -1/3 & 1/3 & 0 \\ 1/3 & 2/3 & -1 \end{bmatrix}$ **9.** $A^{-1} = \begin{bmatrix} -\dfrac{1}{3} & -1 & \dfrac{1}{3} \\ 1 & 1 & 0 \\ -\dfrac{1}{3} & 0 & \dfrac{1}{3} \end{bmatrix}$

11. $A^{-1} = \begin{bmatrix} 0.9 & 0.2 & -0.7 \\ -0.5 & 0 & 0.5 \\ 0.7 & -0.4 & -0.1 \end{bmatrix}$

13. $C^{-1} = \begin{bmatrix} 1 & 0 & 1 & 1 \\ 0 & 1 & 1 & 1 \\ 0 & 0 & 1 & 0 \\ 0 & 0 & 0 & 1 \end{bmatrix}$　　**15.** $X = \begin{bmatrix} 10 \\ 20 \end{bmatrix}$

17. $x = 24, y = 27, z = 14$　　**19.** $x = 1, y = 1, z = 1$
21. $x = 2, y = 1, z = -1$
23. $x_1 = 34, x_2 = 16, x_3 = -40, x_4 = 96$

7.4 Exercises

25. Company X: 225,000; company Y: 45,000
27. 50% Republican, 50% Democrat

29. a. $\begin{cases} x + y + z = 400,000 \\ x + y - z = -100,000 \\ x - \dfrac{1}{2}y = 0 \end{cases}$　**b.** $50,000; $100,000; $250,000

31. $50,000 at 6%; $100,000 at 8%; $250,000 at 10%　　**33.** 47%
from business loans, 27% from auto loans, 32% from home loans
35. a. j　u　s　t　_　d　o　_　i　t
　　　10　21　19　20　27　4　15　27　9　20
b. 124, 52, 156, 59, 124, 35, 168, 69, 116, 49
37. 80, 27, 50, 156, 63, 106, 260, 138, 168, 156, 96, 108, 212, 111, 146
39. Vote early　　**41.** Mind your manners
43. Monday night　　**45.** Answers will vary.

7.5 Skills Check

1. $x = 0, y = 0; x = -3, y = 9$
3. $x = 2, y = 0; x = -4, y = 4$
5. $x = 4, y = 8; x = -4, y = -8$
7. $x = 2, y = 6; x = 6, y = 2$
9. $x = 2, y = 8; x = -5, y = -6$
11. $x = 5, y = 33; x = -8, y = 26.5$

13. $x = 4, y = 3$　　**15.** $x = 1, y = 2.25; x = -10, y = -\dfrac{1}{2}$

17. $x = 2, y = 0; x = 5, y = 0; x = -10, y = -1800$
19. $x = 2, y = 3$　　**21. a.** $x^3 - x^2 - 4 = 0$
b. Synthetic division by $x - 2$ gives $x^2 + x + 2$, which has
no real zeros; thus, 2 is the only real solution to the system.
23. Possible answers: $x = 3, y = 2; x = 2, y = 3.25$;
$x \approx -0.44, y \approx 4.20$;
$x \approx -4.56, y \approx -0.95$

7.5 Exercises

25. $290; 1200 units　　**27.** $4700; 6500 units　　**29.** 18,000 units
and 10,000 units　　**31.** 30 units and 5425 units　　**33.** 15 in. by 12 in.
35. 10 cm by 10 cm by 20 cm　　**37.** 10 yr

Chapter 7 Skills Check

1. $x = 2, y = 2, z = 4$　　**2.** $x = 3, y = -2, z = 1$
3. $x = 1, y = 1, z = -1$　　**4.** $x = 2, y = -1, z = 3$

5. $x = 3, y = 1, z = 2$　　**6.** $x = 1, y = 2, z = -1$
7. $x = 41z - 160$
　　$y = -18z + 70$
　　$z = z$
8. $x = 31z - 110, y = 50 - 14z, z = z$　　**9.** No solution
10. No solution　　**11.** $x = 1, y = 3, z = -2, w = 1$

12. $x = 2, y = 1, z = -2, w = -1$　　**13.** $\begin{bmatrix} -1 & 5 & 2 \\ -1 & 1 & 5 \end{bmatrix}$

14. $\begin{bmatrix} -3 & 1 & 0 \\ -5 & 3 & -1 \end{bmatrix}$　**15.** $\begin{bmatrix} 10 & 15 \\ -5 & 10 \\ 15 & -10 \end{bmatrix}$　**16.** Not \possible

17. $\begin{bmatrix} 4 & 14 & 1 \\ -3 & 11 & -1 \end{bmatrix}$　**18.** $\begin{bmatrix} -13 & 12 & 8 \\ -4 & 1 & 3 \\ 0 & 5 & -1 \end{bmatrix}$

19. $\begin{bmatrix} -4 & -2 \\ -2 & -9 \end{bmatrix}$　**20.** $\begin{bmatrix} 10 & 6 & -6 \\ 9 & 25 & 0 \\ 3 & 15 & 10 \end{bmatrix}$

21. $A^{-1} = \begin{bmatrix} -\dfrac{1}{8} & \dfrac{3}{8} & -\dfrac{3}{8} \\ \dfrac{1}{8} & \dfrac{1}{40} & \dfrac{7}{40} \\ -\dfrac{1}{4} & \dfrac{3}{20} & \dfrac{1}{20} \end{bmatrix}$　**22.** $\begin{bmatrix} \dfrac{4}{3} & \dfrac{1}{3} & -1 \\ -1 & 0 & 1 \\ -\dfrac{2}{3} & -\dfrac{2}{3} & 1 \end{bmatrix}$

23. $\begin{bmatrix} 1 & 0 & -1 \\ -1 & 1 & 1 \\ 2 & -2 & -1 \end{bmatrix}$　**24.** $\begin{bmatrix} -0.4 & 0.2 & 0.2 & 0 \\ -0.8 & 0.4 & -0.6 & 1 \\ 0.6 & 0.2 & 0.2 & -1 \\ 0 & 0 & 0 & 1 \end{bmatrix}$

25. $x = 2, y = 3, z = -1$　　**26.** $x = -16, y = 20, z = -8$
27. $x_1 = 2, x_2 = 4, x_3 = -4, x_4 = 12$
28. $x_1 = -2, x_2 = 0, x_3 = -2, x_4 = 8$
29. $x = 1, y = 0; x = 4, y = 12$
30. Possible answers: $x = 10, y = 20; x = -10, y = 20$;
$x \approx 6.32, y = 50; x \approx -6.32, y = 50$

Chapter 7 Review

31. 2800 at $40, 800 at $60, 400 at $100　　**32.** 2.8 mg of
medication A, 4.2 mg of medication B, 3.5 mg of medication C
33. $350,000 at 12%, $150,000 at 15%, $250,000 at 10%
34. $80,000 at 12%, $40,000 at 16%, $240,000 at 8%
35. $110,000, $185,000, $885,000　　**36.** 5 units of
portfolio I, 10 units of portfolio II, 12 units of portfolio III
37. The first property costs $150,000, the second property
costs $100,000, and the third property costs $125,000.
38. 20 g of food I, 30 g of food II, 40 g of food III
39. 3 passenger planes, 4 transport planes, and 4 jumbo planes
40. If x = tech, y = balanced, z = utility,
$$x = \frac{2800 - z}{3}, \quad y = 200 - \left(\frac{2}{7}\right)z, \quad 0 \le z \le 700$$
41. x = number of units of product A, $x = 252 - 1.2y$;
　　y = number of units of product B, $0 \le y \le 210$;
　　z = number of units of product C, $z = 74$
42. $x = 2z, y = 2000 - 4z$ for $0 \le z \le 500; A$ = twice number
of $C, B = 2000 - 4$ times number of $C, 0 \le C \le 500$

43. a. Exports

$$A = \begin{bmatrix} 97{,}470 & 160{,}923 \\ 97{,}412 & 169{,}924 \\ 110{,}835 & 189{,}880 \\ 120{,}365 & 211{,}899 \\ 133{,}979 & 230{,}656 \end{bmatrix}$$

b. Imports

$$B = \begin{bmatrix} 134{,}616 & 209{,}088 \\ 138{,}060 & 221{,}595 \\ 155{,}902 & 256{,}360 \\ 170{,}109 & 290{,}384 \\ 198{,}253 & 302{,}438 \end{bmatrix}$$

c. $\begin{bmatrix} -37{,}146 & -48{,}165 \\ -40{,}648 & -51{,}671 \\ -45{,}067 & -66{,}480 \\ -49{,}744 & -78{,}485 \\ -64{,}274 & -71{,}782 \end{bmatrix}$ **d.** No   **e.** Canada

44. a. $x_1 + 3050 = x_2 + 2500$
$x_2 + 4100 = x_3 + 2000$
$x_3 + 1800 = x_4 + 3800$
    b. $x_1 = x_4 - 650$
$x_2 = x_4 - 100$
$x_3 = x_4 + 2000$
$x_4 \geq 650$

45. a. $y = 1.155x + 432.329$, x in \$thousand
b. $y = 1.082x + 444.421$, x in \$thousand   **c.** Approximately
\$165,803   **46. a.** $x + y = 10$ **b.** $0.20x + 0.05y = 0.155(10)$
c. 7 cc of 20% and 3 cc of 5%   **47.** 14 units when the price is \$592
48. 20 units and approximately 1252 units

Chapter 8   Special Topics

8.1   Skills Check

1. $y \leq 5x - 4$

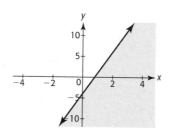

3. $6x - 3y \geq 12$

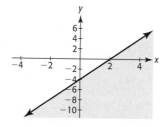

5. $4x - 5y \leq 20$

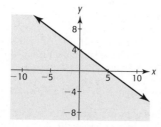

7. C   **9.** D   **11.** Corners: $(0, 0)$, $(0, 5)$, $(4, 0)$, $(3, 2)$
13. Corners: $(0, 5)$, $(1, 2)$, $(2, 0)$
15. Corners: $(0, 5)$, $(1, 2)$, $(3, 1)$, $(6, 0)$

17. $\begin{cases} y \leq 8 - 3x \\ y \leq 2x + 3 \\ y > 3 \end{cases}$

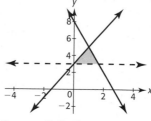

Corners: $(0, 3)$, $(1, 5)$, $(5/3, 3)$

19. $\begin{cases} 2x + y < 5 \\ 2x - y > -1 \\ x \geq 0;\ y \geq 0 \end{cases}$

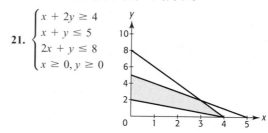

Corners: $(0, 0)$, $(0, 1)$, $(2.5, 0)$, $(1, 3)$

21. $\begin{cases} x + 2y \geq 4 \\ x + y \leq 5 \\ 2x + y \leq 8 \\ x \geq 0,\ y \geq 0 \end{cases}$

Corners: $(3, 2)$, $(4, 0)$, $(0, 2)$, $(0, 5)$

23. $\begin{cases} x^2 - 3y < 4 \\ 2x + 3y < 4 \end{cases}$

$(-4, 4)$ $(2, 0)$

25. $\begin{cases} x^2 - y - 8x \leq -6 \\ y + 9x \leq 18 \end{cases}$

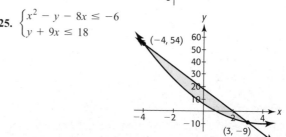

$(-4, 54)$ $(3, -9)$

8.1 Exercises

27. a. $x + y \geq 780$

b.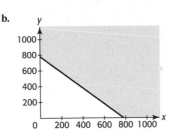

29. a. $240x + 150y \leq 36,000$ **b.**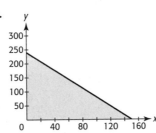

31. a. $\begin{cases} 0.12x + 0.009y \geq 7.56 \\ x + y \geq 100 \\ x \geq 0, y \geq 0 \end{cases}$ **b.** Corners: (0, 840), (60, 40), (100, 0)

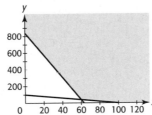

33. $\begin{cases} x + y \geq 780 \\ 78x + 117y \leq 76,050 \\ x \geq 0, y \geq 0 \end{cases}$

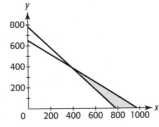

Corners: (780, 0), (975, 0), (390, 390)

35. a. $\begin{cases} 80x + 40y \geq 3200 \\ 20x + 20y \geq 1000 \\ 100x + 40y \geq 3400 \\ x \geq 0, y > 0 \end{cases}$ **b.** Corners: (30, 20), (0, 85), (50, 0), (10, 60)

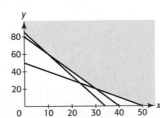

37. $\begin{cases} 4x + 10y \leq 1600 \\ 6x + 7y \leq 1600 \\ x \geq 0, y \geq 0 \end{cases}$

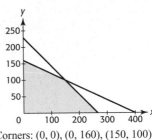

Corners: (0, 0), (0, 160), (150, 100), (266 2/3, 0)

39. a. $\begin{cases} 4x + 6y \leq 480 \\ 2x + 6y \leq 300 \\ x \geq 0, y \geq 0 \end{cases}$ **b.**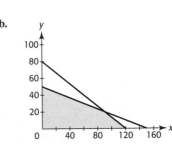

Corners: (90, 20), (0, 50), (120, 0), (0, 0)

41. a. $\begin{cases} x + y \leq 1400 \\ x \geq 500 \\ y \geq 750 \\ x \geq 0, y \geq 0 \end{cases}$ **b.**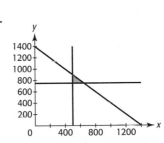

Corners: (500, 750), (500, 900), (650, 750)

8.2 Skills Check

1. Maximum: 382 at (10, 38) minimum: 0 at (0, 0)

3. Maximum: 22 at (4, 3) minimum: 0 at (0, 0)

5. a.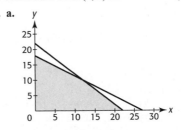

Corners: (0, 0), (0, 18), (22, 0), (12, 10)

b. Maximum is 90 at $x = 0, y = 18$

7. a.

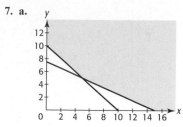

Corners: $(0, 10)$, $(5, 5)$, $(15, 0)$

b. Minimum is 20 at $x = 0$, $y = 10$

9. 140 at $(4, 2)$ **11.** 1200 at all points on the line segment from $(2, 6.5)$ to $(0, 7.5)$ **13.** 140 at $(2, 2)$ **15.** 161 at $(3, 1)$ **17.** 70 at $(0, 7)$

8.2 Exercises

19. 975 Turbo and 0 Tornado models gives maximum profit of $31,200.
21. Maximum daily profit is $396 with 9 Safecut and 6 Safecut Deluxe.
23. a. 100 minutes on TV and 80 minutes on radio   **b.** Maximum profit is $94,000.   **25.** Using assembly line 1 for 50 days and assembly line 2 for 0 days gives a minimum cost of $1,000,000.
27. Profit is maximized at $900,000 with 15 Van Buren and 0 Jefferson models, with 5 Van Buren and 8 Jefferson models, or with 10 Van Buren and 4 Jefferson models.   **29.** Operating facility A for 8 weeks and facility B for 2 weeks gives minimum cost, $160,000.   **31.** 4 servings of diet A and 10 servings of diet B; minimum is .71 oz.

8.3 Skills Check

1. 5, 7, 9, 11, 13, 15   **3.** $10, 5, \dfrac{10}{3}, \dfrac{5}{2}, 2$ **5.** 9, 11, 13
7. 25   **9.** 24, 48, 96, 192   **11.** 2430   **13.** 5, 3, 1, -1
15. 2, 7, 17, 37

8.3 Exercises

17. $10,500   **19. a.** $300 + 60n$ **b.** 360, 420, 480, 540, 600, 660
21. a. First by $2000   **b.** They are the same.   **c.** Years on job; if less than 10 yr, choose first.   **23.** 1050, 1102.50, 1157.63, 1215.51   **25. a.** $20,000   **b.** $50,000 - 10,000n$
c. 40,000, 30,000, 20,000, 10,000, 0   **27.** 320,000   **29.** $\dfrac{1}{2}$ foot
31. $7,231,366.37   **33.** 15 years   **35.** $22,253.46
37. 13, 21, 34, 55   **39.** 1100, 990, 891, 801.90

8.4 Skills Check

1. $\dfrac{364}{27}$ **3.** 205   **5.** 150   **7.** 98,301   **9.** 5115   **11.** $\dfrac{729}{8}$
13. 126   **15.** $\dfrac{73}{12}$ **17.** 3   **19.** Not possible, infinite

8.4 Exercises

21. $2400   **23.** 11   **25. a.** 78   **b.** 156
27. $42,768.57   **29.** 16 cm^3 **31. a.** 5   **b.** 25   **c.** 125; 625
d. Geometric, with $r = 5$ **33.** $2824.30   **35.** 213 ft
37. a. $s_n = 35,000(1 - 0.84^n)$ **b.** $35,000(0.84)^n$
39. $15,992.73

41. n^{th} term: $R(1 + i)^{-n}(1 + i)^{n-1}$

$$s_n = \frac{R(1 + i)^{-n}(1 - (1 + i)^n)}{1 - (1 + i)}$$

$$= R\left[\frac{(1 + i)^{-n} - (1 + i)^0}{-i}\right]$$

$$= R\left[\frac{(1 + i)^{-n} - 1}{-i}\right]$$

$$= R\left[\frac{1 - (1 + i)^{-n}}{i}\right]$$

8.5 Skills Check

1. $10x^4 + 33x^2 + 2x + 12$ **3.** $\dfrac{x^3 + 6}{x^3}$ **5.** 26
7. $0.8125 = \dfrac{13}{16}$ **9.** $y = 12x - 16$ **11. a.** $4(x + h) + 5$
b. $4h$ **c.** 4   **13.** $y = 3 - 2x^2$ **15.** $y = \dfrac{18 - 9x^3}{5}$
17. a. $x = 1$ **b.** Same   **c.** $(1, 4)$ **19. a.** $x = 2$
b. Same   **c.** $(2, -9)$ **21. a.** $x = -2, x = 8$ **b.** $y = 67$ at $x = -2; y = -433$ at $x = 8$ **c.** $x = -2$ **23.** $y = 3x^{1/2}$
25. $y = 2x^{2/3}$ **27.** $y = (x^2 + 1)^{1/3}$ **29.** $y = (x^3 - 2)^{2/3}$
31. $y = x^{1/2} + (2x)^{1/3}$ **33.** $y = \pm\sqrt{3 - 4x}$
35. $y = \dfrac{-1 \pm \sqrt{1 + 24x}}{2}$ **37.** $2; 4x^3 + 5$ **39.** $x^3 + x; \frac{1}{2}$
41. $3x\sqrt{x^2 - 1}$ **43.** $y = 3x^{-1} - 4x^{-2} - 6$
45. $y = (x^2 - 3)^{-3}$ **47.** $f'(x) = -\dfrac{6}{x^4} + \dfrac{4}{x^2} + \dfrac{1}{x}$
49. $f'(x) = \dfrac{8x}{(4x^2 - 3)^{1/2}}$ **51.** $3\log x + 5\log(3x - 4)$
53. $x = 0, x = 1$ **55.** $x = 3, x = \dfrac{3 + \sqrt{69}}{5}, x = \dfrac{3 - \sqrt{69}}{5}$

Chapter 8 Skills Check

1.

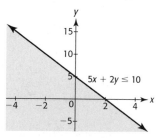

2.

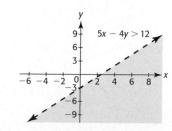

3. $\begin{cases} 2x + y \le 3 \\ x + y \le 2 \\ x \ge 0, y \ge 0 \end{cases}$

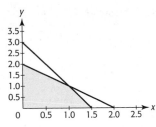

Corners: $(0, 0)$, $(0, 2)$, $(1, 1)$, $(1.5, 0)$

4. $\begin{cases} 2x + y \le 3 \\ x + y \le 2 \\ x \ge 0, y \ge 0 \end{cases}$

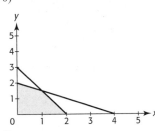

Corners: $(0, 0)$, $(0, 2)$, $(1, 1.5)$, $(2, 0)$

5. $\begin{cases} 2x + y \le 30 \\ x + y \le 19 \\ x + 2y \le 30 \\ x \ge 0, y \ge 0 \end{cases}$

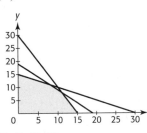

Corners: $(0, 0)$, $(0, 15)$, $(15, 0)$, $(11, 8)$, $(8, 11)$

6. $\begin{cases} 2x + y \le 10 \\ x + 2y \le 11 \\ x \ge 0, y \ge 0 \end{cases}$

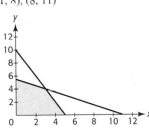

Corners: $(0, 0)$, $(0, 5.5)$, $(3, 4)$, $(5, 0)$

7. $\begin{cases} 15x - x^2 - y \ge 0 \\ y - \dfrac{44x + 60}{x} \ge 0 \\ x \ge 0, y \ge 0 \end{cases}$

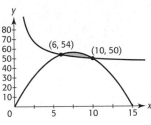

8. $\begin{cases} x^3 - 26x + 100 - y \ge 0 \\ 19x - 20 - y \le 0 \\ x \ge 0, y \ge 0 \end{cases}$

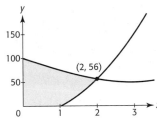

9. Minimum is 14 at $(2, 2)$ **10.** Maximum is 168 at $(12, 7)$.
11. Maximum is 79 at $(8, 11)$.   **12.** Geometric; $r = 6$
13. Arithmetic; $d = 12$ **14.** Geometric; $r = -\dfrac{3}{4}$ **15.** $a_5 = 4$
16. 27   **17.** $S_{10} = -1705$ **18.** 234   **19.** 4
20. $5x^4 - 9x^2 + 6x - 10$ **21.** $x = 0, x = 9, x = -1$
22. $y = x + x^{-2} - 2x^{-3/2} + 3x^{-1/3}$ **23.** $f'(x) = \dfrac{8}{x^3} + \dfrac{5}{x} + 4x$
24. $x = 0, x = -1, x = \dfrac{1}{2}$

Chapter 8 Review

25. \$25,200 profit at 360 units each   **26. a.** Pottstown, 20 days, Ethica, 30 days   **b.** \$1,120,000   **27.** Minimum cost is \$48 with 32 units of feed A and 2 units of feed B.   **28.** 220 leaf blowers and 240 weed wackers   **29.** Maximum profit is \$71,000 with 60 two-bedroom and 20 three-bedroom apartments.   **30.** \$30 million in auto loans and no home equity loans   **31. a.** Job 2   **b.** Job 1
32. Approximately 999.98 mg   **33. a.** 9.22337×10^{18} grains
b. 1.84467×10^{19} grains   **34.** \$26,764.51   **35.** \$24,500.90
36. $8 < x < 52$

index

Note: Page numbers followed by f indicate figures; those followed by t indicate tables.